Air Pollution Science
for the 21ˢᵗ Century

Photograph front cover:

Downtown Hong Kong from Wilson Trail
Malcolm Downes
Olympus OM2N Zuiko 100mm Fuji Sensia II

Air Pollution Science for the 21st Century

Edited by

Jill Austin
Peter Brimblecombe
William Sturges

School of Environmental Sciences, University of East Anglia,
Norwich, NR4 7TJ, UK

2002
Elsevier

Amsterdam – Boston – London – New York – Oxford – Paris
San Diego – San Francisco – Singapore – Sydney – Tokyo

ELSEVIER SCIENCE Ltd.
The Boulevard, Langford Lane
Kidlington, Oxford OX5 1GB, UK

First edition 2002

Library of Congress Cataloging in Publication Data
A catalog record from the Library of Congress has been applied for.

British Library Cataloguing in Publication Data
A catalogue record from the British Library has been applied for.

ISBN: 0 08 044119 X
Series ISSN: 1474-8177

⊗ The paper used in this publication meets the requirements of ANSI/NISO Z39.48-1992 (Permanence of Paper).
Printed in The Netherlands.

Contents

Developments in Environmental Science

Introduction to the Book Series

Environmental pollution has played a critical role in human lives since the early history of the nomadic tribes. During the last millennium industrial revolution, increased population growth and urbanization have been the major determinants in shaping our environmental quality.

Initially primary air pollutants such as sulfur dioxide and particulate matter were of concern. For example, the killer fog of London in 1952 resulted in significant numbers of human fatality leading to major air pollution control measures. During the 1950s, scientists also began to understand the cause and atmospheric mechanisms for the formation of the Los Angeles photochemical smog. We now know that surface level ozone and photochemical smog are a worldwide problem at regional and continental scales, with specific geographic areas of agriculture, forestry and natural resources, including their biological diversity at risk. As studies continue on the atmospheric photochemical processes, air pollutant transport, their atmospheric transformation and removal mechanisms, so is the effort to control the emissions of primary pollutants (sulfur dioxide, oxides of nitrogen, hydrocarbons and carbon monoxide), mainly produced by fossil fuel combustion.

During mid 1970s environmental concerns regarding the occurrence of "acidic precipitation" began to emerge to the forefront. Since then, our knowledge of the adverse effects of air pollutants on human health and welfare (terrestrial and aquatic ecosystems and materials) has begun to rise substantially. Similarly, studies have been directed to improve our understanding of the accumulation of persistent inorganic (heavy metals) and organic (polyaromatic hydrocarbons, polychlorinated biphenyls) chemicals in the environment and their impacts on sensitive receptors, including human beings. Use of fertilizers (excess nutrient loading) and herbicides and pesticides in both agriculture and forestry and the related aspects of their atmospheric transport, fate and deposition; their direct runoff through the soil and impacts on ground and surface water quality and environmental toxicology have become issues of much concern.

In the recent times environmental literacy has become an increasingly important factor in our lives, particularly in the so-called developed nations. Currently the scientific, public and political communities are much concerned with the increasing global scale air pollution and the consequent global climate change. There are efforts being made to totally ban the use of chlorofuorocarbon and organo-bromine compounds at the global scale. However, during this millennium many developing nations will become major forces governing environmental health as their populations and industrialization grow at a rapid pace. There is an on-going international debate regarding policies and the mitigation strategies to be adopted to address the critical issue of climate change. Human health and environmental impacts and risk assessment and the associated cost-benefit analyses, including global economy are germane to this controversy.

An approach to understanding environmental issues in general and in most cases, mitigation of the related problems requires a systems analysis and a multi- and inter-disciplinary philosophy. There is an increasing scientific awareness to integrate environmental processes and their products in evaluating the overall impacts on various receptors. As momentum is gained, this approach constitutes a challenging future direction for our scientific and technical efforts.

The objective of the book series "Developments in Environmental Science" is to facilitate the publication of scholarly works that address any of the described topics, as well as those that are related. In addition to edited or single and multi-authored books, the series also considers conference proceedings for publication. The emphasis of the series is on the importance of the subject topic, the scientific and technical quality of the content and timeliness of the work.

Sagar V. Krupa
Chief Editor, Book Series

Foreword

Acid rain, ozone photochemistry, long-range transport of pollutants, greenhouse gas emissions and aerosols dominated tropospheric air pollution research in the last 30 years of the 20th century. At the start of the 21st century, acid rain is subject to planned improvement in Europe and North America, but is a growing problem in Asia. Ozone pollution is much better understood, but the problem is still with us, and desirable levels are difficult to achieve over continental Europe. The heterogeneous chemistry which is responsible for ozone depletion in the stratosphere is now reasonably well understood, but there is on-going interest in the sources and sinks of CFC (chlorofluorocarbon) replacements in troposphere. There is increasing interest in indoor air quality, and the origin and health implications of atmospheric particles. Perhaps most important on a global perspective, intensive research has not yet determined the relationship between greenhouse gases, aerosols and surface temperature. The climatic implications of these are now more urgent than ever.

In 1998 *Atmospheric Environment* began to encourage the submission of Millennial Reviews on a range of key topics. These papers assessed our knowledge at the end of the 20th century, and looked forward to the 21st century. The Millennial Reviews have been collected together as the main chapters of this book, and the authors of some of the earlier papers have supplied short up-date articles. These are included as short appendices to the main articles.

In 1995 *Atmospheric Environment* also introduced an exciting feature, the New Directions columns. These short articles were written in a journalistic style, and authors were encouraged to speculate and express controversial opinions. Initially these articles were all invited but, as the column became better known, submitted articles became the norm. These are now published at least once a month. In 1999 a selection of Future Directions columns were invited, where the authors were encouraged to look ahead to possible developments in air pollution science at the end of the 21st century. A group of recent New and Future Directions columns have been chosen for this book, to complement the Millennial Reviews. We hope that both the reviews and the thought provoking essays will stimulate the imagination of current researchers, and encourage breadth of vision in future research.

Jill Austin
Peter Brimblecombe
William Sturges

School of Environmental Sciences
University of East Anglia
Norwich
NR4 7TJ, UK

Chapter 1

Urban air quality

Jes Fenger

National Environmental Research Institute, Department of Atmospheric Environment,
Frederiksborgvej 399, DK-4000 Roskilde, Denmark

Abstract

Since 1950 the world population has more than doubled, and the global number of cars has increased by a factor of 10. In the same period the fraction of people living in urban areas has increased by a factor of 4. In year 2000 this will amount to nearly half of the world population. About 20 urban regions will each have populations above 10 million people.

Seen over longer periods, pollution in major cities tends to increase during the built up phase, they pass through a maximum and are then again reduced, as abatement strategies are developed. In the industrialised western world urban air pollution is in some respects in the last stage with effectively reduced levels of sulphur dioxide and soot. In recent decades however, the increasing traffic has switched the attention to nitrogen oxides, organic compounds and small particles. In some cities photochemical air pollution is an important urban problem, but in the northern part of Europe it is a large-scale phenomenon, with ozone levels in urban streets being normally lower than in rural areas. Cities in Eastern Europe have been (and in many cases still are) heavily polluted. After the recent political upheaval, followed by a temporary recession and a subsequent introduction of new technologies, the situation appears to improve. However, the rising number of private cars is an emerging problem. In most developing countries the rapid urbanisation has so far resulted in uncontrolled growth and deteriorating environment. Air pollution levels are here still rising on many fronts.

Apart from being sources of local air pollution, urban activities are significant contributors to transboundary pollution and to the rising global concentrations of greenhouse gasses. Attempts to solve urban problems by introducing cleaner, more energy-efficient technologies will generally have a beneficial impact on these large-scale problems. Attempts based on city planning with a spreading of the activities, on the other hand, may generate more traffic and may thus have the opposite effect.

First published in Atmospheric Environment 33 (1999) 4877–4900

1. Introduction

"Urban air quality" is a vast subject with different socio-economic aspects in different parts of the world – and even within a specific region. This is demonstrated for selected regions. The European Union representing the industrialised western countries is treated in most detail, the eastern part of Europe gives examples of economies in transition and East Asia – specifically China – represents the developing regions. Other regions: North America, Latin America, West Asia and Africa are only briefly discussed – not because their problems are less important, but because the general aspects of urban air pollution and air quality are already covered.

1.1. The historical perspective

Cities are by nature concentrations of humans, materials and activities. They therefore exhibit both the highest levels of pollution and the largest targets of impact. Air pollution is, however enacted on all geographical and temporal scales, ranging from strictly "here and now" problems related to human health and material damage, over regional phenomena like acidification and forest die back with a time horizon of decades, to global phenomena, which over the next centuries can change the conditions for man and nature over the entire globe. In this respect the cities act as sources.

Initially outdoor air pollution was by and large a purely urban phenomenon, and literature as well as historical records testify that the problems were extensive. They may even be underestimated, since generally people were less critical about their living conditions, and they had no means of evaluating long-term impacts of, e.g. carcinogens. Further, many of the records concern aesthetic impacts in the form of smell and soiling, which are not deleterious to health in themselves.

Finally it should be recognised that up to the Second World War many people had an ambivalent attitude towards pollution, which to some extent was perceived as a symbol of wealth and growth. Thus advertisements showed pictures of fuming chimney stacks and cars with visible exhaust – images hardly anyone would cultivate today!

Semi-quantitative evaluations of the early urban air pollution have been attempted in various ways, i.a. via records of material damage and impacts on human health and vegetation. Also simplified dispersion modelling is a possibility, when the consumption of fuels and raw materials within a confined area are reasonably well known (Brimblecombe, 1987).

Some direct measurements of air pollutants were carried out by scientists and amateur enthusiasts in the last century, but systematic and official investigations with continuous time series are of fairly recent date. In England the number of measuring sites was thus not increased substantially before the London smog disaster in 1952 was followed by the Clean Air Act in 1956 (Brimblecombe, 1998).

1.2. Present day records

In most of the industrialised world urban air pollution is now monitored routinely. Since 1974 WHO and UNEP have, within the "Global Environment Monitoring System", collaborated on a project to monitor urban air quality, the so-called GEMS/AIR (UNEP, 1991; WHO/UNEP, 1992; GEMS/AIR, 1996; and a series of related reports). Concentrations of air pollutants in selected countries are also reported yearly by the OECD (1997). A comprehensive presentation of urban air pollution in Europe, based on data from 79 cities in 32 countries (Richter and Williams, 1998) has recently been published by the European Environment Agency.

These data and similar ones give an indication of trends in ambient air quality at national level and in cities. However, one should be cautious when comparing absolute values from different regions. Often the data are based on one or a few monitoring stations, placed at critical sites and thus represent micro-environments. It should also be taken into account that the coverage of stations is different for different countries (Larssen and Hagen, 1997), and that average values can therefore be differently biased.

Information on air pollution in the developing countries and in some countries with economies in transition is limited and longer time-series are very rare. In some cases a general trend in air quality can only be estimated on the basis of uncertain emission inventories. Data presented in the open literature are seldom up to date and normally concern specific cities, which may not be fully representative. In recent years many governmental and private institutions from the industrialised countries (including Denmark) have acted as consultants in developing countries or performed investigations, but not all the efforts are reported in the open literature.

1.3. Air quality indicators

To date nearly 3000 different anthropogenic air pollutants have been identified, most of them organic (including organometals). Combustion sources, especially motor vehicles, emit about 500 different compounds. However, only for about 200 of the pollutants have the impacts been investigated, and the ambient concentrations are determined for an even smaller number.

This complex nature of air pollution, especially with respect to health impacts in cities, has prompted attempts to define the so-called indicators

(Wiederkehr and Yoon, 1998), which condense and simplify the available monitoring data to make them suitable for public reporting and decision makers. OECD (1998) has applied major pollutants measured in a specified way as indicators for the total mix of pollutants (an example is shown later as Fig. 17), but also weighed means of concentrations of several pollutants relative to guideline values has been used.

In another type of indicators OECD (1998) has aggregated monitoring data from various regions (Western Europe, USA, Japan) to demonstrate overall trends in pollution levels. The results of such an exercise should be treated with some caution, but yearly averages appear to represent the general developments reasonably well. Values for the period 1988–1993 and the pollutants SO_2, NO_2, CO and O_3 are shown in later sections.

1.4. Local emissions and global pollution

The growing global consumption of fossil fuels leads to energy-related emissions of carbon dioxide (e.g. Ellis and Tréanton, 1998) and may eventually, via the enhanced greenhouse effect result in climate changes with impacts on all human activities and natural ecosystems. One of the results of the UN-conference on environment and development in Rio de Janeiro in 1992 was an action plan for the attainment of a sustainable global development in the next century – the so-called *Agenda 21*. As a consequence many cities and administrative units in the industrialised world have embarked on local programs, and more than 290 European cities have signed the Aalborg Charter of European Cities and Towns towards Sustainability. Noteworthy in this connection is The International Council for Local Environmental Initiatives (ICLEI, 1998) with the purpose to achieve and monitor improvements in global environmental conditions through cumulative local actions. Although the political attention emphasises climate protection and thus reduction of emission of the ultimate product of combustion – CO_2, the attempts to save energy may also improve the urban air quality (e.g. Pichl, 1998).

2. Global growth and increasing urbanisation

2.1. Population and urbanisation

Just after the Second World War the world population was about 2.5 billion, and in mid-1998 it had more than doubled to 5.9 billion. The more developed countries account now for 1.2 billion and the less developed for 4.7 billion, among these China alone accounts for 1.2 billion. In the same 50 year period the global urbanisation, defined as the fraction of people living in settlements above 2000 inhabitants, has risen from below 30 to 44%. In the more developed countries it is now on the average 73% and in the less developed 36% (Population Reference Bureau, 1998).

There is no indication that these global trends will change in the next decades (Fig. 1). Around year 2000 nearly half of the world population will live in urban areas, and various projections estimate a total population in the order of 10 billion in the middle of the next century with the sharpest rise in the (now) developing regions – and among them most in Africa.

Especially in the developing countries there is a significant migration of people from the countryside to the towns – both because of a mechanisation of farming and opportunities in new industries and public services. This may here lead to a further growth in urbanisation with a factor in the order of 1.5 in the next 50 years (UNEP, 1997). In China alone more than 100 million people are reported to move around in search for work, and, e.g. Beijing appears actively to prevent their permanent settlement.

In Asia, Latin America and Africa this urbanisation has been accompanied by the proliferation of slums and squatter settlements (United Nations, 1997a). In some cases the situation has been aggravated by polluting industries, which have been transferred from industrialised countries with stricter environmental legislation and higher wages.

Regions with high birth rates and immigration are therefore faced with environmental problems due to unplanned urban growth and emerging megacities. In 1950 there were only eight cities with inhabitant numbers above 5 million,

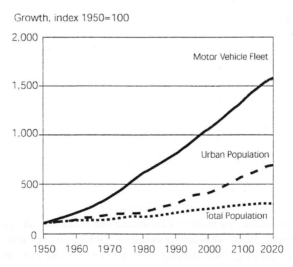

Figure 1. The increase since 1950 of the total world population, the urban population and the number of motor vehicles – excluding motorbikes and three wheelers (UNEP/WHO, 1992).

in 1990 there were about 35, and in year 2000 there will probably be more than 40. About half of them will be situated in East Asia (WHO/UNEP, 1992). Probably more than 20 of the cities will have more than 10 million inhabitants.

On the other hand there has been a tendency in the industrialised world that relatively fewer people live in the inner areas of the cities (OECD, 1995b). This leads to increasing urban travel and expanding road systems.

2.2. *Global energy consumption, production and emissions*

The global number of motor vehicles – excluding 2- and 3-wheelers has increased by a factor of 10 since 1950 (Fig. 1) and is now above 600 million. In addition to this there is now an estimated 80 million motorcycles (OECD, 1995a). In the same period the industrial production has increased by a factor 10 and the global energy consumption by nearly a factor 5 (United Nations, 1997b and earlier reports). Since the major part of the energy has been produced by fossil fuels, and to a minor extent by biofuels, initially without flue gas cleaning, the global emissions of air pollutants have increased correspondingly. Since 1950 the global emission of sulphur oxides has more than doubled, and the emission of nitrogen oxides increased by a factor of 4.

2.3. *Pollution development*

In general the environmental quality in a given country depends upon the average income of the inhabitants (The World Bank, 1992). The availability of safe water and adequate sanitation increases with the income, and so does the amount of municipal waste per capita. The air pollution however, appears initially to increase with the income up to a point and then to decrease (Fig. 2).

Based on more recent data Grossman and Krueger (1995) estimate that the turning point for different pollutants vary, but in most cases comes before a country reaches a per capita income of USD 8000. Although this suggests that global emissions will decrease in the very long run, a continued rapid growth over the next several decades is forecasted (Selden and Song, 1994).

Seen in a time perspective the air pollution in a developing urban area initially increases, goes through a maximum and is then again reduced, when pollution abatement becomes effective (Fig. 3). As it will be demonstrated in later sections, cities in the industrialised western world are in some respects in the last stage of this development, in economies in transition many cities are in the stabilisation stage, whereas in the developing countries the pollution levels are still rising. In the developing countries indoor pollution from biomass combustion for heating and cooking may, however be an even more serious problem with concentrations of particles orders of magnitude higher than the safe levels in WHO guidelines (Smith, 1988).

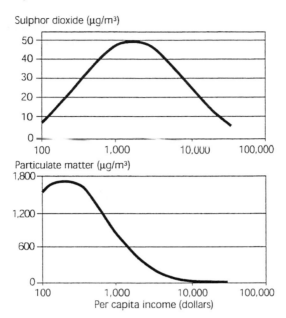

Sulphor dioxide (µg/m³)

Particulate matter (µg/m³)

Per capita income (dollars)

Figure 2. Urban concentrations of SO_2 and particulate matter as a function of national per capita income (based on data from the 1980s. The World Bank, 1992). The curves only show general tendencies, there are marked deviations. Thus in the early 1980s Kuwait had both some of the highest incomes and highest pollution levels (Smith, 1988).

Fully in line with this typical development The World Commission on Environment and Development in its report "Our common future" (The World Commission on Environment and Development, 1987) conceives technological development and rising standard of living as a prerequisite for environmental improvement. Or as Bertholt Brecht has put it: "Erst kommt das Fressen, dann kommt die Moral" – in modern form: "First development and only later pollution control".

3. Pollutants, sources and emissions

The air pollutants can be divided into two groups (Wiederkehr and Yoon, 1998): The traditional *Major Air Pollutants* (MAP, comprising sulphur dioxide, nitrogen dioxide, carbon monoxide, particles, lead and the secondary pollutant ozone) and the *Hazardous Air Pollutants* (HAP, comprising chemical, physical and biological agents of different types). The HAP are generally present in the atmosphere in much smaller concentrations than the MAP, and they appear often more localised, but they are – due to their high specific activity –

8

J. Fenger

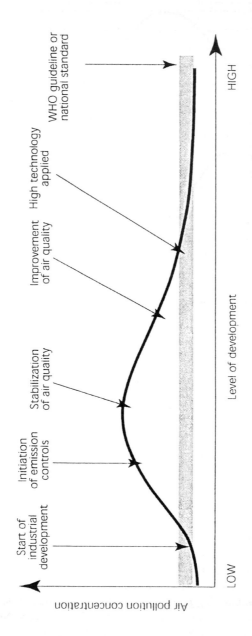

Figure 3. Schematic presentation of a typical development of urban air pollution levels. Depending upon the time of initiation of emission control the stabilisation and subsequent improvement of the air quality may occur sooner or later in the development. (Based on WHO/UNEP, 1992; Mage et al., 1996).

nevertheless toxic or hazardous. Both in scientific investigations and in abatement strategies HAP's are difficult to manage, not only because of their low concentrations, but also because they are in many cases not identified.

3.1. The major pollutants

Sulphur dioxide (SO_2) is the classical air pollutant associated with sulphur in fossil fuels. The emission can be successfully reduced using fuels with low sulphur content e.g. natural gas or oil instead of coal. On large plants in industrialised countries desulphurisation of the flue gas is an established technique.

Nitrogen oxides (NO_x) are formed by oxidation of atmospheric nitrogen during combustion. The main part, especially from cars, is emitted in the form of the nontoxic nitric oxide (NO), which is subsequently oxidised in the atmosphere to the secondary "real" pollutant NO_2. The emissions can be reduced by optimisation of the combustion process (low NO_x burners in power plants and lean burn motors in motor vehicles) or by means of catalytic converters in the exhaust.

Carbon monoxide (CO) is the result of incomplete combustion with motor vehicles as the absolutely dominant source. The emissions can be reduced by increasing the air/fuel ratio, but with the risk of increasing the formation of nitrogen oxides. Most effective reductions are carried out with catalytic converters.

Particulate matter (PM) is not a well-defined entity as, e.g. carbon monoxide. Originally it was determined as soot or "black smoke" for which there is an EC air quality limit value (Edwards, 1998). Later the concept of total suspended particulate matter (TSP) was introduced, but since 1990 size fractionating has been attempted by measurements of PM_{10} (particles with diameter less than 10 µm). Unfortunately the major part of PM_{10} may have a natural origin (e.g., sea spray or desert and soil dust), and it is therefore important also to measure $PM_{2.5}$ or even, when the appropriate technology has been developed, PM_1.

The applications of different concepts and measuring techniques complicate evaluations of the development in pollution levels. To some extent it is possible to establish relationships between the concentrations of fine and coarse particles relevant to epidemiological studies (Wilson and Sue, 1997), but only under well-defined conditions. Measurements from Erfurt in the former DDR thus show that the level of $PM_{2.5}$ has been reduced substantially after the reunification and the subsequent introduction of updated technology. Nevertheless the amount of even smaller particles has increased and so has the total number of particles indicating a change in major sources (Tuch et al., 1997).

A further complication is that the chemical composition of particles is not well known, and that the health impacts may be due to other pollutants adsorbed on them – typically heavy metals or less volatile organic compounds.

The emissions of particles of antropogenic origin can be reduced by use of cleaner fuels, better combustion techniques and a series of filtration or impaction technologies.

Lead (Pb) as an additive to petrol has been phased out in the major part of the industrial world, but is still used in many developing countries and economies in transition, where also emissions from industrial activities play a role.

3.2. The hazardous pollutants

Volatile organic compounds (VOC) as air pollutants are the result of incomplete combustion of fuels or are formed during the combustion – typically in cars, where also evaporation may play a role. Also some industrial processes and the use of solvents result in the emission of VOC. In urban air the most important compounds are benzene and the series of polyaromatic hydrocarbons (PAH) among which some have until recently been unnoticed (Enya et al., 1997), but also, i.a. 1,3-butadiene, ethene, propene and a series of aldehydes have received attention (Larsen and Larsen, 1998).

It is a formal question whether biogenic VOC e.g. from vegetation constitute a pollution, but they must be taken into account in relation to photochemical air pollution.

The removal of lead as an additive has not been completely without side effects. Changes in the mixing of the petrol to increase the octane number may increase the emission of aromatic hydrocarbons i.a. benzene. Benzene concentrations have increased in many urban atmospheres with the introduction of catalytic converters (Richter and Williams, 1998). An alternative additive MTBE (methyl-tert-butyl ether) not only increases the octane number, but also improves the combustion and thus reduces the emissions of carbon monoxide and hydrocarbons. It is however an air pollutant causing both immediate eye and respiratory irritation and long-term risk of cancer. More important may be the contamination of soil and groundwater, especially around petrol filling stations (trans-media pollution). In Denmark MTBE is only used for 98 octane petrol.

Other heavy metals than lead of interest as air pollutants include cadmium, nickel and mercury, all with industrial sources.

3.3. Urban sources of air pollution

Since combustion is the dominant cause of urban air pollution, the various sources emit to a large extent the same pollutants – only in varying proportions. Table 1 indicates the typical relative importance of source categories for emission of the main pollutants. The distribution of course varies, thus e.g. in Eastern Europe SO_2 from space heating play a relatively more important role compared to Western and Southern Europe, and in Southern Europe the contribution from SO_2 from traffic is relatively high due to the use of diesel oil with a high sulphur content.

3.4. Emission inventories

National or regional inventories of emissions as they are carried out e.g. in the form of the European Corinair database (CORINAIR, 1996) are used in international negotiations. They may also for lack of better information suggest general trends in air pollution levels, but dispersion modelling in urban areas requires time and space resolutions relevant to the applied scale. Proxy data for larger areas can be generated on the basis of information on i.a. traffic pattern (Friedrich and Schwarz, 1998), but detailed investigations of individual streets (Berkowicz, 1998) are based on actual traffic counts.

4. From emissions to impacts

In the design of cost effective abatement strategies (e.g., Krupnick and Portney, 1991) it must be realised that the relations between emissions and resulting concentrations are by no means simple. Measurements are still the foundation of our understanding, but application of mathematical modelling, and also

Table 1. Main emission sources and pollutants in air pollution in commercial non industrial cities. The table indicates the relative importance of *urban sources* for the main *urban pollutants*. x: 5–25%; xx: 25–50%; xxx: More than 50%. (Based on Stanners and Bourdeau, 1995)

Source category		Pollutant						
		SO_2	NO_2	CO	TSP	Organic	Pb	Heavy metals[a]
Power generation (Fossil fuel)		xx	x	x				x/xx
Space heating	- coal	xx	x	xx	xx	xx/x		x/xx
	- oil	xx	x					
	- wood				xx	xx/x		
Traffic	- gasoline		xx	xxx		xx	xxx	
	- diesel	x	xx		xx	xx		
Solvents						x		
Industry		x		x	x	x	x	xx/xxx

[a]With the exception of lead (Pb).

of physical modelling in wind tunnels, is of increasing importance in urban air pollution management. As a consequence numerous techniques have been developed for different spatial scales ranging from entire regions down to individual streets. Some models only describe dispersion or have simple reaction schemes; more sophisticated models comprise a large number of interacting reactions. Such models can be further developed to form full decision support systems (e.g., Dennis et al., 1996).

4.1. Dispersion in the urban area

The importance of dispersion was recognised already with the invention of chimneys, and the meteorological conditions played a crucial role in a series of pollution disasters (Brimblecombe, 1987). The dispersion mechanisms have received special interest with the increasing urban traffic in built up areas. Fig. 4 demonstrates the significance of wind speed for the resulting air pollution in a street canyon, where a fairly strong wind ($8 \, \mathrm{m \, s^{-1}}$) is seen to nearly halve the concentration of NO_2 at rush hours.

Measurements and dispersion calculations (e.g. Berkowicz, 1998) have shown that also the wind direction is important; therefore in areas with a dominant wind direction – thus in Denmark from the west – the orientation of the individual streets counts. The overall significance of the climatological conditions is clearly demonstrated in a comparison between the air quality in Copenhagen and Milan. Since the frequency of low wind speeds is considerably higher in Milan than in Copenhagen, Milan has much higher pollution levels for comparable emissions (Vignati et al., 1996).

In more open spaces (parks, squares, residential areas) the pollution levels take the form of an urban background, with increasing impact of more distant sources. Recent applications of mesoscale computer models have demonstrated that also the regional component is important, especially in areas with a complex landscape such as coastal regions; thus studies in the Mediterranean Region and Southern Europe have indicated that in certain periods the urban areas may be significantly affected by sources located hundreds of kilometres away (Kallos, 1998).

4.2. Chemical reactions

During dispersion the pollutants interact chemically (e.g. Finlayson-Pitts and Pitts, 1997) and for the urban atmosphere reactions between nitrogen oxides, organic compounds and ozone are the most important. Photochemical smog with formation of ozone and other oxidants was first recognised in Los Angeles in the mid-1940s as an urban phenomenon related to car exhaust in a subtropical topographically confined region.

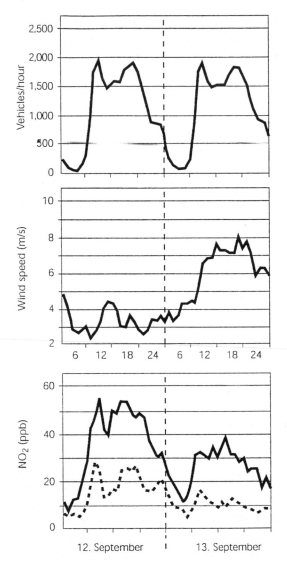

Figure 4. Relations between traffic intensity, wind velocity and NO_2-concentrations on two consecutive weekdays in 1997, measured at Jagtvej, a busy street surrounded with buildings in Copenhagen. The concentrations are measured both at the roadside (upper curve) and at a nearby rooftop (lower curve). (Based on material from Kemp et al., 1998 and related reports.)

It is now observed in most parts of the world, but with distinctly different patterns. In the south of Europe cities like Athens and Rome may experience a "summer smog" of the Los Angeles type, but in many cases it is a large scale phenomenon (Guicherit and van Dop, 1977). In cities in the northern part of Europe the predominant reaction is a reduction of ozone by nitric oxide in car exhaust to form oxygen and nitrogen dioxide. As a result ozone levels are generally lower at ground level in the streets than at roof level or in the surrounding countryside. Urban ozone levels are further higher during weekends with low traffic and may be practically nil during some pollution episodes (Fig. 5). Note also in Fig. 5 that the concentration of NO follows the traffic with rush hours and weekends much more closely than NO_2, the concentration of which is largely determined by the available O_3, supplied from outside the city.

These mechanisms often lead to formation of elevated ozone concentrations downstream from the city (city or urban plume). As a similar phenomenon on a larger-scale elevated ozone concentration in Central Europe due to extended high pressure events can, via long-range transport, also be detected in Northern Europe, where concentrations normally built up over several days – often in parallel with rising temperatures.

Since ozone is a secondary pollutant it can only be regulated via the primary pollutants. Long-range models can demonstrate the effects of ozone levels of changes in the emissions, and they indicate that a concerted international effort is necessary. In computer experiments it has thus been shown that in the hypothetical situation, where all Danish emissions were reduced to zero, the average ozone levels in Denmark would go slightly *up* (Zlatev et al., 1996).

Generally reductions in emissions of hydrocarbons are more effective than reductions of nitrogen oxides, but even with simultaneous reductions of 95% the European vegetation will not be fully protected against damages.

4.3. Formation of particles

Usually particles are grouped in the three so-called modes: ultrafine, fine and coarse (Fig. 6). The ultrafine particles are chemically formed or condensed from hot vapour e.g. from diesel exhaust and coagulate into fine particles (Whitby and Sverdrup, 1980). Defined as having an aerodynamic diameter less than 2.5 μm (UNEP/WHO, 1994), the ultrafine and fine particles, which are predominantly of antropogenic origin are deposited with high probability in the lower parts of the human respiratory system and thus have the largest impact. Coarse particles, on the other hand, are often of natural origin (dust, seaspray, pollen or even insects). Their health impacts are modest, both because they are deposited in the upper airways, and because they may be less toxic. Still, in determinations of total suspended particulate matter (TSP) the coarse particles dominate with their high weight.

In the atmosphere the actual size spectra show quantitative differences with e.g. more pronounced mass peaks for fine particles in urban and suburban sites and larger peaks for coarse particles near sea coasts.

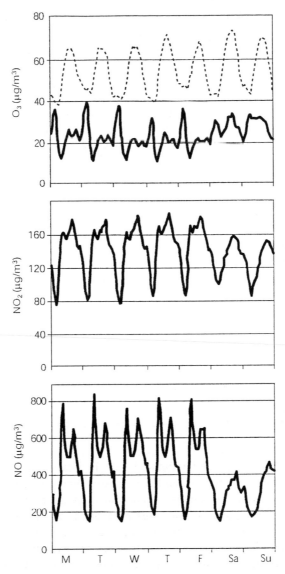

Figure 5. Average weekly variations for ozone and nitrogen oxides at H.C. Andersens Boulevard in Copenhagen with a daily traffic of 60,000 cars. For comparison is shown (dotted) a typical O_3-variation at a rural site. (Based on HLU, 1994.)

4.4. Exposure

Human health is the main concern in the regulation of urban air quality, and it is therefore an important question to which extent people are actually exposed

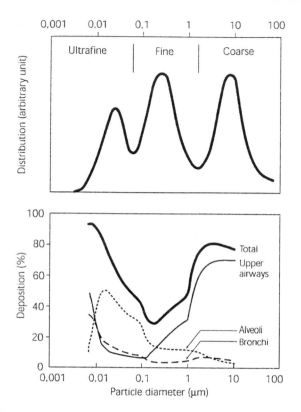

Figure 6. Schematic size distribution of particulate matter in the atmosphere and the corresponding deposition in the human respiratory system. (Based partly on Whitby and Sverdrup, 1980.)

to the measured or calculated pollution levels. In most cases so far the pollution exposure of the population has been assessed through crude assumptions, e.g. that levels observed at a single or a few street monitoring stations are representative for the exposure of the entire population in an urban area. Results of such an evaluation (WHO, 1995) are summarised in Section 6 (Table 2).

Direct monitoring (Fig. 7) demonstrates that the levels a person is exposed to vary drastically during the day, and that the ambient, outdoor air quality, regulated by various limit values do not describe the overall load, even in industrialised countries. Similar plots for, e.g., particles give slightly different patterns, but lead to the same general conclusion.

Still however, a realistic evaluation of population exposures requires more statistical information, where the time and activity pattern of the entire population must be taken into account. In reasonable agreement with Fig. 7 earlier Danish studies (Andersen, 1988) indicate that on the average the population

Table 2. Population exposure to various pollutants in Europe before and after 1985. The table shows how large a fraction (%) of a registered city population is exposed to (living in) concentrations above WHO guidelines for yearly averages (SO_2 and black smoke) or to the lowest level at which long-term exposure is reported to cause respiratory effects (SPM and NO_2); unit: $\mu g\,m^{-3}$ (Based on WHO, 1995)

Pollutant	"Limit value"	Western countries	CCEE countries	USSR	Total Europe
SO_2	50	58.2/14.3	37.3/11.1	77.9/48.1	63.1/20.5
Black smoke	50	25.5/19.1	65.7/58.2	100/25.5	33.4/22.9
SPM	60	91.6/51.8	—/100	—/100	—/61.0
NO_2	60	20.3/23.9	38.5/24.1	19.2/3.7	20.8/19.1

spend only 1 h per day outdoors, 1 h commuting and the remaining 22 h indoors. Ongoing Danish studies (Jensen, 1998) apply a geographic information system (GIS) to combine air pollution data calculated with the OSPM model (Berkowicz, 1998) with population data available from administrative databases. In view of the large fraction of time spend indoors in cities with a cold climate, also the relations between outdoor pollution and related indoor levels (typically about the half), must be taken into account. And then nothing has been said about active and passive smoking, which have large regional and cultural variabilities, but in general appear to have health impacts an order of magnitude above general air pollution!

In the so-called "EXPOLIS" study (Jantunen et al., 1998) personal exposure is determined by monitors and diary records in six European cities (Helsinki, Bilthoven, Prague, Basel, Grenoble and Milan), but detailed results are not yet available.

4.5. Impacts

Urban air pollution has a series of impacts on materials, vegetation (including urban agriculture) and visibility. These impacts depend of course on the relevant levels, but also on other factors, for material damage (Tidblad and Kucera, 1998) thus on temperature and humidity and the possibility of interaction between different components. The main concern, however – and the one which by and large determines the abatement strategies – is its impact on human health and well-being, although the health of urban populations is determined by many factors (Phillips, 1993), which may blur the picture.

Various types of health impacts of the major air pollutants are well established (WHO, 1987/1999), but a series of, notably organic, compounds are not sufficiently investigated. In recent years numerous epidemiological studies of short- and long-term effects of air pollution have shown that fine particles at the present levels are responsible for significant impacts, especially on people

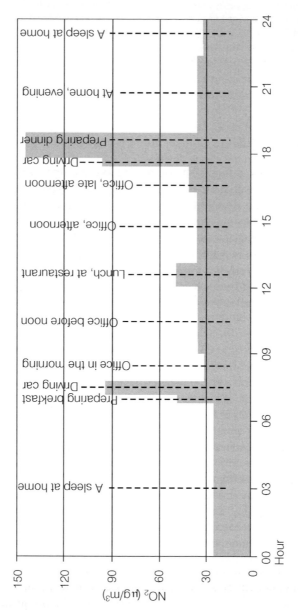

Figure 7. An example of personal exposure with NO_2 during a working day in The United States. Note the peak during preparation of dinner; it of course depends on the means of heating (Sexton and Ryan, 1988).

suffering from respiratory and cardio-pulmonary diseases (Pope et al., 1995). Thus Schwartz et al. (1997) have reported that a $10\,\mu g\,m^{-3}$ increase in $PM_{2.5}$ 2 day mean was associated with a 1.5% increase in total daily mortality, although the actual mechanisms are still not known. For coarse particles (above $10\,\mu m$) no impacts were observed.

4.6. City planning

The impacts of urban air pollution can be mitigated by constructive city planning. The complete separation of industry and habitation, originally envisaged as an environmental improvement and a reasonable solution in a society with heavily polluting industries, is now outdated and only leads to increased commuting traffic and congestion. Attempts to reduce urban driving by various types of economic incentives (green taxes, road pricing, km taxes), parking restrictions and pedestrian streets have had some success, but has often been opposed by trade. It must also be considered that driving restrictions *in* cities may increase recent years growth of big shopping centres, hotels and office buildings *outside* the cities, where they can offer free parking space and other facilities as e.g. child minding, thus often resulting in an increase in total traffic (OECD, 1995b).

The goal now is integrated land use, which minimises transport and thus total urban emissions. Open spaces and parks can here be used to improve the environmental quality especially in residential areas (Fig. 8).

In existing cities the possibilities of restructuring are limited, but the construction of ring-roads, which lead part of the traffic round the city-centre is one of the options (OECD, 1995).

In the industrialised world few cities and urban areas can be constructed from scratch, but when possible new concepts of integrating urban planning, building design and supply of renewable energy should be applied. Also the climate of the city is important, and the influence of buildings and street canyons on solar radiation, shade and wind pattern should be taken into account (Bitan, 1992).

In this planning, which to a large extent is planning of traffic, it must be realised that air pollution is not its only environmental impact, and probably not even its most important. It is estimated (European Commission, 1995) that on the average the external costs of air pollution (not including greenhouse effect) from transport in the EU amount to 0.4% of the GNP, compared to 0.2% from noise, 1.5% from accidents and 2.0% from congestion.

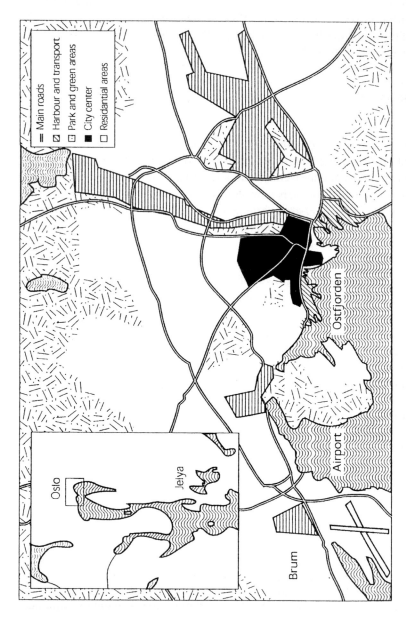

Figure 8. Oslo, the capital of Norway. The map shows a city plan aimed at mixing polluting activities and green areas to improve environment (Grønskei, 1998).

5. National and international legislation

Past experiences have with depressing clarity shown that existing technical possibilities and recommended management practices will not be used unless legally or economically enforced. Air quality is controlled by limit values. Their scientific foundation is experiments on humans or animals and epidemiological investigations. The results are evaluated by the World Health Organization (WHO) and expressed in the form of guidelines (WHO, 1987/1999), which are subsequently used as a main part of the basis for legally binding limit values.

Most countries have established such limit values for the major air pollutants and they use in addition guideline values for a series of other compounds. Most important are the legislation in the European Community and in The United States, which in many cases have served as models for other regions. Also US emission standards, especially for motor vehicles (Faiz et al., 1996) have been used in this way.

5.1. Air quality in the European Union

In the European Union the setting of limit values is a multistep process (Edwards, 1998) with a system of EC directives, the first being adapted in 1980. Since 1996 a framework directive provides a basic structure, and daughter directives, lay down limit values and proscribe dates for attainment, methods for measurements, etc., which are mandatory throughout the territory. These directives will be ratified in the individual memberstates in the form of national legislation. The first directive comprises sulphur dioxide, particulate matter, nitrogen dioxide and lead. Threshold values for ozone for information and warning to the public are also regulated by EU-directives. Other pollutants for which daughter directives are planned include: Benzene, carbon monoxide, polyaromatic hydrocarbons, cadmium, arsenic, nickel and mercury (Edwards, 1998).

5.2. Vehicle emissions

The most direct mean of regulating air quality is of course through regulation of emissions. The EC legislation on vehicle emissions and fuel quality standards has evolved greatly since the first directive in 1970. The early legislation had the dual purpose of reducing pollution and avoiding barriers to trade due to different standards in different member states. It is now giving place for designs aimed at meeting air quality targets. Thus the "Auto Oil I Programme" carried out by the European Commission in conjunction with industry (EU Commission, 1996a) has set up targets for a series of traffic related pollutants and assessed different technologies and fuel quality standards. Thus the target for nitrogen dioxide was a full compliance with the new WHO guideline of $200\,\mu g\,m^{-3}$ as a maximum 1 h average.

By means of models with simplified chemical reactions, the impacts, compared to 1990, on the air quality in seven representative European cities (Athens, Cologne, London, Lyon, Madrid, Milan and The Hague) have been estimated (EC Commission, 1996b). Already agreed measures (i.a. introduction of 3-way catalytic converters) were expected to reduce pollution from vehicles by 40–50% in 2010 compared to 1990. The auto oil programme will increase this reduction to 70% – even for the expected higher traffic.

It appeared that the objectives for carbon monoxide ($10\,mg\,m^{-3}$, 1 h max.), benzene ($10\,\mu g\,m^{-3}$ annual mean) and particulate matter ($50\,\mu g\,m^{-3}$ 24 h average) would be met by year 2010, and also the NO_2 objective should be met in most of the Union in 2010. In some cities like Athens, however further action would be needed. A more stringent target value for benzene of $2.5\,\mu g\,m^{-3}$, preferred by several member states, would be exceeded in all investigated cities except The Hague.

In an ongoing second programme also nontechnical measures will be evaluated. Possibilities for local use in areas with high pollution levels include road pricing, traffic management and scrapping schemes. Further the resulting reductions in pollution levels will be calculated in more detail.

5.3. Stationary sources

EC legislation on industrial air pollution has always taken air quality limit values into account and required the operators to use "Best available technology" not entailing excessive costs. In a Council Directive from 1996 on integrated pollution prevention and control (IPPC) the purpose is to reduce pollution to the environment as a whole, avoiding transfer from one medium to the other (Edwards, 1998).

5.4. The United States Clean Air Act

In the United States the first federal air pollution legislation was enacted in 1955, and in 1970 the administration was transferred to the new US Environmental Protection Agency (EPA). The Clean Air Act was first passed by Congress in 1967 as the Air Quality Act and has later been followed up by amendments in 1990. Under the act the EPA set national air quality standards (NAAQS) for six pollutants: Sulphur dioxide, nitrogen dioxide, carbon monoxide, particulate matter and lead.

The standards are reviewed regularly, and new proposals were put forward in 1997. For ozone the new EPA standard allows no more than 0.08 ppm

(160 µg m^{-3}) as an 8 h average. For particles a separate standard will be set for PM$_{2.5}$ (Brown, 1997).

5.5. Long-range transport and urban air pollution

The United Nations Economic Commission for Europe (UN-ECE), comprising all European countries and Canada and the US has been an important forum for east–west discussions of air pollution. It was therefore also in the ECE that the so-called Geneva Convention on long range, transboundary air pollution was established and undersigned in 1979. In 1983 it had been ratified by a sufficient number of member states and went into force. A series of related protocols set targets for reductions of national emissions of sulphur dioxide, nitrogen dioxide and volatile organic compounds.

These protocols are all aimed at protecting natural systems, and a new multipollutant-multieffect protocol will comprehensively address acidification, eutrofication and photochemical air pollution. Since however, the main part of the relevant emissions take place in urban areas, the necessary reductions have a direct impact on urban air quality. Thus the later decreases in levels of sulphur dioxide in the 1980s are related to the use of natural gas, low sulphur fuel oil and desulphurisation (or alternative sources of energy), necessary to comply with the international agreements. Also attempts to reduce ecological impacts of large scale photochemical air pollution will directly influence urban ozone levels – especially in the North of Europe.

In other parts of the world, notably in East Asia, unregulated local air pollution emissions have increasing regional and transboundary impacts.

6. Europe

6.1. The characteristics of Europe

Europe is a highly urbanised continent with more than 70% of the population living in cities. The population changes in many countries however, are modest and partly due to refugees and migration from east to west. The most extensive comprehensive treatment of Europe's environment up to 1992 was given in the so-called Dobris Assessment (Stanners and Bourdeau, 1995) from the European Environment Agency. It is now updated in a second assessment (EEA, 1998). The general air quality in European cities has improved in recent decades – often in spite of an increase in population density and standard of living – but air pollution is still considered a top priority environmental problem with both urban and large scale impacts. Its special aspects are treated in more detail in several reports from the agency (i.a. Jol and Kielland, 1997; Richter and Williams, 1998).

The European cities differ in various ways, which influence the relations between emissions and resulting pollution levels. Thus infrastructure and town planning determine the emission pattern, and meteorology and topography determine the possibilities of dispersion and transformation (Grønskei, 1998). *Western Europe* is influenced by the predominant westerly wind bringing moist air from the sea, a climate which also favours long-range transport. In the *northern part of Europe* the small amount of sunlight favours persistent inversions with poor dispersion conditions. In *Central and Eastern Europe* high pressures with air stagnation and accumulation of local pollution are frequent. During the summer the climate in the *Mediterranean Region* likewise favours accumulation of local emissions, whereas during the winter large-scale wind systems are more frequent. Formation of photochemical oxidants depends upon sunlight, which in combination with poor dispersion conditions result in frequent episodes during summer.

6.2. The political and economic development

The large resources of coal in England, Central Europe and Eastern Europe have been the primary source of energy during the industrial revolution, but in recent decades oil and gas have been found in the North Sea and in Russia and has been transported through pipelines to centralised regions. In western urban areas the consumption of coal in small units, e.g. used for domestic heating, has therefore been substituted with less polluting fuels resulting in reduced emissions of sulphur dioxide and particles.

In Eastern Europe however, solid fuels are still used in private houses and industries. Many buildings are badly insulated with large potential savings, but combined heat and power production is limited partly due to problems with financing and management. As a consequence in 1990 the energy intensity of economies in Central and Eastern Europe (CEE) was about 3 times higher than in the Western Europe; per unit of GDP emissions of NO_x and SO_2 were more than 4 times higher; and emissions of particles and volatile organic compounds (VOC) considerably higher (Bollen et al., 1996; UNEP, 1997). SO_2-emissions were highest in the northern part of the region (Poland), which depends heavily on indigenous coal and lignite (Adamson et al., 1996), and lower in the southern CEE, where oil and gas are available.

In general it is fair to say that industrial "hot spots" have shifted from Western Europe to the east and southeast, where heavy industry, use of low-quality fuels and outdated production technologies have resulted in high emission levels.

When, nevertheless, emissions have been reduced, it is partly a result of the German reunification in 1990 and the collapse of the Soviet Union in 1991. Already in 1990/1991 distribution problems, ethnic conflicts and mafia-crime in combination with an outdated technology resulted in a drop in the Soviet Russian GDP of 10–15%. And in the period 1990–1996 NO_x-emissions dropped with $1/3$ and SO_2 with $1/4$.

A typical Polish example is Katowice (GEMS/AIR, 1996), where since the late 1980s many older more polluting industrial plants have been closed down. In the period 1988–1992 this has – in combination with introduction of better emission controls – resulted in a fall in emissions of nitrogen oxides and particulate matter of 22 and 41%, respectively. In the same period the air quality was improved with reductions in overall long-term averages of 30 and 44% for SO_2 and NO_x, respectively.

Further substantial reductions in emissions are possible, even when only current Western European practices are applied (Bollen et al., 1996), but it is a serious problem for Central and Eastern Europe to raise funds for economic and technological growth in connection with a transition to market economy. For some cities e.g. Krakow the most effective strategy to improve air quality would be a ban on use of coal – possibly limited to the town centre (Adamson et al., 1996).

The southeastern urban areas have a partly outdated car fleet and are decades behind in the organisation of road traffic. Only recently have efforts been put into the construction of ring highways (i.a. in Budapest and Prague) to reduce unnecessary crossings of the city centres.

6.3. Total emissions

In Europe the total emission of sulphur dioxide steadily increased from about 5 million metric ton in 1880 to a maximum of nearly 60 million in the 1970s only interrupted by the Second World War. It peaked in the mid-1970s, but has now been reduced to less than half. For the traffic-related pollutants nitrogen oxides, carbon monoxide and volatile organic compounds an increase has only recently been reversed (10–15%) by the introduction of three way catalytic converters (TWC)(EMEP/MSC-W, 1998 and related reports).

European road traffic currently accounts for the main part of the total CO-emissions, more than half of the NO_x-emissions and a third of the emissions of volatile organic compounds. The emission of sulphur dioxide from traffic is of minor importance; in western and northern cities it accounts for only about 5%; in southern cities, where diesel oil have a higher sulphur content, it accounts for 14%.

It is estimated that in 2010 the pollution from traffic with CO, VOC and NO_x within the EU-region will be reduced drastically in spite of the expected growth

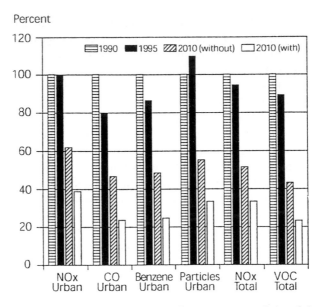

Figure 9. Expected impacts of the European auto oil programme on emission of air pollutants in the European Union and specifically in urban areas. The diagram shows calculated emissions in 1990, 1995 and projected emissions in 2010 without and with the programme in force. (Based on EC Commission, 1996a.)

in traffic (Fig. 9). On the other hand, total emissions within the European area covered by the European Monitoring and Evaluation Programme, comprising a series of eastern countries (EMEP/MSC-W, 1998) are not expected to be reduced much in 2010 compared to 1980: VOC and NO_x only about 10% and CO about 18%. In comparison SO_2 will be reduced by 53%.

6.4. Smoke

The most visible air pollutant is black smoke, which acts as a precursor for fogs (an important ingredient in early English detective novels). The drastic improvements in urban air quality in the 1960–1970s, partly brought about by a change from coal to less polluting fuels for domestic heating partly by the closing down of polluting industry, has also resulted in a marked reduction in incidence of fogs, shown in Fig. 10 for Lincoln in UK.

Most other West European cities show similar downward trends in particulate matter, but the lack of chemical analysis and size fractionation preclude more than a qualitative evaluation of the health impacts.

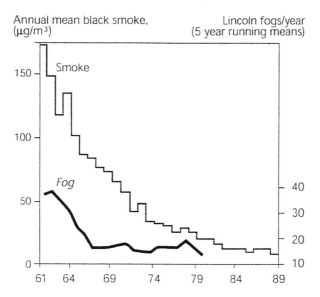

Figure 10. The reduction of annual mean black-smoke concentrations and the number of fogs per year (5 yr running means) in Lincoln, UK (Eggleston et al., 1992).

6.5. Sulphur

In many West European cities also the severe pollution with sulphur dioxide is a thing of the past. Thus in Copenhagen the yearly average of the SO_2 concentration was around 1970 about $80 \, \mu g \, m^{-3}$ and during the winter about $120 \, \mu g \, m^{-3}$. Today it is reduced to a fraction, well below the WHO guideline of $50 \, \mu g \, m^{-3}$. In provincial towns it is even lower (Fig. 11).

It appears though that the reduction is most evident for medium and average values, which are determined by the contributions from local sources, whereas there still are significant peak values (95- and 98-percentiles), identified as being due to long-range transport from Central and Eastern Europe. Consistent with this the reduction of the levels of particulate sulphur (sulphate) show only a slight downward trend up to 1996. In 1997 levels for all types of sulphur pollution were very low, but so far it has not been established, whether it has been due to reduced emissions or special meteorological conditions.

This development has had several causes. An important aspect is the Geneva Convention on transboundary air pollution, which resulted in reduction of total emissions, but also a widespread transition from individual to district space heating produced on large units with high stacks (often as combined heat and power production) has played a role. A further factor is that the oil crisis in the beginning of the 1970s resulted in better insulation of buildings. Thus in

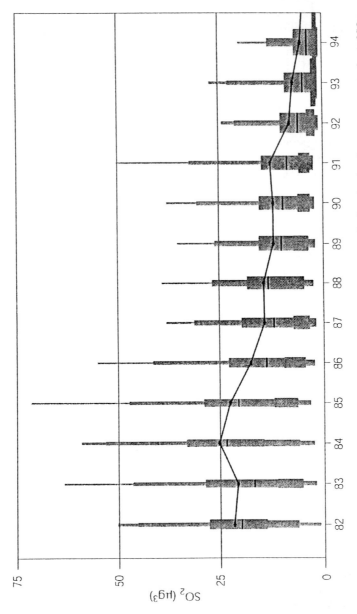

Figure 11. Trends for annual 98-, 95-, 75-, 50-, 25- and 5-percentiles, average and minimum value based on hourly average concentrations of SO_2 measured in Aalborg, a major provincial town in Denmark. The widths of the bars indicate the number of observations of a given concentration (Kemp et al., 1998).

Denmark the consumption of energy used for space heating was reduced by about 30% in the period 1972–1982 – in spite of a simultaneous increase in area of about 20% (ENS, 1997).

In 1990, 10 European cities observed exceedances of the long-term WHO-AQG for SO_2 of $50\,\mu g\,m^{-3}$; in 1995 it was only in Katowice and Istanbul. The short-term guideline of $125\,\mu g\,m^{-3}$ daily average is, however, still exceeded for a few days per year in many countries.

6.6. Nitrogen oxides

In recent years urban air pollution with nitrogen dioxide has shown a downward trend in most European cities, although the short-term WHO guideline (corresponding to $200\,\mu g\,m^{-3}$ as maximum hourly value) is exceeded in cities (EEA, 1998).

The situation is complicated by chemical reactions in the urban atmosphere. Thus since the introduction of catalytic converters on all Danish petrol cars registered after 1990 there has been a marked reduction in the levels of NO in Danish cities. The levels of NO_2, however are reduced much less, the median in Copenhagen about 20%, reflecting that the emissions of NO is not the limiting factor, but rather the available O_3 (Kemp et al., 1998).

6.7. Carbon monoxide

Urban concentrations of carbon monoxide have likewise been reduced since 1990, although exceedances of the 8-h WHO guideline have been reported from many cities (EEA, 1998).

6.8. Lead

In countries that have reduced or eliminated lead in petrol the lead levels have substantially declined (EEA, 1998), especially in countries with few or no lead-emitting industries. Since car exhaust is emitted at low height, and often in street canyons, there has been a close correspondence with the lead concentrations in urban air. In Denmark, where lead from petrol in 1977 accounted for 90% of the national lead emissions, the problem has virtually disappeared. Remaining low lead concentrations of about $20\,ng\,m^{-3}$ are essentially due to long-range transport (Fig. 12).

In the early 1980s, 5% of Europe's urban population in cities with reported lead levels were exposed to more than the WHO guideline of $0.5\,\mu g\,m^{-3}$. At the end of the decade levels above the guideline value were no longer reported from the western countries, but were still found in Eastern Europe, notably in Romania and Bulgaria near large uncontrolled metal industries (WHO, 1995).

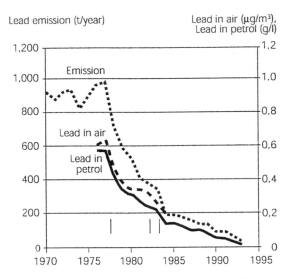

Figure 12. Annual average values for the total Danish lead emissions 1969–1993, the lead pollution in Copenhagen since 1976, and the average lead content in petrol sold in Denmark (Jensen and Fenger, 1994). The dates of tightening the restrictions on lead content are indicated with bars. Lead concentrations for the recent years can be found in Kemp et al. (1998). In 1997 they were below $20\,\text{ng m}^{-3}$.

6.9. Volatile organic compounds

Recently published (EEA, 1998) levels of benzene range from a few to more than $50\,\mu\text{g m}^{-3}$ with the highest values normally found near streets with high traffic. Of the reporting cities only Antwerp did not observe exceedances of the WHO-AQG corresponding to $2.5\,\mu\text{g m}^{-3}$ as yearly average.

In the 1960s the annual average concentrations of BaP was above $100\,\text{ng m}^{-3}$ in several European cities (WHO, 1987/1999). In most developed countries improved combustion technology, change of fuels and catalytic converters on motor vehicles have reduced the urban levels to $1–10\,\text{ng m}^{-3}$. Still, however, the urban air pollution with potentially carcinogenic species is not satisfactorily understood.

6.10. Ozone

European ozone levels appear to have increased from about $20\,\mu\text{g m}^{-3}$ around 1900 to now about the double with the most rapid rise between 1950 and 1970 concurrent with the rise in emissions of primary precursors (Volz and Kley, 1988). In the outskirts of Paris the early ozone levels were about $20\,\mu\text{g m}^{-3}$, but in the centre only $3–4\,\mu\text{g m}^{-3}$. Since at that time there were no nitric oxide from motor vehicles it is assumed, that O_3 was reduced by SO_2 or NH_3 (Anfossi and Sandroni, 1997).

Generally the present European ozone concentrations increase from northwest towards southeast, and summer smog with high ozone concentrations occurs in many European countries. As an urban phenomenon it is most serious in Athens and Barcelona with concentrations up to $400\,\mu g\,m^{-3}$, but also Frankfurt, Krakow, Milan, Prague and Stuttgart are affected.

An EU ozone directive contains a threshold value for information to the population of $180\,\mu g\,m^{-3}$ and for warning of $360\,\mu g\,m^{-3}$. In the Nordic countries the level for information is seldom exceeded – and the level for warning never.

Implementation of the VOC-protocol is expected to result in a 40–60% reduction in high peak values and 1–4% in annual average O_3-concentrations (EEA, 1995).

6.11. OECD trend indicators

The OECD (1998) uses average values for typical sites and major air pollutants as indicators for the development of air quality. In Western European cities in the period 1988–1993 the trends were (with approximate average values in 1993 in parentheses):

SO_2, annual average		
Traffic sites	$(18\,\mu g\,m^{-3})$	37% decrease
Residential areas	$(22\,\mu g\,m^{-3})$	22% decrease
NO_2, annual average		
Traffic sites	$(48\,\mu g\,m^{-3})$	10% decrease
Residential areas	$(35\,\mu g\,m^{-3})$	12% decrease
CO, annual max. 8 h average		
Traffic sites	$(8.3\,mg\,m^{-3})$	unchanged
Residential areas	$(5.7\,mg\,m^{-3})$	unchanged

6.12. Overview of population exposure in Europe

It is not only the pollution levels as such, which determine the potential health impacts, but also the extent to which people are exposed to them. Table 2 summarises the results of a WHO (1995) study, where a series of data for the period 1976 to 1990 from European urban areas with populations above 50,000 were pooled in two groups: up to and after 1985. Notwithstanding the limited and not fully representative data it appears that there has – taken as a whole – been a substantial general improvement for SO_2 and a somewhat smaller, and varying, improvement for particles. Concerning nitrogen dioxide there is an increase in the western countries and decreases in the eastern – probably reflecting an increasing traffic and an economic recession, respectively.

Also the number of peak values has decreased, thus the total population experiencing episodes exceeding $250\,\mu g\,m^{-3}$ SO_2 decreased during the 1980s from 71 to 33% in the western countries and from 74–51% in Russia (WHO, 1995). In spite of the general improvement of European air quality however, the WHO short-term air quality guidelines for SO_2 and TSP are often violated during winter-type smog, where the highest exceedances are observed in Central European cities (Fig. 13).

7. North America

The two countries in North America: The United States and Canada are among the wealthiest in the world both with respect to natural resources and production. This has formerly, especially in the US, lead to serious urban pollution.

7.1. United States

Many cities in the early industrialised United States were like in Europe characterised by heavy smoke and have subsequently gone through the typical development shown in Fig. 3. As an example (Davidson, 1979) Pittsburgh had pollution problems already in 1804, and they by and large increased until the first effective smoke control in the late 1940s after the city experienced severe pollution due to the Second World War steel production.

As a general measure of the development in the period 1970–1993 in air pollution the overall emissions of carbon monoxide, volatile organic compounds and particulate matter have been reduced by 24, 24 and 78%, respectively. A decline in emissions of nitrogen oxides from vehicles has been offset by increased electricity generation to a resulting increase of 14%, but since the emissions are generally not in urban areas, it has not prevented a reduction of NO_2 levels in cities. Lead as an air pollutant has virtually disappeared, but toxic chemicals are still a problem.

Los Angeles still has the largest ozone problem in the US. In 1990 the highest 1 h average was $660\,\mu g\,m^{-3}$. In the US as such ozone concentrations showed a significant decrease of 30% from 1988–1993 in urban residential areas both as an average and in the most polluted cities. With the new ozone standard of 0.08 ppm (Brown, 1997) many counties will find it hard to comply (Fig. 14).

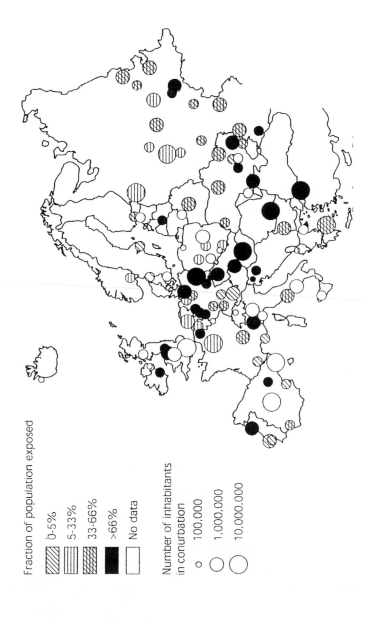

Figure 13. The present exposure of inhabitants in larger European conurbation's to exceedances of short-term WHO air quality guidelines for SO_2 and/or particulate matter (EEA, 1998).

Fraction of population exposed

0-5%
5-33%
33-66%
>66%
No data

Number of inhabitants in conurbation

○ 100,000
○ 1,000,000
○ 10,000,000

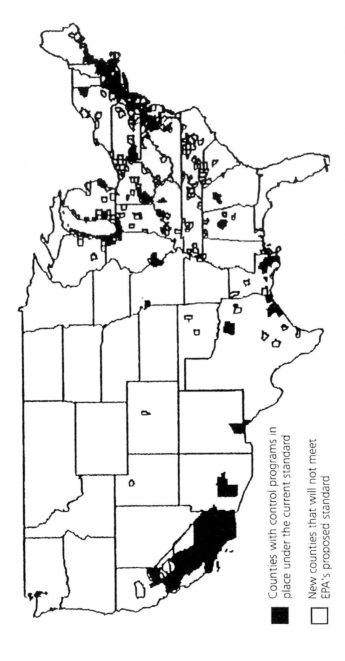

Figure 14. Counties not meeting EPA's ozone proposal standard, 8 h average 3rd maximum, 0.08 ppm. Based on 1993–1995 data (EPA Office of Air and Radiation World Wide Web site at http:ttmwww.rtpnc.epa.gov/naaqspro/).

7.2. OECD trend indicators for urban air quality in the US

According to the OECD (1998) the trends for air quality in cities in the US in the period 1988–1993 were (with approximate average values in 1993 in parentheses):

SO_2, annual average		
Traffic sites	$(20 \, \mu g \, m^{-3})$	23% decrease
Residential areas	$(20 \, \mu g \, m^{-3})$	23% decrease
NO_2, annual average		
Traffic sites	$(42 \, \mu g \, m^{-3})$	11% decrease
Residential areas	$(40 \, \mu g \, m^{-3})$	16% decrease
CO, annual max. 8 h average		
Traffic sites	$(6.8 \, mg \, m^{-3})$	22% decrease
Residential areas	$(6.4 \, mg \, m^{-3})$	25% decrease

7.3. Canada

In Canada (UNEP, 1997) emissions of sulphur dioxide and particulate matter have been reduced significantly since the early 1970s, and lead has virtually disappeared. Still however, some central Canadian cities experience unacceptable air quality with high levels of ozone and particulate matter, especially during the summer.

8. East Asia

This region contains three of the worlds largest countries (China, India and Indonesia), several minor landlocked states and a series of archipelagic states (including the highly industrialised Japan). The region is 35% urbanised and contains about half of the largest cities in the world. They therefore also represent all stages of pollution development. Urbanisation is not restricted to the continent and the major archipelagic states, but is also seen as in-migration to the main island on small island states as the Maldives. Urban congestion and air pollution is seen as a high-priority problem in many countries such as China, India, Pakistan, Indonesia, Philippines and Thailand (UNEP, 1997).

8.1. The general pollution with sulphur and particles

Taken as a whole the rapid growth of energy use, combined with extensive use of coal in most of Asia has resulted in a drastic increase in emissions of sulphur dioxide. Attempts to solve *local* problems by installing taller stacks only transformed them into an extensive *regional* pollution. The acidification phenomena, known from North America and Europe are now emerging. A network comprising 45 locations throughout Asia (Carmichael et al., 1995) has demonstrated annual average levels of SO_2 above $20\,\mu g\,m^{-3}$, especially in China, with the highest concentrations (about $60\,\mu g\,m^{-3}$) in the industrial area in Luchongguan, Guiyang.

In China coal is the main source of energy, in 1995 accounting for 76% and resulting in a sulphur dioxide emission of 21,000 kt. On the basis of projected energy consumption and available desulfurisation investment it is estimated (Wang and Wang, 1996) that the emission in 2020 will reach 31,780 kt or an 80% increase from the 1990 level. This will increase the emerging acidification in the southern part of China, and will undoubtedly also influence the urban air quality.

In the important agricultural and industrial region of the Jiangsu Province and Shanghai Municipality the SO_2 emissions are already high and are projected to double by the year 2010 (Chang et al., 1998). Model calculations demonstrate that in large regions the WHO-guideline for long-term exposure $(40–60\,\mu g\,m^{-3})$ is exceeded and in some regions even the 1 h guideline $(350\,\mu g\,m^{-3})$. Without drastic measures the short-term guideline will by 2010 be exceeded in large part of the province for more than 5% of the time. In line with this the new version of Atmospheric Pollution Control Law passed by the National People's Congress in 1995 calls for emission reductions on power plants and other large coal users based on a permit system and emission taxes (Chang et al., 1998).

Aerosol analysis 1987–1992 by a privately established network comprising five cities in China (Beijing, Chengdu, Bautou, Lanzhou, Urimuqi), one in Korea (Seoul) and one in Japan (Tokyo) (Hashimoto et al., 1994) demonstrated TSP-levels for Chinese cities up to averages above $500\,\mu g\,m^{-3}$ (Lanzhou) or seven times higher than in Tokyo to be compared with a tentative WHO-guideline of $120\,\mu g\,m^{-3}$ as a 1 h average. Lower values in Seoul (around $70\,\mu g\,m^{-3}$) showed a gradual increase.

Some of the highest reported levels of TSP, compared to what is seen in e.g. European cities, should however be evaluated critically, since not all is of human origin; thus high concentrations observed in Beijing in the December–April period are partly due to sand storms from the northern desert.

A further complication, with a different relation to abatement strategies, is forest fires, which are generally blamed for causing smog over major cities in the ASEAN region (Hassan et al., 1997). In Malaysia it has been confirmed that the high concentration of SPM during haze episodes are largely caused by transboundary impacts from intense biomass burning, not by build-up of local pollution (NERI, 1998).

8.2. Traffic

Urban transport is an increasing problem, which has been treated in a series of reports from The World Bank (Walsh, 1996; Walsh and Shah, 1997). A special aspect is that Asia is responsible for the main part (about 90%) of the worlds motorbike production, and China alone accounts for about 40% (Fig. 15).

Asia is also responsible for most of the motorbike use – partly because motorbikes constitute the cheapest mean of individual motorised transportation for the expanding working class, partly because many Asian cities are too crowded to allow a drastic expansion of the car fleet. In Beijing motorcycles accounted for 27% of the vehicle fleet in 1992 and in Guangzhou for 65%. This results in comparatively large emissions in relation to the fuel consumption, since most motorcycles have two stroke engines with poor pollution characteristics.

8.3. Examples of Asian cities

Taiwan is characterised by limited land and rapid economic growth, consequently the environmental stress is serious. In the early 1990s the monthly SO_2 averages in *Taipei City* were reported to be about 30 ppb ($80 \, \mu g \, m^{-3}$), but a gradual decrease is expected concurrent with a reduction of permissible sulphur contents in fuels. Thus the limit for diesel fuel is scheduled to be reduced to 0.05% in 1999. Also lead in petrol is being phased out (Fang and Chen, 1996).

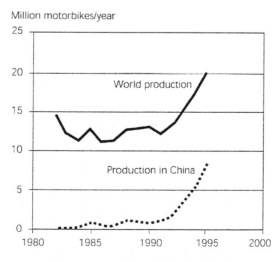

Figure 15. Annual motorbike production 1992–1995 in the World (upper curve) and in China (lower curve). (Based on material compiled by G. Gardner in Brown et al., 1998b.)

In *Beijing* the air quality is in a transition from coal burning caused problems to traffic exhaust related pollution (Zhang et al., 1997), and vehicle emissions are projected to double within the next two decades unless drastic strategies both to lower actual driving and emissions per km are employed (Fig. 16).

The SO_2 levels have been fairly constant around $100 \mu g\, m^{-3}$ as annual average since the early 1980s with a slight recent decrease, and TSP has been reduced from about 500 to $400 \mu g\, m^{-3}$, but the NO_x levels have increased drastically from about 70 to $170 \mu g\, m^{-3}$. Also general O_3 levels and frequency of exceedance of the $80 \mu g\, m^{-3}$ national standard have been increasing (Zhang et al., 1997) with maximum concentrations above $300 \mu g\, m^{-3}$ (Zhang and Xie, 1998). Similar values have been observed in the Guangzhou area (Zhang et al., 1998).

Shanghai is the commercial centre of China with an extensive industry. Data from the late 1980s (WHO/UNEP, 1992) demonstrate levels of both SO_2 and TSP well above WHO guidelines. Shanghai has the highest cancer mortality in China and the mortality for male lung cancer doubled from 1963 to 1985.

In *Ho Chi Minh City* in Vietnam the air quality has been deteriorating i.a. due to the increasing vehicular traffic. Although the fuel consumption per capita is still low, a series of factors as narrow overcrowded streets and outdated cars result in high pollution levels. Daily average values of NO_2 are between 50 and $250 \mu g\, m^{-3}$ with hourly peaks, which can exceed $700 \mu g\, m^{-3}$ (Duc, 1998). The average lead concentration measured on top of an eight-storey building in a residential area was found to about $180\, ng\, m^{-3}$ for the period 1993–1994 (Hien et

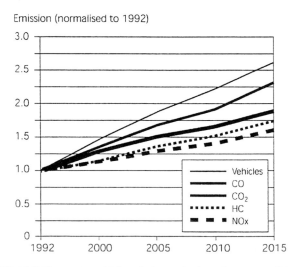

Figure 16. Official Beijing motor vehicle emission projections. Summer conditions. More likely projections result in about the double increase (Walsh, 1996).

al., 1997). Some contributions from open burning of refuse and firecracker discharges (sic!) can not be excluded, but lead from traffic is considered the main source. From 1995 the maximum permitted lead content in petrol is $0.4\,g\,l^{-1}$. Typical concentrations of TSP were $100\,\mu g\,m^{-3}$.

In a series of major cities, where energy production is based on gas or low sulphur coal (e.g. *Bombay, Calcutta, Bangkok*) SO_2 with average levels about $30\,\mu g\,m^{-3}$ is not a serious problem. In many cases, however TSP levels are above WHO guidelines (WHO/UNEP, 1992).

The capital of Japan, *Tokyo*, is an encouraging example of an industrial megacity, where air pollution is controlled. In the 1960s it was heavily polluted due to coal combustion and insufficient emission control. Concentrations peaked in the late 1960s with annual mean values for SO_2 and TSP up to 200 and $400\,\mu g\,m^{-3}$ respectively (Komeiji et al., 1990). By switching the major fuel consumption from coal to oil and installation of dust collectors the annual mean values of SO_2 and TSP were brought well below WHO guidelines already in the 1980s (WHO/UNEP, 1992), and as a spectacular example the reduction of particulate emissions in the 1970s (Kurashige and Miyashita, 1998) has resulted in a corresponding increase in visibility.

As in the western world stricter control of vehicle emissions has been counteracted by a growth in traffic and the levels of traffic related pollution have merely been stabilised. However the ozone levels have generally been halved since 1970, when the yearly average was about $80\,\mu g\,m^{-3}$.

8.4. OECD trend indicators for urban air quality in Japan

Japan is in many respects not typical for East Asia, but more like the western countries. According to the OECD (1998) the trends for air quality in Japanese cities in the period 1988–1993 were (with approximate average values in 1993 in parentheses):

SO_2, annual average		
Traffic sites	$(18\,\mu g\,m^{-3})$	20% decrease
Residential areas	$(22\,\mu g\,m^{-3})$	17% decrease
NO_2, annual average		
Traffic sites	$(50\,\mu g\,m^{-3})$	unchanged
Residential areas	$(35\,\mu g\,m^{-3})$	unchanged
CO, annual max. 8 h average		
Traffic sites	$(5.5\,mg\,m^{-3})$	26% decrease
Residential areas	$(4.3\,mg\,m^{-3})$	25% decrease

In urban residential areas the average concentrations of oxidants increased by about 5% from 1988 to 1993, while the 95-percentile decreased.

9. Other developing regions

9.1. Latin America

As demonstrated in a series of case studies (Onursal and Gautam, 1997) Latin America and the Caribbean is the most urbanised region in the developing world with a rapidly increasing vehicle fleet, which is the dominant source of air pollution. In Mexico City e.g. it accounts for 99% of the CO, 54% of the hydrocarbons and 70% of NO_x. Leaded gasoline is still permitted in most countries and even in 1995 constituted the entire sale in Venezuela. In other countries it is being phased out; thus in Mexico City the lead concentrations decreased 80% in the period 1990–1994.

Most of the air pollution occurs in major urban centres. In 1994, 43 of them had more than 1 million inhabitants. Often the situation is aggravated by the cities (e.g. Mexico City and Santiago de Chile) being situated in valleys surrounded by mountains.

Not surprisingly, the most critical air pollutants in Mexico City are ozone, NO_2, VOC and PM. The ambient ozone concentrations have concistently exceeded the Mexican 1 h standard of $220\,\mu g\,m^{-3}$. The highest value ever recorded (in 1992) was $955\,\mu g\,m^{-3}$.

In Santiago the most critical air pollutant is particulates, especially in the colder period (Apr.–Sep.). The principal source being a large number of poorly maintained diesel busses (GEMS/AIR, 1996). The concentration of TSP is among the highest in any urban area in the world. In 1995 it reached a 1 h mean of $621\,\mu g\,m^{-3}$ to be compared with the Chilean standard of $260\,\mu g\,m^{-3}$.

Sao Paulo in Brazil is the third of the three most polluted cities in the study. There has been some success in attempts to control emissions from the rapidly growing industry, and the SO_2 concentrations have been reduced substantially during the 1980s (WHO/UNEP, 1992), but ambient air quality standards for all traffic relevant pollutants are exceeded.

9.2. West Asia

In the recent two decades West Asia has been radically transformed with an urban growth rate above 4%. Although the most pressing urban problem seems to be waste management, also air pollution is emerging (UNEP, 1997). In many cases protective trade regimes and lack of environmental regulations have prevented adequate substitution of outdated polluting industries. Fuels with high sulphur content and old inefficient cars using leaded petrol have exacerbated urban air pollution. Recently however, the situation has been improving, thus in Oman new industries will be regulated with environmental standards.

9.3. Africa

For the larger African cities Cairo, Alexandria, Nairobi and Johannesburg air pollution has been monitored for some years and there is an increasing awareness of the need for air quality management (WHO/UNEP, 1992; GEMS/AIR, 1996).

Urbanisation is however increasing rapidly all over Africa and especially in the least developed countries with up to 5% per year, the driving force being a mixture of population growth, natural disasters, and armed ethnic conflicts. Most African cities have been unable to keep pace with this development and lack adequate industrial and vehicle pollution control.

Much of the urban population growth is in coastal cities i.a. in the Mediterranean area. So far the general air pollution appears to be modest, but urban problems are emerging. In most countries however, emission inventories are nearly non-existent, pollution is neither monitored nor controlled, and there are no long-term records of pollution levels and impacts (UNEP, 1997).

10. Conclusion

Urban air pollution and its impact on urban air quality is a world-wide problem. It manifest itself differently in different regions depending upon the economical, political and technological development, upon the climate and topography, and last – but not least – upon the nature and quality of the available energy sources. Nevertheless a series of general characteristics emerges.

10.1. From space heating to traffic

Originally urban air pollution was a strictly local problem mainly connected with space-heating and early industry – and not far from being considered unavoidable or even a symbol of growth and prosperity. The situation in the industrialised western world has in some respects proved this viewpoint outdated. Emissions from industry and space heating are by and large controllable, but the urban atmosphere is now in most cities dominated by traffic emissions with documented impacts on human health. Thereby the attention has been shifted from sulphur dioxide and soot to nitrogen oxides, the whole spectrum of organic compounds and particles of various sizes and composition, which are reported to be carcinogenic and/or cause a significant reduction of life through respiratory and cardiovascular diseases. These pollutants require much more detailed investigations both in the form of chemical analysis and computer modelling.

In principle the control of sulphur emission is relatively straightforward, when it is related to power production on large plants, which can be compelled to use clean fuels and equipped with proper cleaning technology. Traffic emissions are more difficult to control, since they, in the nature of things, arise from small units. Meeting the increasing stringent air pollution targets is therefore not an easy task. According to the conclusion of the "auto oil programme" even with the maximum technical package introduced in the EU not all cities will be able to comply.

The situation in developing countries is mixed. In some major cities in Asia the sulphur emissions have been brought under control, i.a. via a transition from coal to natural gas, but especially in China a rapid growth in energy production based on coal has resulted in increasing sulphur pollution both on an urban and a regional scale. Also in the developing countries, however and in some economies in transition (including Eastern Europe) traffic is becoming *the* problem. The expectation of private cars in these countries is understandable, but regrettable. This will in the future be a challenge in town planning.

As a demonstration that the traffic-related nitrogen dioxide in many cities exceed the WHO air quality guideline is shown (Fig. 17) a bar chart for peak statistics comprising more than 60 cities all over the world. In fact very few never experience exceedenses.

10.2. Regional impacts on urban air pollution

The interactions between the cities and their surroundings are becoming increasingly important. With the expanding and often merging urban areas and the diminishing sulphur emissions in the cities proper, the pollution levels can to a large extent be determined by long-range transport. The same applies to lead pollution in countries, where lead has been removed from petrol. Another example is photochemical air pollution, which in many cases is a large-scale phenomenon, where emissions and atmospheric chemical reactions in one country may influence urban air quality in another.

10.3. The cities as sources of pollution

A more far reaching problem is the city as a source of pollution. In the past, local problems were attacked by dispersing pollutants from high stacks, but this only resulted in a transfer to a larger geographical scale in the form of acidification and other transboundary phenomena. Now long-lived greenhouse gases, and especially carbon dioxide, threatens the global climate – irrespective

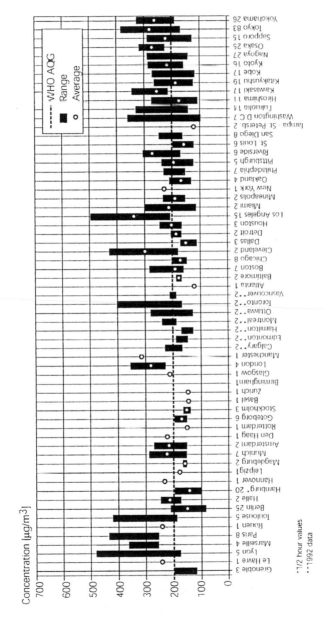

Figure 17. Urban peak statistics for nitrogen dioxide with averages and ranges of 1-h values in 1993. The numbers after the city name indicate the number of monitoring stations (Wiederkehr and Yoon, 1998).

of the origin. This problem can only be solved by a general reduction in the net emissions (Brown et al., 1998a).

In the industrialised countries the development in technology and legislation to protect air and water quality in many cases resulted in improved energy efficiency and emission reductions, although some means of improving urban air quality, such as catalytic converters, which consume energy and emit nitrous oxide (another greenhouse gas) are contrary to this goal. On a global basis, the growing population and its demand for higher material standard of living have so far counteracted any reductions in total emissions. Therefore the responsibility for the future is both national and global with many actors comprising national environmental agencies, international organisations, The World Bank, etc.

Long-term abatement however, has been intensified in recent years. An example is the ICLEI initiative with the purpose to achieve and monitor improvements in global environmental conditions through cumulative local actions.

10.4. A comprehensive approach

The realisation that traffic is rapidly becoming *the* urban air quality problem both in industrialised and developing countries calls for comprehensive solutions, where traffic-related air pollution is seen in connection with other impacts of traffic as noise, accidents, congestion and general mental stress. As a consequence technological improvements in the form of less polluting vehicles are not sufficient. Also support of infrastructures, where the need of transport is minimised, and where use of public means of transport dominate over individual private cars and motorbikes should be encouraged. Unfortunately most attempts of control will be perceived as a restriction of the individual freedom, and they are frequently met with outspoken opposition. Obviously a change in attitude is called for.

11. Note to references

The literature relevant to urban air quality is vast, and the references in this paper are only given as typical examples. The present review however, contains material from a recently published textbook (Fenger, Hertel and Palmgren, 1998), where further references on specific subjects can be found. Some of the presented data have been used by different authors and are compiled from various sources. Not all references are therefore primary, but they are a key to a full documentation.

Acknowledgements

The author thanks the colleges, who have provided material and offered advice during the preparation of this article – and especially my co-editors on our book "Urban air pollution – European aspects" (Ole Hertel and Finn Palmgren), the librarians at the our institute (Tove Juul Hansen and Kit Andersen) and the lithographic artists (Britta Munter and Beatrix Rauch).

References

Adamson, S., Bates, R., Laslett, R., Pototschnig, A., 1996. Energy use, air pollution, and environmental policy in Krakow. Can economic incentives really help? World Bank Technical Paper no. 308. The World Bank, Washington, DC.

Anfossi, D., Sandroni, S., 1997. Ozone levels in Paris one century ago. Atmospheric Environment 31, 3481–3482.

Andersen, D., 1988. The everyday life of the Danish population 1987 (In Danish). Institute of Social Science, Copenhagen.

Berkowicz, R., 1998. Street scale models. In: Fenger, J., Hertel, O., Palmgren, F. (Eds.), Urban Air Pollution, European Aspects. Kluwer Academic Publishers, Dordrecht, pp. 223–251.

Bitan, A., 1992. The high climatic quality city of the future. Atmospheric Environment 26B, 313–329.

Bollen, J.C., Hettelingh, J.-P., Maas, R.J.M., 1996. Scenarios for economy and environment in Central and Eastern Europe. RIVM Report no. 481505002. Netherlands Institute of Public Health and the Environment.

Brimblecombe, P., 1987. The Big Smoke. A History of Air Pollution in London Since Medieval Times. Methuen, London.

Brimblecombe, P., 1998. History of urban air pollution. In: Fenger, J., Hertel, O., Palmgren, F. (Eds.), Urban Air Pollution, European Aspects. Kluwer Academic Publishers, Dordrecht, pp. 7–20.

Brown, K.S., 1997. A decent proposal? EPA's new clean air standards. Environmental Health Perspectives 105, 378–383.

Brown, L.R. et al., 1998a. State of the World. A Worldwatch Institute Report on Progress Toward a Sustainable Society. W.W. Norton and Company, New York.

Brown, L.R., Renner, M., Flavin, C., 1998b. Vital signs 1998. The Environmental Trends that are Shaping Our Future. World Watch Institute, W.W. Norton and Company, New York.

Carmichael, G.R., Ferm, M., Adikary, S., Ahmad, J., Mohan, M., Hong, M.S., Chen, L., Fook, L., Liu, C.M., Soedomo, M., Tran, G., Suksomsank, K., Zhao, D., Arndt, R., Chen, L.L., 1995. Observed regional distribution of sulfur dioxide in Asia. Water, Air and Soil Pollution 85, 2289–2294.

Chang, Y.S., Arndt, R.L., Calori, G., Carmichael, G.R., Streets, D.G., Su, H.P., 1998. Air quality impacts as a result of changes in energy use in China's Jiangsu province. Atmospheric Environment 32, 1383–1395.

CORINAIR, 1996. Atmospheric Emission Inventory Guidebook. European Environment Agency, Copenhagen.

Davidson, C.I., 1979. Air pollution in Pittsburgh: a historical perspective. Air Pollution Control Association 29, 1035–1041.

Dennis, R.L., Byon, W.D., Novak, J.H., Galuppi, K.J., Coats, C.J., 1996. The next generation of integrated air quality modelling: EPA's models–3. Atmospheric Environment 30, 1925–1938.

Duc, H.N., 1998. Some Aspects of Air Quality in Ho Chi Minh City, Vietnam. Environment Protection Authority NSW.

Edwards, L., 1998. Limit values. In: Fenger, J., Hertel, O., Palmgren, F. (Eds.), Urban Air Pollution, European Aspects. Kluwer Academic Publishers, Dordrecht, pp. 419–432.

EEA, 1998. Europe's Environment. The Second Assessment. European Environment Agency, Copenhagen.

Eggleston, S., Hackman, M.P., Heyes, C.A., Irwin, J.G., Timmis, R.J., Williams, M.L., 1992. Trends in urban air pollution in The United Kingdom during recent decades. Atmospheric Environment 26B, 227–239.

Ellis, J., Tréanton, K., 1998. Recent trends in energy-related CO_2 emissions. Energy Policy 26, 159–166.

EMEP/MSC-W, 1998. Transboundary acidifying air pollution in Europe, Norwegian Meteorological Institute, Oslo. Research Report no. 66.

ENS, 1997. Energy Statistics (in Danish). Danish Energy Agency, Copenhagen.

Enya, T., Suzuki, H., Watanabe, T., Hirayama, T., Hisamatsu, Y., 1997. 3-Nitrobenzanthone, a powerfull bacterial mutagen and suspected human carcinogen found in diesel exhaust and airborne particles. Environmental Science and Technology 31, 2772–2776.

European Commission, 1995. Towards fair and efficient pricing in transport. COM (95) 691.

EU Commission, 1996a. The European Auto Oil Programme. A Report by the Directorate Generals for Industry; Energy; and Environment. Civil Protection & Nuclear Safety of the European Commission.

EU Commission, 1996b. Air quality report of The European Auto Oil Programme. Report of Sub Group 2.

Faiz, A., Weaver, C.S., Walsh, M.P., 1996. Air pollution from motor vehicles. Standards and technologies for controlling emissions. The World Bank, Washington, DC.

Fang, S.-H., Chen, H.-W., 1996. Air quality and pollution control in Taiwan. Atmospheric Environment 30, 735–741.

Fenger, J., Hertel, O., Palmgren, F., 1998. Urban air Pollution, European Aspects. Kluwer Academic Publishers, Dordrect.

Finlayson-Pitts, B., Pitts, J.N., Jr., 1997. Tropospheric air pollution, ozone, airborne toxics, polycyclic aromatic hydrocarbons, and particles. Science 276, 1045–1052.

Friedrich, R., Schwarz, U.-B., 1998. Emission inventories. In: Fenger, J., Hertel, O., Palmgren, F. (Eds.), Urban Air Pollution, European Aspects. Kluwer Academic Publishers, Dordrecht, pp. 93–106.

GEMS/AIR, 1996. Air quality management and assessment in 20 major cities. United Nations Environment Programme and World Health Organization.

Grossman, G.M., Krueger, A.B., 1995. Economic growth and the environment. The Quarterly Journal of Economics CX, 353–377.

Grønskei, K.E., 1998. Europe and its Cities. In: Fenger, J., Hertel, O., Palmgren, F. (Eds.), Urban Air Pollution, European Aspects. Kluwer Academic Publishers, Dordrecht, pp. 21–32.

Guicherit, R., van Dop, H., 1977. Photochemical production of ozone in Western Europe (1971–1975) and its relation to meteorology. Atmospheric Environment 11, 145–155.

Hashimoto, Y., Sekine, Y., Kim, H.K., Chen, Z.L., Yang, Z.M., 1994. Atmospheric fingerprints of East Asia, 1986–1991. An urgent record of aerosol analysis by The Jack Network. Atmospheric Environment 28, 1437–1445.

Hassan, H.A., Taha, D., Dahalan, M.P., Mahmud, A. (Eds.), 1997. Transboundary pollution and the Sustainability of Tropical Forests. Asean Institute of Forest Management, Kuala Lumpur.

HLU, 1994. Air Quality in the Copenhagen Area (in Danish). Environmental Protection Agency Copenhagen, Copenhagen.

Hien, P.D., Binh, N.T., Ngo, N.T., Ha, V.T., Truong, Y., An, N.H., 1997. Monitoring lead in suspended air particulate matter in Ho Chi Minh City. Atmospheric Environment 31, 1073–1076.

ICLEI, 1998. Recent publications. The International Council for Local Environmental Initiatives. Http://www.iclei.org/iclei/icleipub.htm.

Jantunen, M.J., Hänniken, O., Katsouyanni, K., Knöppel, H., Kuenzli, N., Lebret, E., Maroni, M., Saarela, K., Srám, R., Zmirou, D., 1998. Air pollution exposure in European cities. The "*EX-POLIS*" study. Journal of Exposure Analysis and Environmental Epidemiology 8, 495–518.

Jensen, F.P., Fenger, J., 1994. The air quality in Danish urban areas. Environmental Health Perspectives 102 (Suppl. 4), 55–60.

Jensen, S.S., 1998. Mapping human exposure to traffic air pollution using GIS. Journal of Hazardous Materials 61, 385–392.

Jol, A., Kielland, G. (Eds.), 1997. Air Pollution in Europe 1997. EEA Environmental Monograph No. 4. European Environment Agency, Copenhagen.

Kallos, G., 1998. Regional/mesoscale models. In: Fenger, J., Hertel, O., Palmgren, F. (Eds.), Urban Air Pollution, European Aspects. Kluwer Academic Publishers, Dordrecht, pp. 177–196.

Kemp, K., Palmgren, F., Mancher, O.H., 1998. The Danish air quality monitoring programme. Annual Report for 1997. NERI Technical Report No. 245. National Environmental Research Institute, Roskilde.

Komeiji, T., Aoki, K., Koyama, I., 1990. Trends of air quality and atmospheric deposition in Tokyo. Atmospheric Environment 24A, 2099–2103.

Krupnick, A.J., Portney, P.R., 1991. Controlling urban air pollution: A benefit-cost assessment. Science 252, 522–528.

Kurashige, Y., Miyashita, A., 1998. How many days can Mt. Fuji and the Tokyo Tower be seen from the Tokyo suburban area. Journal of Air and Waste Management Association 48, 763–765.

Larsen, J.C., Larsen, P.B., 1998. Chemical carcinogens. In: Hester, R.E., Harrison, R.M. (Eds.), Air Pollution and Health. Issues in Environmental Science and Technology, Vol. 10. The Royal Society of Chemistry, Cambridge, pp. 33–56.

Larssen, S., Hagen, L.O., 1997. Air pollution monitoring in Europe – Problems and Trends. Topic report 26. 1996. European Environment Agency, Copenhagen.

Mage, D., Ozolins, G., Peterson, P., Webster, A., Orthofer, R., Vandeweerd, V., Gwynne, M., 1996. Urban air pollution in megacities of the World. Atmospheric Environment 30, 681–686.

NERI, 1998. The origin, formation and composition of aerosol haze in Malaysia. Unpublished project report. National Environmental Research Institute, Roskilde, Denmark.

OECD, 1995a. Motor vehicle pollution. Reduction strategies beyond 2010. Organisation for Economic Co-operation and Development, Paris.

OECD, 1995b. Urban travel and sustainable development. Organisation for Economic Co-operation and Development, Paris.

OECD, 1997. OECD Environmental Data. Compendium 1997. Organisation for Economic Co-operation and Development, Paris.

OECD, 1998. Advanced air quality indicators and reporting. Organisation for Economic Co-operation and Development, Paris.

Onursal, B., Gautam, S.P., 1997. Vehicular air pollution. Experiences from seven Latin American urban centers. World Bank Technical Paper no. 373. The World Bank, Washington, DC.

Phillips, D.R., 1993. Urbanization and human health. Parasitology 106, S93–S107.

Pichl, P. (Ed.), 1998. Exchange of Experiences Between European Cities in the Field of Clean Air Planning, CO_2 Reduction and Energy Concepts. Umwelt Bundes Amt, Berlin.

Pope III, C.A., Dockery, D.W., Schwartz, 1995. Review of epidemiological evidence of health effects of particulate air pollution. Inhalation Toxicology 7, 1–18.

Population Reference Bureau, 1998. World population data sheet. Demographic data and estimates for the countries and regions of the World. Population Reference Bureau, Washington, DC.

Richter, D.U.R., Williams, W.P., 1998. Assessment and management of urban air quality in Europe. EEA Monograph no. 5. European Environment Agency, Copenhagen.

Schwartz, J., Dockery, D.W., Neas, L.M., 1997. Is daily mortality associated specifically with fine particles? Journal of air and Waste Management Association 46, 927–939.

Selden, T.M., Song, D., 1994. Environmental quality and development: is there a Kuznets curve for air pollution emissions? Journal of Environmental Economics and Management 27, 147–162.

Sexton, K., Ryan, P.B., 1988. Assessment of human exposure to air pollution: methods, measurements and models. In: Watson, A.Y., Bates, R.R., Kennedy, D. (Eds.), Air Pollution, the Automobile, and Public Health. National Academy Press, Washington, DC, pp. 207–238.

Smith, K.R., 1988. Air pollution. Assessing total exposure in developing countries. Environment 30, 16–20, 28–30, 33–35.

Stanners, D., Bourdeau, P. (Eds.), 1995. Europe's Environment: The Dobris Assessment, European Environment Agency. Office for Publications of the European Communities, Luxemburg.

The World Bank, 1992. World Development Report 1992. Development and the Environment. Oxford University Press, New York.

The World Commission on Environment and Development, 1987. Our Common Future. Oxford University Press, Oxford.

Tidblad, J., Kucera, V., 1998. Materials damage. In: Fenger, J., Hertel, O., Palmgren, F. (Eds.), Urban Air Pollution, European Aspects. Kluwer Academic Publishers, Dordrecht, pp. 343–361.

Tuch, Th., Heyder, J., Heinrich, J., Wichmann, H.E., 1997. Changes of the particle size distribution in an Eastern German city. Presented at the 6th International Inhalation Symposium, Hannover Medical School.

UNEP, 1991. Urban air pollution. UNEP/GEMS Environment Library No 4. United Nations Environment Programme, Nairobi.

UNEP, 1997. Global Environment Outlook. United Nations Environment Programme and Oxford University Press.

UNEP/WHO, 1994. GEMS/AIR Methodology Reviews, Vol. 3: Measurement of suspended particulate matter in ambient air. WHO/EOS/94.3, UNEP/GEMS/94.A.4. UNEP, Nairobi.

United Nations, 1997a. Report of the United Nations Conference on Human Settlements (Habitat II), Istanbul, 3–14 June 1996. United Nations, New York.

United Nations, 1997b. Statistical Yearbook 1995. United Nations, New York.

Vignati, E., Berkowicz, R., Hertel, O., 1996. Comparison of air quality in streets of Copenhagen and Milan, in view of the climatological conditions. Science of the Total Environment 189/190, 467–473.

Volz, A., Kley, D., 1988. Evaluation of the Montsouris series of ozone measurements made in the nineteenth century. Nature 332, 240–242.

Walsh, M.P., 1996. Motor vehicle pollution control in China: an urban challenge. In: Stares, S., Zhi, L. (Eds.), China's Urban Development Strategy. World Bank Discussion Paper No. 353. The World Bank, Washington, DC, pp. 105–151.

Walsh, M., Shah, J.J., 1997. Clean fuels for Asia. Technical options for mowing toward unleaded gasoline and low-sulfur diesel. World Bank Technical Paper no. 377. The World Bank, Washington, DC.

Wang, W., Wang, T., 1996. On acid rain formation in China. Atmospheric Environment 30, 4091–4093.

Whitby, K.T., Sverdrup, G.M., 1980. California aerosols: their physical and chemical characteristics. Advances in Environmental Science and Technology 9, 477–517.

WHO/UNEP, 1992. Urban air pollution in megacities of the World. World Health Organization and United Nations Environment Programme, Blackwell, Oxford. (In short form as Mage et al., 1996.)

WHO, 1995. Concern for Europe's tomorrow: Health and environment in the WHO European region. WHO European Centre for Environment and Health, Wissenschaftliche Verlagsgesellschaft mbH, Stuttgart.

WHO, 1987/1999. Air quality guidelines for Europe. World Health Organization, Copenhagen. A revised version is being prepared and may be published in 1999.

Wiederkehr, P., Yoon, S.-J., 1998. Air quality indicators. In: Fenger, J., Hertel, O., Palmgren, F. (Eds.), Urban Air Pollution, European Aspects. Kluwer Academic Publishers, Dordrecht, pp. 403–418.

Wilson, W.E., Suh, H.H., 1997. Fine particles and coarse particles: concentration relationships relevant to epidemiologic studies. Journal of Air and Waste Management Association 47, 1238–1249.

Zhang, Y., Xie, S., 1998. The Vehicular Exhaust Pollution in Some China Big Cities. Center of Environmental Sciences, Peking University.

Zhang, Y., Shao, M., Hu, M., 1997. Air Quality in Beijing and its transition from coal burning caused problems to traffic exhaust related pollution. Air and Waste Management Association, Toronto, June 8–13, 1997.

Zhang, Y., Xie, S., Zeng, L., Wang, H., Shao, M., Yu, K., Zhu, C., Pan, N., Wang, B., 1998. The traffic emission and its impact on air quality in Guangzhou area. Center of Environmental Sciences, Peking University.

Zlatev, Z., Fenger, J., Mortensen, L., 1996. Relationships between emission sources and excess ozone concentrations. Computers in Mathematical Applications 32, 101–123.

Appendix

Neither environmental science nor management progresses as fast as some journalists and politicians seem to believe. And the global situation of urban air quality has not changed significantly since the publication in 1999 of the Millennium Review on Urban Air Quality.

In the industrialised world, the classical pollutants (soot and sulphur dioxide) from heating systems are by and large under control and – in spite of widespread use of catalytic converters and other measures – traffic emissions are now the main problem. This applies especially to particulate matter, which for a foreseeable future will remain a significant urban air quality problem (European Commission 2000). A similar development is appearing in economies in transition, and it can be anticipated in major cities in the developing countries.

Some aspects will, however, in the coming years receive increasing attention, partly as a result of the development of better analytical techniques, partly due to a growing awareness of the importance of comprehensive studies and abatement policies.

1. New pollutants

In general the *concentration* of major pollutants in the environment are decreasing, but the *number* of pollutants being detected is increasing. Today some hundred thousand chemical compounds must be evaluated internationally. Many of them, mostly volatile organic compounds (VOC), may be present in the urban atmosphere due to vehicle exhaust or industrial emissions, but so far only a few indicator compounds have been studied in detail.

A new EU (European Union) directive (1999/30EC; 22 April 1999) includes limit values for particulate matter (PM_{10}) and obligations for the member states to collect data on $PM_{2.5}$. So far however, WHO (World Health Organisation) has not recommended a limit value for particulate matter.

Studies of fine and ultrafine particles from traffic are therefore intensified. Not only because of their health impacts, but also because their emission has been reported to *increase*, even if the total mass of particles emitted from vehicles *decreases*. A new technique based on measurements with a Differential Mobility Analyser (DMA) followed by factor analysis and constrained linear receptor modelling has been developed (Wåhlin et al., 2000). It allows field identification of particle spectre from diesel and petrol fuelled vehicles. It appears that although diesel cars are the main source of very fine particles, petrol driven cars also contribute with generally larger particles.

2. Population exposure and health impacts

Air quality management has so far mainly been carried out on the basis of measurements and model results of ambient concentrations of pollutants.

However, direct personal monitoring demonstrates that the levels *an individual* is exposed to can vary drastically during the day (e.g. Zhiqiang et al., 2000), and that the ambient, outdoor air quality does not adequately describe the actual exposure in sufficient detail. Many tools are available (Hertel et al., 2000) including GPS (Global Positioning System) to trace the movements of test persons, but still some mechanisms are poorly understood. Thus, for example, people appear to stir up "personal clouds" of particle-laden dust from their surroundings. They may therefore experience exposure to fine particles about 60% greater than classical monitoring may suggest (Renner, 2000).

In view of the large fraction of time spent indoors, especially in cities with a cold climate, the relations between outdoor pollution and related indoor levels of pollutants from outdoor sources must also be taken into account. That is not always simple. Typically indoor levels are only about half the outdoor levels, but for benzene this was not reflected convincingly in the *personal exposure* in a series of European Cities (Cocheo et al., 2000). In some cases the population exposures were found to exceed the average ambient urban concentration. It is speculated that the reason may be that people are generally outdoors, when the ambient levels are high, and indoors, when they are low, and that the indoor environment may store pollution with an outdoor origin (a sort of "flywheel effect").

Notwithstanding an extensive documentation (see e.g. Künzli et al., 2000 for a European assessment) some important problems concerning relations between exposure and impacts remain to be satisfactorily elucidated. First of all the statistical procedures are debated. Thus it is a question how to relate daily mortality with longer term mortality effects (McMichael et al., 1998) and especially how to filter out a possible "harvesting effect" (Zeger et al., 1999). Secondly the assessment of human exposure is still being developed (Mage et al., 1999). In addition, it must be admitted that epidemiological studies provide little (or no) information on the underlying impact mechanisms, which are still poorly understood (Brunekreef, 1999). For example, it is not obvious to what extent the particulate matter *per se* is important and what is the role of the chemical composition (e.g. PAH). This may influence the future design and application of exhaust cleaning devices. For an extensive critical review see Lightly et al. (2000).

3. Cities in a global context

In its 3rd assessment report The Intergovernmental Panel on Climate Change (IPCC, 2001) applies a more holistic approach to human induced climate change than in its previous reports. It is now explicitly acknowledged that some changes can not be avoided and various means of adaptation are discussed.

In this context the interplay of various environmental loads should be considered. In the long run climate changes may influence urban air quality via changes in temperature, precipitation, wind pattern and solar radiation. Thus for example, formation and dispersion of photochemical pollution may be enhanced and allergens and irritants may have larger impacts. On the other hand the mix of pollutants may change with less need for space heating. More important is undoubtedly the attempt of general phasing out of fossil fuels. To the extent that this is successful many problems with urban air quality may be solved.

4. Conclusion

The conclusion can not be better expressed than in a report on the recently concluded Auto-Oil II project (European Commission 2000):

- There is a need for improved knowledge on the links between emission targets and air quality requirements.
- A really cost-effective policy package will require an integrated approach across sources, pollutants and measures.

References

Brunekreef, B., 1999. All but quiet on the particulate front. Am. J. Respir. Crit. Care Med. 159, 354–356.

Chocheo, V. et al., 2000. Urban benzene and population exposure. Nature 40, 141.

European Commission, 2000. Communication on Auto-oil II programme review. Europe Environment supplement to n° 578.

Hertel, O., de Leeuw, F.A.A.M., Raaschou-Nielsen, O., Jensen, S.S., Gee, D., Herbarth, O., Pryor, S., Palmgren, F., Olsen, E., 2000. Human Exposure to Outdoor Air Pollution. IUPAC recommendation. Approved by the IUPAC commissions on Toxicology and Atmospheric Chemistry and submitted for publication in PAC.

IPCC (Intergovernmental Panel on Climate Change), 2001. 3rd. Assessment. To be released in spring 2001.

Künzli, N. et al., Public-health impacts of outdoor and traffic-related air pollution: a European assessment. Lancet 356, 795–801.

Lightly, J.S., Veranth, J.M., Sarofim, A.F. 2000. Combustion aerosols: Factors governing their size and composition and implications to human health. J. Air & Waste Manag. Ass. 50, 1565–1618.

Mage, D., Wilson, W., Hasselblad, V., Grant, L., 1999. Assessment of human exposure to ambient particulate matter. J. Air & Waste Manag. Ass. 49, 1280–1291.

McMichael, A.J., Anderson, H.R., Brunekreef, B., Cohen, A.J., 1998. Inappropriate use of daily mortality analyses to estimate longer-term mortality effects of air pollution. International Epidemiological Association 27, 450–453.

Renner, R., 2000. Pollution monitoring should get personal, scientists say. Environ. Sei. Technol./News 64–65A.

Wåhlin, P., Palmgren, F., Dingenen, R. van, 2000. Experimental studies of ultrafine particles in streets and the relationship to traffic. Atmospheric Environment in press.

Zeger, S.L., Dominici, F., Samet, J., 1999. Harvesting-resistant estimates of air pollution effects on mortality. Epidemiology 10, 171–175.

Zhiqiang, Q., Siegmann, K., Keller, A., Matter, U., Sherrer, L., Siegmann, H.C., 2000. Nanoparticle air pollution in major cities and its origin. Atmospheric Environment 34, 443–451.

Air Pollution Science for the 21st Century
J. Austin, P. Brimblecombe and W. Sturges, editors
© 2002 Elsevier Science Ltd. All rights reserved.

Chapter 2

New Directions: Sustainability in strategic air quality planning

Michael E. Chang

*Center for Urban and Regional Energy, Georgia Institute of Technology, Atlanta,
GA 30332-0595, USA
E-mail: chang@eas.gatech.edu*

Sustainability is often defined as meeting the needs of the present without compromising the ability of future generations to meet their needs. Rather than a shortsighted approach to distribution of capital and opportunity, sustainability emphasizes equity over much longer, intergenerational time scales. While still a relatively new paradigm, support for the concepts inherent to sustainability is spreading. Current policies for meeting and maintaining ambient air quality standards, however, often fail to recognize these precepts. In the US, the Clean Air Act and its amendments have traditionally provided a planning window on the order of only five to ten years, with a maximum of 20. As will be discussed here, the relatively near-team provisions of these policies compound ignorance of the effects of current actions on future opportunities, which could, in turn, preclude a region from ever attaining clean air under current or future standards.

Consider two hypothetical and competing opportunities for improving air quality: A and B. Further assume that the choice between the two is discrete – one may choose only one of the options. If implemented, the effectiveness of each opportunity is illustrated in Fig. 1. Opportunity A rapidly attains the current air quality standard and is able to maintain it in perpetuity. Opportunity B is also able to meet the current standard, but at a much later date, and only after an initial increase in air pollution. Unlike opportunity A however, B continues to decrease air pollution beyond the current standard. If decision-makers are forced to attain the current air quality standard in five or ten years, then clearly A will be selected. Only it meets the goal of attainment. The selection of A however, also presumes that current air quality standards will not change. This presumption fails the test of sustainability. Just as the current generation is debating air quality standards, it is reasonable to expect that future generations will also revise the standards to reflect their contemporary values. The choice

First published in Atmospheric Environment 34 (2000) 2495–2496

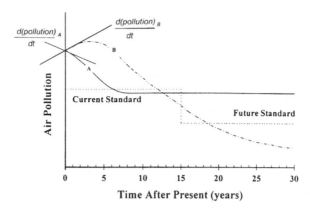

Figure 1.

to pursue A, however, precludes future generations from reasonably attaining a more strict air quality standard. On the other hand, B preserves the opportunity to lower the standard in the future, and even if this is not desired, it provides other opportunities by providing more pollution allowances under the cap.

Short planning horizons demand decisions be made at the margin. That is, the first derivaties d(pollution)/dt at time $t = current$ are used to determine the most appropriate action. The option that is selected for implementation is the one that provides the greatest immediate decrease in pollution (or more likely, greatest pollution decrease per unit cost). The effect of the action on longer-term future opportunities is discounted since the future is over the planning horizon. Sustainability on the other hand, demands a more holistic approach and, rather than discount the future consequences, it discounts those options that fail to preserve future opportunity.

Atlanta, Georgia is a large urban center located in the southeastern United States. The US EPA designated Atlanta nonattainment for ground-level ozone in 1978. Since then, the state of Georgia has prepared and implemented five separate plans to improve air quality (1979, 1985, 1987, 1994, and 1998). None of these have extended beyond a five-year planning horizon, and none have resulted in attainment. The Atlanta region's primary sources of anthropogenic ozone precursors are the transportation sector and coal-fixed electric power production. Regional transportation plans however, because they necessitate huge investments in capital, are prescribed 20 years in advance. They also persist for multiple generations once they are built. The fundamental elements of Atlanta's current urban expressway were designed in 1946 and building was completed in 1970. Electric power plants are designed, constructed, and operated across similar scales. Of the larger coal burning facilities near Atlanta, the newest plant began commercial operations in 1976. It is this long-in-the-making and persistent urban paradigm that created Atlanta's nonattainment problem. A strategy that employs innovative technology and clever policy may make attainment possible in Atlanta while preserving the paradigm, but this has not yet been realized (in Atlanta, the last two air quality plans could not envision a scenario in which attainment is even possible). Further, the short-sighted policies that guide actions may be reinforcing the nonattainment paradigm rather than shifting it to one in which attainment is more readily achieved. When Georgia's regulators require a power plant near Atlanta to invest in expensive technologies (e.g. NO_x RACT in 1994) to reduce marginally its emissions of ozone precursors (plan A in the discussion above), they also extend the commercial service life of this facility. The operator must continue to run the facility and earn a return on investment at least until the newly acquired capital costs (needed to implement the control technologies) are fully amortized. If however, future generations wish to shut down the power plant (plan B) for air quality or other reasons, the opportunity to do this is lost or at least delayed. Just as creating new power plants takes time and significantly alters the paradigm, shutting one down must take considerable time as well. New capacity or effective conservation matters must be in place before operations can be terminated. Because the benefits of these measures may not be realized before the deadlines that existing air quality policies mandate for clean air however, they are not explored in the air quality planning process.

The argument presented here is in no way intended to discredit the seeking of near-team solutions to air quality problems. Current generations are just as entitled to clean air as future generations. Furthermore, many solutions may in fact provide benefits now and into the future. The challenge is to consider the effect of near-team actions on future opportunities. To do this, policies must provide incentives for planners and decision-makers to seek longer planning horizons. This in turn will expand the set of opportunities available to improve air quality. The Clean Air Act and its amendments have improved air quality largely through a mandated series of actions planned and implemented over five to ten year periods. Now after nearly three decades of concerted, yet fragmented effort, and marginal effectiveness, one can only speculate if better performance might have been achieved if only long-term opportunities would have been more thoughtfully considered. Finally, while the argument presented here is strictly concerned with air quality, the reasoning is universal. Sustainability is relevant in the context of any system – climate, water resources, the economy, etc. – and while a truly holistic perspective may be difficult to attain, it is no less necessary if intergenerational progress is expected.

Air Pollution Science for the 21st Century
J. Austin, P. Brimblecombe and W. Sturges, editors
© 2002 Elsevier Science Ltd. All rights reserved.

Chapter 3

Indoor air quality and health

A.P. Jones

School of Environmental Sciences, University of East Anglia, Norwich, Norfolk, NR4 7TJ, UK
E-mail: a.p.jones@uea.ac.uk

Abstract

During the last two decades there has been increasing concern within the scientific community over the effects of indoor air quality on health. Changes in building design devised to improve energy efficiency have meant that modern homes and offices are frequently more airtight than older structures. Furthermore, advances in construction technology have caused a much greater use of synthetic building materials. Whilst these improvements have led to more comfortable buildings with lower running costs, they also provide indoor environments in which contaminants are readily produced and may build up to much higher concentrations than are found outside. This article reviews our current understanding of the relationship between indoor air pollution and health. Indoor pollutants can emanate from a range of sources. The health impacts from indoor exposure to combustion products from heating, cooking, and the smoking of tobacco are examined. Also discussed are the symptoms associated with pollutants emitted from building materials. Of particular importance might be substances known as volatile organic compounds (VOCs), which arise from sources including paints, varnishes, solvents, and preservatives. Furthermore, if the structure of a building begins to deteriorate, exposure to asbestos may be an important risk factor for the chronic respiratory disease mesothelioma. The health effects of inhaled biological particles can be significant, as a large variety of biological materials are present in indoor environments. Their role in inducing illness through immune mechanisms, infectious processes, and direct toxicity is considered. Outdoor sources can be the main contributors to indoor concentrations of some contaminants. Of particular significance is Radon, the radioactive gas that arises from outside, yet only presents a serious health risk when found inside buildings. Radon and its decay products are now recognised as important indoor pollutants, and their effects are explored. This review also considers the phenomenon that has become known as Sick Building Syndrome (SBS), where the occupants of certain affected buildings repeatedly describe a complex range of vague and often subjective health complaints. These are often attributed to poor

First published in Atmospheric Environment 33 (1999) 4535–4564

air quality. However, many cases of SBS provide a valuable insight into the problems faced by investigators attempting to establish causality. We know much less about the health risks from indoor air pollution than we do about those attributable to the contamination of outdoor air. This imbalance must be redressed by the provision of adequate funding, and the development of a strong commitment to action within both the public and private sectors. It is clear that meeting the challenges and resolving the uncertainties associated with air quality problems in the indoor environment will be a considerable undertaking.

1. Introduction

Until recently, the health effects of indoor air pollution have received relatively little attention from the scientific community. Prior to the 1970s, problems with indoor air quality in residences and the non-industrial workplace were occasionally investigated, but the level of interest was low (Stolwijk, 1992). Even today, the bulk of public concern continues to be directed at the health impacts of outdoor pollution. Numerous studies suggest that most members of the public perceive the risks from poor quality outdoor air as being substantially higher than those from indoor contamination (LHEA, 1997). These perceptions are formed despite the fact that, in developed societies, many people pass most of their time indoors. For example, a recent investigation of time budgets amongst US residents found that, on average, individuals spent 88% of their day inside buildings, and 7% in a vehicle. Only 5% of participants'time was actually spent outside (Robinson and Nelson, 1995). Whilst this does not per se mean that indoor exposures will produce more harmful health effects, the evidence is that indoor concentrations of many pollutants are often higher than those typically encountered outside.

A preoccupation with the relationship between outdoor air quality and health is understandable. Particularly in more economically developed and rapidly developing countries, recent and unprecedented changes in lifestyles and environmental quality have meant that an increasing number of people are exposed to the contaminants of urban air (Lipfert, 1997). There is certainly evidence that, especially amongst more vulnerable members of society, outdoor pollutants may pose a real health risk (Burr, 1995). Furthermore, the frequently invisible nature of indoor air quality problems must be contrasted with the often only too obvious photochemical and particulate substances that characterise outdoor contamination.

Historically, problems with indoor air were unquestionably much more apparent than they are today. Soot found on the ceilings of prehistoric caves provides ample evidence of the high levels of pollution that was associated with inadequate ventilation of open fires (Spengler and Sexton, 1983). Whilst chimneys first began to appear in European homes in the late twelfth century, most large medieval houses still had a central hearth in the great hall, ventilated by a louvre in the roof. It was only during the sixteenth century that chimney stacks and fireplaces against walls came into general use (Brimblecombe, 1987; Burr, 1997). The blackened roof timbers in many buildings that predate these innovations bear testimony to the severe pollution problems that their inhabitants faced.

Today, concern over the health effects of poor quality indoor air is increasing. Despite the fact that the vast majority of buildings exhibit no immediately apparent problems, a wide spectrum of symptoms and illnesses are attributed to non-industrial indoor air pollution (Redlich et al., 1997). Indeed, problems associated with shortcomings in indoor air may be one of the most common environmental health issues most doctors now face (Seltzer, 1995).

Over recent decades, there have been many changes in the way buildings are constructed and operated. To some extent, modifications in building design have been driven by the need for increased energy efficiency, largely brought about by higher fuel costs since the 1970s oil crisis (Jones, 1998). Modern homes and offices are much better insulated than was previously the case. Houses with ventilation rates as low as 0.2 to 0.3 air exchanges h^{-1} are now widespread (Platts-Mills et al., 1996). In older properties, particularly those with open fireplaces, ventilation rates above 1 air exchange h^{-1} are more common. Improved insulation has been accompanied by numerous other modifications to the management of indoor environments, and advances in construction technology have led to a much greater use of synthetic building materials (D'Amato et al., 1994). All these changes have undoubtedly meant that buildings are more comfortable. However, they have also provided an environment in which airborne contaminants are readily produced and may build up to substantially higher concentrations than are typically encountered outside (Teichman, 1995).

Indoor air pollutants emanate from a range of sources (Table 1). They are emitted by the fabric of buildings, and may also be a by-product of the activities that are undertaken within them. Sources can be broadly classified as being associated with the activities of building occupants and other biological sources, the combustion of substances for heating or fuel, and emissions from building materials. For some contaminants, infiltration from outside, either through water, air, or soil, can also be a significant source.

The concentration of a pollutant indoors depends on the relationship between the volume of air contained in the indoor space, the rate of production or release of the pollutant, the rate of removal of the pollutant from the air via reaction or settling, the rate of air exchange with the outside atmosphere, and the outdoor pollutant concentration (Maroni et al., 1995). However, actual human exposures are often difficult to quantify. This is largely because the behaviour and activity patterns of individuals can strongly affect their levels of

Table 1.   Major indoor pollutants and emission sources[a]

Pollutant	Major emission sources
Allergens	House dust, domestic animals, insects
Asbestos	Fire retardant materials, insulation
Carbon dioxide	Metabolic activity, combustion activities, motor vehicles in garages
Carbon monoxide	Fuel burning, boilers, stoves, gas or kerosene heaters, tobacco smoke
Formaldehyde	Particleboard, insulation, furnishings
Micro-organisms	People, animals, plants, air conditioning systems
Nitrogen dioxide	Outdoor air, fuel burning, motor vehicles in garages
Organic substances	Adhesives, solvents, building materials, volatilisation, combustion, paints, tobacco smoke
Ozone	Photochemical reactions
Particles	Re-suspension, tobacco smoke, combustion products
Polycyclic aromatic hydrocarbons	Fuel combustion, tobacco smoke
Pollens	Outdoor air, trees, grass, weeds, plants
Radon	Soil, building construction materials (concrete, stone)
Fungal spores	Soil, plants, foodstuffs, internal surfaces
Sulphur dioxide	Outdoor air, fuel combustion

[a]Taken from Spengler and Sexton (1983).

exposure (Harrison, 1997). Results from the TEAM (Total Exposure Assessment Methodology) studies undertaken by the US Environmental Protection Agency (EPA) during the 1980s consistently show that personal exposures to many pollutants can markedly exceed those anticipated from concentrations in ambient air (Wallace, 1991a). This phenomenon has become known as the 'personal cloud effect' (Furtaw et al., 1996).

Problems in quantifying exposure also arise because many of the advanced technologies that have been developed for measuring outdoor pollution are not suitable for indoor use. In part, this is due to considerations of cost, size, and the amount of air they displace (Maroni et al., 1995). Although miniaturised measuring devices have been developed for indoors, they often only record average pollutant concentrations over several hours or days. This can create uncertainties, especially for pollutants with health effects thought to be related to short-term extreme exposures. In the case of recording pollutants from sources such as micro-organisms an opposite problem can apply, whereby monitoring programmes must be operational for a very long time period to provide accurate estimates of exposure (Blomquist, 1994).

Because of the difficulties in assessing public exposure to indoor air pollution, coupled, perhaps, with the fact that the resurgence of scientific interest into the links between indoor air quality and health is still relatively recent, we know much less about the health risks from polluted indoor air than we do about those associated with outdoor contamination. Indeed much of what has been learnt about the health impacts of indoor air comes from studies of the outdoor environment. Exposure to indoor toxicants can potentially lead to a variety of adverse health outcomes (Bascom et al., 1995). The likelihood that an individual will become ill from the presence of a contaminant depends upon factors such as the individual's sensitivity to that contaminant, the contaminant concentration, the current state of their psychological and physical health, and the duration and frequency of exposure (Seltzer, 1997). Indoor air pollutants have the potential to cause transient morbidity, disability, disease, and even death in extreme cases (Berglund et al., 1992). Recent research into these health outcomes has involved human, animal, and in vitro studies (Maroni et al., 1995).

Human studies entail the observation and measurement of symptoms and other effects in individuals exposed to pollutants, either routinely (as in epidemiological studies) or in experimental circumstances. The main advantage of studies of routine situations is that the conditions under which exposure takes place are realistic, although the power of these investigations is sometimes insufficient to signify the likely causality of associations. Experimental exposures have the advantage that exposure conditions and subject selection may be determined by the investigator. However, they are only suitable for studying slight, reversible, short-term effects in healthy or only moderately ill individuals.

Animal studies involve the assessment of the health effects of exposures in laboratory animals. Their main limitation is that it is necessary to extrapolate the findings to humans. Such extrapolations are fraught with difficulties, particularly as results are often based on much higher doses than would be commonly encountered in the indoor environment.

In in vitro research, the effects of pollutants on cell or organ cultures are examined. They have cost advantages over animal studies and are quick to undertake, but there are problems associated with the use of their findings to predict effects on whole organisms.

It is the aim of this review to draw together the results from these investigations and present a picture of our current understanding of the relationship between indoor air pollution and health. Most of the evidence presented will be taken from epidemiological studies. Animal and in vitro research has provided many important findings. However, the conditions of these studies may be rather removed from those that typically prevail when individuals are exposed to pollutants indoors. Major progress has recently been achieved in the field of developing guidelines for volatile organic compounds from bioassays (ECJRA, 1997), but their use for many substances is less well developed. Hence, this is not the place for a detailed consideration of their findings. Neither will research into the health impacts of occupational exposures to pollutants be discussed.

Such exposures are usually associated with high contaminant doses, and the health outcomes are often severe. Hence, the aetiology and mechanisms of industrial exposures are sufficiently different from those commonly encountered in the domestic or office environment to warrant consideration elsewhere.

In this account, pollutants are considered according to whether they arise from biological or non-biological sources. Of course, dividing them in this way provides a simplistic picture of true patterns of exposure; in reality individuals are often concurrently exposed to a wide variety of contaminants from a range of different sources. In some situations, they may react with each other, leading to patterns and levels of exposure that are rather different from the apparent sum of their parts. However, the use of a simplified division is necessarily adopted for the sake of clarity.

Also considered in this text is the phenomenon that has become known as 'sick building syndrome'(SBS), where the occupants of some buildings repeatedly describe a complex range of vague and often subjective health complaints (Horvath, 1997). Since the early 1970s, numerous outbreaks of SBS have been reported. These have usually been in offices, but sometimes schools, hospitals, homes for the elderly, or apartments have been involved. The symptoms of SBS are often relatively minor, non-specific, and common amongst the general population, but are more frequent amongst the occupants of 'sick' buildings (Lahtinen et al., 1998). Despite their minor nature, SBS symptoms may have a greater impact on public health and cost to the economy than some major diseases due to widespread absenteeism and lowered productivity amongst affected workers (Wallace, 1997). For example, it has been estimated that the annual 'cost' of headaches amongst the employees of the US Environmental Protection Agency may be as high as $2 million (Wallace, 1997).

Sick building syndrome is normally distinguished from the term 'building related illness' (BRI) (Ryan and Morrow, 1992). Whilst BRI has a known aetiology, many studies of SBS have failed to determine any specific factor that may explain the health complaints (Lahtinen et al., 1998). Outbreaks of SBS have been known to occur in older, naturally ventilated buildings. However they are more common in modern energy-efficient 'airtight' buildings, in particular those served by mechanical heating, ventilation, and air-conditioning (HVAC) systems (Redlich et al., 1997). This observation has led many to conclude that SBS must in some way be associated with indoor air quality problems.

2. Pollution from non-biological sources

It is important to note that outdoor sources may be the main contributors to indoor concentrations of a number of non-biological pollutants commonly found in indoor air. This is especially the case for contamination in buildings situated in urban areas and close to industrial zones or streets with heavy traffic. The factors which determine the contribution of outdoor pollution to indoor air quality include the type of ventilation in use (natural or forced), the ventilation rate (air changes per hour), and the nature of the contaminants in question (Wanner, 1993). The major outdoor sources of important indoor air pollutants are given in Table 2. The USA EPA TEAM studies have shown that reactive gasses such as ozone tend to occur at lower concentrations indoors than outdoors because they react rapidly with indoor surfaces (Wallace, 1987). Non-reactive gases may accumulate indoors and exposures there may be greater than outside.

Where meaningful, typical indoor–outdoor concentration ratios of pollutants are provided. However, to discuss the full impact of outdoor pollutants on indoor air quality would require a detailed consideration of the chemistry and processes that operate within the outdoor environment. This is not the place for such a discussion, although the health effects of one outdoor pollutant, radon, are considered. Although radon arises from outside, it merits consideration as it only presents a significant risk to health when found inside buildings.

Heating and cooking are often essential indoor activities but may produce smoke and gasses that present a problem of disposal. This difficulty is greatest in colder climates, where it is necessary to retain heat at the same time as removing combustion by-products (Burr, 1997).

A wide range of pollutants are associated with combustion. In the past, a high reliance on wood and coal for the production of heat meant that the indoor pollution profile was dominated by smoke from these sources (Wallace, 1996). In some countries, especially those of eastern Europe and China, the use of these fuels still predominates (Lan et al., 1993). In much of western

Table 2. Outdoor sources of major indoor air pollutants

Pollutant	Percentage of emissions associated with industry[a]	Percentage of emissions associated with transport[a]
Benzene	32	65
Carbon monoxide (CO)	3	90
Lead (Pb)	31	60
Oxides of nitrogen (NO_x)	38	49
Particulates (PM_{10})	56	25
Sulphur dioxide (SO_2)	90	2
Volatile organic compounds (VOCs)	52	34
Ozone (O_3)	Arises from atmospheric chemical reactions	

[a]Figures based on UK estimates (DOE, 1997).

Europe and the USA, coal and wood have largely been replaced by natural gas and electricity (Lambert, 1997), and approximately one-half of all American homes now use gas for cooking and heating (Hines et al., 1993). If properly vented to the outside, most appliances are typically not major contributors to indoor air quality problems. However, the use of unvented appliances or those that are malfunctioning or improperly installed can generate severe problems (Nagda et al., 1996).

Cigarette smoking is an important source of indoor combustion related pollution. Tobacco smoke is an aerosol containing several thousand substances that occur as particles, vapours, and gases. Exposure to cigarette smoke has been associated with a wide range of acute and chronic health impacts. Although the prevalence of smoking in western societies is much lower than it was in the 1950s and 1960s, tobacco smoke still contributes a significant proportion of the total indoor pollutant dose for many individuals.

Materials that comprise the fabric of buildings are another important source of non-biological pollutants. A large number of the chemical compounds which are found in indoor air originate from paints, varnishes, solvents, and wood preservatives used in buildings (ECJRC, 1997). Furthermore, deteriorating materials that comprise the fabric of a building can become friable and release contaminants into the air. Pollutants from these sources are often difficult to quantify because they are present in relatively low concentrations, and their sources are diffuse (Wanner, 1993). An additional complication arises where levels of contamination are mediated by climatic factors such as temperature and relative humidity. Hence, possible health risks of contaminants from building materials depend very much on the nature and concentration of the pollutants involved.

The following sections consider the typical concentrations and important health effects of the range of non-biological pollutants from the various sources that are most commonly encountered in the indoor environment.

2.1. Asbestos

Asbestos is a generic term that applies to a group of impure hydrated silicate minerals which occur in various forms, and are incombustible and separable into filaments (Maroni et al., 1995). Until recently, asbestos was widely used by the building industry for the production of many different products. It was particularly valued for its electrical and thermal insulating properties, and was heavily utilised in pipe and boiler insulation, cement-board, thermal tiles, paint, and wallpaper (Bigonon et al., 1989). Concern over the health effects of asbestos exposure has led to legislation being introduced to prohibit its use in many countries. Hence, most asbestos-related air quality problems are associated with the management of existing asbestos installations.

Accurate assessment of exposure to asbestos is particularly difficult because most airborne fibres are too small to be counted by optical microscopy, and hence scanning and transmission electron microscopy must be used to analyse samples (Hines et al., 1993). Because of this, relatively little information on typical indoor concentrations is available (Gaensler, 1992). In one of the few studies that have been undertaken, Lee et al. (1992) examined exposure to airborne asbestos in 315 public, commercial, residential, school, and university buildings in the USA. The average concentration of asbestos was 0.02 structures ml^{-1} of air, whilst the average concentration of fibres greater than 5 μm in length was 0.00013 fibres ml^{-1}. However, it is important to note that, in 48% of samples, no asbestos was detected. In general, asbestos-containing materials within buildings that are in good repair are unlikely to lead to exposure of occupants to concentrations higher than ambient levels. A recent literature review commissioned by the US EPA cost 4 million dollars and suggested that well-maintained asbestos in public buildings posed little risk to office workers (Hines et al., 1993).

Acute exposure to asbestos can cause skin irritation (Spengler and Sexton, 1983). However, the most serious health effects from asbestos are lung cancer, mesothelioma (cancer involving a proliferation of mesothelial cells) and asbestosis (a slowly developing but lethal fibrosis of the lung) (McDonald, 1991). Recent data suggest that the fibres must remain in the respiratory tract for approximately one year to produce effects, and those with a diameter of less than 1 μm and a length greater than 5–10 μm are believed to be particularly dangerous. There is generally a latent period of 20–50 yr between first exposure to asbestos, and the clinical manifestation of tumours (Doll and Peto, 1985).

Whilst the relationship between asbestos exposure and the incidence of mesothelioma and asbestosis is well documented, clarifying the relationship with lung cancer is complicated, and is still a controversial area (Huncharek, 1994). It is accepted that exposure to asbestos and tobacco smoke from active smoking has a marked synergistic effect. It is important to note that the majority of population risk from asbestos arises from industrial exposures. However, cases of disease associated with non-industrial exposures do occur. For example, Roggli and Longo (1991) highlight the instance of a teacher's aid with pleural mesothelioma who was for many years exposed to asbestos-containing acoustic plaster in a school building.

2.2. Carbon dioxide

Carbon dioxide (CO_2) is a colourless, odourless gas. Humans continuously exhale CO_2 formed in the body during metabolic processes and, where fuel is not being burnt, these emissions comprise the greatest contribution to indoor concentrations (Wanner, 1993). Carbon dioxide is also the main combustion product from gas, kerosene, and wood or coal fuelled appliances and these can represent significant sources when in operation (Moriske et al., 1996). Typical indoor CO_2 concentrations range between 700 and 2000 ppm (approximately 3657 mg^{-3}) but can exceed 3000 ppm (5486 mg^{-3}) during the use of unvented appliances (Arashidani et al., 1996).

Carbon dioxide is a simple asphyxiant, and can also act as a respiratory irritant (Maroni et al., 1995). However, although indoor to outdoor ratios of the gas are typically in the range 1–3 for most environments, exposure to an extremely high CO_2 concentration (above 30,000 ppm or 54860 mg^{-3}) is required before significant health problems are likely (NASA, 1973). At moderate concentrations, CO_2 can cause feelings of stuffiness and discomfort, an effect noted by Pettenkofer as far back as 1858 (Pettenkofer, 1858). Respiration can be slightly affected at levels above 15,000 ppm (27430 mg^{-3}). Exposures above 30,000 ppm can lead to headaches, dizziness, and nausea (Schwarzberg, 1993). Yang et al. (1997) found that these concentrations also affect perception of motion. This may be because CO_2 has been shown to moderate the activity of cells within the visual cortex.

2.3. Carbon monoxide

Carbon monoxide (CO) is a toxic odourless gas produced by the incomplete combustion of fuel (IEH, 1996). Water heaters, gas or coal heaters, and gas stoves are all indoor sources of CO (Gold, 1992). High indoor levels of CO can also result from the entry of outdoor vehicle exhausts into the ventilation system of a building. This can be a particular risk in buildings that have delivery areas where vehicles are parked with their motors running. In addition, tobacco smoking is an important source of transitory indoor CO pollution (Sterling, 1991), and the burning of charcoal briquettes and the use of gasoline-powered electrical generators during disruptions to electrical services can present temporary problems (Houck and Hampson, 1997). Another source of CO is methylene chloride, a substance used as a paint stripper. It is metabolised within the body to form CO, and its use can lead to a significant dose in even well ventilated rooms (Gold, 1992).

In the absence of emission sources, concentrations of CO in indoor environments are generally lower than those outdoors. In a US Environmental Protection Agency (EPA) survey of the personal CO exposures of 1000 non-smoking adults, residential exposures were generally low, ranging from 2 to 4 ppm (approximately 2.3–4.7 µg m^{-3}) (Akland et al., 1985). Where gas stoves are in operation, hourly concentrations of CO are generally around 6 ppm (6.9 µg m^{-3}), and do not often exceed 12 ppm (13.8 µg m^{-3}) (Samet et al., 1987).

The toxic properties of CO are largely associated with its high affinity for oxygen-carrying proteins such as haemoglobin and myoglobin (Coultas and Lambet, 1991). Because its affinity for haemoglobin is approximately 200 times greater than that of oxygen, CO displaces oxygen, forming carboxy-haemoglobin (COHb), lowering the oxygen carrying capacity of the blood, and producing a left shirt in the oxyhaemoglobin dissociation curve (Roughton and Darling, 1994). Carbon monoxide can also interfere with oxygen diffusion into cellular mitochondria, and interfere with intracellular oxidation (Gold, 1992).

The health effects of exposure to CO are generally described relative to COHb levels (Madany, 1992). In non-smoking individuals unexposed to environmental CO, blood COHb levels are usually around 0.5% (Lambert, 1997). Various symptoms of neuropsychological impairment have been associated with acute low-level exposures. Amitai et al. (1998) found that subjects exposed to CO from residential stoves for up to 2.5 h showed declines in their learning and planning abilities, as well as a drop in their attention and concentration spans. Chronic exposure (at 10–30% COHb) often produces symptoms that are easily misdiagnosed or overlooked, such as headache, fatigue, dizziness, and nausea (Stewart et al., 1970). There is evidence from animal studies that some foetal damage may occur from maternal exposure to CO at these levels (Longo, 1977).

Carbon monoxide poisoning has its most toxic acute effects on the organs with high oxygen requirements; the heart and brain. Hence, individuals with ischaemic heart disease are at particularly high risk (USEPA, 1991). At moderate concentrations of CO, adverse cardiovascular effects may be observed amongst susceptible individuals (Dahms et al., 1993). Amongst 20 non-smoking men with ischaemic heart disease, Lambert (1994) found that the probability of occurrence of an episode of myocardial ischemia was 2.1 times higher at COHb levels of 2% relative to those below 1%.

Although non-reversible impairment is relatively rarely associated with CO exposure, the greatest danger of serious CO poisoning comes from faulty combustion appliances, or those with blocked or malfunctioning external vents (Howell et al., 1997). In such situations, COHb levels of between 50 and 60% can result in fainting and convulsions, whilst higher exposures can lead to coma and death. In England and Wales, there are on average around 60 deaths annually associated with accidental CO poisoning (Burr, 1997), and similar rates have been observed in the USA (Cobb and Etzel, 1991). Individuals who survive acute CO poisoning may still exhibit neurological and psychological symptoms many weeks or months after exposure, particularly if a period of unconsciousness has occurred (Choi, 1983).

2.4. Formaldehyde

Formaldehyde is the most widespread aldehyde found in the environment. Although it is a volatile compound, it is not detected by the gas chromatographic methods commonly applied to VOC analysis, and hence is often considered separately (Maroni et al., 1995).

At normal room temperatures, formaldehyde is a colourless gas with a pungent odour. The primary sources of formaldehyde are building materials such as particleboard, medium-density fibreboard, plywood, resins, adhesives, and carpeting (Hines et al., 1993). It is also used in the manufacture of urea formaldehyde foam insulation (UFFI) which is injected into wall cavities to supplement the insulation in existing buildings. However, because of health concerns, UFFI is little used nowadays.

As with all volatile organic compounds, the concentration of formaldehyde within a given indoor space will be very dependent upon the presence of important emission sources. The background concentration of formaldehyde in outdoor air is generally lower than 0.1 ppm (Maroni et al., 1995). Indoors, the rate of emission of formaldehyde varies according to conditions of temperature and humidity. Indoor formaldehyde concentrations usually exceed those observed outdoors. In an early study of 23 Danish homes by Anderson et al. (1975), the average formaldehyde concentration was 0.5 ppm (0.6 mg m^{-3}), with a range from 0.07 to 1.9 ppm (0.08–2.28 mg m^{-3}). Similar findings have subsequently been reported in Germany by Prescher and Jander (1987), in Finland by Niemala et al. (1985) and in the USA by Breysse (1984).

Adverse health effects from formaldehyde exposure may arise from inhalation, or direct contact. A range of acute health impacts have been attributed to the substance (Table 3). Exposure to concentrations of less than 1 ppm (1.2 mg m^{-3}) may result in sneezing, coughing, and minor eye irritation, although these symptoms often rapidly subside after the start of the exposure (Koeck et al., 1997). Numerous studies show that formaldehyde vapour is also an irritant of the skin (Eberlein-König et al., 1998) and the respiratory tract (Bardana and Montanaro, 1991).

There is conclusive evidence that formaldehyde is an animal carcinogen (Morgan, 1997). In the 1980s, a number of occupational research projects were carried out to address the potential carcinogenicity of formaldehyde in humans (e.g. Wong, 1983). All subjects were workers exposed to high concentrations of formaldehyde, but none of the studies found strong evidence of a cancer risk. One of the few pieces of evidence of a risk to humans from typical indoor exposures comes from the work of Vaughan et al. (1986) who reported a significant correlation between formaldehyde exposure and nasopharyngeal cancer in mobile home residents.

Table 3. Acute health effects from formaldehyde exposure

Formaldehyde concentration (ppm)	Observed health effects
< 0.05	None reported
0.05–1.5	Neurophysiologic effects
0.05–1.0	Odour threshold limit
0.01–2.0	Irritation of eyes
0.10–25	Irritation of upper airway
5–30	Irritation of lower airway and pulmonary effects
50–100	Pulmonary edema, inflammation, pneumonia
> 100	Coma, death

Sources: Hines et al. (1993).

2.5. Nitrogen dioxide

Nitrogen dioxide (NO_2) is a water-soluble red to brown gas with a pungent acrid odour. It is formed from the combination of nitrogen and oxygen during combustion at high temperatures (Maroni et al., 1995). Hence, the production of NO_2 is particularly associated with the operation of gas appliances, kerosene heaters, and woodburning stoves, as well as the smoking of cigarettes. Additionally, outdoor air can act as an important source for indoor NO_2 pollution (Chan et al., 1990) although, in many areas of the USA and Europe, ambient outdoor NO_2 levels are relatively low.

There have been numerous studies of indoor concentrations of NO_2. In the absence of emission sources, levels generally correlate well with those observed outdoors (Monn et al., 1998). In homes with gas cooking stoves in Albuquerque, New Mexico, Lambert et al. (1993a) found that 2-week average NO_2 levels were 21 ppb (39 μg m^{-3}) in bedrooms and 34 ppb (63 μg m^{-3}) in kitchens. In comparison, bedroom concentrations in homes using electric stoves averaged just 7 ppb (13 μg m^{-3}). On average, normal use of a vented gas cooking range adds 25 ppb (47 μg m^{-3}) of NO_2 to the background concentration in a home (Samet et al., 1987). In homes with unvented kerosene space heaters, 1-week average concentrations exceeding 45 ppb (84 μg m^{-3}) have been observed (Leaderer et al., 1986). One-week average levels of greater than 50 ppb (94 μg m^{-3}) have been reported in homes with unvented gas space heaters (Ryan et al., 1989).

Transient peak concentrations of NO_2 during the operation of appliances can greatly exceed average measurements. Whilst cooking with a gas range, peak levels in the kitchen may be as high as 400–1000 ppb (752–1880 μg m^{-3}) (Spengler et al., 1981). As individuals operating gas ranges often stay close to the appliance, personal exposures may be even greater.

Nitrogen dioxide is an oxidising agent that can be very irritating to the mucous membranes of the lung (Spengler, 1993). It is highly soluble in water, and a large proportion of inhaled NO_2 is removed in the respiratory tract (Lambert, 1997). It is thought to combine with water in the lungs to form nitric acid (HNO_3) and may react with lipids and proteins to form nitrite anions and hydrogen ions (Postlethwait and Bidani, 1990). Nitrogen dioxide within the airways is also converted into vapour phase nitrous acid (HONO) via heterogeneous reactions involving water vapour, invoking the formation of free oxygen radicals and lipo-peroxidation (Spicer et al., 1993). Whilst it has been suggested that oxidant injury is the principal mechanism by which NO_2 damages the lung, substantial uncertainty remains (Lambert, 1997).

Evidence from experimental research suggests that exposure to NO_2 may increase respiratory infections, and adversely affect lung function (Frampton et al., 1991). However, there is rather little evidence from epidemiological studies that exposure to NO_2 has deleterious health effects amongst the majority of the population. Two important projects have discovered possible associations. In Italy, Viegi et al. (1992) found the use of bottled gas for cooking was associated with increased reporting of cough and phlegm in males. In England, Jarvis et al. (1996) found that a general population sample of females who reported they used mainly gas for cooking were more likely to report respiratory symptoms in the 12 months prior to survey. These women were also found to have reduced lung function and increased airway obstruction. No effects were observed amongst males, suggesting women may be more susceptible to NO_2, or are exposed to higher concentrations.

There is evidence that the health effects of NO_2 may be greater amongst certain vulnerable population subgroups, such as children and asthmatics (Li et al., 1994). In the Harvard Six Cities Study, involving 10,106 children aged 6–10 yr, serious illness in infancy was more common amongst infants from homes with gas cooking (Ware et al., 1984). Recent meta-analyses suggest that the risk associated with NO_2 exposure is probably only significantly increased for children aged over 2 yr (Hasselblad et al., 1992). In Australia, Pilotto et al. (1997) monitored NO_2 exposures amongst 388 children aged between 6 and 11 yr. They found hourly peak levels of 80 ppb ($150 \mu g \ m^{-3}$) and above were associated with significant increases in the reporting of sore throats, colds, and absences from school, and they concluded that it was important to consider short-term peak exposures. However, in another study, Samet et al. (1993) found no association between personal exposures and symptoms amongst infants, and Brunekreff et al. (1990) failed to find an association between NO_2 and the pulmonary function of a sample of children.

Amongst asthmatic women and children, Goldstein et al. (1988) examined the relationship between mean 48 h NO_2 concentrations in the subjects' kitchens, and their spirometric lung function. They found that exposure to NO_2 levels ranging between 300 and 800 ppb (564 and $1504 \mu g \ m^{-3}$) was associated with a reduction in lung capacity (measured by FEV) of the order of 10%.

Salome et al. (1996) also found that the experimental exposure of 600 ppb (1128 µg m^{-3}) of NO_2 over a period of one hour was associated with a slight increase in airway hyperresponsiveness amongst a sample of 20 asthmatics.

Exposure to NO_2 may act as a trigger for asthma in one of two ways (Jones, 1997). One possibility is that the pollutant causes a direct effect on the lungs by inflicting toxic damage. Another is that it may irritate and sensitise the lungs, making individuals more susceptible to allergic response upon contact with indoor allergens. Evidence for the sensitisation mechanism comes from a study by Tunnicliffe et al. (1994); in asthmatics previously exposed to 400 ppb (752 µg m^{-3}) NO_2 for a period of 1 h, lung function dropped by 19% after the inhalation of house dust mite allergen. This was compared to a reduction of just 14% in those exposed only to air.

2.6. Sulphur dioxide

Sulphur dioxide (SO_2) is produced by the oxidation of sulphur impurities during the burning of coal and other fuels that contain sulphur (Burr, 1997). It is a colourless gas with a strong pungent odour that can be detected at about 0.5 ppm (0.9 mg m^{-3}). It is readily soluble in water and can be oxidised within airborne water droplets (Maroni et al., 1995). Sulphur dioxide levels are generally lower indoors than outdoors, and indoor/outdoor concentration ratios between 0.1 and 0.6 are commonly observed in buildings without indoor sources. As a result of reductions in emissions, annual mean levels of ambient SO_2 in most major cities in Europe and the USA are below 20 ppb (52 µg m^{-3}). However indoor SO_2 concentrations can be higher for homes with kerosene heaters and poorly vented gas and coal appliances. Leaderer et al. (1993) recorded average concentrations of 30 ppb (78 µg m^{-3}) in a study of 33 homes with kerosene space heaters in Connecticut, USA. Mean values of 57 ppb (149 µg m^{-3}) have been measured in homes equipped with both kerosene heaters and gas stoves (Leaderer et al., 1984).

From a health effects viewpoint, two substances are important; the SO_2 itself, and the acid aerosols that may result from its oxidation in the atmosphere. Absorption of SO_2 in the mucous membranes of the nose and upper respiratory tract occurs as a result of its aqueous solubility. The deposition pattern of acid aerosols within the respiratory tract will be dependent upon the size distribution of the droplets and the level of humidity indoors (Maroni et al., 1995).

Whilst evidence from animal experiments and occupational exposures suggests that exposure to extreme concentrations of SO_2 and acid aerosols can precipitate an acute reduction in lung function (Islam and Ulmer, 1979), there are relatively few indications of short-term health impacts associated with typical indoor concentrations. However, long term exposures to indoor SO_2 may be associated with elevated reporting of chronic respiratory complaints. In the UK, Burr et al. (1981) found that residents of South Wales who had open coal fires were more likely than others to suffer from breathlessness and wheezing. They speculated that a high death rate from respiratory conditions amongst miners wives may have been associated with exposure to SO_2 from the burning of concessionary coal, although particles may have also played a part. Recent studies in China, where the domestic burning of coal is still widespread, have also associated exposure to SO_2 with impaired lung function, and a range of other respiratory symptoms (Qin et al., 1993; Jin et al., 1993).

2.7. Radon

Radon is an inert radioactive gas that arises directly from the decay of radium-226 contained in various minerals (Lyman, 1997). It has a half-life of 3.82 d. As radon undergoes further radioactive decay, it produces a series of short-lived radioisotopes, known as radon daughters or progeny (Table 4). Radon itself is inert and causes little damage as most of it is exhaled in the breath. However, the progeny, Po-218 and Po-214 are electrically charged and can be inhaled either directly or through their attachment to airborne particles (Cohen, 1998). Once inhaled, they tend to remain in the lungs, where they may eventually cause cancer (Polpong and Bovornkitti, 1998). As activities such as the smoking of cigarettes can lead to considerably elevated levels of airborne particles, smokers are at particular risk from the inhalation of radon progeny (Hampson et al., 1998). Indeed, the US EPA has estimated that the cancer risk from radon for smokers is as much as 20 times the risk for individuals who have never smoked (US EPA, 1992).

In the past, contamination of air by radon and the subsequent exposure to radon daughters were believed to be a problem only for uranium and phosphate miners. However, it has recently been recognised that homes and buildings far away from uranium or phosphate mines can exhibit high concentrations of radon. As a consequence, radon and radon progeny are now recognised as important indoor pollutants (Létourneau, 1997).

Radon formed in rocks and soils is released into the surrounding air. Typical rates of radon release from soils throughout the world range from about 0.0002 to 0.07 bequerels (Bq) m^{-3} s^{-1}. Production rates from any soil are very dependent upon the geological characteristics of the soil itself and its underlying geological strata (Lévesque et al., 1997). Porous soils overlaying uranium rich alum shales, granites and pegmatites are a particularly high risk for radon, while gas-impermeable soils consisting of fine sand, silt, and moist clay present a low risk (IARC, 1988).

Outdoors, radon emanating from the ground is quickly dispersed, and concentrations never reach levels that may be a threat to health. However, inside

Table 4. The uranium-238 decay series

Nuclide	Half-life	Decay particle
U-238	4.5×10^9 yr	Alpha
Th-234	24 d	Beta
Pa-234	1.2 min	Beta
U-234	2.5×10^5 yr	Alpha
Th-230	8×10^4 yr	Alpha
Ra-226	1620 yr	Alpha
Radon-222	3.8 d	Alpha
Po-218	3 min	Alpha
Pb-214	27 min	Beta
Bi-214	20 min	Beta
Po-214	160μ	Alpha
Pb-210	22 yr	Beta
Bi-210	5 d	Beta
Po-210	138 d	Alpha
Pb-206	Stable	-

Source: Lyman (1997).

confined areas, low rates of air exchange can result in a build-up of radon and its daughters to concentrations tens of thousands of times higher than those observed outside (Wanner, 1993). Radon concentrations within a building depend very much on both the concentration of radon in the soil surrounding the structure, and the presence of entry points that allow the gas to infiltrate from outside (Jedrychowski et al., 1995). Some of the common entry points of radon into buildings include foundation joints, cracks in floors and walls, drains and piping, electrical penetrations, and cellars with earth floors (Nielson et al., 1997).

In the USA, Nero et al. (1986) summarised the results of 19 studies of indoor radon concentrations, covering 552 single-family homes. They found the mean indoor concentration was 56 Bq m^{-3}. As part of the more recent US National Residential Radon Survey, Marcinowski et al. (1994) estimated an annual average radon concentration of 46.3 Bq m^{-3} in US homes. They also calculated that around 6% of homes had radon levels greater than the US EPA action level for mitigation of 148 Bq m^{-3}. Outside the USA, Albering et al. (1996) found a much higher average concentration of 116 Bq m^{-3} in 116 homes in the township of Visé in a radon prone area in Belgium. In Italy, Bochicchio et al. (1996) reported an average concentration of 75 Bq m^{-3} in a sample of 4866 dwellings, and observed concentrations exceeding 600 Bq m^{-3} in 0.2% of homes. Yun et al. (1998) recently undertook one of the relatively few studies of radon concentrations in the office environment. In 94 Hong Kong office buildings they recorded radon concentrations similar to those that have been observed in domestic situations, with a mean of 51 Bq m^{-3}.

Radon exposure has been linked to lung carcinogenesis in both human and animal studies. It has also been associated with the development of acute myeloid and acute lymphoblastic leukaemia. However, the estimation of health risks from residential radon is extremely complex, and encompasses many uncertainties. Studies of smoking and non-smoking uranium miners indicate that radon is a substantial risk factor for lung cancer at high concentrations. Based on data on dose–response relationships amongst miners, it is estimated that between 5 and 15% of lung cancer deaths might be associated with exposure to residential radon (Steindorf et al., 1995). The relevance of data from mines to the lower-exposure home environment is often questioned (Lubin et al., 1997). Nevertheless, a meta-analysis of eight epidemiological studies undertaken by Lubin and Boice (1997) found that the dose–response curve associated with domestic radon exposures was remarkably similar to that observed amongst miners.

Ecological (geographical) study designs have been adopted by a number of recent epidemiological investigations into the health risks associated with non-industrial radon exposures. Lucie (1989) reported positive county level correlations between radon exposure and mortality from acute myeloid leukaemia in the UK, and Henshaw et al. (1990) found that mean radon levels in 15 counties were significantly associated with the incidence of childhood cancers and, specifically, all leukaemias. However, these reports have been met with considerable criticism because ecological designs can suffer from serious limitations (Wolff, 1991). In particular, the effects of migration are often difficult to account for, information on potential confounding variables can be unavailable, and estimates of exposure for populations of large areas may differ greatly from actual individual doses. Recent more refined ecological analyses, such as that undertaken by Etherington et al. (1996), have reported no association between indoor radon exposure and the occurrence of cancer.

An alternative to ecological analyses is the case-control study design, where radon exposures amongst individuals with cancer are compared to those of control subjects free from the disease. Most case-control studies have reported a small, but significant, association between radon exposure and lung cancer mortality (Pershagen et al., 1994). For example, in a recent examination of over 4000 individuals in Sweden, Lagarde et al. (1997) estimated that there is an excess relative risk of contracting lung cancer of between 0.15 and 0.20 per $100\,\text{Bq m}^{-3}$ increase in radon exposure.

2.8. Respirable particles

Smoke from the burning of wood and fossil fuels produces an extremely complex mixture of pollutants, both in physical and chemical characteristics, and toxicological properties (Lambert, 1997). One of the primary constituents of the smoke are respirable particles (Cooper, 1980). These are aerosols that are of a small enough diameter to enter and remain in the lung. Many are around 6–7 µm, or less in diameter (Martonen et al., 1992). The particulate mater comprises a mix of organic and inorganic substances including aromatic hydrocarbon compounds, trace metals, nitrates, and sulphates (Maroni et al., 1995). In developing countries, exposure to smoke is arguably the greatest indoor pollution problem, as in many buildings the burning of wood, charcoal, crop residues, or animal dung is often undertaken without adequate ventilation (Gold, 1992). For example, in India, Pandey et al. (1989) found that airborne particle concentrations, measured during cooking, were as high as 21,000 µg m^{-3}. Such figures are not typical of indoor exposures in the developed world.

In most developed societies, a movement away from the use of coal and wood to provide heat has meant that indoor particulate pollution associated with combustion is rather lower than in the past. However, the popularity of woodburning stoves has increased in both Europe and the USA in recent years (Samet, 1990). Whilst particulate pollution from modern airtight stoves is much lower than was previously common with open fires, problems can still occur during start-up, stoking, and reloading.

In the absence of significant indoor sources, indoor to outdoor ratios of respirable particles are generally slightly below unity (Wallace, 1996; Janssen et al., 1998). However, measurement techniques routinely used in reported studies are often not able to differentiate combustion related particles from those from other sources (Maroni et al., 1995). Given the heterogeneous nature of particulate material, this makes it impossible to formulate a direct relationship between exposure to specific compounds in the aerosol and health effects.

As part of the Harvard Six-City Study, Neas et al. (1994) recorded mean annual $PM_{2.5}$ (particles less than 2.5 µm in diameter) concentrations in the order of 17 µg m^{-3} in a sample of 470 non-smoking homes. The presence of a smoker can greatly add to indoor concentrations, and is discussed elsewhere in this review. In another sample of homes, Traynor et al. (1986) reported indoor concentrations of respirable particles slightly above background (up to 30 µg m^{-3}) during the use of airtight woodstoves and substantially higher with non-airtight stoves (200–1900 µg m^{-3}). When not in operation, homes with woodburning stoves have on average about 4 µg m^{-3} higher indoor particulate concentrations than homes without the appliances.

It is important to note that personal exposures to particles are often higher than ambient indoor concentrations. This is particularly the case during the daytime. For example, Clayton et al. (1993) observed that daytime mean personal PM_{10} (particles less than 10 µm in diameter) exposure in a sample of individuals was more than 50% higher than either indoor or outdoor levels. Similar findings have recently been reported by Janssen et al. (1998).

In general, the health impacts of outdoor exposures to respirable particles are rather better studied than those associated with indoor exposures (see Ostro and Chestnut, 1998). Hence, the most extensive data on the health effects from exposure to particles are derived from epidemiological studies of outdoor air. Given that total exposure to particles is greater indoors than outside by virtue of the time spent indoors, a good deal of the apparent effects of outdoor particles probably occurs due to exposure indoors. Whilst in vitro and animal experiments suggest that emissions from coal or woodburning stoves are mutagenic, epidemiological evidence from studies in the indoor environment is limited to a relatively small number of reports (Marbury, 1991).

Irritant effects from a range of inhaled particles may result in airway constriction. There is evidence that woodsmoke may be associated with respiratory illness, particularly amongst vulnerable groups such as children and those with pre-existing chronic respiratory disease. Honickey et al. (1985) examined respiratory symptoms amongst 62 children in Michigan, USA. They found 84% of children in homes with woodburning stoves recounted at least one severe symptom, compared to only 3% of children in homes without a stove. As part of the Six-City Study, Dockery et al. (1993) found an odds ratio of 1.32 (95% CI 0.99–1.76) for respiratory illness in households with woodburning stoves in comparison with those using other sources of heating. Koenig et al. (1993) reported that infants exposed to woodsmoke were more likely to recount asthma symptoms, and Abbey et al. (1998) observed a reduction in lung function in non-smokers exposed to high concentrations of indoor particles over a period of 20 yr.

Polycyclic aromatic hydrocarbons (PAHs) in woodsmoke are of particular concern because of their carcinogenic potential. These fat-soluble compounds are produced as a result of incomplete combustion, and include a large number of organic molecules which contain two or more benzene rings (Maroni et al., 1995). Once airborne, PAH compounds can be absorbed onto particles and inhaled into the lungs. It is thought that their carcinogenic properties arise as a result of their metabolism within the body (Sisovic et al., 1996). Evidence for this comes from studies such as that undertaken by Mumford et al. (1995). Amongst residents of Xuan Wei county, China, they found that urinary concentrations of PAH metabolites were highest in villages with correspondingly elevated lung cancer death rates.

2.9. *Tobacco smoke*

Although the health effects of smoking have been recognised for a number of decades, it is only relatively recently that concerns have been directed towards the inhalation of environmental tobacco smoke (ETS) by the non-smoking population. Tobacco smoke is an aerosol containing several thousand substances that are distributed as particles, vapours, and gasses (Gold, 1992). Environmental tobacco smoke is possibly one of the most important indoor air pollutants in homes and offices (Wanner, 1993). This is because a substantial proportion of the population are exposed to high concentrations of ETS on a regular basis. In 1990, it was estimated that approximately 50 million US citizens (corresponding to 26% of the adult population) were smokers (Rando et al., 1997), and around 70% of all children may be living in homes with at least one parent who smokes (Weiss, 1986).

The constituents of ETS are commonly sub-divided into those associated with sidestream and mainstream smoke. Mainstream (MS) smoke is that which is directly exhaled from the smoker, whilst sidestream smoke (SS) is emitted from the smouldering tobacco between puffs (Maroni et al., 1995). As information has become known about mainstream and sidestream smoke generation and its chemical composition, many irritants and carcinogens have been identified (Rando et al., 1997). By inhaling ETS, non-smokers are exposed to most of the toxins inhaled by active smokers, as well as some additional substances. For example, N-nitrosodimethylamine, a potent animal carcinogen, is emitted in quantities 20–100 times higher in SS than in MS smoke (Guerin et al., 1992), and a 'passive' smoker at a distance of 50 cm from a cigarette may inhale more than 10 times the amount of carbonyl compounds actively taken up by the smoker (Schlitt and Knöppel, 1989).

Some of the more common compounds found in ETS are outlined in Table 5. However, the exact composition of the smoke is known to be very dependent upon the type of tobacco being consumed, its packing density, the composition of the wrapping paper, and the puffing rate of the smoker (Hines et al., 1993).

Because of the complex nature of ETS, measurements of respirable particles are often used as indicators of indoor concentrations. Reported cigarette smoke particle sizes have varied in the range from 0.1 to 1.5 μm (Chen et al., 1990). Spengler et al. (1981) observed respirable particulate concentrations in 80 homes over several weeks. They found that a smoker of one pack a day contributed around 20 μg m^{-3} to 24 h indoor concentrations. However, because cigarettes are not uniformly smoked throughout the day, the authors concluded that short-term particulate concentrations of 500–1000 μg m^{-3} were likely when cigarettes were actually ignited. Over a period of 3 months, Leaderer et al. (1990) investigated indoor aerosol mass concentrations in a sample of almost 400 houses. They found that homes with smokers exhibited mass concentrations that were approximately three times higher than those observed in non-smoking residences.

As well as being an important pollutant of the domestic environment, ETS can be a serious problem in the workplace. In a report on tobacco smoke discomfort published in the USA (Anonymous, 1992) 43.5% of non-smoking employees recounted at least some discomfort from ETS at their place of work.

Table 5. Emission factors for mainstream and sidestream smoke

Substance	Mainstream (μg per cigarette)	SS/MS ratio	Calculated sidestream (μg per cigarette)[a]
Carbon dioxide	10,000–80,000	8.1	81,000–640,000
Carbon monoxide	500–26,000	2.5	1200–65,000
Oxides of nitrogen	16–600	4.7–5.8	80–3500
Ammonia	10–130	44–73	400–9500
Hydrogen cyanide	280–550	0.17–0.37	48–203
Formaldehyde	20–90	51	1000–4600
Acrolein	10–140	12	100–1700
N-nitrosodimethylamine	0.004–0.18	10–830	0.04–149
Nicotine	60–2300	2.6–3.3	160–7600
Total particulate	100–40,000	1.3–1.9	130–76,000
Phenol	20–150	2.6	52–390
Catechol	40–280	0.7	28–196
Naphthalene	2.8	16	45
Benzo(a)pyrene	0.008–0.04	2.7–3.4	0.02–0.14
Aniline	0.10–1.20	30	3–36
2-Naphthylamine	0.004–0.027	39	0.02–1.1
4-Aminobiphenyl	0.002–0.005	31	0.06–0.16
N-nitrosonornicotine	0.2–3.7	1–5	0.02–18

Source: Rando et al. (1997).
[a]Calculated from ratio of SS to MS.

Occupational groups particularly at risk from ETS include those such as bartenders and waiters who work in a traditionally smoky environment. Lambert et al. (1993b) recorded a median respirable concentration of 53.2 μg m^{-3} in a study of 7 restaurants, whilst a median concentration of 355 μg m^{-3} has been noted in a billiard lounge (Rando et al., 1997).

The most common acute health effects associated with exposure to ETS are eye, nose, and throat irritation (Maroni et al., 1995). For some people, eye tearing can be so intense as to be incapacitating. Smoke also contains substances that can activate the immune system, and approximately half of allergy prone individuals react to various extracts of tobacco leaf or smoke in skin tests (Maroni et al., 1995).

Exposure to ETS has been particularly associated with the exacerbation of symptoms in asthmatics (Jones, 1998). In Italy, parental smoking and the self-reported prevalence of asthma symptoms was examined amongst 3239 children by Forastiére et al. (1992). They found that the prevalence of asthma symptoms was significantly elevated in children whose mother smoked. Similarly, Gortmaker et al. (1982) found the prevalence of severe asthma was 2.2% amongst a sample of children with a smoking parent, compared to only 1.1% amongst children whose parents did not smoke. There is some evidence that exposure to

ETS may be important in actually inducing the onset of asthma in very young children. On the Isle of Wight, UK, Arshad et al. (1992) reported an odds ratio for asthma at 12 months of 3.33 (95% CI 0.8–14.6) if one parent smoked, compared to 11 (95% CI 2.5–48.2) if both were smokers.

Strong evidence is available for establishing a link between acute childhood lower respiratory tract illnesses and exposure to ETS at home (Somerville et al., 1988). Furthermore, passive smoking may have a significant effect on the level and growth of lung function in children (NRC, 1986) and the onset of chronic obstructive pulmonary disease in later life. Smoke-exposed children also have increased rates of hospitalisation; as far back as 1974, Harlap and Davies observed elevated rates of hospital admission for bronchitis, pneumonia, and bronchiolitis amongst the infants of smokers (Harlap and Davies, 1974). An important current field of research is investigating whether the greatest effects of environmental tobacco smoke on the lung are established prebirth, or are associated with exposures during the first few years of life (Gold, 1992).

The strongest evidence for the carcinogenic effects of ETS comes from research into cases of lung cancer. Based on the results from a case-control study of 191 individuals in the USA, Janerich et al. (1990) concluded that approximately 17% of lung cancers amongst non-smokers can be attributed to exposure to ETS during early life. The risks associated with exposure appear to be particularly high amongst women; a multi-centre national study of lung cancer in lifetime non-smokers found an increased cancer risk of approximately 30% in women whose husbands smoked (Rando et al., 1997). Although it is difficult to determine causality, there is evidence that cancer in sites other than the lung may be associated with passive smoking. Fukuda and Shibata (1990) discovered an association between ETS exposure and cancer of the sinus, and a relationship with the incidence of brain tumours has also been reported (Ryan et al., 1992).

2.10. Volatile organic compounds

Any chemical compound that contains at least one carbon and a hydrogen atom in its molecular structure is referred to as an organic compound. Organic compounds are further classified into various categories which include volatile organic compounds (VOCs), semi-volatile organic compounds (SVOCs) and non-volatile organic compounds (NVOCs). Volatile organic compounds are defined as having a boiling point that ranges between 50°C and 260°C (Maroni et al., 1995). Their low boiling point means that they will readily off-gas vapours into indoor air. However, the fact that materials containing them exhibit the desirable characteristics of good insulation properties, economy, fire-resistance,

Table 6.   Sources of common volatile organic compounds in indoor air

Sources	Examples of typical contaminants
Consumer and commercial products	Aliphatic hydrocarbons (*n*-decane, branched alkanes), aromatic hydrocarbons (toluene, xylenes), halogenated hydrocarbons (methylene chloride), alcohols, ketones (acetone, methyl ethyl ketone), aldehydes (formaldehyde), esters (alkyl ethoxylate), ethers (glycol ethers), terpenes (limonene, alpha-pinene).
Paints and associated supplies	Aliphatic hydrocarbons (*n*-hexane, *n*-heptane), aromatic hydrocarbons (toluene), halogenated hydrocarbons (methylene chloride, propylene dichloride), alcohols, ketones (methyl ethyl ketone), esters (ethyl acetate), ethers (methyl ether, ethyle ether, butyl ether).
Adhesives	Aliphatic hydrocarbons (hexane, heptane), aromatic hydrocarbons, halogenated hydrocarbons, alcohols, amines, ketones (acetone, methyl ethyl ketone), esters (vinyl acetate), ethers.
Furnishings and clothing	Aromatic hydrocarbons (styrene, brominated aromatics), halogenated hydrocarbons (vinyl chloride), aldehydes (formaldehyde), ethers, esters.
Building materials	Aliphatic hydrocarbons (*n*-decane, *n*-dodecane), aromatic hydrocarbons (toluene, styrene, ethylbenzene), halogenated hydrocarbons (vinyl-chloride), aldehydes (formaldehyde), ketones (acetone, butanone), ethers, esters (urethane, ethylacetate).
Combustion appliances	Aliphatic hydrocarbons (propane, butane, isobutane), aldehydes (acetaldehyde, acrolein).
Potable water	Halogenated hydrocarbons (1,1,1-trichloroethane, chloroform, trichloroethane).

Source: Maroni et al., 1995.

and ease of installation, means that their use in construction projects is widespread (Burton, 1997).

Table 6 lists sources of common VOCs found indoors. Out of a total of more than 900 chemical and biological substances that have been identified in indoor air, more than 350 VOCs have been recorded at concentrations exceeding 1 ppb (Brooks et al., 1991). The typical concentrations of some of the most frequently encountered compounds in the home are summarised by Table 7.

Indoor concentrations of VOCs have been well quantified; they are mostly considerably below the odour threshold but often exceed outdoor levels by up to 5 times (Wallace, 1991a). One of the earliest studies was undertaken by Mølhave (1979). He targeted 29 chemicals in 14 office buildings in Denmark. Most VOCs identified were alkylbenzenes and had concentrations in the range of 0.03–$2.8\,\mu g\ m^{-3}$. Shah and Singh (1988) found that levels of the majority

Table 7. Concentration ($\mu g\ m^{-3}$) and distribution of selected VOCs found in indoor air

Pollutant	Concentration Percentile				Mean
	10th	50th	90th	98th	
Benzene	2	10	20	30	10
Toluene	30	65	150	250	80
n-Decane	3	10	50	90	20
Limonene	2	15	70		30
o-Xylene	3	5	10		10
1,1,1-Trichloroethane	2	5	20		10
p-Dichlorobenzene	1	5	20		
1,2,4-Trimethylbenzene		5	20		10
m- and *p*-Xylene	10	20	40		20
Undecane	3	5	25		10
1,3,5-Trimethylbenzene		2	5		5
Dichloroethane		< 10	< 10	600	
Trichloroethane	1	5	20	30	

Source: IEH, 1996.

of a sample of 66 indoor VOCs ranged from 0.4 to $4\,\mu g\ m^{-3}$ and Brown et al. (1994) reported that concentrations of most VOCs they studied were below $5\,\mu g\ m^{-3}$.

In any given environment, the concentration of individual VOCs will be very variable and depend upon the presence or absence of an extremely wide range of potential emission sources. Hence, many publications report levels of 'Total Volatile Organic Compounds' (TVOCs) rather than individual values, although this of little help in determining the toxicological properties of substances (ECJRC, 1997). A large study of TVOC concentrations in a sample of 179 UK homes found that the mean concentration of all readings in the rooms measured was 200–$500\,\mu g\ m^{-3}$ (Brown and Crump, 1996). The maximum recorded concentration was $11,401\,\mu g\ m^{-3}$ in a living room. Mean concentrations were highest in main bedrooms, and lowest in second bedrooms. Similar findings have been reported by research in Denmark (Wolkoff et al., 1991), Germany, (Adlkofer et al., 1993) Sweden (Norbäck et al., 1993), and the USA (Hartwell, 1987).

It is important to note that average VOC concentrations may give an inaccurate indication of personal exposures (Rodes, 1991). This is because individuals are often situated close to emission sources, and may also be subjected to pollutants emitted from substances on their person. As an illustration of this, Wallace (1991a) reported the results of measurements of personal exposure to 25 VOCs among 51 Los Angeles Residents. Whilst ambient indoor maxima ranged from 10 to 100 µg m^{-3}, personal maxima were generally between 100 to 1000 µg m^{-3}.

VOC concentrations may be much higher than typical ambient levels in newly constructed buildings, or those in which building work or decoration has recently taken place. This is because many VOCs will off-gas a significant proportion of their volume in a relatively short time, and hence their concentrations will decline rapidly and exponentially. Because of this, builders and decorators may receive particularly high doses (Wieslander et al., 1997). For example, Wallace et al. (1991), found that breath concentrations of decane in a subject being studied increased by a factor of 100 (from 2.9–290 µg m^{-3}) after they were engaged in painting and the use of solvents.

Exposure to VOCs can result in both acute and chronic health effects. It is possible that asthmatics and other individuals with prior respiratory complaints may be particularly susceptible to low-dose VOC exposures. In a recent study, Norbäck et al. (1995) reported a positive association between levels of VOCs and the prevalence of nocturnal breathlessness amongst 88 Swedish asthmatics aged between 20–45 yr. However, most information on VOC toxicity has been established from animal and experimental studies at high concentrations, as levels in the majority of indoor environments are well below those required to demonstrate measurable health impacts.

At high concentrations, many VOCs are potent narcotics, and can depress the central nervous system (Maroni et al., 1995). Exposures can also lead to irritation of the eyes and respiratory tract, and cause sensitisation reactions involving the eyes, skin, and lungs. Mølhave (1991) reported complaints of unpleasant mucous membrane irritation amongst individuals exposed to a mix of 22 VOCs at a concentration of 8 µg m^{-3}. Because of the similarity of these symptoms, exposure to VOCs has frequently been attributed as a cause of sick building syndrome, discussed later in this text. To back this up, a number of studies have reported a strong association between mucous membrane irritation, central nervous system symptoms, and total exposure to VOCs amongst office workers (Hodgson et al., 1991).

At extreme concentrations, some VOCs may result in impaired neurobehavioral function (Burton, 1997). In an experimental study, Otto et al. (1992) noted that subjects exposed to a concentration of 22 VOCs at 25 µg m^{-3}, reported symptoms of headache, drowsiness, fatigue, and confusion. At concentrations as high as 188 µg m^{-3}, VOCs such as toluene may cause symptoms of lethargy, dizziness, and confusion. These may progress to coma, convulsions, and possibly death at levels in excess of 35,000 µg m^{-3} (Sandmeyer, 1982). However, these concentrations have never been recorded in the non-industrial environment.

Exposure to high concentrations of several VOCs commonly found in indoor air have been associated with cancers in laboratory animals. Based on laboratory evidence, Wallace (1991b) estimated that typical concentrations of seven VOCs exceeded the 1×10^{-6} risk of cancer by at least a factor of 10. These include benzene, vinylidine chloride, p-dichlorobenzene, chloroform, ethylene dibromide, methylene chloride and carbon tetrachloride. However, it should be noted that many risk calculations involve an assumption of a linear relationship of the dose–response curve when extrapolating from high to low exposures, and hence they are subject to much uncertainty (Stolwijk, 1991).

One recent theory proposes that the products of chemical reactions involving VOCs may be more important than direct exposure to the VOCs themselves (Wolkoff et al., 1997). This proposition is based on the fact that the majority of sick building syndrome studies have recorded concentrations of VOCs at considerably lower levels than those required to induce symptoms. Whilst there is an assumption here that sick building syndrome must be associated with exposure to VOCs, there is increasing evidence that many chemical reactions take place on surfaces and in air in the indoor environment (Reiss et al., 1995). Results from epidemiological research suggests that reactions between indoor ozone and VOCs may produce irritant substances that could cause SBS symptoms (Groes et al., 1996). Reactions involving nitrogen dioxide (Grosjean et al., 1992), and particles (Schneider et al., 1994) may also be important.

3. Pollution from biological sources

Whilst discussions of indoor air pollution frequently concentrate on chemical pollutants, the health effects of inhaled biological particles should not be overlooked, as a large variety of biological material is present in indoor environments (Montanaro, 1997). Biological agents can cause disease through atopic mechanisms, infectious processes, or direct toxicity. House dust in carpets, on sofas, and in air ducts is a major source of a range biological allergens (Lewis et al., 1994). The growth of moulds is not only aesthetically unpleasant, but can pose serious health problems, and some indoor environments provide ideal conditions for the maintenance of populations of harmful viruses and bacteria. The health effects of these agents are discussed below.

3.1. Indoor biological allergens

Table 8 summarises the main biological allergens found in indoor air. Probably the best studied source of indoor allergens is the house dust mite. In temperate climates, ten species of mites can be found from the genera *Dermatophagoides, Euroglyphus, Malayoglyphus, Hirstia,* and *Sturnophagoides* (D'Amato et al., 1994). D. *pteronyssinus* and E. *Maynei* mites appear to comprise around 95%

Table 8.   Common indoor allergic agents

Source	Genus	Species	Allergen
Dust-mite	*Dermatophagoides*	*pteronyssinus*	*Der p*
	Dermatophagoides	*farinae*	*Der f*
	Euroglyphus	*maynei*	*Eur m*
	Hirstia	*domicola*	*Hir d*
	Lepidoglyphus	*destructor*	*Lep d*
	Malayoglyphus	*intermedius*	*Mal I*
	Malayoglyphus	*carmelitus*	*Mal C*
	Sturnophagoides	*brasiliensis*	*Stu b*
Cat	*Felis*	*domesticus*	*Fel d*
Dog	*Canis*	*familiaris*	*Can f*
Rodent	*Mus*	*musculus*	*Mus m*
	Rattus	*norvegicus*	*Rat n*
Cockroach	*Blattela*	*germanica*	*Bla g*
	Periplanetta	*americana*	*Per a*
Fungi	*Alternaria*	*alternato*	*Alt a*
	Aspergillus	*fumigatus*	*Asp f*
	Cladosporium	*herbarium*	*Cla h*

of all mites found in a typical house (Hart and Whitehead, 1990). In tropical climates the mite *Blomia tropicalis* is also commonly found. Undoubtedly the most researched mite is the common house dust mite *D. pteronyssinus*. They measure 250–350 µm in size, and grow from egg to adult in around 25 d (D'Amato et al., 1994). Most mite populations prefer constant temperatures of around 25°C and a relative humidity of between 70 and 80% (Arlain et al., 1990; Salerno et al., 1992). They do not survive well in cool, dry conditions. Mites inhabit a range of soft furnishings, including sofas, fabrics, carpets, sheets, duvets, pillows, and mattresses.

The droppings from the mite are a primary source of indoor antigens (Kaliner and White, 1994). Mite faeces are encased in a coating of intestinal enzymes including a protein which is a strong allergen. Although over 10 different antigens have been characterised for *D. pteronyssinus* so far, the first to be identified, *Der p* 1, is the most commonly reported.

There may be up to 100,000 mite faecal particles, ranging between 10 and 40 µm in diameter, in a gram of house dust (Platts-Mills et al., 1991). The size of the particles, similar to that of pollen grains, means they do not remain airborne for long, although concentrations in disturbed rooms can be more than 1000 times higher than those observed in undisturbed environments (Kalra et al., 1990). A concentration of mite allergen above 2 µg *Der p* 1 g^{-1} (equivalent to 100 mites) of dust appears to represent a significant factor for mite sensitisation (WHO, 1995), and 10 µg *Der p* 1 g^{-1} increases the risk of triggering an acute or severe asthma attack in mite allergic individuals (IEH, 1996).

Typical *Der p* 1 concentrations are below $5 \mu g \ g^{-1}$ of dust (Verhoeff, 1994). However, the actual concentration of mite antigen in a home will strongly depend upon a range of factors, and is particularly influenced by climatic variables. This is illustrated by the work of Friedman et al. (1992). They reported mean concentrations of *Der p* 1 and *Der f* 1 as being 3.98 and $8.17 \mu g \ g^{-1}$ of dust in a sample of 15 New England homes during the month of June. In September the corresponding mean concentrations had risen to 21.43 and $14.53 \mu g \ g^{-1}$.

Given that inhalation is the main route of exposure for the majority of allergens in indoor air, it is unsurprising that most research has concentrated on the effects on atopic (allergic) respiratory diseases, and in particular asthma. Over 70 yr ago, Ancona (1923) made the first documented observation that mites may be a cause of respiratory symptoms. Today, the evidence suggests that exposure to airborne mite allergen will exacerbate symptoms in up to 85% of asthmatics (Platts-Mills and Carter, 1997). Studies of the effects of dust mite allergen exposure are numerous. Recently, Björnsson et al. (1995) reported that Swedish subjects inhabiting homes with large mite populations were more likely to report nocturnal breathlessness and other asthma-related symptoms. Peat et al. (1994) also found an association between exposure to dust mite *Der p* 1 and wheezing in the previous year amongst Australian children.

As well as being a major trigger of attacks of asthma, there is some evidence that, particularly amongst infants, exposure to mite allergen may induce the onset of the condition in previously healthy individuals (Tariq et al., 1998). In an important investigation, Arshad et al. (1992) studied the role of early exposure to dust-mite allergens in the development of allergic disorders amongst infants on the Isle of Wight. One hundred and twenty infants were randomly allocated to prophylactic and control groups. In the prophylactic group, the infants' bedrooms and living rooms were treated to control mite populations. In the control group, there were no interventions. After 12 months, the odds ratio for asthma was 4.13 (95% CI 1.1–15.5) amongst the control group. After 24 months, the excess number of cases amongst the control group had fallen, but still remained (Hide et al., 1994).

Like dust mites, cockroaches have been associated with manifestations of symptoms in individuals with allergic asthma, and up to 60% of asthmatics test positive to cockroach allergen in the USA (Kuster, 1996). Although cockroaches are generally indigenous to warm tropical climates, some species are able to thrive elsewhere due to the presence of central heating. The most commonly found cockroach in homes in Europe is the German Cockroach, *Blattella Germanica*, from which at least 5 antigens have been isolated (*Bla g* 1–5). In the USA, the American Cockroach, *Periplaneta americana* is ubiquitous. Sources of cockroach allergens have been identified in body parts, as well as faecal extracts (Musmand et al., 1995).

Van Wijnen et al. (1997) studied concentrations of *Bla g* 1 allergen in 46 homes in Amsterdam. It was detected in over 44% of homes, with mean concentrations ranging between 1.3 and 11 ng g^{-1} of dust. The highest level recorded in a single sample was 3899 ng g^{-1}. In general, concentrations were elevated in rooms with textile floor coverings. Similar findings were made in the USA by Rosenstreich et al. (1997). They measured concentrations of the allergen in dust taken from 476 homes situated in various inner-city locations. Interestingly, they found that asthmatic children allergic to cockroaches were three times more likely to be hospitalised for their asthma if they lived in a home with a large cockroach population. The authors concluded that the problems of cockroach sensitisation might be particularly severe amongst the residents of poor quality inner-city housing, as these homes provide an ideal environment for cockroach populations.

Domestic cats and dogs are a further important source of allergens in indoor air. This is particularly so in colder climates, which are less favourable to dust mites. Cat allergen (*Fel d*) is found in saliva, skin and dander. Particles of the primary cat allergen, *Fel d* 1, are often small, and hence remain airborne for many hours (Luczynska et al., 1990). *Fel d* 1 concentrations in domestic dust can exceed 10 µg g^{-1} and airborne levels can vary between 2 and 20 ng m^{-3} (Luczynska, 1994). As with cat allergens, dog allergens are mostly found in saliva and dander. The major dog allergen has been termed *Can f* 1 (Montanaro, 1997). Concentrations of *Can f* 1 in dust are often similar to *Fel d* 1, although *Fel d* 1 can remain airborne longer due to its smaller particle size (Puerta et al., 1997).

Because *Fel d* and *Can f* are readily transported by becoming attached to clothing, exposure to these allergens can even be a problem in public places such as schools, trains, and hospitals, where pets are not allowed. Custovic et al. (1998) recently examined concentrations of mite, cockroach, dog, and cat allergens in a sample of 14 British hospitals. Although levels of mite and cockroach allergen were low, high concentrations of *Fel d* 1 (Mean 22.9 µg g^{-1}, range 4.5–58) and *Can f* 1 (Mean 21.6 µg g^{-1}, range 4–63) were found in upholstered chairs. Out of a total of 10 sampling days, airborne *Can f* 1 was detected on every occasion (range 0.12–0.56 ng m^{-3}), whilst airborne *Fel d* 1 was recorded on 7 d (range 0.09–0.22 ng m^{-3}).

Recently, it has been proposed that concentrations of dog and cat allergens over 8 µg g^{-1} in house dust represent a threshold for allergic sensitisation (D'Amato et al., 1994). Because of the generally small particle size, dog and cat allergen readily enters the lung and can produce a rapid and severe asthmatic response. For example, Pollart et al. (1991) found 30 out of 188 asthmatic patients admitted to a hospital emergency department were allergic to cats, compared to only 1 out of 202 control asthmatics who did not require emergency treatment.

3.2. *Fungi, bacteria and viruses*

Microorganisms are an important form of biological pollution in the indoor environment. A large number of species of fungi and bacteria are found indoors, where they are associated with the presence of organic matter (e.g. wall coatings, wood, foodstuffs) (IEH, 1996). The outdoor air is one of the major sources of fungi and bacteria in indoor environments, particularly during the summer and autumn (Wanner et al., 1993). It is also well documented that high levels of humidity favour fungal growth (Sterling and Lewis, 1998). They are frequently found in homes that contain damp conditions, especially those with structural faults, or basements and underfloor crawl spaces (Montanaro, 1997).

A major difficulty to the sampling of viable airborne fungi is the large variability with time, even over short periods (IEH, 1996). A particular problem arises as temporal variations in airborne concentrations are often much greater in magnitude than variations between homes (Flannigan and Miller, 1994). Despite these difficulties, indoor concentrations of bacteria and fungi have been relatively well studied.

In the UK, Hunter et al. (1996) used air filtration to monitor 163 homes for the presence of fungi and bacteria over the period November 1990 to December 1992. The geometric mean count was 234 colony forming units CFU m^{-3} air for fungi, and 365.6 CFU m^{-3} air for bacteria. In a more intensive study of 35 of the houses, mean counts were 912 and 818 CFU m^{-3} air for fungi in living rooms and bedrooms respectively, compared to 917 and 933 CFU m^{-3} air for bacteria. *Penicillium* was the most frequently isolated fungus, found in 53% of samples. It was followed by *Cladosporium* and *Aspergillus*. The dominant bacteria were *Bacillus*, followed by *Staphylococcus* and *Micrococcus*. In Norway, Dotterud et al. (1995) found that *Penicillium* was the most common microfungus in homes and schools, followed by *Aspergillus*, *Cladosporium*, and *Mucor*. In a study in Japan, Takahashi (1997) reported that concentrations of fungal CFU ranged from between 13 to 3750 CFU m^{-3} in samples of air from indoor environments. They also found that concentrations were strongly correlated with indoor temperature and relative humidity, as well as outdoor climatic factors. Macher et al. (1991) measured rather less extreme spore concentrations, of the order of 198 CFU m^{-3}, in an apartment in the USA.

Exposure to airborne bacteria and fungi is associated with a number of well-defined diseases, as well as various less well-defined symptoms (Peat et al., 1998). Table 9 summarises some of the main reported health impacts. Although many studies use indoor dampness as a surrogate measure of microorganism concentrations, some direct measurements have been attempted. In the UK, Platt et al. (1989) examined the relationship between fungal spore counts and self-reported symptoms amongst the occupants of almost 600 homes. They found higher numbers of CFU m^{-3} were associated with an elevated preva-

Table 9.   Diseases and disease syndromes associated with exposure to bacteria and fungi

Disease/Syndrome	Examples of causal organisms cited
Rhinitis (and other upper respiratory symptoms)	*Alternaria, Cladosporium, Epicoccum*
Asthma	Various aspergilli and penicillia, *Alternaria, Cladosporium, Mucor, Stachybotrys, Serpula* (dry rot)
Humidifier fever	Gram-negative bacteria and their lipopolysaccharide endotoxins, Actinomycetes and fungi
Extrinsic allergic alveolitis	*Cladosporium, Sporobolomyces, Aureobasidium, Acremonium, Rhodotorula, Trichosporon, Serpula, Penicillium, Bacillus*
Atopic dermatitis	*Alternaria, Aspergillus, Cladosporium*

Source: IEH, 1996.

lence of reported wheeze and fever in children, and high blood pressure, and breathlessness in adults. Also in the UK, Potter et al. (1991) detected allergic responses to nine different fungal allergens in a survey of 2000 patients with allergic respiratory disease. In Sweden, Wickman et al. (1992) found that *Penicillium*, *Alternaria*, and *Cladosporium* moulds were more common in homes of children with allergies, and Neas et al. (1996) reported that morning lung function was inversely associated with *Epicoccum* and *Cladosporium* spore concentrations amongst a panel of 108 children. Contrary to many findings, Verhoff et al. (1994) were unable to find any relationship between microbial contamination and the directly measured pulmonary function of a sample of children living in 60 homes in the Netherlands. Although levels of CFU were higher in dust from mattresses in damp homes, they were not associated with objective measures of lung function amongst the infants.

Interestingly, the most recent research on moulds and fungi suggests that allergic reactions may not be the most important factor in the development of respiratory symptoms associated with exposure to spores (Howden-Chapman et al., 1996). Mycotoxins are toxic compounds produced naturally by many fungi (Hendry and Cole, 1993). They induce a wide range of acute and chronic systemic effects that cannot be attributed to fungal growth within the host. Samson et al. (1994) provide evidence that mycotoxins emitted by fungi are readily absorbed through the membranes of the respiratory tract, and it could be that their presence in the lung affects the immune system, precipitating the onset of symptoms.

Most bacterial and viral infections that spread within buildings are transmitted from human to human by airborne droplets (Ayars, 1997). More often than not, the building is an 'innocent bystander' and plays no role in harbouring populations of infectious agents, other than by providing a living environment for infected individuals. However, there are examples of serious bacterial and viral infections where the building itself can act as a potential reservoir (Burrel, 1991).

Probably the best-known infection associated with the indoor environment is Legionnaire's disease. It is caused by the bacteria *Legionella pneumophila*. It was first recognised at a legionnaire's convention in Philadelphia in 1976. One hundred and eighty four legionnaires attending the symposium developed symptoms of the disease, and 29 died (Fraser et al., 1977). The disease has an incubation period of 2–12 d, and an attack rate of 1–7%. Typical symptoms include malaise and headache, followed by dry cough, chest pain, diarrhoea, and altered mental status (Ayars, 1997). Since that epidemic, numerous other outbreaks and a host of other *Legionella* species have been identified (Hanrahan et al., 1987).

Legionella bacteria thrive in a warm and humid environment, and a common denominator in most outbreaks is a source of water in or around the building. Typical reservoirs of the bacteria include air humidifiers, air conditioning cooling towers, warm water supplies, shower heads and plumbing systems (Hines et al., 1993). The mechanisms by which *Legionella* reach the lungs were, until recently, thought to be via the inhalation of aerosols. However, recent evidence suggests that many cases may be associated with the aspiration of potable water (Yu, 1993).

An influenza illness known as Pontiac fever has also been identified, and a number of outbreaks have been documented. It is caused by *Legionella* organisms, but is more benign than Legionnaire's disease (Glick et al., 1978). Q fever is another bacterial infection that can spread in buildings independent of humans (Ayars, 1997), and the bacterium that causes tuberculosis can also be transmitted via air-management systems in closed environments. One of the most studied examples occurred in 1965 aboard a US naval vessel when 7 crew members developed tuberculosis that was spread via the ship's ventilation system (Houk et al., 1968).

Although outbreaks of bacterial infections are more commonly associated with indoor environments, there are also a number of serious viral diseases that can be harboured within buildings. Included are the often deadly but rare viruses that produce haemorrhagic fevers such as the Marburg virus, Ebola virus, and Lassa fever (Ayars, 1997). Many of these agents appear to be spread by aerosolised animal products, particularly urine, and hence may pose particular problems in buildings with rodent infestations.

4. Sick building syndrome

Over the past two decades, numerous field studies on indoor air quality and the sick building syndrome (SBS) have been conducted, mostly in office environ-

Table 10. Common symptoms of sick building syndrome

- Headache and nausea
- Nasal congestion (runny/stuffy nose, sinus congestion, sneezing)
- Chest congestion (wheezing, shortness of breath, chest tightness)
- Eye problems (dry, itching, tearing, or sore eyes, blurry vision, burning eyes, problems with contact lenses)
- Throat problems (sore throat, hoarseness, dry throat)
- Fatigue (unusual tiredness, sleepiness, or drowsiness)
- Chills and fever
- Muscle pain (aching muscles or joints, pain or stiffness in upper back, pain or stiffness in lower back, pain or numbness in shoulder/neck, pain or numbness in hands or wrists
- Neurological symptoms (difficulty remembering or concentrating, feeling depressed, tension, or nervousness)
- Dizziness
- Dry skin

Source: Wallace (1997).

ments. The symptoms of SBS are usually non-specific, and are often somewhat particular to the building being occupied by the workers (Table 10). Amongst employees, symptoms often worsen during working hours, and lessen or disappear after leaving the building. It may be that SBS does not warrant separate attention from the rest of the discussion of indoor air pollutants, as it is unlikely that its mechanisms are much different from those of some well defined pollution problems. However, the study of cases of SBS provides a valuable insight into the problems faced by investigators attempting to find evidence of causal relationships between any indoor air pollution and health.

For a time, exposure to VOCs was thought to be the major cause of SBS. This view was based on the results of experimental chamber studies in which persons exposed to mixtures of VOCs exhibited symptoms commonly associated with SBS (Mølhave et al., 1986). Some epidemiological studies have also supported the hypothesis that VOCs may be important. In a longitudinal study of a 'sick' library building, Berglund et al. (1990) found an association between the reporting of SBS symptoms and temporal variations in VOC concentrations. In a cross-sectional study of SBS symptoms amongst 147 office workers, Hodgson et al. (1991) found that VOC concentrations in the breathing zone of the building occupants were good predictors of mucous membrane irritation, and central nervous system complaints. However, despite these findings, a large number of studies have been unable to find any association between VOC exposures and SBS outbreaks. Indeed, one recent large-scale project actually reported a negative relationship (Sundell et al., 1993). Therefore it appears that, in many cases, VOCs may not be primarily to blame.

Some attention has recently been focused on the role that building ventilation systems may play in SBS (Bourbeau et al., 1997). There is much debate as to whether ventilation may actually be a cure or cause of SBS. Certainly, hot stuffy air is a common complaint in SBS studies, suggesting that efficient ventilation is desirable (Wallace, 1997). However, the exact role of building ventilation in this phenomenon is not well understood, and the results from studies that have been undertaken are often conflicting (Mendell, 1994). Although many previous reports have found reduced symptom prevalence to be associated with increased outside air ventilation (e.g. Nagda et al., 1991; Sundell et al., 1993) a more recent double-blind experiment by Jaakkola et al. (1994) found no significant difference in symptom prevalence amongst the occupants of buildings with ventilation rates between 13 and 42 cfm p^{-1} of outside air.

In an important meta-analysis of six previous studies, Mendell and Smith (1990) examined the relationship between symptom reporting and ventilation provision amongst office workers. Compared with those in naturally ventilated buildings, workers in air-conditioned offices consistently reported an increased prevalence of work related headache (OR = 1.3–3.1), lethargy (OR = 1.4–5.1), and upper respiratory/mucus membrane symptoms (OR = 1.3–4.8). Interestingly, the provision of mechanical ventilation without air conditioning was not associated with higher symptom prevalence.

The fact that some ventilation systems may themselves contribute to the concentrations of indoor pollutants is a possible explanation for the finding of Mendell and Smith (1990). Sources might include bacterial contamination, fungal mycotoxins, antiseptic agents and pesticides, or VOCs from air filtration systems or maintenance ducts. Harrison et al. (1987) found that, compared with the inhabitants of 8 naturally ventilated buildings, the occupants of 19 buildings with mechanical ventilation reported significantly higher frequencies of eye and nose irritation, headaches, attacks of lethargy, and dry skin. However, a number of more recent studies (e.g., Sundell, 1994) have failed to confirm these findings, and whilst recent research by Vincent et al. (1997) did find some evidence of an association between the mode of building ventilation and SBS symptoms amongst Parisian office workers, the authors concluded that ventilation only explained a relatively small proportion of non-specific symptoms.

Despite the voluminous research undertaken over the last two decades to investigate the reasons for SBS outbreaks, it has been estimated that no specific cause has been identified in over 75% of cases (Rothman and Weintraub, 1995). This may, in part, be due to methodological considerations. One problem is that currently available pollutant sampling techniques are costly and suffer from some technical limitations. This has often meant that only relatively limited data on the actual concentrations of pollutants that may be important in SBS have been available (Bardana, 1997). This is particularly the case for studies of VOCs, where many investigations have measured concentrations of only a defined subset of a limited number of compounds. Potentially important VOCs such as aldehydes and carboxylic acids of low molecular weight are rarely analysed quantitatively (Wolkoff et al., 1997).

Another methodological limitation that has hampered many studies is the lack of a standardised system for the reporting and diagnosis of symptoms. Numerous different questionnaires have been used to elicit information on SBS symptoms, meaning that measures of symptom prevalence are often not comparable between studies. Furthermore, the lack of standards for objective findings or diagnostic criteria, coupled with the non-specific nature of many symptoms, may lead to the potential introduction of bias by respondents who already have strong beliefs concerning the origin of their illness (Burton, 1997).

Whilst perceived sensory irritation is often a primary determinant in the assessment of the quality of indoor air, there is an increasing recognition that any stimulus–response may be greatly influenced by the complex environment surrounding the exposure, which can include the social context or the perceiver's mental state (Dalton, 1996). To illustrate this, Dalton et al. (1997) exposed three groups of individuals to 800 ppm of acetone for a 20 min period. Before the exposure, the groups were given different information about the consequences of long-term exposure to the substance. The study found that subjects given a positive characterisation of the consequences of exposure perceived significantly less odour and irritation during exposure than did subjects given a negative or neutral characterisation ($R^2 = 0.72$, $p < 0.001$).

Whilst a reliance on the self-reporting of symptoms for the assessment of indoor air quality might be problematic, there are uncertainties in formulating more objective measures. For example, Hempel-Jorgensen et al. (1998) examined the relationship between self-reported symptoms, cytological changes, and conjunctival hyperaemia (redness) in the eyes of subjects exposed to mixtures of VOCs. Whilst perceived irritation and hyperaemia did increase with exposure, cytological changes in samples of eye fluid were unrelated to both the level of exposure and the reporting of perceived irritation.

Methodological issues aside, it may be that the likely causal mechanisms that underlie most outbreaks of SBS have not yet been correctly identified. The theory discussed earlier that reactions involving VOCs and substances such as ozone may be more important than direct exposure to the VOCs themselves could be significant (Wolkoff et al., 1997). If reactive mechanisms do prove to be one cause SBS, the prudent siting of electrostatic office equipment (such as photocopiers, and laser printers) that produce these substances may be particularly significant in preventing outbreaks (Wolkoff et al., 1997).

Another possibility that has been suggested is that the symptoms of SBS may, in some part, be associated with the syndrome of multiple chemical sensitivity (MCS). Proponents of MCS believe that symptoms may be produced by exposure to many chemically distinct substances at very low doses (Nethercott, 1996). The concept of MCS has largely evolved out of concern over food allergies during the 1960s (Shorter, 1997), and recent research by Meggs et al. (1996) suggests that around one-third of US citizens consider themselves to be suffering from MCS. The existence of MCS could explain why outbreaks of SBS have been recorded in buildings with extremely low concentrations of individual pollutants. However, MCS is not generally recognised by traditional medicine and a number of recent studies have failed to identify any clinical mechanism that might underlie it (Wolf, 1996). For example, Simon et al. (1993) were unable to detect any immunological differences between 34 control subjects and a sample of 41 patients reporting MCS. Interestingly, the authors did, however, find that MCS patients were more likely to show symptoms of anxiety and depression. Hence it has been argued that MCS may have a much stronger psychological than clinical basis (Shorter, 1997).

The debate over MCS highlights the importance that psychological factors may play in SBS. The symptoms commonly associated with SBS also characterise the diagnostic criteria for generalised anxiety disorder and panic disorder. Furthermore, individual characteristics and non-toxicological job related factors often appear to be important correlates of SBS (Burton, 1997).

Research on occupational stress has repeatedly shown that factors such as work overload, role ambiguity, and low status can be important in determining the health and well-being of employees (Kivimäki, 1996). In their study of the significance of psychosocial factors on the prevalence of SBS symptoms amongst 450 Swedish office workers, Erikssonm and Höög (1993) found that the combination of a high workload, a perception of low control of the personal environment and workload, and lack of support from supervisors generated an increased risk of symptom reporting. In an extensive study of 4900 office employees in the USA, Wallace et al. (1993) also found that a heavy workload and conflicting demands in work were associated with headaches, eye irritations, dizziness, and nasal and chest discomfort. Amongst women, the researchers found that low educational status was also significant. Letz (1990) has produced a theoretical framework that explains how psychological mechanisms may explain the observed increases in the number of outbreaks of SBS in the last two decades. It is outlined in Fig. 1.

It seems, then, that SBS probably has a multi-factorial aetiology, in which chemical, physical, biological, and psychosocial factors all interact to produce symptoms and discomfort (Baker, 1989). Psychosocial factors may act as modifiers of the responses of individuals to chemical, physical, and biological challenges in the indoor environment (Lahtinen et al., 1998). It seems probable that work stress functions as a mediating factor around environmental symptoms, possibly by altering the body's sensitivity to perceived physical demands and

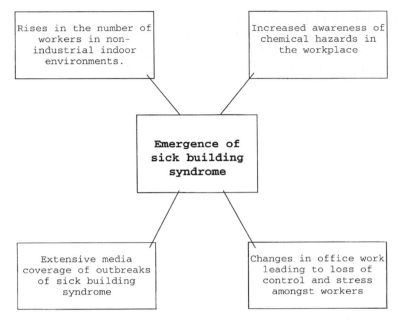

Figure 1. Possible socio-psychological factors responsible for the recent emergence of sick building syndrome.

toxic exposures (Hedge et al., 1992). However, despite the multitude of theories that have been proposed, the exact processes of SBS are still unknown, and many uncertainties remain.

5. Discussion

Differences in factors such as climatic conditions, lifestyles, and construction habits mean that indoor environments vary from region to region, and country to country. However, despite this variation, many industrialised societies are increasingly facing a growing number of very similar indoor air quality problems. Although considerable advances have been made in our knowledge and understanding of the sources of indoor air pollution, progress on the definition of the effects on human health and welfare has been on a much smaller scale (Stolwijk, 1992).

There is undoubtedly a need for more high-quality research to be undertaken to investigate further the links between indoor air quality and health. At present, projects concerning indoor air receive only around one tenth of the funding allocated to those studying the health effects of outdoor pollution (Seltzer, 1995). Given the evidence for the health impacts of indoor pollutants, this imbalance clearly needs to be addressed. Future research must fully acknowledge the complex nature of the indoor environment. Indoor air quality is the result of an intricate series of interactions involving many indoor and outdoor ventilation, microbiological, toxicological, and physical systems. Their interdependence may lead to the instigation of a multitude of synergistic processes. Hence the current paradigm, of categorising air quality problems according to single independent fields of expertise, will probably not provide solutions to most of unanswered questions.

Protocols must be developed and validated for analytical techniques that can consider, not just individual classes of pollutants, but a whole range of combinations and mixtures. It must be recognised that research limited to knowledge of a single class of pollutant may produce interesting results in a laboratory situation, but is unlikely to explain accurately the human health problems that will result from several interacting contaminants (Seltzer, 1995). All of the required advances will demand more refined methods for measuring and characterising pollutant emissions and concentrations. At present, techniques for measuring the complex combinations of pollutants that exist in many indoor environments are not well developed. In the future, it is important that this information is available in a useable form that will facilitate study, and aid the development of pollution mitigation strategies (Farland, 1991).

There is also a need for research to refine our understanding and measurement of health impacts. A lack of knowledge has led to particular difficulties in identifying many of the causes of sick building syndrome. Studies of sick buildings also highlight the need to consider mechanisms that lie outside the traditional toxicological dose–response relationships. Among the factors that should be encompassed in the development of future methodologies are the impact of environmental variables such as temperature and lighting, as well as psychological indicators that include job dissatisfaction, controllability, and stress.

As the health effects of indoor air pollution become better understood, the legal and regulatory aspects of the issue will require careful consideration. Nowadays, indoor air quality and pollution are becoming concerns not just of scientists, but also of the legal community.

The legal community's focus is on efforts to control the quality of indoor air through the passage of legislation and the development of regulations, as well as to impose liability upon those who are allegedly responsible for the indoor environment (Tokarz and Potterfield, 1994). It is important to remember that we live in an increasingly litigious society, and the number of claims from workers who feel their health has been damaged by the contamination of their surroundings is increasing.

With the introduction of the US Clean Air Act in 1970, the US EPA was faced with the task of setting national ambient air quality standards for a set of common outdoor air pollutants. No consideration of the health risks of indoor contamination was included. The Act was substantially revised in 1990, but still does not provide the EPA with clear authority to address indoor pollution. Complicating the public-policy issue is the shortage of solid data. Within the United Kingdom, the National Environmental Protection Plan, published in 1996, identifies indoor air quality as a key area for action (DOE, 1996), but provides no mandatory regulations. Indeed, outside the Scandinavian counties, Holland, and Germany, policy on indoor air quality issues is generally poorly developed in Europe as a whole (Harrison, 1997).

The introduction of new legislation may go some way towards reducing levels of contamination. However, it is important to remember that simple interventions by home-owners and building operators can also have major impacts. Measures such as the avoidance of smoking indoors, the deployment of dehumidifiers in damp environments, and the use of natural ventilation are important. Here, education will be critical. Good education programmes are the key. The US Master Home Environmentalist Program (MHEP) is an example of a public education programme designed to reduce exposures amongst homeowners by providing personal contact with experts in the home (Roberts and Dickey, 1995). Volunteers operating the scheme provide free home assessments and follow-ups to families wishing to reduce their exposure to indoor air pollutants, house dust, and chemicals. Programmes such as this may be one of the most efficient ways to teach control methods to motivated individuals.

It might be necessary to re-evaluate the way in which buildings are designed and constructed in the future. Whilst problems with existing homes and offices are often difficult to overcome, architects and builders may need to be educated so as to create less contaminated structures. Simple measures, such as the selection of building materials and indoor fittings on the principle that the emission of pollutants such as VOCs and formaldehyde should be as low as reasonably achievable may prove beneficial. More radical ideas, such as a change of attitude with respect to the energy efficiency of buildings may also be required. Increased fuel costs could be offset against potential reductions in the considerable financial burden from the treatment of both acute and chronic indoor pollution related diseases. In these areas, a policy shift from a reactive to a proactive mode with respect to building regulations may be required. However, improved data on the links between building design and health will have to precede the development and enforcement of any new construction legislation.

Meeting the challenges that are posed by indoor air pollution will take some time. A strong commitment will be required within both the public and private sectors, and the provision of adequate funding will be essential. In the meantime, we must manage our indoor environments using the best scientific advice available. Our accumulating knowledge of the sources, exposures, and health impacts of indoor pollutants will need to be put to good use to ensure that future indoor environments provide the healthiest possible conditions.

References

Abbey, D.E., Burchette, R.J., Knutsen, S.F., McDonnell, W.F., Lebowitz, M.D., Enright, P.L., 1998. Long-term particulate and other air pollutants and lung function in non-smokers. American Journal of Respiratory Critical Care Medicine 158 (1), 289–298.

Adlkofer, F., Angerer, J., Ruppert, T., Scherer, G., Tricker, A.R., 1993. Determination of benzene exposure from occupational and environmental sources. In: (Anon) Volatile Organic Compounds in the Environment. Indoor Air International, Rotherfleet, pp. 511–518.

Akland, G.G., Hartwell, T.D., Johnson, T.R., Whitmore, R.W., 1985. Measuring human exposure to carbon monoxide in Washington, DC, and Denver, Colorado, during the winter of 1982–1983. Environmental Science and Technology 19 (10), 911–918.

Albering, H.J., Hoogewerff, J.A., Kleinjans, J.C., 1996. Survey of ^{222}Rn concentrations in dwellings and soils in the Dutch Belgian border region. Health Physics 70 (1), 64–69.

Amitai, Y., Zlotogorski, Z., Golan-Katzav, V., Wexler, A., Gross, D., 1998. Neuropsychological impairment from acute low-level exposure to carbon monoxide. Archives of Neurology 55 (6), 845–848.

Ancona, G., 1923. Asma epidemica da pediculoides ventricosus. Policlinico Sez Med 30 (1), 45–70.

Anderson, I., Lundquist, G.R., Molhave, L., 1975. Indoor air pollution due to chipboard used as a construction material. Atmospheric Environment 9, 1121–1127.

Anonymous, 1992. Discomfort from environmental tobacco smoke among employees at worksites with minimal smoking restrictions. US Morbidity and Mortality Weekly Report 41 (20), 351–354.

Arashidani, K., Yoshikawa, M., Kawamoto, T., Matsuno, K., Kayama, F., Kodama, Y., 1996. Indoor pollution from heating. Industrial Health 34 (3), 205–215.

Arlain, L.G., Rapp, C.M., Ahmed, S.G., 1990. Development of Dermatophagoides pteronyssinus. Journal of Medical Entomology 27 (6), 1035–1040.

Arshad, S.H., Matthews, S., Gant, C., Hide, D.W., 1992. Effect of allergen avoidance on development of allergic disorders in infancy. Lancet 339, 1493–1497.

Ayars, G.H., 1997. Biological agents and indoor air pollution. In: Bardana, E.J., Montanaro, A. (Eds.), Indoor Air Pollution and Health. Marcel Dekker, New York, pp. 11–60.

Baker, D.B., 1989. Social and organisational factors in office building-associated illness. Occupational Medicine 4 (4), 607–624.

Bardana, E.J., 1997. Sick building syndrome — a wolf in sheep's clothing. Annals of Allergy and Asthma Immunology 79 (4), 283–293.

Bardana, E.J., Montanaro, A., 1991. Formaldehyde: an analysis of its respiratory, cutaneous, and immunologic effects. Annals of Allergy 66 (6), 441–452.

Bascom, R., Kesavanathan, J., Swift, D.L., 1995. Human susceptibility to indoor contaminants. Occupational Medicine 10 (1), 119–132.

Berglund, B., Brunekreef, B., Knoppel, H., Lindvall, T., Maroni, M., Mølhave, L., Skov, P., 1992. Effects of indoor air pollution on human health. Indoor Air 2 (1), 2–25.

Berglund, B., Jöhansson, I., Lindvall, T.H., Lundin, L., 1990. A longitudinal study of perceived air quality and comfort in a sick library building. In: Proceedings of the Fifth International Conference on Indoor Air Quality and Climate, Vol. 3. Canada Mortgage and Housing Corporation, Ottawa, Canada.

Bigonon, J., Peto, J., Saracci, R. (Eds.), 1989. Non-occupational Exposure to Mineral Fibres. International Agency for Research on Cancer, Lyon.

Björnsson, E., Nörback, D., Janson, C., 1995. Asthmatic symptoms and indoor levels of microorganisms and house dust mites. Clinical and Experimental Allergy 25 (5), 423–431.

Blomquist, G., 1994. Sampling of biological particles. Analyst 119 (1), 53–56.

Bochicchio, F., Campos-Venuti, G., Nuccetelli, C., Piermattei, S., Risica, S., Tommasino, L., Torri, G., 1996. Results of the representative Italian national survey on radon indoors. Health Physics 71 (5), 741–748.

Bourbeau, J., Brisson, C., Allaire, S., 1997. Prevalence of the sick building syndrome symptoms in office workers before and six months and three years after being exposed to a building with an improved ventilation system. Occupational and Environmental Medicine 54 (1), 49–53.

Breysse, P.A., 1984. Formaldehyde levels and accompanying symptoms associated with individuals residing in over 1000 conventional and mobile homes in the state of Washington. Indoor Air 3, 403–408.

Brimblecombe, P., 1987. The Big Smoke: a History of Air Pollution in London Since Medieval Times. Routledge, London.

Brooks, B.O., Utter, G.M., DeBroy, J.A., Schimke, R.D., 1991. Indoor air pollution: an edifice complex. Clinical Toxicology 29 (3), 315–374.

Brown, S.K., Sim, M.R., Abramson, N.J., Gray, C.N., 1994. Concentrations of volatile organic compounds in indoor air–a review. Indoor Air 4 (1), 123–134.

Brown, V.M., Crump, D.R., 1996. Volatile organic compounds. In: Berry, R.W., Brown, V.M., Coward, S.K.D. et al. (Eds.), Indoor Air Quality in Homes, the Building Research Establishment Indoor Environment Study, part 1. Construction Research Communications, London.

Brunekreef, B., Houthuijs, D., Dijkstra, L., Boleij, J.S., 1990. Indoor nitrogen dioxide exposure and children's pulmonary function. Journal of the Air and Waste Management Association 40 (9), 1252–1256.

Burr, M.L., 1997. Health effects of indoor combustion products. Journal of the Royal Society of Health 117 (6), 348–350.

Burr, M.L., 1995. Pollution: does it cause asthma? Archives of Disorders in Childhood 72 (5), 377–387.

Burr, M.L., St-Leger, A.S., Yarnell, J.W.G., 1981. Wheezing, dampness, and coal fires. Community Medicine 3 (3), 205–209.

Burrel, R., 1991. Microbiological agents as health risks in indoor air. Environmental Health Perspectives 95 (11), 29–34.

Burton, B.T., 1997. Volatile organic compounds. In: Bardana, E.J., Montanaro, A. (Eds.), Indoor Air Pollution and Health. Marcel Dekker, New York, pp. 127–153.

Chan, C.C., Yanagisawa, Y., Spengler, J.D., 1990. Personal and indoor/outdoor nitrogen dioxide exposure assessments of 23 homes in Taiwan. Toxicology and Industrial Health 6 (1), 173–182.

Chen, B.T., Nanenyi, J., Yeh, H.C., Mauderly, J.L., Cuddihy, R.G., 1990. Physical characterisation of cigarette-smoke aerosol generated from a walton smoke machine. Aerosol Science and Technology 12 (2), 364.

Choi, I.S., 1983. Delayed neurological sequelae in carbon monoxide intoxication. Archives of Neurology 40 (7), 433–435.

Clayton, C.A., Perritt, R.L., Pellizzari, E.D., Thomas, K.W., Whitmore, R.W., Özkaynak, H., Spengler, J.D., Wallace, L.A., 1993. Particle total exposure assessment methodology (PTEAM) study: distributions of aerosol and elemental concentrations in personal, indoor, and outdoor air samples in a Southern California community. Journal of Exposure Analysis and Environmental Epidemiology 3 (2), 227–250.

Cobb, N., Etzel, R.A., 1991. Unintentional carbon-monoxide related deaths in the United States, 1979 through 1988. Journal of the American Medical Association 266 (5), 659–663.

Cohen, B.S., 1998. Deposition of charged particles on lung airways. Health Physics 74 (5), 554–560.

Cooper, J.A., 1980. Environmental impact of residential wood combustion emissions and its implications. Journal of the Air Pollution Control Association 30, 855–861.

Coultas, D.B., Lambet, W.E., 1991. Carbon monoxide. In: Sanet, J.M., Spengler, J.D. (Eds.), Indoor Air Pollution: A Health Perspective. Johns Hopkins University Press, Baltimore, MD, pp. 187–208.

Custovic, A., Fletcher, A., Pickering, C.A., 1998. Domestic allergens in public places III: house dust mite, cat, dog and cockroach allergens in British hospitals. Clinical and Experimental Allergy 28 (1), 53–59.

Dalton, P., 1996. Odor perception and beliefs about risk. Chemical Senses 21, 447–458.

Dalton, P., Wysocki, C.J., Brody, M.J., Lawley, H.J., 1997. The influence of cognitive bias on the perceived odor, irritation, and health symptoms from chemical exposure. International Archives of Occupational and Environmental Health 69, 407–417.

D'Amato, G., Liccardi, G., D'Amato, M., 1994. Environment and the development of respiratory allergy. II: Indoors. Monaldi Archive of Chest Disorders 49 (5), 412–420.

Dahms, T.E., Younis, L.T., Wiens, R.D., Zarnegar, S., Byers, S.L., Chaitman, B.R., 1993. Effects of carbon monoxide exposure in patients with documented cardiac arrhythmias. Journal of the American College of Cardiology 21 (2), 442–450.

Dockery, D.W., Pope, C.A., Xu, X., Spengler, J.D., Ware, J.H., Martha, E.F., Ferris, B.G., Spezier, F.E., 1993. An association between air pollution and mortality in six U.S. cities. The New England Journal of Medicine 329 (24), 1753–1759.

DOE (Department of the Environment), 1997. The United Kingdom National Air Quality Strategy. H.M. Stationary Office, London.

DOE (Department of the Environment), 1996. The United Kingdom National Environmental Health Action Plan. H.M. Stationary Office, London.

Doll, R., Peto, J., 1985. Asbestos: Effects on Health of Exposure to Asbestos. H.M. Stationary Office, London.

Dotterud, L.K., Vorland, L.H., Falk, E.S., 1995. Viable fungi in indoor air in homes and schools in the Sor-Varanger community during winter. Paediatric Allergy and Immunology 6 (4), 181–186.

Eberlein-König, B., Przybilla, B., Kühnl, P., Pechak, J., Gebefügi, I., Klenschmidt, J., Ring, J., 1998. Influence of airborne nitrogen dioxide or formaldehyde on parameters of skin function and cellular activation in patients with atopic eczema and control subjects. Journal of Allergy and Clinical Immunology 101 (1), 141–143.

ECJRC (European Commission Joint Research Centre), 1997. Total volatile organic compounds (TVOC) in indoor air quality investigations. European Commission, Luxembourg.

Erikssonm, N., Höög, J., 1993. The office illness project in northern Sweden, the significance of psychosocial factors for the prevalence of the 'sick building-syndrome'. A case reference study. In: Proceedings of the Sixth International Conference on Indoor Air Quality and Climate, Vol. 1. Helsinki, Finland, pp. 333–337.

Etherington, D.J., Pheby, D.F., Bray, F.I., 1996. An ecological study of cancer incidence and radon levels in south west England. European Journal of Cancer 32 (7), 1189–1197.

Farland, W.H., 1991. Future directions and research needs. Environmental Health Perspectives 95, 131–133.

Flannigan, B., Miller, J.D., 1994. Health implications of fungi in indoor environments–an overview. In: Sampson, R.A., Verhoeff, A.P., Flannigan, B., Flannigan, M.E., Adan, O., Hoekstra, E. (Eds.), Health Implications of Fungi in Indoor Environments. Elsevier, Amsterdam.

Forastiére, F., Corbo, G.M., Michelozzi, P., Pistelli, R., Agabiti, N., Brancato, G., Ciappi, G., Perucci, C.A., 1992. Effects of environment and passive smoking on the respiratory health of children. International Journal of Epidemiology 21 (1), 66–73.

Frampton, M.W., Morrow, P.E., Cox, C., Gibb, F.R., Speers, D.M., Utell, M.J., 1991. Effects of nitrogen dioxide exposure on pulmonary function and airway reactivity in normal humans. American Review of Respiratory Disorders 143 (3), 522–527.

Fraser, D.W., Tsai, T.R., Orestein, W., 1977. Legionnaire's disease: description of an epidemic of pneumonia. New England Journal of Medicine 297 (22), 1187–1197.

Friedman, F.M., Friedman, H.M., O'Connor, G.T., 1992. Prevalence of dust mite allergens in homes and workplaces of the Upper Connecticut river valley of New England. Allergy Proceedings 13, 259–262.

Fukuda, K., Shibata, A., 1990. Exposure–response relationships between woodworking, smoking, or passive smoking, and squamous cell neoplasms of the maxillary sinus. Cancer Causes and Control 1 (2), 165–168.

Furtaw, E.J., Pandian, M.D., Nelson, D.R., Behar, J.V., 1996. Modelling indoor air concentrations near emission sources in imperfectly mixed rooms. Journal of the Air and Waste Management Association 46 (9), 861–868.

Gaensler, E.A., 1992. Asbestos exposure in buildings. Clinics in Chest Medicine 13 (2), 231–242.

Glick, T.H., Gregg, M., Berman, B., 1978. Pontiac fever: an epidemic of unknown etiology in a health department. I. Clinical and epidemiological aspects. American Journal of Epidemiology 107 (2), 149–160.

Gold, D.R., 1992. Indoor air pollution. Clinics in Chest Medicine 13 (2), 215–229.

Goldstein, I.F., Andrews, L.R., Hartel, D., 1988. Assessment of human exposure to nitrogen dioxide, carbon monoxide, and respirable particles in New York inner-city residences. Atmospheric Environment 22 (10), 2127–2139.

Gortmaker, S.L., Walker, D.K., Jacobs, F.H., Ruch-Ross, H., 1982. Parental smoking and the risk of childhood asthma. American Journal of Public Health 72 (6), 574–579.

Groes, L., Pejtersen, J., Valbjørn, O., 1996. Perceptions and symptoms as a function of indoor environmental factors and building characteristics in office buildings. In: Proceedings of the Sixth International Conference on Indoor Air Quality and Climate, Vol. 4. Nagoya, Japan, pp. 237–242.

Grosjean, D., Williams, E.L., Seinfeld, J.H., 1992. Atmospheric oxidation of selected terpenes and related carbonyls: gas-phase carbonyl products. Environmental Science and Technology 26 (8), 1526–1533.

Guerin, M.R., Jenkins, R.A., Tomkins, B.A., 1992. The Chemistry of Environmental Tobacco Smoke: Composition and Measurement. Lewis Publishers, Chelsea, MI.

Hampson, S.E., Andrews, J.A., Lee, M.E., Foster, L.S., Glasgow, R.E., Lichtenstein, E., 1998. Lay understanding of synergistic risk: the case of radon and cigarette smoking. Risk Analysis 18 (3), 343–350.

Hanrahan, J.P., Morse, D.L., Scharf, V.B., 1987. A community hospital outbreak of legionellosis. Transmission by potable hot water. American Journal of Epidemiology 125 (4), 639–649.

Harlap, S., Davies, A.M., 1974. Infant admissions to hospital and maternal smoking. Lancet 1 (857), 529–532.

Harrison, J., Pickering, A.C., Finnegan, M.J., Austick, P.K.C., 1987. The sick building syndrome–further prevalence studies and investigations of possible causes. In: Proceedings of the Fourth International Conference on Indoor Air Quality and Climate. Institute for Water, Soil, and Air Hygiene, Berlin, pp. 487–491.

Harrison, P.T.C., 1997. Health impacts of indoor air pollution. Chemistry and Industry 17, 677–681.

Hart, B.L., Whitehead, L., 1990. Ecology of house dust mites in Oxfordshire. Clinical and Experimental Allergy 20, 203–209.

Hartwell, T.D., Pellizzari, E.D., Perritt, R.L., 1987. Results from the Total Exposure Assessment Methodology (TEAM) study in selected communities in Northern and Southern California. Atmospheric Environment 21, 1995–2004.

Hasselblad, V., Eddy, D.M., Kotchmar, D.J., 1992. Synthesis of environmental evidence: nitrogen dioxide epidemiology studies. Journal of the Air and Waste Management Association 42 (5), 662 671.

Hedge, A., Erickson, W.A., Rubin, G., 1992. Effects of personal and occupational factors on sick building syndrome reports in air-conditioned offices. In: Quick, J.C., Murphy, L.R., Hurrell, J.J. (Eds.), Work and Well-being: Assessment and Interventions for Occupational Mental Health. American Psychological Association, Washington, DC, pp. 286–298.

Hempel-Jorgensen, A., Kjaergaard, S.K., Molhave, L., 1998. Cytological changes and conjunctival hyperemia in relation to sensory eye irritation. International Archives of Occupational and Environmental Health 71 (4), 225–235.

Hendry, K.M., Cole, E.C., 1993. A review of mycotoxins in indoor air. Journal of Toxicology and Environmental Health 38 (2), 183–198.

Henshaw, D.L., Eatough, J.P., Richardson, R.B., 1990. Radon as a causative factor in induction of myeloid leukaemia and other cancers. Lancet 335, 1008–1012.

Hide, D.W., Matthews, S., Matthews, L., Stevens, M., Ridout, S., Twiselton, R., Gant, C., Arshad, S.H., 1994. Effect of allergen avoidance in infancy on allergic manifestations at age 2 years. Journal of Allergy and Clinical Immunology 93 (5), 842–846.

Hines, A.L., Ghosh, T.K., Loyalka, S.K., Warder, R.C. (Eds.), 1993. Indoor Air – Quality and Control. Prentice-Hall, Englewood Cliffs, NJ.

Hodgson, M.J., Frohlinger, J., Permar, E., Tidwell, C., Traven, N.D., Olenchock, S.A., Karpf, M., 1991. Symptoms and micro-environmental measures in non-problem buildings. Journal of Occupational Medicine 33 (4), 527–533.

Honickey, R.E., Osborne, J.S., Akpom, C.A., 1985. Symptoms of respiratory illness in young children and the use of wood-burning stoves for indoor heating. Pediatrics 75 (3), 587–593.

Horvath, E.P., 1997. Building-related illness and sick building syndrome: from the specific to the vague. Cleveland Clinical Journal of Medicine 64 (6), 303–309.

Houck, P.M., Hampson, N.B., 1997. Epidemic carbon monoxide poisoning following a winter storm. Journal of Emergency Medicine 15 (4), 469–473.

Houk, V.N., Baker, J.H., Sorensen, K., Kent, D.C., 1968. The epidemiology of tuberculosis infection in a closed environment. Archives of Environmental Health 16 (1), 26–35.

Howden-Chapman, P., Isaacs, N., Crane, J., Chapman, R., 1996. Housing and health: the relationship between research and policy. International Journal of Environmental Health Research 6 (2), 173–185.

Howell, J., Keiffer, M.P., Berger, L.R., 1997. Carbon monoxide hazards in rural Alaskan homes. Alaskan Medicine 39 (1), 8–11.

Huncharek, M., 1994. Asbestos and cancer: epidemiological and public health controversies. Cancer Investigations 12 (2), 214–222.

Hunter, C.A., Hull, A.V., Higham, D.F., Grimes, C.P., Lea, R.G., 1996. Fungi and bacteria. In: Berry, R.W., Brown, V.M., Coward, S.K.D. et al. (Eds.), Indoor Air Quality in Homes, the Building Research Establishment Indoor Environment Study, part 1. Construction Research Communications, London.

IARC (International Agency for Research on Cancer), 1988. IARC Monographs on the Evaluation of the Carcinogenic Risk of Chemicals to Humans, Vol. 43: Man-made Mineral Fibres and Radon. International Agency for Research on Cancer, Lyon.

IEH (Institute for Environment and Health), 1996. IEH assessment on indoor air quality in the home. Institute for Environment and Health, Leicester, UK.

Islam, M.S., Ulmer, W.T., 1979. Threshold concentrations of SO_2 for patients with oversensitivity of the bronchial system. Wissenschaft und Umwelt 1 (1), 41–47.

Jaakkola, J.J.K., Tuomaala, P., Seppänen, O., 1994. Air recirculation and sick building syndrome: a blinded crossover trial. American Journal of Public Health 84, 422–428.

Janerich, D.T., Thompson, W.D., Varela, L.R., 1990. Lung cancer and exposure to tobacco smoke in the household. New England Journal of Medicine 323 (10), 632–636.

Janssen, N.A., Hoek, G., Brunekreef, B., Harssema, H., Mensink, I., Zuidhof, A., 1998. Personal sampling of particles in adults: relation among personal, indoor, and outdoor air concentrations. American Journal of Epidemiology 147 (6), 537–547.

Jarvis, D., Chinn, S., Luczynska, C., Burney, P., 1996. Association of respiratory symptoms and lung function in young adults with the use of domestic gas appliances. Lancet 1 (8999), 426–431.

Jedrychowski, W., Flak, E., Wesolowski, J., Liu, K.S., 1995. Relation between residential radon concentrations and housing characteristics. The Cracow Study. Central European Journal of Public Health 3 (3), 150–160.

Jin, H., Zheng, M., Mao, Y., Wan, H., Hang, Y., 1993. The effect of indoor pollution on human health. In: Jantunen, M., Kalliokoski, P., Kukkonen, E., Saarela, K., Seppänen, A., Vuorelma, H. (Eds.), Proceedings of the Sixth International Conference on Indoor Air Quality and Climate. Helsinki, Finland, pp. 477–482.

Jones, A.P., 1998. Asthma and domestic air quality. Social Science and Medicine 47 (6), 755–764.

Jones, A.P., 1997. The main causative factors of asthma, available information on population exposure to those factors, and current measures of control. In: Asthma in Europe. European Federation of Asthma and Allergy Associations, Holland, pp. 207–301.

Kaliner, M.A., White, M.V., 1994. Asthma: causes and treatment. Complimentary Therapy 20 (11), 645–650.

Kalra, S., Owen, S.J., Hepworth, J., Woodcock, A., 1990. Airborne house dust mite antigen after vacuum cleaning (Letter). Lancet 336 (8712), 449.

Kivimäki, M., 1996. Stress and Personality Factors. In: Specifications of the Role of Test Anxiety, Private Self-Consciousness, Type A Behaviour Pattern, and Self-Esteem in Relationship Between Stressors and Stress Reactions. Finnish Institute of Occupational Medicine, Helsinki.

Koeck, M., Pichler-Semmelrock, F.P., Schlacher, R., 1997. Formaldehyde – study of indoor air pollution in Austria. Central European Journal of Public Health 5 (3), 127–130.

Koenig, J.Q., Larson, T.V., Hamley, Q.S., Rebolledo, V., Dumler, K., Checkoway, H., Wang, S.Z., Lin, D., Pierson, W.E., 1993. Pulmonary lung function in children associated with fine particulate matter. Environmental Research 63 (1), 26–38.

Kuster, P.A., 1996. Reducing the risk of house dust mite and cockroach allergen exposure in inner-city children with asthma. Pediatric Nursing 22, 297–299.

Lagarde, F., Pershagen, G., Akerblom, G., Axelson, O., Bäverstam, U., Damber, L., Enflo, A., Svartengren, M., Swedjemark, G.A., 1997. Residential radon and lung cancer in Sweden: risk analysis accounting for random error in the exposure assessment. Health Physics 72 (2), 269–276.

Lahtinen, M., Huuhtanen, P., Reijula, K., 1998. Sick building syndrome and psychosocial factors – a literature review. Indoor Air (Suppl. 4), 71–80.

Lambert, W.E., 1994. Urban exposures to carbon monoxide and myocardial ischemia in men with ischemic heart disease. Ph.D. Thesis. University of California.

Lambert, W.E., 1997. Combustion pollution in indoor environments. In: Bardana, E.J., Montanaro, A. (Eds.), Indoor Air Pollution and Health. Marcel Dekker, New York, pp. 83–103.

Lambert, W.E., Samet, J.M., Hunt, W.C., Skipper, B.J., Schwab, M., Spengler, J.D., 1993a. Nitrogen dioxide and respiratory illness in infants – part II: assessment of exposure to nitrogen dioxide (research report 58). Health Effects Institute, Cambridge, MA.

Lambert, W.E., Samet, J.M., Spengler, J.D., 1993b. Environmental tobacco smoke concentrations in no-smoking and smoking sections of restaurants. American Journal of Public Health 83 (9), 1339–1341.

Lan, Q., Chen, W., Chen, H., He, X.Z., 1993. Risk factors for lung cancer in non-smokers in Xuanwei County of China. Biomedical and Environmental Science 6 (2), 112–118.

Leaderer, B.P., Stowe, M., Li, R., Sullivan, J., Koutrakis, P., Wolfson, M., Wilson, W., 1993. Residential levels of particle and vapor phase acid associated with combustion sources. In: Jantunen, M., Kalliokoski, P., Kukkonen, E., Saarela, K., Seppänen, A., Vuorelma, H. (Eds.), Proceedings of the Sixth International Conference on Indoor Air Quality and Climate. Helsinki, Finland, pp. 147–152.

Leaderer, B.P., Koutrakis, P., Briggs, S.L.K., Rizzuto, J., 1990. Measurement of toxic and related air pollutants. In: Proceedings of the EPA/Air and Waste Management Association International Symposium (VIP-17). United States Environmental Protection Agency, Washington, DC, pp. 567.

Leaderer, B.P., Zagraniski, R.T., Berwick, M., Stolwijk, J.A.J., 1986. Assessment of exposure to indoor air contaminants from combustion sources: methodology and application. American Journal of Epidemiology 124 (2), 275–289.

Leaderer, B.P., Stolwijk, J.A.J., Zagraniski, R.T., Qing-Shan, M., 1984. A field study of indoor air contaminant levels associated with unvented combustion sources. In: Proceedings of the 77th Annual Meeting of the Air Pollution Control Association, San Francisco, CA.

Lee, R.J., Van Orden, D.R., Corn, M., Crump, K.S., 1992. Exposure to airborne asbestos in buildings. Regulatory Toxicology and Pharmacology 16 (1), 93–107.

Letz, G.A., 1990. Sick building syndrome: acute illness among office workers – the role of building ventilation, airborne contaminants, and work stress. Allergy Proceedings 11 (3), 109–116.

Lévesque, B., Gauvin, D., McGregor, R., Martel, R., Gingras, S., Dontigny, A., Walker, W.B., Lajoie, P., Létourneau, E., 1997. Radon in residences: influences of geological and housing characteristics. Health Physics 72 (6), 907–914.

Lewis, R.G., Fortmann, R.C., Camann, D.E., 1994. Evaluation of methods for monitoring the potential exposure of small children to pesticides in the residential environment. Archives of Environmental and Contamination Toxicology 26 (1), 37–46.

LHEA (London Health Education Authority), 1997. What people think about air pollution, their health in general, and asthma in particular. Health Education Authority, London.

Li, Y., Powers, T.E., Roth, H.D., 1994. Random effects linear regression meta-analysis models with application to nitrogen dioxide health effects studies. Journal of the Air and Waste Management Association 44 (3), 261–270.

Lipfert, F.W., 1997. Air pollution and human health: perspectives for the '90s and beyond. Risk Analysis 17 (3), 137–146.

Longo, L.D., 1977. The biological effects of carbon monoxide on the pregnant woman, fetus, and new-born infant. American Journal of Obstetrics and Gynaecology 129 (1), 69–103.

Lubin, J.H., Boice, J.D., 1997. Lung cancer risk from residential radon: meta-analysis of eight epidemiological studies. Journal of the National Cancer Institute 89 (1), 49–57.

Lubin, J.H., Tomásek, L., Edling, C., Hornung, R.W., Howe, G., Kunz, E., Kusiak, R.A., Morrison, H.I., Radford, E.P., Samet, J.M., Tirmarche, M., Woodward, A., Yao, S.X., 1997. Estimating lung cancer mortality from residential radon using data for low exposures of miners. Radiation Research 147 (2), 126–134.

Lucie, N.P., 1989. Radon and leukaemia. Lancet 2 (8654), 99–100.

Luczynska, C.M., 1994. Risk factors for indoor allergen exposure. Respiratory Medicine 88 (10), 723–729.

Luczynska, C.M., Li, Y., Chapman, M.D., Platts-Mills, T.A.E., 1990. Airborne concentrations and particle size distribution of allergen derived from domestic cats: measurements using a cascade impactor, liquid impinger and a two-site monoclonal antibody assay for Fel d I. American Review of Respiratory Disorders 141 (2), 361–367.

Lyman, G.H., 1997. Radon. In: Bardana, E.J., Montanaro, A. (Eds.), Indoor Air Pollution and Health. Marcel Dekker, New York, pp. 83–103.

Macher, J.M., Huang, F.Y., Flores, M., 1991. A two-year study of microbiological indoor air quality in a new apartment. Archives of Environmental Health 46 (1), 25–29.

Madany, I.M., 1992. Carboxyhemoglobin levels in blood donors in Bahrain. Science of the Total Environment 116 (1), 53–58.

Marbury, M.C., 1991. Wood Smoke. In: Samet, J.M., Spengler, J.D. (Eds.), Indoor Air Pollution: A Health Perspective. John Hopkins University Press, Baltimore, MD, pp. 209–222.

Marcinowski, F., Lucas, R.M., Yeager, W.M., 1994. National and regional distributions of airborne radon concentrations in US homes. Health Physics 66 (6), 699–706.

Maroni, M., Seifert, B., Lindvall, T. (Eds.), 1995. Indoor Air Quality – a Comprehensive Reference Book. Elsevier, Amsterdam.

Martonen, T.B., Katz, I., Fults, K., Hickey, A.J., 1992. Use of analytically defined estimates of aerosol respirable fraction to predict lung deposition. Pharmaceutical Research 9 (12), 1634–1639.

McDonald, J.C., 1991. An epidemiological view of asbestos in buildings. Toxicology and Industrial Health 7 (5–6), 187–193.

Meggs, W.J., Dunn, K.A., Bloch, R.M., Goodman, P.E., Davidoff, A.L., 1996. Prevalence and nature of allergy and chemical sensitivity in a general population. Archives of Environmental Health 51 (4), 275–282.

Mendell, M.J., Fine, L., 1994. Building ventilation and symptoms – where do we go from here? (editorial). American Journal of Public Health 84 (3), 346–348.

Mendell, M.J., Smith, A.H., 1990. Consistent pattern of elevated symptoms in air-conditioned office buildings: a reanalysis of epidemiologic studies. American Journal of Public Health 80 (10), 1193–1199.

Mølhave, L., 1991. Indoor climate, air pollution, and human comfort. Journal of Exposure Analysis and Environmental Epidemiology 1 (1), 63–81.

Mølhave, L., 1979. Indoor air pollution due to building materials. In: Proceedings of the First International Indoor Climate Symposium, Copenhagen, Denmark, p. 89.

Mølhave, L., Bach, B., Pedersen, O.F., 1986. Human reactions to low concentrations of volatile organic compounds. Environment International 12 (1–4), 167–175.

Monn, C., Brändli, O., Schindler, C., Ackermann-Liebrich, U., Leuenberger, P., 1998. Personal exposure to nitrogen dioxide in Switzerland: SAPALDIA team Swiss study on air pollution and lung diseases in adults. Science of the Total Environment 215 (3), 243–251.

Montanaro, A., 1997. Indoor allergens: description and assessment of health risks. In: Bardana, E.J., Montanaro, A. (Eds.), Indoor Air Pollution and Health. Marcel Dekker, New York, pp. 201–214.

Morgan, K.T., 1997. A brief review of formaldehyde carcinogenesis in relation to rat nasal pathology and human health risk assessment. Toxicologic Pathology 25 (3), 291–307.

Moriske, H.J., Drews, M., Ebert, G., Menk, G., Scheller, C., Schöndube, M., Konieczny, L., 1996. Indoor air pollution by different heating systems: coal burning, open fireplace and central heating. Toxicology Letters 88 (1–3), 349–354.

Mumford, J.L., Li, X., Hu, F., Lu, X.B., Chuang, J.C., 1995. Human exposure and dosimetry of polycyclic aromatic hydrocarbons in urine from Xuan Wei, China with high lung cancer mortality associated with exposure to unvented coal smoke. Carcinogenesis 16 (12), 3031–3036.

Musmand, J.J., Horner, W.E., Lopez, M.M., Lehrer, S.B., 1995. Identification of important allergens in German cockroach extracts by sodium dodecylsulphate-polyacrylamide gel electrophoresis and Western blot analysis. Journal of Allergy and Clinical Immunology 95, 877–885.

Nagda, N.L., Koontz, M.D., Billick, I.H., Leslie, N.P., Behrens, D.W., 1996. Causes and consequences of backdrafting of vented gas appliances. Journal of the Air and Waste Management Association 46 (9), 838–846.

NASA (National Aeronautics and Space Administration), 1973. Bioastronautics Data Book – SP 3006. NASA, Washington, DC.

Neas, L.M., Dockery, D.W., Burge, H., Koutrakis, P., Speizer, F.E., 1996. Fungus spores, air pollutants, and other determinants of peak expiratory flow rate in children. American Journal of Epidemiology 143 (8), 797–807.

Neas, L.M., Dockery, D.W., Ware, J.H., Spengler, J.D., Ferris, B.G., Speizer, F.E., 1994. Concentration of indoor particulate matter as a determinant of respiratory health in children. American Journal of Epidemiology 139 (11), 1088–1099.

Nero, A.V., Schwehr, M.B., Nazaroff, W.W., Revzan, K.L., 1986. Distribution of airborne radon-222 concentrations in US homes. Science 234 (4779), 992–997.

Nethercott, J.R., 1996. Whither multiple chemical sensitivities?. American Journal of Contact Dermatology 7 (4), 199–201.

Nielson, K.K., Rogers, V.C., Holt, R.B., Pugh, T.D., Grondzik, W.A., de Meijer, R.J., 1997. Radon penetration of concrete slab cracks, joints, pipe penetrations, and sealants. Health Physics 73 (4), 668–678.

Niemala, R., Vaino, H., 1985. Formaldehyde exposure in work and the general environment. Scandinavian Journal of Work and Environmental Health 7 (1), 95–100.

Norbäck, D., Björnsson, E., Janson, C., Widstrom, J., Boman, G., 1995. Asthma and the indoor environment: the significance of emission of formaldehyde and volatile organic compounds from newly painted indoor surfaces. Occupational and Environmental Medicine 52 (6), 388–395.

Norbäck, D., Björnsson, E., Widström, J., 1993. Asthma symptoms in relation to volatile organic compounds (VOCs) and bacteria in dwellings. In: (Anon) Volatile Organic Compounds in the Environment. Indoor Air International, Rotherfleet, pp. 377–386.

NRC (National Research Council), 1986. Environmental Tobacco Smoke: Measuring Exposures and Assessing Health Effects. National Academy Press, Washington, DC.

Ostro, B., Chestnut, L., 1998. Assessing the health benefits of reducing particulate matter air pollution in the United States. Environmental Research 76 (2), 94–106.

Otto, D., Hundell, H., House, D., Mølhave, L., Counts, W., 1992. Exposure of humans to a volatile organic mixture. I. Behavioural assessment, Archives of Environmental Health 47 (1), 23–30.

Pandey, M.R., Smith, K.R., Boleij, J.S.M., 1989. Indoor air pollution in developing countries and acute respiratory infection in children. Lancet 1 (8635), 427–429.

Peat, J.K., Dickerson, J., Li, J., 1998. Effects of damp and mould in the home on respiratory health: a review of the literature. Allergy 53, 120–128.

Peat, J.K., Tovey, E., Gray, E.J., Mellis, C.M., Woolcock, A.J., 1994. Asthma severity and morbidity in a population sample of Sydney schoolchildren. Australian and New Zealand Journal of Medicine 24 (3), 270–276.

Pershagen, G., Akerblom, G., Axelson, O., Clavenson, B., Damber, L., Desai, G., Enflo, A., Lagarde, F., Mellander, H., Svartengren, M., Swedjemark, G.A., 1994. Residential radon exposure and lung cancer in Sweden. New England Journal of Medicine 330 (3), 159–164.

Pettenkofer, M.S., 1858. Über den luftwechel in Wohngebauden. Cottashe Buchhandlung, Munich.

Pilotto, L.S., Douglas, R.M., Attewell, R.G., Wilson, S.R., 1997. Respiratory effects associated with indoor nitrogen dioxide exposure in children. International Journal of Epidemiology 26 (4), 788–796.

Platt, S.D., Martin, C.J., Hunt, S.M., Lewis, C.W., 1989. Damp housing, mould growth, and symptomatic health state. British Medical Journal 298, 1673–1678.

Platts-Mills, T.A.E., Carter, M.C., 1997. Asthma and indoor exposure to allergens. New England Journal of Medicine 336 (19), 1382–1384.

Platts-Mills, T.A.E., Woodfolk, J.A., Chapman, M.D., Heymann, P.W., 1996. Changing concepts of allergic disease: the attempt to keep up with real changes in lifestyles. Journal of Allergy and Clinical Immunology 98 (Suppl.), S297–S306.

Platts-Mills, T.A.E., Ward, G.W., Sporik, R., Gelber, L.E., Chapman, M.D., Heymann, P.W., 1991. Epidemiology of the relationship between exposure to indoor allergens and asthma. International Archives of Allergy and Applied Immunology 94 (1–4), 339–345.

Pollart, S.M., Smith, T.F., Morris, E.C., Gelber, L.E., Platts-Mills, T.A.E., Chapman, M.D., 1991. Environmental exposure to cockroach allergen: analysis with monoclonal antibody based enzyme immunoassays. Journal of Allergy and Clinical Immunology 87 (2), 505–510.

Polpong, P., Bovornkitti, S., 1998. Indoor Radon. Journal of the Medical Association of Thailand 81 (1), 47–57.

Postlethwait, E.M., Bidani, A., 1990. Reactive uptake governs the pulmonary air space removal of inhales nitrogen dioxide. Journal of Applied Physiology 68 (2), 594–603.

Potter, P.C., Juritz, J., Little, F., McCaldin, M., Dowdle, E.B., 1991. Clustering of fungal-allergen specific IgE antibody responses in allergic subjects. Annals of Allergy 66 (2), 149–153.

Prescher, K.E., Jander, K., 1987. Formaldehyde in indoor air. Bundesgesundheitsblatt 30, 273–278.

Puerta, L., Fernandez-Caldas, E., Lockey, R.F., 1997. Indoor allergens: detection and environmental control measures. In: Bardana, E.J., Montanaro, A. (Eds.), Indoor Air Pollution and Health. Marcel Dekker, New York, pp. 215–229.

Qin, Y.H., Zhang, X.M., Jin, H.Z., Liu, Y.Q., Fan, D.L., Cao, Z.J., 1993. Effects of indoor air pollution on respiratory illness of school children. In: Jantunen, M., Kalliokoski, P., Kukkonen, E., Saarela, K., Seppänen, A., Vuorelma, H. (Eds.), Proceedings of the Sixth International Conference on Indoor Air Quality and Climate. Helsinki, Finland, pp. 477–482.

Rando, R.J., Simlote, P., Salvaggio, J.E., Lehrer, S.B., 1997. Environmental tobacco smoke: measurement and health effects of involuntary smoking. In: Bardana, E.J., Montanaro, A. (Eds.), Indoor Air Pollution and Health. Marcel Dekker, New York, pp. 61–82.

Redlich, C.A., Sparer, J., Cullen, M.R., 1997. Sick-building syndrome. Lancet 349 (9057), 1013–1016.

Reiss, R., Ryan, P.B., Koutrakis, P., Tibbetts, S.J., 1995. Ozone reactive chemistry on interior latex paint. Environmental Science and Technology 29 (8), 1906–1912.

Roberts, J.W., Dickey, P., 1995. Exposure of children to pollutants in house dust and indoor air. Reviews of Environmental Contamination and Toxicology 143 (1), 59–78.

Robinson, J., Nelson, W.C., 1995. National Human Activity Pattern Survey Data Base. United States Environmental Protection Agency, Research Triangle Park, NC.

Rodes, C.E., Kamens, R.M., Wiener, R.M., 1991. The significance and characteristics of the personal activity cloud on exposure assessments for indoor contaminants. Indoor Air 2, 123–145.

Roggli, V.L., Longo, W.E., 1991. Mineral fibre content of lung tissue in patients with environmental exposures: household contacts vs. building occupants. Annals of the New York Academy of Science 643, 511–518.

Rosenstreich, D.L., Eggleston, P., Kattan, M., 1997. The role of cockroach allergy and exposure to cockroach allergen in causing morbidity amongst inner-city children with asthma. New England Journal of Medicine 336, 1356–1363.

Rothman, A.L., Weintraub, M.I., 1995. The sick building syndrome and mass hysteria. Neurologic Clinics 13 (2), 405–412.

Roughton, F.J.W., Darling, R.C., 1994. The effect of carbon monoxide on the oxyhemoglobin dissociation curve. American Journal of Physiology 141 (1), 17–31.

Ryan, C.M., Morrow, L.A., 1992. Dysfunctional buildings or dysfunctional people: an examination of the sick building syndrome and allied disorders. Journal of Consulting and Clinical Psychology 60 (2), 220–240.

Ryan, P.B., Lee, M.W., North, B., McMichael, A.J., 1992. Risk factors for tumours of the brain and meninges: results from the Adelaide adult brain tumour study. International Journal of Cancer 51 (1), 20–27.

Ryan, P.B., Hemphill, C.P., Billick, I.H., Nagda, N.L., Koontz, M.D., Fortmann, R.C., 1989. Estimation of nitrogen dioxide concentrations in homes equipped with unvented gas space heaters. Environment International 15 (1–6), 551–556.

Salerno, M.S.R., Huss, K., Huss, E.W., 1992. Allergen avoidance in the treatment of dust-mite allergy and asthma. Nursing Practitioner 17 (1), 53–65.

Salome, C.M., Brown, N.J., Marks, G.B., Woolcock, A.J., Johnson, G.M., Nancarrow, P.C., Quigley, S., Tiong, J., 1996. Effect of nitrogen dioxide and other combustion products on asthmatic subjects in a home-like environment. European Respiratory Journal 9 (5), 910–918.

Samet, J.M., 1990. Environmental controls and lung disease. American Review of Respiratory Disorders 142 (6), 915–939.

Samet, J.M., Lambert, W.E., Skipper, B.J., Cushing, A.H., Hunt, W.C., Young, S.A., McLaren, L.C., Schwab, M., Spengler, J.D., 1993. Nitrogen dioxide and respiratory illnesses in infants. American Review of Respiratory Disorders 148 (8), 1258–1265.

Samet, J.M., Marbury, M.C., Spengler, J.D., 1987. Health effects and sources of indoor air pollution: part 1. American Review of Respiratory Disorders 136 (6), 1486–1508.

Samson, R.A., Flannigan, B., Flannigan, M.E., Verhoeff, A.P., Adan, O.C.G., Hoekstra, E.S., 1994. Health indicators of fungi in indoor environment. In: Air Quality Monographs, Vol. 2. Elsevier, Amsterdam.

Sandmeyer, E.E., 1982. Aromatic hydrocarbons. In: Clayton, G.D., Clayton, F.E. (Eds.), Patty's Industrial Hygiene and Toxicology, Vol. 2, 3rd Edition. Wiley, New York, pp. 3253–3431.

Schlitt, H., Knöppel, H., 1989. Carbonyl compounds in mainstream and sidestream tobacco smoke. In: Bieva, C.J., Courtois, Y., Govaerts, M. (Eds.), Present and Future of Indoor Air Quality. Excerpta Medica, Amsterdam, pp. 197–206.

Schneider, T., Bohgard, M., Gudmundsson, A., 1994. A semiempirical model for particle deposition onto facial skin and eyes. Role of air currents and electric fields. Journal of Aerosol Science 25 (3), 583–593.

Schwarzberg, M.N., 1993. Carbon dioxide level as migraine threshold factor: hypothesis and possible solutions. Medical Hypotheses 41 (1), 35–36.

Seltzer, J.M., 1997. Sources, concentrations, and assessment of indoor pollution. In: Bardana, E.J., Montanaro, A. (Eds.), Indoor Air Pollution and Health. Marcel Dekker, New York, pp. 11–60.

Seltzer, J.M., 1995. Effects of the indoor environment on health. Occupational Medicine 10 (1), 26–45.

Shah, J.J., Singh, H.B., 1988. Distribution of volatile organic compounds in outdoor and indoor air. Environmental Science and Technology 22 (12), 1381–1388.

Shorter, E., 1997. Multiple chemical sensitivity: pseudodisease in historical perspective. Scandinavian Journal of Work and Environmental Health 23 (Suppl. 3), 35–42.

Simon, G.E., Daniell, W., Stockbridge, H., Claypoole, K., Rosenstock, L., 1993. Immunologic, psychological, and neuropsychological factors in multiple chemical sensitivity. A controlled study. Annals of International Medicine 119 (2), 97–103.

Sisovic, E., Fugas, M., Sega, K., 1996. Assessment of human inhalation exposure to polycyclic aromatic hydrocarbons. Journal of Exposure Analysis and Environmental Epidemiology 6 (4), 439–447.

Somerville, S.M., Rona, R.J., Chinn, S., 1988. Passive smoking and respiratory conditions in primary school children. Journal of Epidemiology and Community Health 42 (2), 105–110.

Spengler, J.D., 1993. Nitrogen dioxide and respiratory illnesses in infants. American Review of Respiratory Disorders 148 (5), 1258–1265.

Spengler, J.D., Sexton, K., 1983. Indoor air pollution: a public health perspective. Science 221 (4605), 9–17.

Spengler, J.D., Dockery, D.W., Turner, W.A., Wolfson, J.M., Ferris, B.G., 1981. Long-term measurements of respirable particles, sulphates, and particulates inside and outside homes. Atmospheric Environment 15 (1), 23–30.

Spicer, C.W., Kenny, D.V., Ward, G.F., Billick, I.H., 1993. Transformations, lifetimes, and sources of NO_2, HONO, and HNO_3 in indoor environments. Journal of the Air and Waste Management Association 43 (11), 1479–1485.

Steindorf, K., Lubin, J., Wichmann, H.E., Becher, H., 1995. Lung cancer deaths attributable to indoor radon exposure in West Germany. International Journal of Epidemiology 24 (3), 485–492.

Sterling, D.A., Lewis, R.D., 1998. Pollen and fungal spores indoor and outdoor of mobile homes. Annals of Allergy and Asthma Immunology 80 (3), 279–285.

Sterling, T.D., 1991. Concentrations of nicotine, RSP, CO and CO_2 in non-smoking areas of offices ventilated by air recirculated from smoking designated areas. American Industrial Hygiene Association Journal 52 (10), 564–565.

Stewart, R.D., Peterson, J.E., Baretta, E.D., Bachand, R.T., Hosko, M.J., Herrmann, A.A., 1970. Experimental human exposure to carbon monoxide. Archives of Environmental Health 21 (2), 154–164.

Stolwijk, J.A., 1992. Risk assessment of acute health and comfort effects of indoor air pollution. Annals of the New York Academy of Sciences 641, 56–62.

Stolwijk, J.A., 1991. Assessment of population exposure and carcinogenic risk posed by volatile organic compounds in indoor air. Risk Analysis 10 (1), 49–57.

Sundell, J., 1994. On the association between building ventilation characteristics, some indoor environmental exposures, some allergic manifestations, and subjective symptom reports. Indoor Air (Suppl. 2), 1.

Sundell, J., Andersson, B., Andersson, K., Lindvall, T., 1993. Volatile organic compounds in ventilating air in buildings at different sampling points in the buildings and the relationship with the prevalence of occupant symptoms. Indoor Air 2 (1), 82–93.

Takahashi, T., 1997. Airborne fungal colony-forming units in outdoor and indoor environments in Yokohama. Japan. Mycopathologia 139 (1), 23–33.

Tariq, S.M., Matthews, S.M., Hakim, E.A., Stevens, M., Arshad, S.H., Hide, D.W., 1998. The prevalence of and risk factors for atopy in early childhood: a whole population birth cohort study. Journal of Allergy and Clinical Immunology 101 (5), 587–593.

Teichman, K.Y., 1995. Indoor air quality: research needs. Occupational Medicine 10 (1), 217–227.

Tokarz, A.P., Potterfield, P.M., 1994. Regulatory and legal aspects of indoor air quality and pollution. Indoor Air Pollution 14 (3), 679–691.

Traynor, G.W., Apte, M.G., Carruthers, A.R., 1986. Indoor Air Pollution Due to Emission from Woodburning Stoves. Lawrence Berkeley Laboratories, Berkeley, CA.

Tunnicliffe, W.S., Burge, P.S., Ayres, J.G., 1994. Effect of domestic concentrations of nitrogen dioxide on airway responses to inhaled allergen in asthmatic patients. Lancet 344 (8939–8940), 1733–1736.

US EPA (United States Environmental Protection Agency), 1992. Technical support document for the 1992 citizen's guide to radon (EPA 400-R-92-011). US Environmental Protection Agency, Washington, DC.

US EPA (United States Environmental Protection Agency), 1991. Air Quality Criteria for Carbon Monoxide. US Environmental Protection Agency, Washington, DC.

Vaughan, T.L., Strader, C., Davis, S., Daling, J.R., 1986. Formaldehyde and cancers of the pharynx, sinus, and nasal cavity: II. Residential exposures. International Journal of Cancer 38 (5), 685–688.

Verhoeff, A.P., 1994. Home Dampness, Fungi and House Dust Mites, and Respiratory Symptoms in Children. Posen & Looijen b.v., Wageningen, Germany.

Viegi, G., Carrozzi, L., Paoletti, P., Vellutini, M., DiViggiano, E., Baldacci, S., Modena, P., Pedreschi, M., Mammini, U., di-Pede, C., 1992. Effects of the home environment on respiratory symptoms of a general population sample in middle Italy. Archives of Environmental Health 47 (1), 64–70.

Vincent, D., Annesi, I., Festy, B., Lambrozo, J., 1997. Ventilation system, indoor air quality, and health outcomes in Parisian modern office workers. Environmental Research 75 (2), 100–112.

van Wijnen, J.H., Verhoeff, A.P., Mulder-Folkerts, D.K.F., Brachel, H.J.H., Schou, C., 1997. Cockroach allergen in house dust. Allergy 52 (4), 460–464.

Wallace, L.A., 1991a. Personal exposure to 25 volatile organic compounds. EPA's 1987 team study in Los Angeles. California. Toxicology and Industrial Health 7 (5–6), 203–208.

Wallace, L.A., 1991b. Comparison of risks from outdoor and indoor exposure to toxic chemicals. Environmental Health Perspectives 95 (1), 7–13.

Wallace, L.A., 1996. Indoor particles: a review. Journal of the Air and Waste Management Association 46 (2), 98–126.

Wallace, L.A., 1997. Sick building syndrome. In: Bardana, E.J., Montanaro, A. (Eds.), Indoor Air Pollution and Health. Marcel Dekker, New York, pp. 83–103.

Wallace, L.A., Nelson, C.J., Highsmith, R., Dunteman, G., 1993. Association of personal and workplace characteristics with health, comfort, and odor: a survey of 3948 office workers in three buildings. Indoor Air 3 (1), 193–205.

Wallace, L.A., Nelson, W.C., Ziegenfus, R., Pellizzari, E., 1991. The Los Angeles TEAM study: Personal exposures, indoor–outdoor air concentrations, and breath concentrations of 25 volatile organic compounds. Journal of Exposure Analysis and Environmental Epidemiology 1 (2), 37–72.

Wallace, L.A., 1987. The Total Exposure Assessment Methodology (TEAM) Study: Summary and Analysis. US Environmental Protection Agency, Washington, DC.

Wanner, H.U., 1993. Sources of pollutants in indoor air. IARC Scientific Publications 109, 19–30.

Ware, J.H., Dockery, D.W., Spiro, A., Speizer, F.E., Ferris, B.G., 1984. Passive smoking, gas cooking, and respiratory health of children living in six cities. American Review of Respiratory Disorders 129 (3), 366–374.

Weiss, S.T., 1986. Passive smoking and lung cancer: what is the risk? American Review of Respiratory Disorders 133 (1), 1–3.

WHO (World Health Organisation), 1995. Global Strategy of Asthma Management and Prevention. World Health Organisation, Geneva.

Wickman, M., Gravesen, S., Nordvall, S.L., Pershagen, G., Sundell, J., 1992. Indoor viable dust-bound microfungi in relation to residential characteristics, living habitats, and symptoms in atopic and control children. Journal of Allergy and Clinical Immunology 89 (3), 752–759.

Wieslander, G., Norbäck, D., Björnsson, E., Janson, C., Boman, G., 1997. Asthma and the indoor environment: the significance of emission of formaldehyde and volatile organic compounds from newly painted indoor surfaces. International Archives of Occupational and Environmental Health 69 (2), 115–124.

Wolf, C., 1996. Multiple chemical sensitivity (MCS) – the so-called chemical multiple hypersensitivity. Versicherungsmedizin 48 (5), 175–178.

Wolff, S.P., 1991. Leukaemia risks and radon. Nature 352 (6333), 288.

Wolkoff, P., Clausen, P.A., Jensen, B., Nielsen, G.D., Wilkins, C.K., 1997. Are we measuring the relevant indoor pollutants? Indoor Air 7 (1), 92–106.

Wolkoff, P., Clausen, P.A., Neilsen, P.A., Mølhave, L., 1991. The Danish twin apartment study; part 1: formaldehyde and long-term VOC measurements. Indoor Air 4, 478–490.

Wong, O., 1983. A epidemiologic mortality study of a cohort of chemical workers potentially exposed to formaldehyde, with a discussion on SMR and PMR. In: Gibson, J.E. (Ed.), Formaldehyde Toxicity. Hemisphere Publishing Corporation, New York, NY.

Yang, Y., Sun, C., Sun, M., 1997. The effect of moderately increased CO_2 concentration on perception of coherent motion. Investigative Ophthalmology and Visual Science 38 (4), 1786.

Yu, V.L., 1993. Could aspiration be the major mode of transmission for Legionella? American Journal of Medicine 95 (1), 13–15.

Yun, K.N., Young, E.C., Stokes, M.J., Tang, K.K., 1998. Radon properties in offices. Health Physics 75 (2), 159–164.

Appendix

Since the completion of the Millennium Review, there have been a number of important references published in the areas of respiratory health (and in particular the effects of passive smoking on infants), and radon and cancer risk.

1. Respiratory health

A series of ten articles reviewing the health effects of passive smoking, largely written by Cook and Strachan, but with contributions from Anderson, Carey, and Coultas, have been published in the journal *Thorax*. All except one have focussed on the effects of parental smoking on children's health. The findings of these papers have been based on the results of a systematic review and quantitative meta-analysis of the existing literature.

In the first article of the series, Strachan and Cook (1997) examined the effects of parental smoking on lower respiratory tract illness in infancy and early childhood. When using a broad definition of symptoms that included wheeze, bronchitis, bronchoilitis or pneumonia, they found an odds ratio of 1.57 (95% CI 1.42–1.74) for the presence of symptoms in infants aged under two years if either parent smoked. If the mother smoked, the odds ratio was raised to 1.72 (95% 1.55–1.91). The elevated maternal influence may be explained by the children spending more time with their mothers, but could also suggest that prenatal exposure to the products of smoking might compromise lung development and hence place infants at increased risk of contracting illnesses.

Cook and Strachan (1997) assessed the relationship between parental smoking and the prevalence of asthma and respiratory symptoms in school-aged children. They found an odds-ratio of 1.21 (95% CI 1.10–1.34) for cases where either parent smoked. This increased to 1.50 (95% CI 1.29–1.73) where both parents were smokers, illustrating the probable importance of environmental exposures. For either parent smoking, elevated symptom prevalence was also detected for the prevalence of wheeze (odds ratio = 1.24, 95% CI 1.17–1.31), cough (odds ratio = 1.40, 95% CI 1.27–1.53), phlegm production (odds ratio = 1.35, 95% CI 1.13–1.62), and breathlessness (odds ratio = 1.31, 95% CI 1.08–1.59).

Interestingly, some recent studies that have investigated the relationship between current parental smoking and the prevalence of respiratory symptoms in large samples of infants, have reported conflicting results. Research by Hu et al. (1997) in Chicago found that wheezing was inversely associated with current maternal smoking, whilst Forsberg et al. (1997) detected an inverse association between current smoking in the home and the prevalence of asthma attacks and dry cough. It may be that these negative associations are due to the avoidance of smoking amongst the parents of symptomatic children (Cook and Strachan, 1999). Further research is undoubtedly required in this area.

The link between common lower respiratory tract illnesses in infancy and asthma in later childhood is controversial. The *Thorax* review series presented evidence to suggest that parental smoking was more influential as a cause of early "wheezy bronchitis" than of later onset asthma. However, recent Norwegian research (Siersted et al., 1998) suggests that teenagers with asthmatic symptoms are less likely to receive a diagnosis of asthma if their parents smoke. This raises the possibility that the association between Environmental Tobacco Smoke (ETS) and asthma may have been underestimated by studies relying of physician diagnosed symptoms.

In contrast to much of the previous literature, Strachan and Cook (1998a) did not find evidence of a positive association between allergic sensitisation and parental smoking, either before or after birth. This may be because many studies included asthma and wheezing as measures of sensitisation, and these might be related to exposure to ETS by mechanisms other than allergy. In their review of the relationship between exposure to ETS and the natural history and severity of asthma and wheezing, Strachan and Cook (1998b), also found inconsistent results amongst both case-control and longitudinal studies. Early prognosis appeared to be worse if parents smoked, whereas the persistence of symptoms into the teens and twenties was less common in the children of smokers. However, a meta-analysis of the relationship between bronchial re-activity, as assessed by challenge test and exposure to (largely maternal) ETS was more consistent, suggesting a small but statistically significant increase in bronchial hyper-responsiveness amongst the children of smoking mothers (Odds ratio = 1.29, 95% CI 1.10–1.50) (Cook and Strachan, 1998).

Cook et al. (1998) examined the relationship between parental smoking and infant's spirometric index scores. They concluded that maternal smoking was generally associated with small but statistically significant deficits in spiro-metric indices (for example a 0.9% (95% CI −1.2–0.7) reduction in forced expiratory volume in one second). The authors concluded that this was almost certainly a causal relationship, and that much of the effect might be due to maternal smoking during pregnancy which appears to have a particularly large influence on neonatal lung mechanics.

Anderson and Cook (1997) reviewed the evidence for a relationship between sudden infant death syndrome (SIDS) and passive smoking, and reported an odds ratio of 2.13 (95% CI 1.86–2.43) for maternal smoking. An effect of this magnitude is unlikely to be due to residual confounding. Based on the limited available evidence where mothers claimed to be non-smokers during pregnancy, it seems that postnatal exposure may play the most important role.

Strachan and Cook (1998c) reviewed studies of the relationship between parental smoking, and infant middle ear disease and adenotonsillectomy. Where either parent smoked, they reported odds ratios of 1.48 (95% CI 1.08–2.04) for recurrent otitis media, 1.38 (95% CI 1.23–1.55) for middle ear effu-sion, and 1.21 (95% CI 0.95–1.53) for referral for glue ear. All these results suggest that exposure to ETS may be a significant risk for the development of middle ear problems.

In summarising the findings from their meta-analyses, Cook and Strachan (1999) compared their results with those of a review undertaken by the Cal-ifornian Environmental Protection Agency (CEPA) (Dunn and Zeise, 1997). Many of the conclusions were remarkably consistent, although the CEPA in-terpreted the inconsistent data on allergic sensitisation, as providing evidence that exposure to ETS may be a risk. The CEPA study did evaluate the effects of ETS on cystic fibrosis, a condition not covered in the *Thorax* articles. It reported that hospital admissions for cystic fibrosis exacerbations were signifi-cantly related to parental smoking in 3 of the 4 studies reviewed, and that ETS exposure also appeared to be linked to other measures of disease severity.

In the only article in the *Thorax* series that is not concerned with infants, Coultas (1998) examined the relationship between passive smoking and the risk of adult asthma and chronic obstructive pulmonary disease (COPD). Al-though the available literature is limited by difficulties in measuring a dose-response relationship, the survey found evidence that adults regularly exposed to ETS at home or in the workplace appear to have a 40–60% increased risk of asthma compared to non-exposed individuals. A weak relationship was also detected between passive smoking and the occurrence of COPD, although the literature contains much variability in the adopted definitions of COPD.

In an important article published in *Allergy*, Garrett et al. (1999) examined the relationship between indoor exposure to formaldehyde and the prevalence of sensitisation to common aero-allergens amongst a sample of 148 Australian infants aged between 7–14 years. Formaldehyde measurements were taken with passive samplers on four occasions between March 1994 and February 1995. A respiratory questionnaire was completed, and skin prick tests were performed.

The median indoor formaldehyde level was $15.8\,\mu\text{g m}^{-3}$. Marginally higher mean formaldehyde levels were recorded in the bedrooms of atopic compared to non-atopic children ($19\,\mu\text{g m}^{-3}$ versus $16.4\,\mu\text{g m}^{-3}$, $p = 0.06$). The age-sex adjusted odds ratio for atopy with an increase in bedroom formaldehyde levels of $10\,\mu\text{g m}^{-3}$ was 1.40 (95% CI 0.99–2.04), and this remained after controlling for passive smoking, the presence of pets, and concentrations of other indoor air pollutants. There were significant differences in the average number of positive skin prick tests between formaldehyde exposure levels (1.2 positive tests for $< 20\,\mu\text{g m}^{-3}$ versus 4 positive tests for $> 50\,\mu\text{g m}^{-3}$, $p = 0.004$). The average relative allergen wheal size for the largest allergen wheal was also associated with formaldehyde concentrations (0.4 allergen wheal ratio for $< 20\,\mu\text{g m}^{-3}$ versus 1.2 allergen wheal ratio for $> 50\,\mu\text{g m}^{-3}$). The findings of Garrett et al. (1999) suggest that low-level exposure to formaldehyde may increase the risk of allergic sensitisation to common aeroallergens in children. The mechanism could be associated with formaldehyde-induced changes to the upper respiratory tract, possibly via increased permeability of the respiratory epithelial layer or suppression of mucosal immune defences.

2. Radon and cancer risk

There has been an interesting debate in the literature concerning the relationship between population-level residential radon exposures and the prevalence of lung cancer. The debate largely arises from the publication of a report by Cohen (1997) of a strong negative relationship between residential radon measurements and US county rates of male and female lung cancers. This direction of relationship is contrary to what might be expected from studies on miners.

Goldsmith (1999) examined Cohen's findings, and claimed a number of flaws in the original study design. In particular, they argued that Cohen's results are confounded by population density (residential radon levels are higher in suburban areas leading to a negative association with population density, whilst population density is positively associated with county-level lung cancer prevalence).

Using a more sophisticated approach to that of Cohen, Bogen (1998) fitted a cytodynamic two-stage (CD2) cancer model to age specific lung cancer mortality data and estimated radon-exposure in white females aged 40+ years who resided in 2821 US counties between 1950–1954. The results of this model suggest that residential radon exposure may have a non-linear U-shaped relation to lung cancer mortality risk, and hence linear extrapolations could be inaccurate. The presence of a U-shaped relationship has also been detected in field measurements by Tóth et al. (1998), in their study of residential radon exposure and lung cancer risk amongst women in Mátraderecske, Hungary. They found the lowest cancer risk was actually amongst women living in medium-high radon concentrations of between 110 and 165 Bq m^{-3}. Bogen (1998) postulates that the mechanism of such a relationship may be associated with the influence of radiation exposure on the growth kinetics of pre-malignant foci. In particular the competition between extinction (via alpha-induced cell killing) within pre-existing pre-malignant foci, and the induction of new pre-malignant foci might be important.

References

Anderson, H.R., Cook, D.G., 1997. Health effects of passive smoking 2. Passive smoking and sudden infant death syndrome. Review of the epidemiological evidence. Thorax 52, 1003–1009.

Bogen, K.T., 1998. Mechanistic model predicts a U-shaped relation of radon exposure to lung cancer risk reflected in combined occupational and US residential data. Human and Experimental Toxicology 17, 691–696.

Cohen, B.L., 1997. Lung cancer rate vs. mean radon level in US counties of various characteristics. Health Physics 72, 114–119.

Cook, D.G., Strachan, D.P., 1997. Health effects of passive smoking 3. Parental smoking and respiratory symptoms in school-children. Thorax 52, 1081–1094.

Cook, D.G., Strachan, D.P., 1998. Health effects of passive smoking 7. Parental smoking, bronchial reactivity and peak flow variability in children. Thorax 53, 295–301.

Cook D.G., Strachan, D.P., 1999. Health effects of passive smoking 10. Summary of effects of parental smoking on the respiratory health of children and implications for research. Thorax 54, 357–366.

Cook, D.G., Strachan, D.P., Carey, I.M., 1998. Health effects of passive smoking 9. Parental smoking and spirometric indices in children. Thorax 53, 884–893.

Coultas, D.B., 1998. Health effects of passive smoking 8. Passive smoking and risk of adult asthma and COPD: an update. Thorax 53, 381–387.

Dunn, A., Zeise, L., 1997. Health effects of exposure to environmental tobacco smoke. California Environmental Protection Agency, USA.

Forsberg, B., Pekkanen, J., Clench-Aas J., et al., 1997. Childhood asthma in four regions of Scandinavia: risk factors and avoidance effects. International Journal of Epidemiology 26, 610–619.

Garrett, M.H., Hooper, M.A., Hooper, B.M., Rayment, P.R., Abramson, M.J., 1999. Increased risk of allergy in children due to formaldehyde exposure in homes. Allergy 54, 330–337.

Goldsmith, J.R., 1999. The residential radon-lung cancer association in US counties: a commentary. Health Physics 76 (5), 553–557.

Hu, F.B., Persky, V., Flay, B.R. et al., 1997. Prevalence of asthma and wheezing in public schoolchildren: association with maternal smoking during pregnancy. Annals of Allergy and Asthma Immunology 79, 80–84.

Siersted, H.C., Boldsen, J., Hansen, H.S. et al., 1998. Population based study of risk factors for under-diagnosis of asthma in adolescence: Odense schoolchild study. BMJ 316, 651–655.

Strachan, D.P., Cook, D.G., 1997. Health effects of passive smoking 1. Parental smoking and lower respiratory illness in infancy and early childhood. Thorax 52, 905–914.

Strachan, D.P., Cook, D.G., 1998c. Health effects of passive smoking 4. Parental smoking, middle ear disease and adenotonsillectomy in children. Thorax 53, 50–56.

Strachan, D.P., Cook, D.G., 1998a. Health effects of passive smoking 5. Parental smoking and allergic sensitisation in children. Thorax 53, 117–123.

Strachan, D.P., Cook, D.G., 1998b. Health effects of passive smoking 6. Parental smoking and childhood asthma: longitudinal and case-control studies. Thorax 53, 204–212.

Tóth, E., Lázár, I., Selmeczi, D., Marx, G., 1998. Lower cancer risk in medium-high radon. Pathology Oncology Research 4 (2), 125–129.

Chapter 4

Exposure assessment of air pollutants: a review on spatial heterogeneity and indoor/outdoor/personal exposure to suspended particulate matter, nitrogen dioxide and ozone

Christian Monn

*Institute of Hygiene and Applied Physiology, ETH-Zürich, Clausiusstrasse 25,
8092 Zürich, Switzerland
E-mail: monn@iha.bepr.ethz.ch*

Abstract

This review describes databases of small-scale spatial variations and indoor, outdoor and personal measurements of air pollutants with the main focus on suspended particulate matter, and to a lesser extent, nitrogen dioxide and photochemical pollutants. The basic definitions and concepts of an exposure measurement are introduced as well as some study design considerations and implications of imprecise exposure measurements. Suspended particulate matter is complex with respect to particle size distributions, the chemical composition and its sources. With respect to small-scale spatial variations in urban areas, largest variations occur in the ultrafine (< 0.1 μm) and the coarse mode ($PM_{10-2.5}$, resuspended dust). Secondary aerosols which contribute to the accumulation mode ($0.1–2$ μm) show quite homogenous spatial distribution. In general, small-scale spatial variations of $PM_{2.5}$ were described to be smaller than the spatial variations of PM_{10}. Recent studies in outdoor air show that ultrafine particle number counts have large spatial variations and that they are not well correlated to mass data. Sources of indoor particles are from outdoors and some specific indoor sources such as smoking and cooking for fine particles or moving of people (resuspension of dust) for coarse particles. The relationships between indoor, outdoor and personal levels are complex. The finer the particle size, the better becomes the correlation between indoor, outdoor and personal levels. Furthermore, correlations between these parameters are better in longitudinal analyses than in cross-sectional analyses. For NO_2 and O_3, the air chemistry is important. Both have considerable small-scale spatial variations within urban areas. In the absence of indoor sources such as gas appliances, NO_2 indoor/outdoor relationships are strong. For ozone, indoor levels are quite small. The study hypothesis largely determines the choice of a specific concept in exposure assessment,

First published in Atmospheric Environment 35 (2001) 1–32

i.e. whether personal sampling is needed or if ambient monitoring is sufficient. Careful evaluation of the validity and improvements in precision of an exposure measure reduce error in the measurements and bias in the exposure–effect relationship.

1. Introduction, definitions and concepts

1.1. Content and objective

The objective of this review is to describe exposure assessment techniques and to discuss databases of outdoor, indoor, and personal exposure levels of the following main air pollutants: suspended particulate matter (SPM), nitrogen dioxide (NO_2) and photochemical oxidants. The focus is on long-term exposure but also some important aspects of short-term exposure are addressed. Special emphasis is placed on suspended particulate matter because of its increasing significance in health-effect issues. A review on bioaerosols associated with particulates which may have a strong allergenic and inflammatory potential is also included. A short introduction outlines the definitions, concepts, some important aspects of study designs and the effects from errors in the measurements.

1.2. Definitions and concepts

The basic concepts used in exposure assessments were developed in the early 1980s by Duan (1982) and Ott (1982). Their introduction of the term "human exposure" (more simply exposure) emphasises that the human being is the most important receptor of pollutants in the environment. Ott (1982) elaborated a system of definitions for the term "exposure" and defined exposure as "an event that occurs when a person comes in contact with the pollutant". This is a definition of an instantaneous contact between a person i (or a group of persons) and a pollutant with concentration c, at a particular time t. This definition refers to a contact with a pollutant, but it is not necessary that the person inhales or ingests the pollutant. When the duration of exposure is also taken into consideration, the result is an "integrated exposure", calculated by integrating the concentration over time (t_a) (units: ppmh or $\mu g\,h\,m^{-3}$) (Fig. 1). These units, however, are uncommon and the calculation of such an integral is, in most cases, not possible. More easily understood is the term "average exposure"; it can be calculated by dividing the integrated exposure by the specified time and has the unit of mass in a volume of air (Fig. 1). In a pragmatic approach, average exposure is simply deduced by averaging the pollutant concentration over the specified period (e.g. expressed as annual mean). In air pollution epidemiology, the unit "concentration" is most commonly used. It refers to the

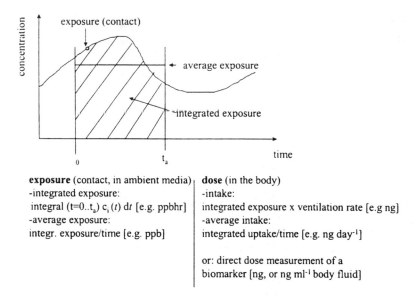

exposure (contact, in ambient media) | dose (in the body)
-integrated exposure: | -intake:
integral (t=0..t_a) c_i (t) dt [e.g. ppbhr] | integrated exposure x ventilation rate [e.g ng]
-average exposure: | -average intake:
integr. exposure/time [e.g. ppb] | integrated uptake/time [e.g. ng day^{-1}]
 |
 | or: direct dose measurement of a
 | biomarker [ng, or ng ml^{-1} body fluid]

Figure 1. Definitions of exposure and dose (Ott, 1982; Sexton and Ryan, 1988; NRC, 1991).

"average exposure" to which the population has been exposed over a specific time. Besides mean concentrations, the use of other statistical parameters (e.g. 95th percentiles, median, frequency of exceedance of a certain value) is also common.

The definitions of exposure described above refer to levels of pollutants in the ambient media. However, once the pollutant has crossed a physical boundary (e.g. skin, alveolar epithelial cells), the concept of "dose" is used (Ott, 1982). "Dose" is the amount of material absorbed or deposited in the body for an interval of time and is measured in units of mass (or mass per volume of body fluid in a biomarker measurement) (Fig. 1). Dose can be determined as internal dose or as a biologically effective dose (NRC, 1991). When data on exposure values are available, an "intake" (also called "potential dose" which assumes total absorption of the contaminant (NRC, 1991)), can be calculated by multiplying the integrated exposure with the volume of air exchanged in the lung per specified time (unit: mass). "Average intake" (in analogy to average exposure) or "dose rate" can be deduced by dividing intake by time (unit: mass × time^{-1}). In Section 2.1.2 the difficulties of relating ambient levels with levels of biomarkers will be addressed.

A comprehensive exposure assessment is part of a risk assessment that evaluates the relationship between the source of a pollutant and its health effect (Ott, 1990). The link between a pollutant emission and a particular target in the body consists of a sequence of events:

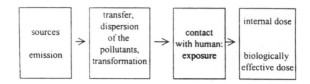

Figure 2. "Sequence of exposure" from source to biologically effective dose.

Fig. 2 shows a simplified version from Lioy (1990) and describes the route of a contaminant from its source into the body. In general, this scheme is valid for all environmental media (water, air, food). Air pollutants are dispersed ubiquitously, and contact between the target organ (airway system) of a human and the pollutant takes place continuously. In the comprehensive concept of "Total Exposure Assessment" (Ott, 1990; Lioy, 1990), all routes into the body have to be considered: contact with soil, water, food and air.

Conceptual approaches to measuring exposure can be classified according to their potential agreement with a "perfectly precise" personal exposure measurement. Table 1 depicts such a classification hierarchy based on the characterisation by Lioy (1995). Personal exposure measurement obviously reflects the individuals' exposure levels best, whereas qualitative methods provide less precise estimates.

Representative epidemiological studies have to include a large number of study individuals from the general population. Collecting exposure data from all these individuals is an expensive undertaking, so that a geographical clustering of the population is often used. In such studies, exposure data are collected

Table 1. Classification of exposure measurements with respect to true personal measurement

	Environmental study: examples	Occupational study: examples	Agreement with personal exposure
Direct/internal	Internal dose/biologically effective dose	Internal dose/biologically effective dose	Perfect
Direct/external	Personal measurement	Personal measurement	Very good
Indirect/external	Area measurement	Stationary measurement at workplace	Moderate–good
	Quantitative surrogate: e.g. distance to street	Quantitative surrogate: e.g. contact with chemicals	Moderate
	Qualitative data by questionnaire: e.g. high/medium/low pollution	Qualitative data by questionnaire: frequency of exposure, estimates high/low	Poor–moderate
	Qualitative data, categorical: polluted/unpolluted	Qualitative data: exposure: yes/no	Poor

by a central monitor and attributed to the residents of the cities in question. On the other hand, collecting exposure data on an individual level has the advantage of assessing frequency distributions in order to reveal whether part of the population is exposed to levels much higher or lower than the average level (Sexton and Ryan, 1988). Krzyzanowski (1997) stressed other key elements of exposure assessment, such as representative sampling, the control of confounding factors (factors which are related to the exposure and the health variable) and the appropriate averaging time. A further difficulty in an exposure assessment for air pollutants is related to the fact that pollutants are present as mixtures. Therefore assessment of exposure has to rely on measurements of markers in these mixtures (Leaderer et al., 1993). Markers, also called indicators, should be unique to the mixture's sources; they should be readily detectable in air at low concentrations and present in a consistent ratio to other components (Leaderer, 1993). Examples of such indicators are NO_2, O_3, airborne particles, metabolites in biological specimens, or variables obtained from questionnaires (e.g. contact to sources).

2. Methodology

Personal exposure measurements can be performed *directly* or *indirectly* (Ott, 1982). In the direct approach exposure levels are determined on an individual (by using a personal sampler or a biological marker); in the indirect approach exposure levels are either measured stationarily or determined by models (Ott, 1982; Lioy, 1995).

The evaluation of a method has to consider method-inherent criteria. These criteria are sensitivity, precision, accuracy, selectivity and detection limit. Besides these criteria, cost, and applicability are important factors in the choice of a particular method. WHO brochures summarise quality assurance (QA) and quality control (QC) protocols (Series GEMS/AIR handbook series, 1994), which are also important to warrant high-quality measurements. Table 2 lists the most important techniques used in air pollution studies. In addition the PAS sensor for PAH-loaded airborne fine particles is added (Burtscher and Siegmann, 1993, 1994). For the separation between gas- and particle-phase the so-called denuders have been developed (Koutrakis et al., 1988, 1989; Possanzini et al., 1983).

2.1. Direct measurements

2.1.1. Personal sampling

Passive samplers are the most widespread and easily used devices employed in personal sampling. They rely on the principle of the passive diffusion of

Table 2. Techniques for measuring air pollutants (note: continuous refers to a response signal within a few seconds to minutes) (Harrison and Perry, 1986; Finlayson-Pitts and Pitts, 1986; Williams, 1995; Willeke and Baron, 1993; Chow, 1995; Wijnand, 1996)

Pollutants	Measurement techniques	Time resolution
Gases: SO_2, NO_x, O_3, CO	Fluorescence, chemiluminescence photometry, non-dispersive infra red	Continuous, minutes
Particles, TSP, PM_{10}, $PM_{2.5}$	Gravimetry (size fraction: impaction, cyclone)	One day, hours
	Beta meter	Integrated, day, hours
	Tapered element	Continuous, minutes
	Nephelometer	Continuous, minutes
	Photoelectric aerosol sensor (PAS)	Continuous, minutes
Gases (personal)	Passive sampler (NO_2, O_3)	Integrated, days
	Electrochemical sensor (CO)	Continuous, minutes
Particles (personal)	Size fraction: impaction, cyclone gravimetry	Integrated: hours, one day
	Light scattering	Continuous, minutes
	Photo-emission sensor (PAS)	Continuous, minutes
Bioaerosols	Slit sampler	Integrated, minutes

a gas and the concentration in air can be calculated according to Fick's law of diffusion (Palmes et al., 1976). The samplers are straightforward for personal sampling as they are light, do not need electricity, and can be easily fixed to outer clothing. They exist as tubes or small personal badges. Passive samplers can also be used to take stationary measurements in outdoor and indoor settings. The most established method is the passive sampler for NO_2 (Palmes et al., 1976; Yanagisawa and Nishimura, 1982; Hangartner, 1990). The sampling time is usually from a few days up to one week, depending on concentration. Passive samplers also exist for CO, SO_2, VOC, O_3, formaldehyde and ammonia (Lee et al., 1992; McConnaughey et al., 1985; Shields and Weschler, 1987; Weschler et al., 1990; Koutrakis et al., 1993; Monn and Hangartner, 1990). Brown (1993) reviewed state-of-the art passive samplers for ambient monitoring. The precision of NO_2 tubes is quite good; the error is in the range of 5–10%. For example, in the Swiss Study on Air Pollution and Lung Diseases in Adults (SAPALDIA), the coefficient of determination between the continuous monitors and passive samplers ranged between 0.69 and 0.93. On average, concentrations of duplicates were within 5%. The problem for passive devices is, however, the lack of accuracy, i.e. the agreement with a reference method. The O_3 tube (Monn and Hangartner, 1990; Hangartner et al., 1996), for example, has acceptable precision (variation <5–10%) but the agreement with a continuous monitor is not always very good.

Real-time personal monitors for gases are also available, but the detection limit is often too high, so that their use is limited to occupational settings only. A personal monitor for CO has also been shown to be practical in ambient conditions (Ott et al., 1986; Jantunen, 1998). Personal monitors for particles exist with different cut-offs and size fractions, such as $PM_{2.5}$, PM_{10} or multistage samplers (analysis by gravimetry, collection of particles on filters). Real-time monitors, based on the principle of light scattering provide particle numbers in a volume of air (usually for particles with a $d_{ae} > 300$ nm). A real-time personal sampler also exists for the detection of combustion aerosols (coated with PAH), using the principle of a photoemission aerosol sensor (PAS) (Burtscher and Siegmann, 1993, 1994). The detection signal is related to levels of ultrafine combustion particles <1 µm.

2.1.2. Biological markers

Biological markers can be grouped into markers of exposure and markers of effects. A marker of an effect is generally a pre-clinical indicator of abnormalities which can also comprise a medical diagnosis (e.g. decreased lung function) (Grandjean, 1995). A marker of exposure reflects the concentration of the analyte which has passed a human boundary. According to Hulka and Wilcosky (1988) and Hulka (1990) biological markers have to fulfil the following purposes: elucidation of pathogenic mechanism, improvement in aetiological classification and the recognition of early effects. Grandjean (1995) further discussed the importance of "susceptibility-biomarkers" because the variability between individuals can be large due to heterogeneities of enzymes and genetic factors. Genetic polymorphism for enzymes will be an important factor in future epidemiological studies (e.g. for oxidative scavengers). Valid biological markers are those which have biological relevance, known pharmacokinetics, temporal relevance and defined background variability (Schulte and Talaska, 1995).

Biological markers can be collected from breath, urine, hair, nails, nasal lavage or, in a more difficult procedure, from blood or fluids from bronchoalveolar lavage. The use of biomarkers is most widespread in occupational studies with known specific exposures (e.g. solvents) (Lowry, 1995). The advantage of the use of a biological marker is that exposure is integrated over time and that all exposure pathways are included. This is, on the other hand, also a shortcoming as it is not possible to differentiate between exposure pathways anymore. For example, dermal exposure to pesticides (and some VOCs) can be very important, but little information on intake rates through the skin is available when looking only at the biological markers. Lioy (1990) described the parameters required to calculate and interpret internal dose relationships with ambient exposures. Data on kinetics, half-life, solubility, excretion and metabolic transformation are needed. Models can be used to elucidate the relationship between concentration in air and level in the body. Such models rely on sophisticated pharmacokinetic data and are presently still in the stage of development (Georgopoulos et al., 1997).

To assess the usability of a biological marker it is important to know its half-life in the body and its temporal variability. If the elimination rate of the contaminant is small and a biological marker is accumulated in the target tissue, then the measurement of the biological marker offers a significant advantage over external measurements (Rappaport et al., 1995). The key issue for the promotion of biological markers lies in the validation of the relationship between concentration in air and concentration of the marker in the body (Rappaport et al., 1995). A shortcoming is that biological markers reflect recent exposure only; therefore, they can only be used for prospective studies on acute effects.

Table 3 shows some examples of biological markers of effect and exposure. One example of an exposure marker is lead in blood. Pb is also retained in the bone marrow and has a rather long half-life in the body. Wallace et al. (1988, 1989) used VOCs in exhaled breath to assess personal exposure to VOCs. It corresponded well to previous exposure to VOCs. DNA- and protein-adducts can be used as markers for exposure to complex particle mixture which contain PAHs (diesel, tobacco smoke, coal emissions). Such adducts imply a carcinogenic and mutagenic potential (Mumford et al., 1996; Lewtas et al., 1993; Meyer and Bechtold, 1996). Yanagisawa et al. (1988) used hydroxy-proline as a biological marker for exposure to NO_2 and tobacco smoke in urinary samples. In urinary samples, secretion of creatinine is rather stable, so that the ratio between hydroxy-proline and creatinine can be used as a marker for exposure. This ratio mainly results from personal NO_2 exposure and smoking levels (active and passive). The ratio was also elevated in persons living near major roads. Nasal and bronchoalveolar lavage have been used in studies on effects of O_3 (Graham and Koren, 1990). Albumine, neutrophils, eosinophiles and cytokines can be used as markers of effects after exposure to pollutants.

Table 3.   Examples of biological markers of exposure and effect (Lewtas et al., 1993; Wallace et al., 1988, 1989; Graham and Koren, 1990)

Type of marker	Examples of markers
Biological marker of exposure	Lead in blood, VOC in exhaled air, DNA/protein adducts, chemicals (DDT, PCBs) in mothers' milk, hydroxy-proline
Biological marker of effect	Chromosome aberration, lung function change mediators, cytokines

2.2. Indirect measurements

2.2.1. Ambient measurements

In many epidemiological studies exposure data are obtained from ambient monitoring networks. In these studies, people living in defined areas (e.g. in a particular city) are assigned to the same pollution concentration. In this type of study, the units of analysis are populations or groups of people rather than individuals (ecological analysis, according to Last, 1988). Ambient monitoring networks have been established all over Europe and the USA by national institutions or local councils. They are equipped with on-line monitors providing continuous data with sufficient time resolution (half-hourly values). The accuracy and precision of these monitors is generally good (within 5–10%). Running such a network is expensive, especially the implementation of quality assurance and quality control procedures. The quality control includes several steps from the calibration with independent standards to internal plausibility check, technical controls, and data acquisition control (EMPA, 1994). Ring calibration, where an external monitoring van moves around from site to site, is used to check the agreement between these sites.

2.2.2. The use of microenvironments (MEs)

An indirect way of assessing personal exposure is to use a microenvironmental model. In daily life, people move around and thus are exposed to various levels of pollutants in various locations. Duan (1982) introduced the term "microenvironments" (MEs), which is defined as a "chunk of air space with homogeneous pollutant concentration". Such microenvironments can either represent outdoor locations (e.g. in front of the home) or indoor locations (bedroom, kitchen, etc.). Mage (1985) defined MEs as a volume in space, during a specific time interval, during which the variance of concentration within the volume is significantly less than the variance between that ME and its surrounding MEs.

Selective measurements in MEs and a time-activity/time-budget questionnaire are used for the estimation of personally encountered pollution levels, calculated as integrated dose or concentration in a cubic meter of air. The total average exposure (X) can be defined as

$$X = \sum X_i t_i / \sum t_i,$$

where X_i is the total exposure in the ith ME, visited in sequence by the person for a time interval t_i (Mage, 1985).

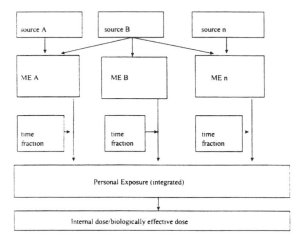

Figure 3. The concept of calculating personal exposure using time-activity data and pollutant levels in microenvironments (ME).

For J different microenvironments, integrated personal exposure can be calculated as follows:

$$X = \sum_{j=1}^{J} C_{iJ} T_{ij} / \sum t_{ij}.$$

C_{ij} is the concentration in microenvironment j in which the individual remains during period t_{ij}.

Fig. 3 shows the method for calculating personal exposure levels. Different emission sources contribute to pollution levels in different MEs. The time fraction spent in each ME allows calculation of integral personal exposure levels (= the sum from all MEs). Such approaches are based on easy-to-use and reliable time-activity diaries.

It is demanding and expensive to obtain refined data of dozens of MEs in large-scale epidemiological surveys. The most feasible approach is to use "group-MEs", where similar MEs are aggregated into ME types (e.g. indoor and outdoor MEs). Stock et al. (1985) used personal-activity profiles and household characteristics to partition the locations into seven broad microenvironments. Three of them were indoors, two outdoors and two in transportation modes. The following list shows the main microenvironments which cover most of the daily activities of adult persons. This list is similar to that of the EXPOLIS study (Jantunen et al., 1998):

- outdoor (at home, in study region),
- indoor home (kitchen, bedroom, living room),

- in transit (in car, train, bus, as pedestrian),
- others (shopping malls, restaurants, theatres, indoor sports),
- workplace.

2.2.3. Further models and the use of questionnaires

Models exist for a broad range of mathematical descriptions to predict the exposure of individuals or of populations. Generally, models can be grouped into physical (or deterministic) and statistical (or stochastic) models (Sexton and Ryan, 1988). Some models rely on both physical–chemical knowledge and incorporate statistical approaches (hybrid models). Physical models are based on mathematical equations, describing known physical/chemical mechanisms in the atmosphere. Statistical models are based on measured data and explanatory variables. For outdoor pollutants, sophisticated dispersion models (e.g. Gauss models) which incorporate meteorological variables and chemical processes, have been developed. They can be used to predict outdoor spatial and temporal behaviour of pollutants (Hanna et al., 1982). Their most important shortcoming is the need for detailed emission inventories, which are in most cases unavailable. Moreover, the application of dispersion models in epidemiology is sparse to date. Ihrig et al. (1998) demonstrated the usefulness of a combination of an atmospheric dispersion model with a geographical information system (GIS). In this approach, associations were detected between exposure to arsenic and stillbirths in Texas. In another study, McGraven et al. (1999) used a regional atmospheric transport model which incorporated spatially varying meteorology and environmental parameters for exposure calculations of beryllium in order to find association with lifetime cancer incidences. In these two examples, distinct air toxins with known sources were used. For air pollution mixtures with a variety of sources, the use of such models may be more difficult. Other models such as receptor models, based on mass conservation or with multivariate approaches, have found wide applications in source apportionment (Hopke, 1985).

Questionnaires are important tools for assessing exposure. They can be used to identify contact with emission sources and frequencies of contacts with potential sources (e.g. in a household) (Lebowitz et al., 1989). This is especially important for the identification of contacts to indoor sources which do not reflect the same mixtures than outdoor sources (e.g. NO_2 from gas cooking, $PM_{2.5}$ from tobacco smoke).

In microenvironmental models, questionnaires are used to obtain data on time-budget and time-activity patterns and they are essential in assessing long-term exposure to pollutants in retrospective studies (e.g. Künzli et al., 1997). Questionnaires can be used to assess the perception of traffic near the home, representing a surrogate for the traffic intensity and hence pollution levels in air. The most important advantage of questionnaires is their low cost. However, as with all other methods, validation studies have to be performed to test reliability and validity.

3. Validity, errors, precision of exposure data and spatial variation

3.1. Validity

The validity of a measurement can be defined as "the degree to which a measurement measures what it purports to measure" (Seifert, 1995). The validity of an exposure measurement can also be defined as the capacity to measure the "true" personal exposure (Armstrong et al., 1992). Seifert (1995) defines three important aspects of validity:

- The "content-validity" implies that all important contacts which could cause the effect have to be considered. It further implies that all locations of potential contacts have to be incorporated into the exposure measurement (e.g. all microenvironments).
- The "criterion-validity" is the "extent to which the measurement correlates with the phenomenon under study". In practice, the criterion-validity can be determined by comparing personal exposure values with values obtained from surrogate measurements. The coefficient of determination ρ^2 (later called "validity coefficient") is a direct measure of the validity of the surrogate method.
- A further aspect is the "construct-validity", which is "the extent to which the measurement corresponds to the theoretical concepts concerning the phenomenon under study". This requires the evaluation of the biologically relevant exposure (which is often unknown) in relation to the type of measurements used.

For a critical review of a study, all three definitions have to be considered. In practice, however, only the criterion-validity can be quantified, for example by comparing the personal exposure measurement with a surrogate measurement.

Studies on long-term exposure to pollutants encounter further difficulties and potential lack of validity, as historical data on life-long exposure are scarce. Data from actual measurements (e.g. an annual mean) are often used as surrogates for long-term exposure. The only way to validate actual measurements reflecting a long-term trend is to use emission inventories. This information, however, is not easily accessible, and, most often, not very detailed in its geographical resolution. In addition, time-activity patterns over long-term periods may influence the exposure. Künzli et al. (1997) evaluated a more refined method for assessing long-term exposures to ozone in California. Retrospective exposures to O_3 were estimated using available O_3 data and questionnaires on time-activity/budgets. From questions on the total time spent outdoors, the location of residence, and the monthly measured O_3 values at these sites, the integrated exposure over lifetime was calculated.

3.2. Types of measurement errors

Errors in measurement can occur as *systematic* or as *random* errors (Ahlbom and Norell, 1990). A systematic error can be defined as the mean of measured values minus the true value (Profos and Pfeifer, 1994). It occurs, for example, when a calibration procedure is based on a false standard. For exposure–effect relationships, a systematic error is not critical as it does not lead to a bias; it only shifts the regression line up or downwards. Moreover, if the systematic error is known, data can easily be corrected. Random errors are critical as they lead to a bias in the exposure–effect relationship, in most cases to an attenuation of the true effect (Armstrong et al., 1992). Random errors can occur at all stages of measurements (e.g. for particle measurements: during weighing, air flow variability, and erroneous use of exposure time). The extent of these errors are not predictable a priori; these errors can only be quantified from repeated measurements (e.g. by a Gaussian density curve).

These errors may occur as non-differential errors or as differential errors (Ahlbom and Norell, 1990). In the case of non-differential errors, the extent of the error is the same in the case and control group (in a case–control study), or does not vary over an entire range of exposure in a study using continuous exposure variables. In the case of differential errors, the errors deviate between the case and control group or are not the same over an entire exposure range in a study using continuous variables. Both differential and non-differential errors are critical as they distort the exposure–effect relationship (Armstrong et al., 1992). Theories on errors in measurements and their effects were published by Cochran (1968), Armstrong (1990), Armstrong et al. (1992) and Ahlbom and Steineck (1992). Most of the studies refer to biases occurring in case–control analysis. For environmental studies, analyses of dose–response relationships based on exposure data from individuals (or geographically clustered data) are more relevant.

3.3. Precision of exposure measurement and attenuation

For exposure assessment, precision of exposure measurement is related to the technical and analytical properties of the instrument and to the variation of the air pollutants in time and space (Armstrong et al., 1992). This is different from a "measurement error" in the narrow sense which is based on an error in sampling and analysis only (Brunekreef et al., 1987). Precision and error discussed here are related to the error variance determined by sampling at different points in space and time. The variability of the pollutant concentration in time and space is often larger than the technically inherent precision of the instrument; an assessment of precision in this wider sense allows quantification of attenuation in an exposure–effect relationship.

3.3.1. Validity coefficient and attenuation

Armstrong et al. (1992) presented a model to quantify the extent of attenuation when multiple exposure values of individuals are available. This model refers to an exposure–effect model with only one exposure variable. The estimated (observed) β of an exposure–effect relationship is related to the true β_T by a factor ρ_{TX}^2 [1] (Allen and Yen, 1979). ρ_{TX}^2 is called the "validity coefficient"; in a "perfect" measurement ρ_{TX}^2 reaches 1 and no attenuation is observed.

$$\beta = \rho_{TX}^2 \beta_T, \tag{1}$$

where β is the expected (observed) effect estimate; β_T: true effect and ρ_{TX}^2 the validity coefficient.

3.3.2. Calculation of the validity coefficient ρ_{TX}^2

ρ_{TX}^2 is the coefficient of determination between the true (T) and the measured exposure value (X). When two data sets of exposure measurement in individuals are available, e.g. one from surrogate measurements (X) (e.g. indoor home levels) and another from personal ($\approx$ "true") measurements (T), ρ_{TX}^2 can be calculated by a regression analysis (coefficient of determination $= \rho_{TX}^2$).

In another case, where multiple exposure measurements from each individual are available, Cochran (1968), Allen and Yen (1979) and Armstrong et al. (1992) proposed a model which allows direct calculation of ρ_{TX}^2:

$$\rho_{TX}^2 = 1 - \frac{\sigma_E^2}{\sigma_X^2} = \frac{\sigma_T^2}{\sigma_X^2}, \tag{2}$$

where σ_X^2 is the total variance, σ_T^2 the true error variance ("between-subject" variance) and σ_E^2 the subject error variance (error component, "within-subject" variance).

In a study with multiple measurements from each individual, ρ_{TX}^2 can be calculated using σ_E^2 and σ_T^2:

$$\rho_{TX}^2 = \frac{1}{1 + \sigma_E^2/\sigma_T^2} = \frac{1}{1 + \lambda}. \tag{3}$$

The ratio λ between the subject error variance σ_E^2 and the true variance σ_T^2 is defined as the variance ratio (λ) or the relative precision of a measurement, as it reflects the ratio between the variance of an individual relative to the variance of the true exposure in the study population. Having obtained a value for λ, the potential attenuation can be calculated. Table 4 shows some calculations for the expected attenuation with increasing variance ratios: when λ reaches unity, an attenuation of 50% has to be expected.

Examples of calculated λ from field measurements of NO_2 are shown in Table 5. Variance ratios from indoor, outdoor and personal values from the City of Basle are shown in the first three rows (≈ 70 individuals, repetition: 3–4 times). The next three rows show variance ratios from different locations indoors from Holland (Brunekreef et al., 1987). The largest variability was found in NO_2 outdoor measurements.

In practice, not all the assumptions (no correlation between true value and error, the same magnitude of the error over the whole exposure range) of Armstrong's theory are fulfilled. At higher concentrations, the error is often proportional to the value. In addition, normal distributions for T and E cannot

Table 4. Attenuation for different variance ratios (according to Eq. (3))

λ	0.1	0.2	0.4	0.6	0.8	1
ρ_{TX}^2	0.91	0.83	0.71	0.63	0.55	0.5

Table 5. Estimates for lambda for NO_2 indoor, outdoor and personal values and also in indoor locations

	λ
Indoor NO_2, full year[a]	0.33
Outdoor NO_2, full year[a]	0.71
Personal NO_2, full year[a]	0.41
Kitchen, March–May[b]	0.18
Living room, March–May[b]	0.22
Bedroom, March–May[b]	0.41

[a]Monn et al. (1998).
[b]Brunekreef et al. (1987).

always be guaranteed. Despite that, Armstrong's theory provides a useful tool for estimating effects of measurement errors. A further shortcoming is that this concept applies to a regression model with a single variable only. In most cases, however, more than one exposure variable is included in a model, which again can be subject to errors. The correlation among the exposure variables and also errors in confounding variables will have an influence on the bias in the regression coefficients.

3.4. Grouped data (Berkson case) and the "ecological" analysis

In the previous section, a model for estimating attenuation was presented for random errors in exposure variables for individuals. In this case (so-called "classical random" errors), the model assumes independence between the error E_i and the observed value T_i (true value) and the bias can be quantified by the validity coefficient (Armstrong, 1990). The following cases need more detailed discussion: In the so-called Berkson case, group averages are used instead of individual values to estimate the regression coefficients (Table 6) (Armstrong, 1990). Cochran (1968) described the Berkson case as a case where measured values (X) are set at certain preassigned levels; with errors in the measurements present, the actual amount of the true values (T) will not be exactly equal to the pre-assigned levels but will vary about these values. It is assumed that the error term E' has a constant variance, a mean value of zero and is independent of the observed value $X_{\text{avg cat}}$.

The implication of the Berkson case is that random errors in the exposure variables do not lead to a bias in an exposure–effect model (Cochran, 1968; Armstrong, 1990). However, the confidence intervals become larger with increasing random errors. Lebret (1990) empirically tested effects of random errors in such cases and confirmed that no attenuation in the exposure–effect model occurs. In addition, he demonstrated that in multiple regression models the effect of errors in the measurement may not always produce an underestimation but may also produce an overestimation of the true effect. In multiple regression models with more than one exposure and confounding variable, the

Table 6. Description of types of errors and effects

	Description of random error	Effect
Individual data	Classical random error model: $X_I = T_i + E_i$	Bias, ρ^2_{TX}
Berkson case	$T_i = X_{\text{av cat}} + E'$	No bias
Ecological cases		
One stationary site	Random error model	Bias, ρ_{TX}
Group averages	Berkson case	No bias

extent of distortion was more pronounced in a variable with a strong exposure–effect relationship than in a variable with poor correlation with the effect variable. Another important finding was that an increase in the study size did not remove the bias in the exposure–effect relationship.

The so-called "ecological" studies in most cases, are designed as "semi-individual" studies (health variables are available from each study individual; exposure variables are available as average pollution levels over the city of residence) (Künzli and Tager, 1997). The following situations have to be distinguished (Table 6, last lines):

- Exposure data available from all study individuals (or from a representative sample): an average exposure level over each area can be calculated. The case refers to a Berkson case and no bias in the exposure–effect regression is to be expected.
- Data available from a single fixed site ambient monitor in each area only:
  - If the monitor reflects a population-based average level in each area, the case refers to the above-mentioned case and no bias is to be expected. This case, however, is not very realistic as fixed site monitors are mainly set up for the surveillance of air quality standards.
  - If the monitor does not reflect average population exposure, a bias is to be expected. Given a large number of study sites and an occurrence of these errors in a random fashion, the bias can be estimated according to Armstrong's model, if information on these errors is available. Given a small number of study sites, however, it is very likely that errors at the leverage points distort the exposure–effect relationship; the effect might be an under- or an overestimation. Brenner et al. (1992) pointed out that in ecological analysis based on dichotomous variables, the effect of non-differential misclassification is an overestimation of the exposure–effect association. A general discussion on ecological studies is out of the focus of this review, but it has to be noted that they need careful control of covariates and that an increase of the number of regions does not particularly cancel out biases in covariates' distributions (Greenland and Robins, 1992).

The latter cases show that information on the representativity of the fixed site monitor is needed in order to eliminate or reduce bias. Within-area spatial variability and the position of the monitor within the area have to be assessed. A study design using a combination of passive samplers (or spatial random samples with continuous monitors) with a fixed site monitor is most useful. The fixed site monitor assesses the temporal course (e.g. seasonal variation) of the pollutants and the passive samplers assess the spatial variability (during seasons).

3.5. Study design considerations and cost efficiency

When exposure data are related to health outcomes, effect estimates for health impacts are determined. An important goal of a study design is therefore to reduce the variance of the effect estimate (i.e. to obtain small confidence intervals). There are three factors which influence this variance: the range of exposure, the number of observations and the validity coefficient (statistically named the coefficient of determination) (Armstrong, 1990). Eq. (4) shows the relationship between the variance of the effect estimate (beta) and these three factors. For the reduction of the variance improvements can be made by increasing the number of study sites (or the number of study persons in an individual analysis), by choosing sites that reflect large pollution differences or by choosing an exposure measurement with good validity (ρ^2):

$$\text{var}(\beta) = \frac{\text{var(error)}}{N\rho^2 \, \text{var(exp)}}, \tag{4}$$

where var(β) is the variance of the effect estimate, var(error) the variance of the exposure-health model, N the sampling size (e.g. number of subjects, number of sites, or number of days in a day-to-day analysis), ρ^2 the validity coefficient (coefficient of determination between "true" and "approximate" exposure) and var(exp) the variance of exposure (range of exposure concentrations between study sites).

3.6. Introduction to spatial variations

Most health-effects studies are based on exposure data from one "central" monitor. For uniformly distributed pollutants the choice of the site is uncritical, however, for most air pollutants considerable spatial variations in concentration levels occur.

An important factor for the spatial variation of a primary air pollutant is the geographical distribution and the type of the emission sources (e.g. line source, point source). After the emission takes place, inert pollutants (e.g. CO) simply disperse and a concentration gradient with increasing distance to the source develops (Seinfeld, 1986). For chemically reactive pollutants such as NO, a steeper concentration gradient than for inert pollutants develops (Seinfeld, 1986). In contrast, the formation of secondary pollutants (e.g. ammonium sulphate, ammonium nitrate, ozone) is a large-scale phenomenon and these pollutants have quite uniform spatial distributions (US-EPA, 1996a; Spengler et al., 1990). An exception to this occurs for the reactive gas ozone in the vicinity of other reactive species (e.g. depletion of O_3 by NO along traffic arteries) (BUWAL, 1996). For primary suspended particulate matter, physical processes such as sedimentation and coagulation are the important factors for causing spatial heterogeneity (Hinds, 1982). Furthermore, diffusion and transport of pollutants are determined by atmospheric conditions such as wind speed, vertical temperature gradient, and solar radiation (Seinfeld, 1986). The time-scale of the small-scale spatial variation may also be important; the size of short-term (e.g. within minutes) spatial fluctuations is different from spatial fluctuations in annual means. Brimblecombe (1986) established the relationship between the half-life and the spatial coefficient of variation of several gases on a global scale: for highly reactive species (e.g. radicals) the biggest CVs were observed, for inert gases (e.g. O_2) the CVs were smallest. Table 7 shows estimates of spatial coefficients of variations at more than 25 rural and urban monitoring sites in Switzerland (based on annual means considered to represent long-term data). Data represent mid-range spatial variations of an area of about 200×100 km^2. An estimate for an OH radical, which has the largest spatial variation, is added (from Brimblecombe, 1986). NO, known as a reactive gas (Atkinson, 1990), had the highest CV of all gases. As this analysis was based on annual means, the reactive O_3 did not exhibit a stronger spatial variation than other gases, as it would do in a short-term analysis. Least spatial variation was observed for total suspended particles and for PM_{10}. The spatial variability of the inert gases CO_2 and O_2 is extremely small. All information in Table 7 is based on mid-scale spatial variation but may also be used for

Table 7. Spatial coefficients of variation, based on 25 sites in Switzerland in non-alpine regions in 1993 (PM_{10}, 15 sites only) (data from annual mean levels)

Species	Spatial CV[a]
OH radical[b]	10
NO	1.134
SO_2	0.608
CO	0.569
NO_2	0.525
O_3	0.432
TSP	0.336
PM_{10}	0.205
CO_2[b]	10^{-2}
O_2[b]	10^{-5}

[a]CV: coefficient of variation: standard deviation divided by mean level.
[b]Estimates for highly reactive and inert gases from Brimblecombe (1986).

small-scale spatial variation estimates: the magnitude might be different but the ranking between the pollutants might be similar.

4. Databases of exposure measurements

4.1. Airborne particles

Air pollution by suspended particulate matter has received much attention over the last decade due to its strong association with health parameters (e.g. Dockery et al., 1993; Schwartz, 1994).

Aerosols, by definition, comprise liquid or solid particles in a continuum of surrounding air molecules. Whitby and Sverdrup (1980) proposed the terms nucleation mode ($d_{ae} < 0.1$ μm), accumulation mode (0.08–1 μm) and coarse mode (>1.3 μm) for various size ranges. Size fractions usually refer to the aerodynamic diameter (d_{ae}) which is defined as the diameter of a sphere of unity density ($1\ g\,cm^{-3}$) which has the same terminal settling velocity in air as the particle under consideration.

4.1.1. Size distribution

A bi-modal size distribution is very common in ambient urban aerosols (Hinds, 1982). The peak in the larger size range (8–15 μm) is derived from emissions from natural sources (e.g. from wind-blown dust); the peak in the lower size range (1–2 μm) originates from anthropogenic processes (fuel combustion emissions), gas-to-particle conversions and secondary formation of particles. The largest number of particles is found in the range less than 0.1 μm (Aitken particles in the nuclei mode). The greatest *surface area* is in the accumulation mode (0.08 to 1–2 μm). The highest *volume* (or mass) is found in the accumulation mode and also in the mode between 5 and 20 μm (Finlayson-Pitts and Pitts, 1986).

Over the last decade, measurements of total suspended particulates (TSP) have been replaced by total thoracic particles (particles smaller than 10 μm, PM_{10}) and also, more recently, by fine particles (particles smaller than 2.5 μm, $PM_{2.5}$) in the USA. Air quality standards for PM_{10} were implemented in the USA in the 1980s, and are now being set in Europe. A controversial discussion on these cut-points (10, 2.5 μm) took place in the USA as they were defined according to the available technology (US-EPA, 1996a). The choice of 2.5 μm is a technological compromise; it is good for the separation between anthropogenic and natural particles, but Wilson and Suh (1997) also stated that some natural particles also occur in the size range smaller than 2.5 μm. A further reduction of the cut-off is also in discussion (e.g. PM_1 or smaller for diesel particles).

The mass relationship between $PM_{2.5}$ and PM_{10} was determined in the PTEAM study (Clayton et al., 1993). $PM_{2.5}/PM_{10}$ ratios outdoors were around 0.49 during the day, and 0.55 during the night. In the Six City study (Dockery et al., 1993), the average ratio between fine ($PM_{2.5}$) and inhalable particles (PM_{10}) at the study sites was between 0.47 and 0.63. US-EPA (1996a) reviewed data on $PM_{2.5}/PM_{10}$ relationships for most of the US regions. These ratios varied between 0.71 (Philadelphia) and 0.29 (El Centro, CA), indicating a great spatial and seasonal variability. Ratios between fine and coarse particles are largely determined by the amount of coarse material, mainly by resuspended dust. Brook et al. (1997a, b) investigated the relationship between TSP, PM_{10}, $PM_{2.5}$ and inorganic constituents in Canada. The $PM_{2.5}$ fraction accounted for 49% of the PM_{10} mass and PM_{10} accounted for 44% of TSP. The variability between sites was high and the $PM_{2.5}/PM_{10}$ ratio varied between 0.36 and 0.65. The daily variability of the $PM_{2.5}$ mass correlated with the daily variation in the PM_{10} mass. At urban sites, which are influenced by heavy traffic and at the prairie site the fraction $PM_{10-2.5}$ dominated over the $PM_{2.5}$ fraction. Urban areas have higher PM_{10} concentrations than rural areas; the coarse size fraction ($PM_{10-2.5}$) has been identified as the cause of these differences.

The PM_{10}/TSP relationships were assessed on weekly means in the SAPALDIA study (Monn et al., 1995). TSP was collected with high-volume, PM_{10} with Harvard low-volume samplers (Marple et al., 1987). The average PM_{10}/TSP ratios for the whole year ranged from 0.57 to 0.74. PM_{10}/TSP ratios in the highly polluted urban regions (Geneva and Lugano) were found to be 0.75. Current measurements with high-volume PM_{10} devices indicate PM_{10}/TSP ratios around 0.75–0.9 (unpublished data). At rural and suburban sites, the ratios were between 0.57 and 0.62.

4.1.2. Small-scale spatial variation

The terminal settling velocity is an important factor for the spatial variation of suspended particulate matter (Hinds, 1982; Willeke and Baron, 1993). For very fine particles such as a particle of 1 μm d_{ae} the settling velocity is 8.65×10^{-5} cm s^{-1}, for a particle of 1 μm 3.48×10^{-3} and for a particle of 10 μm in size 3.06×10^{-1} cm s^{-1}. Primary particles (Aitken particles <0.1 μm) and large particles (e.g. >10 μm) are expected to have a bigger spatial variability than particles in the accumulation mode (0.1–1 μm). As mentioned above, coagulation (for very fine particles <0.01 μm) and gravitational settling (for particles >1 μm) are the underlying mechanisms which cause spatial heterogeneity (Hinds, 1982). Preliminary experimental field data, however, did not confirm the expected ranking between TSP, PM_{10} and $PM_{2.5}$. Interestingly, the variance in the PM_{10} concentration was smaller than in $PM_{2.5}$. The distribution of $PM_{2.5}$ might be uniform in situation where secondary formation is important; in cities, however, with large emissions from heavy duties (diesel exhaust), $PM_{2.5}$ can also exhibit significant spatial variation.

Spengler et al. (1981) published data from the Six City study on the within-area variability of $PM_{3.5}$. With the exception of one site (extreme pollution levels derived from a single source in Steubenville), the spatial variation within the study sites was found to be small. In the US-PTEAM study performed in Riverside CA, outdoor levels of $PM_{2.5}$ and PM_{10} at different homes were in good agreement with the central monitoring site, indicating a homogeneous spatial distribution (Clayton et al., 1993). Correlations between outdoor (i.e. back yard) levels of $PM_{2.5}$ and levels at the central monitor were very high (0.96 overnight and 0.92 during the day). For PM_{10}, the correlations were also found to be high (0.93 overnight and 0.9 during the day) (Clayton et al., 1993; Wallace, 1996).

Burton et al. (1996) assessed the spatial variation within Philadelphia. The spatial variation was small for $PM_{2.5}$ but larger for PM_{10}. Spatial correlations for $PM_{2.5}$ were found to be near 0.9–1 and around 0.8 for PM_{10}. In a study from Ito et al. (1995) in Chicago and Los Angeles, the spatial correlations for PM_{10} were around 0.7–0.8. Kingham et al. (2000) investigated small-area variations of pollutants within the area of Huddersfield, UK. Spatial variations of pollutants were only modest and there was no association to distance from roads. Absorbance measurements of fine particles provided the best general marker of traffic-related pollutants (diesel exhaust). Furthermore, the indoor/outdoor correlations were best for these absorbance measurements of fine particles, indicating that an outdoor measurement of the absorbance of fine particulates is a useful measure of exposure to traffic-related pollutants.

Roorda-Knape et al. (1998) measured pollution levels near motorways in Holland. Black smoke and NO_2 levels declined with distance from the roadside; for PM_{10}, $PM_{2.5}$ and benzene, however, no concentration gradient was observed. The contribution of the coarse particle fraction ($PM_{10-2.5}$) on PM_{10} can be important. In Holland, where the PM_{10} levels were described to be quite uniformly distributed, a recent study indicated that the coarse fraction has considerable influence on the PM_{10} levels and that it causes spatial variations in PM_{10} (Janssen et al., 1999).

Blanchard et al. (1999) studied the spatial variation of PM_{10} concentrations within the San Joaquin Valley in California. PM_{10} levels varied by 20% over distances from 4 to 14 km from the core sites. Local source influence was observed to affect sites over distances of less than 1 km, but primary particulate emissions were transported over urban and sub-regional scales of approximately 10–30 km, depending on season. Gas-phase precursors of secondary aerosols were transported over distances of more than 100 km. This indicates that primary particles affect local-scale areas whereas secondary particles affect wide-range areas. Magliano et al. (1999) also found uniform spatial distributions of secondary ammonium nitrate (in fall and winter). Site-to-site variations were determined by differences in geological contributions in the autumn and due to carbonaceous sources in the winter.

Harrison and Deacon (1998) suggested that the number of monitors has to be large in order to cover the spatial variability in cities. With few monitors only, a quite general overview of the pollution climate can be obtained. As the correlations of the temporal concentrations profiles of different monitors are high, only a low-density network is needed for assessing short-term fluctuations.

An example of a study on small-scale spatial variations of PM_{10} is shown in Fig. 4. The spatial variability of PM_{10} (and NO_2) near a road in the city of Zürich was studied during one winter and one summer period in 1994/1995 (Monn et al., 1997a). PM_{10} and NO_2 levels were measured at different distances from the road. The measuring sites were positioned at 15 m (B), 50 m (C) and 80 m (D), and two meters above ground (m.a.g.) except for D: 6 m.a.g. During the summer period, an additional site was located at pedestrian level (1.8 m.a.g.) directly at the road (F). Because of the small difference between sites C and D during the winter period, site C was not used in summer. One site was installed on the roof of a house (20 m.a.g.) in order to investigate the vertical distribution. Fig. 4a shows the horizontal PM_{10} concentration profiles at the distances indicated. An almost parallel shift between the seasons was observed. The largest difference in concentrations occurred between the site closest to the road (A) and the first site at 15 m (B). The spatial variation at sites further away was very small indicating good horizontal mixing.

The vertical distribution (Fig. 4b) shows a similar pattern to that of the horizontal gradient. Levels during winter were higher than in the summer, and the

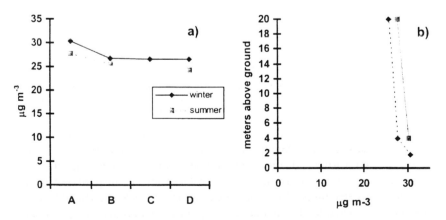

Figure 4. Horizontal (a) and vertical (b) distribution of PM_{10} in the vicinity of a road. (Monn et al., 1997a) (A = 2, B = 15, C = 50, D = 80 m from the road).

shift was almost parallel. The PM_{10} levels at the pedestrian level (F) were by far the highest of all sites. The levels at the upper vertical point 20 m.a.g., on the roof of the building, were similar to the levels 15 and 50 m away from the road, indicating urban "background" levels. Season, weekday and precipitation did not have a significant effect on the spatial CV in this study.

A similar investigation was made by Bullin et al. (1985), distinguishing between fine ($<1 \mu m$) and coarse particles ($>1 \mu m$). For fine particles, concentrations along the horizontal and vertical sites were almost the same. For coarse particles, increased traffic-related aerosols were detected near the road. These findings, however, are different from Roorda's (1998) findings, which did not observe a concentration gradient for PM_{10} and $PM_{2.5}$ with distance from the roadside. Chen and Mao (1998) observed large small-scale horizontal and vertical variations of PM_{10} in Taipei. The vertical profile showed a decrease in PM_{10} concentrations by 58% between the 2nd and 7th floor of a house, but no further decrease between the 7th and the 14th floor. Between a main and a side street in the city, large differences in PM_{10} levels were observed. Micallef and Colls (1998) and Colls and Micallef (1999) measured a vertical PM_{10} concentration profile over the first 3 m in a street canyon. At 0.8 m.a.g., PM_{10} levels were about 35% higher than at 2.8 m.a.g. For inhalable particles, this difference was 12%. Rubino et al. (1998) investigated a vertical profile of PM_{10} at a tower building and observed a decrease of PM_{10} concentrations by 20% between ground level and 80 m.a.g.

Kinney et al. (2000) investigated the small-area variation of $PM_{2.5}$, diesel exhaust particles (DPE) and elemental carbon (EC) in Harlem, New York. Site-to-site variations for $PM_{2.5}$ were only modest, however, EC contractions varied four-fold across sites. These spatial differences were associated with bus and truck counts. Local diesel sources created the spatial variation in sidewalk concentrations of DPE.

Recent studies investigated the distribution of ultrafine particles ($<0.1 \mu m$). Keywood et al. (1999) studied the distribution of $PM_{2.5}$, PM_{10} and ultrafine particles in six Australian cities. The $PM_{2.5}$ fraction dominated the variation in PM_{10}. PM_{10} and $PM_{2.5}$ correlated with each other but the correlation between the coarse fraction ($PM_{10-2.5}$) and PM_{10} was poor. An important finding was a lack of correlation between $PM_{2.5}$ and PM_{10} with ultrafine mass data as well with ultrafine particle number concentrations. This indicates that the former two cannot be used as surrogates for ultrafine particles. Junker et al. (2000) investigated the spatial and diurnal fluctuations of different parameters for particles within the urban area of Basle. Day profiles for ultrafine particle number concentrations, determined by a scanning mobility particle sizer (SMPS), were more closely related to the number of heavy-duty vehicles than to the number of light-duty vehicles. Diesel exhaust is a strong source of particles in the ultrafine mode, this fact is also reflected in the spatial variance of these particles: the site exposed to heavy-duty traffic had two to four times higher particle number concentrations than a background urban site and a residential site, respectively. In a study from Harrison et al. (1999), the ratio between particle number concentrations and PM_{10} was higher at a traffic-influenced site than at a nearby background location. The particle size profile, determined by SMPS, showed a clear difference between roadside and background location with an additional mode in the roadside sample below 10 nm diameter. Measurement of particle number gave the clearest indication of road-traffic emissions and, in contrast to other studies, the correlations between particle numbers and PM_{10} were significant and moderate. The diurnal variation of PM_{10}, particle number counts and Fuchs surface area showed the same general patterns, however, particles number counts gave the clearest indication of road-traffic emissions.

4.1.3. Indoor, outdoor and personal exposure

The largest databases on indoor, outdoor and personal suspended particulates levels are from the Six City study (and related studies) and the PTEAM study (Dockery et al., 1993; Clayton et al., 1993). Not included here are studies related only to indoor air quality.

4.1.3.1. Relationship between indoor and outdoor levels and indoor sources. Earliest data on indoor/outdoor ratios of TSP were presented by Yocom (1982). Outdoor air has been identified as an important source of indoor particulates in homes without apparent indoor sources. In summer, indoor TSP levels were found to be higher than in winter indicating the importance of the ventilation rate. Indoor/outdoor ratios were observed to range between 0.2 and 3.5. In air-conditioned rooms with highly efficient dust filters, indoor/outdoor rates were much lower 0.1–0.3. In Yocom's (1982) review the importance of the difference in the chemical composition of indoor and outdoor particles was emphasised.

In all of the further studies, smoking has been identified as the most important source for indoor particle concentrations (Dockery and Spengler, 1981b; Sheldon et al., 1989; Leaderer, 1990; Santanam et al., 1990; Quackenboss et al., 1991; Neas et al., 1994; Leaderer et al., 1994). Neas et al. (1994) assessed a clear dose–response relationship between $PM_{2.5}$ levels in air and cigarettes smoked. Investigations of the influence of other sources were presented by Quackenboss et al. (1991), Sheldon et al. (1989), Santanam et al. (1990) and Özkaynak et al. (1996b). Gas cooking, vacuum cleaning, dusting and also wood burning were identified as important indoor sources. The influence of gas cooking and kerosene heaters, however, was not confirmed by Leaderer et al. (1994). In a study from Virginia, Leaderer et al. (1999) found strong contribution of kerosene heaters to indoor $PM_{2.5}$, sulphates and acids (H^+) during the winter months.

In Wallace's review (1996) the importance of smoking and cooking was emphasised. Despite the strong effects of these indoor sources, the contribution of outdoor air to indoor PM levels remains significant. Infiltration of outdoor air into homes was estimated to contribute about 70% in naturally ventilated homes and 30% in air-conditioned homes to the indoor levels (Dockery and Spengler, 1981b). In homes without apparent indoor sources, outdoor particles contributed to about 75% of the indoor levels for $PM_{2.5}$ and 66% to the indoor PM_{10} levels in the PTEAM study (Özkaynak et al., 1996b). In homes with important indoor sources (smoking, cooking), outdoor air still contributed to about 55–60% to the indoor PM_{10} and $PM_{2.5}$ levels. Spengler et al. (1981) and Quackenboss et al. (1991) assessed the influence of season on indoor/outdoor relationships and stated that the differences between levels in homes with smokers compared to homes without smokers are stronger during winter than summer, due to reduced ventilation.

With respect to the correlations between indoor and outdoor levels, Dockery and Spengler (1981a), Ju and Spengler (1981) and Sexton et al. (1984) reported rather poor correlations. Indoor sources and differences in the ventilation rates were factors causing these inter-home differences. A further factor, which contributed to indoor PM levels, was that of human activity. Thatcher and Layton (1995) investigated deposition, resuspension and penetration of particles within a residence. The main finding was that the shell of a building did not provide any filtration of airborne particles. Differences in indoor/outdoor ratios in homes without major indoor sources for different size ranges are mainly explained by the difference in the deposition velocities of these particles. Concentrations of fine particles have a lower deposition velocity than coarse particulates. Light household activities such as walking around can increase levels of coarse particles. The resuspension rate increased with increasing particle size. Fine particles, therefore, undergo much less resuspension and deposition, which leads to indoor/outdoor ratios near unity, in the absence of other indoor sources. In a Swiss study investigations of the indoor/outdoor relationship for PM_{10} were performed in seven homes (Monn et al., 1997b). All homes had natural ventilation. The studies took place in the spring–summer periods. In each household, PM_{10} was measured for at least 3 periods for 48 h. Fig. 5 illustrates indoor/outdoor ratios for these homes. The highest indoor/outdoor ratios were found in the two homes with smokers (F, G). The indoor/outdoor ratios in homes with gas appliances were slightly higher than in homes without gas appliances. In the absence of smoking, the factor "activity" (people present in the home) influenced the indoor/outdoor ratios (home C versus A and B). Only in homes with inactive inhabitants did the indoor/outdoor ratio fall below unity (A + B). This figure confirms the importance of smoking and

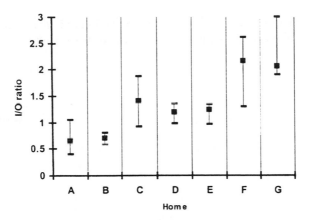

Figure 5. Indoor/outdoor (I/O) ratios for PM_{10} in homes with indoor sources and without obvious indoor source (A, B: no indoor sources, C: high activity of inhabitants; D, E: gas cooking, F, G: smokers) (dot: average, bars: minima and maxima) (Monn et al., 1997b). Reprinted with permission of Elsevier Science Ltd, previously published in *The Science of the Total Environment* 208, 15–21.

gas cooking but also shows the difficulty of the "human activity" factor when explaining indoor/outdoor ratios.

Abt et al. (2000) characterised indoor particle sources in Boston. Cooking, cleaning and the movement of people were identified as the most important indoor particle sources. Cooking (including broiling and baking, toasting and barbecuing) produced particles in a size range between 0.13 and 0.25 μm (volume diameter). Cleaning, moving of people and sautéing produced particles of a diameter between 3 and 4.3 μm. Frying was associated with both, fine and coarse particles.

Indoor/outdoor ratios in homes without apparent indoor sources can be estimated according to Wallace (1996). Indoor/outdoor ratios are a function of the air exchange and particle deposition rate, as the penetration rate is about unity and not different for fine ($PM_{2.5}$) and coarse particles ($PM_{10-2.5}$). Indoor/outdoor (I/O) ratios can be calculated using the equation

$$I/O = a/(a+k),$$

where a is the ventilation rate, and k is the deposition rate. Deposition rates for $PM_{2.5}$ were found to be between 0.4 and 1.0 h^{-1}. For a typical home with an air exchange rate of 0.75 h^{-1}, the indoor/outdoor ratio would be about 0.65 for fine particles and about 0.45 for coarse particles. In only a few homes, however, were such low indoor/outdoor ratios determined, indicating that most indoor environments are not free of indoor sources. In real situations, low indoor/outdoor ratios around 0.2–0.5 were rarely observed; this again indicates the importance of non-apparent indoor sources (Wallace, 1996).

4.1.3.2. Personal exposure. In 1981, Dockery and Spengler (1981a) reported that indoor $PM_{3.5}$ levels were better correlated with personal exposure levels than outdoor levels. The best correlation was observed between indoor sulphate (representing fine particles <1 μm) and personal exposure levels. One of the largest databases on indoor, outdoor and personal particle exposure levels is available from the PTEAM study. From the pilot study, Clayton et al. (1991) concluded that people living in the same household tended to be exposed to similar personal PM levels. In the PTEAM pilot study, the correlations of the concentrations between different rooms of one household were found to be quite high, suggesting that one measuring site per household is adequate. In the PTEAM study, more than 178 non-smoking people participated in a study on indoor, outdoor and personal sampling of PM_{10} and $PM_{2.5}$ (indoor and outdoor only) (Clayton et al., 1993). An important finding was that the personal PM_{10} levels in the daytime sample were higher than the corresponding indoor and outdoor levels. The nighttime personal levels were lower than the daytime personal values. The average personal exposure levels lay between those measured indoors and outdoors. Indoor PM_{10} nighttime values were lower than outdoor values, whereas during the daytime indoor levels were similar to outdoor levels. The correlations for PM_{10} between fixed site monitor levels and outdoor home levels ($r = 0.61$) were higher than for fixed site versus indoor levels ($r = 0.51$) and fixed site versus personal levels ($r = 0.37$). At night all these correlations were higher than during the day ($r = 0.93, 0.59$ and 0.54, respectively). For $PM_{2.5}$, the outdoor home levels during the day were similar to the values at night. The correlations between outdoor fixed site $PM_{2.5}$ levels and residential outdoor levels were good and slightly better at night than during the day. At night the indoor levels were lower than the outdoor levels. The increase in the daytime personal PM_{10} levels was explained by personal activities, mainly by indoor activities such as smoking, vacuuming, dusting, carpeting, cooking, using cloths drier, and spraying. The elemental analysis indicated that the coarse fraction (>2.5 μm) might be responsible for the elevated personal levels. Resuspension of house dust (which comes partly from outdoors) and dust from clothing may largely contribute to the so-called "personal cloud" (Özkaynak et al., 1996b). For personal monitoring Clayton et al. (1993) ruled out a possible sampling bias (in the personal monitor) and skin flakes as a possible source. Özkaynak et al. (1996b) concluded that all elements were elevated in the daytime personal sample, suggesting that both outdoor and indoor particles contributed to the elevated personal levels. With regard to the size range mainly the coarse fraction (>2.5 μm) and only a minor proportion of the fine fraction (<2.5 μm) contributed to the "personal cloud". This find-

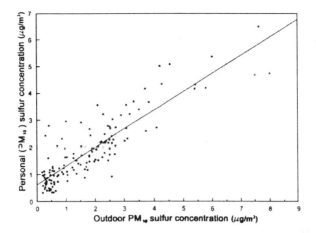

Figure 6. Personal exposure versus outdoor concentrations for sulphur ($R^2 = 0.77$) (Özkaynak et al., 1996b). Reprinted with permission from Nature Publishing Group. First published in *Journal of Exposure Analysis and Environmental Epidemiology* 6, 57–78.

ing was supported by observations of the sulphur levels, mainly found in the fraction below 1 µm, which were not elevated in the personal exposure values compared to those of fixed site monitors. Furthermore, the correlation between outdoor and personal levels for sulphur (representing fine, secondary particles) was good (Fig. 6).

Tamura and Matsumoto (1996) performed a study on personal exposures to PM_2 and PM_{10} in seven elderly non-smoking individuals in Tokyo. Indoor, outdoor and personal measurements were made for eleven 48 h periods. The correlations between the ambient and the personal levels in a cross-sectional analysis were quite high ($r > 0.8$), as was the correlation between the ambient and the indoor levels ($r > 0.8$). This study differed from the American studies with respect to indoor sources; smokers were not present and the reed-mat flooring and habits such as taking shoes off reduced the resuspension of particles. This study confirmed that in the absence of indoor sources the correlation between ambient and personal levels are good, and also that in clean homes without carpets the generation and resuspension of particles is low.

Janssen et al. (1995, 1998) presented results of personal PM_{10} monitoring from non-smoking adults in Amsterdam and from children in Wageningen. Median individual regressions between personal and ambient levels were fairly good (r around 0.6). For the children, parental smoking explained 35% of the variance between the personal and ambient levels. For the adults, dwelling "near a busy road", time spent in traffic and the exposure to ETS explained 75% of the variance between personal and ambient levels. This study confirmed that measurements of outdoor levels are appropriate for estimates of personal levels

Table 8. Potential factors in PM which influence human health

Physical properties[a]	Size mode, number, volume
	Hydrophobicity/philicity
	Electrostatic forces
Chemical composition[b]	Ionic compounds (nitrate, sulphates, acidity)
	Transition metals (e.g. Fe, V, Cr)
	Carbonaceous material (PAH; elemental carbon)
Biological species[c]	Allergens (pollen, fungal spores, glucans)
	Bacterial and bacterial structures (endotoxin)

[a]Yeh et al. (1976), Oberdörster (1995) and Peters et al. (1997).
[b]Spengler et al. (1990), Pritchard et al. (1996) and Lewtas et al. (1993).
[c]Wüthrich et al. (1995) and Rylander (1998).

in follow-up investigations but also that in a cross-sectional analysis the relationship between personal and outdoor levels is poorer. Aggregation of personal exposure levels to daily averages in the study areas, regressed against the daily outdoor averages, also increased the personal–outdoor correlations (Mage and Buckley, 1995; Wallace, 1996). The finer the particles, the better the correlation between the personal and outdoor levels; the best correlation was found for sulphur, representing particles smaller than 1 μm (Wilson and Spengler, 1996).

4.1.3.3. Other potential factors, relevant in an SPM exposure assessment. The particle size is an important factor, as it determines the site of deposition in the respiratory tract (Yeh et al., 1976). Associations between exposure and health are expected to be stronger for fine particles (<2.5 μm) than for coarse particles (>2.5 μm) (Schwartz and Neas, 1996). It is, however, unclear whether the mass load itself, some other physical factors such as the number of particles, the total surface area, or electrostatic factors, or the chemical and biological composition are the causative parameters. In a study from Peters et al. (1997), health effects from fine (0.1–2.5 μm) and ultrafine particles (0.01–0.1 μm) were more strongly attributed to the number of ultrafine particle than to the mass. This single example, while not conclusive, may be a first step in identifying the mechanisms of health effects.

Table 8 summarises parameters of PM which might potentially influence respiratory health and mortality.

4.1.4. Summary

- The spatial distribution of fine particles (<2.5 μm) is relatively uniform; therefore a central monitor for $PM_{2.5}$ may collect a sample representative for a study area. In vicinity to emission sources (e.g. diesel exhaust), however, the PM$_{2.5}$ levels larger than the urban "background" may be observed. For larger particles (the coarse fraction of PM$_{10}$ and TSP) the spatial variation is larger than for fine particles and the collection of representative samples is more critical.

- Ultrafine particle numbers do not strongly correlate with PM$_{10}$ (and PM$_{2.5}$) mass concentrations. Primary traffic-related emissions are not well reflected by mass measurements.
- The fine (<2.5 µm) and the coarse (>2.5 µm) modes represent two different sets of pollutants with different emission sources, different chemical composition and different spatial and temporal behaviour.
- For indoor PM levels, outdoor air is the most important source besides smoking, which is the most important indoor source. The size range influenced by smoking is the fine mode (<1 µm), while other human activities influenced mainly the coarse mode (>2.5 µm).
- Indoor/outdoor ratios of PM$_{10}$ and PM$_{2.5}$ in homes without smokers reach about unity; in homes with smokers, the ratios are larger.
- Indoor PM levels are not well correlated with the corresponding outdoor levels in cross-sectional studies. In follow-up, i.e. day-to-day studies, the correlation between indoor and outdoor concentrations becomes much better.
- The finer the particles, the better the correlation between the personal and outdoor levels.
- Personal PM$_{10}$ levels were generally higher than the corresponding indoor and outdoor levels. The chemical composition of the personal PM$_{10}$ mass mainly indicated contributions from the coarse mode (>2.5 µm).
- The personal, indoor and outdoor sample of PM$_{10}$ have different chemical compositions. A personal sample can even be described as a "confounding factor" in investigations dealing with effects of outdoor particles.
- Up to date, adverse health effects were associated with the measured *mass* concentrations of particles in air. But surface area or number counts may also be relevant. The chemical constituents and biological materials also have to be considered.
- A move towards source-related markers of particles has to be made in order to understand their specific health effects and to develop efficient emission-reduction strategies.

4.2. Bioaerosols and allergens in ambient air

Biological material is present in all ambient aerosols. The biological material consists of proteins, lipids, carbohydrates (starch, cellulose), DNA, and RNA and all these components can be used as unique markers for particles of biological origin (Matthias-Maser, 1998). However, data on the amount of biological material in aerosols are sparse. In studies from Matthias-Maser (1998), the content of primary biological aerosol particles (PBAP) accounted for 10–13% of urban and rural fine aerosol particles, determined by particle counting. The amount of protein accounted for about 2% of the total mass in urban and rural aerosols (Schäppi et al., 1996). In a study from Miguel et al. (1996) biological particles were found to be abundant in paved road dust and in air samples in California. Pollen, pollen fragments, animal dander and moulds were detected in resuspended dust from roads. About 5–12% of the allergenicity of TSP and PM_{10} were attributed to resuspended road dust. In contrast to on-line measurements used for air pollutants, the methods for measuring bioaerosols are more time consuming and difficult. Therefore, large databases on small-scale variations of bioaerosols do not exist. It has to be assumed that the spatial variations are large as local sources of pollen, bacteria and spores strongly influence receptor sites. On the other hand, in a study on personal exposure to pollen, the correlation between personal exposure and fixed site samples was significant with respect to day-to-day variations for people within a city (Riediker et al., 2000).

4.2.1. Natural constituents of the atmosphere

4.2.1.1. Pollen.   Pollen is produced by vascular flowering plants and is the main source of allergens in the atmosphere. The most frequent pollen-induced allergies are caused by grasses (*Poacea, Phleum, Lolium*), ragweed (*Ambrosia*), birch (*Betula*), olive (*Olea*), pellitoria (*Parietaria*), mugwort (*Artemisia*) and cedar pollen (in Japan) (Spieksma, 1995; Wüthrich et al., 1995; Ishizaki et al., 1987). All these pollen are from wind-pollinating plants. The size of pollen is around 15–40 μm and larger. The aerodynamic shape enables the pollen to remain airborne over long distances. In most European countries and the USA, pollen grains are counted routinely within pollen survey networks (Lewis et al., 1983; Emberlin, 1997). The usual instrument for collecting pollen is the Burkard Pollen trap, a volumetric method in which pollen is deposited on a sticky film. The pollen grains are differentiated into species and counted microscopically. Results are given as numbers of pollen grains per cubic meter of air. The time resolution for routine measurement is hours to daily means.

4.2.1.2. Fungi, spores.   The major structures of fungi are filaments and spores. Spores are the main biological airborne material, with a size range from 2 to 10 μm. Fungal spores can be found in large quantities in the atmosphere during summer and autumn with levels up to more than ten thousand spores per m^3 of air (Spieksma, 1995). In outdoor air, *Alternaria* and *Cladosporium* are the most abundant genera in mid-Europe (Flückiger et al., 1998). Spores can be collected by volumetric methods where they are deposited on a culture dish by impaction. The sampling time is short (about 5–15 min) and provides information on a short-term basis. The culture dishes are incubated for 2–5 d (at 20–35° C) and the number of spores is then counted. The results are given as colony forming units (CFU) per m^3 of air. The non-viability of many spores is a drawback of this method. Direct microscopical counting yields a higher number of spores than the use of culture dishes for viable spores (Burge, 1995). It is especially important to determine the allergen content, e.g. with an ELISA test (enzyme-linked-immuno-sorbent assay) as dead spores may still contain active allergenic components (Flückiger et al., 1999). In addition to allergens, fungi also contain antibiotics and mycotoxins. Cell wall components such as D-glucans are also known to induce irritative effects (Fogelmark et al., 1994).

4.2.1.3. Bacterial aerosols. Environmental bacteria in soil and water can be released into the atmosphere by wind, splashing rains and mechanical disturbances. The particle sizes vary from 1 to 50 μm. In indoor settings, the concentration of bacteria in air is mainly due to the presence of humans; most of these bacteria are non-infectious (COST, 1993). To determine the bacteria content in air, techniques similar to those used for fungal spores are employed: a volume of air is drawn through an impactor and particles are deposited on culture dishes. The results are given in CFUs per m^3 of air. The sampling time is between a few and 15 min, providing short-term data. A major distinction is made among Gram-negative, Gram-positive bacteria and actinomycetes. Relevant for human health is the presence of bacterial toxins. Allergenic material can be determined by ELISA and the quantification of toxins is obtained by analytical chemistry.

4.2.1.4. Endotoxins, lipopolysaccharides. Endotoxins are cytoskeletal molecules of the cell wall of Gram-negative bacteria. Lipopolysaccharides (LPS) are the most important and the biologically active component of endotoxins. In occupational studies (e.g. farming and waste industry) exposure to LPS causes inflammation, airway constriction, chest tightness, and induction of nitric oxide reactions (Fogelmark et al., 1994). In indoor settings, e.g. in the sick building syndrome (SBS), inflammation was often related to endotoxins in the air (COST, 1993). Endotoxins were also detected in ambient air particles, where they were responsible for causing pro-inflammatory reactions in macrophages (Becker et al., 1996; Monn and Becker, 1999).

4.2.1.5. Further biological material in aerosols. Latex is a natural rubber from the plant *Hevea braziliensis*. A variety of allergens have been identified recently in latex (Muguerza et al., 1996). Exposure to latex particles is common in certain occupations (dentistry, medicine, etc.) and the increased use of latex devices as a protective barrier against viral infections in the general population raised the incidence of sensitisation observed during the last decade (Vanden-plas et al., 1995). Besides dermal contact, the intake of latex particles through the airways may be important (e.g. inhalation of latex dust from gloves) (Baur, 1995).

Latex has also been detected in ambient air particles (Miguel et al., 1996). Automobile tyre production is the largest application for natural rubber. The fabrication method has changed recently and a higher content of natural rubber is now used. Latex particles have been identified in airborne and deposited particulates in the near vicinity of roads in California (Miguel et al., 1996). For people living near roads with heavy traffic this might constitute a further sources of contact with latex particles. In a study in Switzerland, latex allergens have been identified in ambient coarse (2.5–10 µm) but not in ambient fine (<2.5 µm) particles (Monn and Tenzer, 2000).

In some occupational settings, for example in the handling and storing in-dustry of farm products such as hay, grain, etc., a large amount of allergenic material can be released into the atmosphere (Lacey and Dutkiewicz, 1994).

Excretions from house dust mites, cats, dogs, birds, insects and cockroaches are important sources of indoor allergens in homes (COST, 1993).

Table 9 depicts some of the most important biological species in ambient air. Major allergens from pollen, e.g. from birch trees (Bet V 1), timothy grass (Phl p 5) and rye grass (Lol p 1), and allergenic fungal spores from *Cladosporium* (Cla h 1) and *Alternaria* (Alt a 1) may be responsible for some acute aller-gic reactions. The peptide sequence of pollen allergens shares similarities with some enzymes, recognition structures, and bacterial defence or stress-related proteins (Knox and Suphioglu, 1996a; Swoboda et al., 1994). To date, only a few of the hundreds of potential fungal allergens have been characterised. Also relevant to health effects are cell components such as D-glucans, antibi-otics from fungi and endotoxins from Gram-negative bacteria. Ambient air

Table 9. Summary of important aerosols of biological origin found in ambient air and some examples of associated inflammatory and allergenic material

Biological species	Content	Examples of inflammatory and allergenic species
Plant: pollen grains, leaves	Proteins, lipids, starch, cellulose, carbohydrates	Bet v 1, Phl p 5, Lol p 1
Fungi: spores, mycel	Proteins, lipids, toxins, glucans, an-tibiotics	Alt a 1, Cla h 1, D-glucans
Bacteria: Gram-negative	Proteins, lipids, toxins, lipopolysac-charides	Endotoxins
Rubber from tires	Protein, latex	Hev b 1

samples collected near roads may contain a major allergen from the airborne latex rubber (Hev b 1) resulting from tyre abrasion.

4.2.2. *Interactions between air pollutants and pollen*

Intact pollen grains have a size larger than 10 μm and should, therefore, be found only in sizes larger than 10 μm. However, allergens from pollen were detected in fine particles (<2.5 μm) (Solomon et al., 1983; Rantio-Lehtimaeki et al., 1994; Spieksma, 1990, Spieksma et al., 1995; Schäppi et al., 1996).

How might these allergen-loaded fine particles be produced? A plausible mechanism was suggested by Knox and Suphioglu (1996a). After contact with water, grass pollen undergo osmotic rupturing. This leads to a release of hundreds of small Lol p 5 containing starch-granules (0.6–2.5 μm) in *Lolium perenne* pollen. Cytosolic allergens (such as Lol p 1) are also released into the atmosphere and are free to react with water droplets and other atmospheric particles. Molecular characterisation of these pollen allergens reveals that they are homologous to recognition proteins. Such cysteine-rich molecules tend to bind to other molecules such as carbohydrates, proteins and lipids (Knox and Suphioglu, 1996b). Lol p 1 has been shown to interact with diesel soot particles (Knox et al., 1997). The release of orbiscules of small size (<0.1 μm) may also be a factor (Ong et al., 1995); allergens may also originate from organs other than pollen. Size-fraction allergen studies revealed that antigens were found in periods outside the pollen season (e.g. for Bet v 1) (Schäppi et al., 1996). Antigens transferred to fine particles may have a different half-life in the atmosphere and may remain in the atmosphere longer than larger particles. Bet v 1 was also found in fine particles before pollen counts were positive, indicating either production of Bet v 1 from other plant organs than pollen or lack of sensitivity in the pollen-counting procedure. Schäppi et al. (1997) suggested a mechanism for the production of fine particles containing Bet v 1 which, in contrast to grass pollen, do not show humidity rupturing. Birch pollen grains were seen to sediment on leaves. After a rain-shower, a germination tunnel grew through which allergen-loaded starch particles of small size were released into the atmosphere. Therefore, according to Schäppi's observation, the content of Bet v 1 in fine particles was higher after a rainy period than before.

There are two important mechanisms in the interaction between air pollutants and pollen allergens: air pollutants can induce stress-related allergens in the pollen antheres and air pollutants can modify the surface of pollen and increase the bioavailability of allergens. Breitender and Scheiner (1990) provided evidence for a higher content of Bet v 1 in birch pollen exposed to traffic compared to that of trees in rural areas. Induction of Bet v 2, which is homologous to the stress-related allergen profilin, has also been observed in birch trees exposed to stress (e.g. bacterial infections, drought, pollution, etc.). The direct effects of ozone on the content of Phl p 5 in *Lolium perenne* in Germany were investigated in open-top and closed-top chambers (Masuch et al., 1997). Elevated ozone levels were associated with an elevated allergen content in grass pollen. The effect was dose dependent, but O_3 peak levels were less important than long-term O_3 averages.

In polluted urban areas, pollen surfaces are contaminated by fine particles (Kainka-Staenicke et al., 1988; Behrendt et al., 1992; Schinko et al., 1994). The modification of the pollen surface may alter the availability of allergens although the outer wall of pollen (exine, made of sporopollinin) and the inner wall (intine, made of polysaccharides) are rather inert. For gases, the germinal pore may offer a site of reaction. Behrendt et al. (1997) investigated interactions between pollen and air pollutants in a floating chamber under controlled conditions. Pollen exposed to particles underwent structural changes and increased allergen release was observed. Exposure to SO_2, but not to NO_2, resulted in a reduction of allergen release. Behrendt concludes that air pollutants may modulate the bioavailablity of grass pollen allergens.

The role of these fine particles in promoting allergic symptoms and sensitisation is, however, unknown. It has been suggested that allergen-loaded fine particles affect lower-airway symptoms (e.g. allergic asthma), whereas intact pollen grains cause upper-airway allergic symptoms such as hay fever (Wilson et al., 1973; Suphioglu et al., 1992). During a thunderstorm in Australia, a large number of people suffered from allergic symptoms. It has been suggested (but not measured) that such fine particles may have played an important role in this case (Bellomo et al., 1992). For a similar event in England, a potential influence of a high allergen content in air and of allergen-loaded fine particles has been discussed (Celenza et al., 1996).

Emberlin (1995a, b) reviewed the literature on the interactions between air pollutants and aeroallergens. She concluded that there was a need for more research on aeroallergens and air pollution. This interdisciplinary field requires collaboration between biologists, air pollution scientists, plant physiologists, physiologists and epidemiologists. Her review focused mainly on outdoor conditions, but she pointed out that compounds of biological origin also play an important role in indoor settings.

4.2.3. Summary

- Both, the coarse (>2.5 µm) and the fine fraction of particles (<2.5 µm) contain material of biological origin (bacteria, pollen, fungal spores), but the major part of intact bioaerosols (pollen, bacteria, spores) occurs in the coarse mode.

- In deposited and resuspended dust, particles of biological origin may be abundant. Furthermore, the allergenicity of suspended particles is largely influenced by deposited dust.
- Pollen allergens present in fine particles (<2.5 µm) offer a potential route of entry into the small airways, whereas the target organ of intact pollen (>10 µm) are the mucous membranes of the nose and mouth.
- In addition to pollen allergens, other compounds of biological origin, such as endotoxins, mycotoxins, spores, allergens from pets and mites (both indoors) may significantly affect human health.
- Interactions between anthropogenic pollutants and biological material are not well understood. Gases (e.g. O_3, NO_2) may increase the allergen content in pollen (e.g. by induction of stress-related proteins). Airborne fine particles may adhere to pollen surfaces and modify the bioavailability of allergens. On the other hand, pollen allergens may also adhere to fine particles and produce allergenic fine particles with a potential to penetrate into the alveoli.

4.3. Nitrogen dioxide

4.3.1. Small-scale spatial variation

A large database on small-scale spatial variations of NO_2 was established in SAPALDIA (Martin et al., 1997; Monn et al., 1998). Data on the spatial variability of NO_2 were collected in two phases; in the cross-sectional study (1991) NO_2 was measured stationarily and in the follow-up study (1992–1993) population-based at home with passive samplers. Examples from this study ill be used to show spatial outdoor differences and an approach to deduce a multiple regression model.

An example of stationary measurements at two sites is shown in Fig. 7. Monitoring took place during four weeks in each season. During spring and summer, large spatial variations in NO_2 concentrations were observed, whereas in the autumn and winter periods, the variation was small. During winter, inversion layers prevail on the Swiss Plateau; this situation favours a homogeneous mixing of the pollutants within this layer. The conversion of the primary pollutant NO to NO_2 is slow during winter (Atkinson, 1990); in spring and summer, however, the conversion from NO to NO_2 is much faster. Higher ozone levels and photochemical reactions favour the oxidation of NO. Therefore, a steeper concentration gradient is expected in summer compared with winter periods. In fact, observed levels at the most polluted sites in spring were higher than in winter, although the regional NO_2 levels were higher during winter than during summer. In 1992–1993, a population-based outdoor and indoor/personal monitoring study was performed during a full year period in SAPALDIA. Overall,

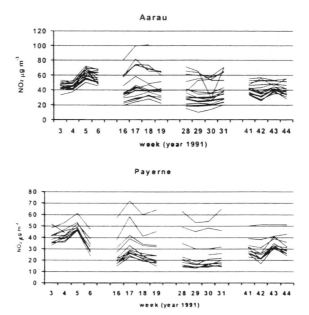

Figure 7. Temporal and spatial concentration profile of NO₂ at Aarau (suburban, 17 sites) and Payerne (village, rural, 12 sites) measured at fixed sites with passive samplers (Hegner, 1994).

about 50 individuals, a random sample of the SAPALDIA population, participated three times during four consecutive weeks in each region. In contrast to the cross-sectional study (1991) a larger number of measurements were obtained and the pollution levels represent randomly population-based outdoor home levels. On the other hand, the number of (temporal) parallel measurement periods was smaller than in 1991 because of the measuring scheme.

Fig. 8 compares the regional (population-based) outdoor annual mean averages (measured with passive samplers) with the annual means of the continuous monitors. At the urban sites, the levels of the central monitors were slightly higher than the population-based outdoor average. This observation is plausible, as in large cities part of the population lives in residential areas outside the city centres (where the central monitors were installed). At the rural sites Payerne and Wald and the Alpine site Montana, the central monitor underestimated the population-based values. At these three sites, the central monitor was located outside the villages and the measured values reflect this fact. A comparison between the technical error (between passive sampler and monitor) and the error due to a "non-representative" sampling shows that the latter can be larger. Such errors in non-representative sampling with respect to population exposures may introduce errors in the exposure-health effect rela-

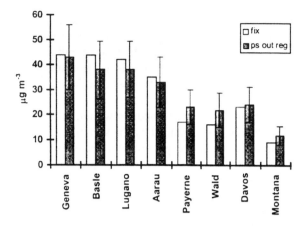

Figure 8. Comparison between ambient NO_2 annual means measured with the fixed site monitors (fix) and the averages obtained from passive sampling measurements outdoors (ps out reg) during 1993 (columns: mean values; bars: average standard deviation of the spatial variation).

tionship. The shown example highlights the need for a careful selection of the central monitors or the use of passive sampler to evaluate the representativity of the central monitor. Based on these results, improved exposure estimates were used in the health-effect analyses of SAPALDIA (Ackermann-Liebrich et al., 1997; Schindler et al., 1998). Based on the spatially refined NO_2 data, Schindler et al. (1998) revealed health effects due to exposure to NO_2 which were hidden in an aggregated analysis.

4.3.2. Indoor/outdoor and personal exposure

Yocom (1982) reviewed indoor/outdoor relationships of NO_2 from early studies. Some of the referred data are from Portage, WI, Boston, Southern California, Holland, Los Angeles and Switzerland (Quackenboss et al., 1986; Leaderer et al., 1986; Drye et al., 1989; Billick, 1990; Noy et al., 1990; Spengler et al., 1994; Monn et al., 1998).

An important study on personal exposures to nitrogen dioxide based on more than 700 individuals in Los Angeles was published by Spengler et al. (1994). The purpose of the "Nitrogen Dioxide Exposure Study" was to quantify the relative contributions of indoor and outdoor sources, as well as factors relating daily activity patterns and human exposure to NO_2. In the main study, which included more than 700 individuals, NO_2 was measured over a 48-h period. In a sub-sample, a microenvironmental study was performed. In addition, these individuals participated more than once in order to quantify the variability between season and within subjects. All the monitoring took place in the bedroom, in outdoor air and on a personal level. Personal levels were higher in homes with gas ranges and pilot lights than in homes with gas ranges only and in homes with electric stoves. Season had a great influence on the outdoor levels, but the influence on the bedroom levels was not substantial. About 40% of the variation in bedroom concentration was explained by outdoor levels. Regression between indoor and outdoor levels yielded slopes of 0.4–0.5 with higher slopes in summer and lower slopes in winter. 60% of the variation in personal exposure levels was explained by the variation in the bedroom level. 51% of the variation in personal exposure could be explained by the outdoor levels. In the SAPALDIA study, outdoor levels explained 33% of the variation in personal levels, and indoor levels about 51% of the personal levels (Monn et al., 1998). The personal exposure showed seasonal and spatial variation. Personal exposure levels were highest during winter months. The spatial patterns in exposure reflected the location of an individual home with respect to the distribution of ambient levels and the intra-urban distribution (zones) of housing characteristics (type of range, pilot light, etc.). The parameters determined to explain personal levels were type of range, season, ambient NO_2 zone, and day of the week. The within-household variation (in some households more than one person was participating) for bedroom or outdoor levels was small. Within-household variation of personal exposure, however, was three times that of the stationary measurement. This suggests that some other factors (such as differences in time-activity pattern) also largely determine the personal exposure levels.

The main findings of these studies, namely that gas stoves, outdoor air and season (ventilation) are the most important parameters for the indoor NO_2 levels, were confirmed in all other studies. Indoor levels were generally better predictors for personal NO_2 exposure levels than outdoor levels. Indoor/outdoor ratios in homes without indoor sources were around 0.4–0.8; and in homes with gas appliances about three times higher (Yocom, 1982; Quackenboss et al., 1986; Leaderer et al., 1986; Ryan et al., 1988a, b; Drye et al., 1989; Billick, 1990; Brunekreef et al., 1990; Schwab et al., 1994; Monn et al., 1998). In homes with gas appliances, attention has to be paid to other compounds, such as HNO_2, which is produced indoors (Brauer et al., 1991). The NO_2-concentration gradient within a household was small in homes using electric appliances but was significant in homes with gas appliances (Ryan et al., 1988a, b; Billick, 1990; Neas et al., 1991). Noy et al. (1990) observed that the correlation between weekly averages and peak levels for personal or stationary indoor measurements was poor. The ratio between the peak and the mean levels was about five. In most of the presented studies, NO_2 levels have been determined with passive samplers exposed for periods lasting from two days up to weeks. This indicates that important information could be lost with this device.

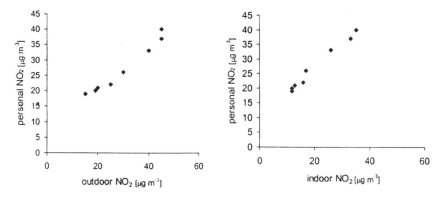

Figure 9. Scatterplot for outdoor/personal ($R^2 = 0.965$ and indoor/personal ($R^2 = 0.983$) NO$_2$ ratios of aggregated data (annual mean estimates) N: indoor = 1501, N: outdoor = 1544, N: personal data = 1494 (Monn et al., 1998).

An example from the SAPALDIA study illustrates indoor/outdoor/personal exposure relationships based on aggregated data. Highly significant correlations ($R^2 > 0.9$) were found for an outdoor versus personal and an indoor versus personal exposure analysis (Fig. 9). The three areas with the highest NO$_2$ outdoor levels (urban) also are the sites with the highest percentage ($>31\%$) of gas users. This might be the reason why the three upper sites do not follow a strict linear line in the outdoor/personal exposure plot. In an analysis of individual single data, however, these correlations were much weaker (indoor/outdoor 0.5, indoor/personal, personal/outdoor <0.3) (Monn et al., 1998).

4.3.3. Summary

- Outdoor small-scale spatial variations of NO$_2$ were observed to be significant, especially during periods of elevated photochemical activity.
- A well-established passive device is available for measuring NO$_2$. The time resolution of the sampler ranges from about one day up to two weeks.
- Indoors, gas stoves are the major emission source of NO$_2$. In homes without indoor sources, outdoor NO$_2$ levels were the strongest source of indoor levels.
- Personal exposure levels for NO$_2$ were better correlated with indoor home levels than with outdoor levels.
- One difficulty in the use of personal and indoor NO$_2$ levels as an indicator is the difference of the pollutant mixture generated indoor and outdoors: outdoor combustion mixtures are different from the mixture of gas stove emissions.

Table 10.   Photochemical pollutants and acids

Parameter	Precursor, reaction partner	I/O ratios approx.[a]
O_3	NO_2, VOC	0.2–0.8
H_2SO_4	SO_2; $HO_2^{\cdot}$	0.86–0.96 (SO_4^{2-})
HNO_3	NO_2; $OH^{\cdot}$	0.1–0.47
HNO_2	NO, $OH^{\cdot}$; heterogeneous	>1–5 (with gas appliance)
H^+	In acids	0.35–0.48
PAN	NO_2, VOC, radicals	Not determined

[a] I/O: estimates for indoor/outdoor ratios, from Brauer et al. (1991, 1995), Yocom (1982), Moschandreas (1981), Lustenberger et al. (1990) and Weschler et al. (1989).

4.4. Photochemical pollutants

4.4.1. Ozone and related photochemical pollutants in the atmosphere

Important photochemical constituents, such as sulphuric acid (H_2SO_4), nitric acid (HNO_3), and peroxy-acetyl-nitrate (PAN) are produced in outdoor air (Atkinson, 1990; Carter et al., 1995; Table 10). Nitrous acid (HNO_2) is an important constituent in ambient air during the night but is also produced by heterogeneous reactions in homes with gas stoves (Brauer et al., 1991). Table 10 shows rough estimates for indoor/outdoor ratios. Indoor levels for O_3, HNO_3, and H^+ were observed to be considerably lower than the corresponding outdoor levels. For H_2SO_4 (determined as SO_4^{2-}), indoor/outdoor ratios reached unity.

Ozone concentrations undergo distinct seasonal and diurnal variations; peak levels are observed on afternoons with intense sunshine in spring and summer (BUWAL, 1996). The photochemical production of O_3 in the troposphere depends on NO_x and VOC emissions (Atkinson, 1990); therefore, highest O_3 levels in rural areas are observed downwind of urban plumes (Altshuller, 1988). The small-scale variation within a study region is dependant on the vicinity of the site to NO emissions: in urban areas, reduced O_3 levels are observed in the vicinity of traffic arteries; in rural areas, O_3 levels are observed to be more or less uniformly distributed (BUWAL, 1996).

Peak values of O_3 were considered to be more relevant for causing adverse health effects than long-term exposures (Lippmann, 1989; Tager, 1993). In considering these findings, air quality standards for O_3 rely on peak values such as 1-h values and 8-h daytime values in the USA and WHO guidelines. In an ecological approach, where the average exposure concentration and also the ranking of the exposure levels between the study sites were important, an inconsistency in the use of different statistical parameters for the description of long-term exposure to O_3 were observed at the SAPALDIA sites (Monn et al., 1999). For SO_2, PM and NO_2, the ranking between the study sites remained the same if percentiles (e.g. 95th or 98th percentiles of $\frac{1}{2}$ values) or annual means were used. For O_3, however, this was not the case. Alpine sites exhibited highest averages but were only in the mid-range for peak values. The reason for these findings can be explained by the fact that two different pathways contribute to O_3 concentration levels in the troposphere: the photochemical reactions, mainly during spring and summer, and the influx of O_3 from the free troposphere, which mainly affects elevated sites (above 1000 m.s.l.) (BUWAL, 1996). A comparison of photochemically produced HNO_3 also revealed that HNO_3 occurred in much lower quantities in the Alps than at urban sites.

Improvements of the assessment of long-term O_3 exposure measures were performed by Künzli et al. (1997). A questionnaire was designed to ask for time activity, time budgets (especially time spent outdoors) and physical activity in the past. Based on measured O_3 levels in two study sites in California, the method enabled a better calculation of life-long exposures than the use of outdoor data only.

4.4.2. Acids

Spengler et al. (1990) reviewed data on acidic compounds in outdoor air and stated that after a change in emission patterns, whereby SO_2 is now largely emitted by high stack power plants, acid sulphates can be transported over large distances. Increasing emissions of NO_x by vehicular traffic has increased the levels of HNO_3 in large urban areas. Acidic aerosols and gases can be neutralised by ammonia, yielding ammonium salts. In the atmospheric surface layer where the amount of NH_3 is higher than in the upper atmosphere, the quantities can be sufficient to neutralise acids. The diurnal variation of acids shows generally higher daytime levels than at night and higher levels during the summer than in winter. Because of higher NH_3 emissions in urban areas (related to the population density) compared to rural areas, the excess of acidity is larger in rural than in urban areas. Acidity can be determined by measuring the strong acids (H^+), H_2SO_4, and HNO_3. Sulphate measurements were also used as surrogates for aerosol acidity, but the two measurements were sometimes weakly correlated. Waldman et al. (1990) found a strong correlation between three sites in the Toronto metropolitan area for aerosols consisting of H^+, NO_3^-, SO_4^{2-}. While for sulphates the peak levels were quite uniformly distributed, strong acidity varied considerably between the sites. Downtown areas had lower aerosol acidity levels than suburban areas. Sulphates and H^+ levels were correlated over time, but considerable differences in the spatial behaviour of these two components were observed (Waldman et al., 1990). Brook and Spengler (1995), Spengler et al. (1996) and Özkaynak et al. (1996a) presented data on ozone and strong acidity in 24 North American communities. Strong acidity was mainly detected in areas close to high sulphur emission areas (western Pennsylvania, eastern Ohio, West Virginia). Low particle strong acidity was found in regions without high sulphur emission (western and midwestern cities). Substantial concentrations of nitric acids were detected in two Californian sites and many sites in the northeast. Suh et al. (1997) studied the distribution of photochemical pollutants in Washington DC. A strong correlation among particulate measurements (PM_{10}, $PM_{2.5}$) with SO_4^{2-} and H^+ was found. H^+ was found to be uniformly distributed across the study area, and a larger spatial variation was observed for coarse particles and NH_3 than for H^+. While sulphur-related acidity is an important problem in parts of the USA, nitrogen species are dominating in Western Europe. In Switzerland and also in other parts of eastern Europe for example, H_2SO_4 levels were found to be negligible (<2 ppb) (Alean-Kirkpatrick, 1993; Brauer et al., 1995; Monn et al., 1998). In western Europe, NO_2-driven chemistry and neutralisation by NH_3 was found to be important. Although SO_2 levels were high in eastern Europe, the conversion to SO_4^{2-} was too slow to produce high amounts of acids. Within a photochemical pollutant mixture, the production of aerosols (<2.5 μm) (e.g. ammonium sulphate) is also an important source of elevated particulate levels (US-EPA, 1996b).

4.4.3. Indoor and personal exposure to photochemical pollutants

Measurements of personal exposure to O_3 and acids have not yet been carried out extensively. Studies from Europe and from the USA (Monn et al., 1993; Brauer and Brook, 1997; Liu et al., 1993; Liu et al., 1997) indicated weak to moderate correlations between outdoor and personal O_3 levels, and also between indoor and personal O_3 levels. However, the correlation coefficients increased for special groups such as children in summer camps or farmers, who spent most of the time outdoors. In such cases, ambient measurements described personal exposure levels much better than in the general working population. Suh et al. (1993) published data on levels of personal exposures to sulphates and aerosol strong acidity. Personal levels of sulphate exposures for children were consistently lower than the measured outdoor values. The same finding was obtained for personal strong acidity levels (H^+). A model was computed in order to calculate personal exposure values by using stationary outdoor sulphate data and information on time spent indoors and outdoors. Indoor levels were modelled using the outdoor levels and a loss-rate for indoor levels (different in air-conditioned than in non-air-conditioned homes). For the calculation of an H^+ concentration model, data on personal NH_3 exposure values were needed, including the reaction rate accounting for a loss due to NH_3 neutralisation. A good model could be computed by using stationary H^+ levels from outdoor air, personal NH_3 levels and time-activity data (time spent indoors and outdoors). However, the accuracy and the precision of this model predicting H^+ personal levels was much lower than for the sulphate model. This is an indication of the large variability in H^+ and NH_3 concentrations. In the validation study it was demonstrated that the interpersonal variability can be very high. It is noteworthy that this model was applied to estimate children's exposure. Children may have a better-defined time-activity pattern with fewer microenvironments than adults, such that the model may not be as accurate and precise for adults. Liu et al. (1997) investigated personal exposures to O_3. Personal exposure levels were not well predicted by outdoor concentrations. An improved model was computed using elevation, distance to stationary monitor and traffic. The low predictive power was due to spatial variability of outdoor O_3 (which is very important in regions with high traffic sources) and also because of errors in time-activity records and in the measurements. Brauer and Brook (1997) studied personal exposures to O_3 and health effects for selected groups (people working indoors, people spending more time outdoors than indoors, such as farm workers). Differences between these groups' exposures to O_3 were associated with time spent outdoors. Leaderer et al. (1999) investigated acids, ammonia and particles in more than 58 homes in Virginia. The contribution of kerosene heaters to increased $PM_{2.5}$, sulphates and acids (H^+) concentrations has already been mentioned. Indoors, nitrous acid levels were higher than outdoors and higher in homes with unvented combustion sources. As ammonia levels are higher indoors than outdoors, the particle acidity was lower indoors than outdoors.

4.4.4. Summary

- The small-scale variation of O_3 within an urban and rural area is dependant on the proximity to NO emissions (e.g. traffic arteries).
- In the assessments of exposure to O_3, short-term parameters are considered to be more important than long-term parameters.
- Besides O_3, a number of other constituents are produced in photochemical reactions such as acids (e.g. H_2SO_4, H^+, HNO_3), radicals and aerosols with different spatial distributions.
- The use of O_3 alone as an indicator of such a mixture is not always adequate, as its spatial and temporal variation differs from that of other species.
- Acidity levels are strongly determined by levels of ammonia, which acts as a neutralising agent.
- Indoor O_3 levels were significantly lower than the corresponding outdoor levels.
- For sulphates, indoor/outdoor ratios reached almost unity.

- Personal exposures to acids showed large interpersonal variations for H^+ and NH_3.

5. Final discussion

In this review, several methods of assessing exposure to air pollutants were reviewed. The main focus was on the small-scale spatial variation of pollutants and the relationship between indoor, outdoor and personal exposure patterns.

Individual personal exposures can be measured directly or indirectly. The direct methods require air pollution measurements on the person (e.g. by carrying a monitoring device), or analyses of biological markers from body fluids (Leaderer et al., 1993). Direct measurements clearly reflect individual personal exposure levels best. With the exception of passive sampling (e.g. for NO_2), measurements of personal exposure are expensive, time consuming and difficult to apply to large study populations. It is important to note that a personal measurement does not a priori provide more valid data than a stationary outdoor measurement, i.e. a personal sample in a study investigating effects from outdoor combustion pollutants is often influenced by sources other than outdoor sources and may thus confound the exposure–effect outcome. Internal dose measurements of biological markers provide the best measurements of personal exposure as they refer directly to the amount of material which has crossed the physical boundary of the body. Understanding of pharmacokinetics is necessary in order to relate the measured biomarkers to ambient pollutant levels. Analysis of most biomarkers is expensive and difficult to carry out in large-scale surveys. In occupational studies, where investigations focus on specific air toxins with known sources, such applications are very useful. In environmental studies on complex mixtures, however, such applications are usually difficult.

An approach to indirect personal exposure measurements is the use of microenvironmental (ME) modelling (Duan, 1982). This approach requires data from selected MEs (e.g. outdoor home, indoor such as bedroom or kitchen, in transit) and the time-activity budgets of the people (Duan, 1982; Duan and Mage, 1997). It is, however, difficult to take all MEs of a persona or population into account; a simpler approach uses data from the most significant MEs only (e.g. outdoor and indoor home levels). A further simplification is to model indoor levels from outdoor levels by incorporating the penetration, ventilation and deposition rates (for particles) and a chemical reaction term (Wallace, 1996; Weschler et al., 1991). In addition to time budgets, people's habits (e.g. smoking, staying outdoors) and their exposure at work have to be taken into account in detailed personal exposure models. The ME method is promising but costly, as the validation of the models, which requires personal measurements, is time consuming. The best application for such ME models will probably lie in risk assessment studies.

In epidemiological studies on air pollutants, indirect measurements are most commonly used and exposure data are collected with fixed site ambient monitors. Measured concentrations derived from these monitors are assigned to people living in these areas. Although, in this "aggregated design", a bias in the exposure–effect outcome may be minimised, an evaluation of the representativity of these "central monitors" has to be preformed. Spatial variations in pollution concentrations within a city may be significant and are most critical for NO_2 and O_3, less critical for PM_{10} and $PM_{2.5}$, but critical again for ultrafine particles (<0.1 μm). The monitors' readings may be influenced by these small-area climates. An assessment of people's average exposure compared to the monitors data is especially important in areas representing "leverage points". Existing deviations between measured fixed site data and average peoples' exposure do not fulfil a Berkson-case and over- or underestimation of effect estimates has to be expected.

To date, the strongest associations between health parameters and air pollution have been observed for mass measurements of small particles below 10 μm (PM_{10}) and particles below 2.5 μm ($PM_{2.5}$) (Dockery et al., 1993). The size and shape of the particles, as well as their physico-chemical properties, determine the depth of inhalation, the extent of exhalation, and the deposition rate in the airways (Yeh et al., 1976). These mass measurements (in $\mu g\,m^{-3}$) of ambient or personal exposure represent a complex mixture which comprises ions, PAHs, salts, acids, metals and also some compounds of biological origin (e.g. pollen, endotoxins). However, the causative agents and mechanism of the observed health effects are not known. For transition metals, PAHs and some compounds of biological origin (e.g. endotoxin) toxicological data are available; possible biochemical pathways for the effects have been proposed but their relationship to ambient pollution is difficult to assess (Lewtas et al., 1993; Pritchard et al., 1996; Becker et al., 1996). It is also well known that allergic symptoms can be attributed to biological particles (pollen, fungal spores, etc.) (Wüthrich et al., 1995). It has been suggested that interactions between anthropogenic air pollutants and natural particles can modify the allergen profiles and promote the formation of allergen-loaded fine particles (Behrendt et al., 1997; Knox et al., 1997). All these findings highlight the complexity of the particles' composition and the difficulties in investigating the causative agents. With respect to exposure parameters it is important to note that personal exposure concentrations for particulate matter do not correlate well with outdoor ambient concentrations (in cross-sectional analyses), and that personal levels were found to be higher than the corresponding outdoor and indoor levels in cross-sectional studies (Clayton et al., 1993). However, when personal levels in a study area were aggregated, the correlation between personal and outdoor values was stronger. This suggests that the variation in personal exposure levels between study persons within a study area was responsible for the poor correlation. In follow-up studies (where multiple samples are collected over time), the correlations between outdoor and personal values were also stronger than in cross-sectional studies (Janssen et al., 1997). For elemental markers of ultrafine particles (traffic-related primary particles, such as Pb and Br), the correlation between personal and outdoor levels is weaker than for markers of secondary particles (e.g. S). This indicates that for the homogenously distributed secondary aerosols, an outdoor measurement is a good approximation to personal exposures and that for the personal ultrafine particle exposure, individual activity pattern and vicinity to sources are important in the short-term analyses (Oglesby et al., 2000). Despite some lack of correlation between personal (PM_{10}) and outdoor values, outdoor fine particle concentrations were strongly associated with mortality and morbidity indicating that outdoor sources (e.g. vehicular emission) emit the toxic entity (Dockery et al., 1993; Schwartz and Neas, 1996).

The complexity of the discussion on the validity of outdoor, indoor or personal exposure markers shows that there is no general recommendation on the use of specific exposure measurements. Depending on the design of the study (e.g. cross-section, longitudinal study, long-term or day-to-day study) and the underlying hypotheses, different approaches have to be chosen.

For a comprehensive risk assessment, the move towards source-attributed particulate concentrations is promising. The type of sources, chemical composition, size and their health effects have to be evaluated in order to take the correct strategy to minimise health risks. In the exposure to pollutants in European cities (EXPOLIS) study such strategies are implemented (Jantunen et al., 1998). The primary goals are to establish a database on microenvironmental concentrations (indoor home, outdoor and occupational), on personal exposure distributions, and on time-activity patterns for a random population in different European cities. In the second step (EXPOLIS-EAS) (Mathys et al., 1999), source identification will be performed based on elemental analyses in order to relate source specific emissions to personal exposures. Such a database will then be used to determine the health risk of specific particulates and their sources and for modelling population exposure distributions and effects of reduction strategies.

Acknowledgements

The author kindly acknowledges the support and collaboration of the following persons and Institutions: Prof. H.-U. Wanner, Prof. Th. Koller, Prof. H. Krueger, the SAPALDIA team and directors (Prof. Ph. Leuenberger, Prof.

U. Ackermann-Liebrich), Dr. Pam Alean (ETH) and Dr. Yasmin Vawda (UK) and Dr. W. Stahel (Seminar for Statistics) for a critical review of the statistics section.

References

Abt, E., Suh, H.H., Allen, G., Koutrakis, P., 2000. Characterization of indoor particle sources: a study conducted in the metropolitan Boston area. Environmental Health Perspectives 108 (1), 35–44.

Ackermann-Liebrich, U., Leuenberger, Ph., Schwartz, J., Schindler, Ch., Monn, Ch. et al., SAPAL-DIA Team, 1997. Lung function and long term exposure to air pollutants in Switzerland. American Journal of Respiratory Critical Care Medicine 155, 122–129.

Ahlbom, A., Norell, S., 1990. Introduction to modern epidemiology. Epidemiology Resources Inc., USA.

Ahlbom, A., Steineck, G., 1992. Aspects of misclassification of confounding factors. American Journal of Industrial Medicine 21, 107–112.

Alean-Kirkpatrick, P., 1993. Temporal fluctuations of atmospheric acidity. Ph.D. Thesis, Institute of Inorganic Chemistry, University of Zürich, Zürich.

Allen, M.J., Yen, W.M., 1979. Introduction to Measurement Theory. Brooks/Cole, Monterey.

Altshuller, A.P., 1988. Some characteristics of ozone formation in the urban plume of St Louis (Missouri, USA). Atmospheric Environment 22 (3), 499–510.

Armstrong, B.G., 1990. The effect of measurement errors on relative risk regressions. American Journal of Epidemiology 132 (6), 1176–1184.

Armstrong, B.K., White, E., Saracci, R., 1992. In: Exposure Measurement Error and its Effect. Principles of Exposure Assessment in Epidemiology. Oxford University Press, Oxford, pp. 49–77.

Atkinson, R., 1990. Gas-phase tropospheric chemistry of organic compounds: a review. Atmospheric Environment 24A (1), 1–41.

Baur, X., 1995. Allergien auf aerogene Latexallergene. Allergologie 18, 568–571.

Becker, S., Soukup, J.M., Gilmour, M.I., Devlin, R.B., 1996. Stimulation of human and rat alveolar macrophages by urban air particulates: effects on oxidant radical generation and cytokine production. Toxicology and Applied Pharmacology 141, 637–648.

Behrendt, H., Becker, W.M., Friedrichs, K.H., Darsow, U., Tomingas, R., 1992. Interactions between aeroallergens and airborne particulate matter. International Archives of Allergy and Immunology 99, 425–428.

Behrendt, H., Becker, W.M., Fritzsche, C., Sliwa-Tomczok, W., Tomczok, J., Friedrichs, K.H., Ring, J., 1997. Air pollution and allergy: experimental studies on modulation of allergen release from pollen by air pollutants. International Archives of Allergy and Immunology 113, 69–74.

Bellomo, R., Gigliotti, P., Treloar, A., Holmes, P., Suphioglu, C., Singh, M.B., Knox, B., 1992. Two consecutive thunderstorm associated epidemics of asthma in the city of Melbourne. The possible role of rye-grass pollen. Medical Journal of Australia 156, 834–837.

Billick, I.H., 1990. Estimation of population exposure to nitrogen dioxide. Toxicology and Industrial Health 6 (2), 325–333.

Blanchard, Ch.L., Carr, E.L., Collins, J.F., Smith, T.B., Lehrman, D.E., Michaels, H.M., 1999. Spatial representativeness and scales of transport during the 1995 integrated monitoring study in California's San Joaquin Valley. Atmospheric Environment 33 (29), 4775–4786.

Brauer, M., Brook, J.R., 1997. Ozone personal exposures and health effects for selected groups residing in the Fraser Valley. Atmospheric Environment 31 (14), 2113–2121.

Brauer, M., Koutrakis, P., Keeler, G.J., Spengler, J.D., 1991. Indoor and outdoor concentrations of inorganic acidic aerosols and gases. Journal of Air and Waste Management Association 41, 171–181.

Brauer, M., Dumyahn, T.S., Spengler, J.D., Gutschmidt, K., Heinrich, J., Wichmann, H.E., 1995. Measurement of acidic aerosol species in eastern Europe: implications for air pollution epidemiology. Environmental Health Perspective 103 (5), 482–488.

Breitender, H., Scheiner, O., 1990. Environmental pollution and pollen allergy – a possible link (abstract). Allergologie 13 (434).

Brenner, H., Savitz, D.A., Jockel, K.H., Greenland, S., 1992. Effects of nondifferential exposure misclassification in ecologic studies. American Journal of Epidemiology 135 (1), 85–95.

Brimblecombe, P., 1986. Air Composition and Chemistry. Cambridge University Press, Cambridge.

Brook, J.R., Wiebe, A.H., Woodhouse, S.A., Audette, C.V., Dann, T.F., Callaghan, S., Piechowski, M., Dabek-Zlotorzynska, E., Dlough, J.F., 1997a. Temporal and spatial relationship in fine particle strong acidity, sulphate, PM_{10}, and $PM_{2.5}$ across multiple Canadian locations. Atmospheric Environment 31 (24), 4223–4236.

Brook, J.R., Dann, T.F., Burnett, R.T., 1997b. The relationship among TSP, PM_{10}, $PM_{2.5}$, and inorganic constituents of atmospheric particulate matter at multiple Canadian locations. Journal of Air and Waste Management Association 42, 2–19.

Brown, R.H., 1993. The use of diffusive passive samplers for monitoring ambient air. Pure and Applied Chemistry 65 (8), 1859–1874.

Brunekreef, B., Noy, D., Clausing, P., 1987. Variability of exposure measurements in environmental epidemiology. American Journal of Epidemiology 125 (5), 892–898.

Brunekreef, B., Houhuijs, D., Dijkstra, L., Boleij, 1990. Indoor nitrogen dioxide exposure and children's pulmonary function. Journal of Air and Waste Management Association 40 (9), 1252–1256.

Bullin, J.A., Bower, S.C., Hinz, M., Moe, R.D., 1985. Aerosols near urban street intersection. Journal of Air Pollution Control Association 35 (4), 355–358.

Burge, H.A., 1995. Bioaerosols in residential environment. Bioaerosol Handbook. CRC Press Inc., Boca Raton, FL, pp. 579–597.

Burton, R.M., Suh, H.H., Koutrakis, P., 1996. Spatial variation in particulate concentrations within Metropolitan Philadelphia. Environmental Science and Technology 30 (2), 400–407.

Burtscher, H., Siegmann, H.C., 1993. Photoemission for in situ analysis of particulate combustion emissions. Water, Air and Soil Pollution 68 (1–2), 125–136.

Burtscher, H., Siegmann, H.C., 1994. Monitoring PAH emissions from combustion processes by photoelectric charging. Combustion Science and Technology 101, 327–332.

BUWAL, 1996. POLLUMET Luftverschmutzung und Meteorologie in der Schweiz. Report No. 63, Federal Office of Environment, Forest and Landscape, Berne.

Carter, W.P.L., Pierce, J.A., Luo, D., Malkina, I.L., 1995. Environmental chamber study of maximum incremental reactivities of volatile organic compounds. Atmospheric Environment 29 (18), 2499–2511.

Celenza, A., Fothergill, J., Kupek, E., Schaw, R.J., 1996. Thunderstorm associated asthma: a detailed analysis of environmental factors. British Medical Journal 7031 (312), 604–608.

Chen, M.L., Mao, I.F., 1998. Spatial variations of airborne particles in metropolitan Taipei. Science of the Total Environment 209 (2–3), 225–231.

Chow, J.C., 1995. Measurement methods to determine compliance with ambient air quality standards for suspended particles. Journal of Air and Waste Management Association 45, 320–382.

Clayton, C.A., Pellizzari, E.D., Wiener, R.W., 1991. Use of a pilot study for designing a large scale probability study of personal exposure to aerosols. Journal of Exposure Analysis Environmental Epidemiology 1 (4), 401–421.

Clayton, C.A., Perritt, R.L., Pellizzari, E.D., Thomas, K.W., Whitmore, R.W., Wallace, L.A., Özkaynak, H., Spengler, J.D., 1993. Particle total exposure assessment methodology (PTEAM) study: distributions of aerosol and elemental concentrations in personal, indoor and outdoor air samples in a Southern Californian community. Journal of Exposure Analysis and Environmental Epidemiology 3 (2), 227–249.

Cochran, W.G., 1968. Errors in measurement in statistics. Technometrics 10 (4), 637–666.

Colls, J.J., Micallef, A., 1999. Measured and modelled concentrations and vertical profiles of airborne particulate matter within the boundary layer of a street canyon. Science of the Total Environment 235 (1–3), 221–233.

COST (Coordination European Scientific and Technique) 613/2, 1993. Biological particles in indoor environments. Report no. EUR 14988 EN, Brussels.

Dockery, D.W., Spengler, J.D., 1981a. Personal exposure to respirable particulates and sulfates. Journal of Air Pollution Control Association 32, 153–159.

Dockery, D.W., Spengler, J.D., 1981b. Indoor–outdoor relationship of respirable sulfates and particles. Atmospheric Environment 15, 335–343.

Dockery, D.W., Pope, C.A., Xu, X., Spengler, J.D., Ware, J.H., Fay, M.E., Ferris, B.G., Speizer, F.E., 1993. An association between air pollution and mortality in six U.S. Cities. New England Journal of Medicine 329 (24), 1753–1759.

Drye, E.E., Özkaynak, H., Burbank, B., Billick, I.H., Baker, Ph.E., Spengler, J.D., Ryan, P.B., Colome, S.D., 1989. Development of models for predicting the distribution of indoor nitrogen dioxide concentrations. Journal of Air and Waste Management Association 39 (9), 1169–1177.

Duan, N., 1982. Models for human exposure to air pollution. Environmental International 8, 305–309.

Duan, N., Mage, D.T., 1997. Combination of direct and indirect approaches for exposure assessment. Journal of Exposure Analysis Environmental Epidemiology 7 (4), 439–470.

Emberlin, J., 1995a. Plant allergens on pauci-micronic airborne particles. Clinical and Experimental Allergy 25 (3), 202–205.

Emberlin, J., 1995b. Interaction between air pollutants and aeroallergens. Clinical and Experimental Allergy 25 (3), 33–39.

Emberlin, J., 1997. In: Kay, A.B. (Ed.), Grass, Tree and Weed Pollens. Allergy and Allergic Diseases, Vol. III. Blackwell, Oxford, pp. 835–857.

EMPA, 1994. Swiss Federal Lab for Materials Testing and Research. Technischer Bericht zum NABEL für Luftfremdstoffe (Technical report of the ambient air pollution network NABEL). Dübendorf.

Finlayson-Pitts, B.J., Pitts, J.N., 1986. Atmospheric Chemistry. Wiley, New York.

Flückiger, B., Monn, Ch., Lüthy, P., Wanner, H.-U., 1998. Hygienic aspects of ground-coupled air systems. Indoor Air 8, 197–202.

Flückiger, B., Bruggmann, D., Monn, Ch., 1999. Measurements of viable spores and fungal allergen concentrations in the homes of allergic patients. Proceedings Indoor Air '99, Vol. I, Edinburgh, pp. 914–919.

Fogelmark, B., Sjostrand, M., Rylander, R., 1994. Pulmonary inflammation induced by repeated beta (1,3) D-Glucan and endotoxin. International Journal of Experimental Pathology 75, 85–90.

Georgopoulos, P.G., Walia, A., Roy, A., Lioy, P.J., 1997. Integrated exposure and dose modeling and analysis system. 1. Formulation and testing of microenvironmental and pharmacokinetic components. Environmental Science and Technology 31 (1), 17–27.

Graham, D.E., Koren, H.S., 1990. Biomarkers of inflammation in ozone-exposed humans. American Review of Respiratory Disease 142, 152–156.

Grandjean, Ph., 1995. Biomarkers in epidemiology. Clinical Chemist 41 (12), 1800–1803.

Greenland, S., Robins, J., 1992. Invited commentary: ecologic studies – biases, misconceptions, and counterexamples. American Journal of Epidemiology 139 (8), 747–760.

Hangartner, M., 1990. Einsatz von Passivsammlern für verschiedene Schadstoffe in der Aussen-luft. Vol. 838. VDI-Aktuelle Aufgaben der Messtechnik in der Luftreinhaltung, VDI Bericht, Heidelberg, pp. 515–526.

Hangartner, M., Kirchner, M., Werner, H., 1996. Evaluation of passive methods for measuring ozone in the European Alps. Analyst 121, 1269–1272.

Hanna, S.R., Briggs, G.A., Hosker Jr., R.P., 1982. Handbook on atmospheric diffusion. US Department of Energy, Report no. DEO/TIC-11223, Technical Information Centre, Oak Ridge, TN.

Harrison, R.M., Perry, R., 1986. Air Pollution Analysis. Chapman & Hall, New York.

Harrison, R.M., Deacon, A.R., 1998. Spatial correlation of automatic air quality monitoring at urban background sites: Implications for network design. Environmental Technology 19 (2), 121–132.

Harrison, R.M., Jones, M., Collins, G., 1999. Measurements of the physical properties of particles in the urban atmosphere. Atmospheric Environment 33 (2), 309–321.

Hegner, H., 1994. Kleinräumige Verteilung von Stickstoffdioxid. Hygiene and Applied Physiology. Ph.D. Thesis 10733, ETH-Zürich, Zürich.

Hinds, W.C., 1982. Aerosol Technology. John Wiley, New York.

Hopke, P.K., 1985. Receptor Modeling in Environmental Chemistry. Wiley Interscience, New York.

Hulka, B.S., 1990. Biological Markers in Epidemiology. Oxford University Press, Oxford.

Hulka, B.S., Wilcosky, T., 1988. Biological Markers in epidemiological research. Archives of Environmental Health 43, 83–89.

Ihrig, M.M., Shalat, S.L., Baynes, C., 1998. A hospital-based case-control study of stillbirths and environmental exposure to arsenic using an atmospheric dispersion model linked to a geographical information system. Epidemiology 9 (3), 290–294.

Ishizaki, T., Koizumi, K., Ikemori, R., Ishiyama, Y., Kushibiki, E., 1987. Studies of the prevalence of Japanese Cedar pollinosis among residents in a densely cultivated area. Annals of Allergy 58, 265–270.

Ito, K., Kinney, P., Thurston, G.D., 1995. Variations in PM_{10} concentrations within two Metropolitan areas and their implication for health effects analyses. Journal of Inhalation Toxicology 7 (5), 735–745.

Janssen, N.A., Hoek, G., Harssema, H., Brunekreef, B., 1995. A relationship between personal and ambient PM. Epidemiology 6 (Suppl.), 45.

Janssen, N.A.H., Harssema, H., Brunekreef, B., 1997. Childhood exposure to PM_{10}: relation between personal, classroom, and outdoor concentrations. Occupational and Environmental Medicine 54, 888–894.

Janssen, N.A., Hoek, G., Brunekreef, B., Harssema, H., Mendink, I., Zuidhof, A., 1998. Personal sampling of particles in adults: relation among personal, indoor and outdoor air concentrations. American Journal of Epidemiology 147 (6), 537–547.

Janssen, L.H.J.M., Buringh, E., van der Meulen, A., van den Hout, K.D., 1999a. A method to estimate the distribution of various fractions of PM_{10} in ambient air in the Netherlands. Atmospheric Environment 33 (20), 3325–3334.

Janssen, N.A., Hoek, G., Brunekreef, B., Harssema, H., 1999b. Mass concentration and elemental composition of PM in classrooms. Occupational and Environmental Medicine 56 (7), 482–487.

Jantunen, M., Hänninen, O., Katsouyanni, K., Knöppel, H., Kuenzli, N., Lebret, E., Maroni, M., Saarela, K., Sram, R., Zmirou, D., 1998. Air pollution exposure in European cities: The EXPOLIS study. Journal of Exposure Analysis and Environmental Epidemiology 8 (4), 495–518.

Jensen, P.A., O'Brien, D., 1993. Industrial Hygiene. Van Nostrand Reinhold, New York, NY.

Ju, C., Spengler, J.D., 1981. Room to room variations in concentration of respirable particles in residences. Environmental Science and Technology 15, 592–596.

Junker, M., Kasper, M., Röösli, M., Camenzind, M., Künzli, N., Monn, Ch., Theis, G., Braun, Ch., 2000. Airborne particle number profiles, particle mass distribution and particle bound PAH concentrations within the city environment of Basle: an assessment of the BRISKA project. Atmospheric Environment 43 (19) 3171–3181.

Kainka-Staenicke, E., Behrendt, H., Friedrichs, K.H., Tomingas, R., 1988. Morphological alterations of pollen and spores induced by airborne pollutants: observations from two differently polluted areas in West Germany. Allergy 43 (Suppl. 7), 57.

Keywood, M.D., Ayers, G.P., Gras, J.L., Gillett, R.W., Cohen, D.D., 1999. Relationships between size segregated mass concentration data and ultrafine particle number concentrations in urban areas. Atmospheric Environment 33 (18), 2907–2913.

Kingham, S., Briggs, D., Elliott, P., Fischer, P., Lebret, E., 2000. Spatial variations in the concentrations of traffic-related pollutants in indoor and outdoor air in Huddersfield, England. Atmospheric Environment 34 (6), 905–916.

Kinney, P.L., Aggrawal, M., Northridge, M.E., Janssen, N.A.H., Shepard, P., 2000. Airborne concentrations of $PM_{2.5}$ and diesel exhaust particles on Harlem sidewalks: a community-based pilot study. Environmental Health Perspectives 108 (3), 213–218.

Knox, B., Suphioglu, C., 1996a. Pollen allergens: development and function. Sex Plant Reproduction 9, 318–323.

Knox, B., Suphioglu, C., 1996b. Environmental and molecular biology of pollen allergens. Trends in Plant Science 1 (5), 156–164.

Knox, R.B., Suphioglu, C., Taylor, P., Desai, R., Watson, H.C., Peng, J.L., Bursill, L.A., 1997. Major grass pollen allergen Lol p 1 binds to diesel exhaust particles: implications for asthma and air pollution. Clinical and Experimental Allergy 27, 246–251.

Koutrakis, P., Wolfson, J.M., Slater, J.L., Brauer, M., Spengler, J.D., 1988. Evaluation of an annular denuder/filter pack system to collect acidic aerosols. Environmental Science and Technology 22 (12), 1463–1468.

Koutrakis, P., Fasano, A.M., Slater, J.L., Spengler, J.D., 1989. Design of a personal annular denuder sampler to measure atmospheric aerosols and gases. Atmospheric Environment 23 (12), 2767–2773.

Koutrakis, P., Wolfson, J.M., Bunyaviroch, A., Froehlich, S., Hirano, K., Mulik, J.D., 1993. Measurement of ambient ozone using a nitrite-coated filter. Analytical Chemistry 65 (3), 209–214.

Krzyzanowski, M., 1997. Methods for assessing the extent of exposure and effects of air pollution. Occupational and Environmental Medicine 54, 145–151.

Künzli, N., Tager, I.B., 1997. The semi-individual study in air pollution epidemiology: a valid design as compared to ecologic studies. Environmental Health Perspectives 105 (10), 1078–1083.

Künzli, N., Lurmann, F., Segal, M., Ngo, L., Balmes, J., Tager, I.B., 1997. Association between lifetime ambient ozone exposure and pulmonary function in college freshmen – results of a pilot study. Environmental Research 72, 8–23.

Lacey, J., Dutkiewicz, J., 1994. Bioaerosols and occupational lung disease. Journal of Aerosol Science 25 (8), 1371–1404.

Last, J.M., 1988. A Dictionary of Epidemiology. Oxford University Press, Oxford.

Leaderer, B.P., 1990. Assessing exposures to environmental tobacco smoke. Risk Analysis 10, 19–26.

Leaderer, B.P., Zagraniski, R.T., Berwick, M., Stolwijk, J.A., 1986. Assessment of exposure to indoor air contaminants from combustion sources: methodology and application. American Journal of Epidemiology 124 (2), 275–289.

Leaderer, B.P., Lioy, P.J., Spengler, J.D., 1993. Assessing exposure to inhaled complex mixtures. Environmental Health Perspectives 101 (S4), 167–174.

Leaderer, B.P., Koutrakis, P., Briggs, S.L., Rizzuto, J., 1994. The mass concentration and elemental composition of indoor aerosols in Suffolk and Onondaga counties, New York. Indoor Air 4, 23–34.

Leaderer, B.P., Naeher, L., Jankum, Th., Balenger, K., Holford, Th.R., Toth, C., Sullivan, J., Wolfson, J.M., Koutrakis, P., 1999. Indoor, outdoor, and regional summer and winter concentrations of PM_{10}, $PM_{2.5}$, SO_4^{2-}, H^+, NH^{4+}, NO^{3-}, NH_3, and nitrous acid in homes with and without kerosene space heaters. Environmental Health Perspectives 107 (3), 223–231.

Lebowitz, M.D., Quackenboss, J.J., Soczek, M.L., Kollander, M., Colome, S., 1989. The new standard environmental inventory questionnaire for estimation of indoor concentration. Journal of Air Pollution Control Association 39, 1411–1419.

Lebret, E., 1990. Errors in exposure measures. Toxicology and Industrial Health 6 (5), 147–156.

Lee, K., Yanasigawa, Y., Hishinuma, M., Spengler, J.D., Billick, I.H., 1992. A passive sampler for measurement of carbon monoxide using solid adsorbent. Environmental Science and Technology 26 (4), 697–702.

Lewis, W.H., Vinay, P., Zenger, V.E., 1983. Airborne and Allergenic Pollen in North America. The John Hopkins University Press, Baltimore.

Lewtas, J., Mumford, J., Everson, R.B., Hulka, B., Wilcosky, T., Kozumbo, W., Thompson, C., George, M. et al., 1993. Comparison of DNA adducts for exposure to complex mixtures in various human tissues and experimental systems. Environmental Health Perspectives 99, 89–97.

Lioy, P.J., 1990. Assessing total human exposure to contaminants. Environmental Science and Technology 24 (7), 938–945.

Lioy, P.J., 1995. Measurement methods for human exposure analysis. Environmental Health Perspective 103 (Suppl. 3), 35–43.

Lippmann, M., 1989. Health effects of ozone. A critical review. Journal of Air Pollution Control Association 39 (5), 672–695.

Liu, L.J., Koutrakis, P., Suh, H.H., Mulik, J.D., Burton, R.M., 1993. Use of personal measurements for ozone exposure assessment: a pilot study. Environmental Health Perspectives 101 (4), 318–324.

Liu, L.J., Delfino, R., Koutrakis, P., 1997. Ozone exposure assessment in a Southern Californian community. Environmental Health Perspectives 105 (1), 58–65.

Lowry, L.K., 1995. Role of biomarkers of exposure in the assessment of health risks. Toxicology Letters 77, 31–38.

Lustenberger, J., Monn, Ch., Wanner, H.-U., 1990. Measurements of ozone indoor and outdoor concentrations with passive sampling devices. Proceedings of Indoor Air '90, Toronto, Canada, Vol. 2, pp. 555–560.

Mage, D.T., 1985. Concepts of human exposure assessment for airborne particulate matter. Environmental International 11, 407–412.

Mage, D.T., Buckley, T.J., 1995. The relationship between personal exposures and ambient concentrations of particulate matter. 88th meeting of the Air and Waste Management Association, San Antonio TX. Paper 95-MP18.01.

Magliano, K.L., Hughes, V.M., Chinkin, L.R., Coe, D.L., Haste, T., Kumar, N., Lutman, F.W., 1999. Spatial and temporal variations in PM_{10} and $PM_{2.5}$ source contributions and comparison to emissions during the integrated monitoring study. Atmospheric Environment 33 (29), 4757–4773.

Marple, V.A., Rubow, K.L., Turner, W., Spengler, J.D., 1987. Low Flow Sharp Cut Impactors for indoor air sampling: design and calibration. Journal of Air Pollution Control Association 37, 1303–1307.

Martin, B., Ackermann, U., Leuenberger, Ph., Künzli, N., Zemp, E., Keller, R., Zellweger, J.-P., Wüthrich, B. et al., 1997. SAPALDIA: Methods and participation in the cross sectional part of the Swiss study on air pollution and lung function in adults. Sozial und Präventivmedizin 42, 67–84.

Masuch, G., Müsken, H., Bergmann, K.-Ch., Wahl, R., 1997. Einfluss von Ozon auf den Gehalt an Phl p 5 in Pollen und Pflanzenbestandteilen von Lolium Perenne. 4. Europ. Pollenflug Seminar, Bad Lippspringe 28.2.–2.3.

Mathys, P., Oglesby, L., Stern, W. et al., 1999. Traffic related $PM_{2.5}$ efficiently penetrate form outdoor to indoor environments (EAS-EXPOLIS). Epidemiology 9, p. 50 (abstract).

Matthias-Maser, S., 1998. Primary biological aerosol particles: their significance, sources, sampling methods and size-distribution in the atmosphere. In: Harrison, R.M., Van Grieken, R. (Eds.), Atmospheric Aerosols. Wiley, New York, pp. 349–368.

McConnaughey, P.W., McKee, E., Pritts, I.M., 1985. Passive colorimetric dosimeter tubes for ammonia, carbon monoxide, carbon dioxide, hydrogen sulfide, nitrogen dioxide and sulfur dioxide. American Industrial Hygiene Association Journal 46, 357–362.

McGraven, P.D., Rood, A.S., Till, J.E., 1999. Chronic beryllium disease and cancer risk estimates with uncertainty for beryllium released to the air from the Rocky Flats Plant. Environmental Health Perspectives 107 (9), 731–744.

Meyer, M.J., Bechtold, W.E., 1996. Protein adduct biomarkers: state of the art. Environmental Health Perspectives 104 (Suppl. 5), 879–881.

Micallef, A., Colls, J.J., 1998. Variation in airborne particulate matter concentration over the first three metres from ground in a street canyon: implications for human exposure. Atmospheric Environment 32 (21), 3795–3799.

Miguel, A.G., Cass, G.R., Weiss, J., Glovsky, M.M., 1996. Latex Allergens in Tire Dust and Airborne Particles. Environmental Health Perspectives 104 (11), 1180–1185.

Monn, Ch., Hangartner, M., 1990. Passive Sampling for Ozone. Journal of Air and Waste Management Association 40 (3), 357–358.

Monn, Ch., Becker, S., 1999. Cytotoxicity and induction of pro-inflammatory cytokines from human monocytes exposed to fine ($PM_{2.5}$) and coarse particles (PM_{10}–$PM_{2.5}$) in outdoor and indoor air. Toxicology and Applied Pharmacology 155, 245–252.

Monn, Ch., Tenzer, A., 2000. Latex – a peculiar component of airborne particles? Aerobiologia, in press.

Monn, Ch., Frauchiger, P., Wanner, H.U., 1993. Exposure assessment for nitrogen dioxide and ozone. Proceedings of Indoor Air '93, Helsinki, Finland, Vol. 3, pp. 319–323.

Monn, Ch., Brändli, O., Schäppi, G., Ackermann, U., Leuenberger, Ph., SAPALDIA Team, 1995. Particulate Matter 10 and Total Suspended Particulates in Urban, Rural and Alpine Air in Switzerland. Atmospheric Environment 29 (19), 2565–2573.

Monn, Ch., Carabias, V., Junker, M., Waeber, R., Karrer, M., Wanner, H.-U., 1997a. Small-scale Spatial Variability of Particulate Matter <10 mm and Nitrogen Dioxide. Atmospheric Environment 31 (15), 2243–2247.

Monn, Ch., Fuchs, A., Högger, D., Kogelschatz, D., Roth, N., Wanner, H.U., 1997b. Relationship between indoor and outdoor concentrations of particulate matter PM_{10} and fine particles $PM_{2.5}$. Science of the Total Environment 208, 15–21.

Monn, Ch., Schindler, Ch., Brändli, O., Ackermann, U., Leuenberger, Ph., SAPALDIA Team, 1998. Personal exposure to nitrogen dioxide. Science of the Total Environment 215, 243–251.

Monn, Ch., Defila, C., Alean, P., Peeters, A., Künzli, N., Ackermann, U., Leuenberger, Ph., SAPALDIA Team, 1999. Air pollution, climate and pollen comparison in urban, rural and alpine regions in Switzerland (SAPALDIA-study). Atmospheric Environment 33, 2411–2416.

Moschandreas, D.J., 1981. Exposures to pollutants and daily time budgets of people. New York Academy of Medicine 57, 845–859.

172   *C. Monn*

Muguerza, J., Capo, C., Porri, F., Jacob, J.L., Mege, J.L., Verloet, D., 1996. Latex allergy: allergen identification in *Havea Braziliensis* fractions by immunoblotting. Clinical and Experimental Allergy 26 (10), 1177–1181.

Mumford, J.L., Williams, K., Wilcosky, T.C., Everson, R.B., Young, T.L., Santella, R.M., 1996. A sensitive color ELISA for detecting polycyclic aromatic hydrocarbon–DNA adducts in human tissues. Mutation Research 359, 171–177.

Neas, L.M., Dockery, D.W., Ware, J.H., Spengler, J.D., Speizer, F.E., Ferris, B.G., 1991. Association of indoor nitrogen dioxide with respiratory symptoms and pulmonary function in children. American Journal of Epidemiology 134 (2), 204–219.

Neas, L.M., Dockery, D.W., Ware, J.H., Spengler, J.D., Ferris, B.G., Speizer, F.E., 1994. Concentrations of indoor particulate matter as a determinant of respiratory health in children. American Journal of Epidemiology 139, 1088–1099.

Noy, D., Brunekreef, B., Boleu, J.S., Houthujis, D., De Koning, R., 1990. The assessment of personal exposure to nitrogen dioxide in epidemiological studies. Atmospheric Environment 24A (12), 1903–1909.

NRC (National Research Council), Lioy, P.J. (Chairman), 1991. Human exposure assessment for airborne pollutants. Washington, DC.

Oberdörster, G., 1995. Lung particle overload: implications for occupational exposures to particles. Regulatory Toxicology and Pharmacology 21 (1), 123–135.

Oglesby, L., Künzli, N., Röösli, M., Braun-Fahrländer, C., Mathys, P., Stern, W., Jantunen, M., Kousa, A., 2000. Validity of ambient levels of fine particles as surrogate for personal exposure to outdoor air pollution. Journal of Air and Waste management Association, in press.

Ong, E.K., Singh, M.B., Knox, R.B., 1995. Aeroallergens of plant origin: molecular basis and aerobiological significance. Aerobiologia 11, 219–229.

Ott, W.R., 1982. Concepts of human exposure to air pollution. Environmental International 7, 179–196.

Ott, W.R., 1990. Total human exposure: basic concepts, EPA field studies and future research needs. Journal of Air and Waste Management Association 40 (7), 966–975.

Ott, W., Rodes, L.E., Drago, R.J., Williams, C., Brumann, F.J., 1986. Automated data-logging personal exposure monitors for carbon monoxide. Journal of Air Pollution Control Association 36, 883–886.

Özkaynak, H., Xue, J., Zhou, H., Spengler, J.D., Thurston, G.D., 1996a. Intercommunity differences in acid aerosol H^+/SO_4^{2-} – ratios. Journal of Exposure Analysis and Environmental Epidemiology 6 (1), 35–55.

Özkaynak, H., Xue, J., Spengler, J., Wallace, L., Pellizzari, E., Jenkins, P., 1996b. Personal exposure to airborne particles and metals: results from the Particle Team Study in Riverside, CA. Journal of Exposure Analysis and Environmental Epidemiology 6, 57–78.

Palmes, E.D., Gunnison, A.F., DiMatto, J., Tomczyk, C., 1976. Personal sampler for nitrogendioxide. Journal of American Industrial Hygiene Association 10 (37), 570–577.

Peters, A., Wichmann, H.E., Tuch, T., Heinrich, J., Heyder, J., 1997. Respiratory effects are associated with the number of ultrafine particles. American Journal of Respiratory and Critical Care Medicine 155 (4), 1376–1383.

Possanzini, M., Febo, A., Liberti, A., 1983. New design of a high-performance denuder for sampling of atmospheric pollutants. Atmospheric Environment 17 (12), 2605–2610.

Pritchard, R.J., Ghio, A.J.J., Lehmann, J.R., Winsett, D.W., Tepper, J.S.P.P., Gilmour, M.I., Dreher, K.L., Costa, D.L., 1996. Oxidant generation and lung injury after particulate air pollutants exposure increase with the concentrations of associated metals. Inhalation Toxicology 8, 457–477.

Profos, P., Pfeifer, T., 1994. Handbuch der industriellen Messtechnik. R. Oldenbourg Verlag, München, Wien.

Quackenboss, J.T., Spengler, J.D., Kanaek, M.S., Letz, R., Duffy, C.P., 1986. Personal exposure to NO₂: relationship to indoor/outdoor air quality and activity patterns. Environmental Science and Technology 20 (8), 775–783.

Quackenboss, J.J., Krzyzanowski, M., Lebowitz, M.D., 1991. Exposure assessment approaches to evaluate respiratory health effects of particulate matter and nitrogen dioxide. Journal of Exposure Analysis and Environment Epidemiology 1 (1), 83–107.

Rantio-Lehtimaeki, A., Viander, M., Koivikko, A., 1994. Airborne birch pollen antigens in different particle sizes. Clinical and Experimental Allergy 24, 23–28.

Rappaport, S.M., Symanski, E., Yager, J.W., Kupper, L.L., 1995. The relationship between environmental monitoring and biological markers in exposure assessment. Environmental Health Perspectives 103 (Suppl. 3), 49–53.

Riediker, M., Koller, T., Monn, Ch., 2000. Differences in sizes elective aerosol sampling for pollen allergen detection using high-volume cascade impactors. Clinical and Experimental Allergy 30 (6), 867–873.

Roorda-Knape, M.C., Janssen, N., De-Hartog, J.H., Van-Vliet, P., Harssema, H., Brunekreef, B., 1998. Air pollution from traffic in city districts near major motorways. Atmospheric Environment 32 (11), 1921–1930.

Rubino, F.M., Floridia, L., Tavazzani, M., Fustinoni, S., Gianpiccolo, R., Colombi, A., 1998. Height profile of some air quality markers in the urban atmosphere surrounding a 100 m tower building. Atmospheric Environment 32 (20), 3569–3580.

Ryan, P.B., Soczek, M.L., Spengler, J.D., Billick, I.H., 1988a. The Boston NO₂ characterisation study I: Preliminary evaluation of the survey methodology. Journal of Air Pollution Control Association 38, 22–27.

Ryan, P.B., Soczek, M.L., Treitman, R., Spengler, J.D., 1988b. The Boston residential NO₂ characterization study II: Survey methodology and population concentration estimates. Atmospheric Environment 22 (10), 2115–2125.

Rylander, R., 1998. Microbial cell wall constituents in indoor air and their relation to disease. Indoor Air 8 (Suppl.), 59–65.

Santanam, S., Spengler, J.D., Ryan, P.B., 1990. Particulate matter exposure estimates from an indoor–outdoor source apportionment. Proceedings of Indoor Air '90, Toronto, Canada, Vol. 2, pp. 583–588.

Schäppi, G.F., Monn, Ch., Wüthrich, B., Wanner, H.-U., 1996. Direct determination of allergens in ambient aerosols: methodological aspects. International Archives of Allergy and Immunology 110, 364–370.

Schäppi, G.F., Taylor, Ph.E., Staff, I.A., Suphioglu, C., Knox, B., 1997. Source of Bet v 1 loaded inhalable particles from birch revealed. Sexual Plant Reproduction 10, 315–323.

Schindler, Ch., Ackermann-Liebrich, U., Leuenberger, Ph., Monn, Ch. et al., SAPALDIA Team, 1998. Association between Lung Function and Estimated average Exposure to NO₂ in eight areas of Switzerland (SAPALDIA). Epidemiology 9 (4), 405–411.

Schinko, H.A.E., Medinger, W., Hager, W., 1994. Pollen, Pollenallergene und partikuläre Luftschadstoffe-Aspektewandel. Allergologie 17 (11), 514–525.

Schulte, P.A., Talaska, G., 1995. Validity criteria for the use of biological markers of exposure to chemical agents in environmental epidemiology. Toxicology 101 (1–2), 73–78.

Schwab, M., McDermott, Spengler, J.D., Samet, J.M., Lambert, W.E., 1994. Seasonal and yearly patterns of indoor nitrogen dioxide levels: data from Albuquerqu, New Mexico. Indoor Air 4, 8–22.

Schwartz, J., 1994. Air pollution and mortality: A review and meta analysis. Environmental Research 64, 36–52.

Schwartz, J.D.D., Neas, L.M., 1996. Is daily mortality associated specifically with fine particles? Journal of Air and Waste Management Association 46 (10), 927–939.

Seifert, B., 1995. Validity criteria for exposure assessment methods. Science of the Total Environment 168 (2), 101–107.

Seinfeld, J.H., 1986. Atmospheric Chemistry and Physics of Air Pollution. Wiley, New York.

Sexton, K., Ryan, P.B., 1988. In: Assessment of Human Exposure to Air Pollution: Methods, Measurements and Models. Air Pollution, the Automobile and Public Health. Health Effect Institute. National Academy Press, Washington, DC, pp. 207–238.

Sexton, K., Spengler, J.D., Treitman, R.D., 1984. Personal exposure for respirable particles: a case study in Waterbury, Vermont. Atmospheric Environment 21 (8), 1385–1398.

Sheldon, L.S., Hartwell, T.D., Cox, B.G., Sickles, II., Pellizzari, E.D., Smith, M.L., Peritt, R.L., Jones, S.M., 1989. An investigation of infiltration and indoor air quality. New York State Energy Research and Development Authority, Albany, NY.

Shields, H.C., Weschler, C.J., 1987. Analysis of ambient concentrations of organic vapors with a passive sampler. Journal of Air Pollution Control Association 37 (9), 1039–1045.

Solomon, W.R., Burge, H.A., Muilenberg, M.L., 1983. Allergen carriage by atmospheric aerosol. I: Ragweed pollen determinants in smaller micron fractions. Journal of Allergy and Clinical Immunology 72, 443–447.

Spengler, J.D., Dockery, D.W., Turner, W.A., Wolfson, J.M., Ferris, B.G., 1981. Long-term measurements of respirable sulfates and particles inside and outside home. Atmospheric Environment 15, 23–30.

Spengler, J.D., Brauer, M., Koutrakis, P., 1990. Acid air and health. Environmental Science and Technology 24, 946–956.

Spengler, J.D., Schwab, M., Ryan, P.B., Colome, S., Wilson, A.L., Billick, I., Becker, E., 1994. Personal exposure to nitrogen dioxide in the Los Angeles Basin. Journal of Air and Waste Management Association 44, 39–47.

Spengler, J.D., Koutrakis, P., Dockery, D.W., Raizenne, M., Speizer, F.E., 1996. Health effects of acid aerosols on North American children: air pollution exposures. Environmental Health Perspectives 104 (5), 492–499.

Spieksma, F.Th.M., 1990. Evidence of grass-pollen allergenic activity in the smaller micronic atmospheric aerosol fraction. Clinical and Experimental Allergy 20, 273–280.

Spieksma, F.Th.M., 1995. Aerobiology of inhalatory allergen carriers. Allergology Et Immunopathology 23 (1), 20–23.

Spieksma, F.Th.M., Nikkels, B.H., Dijkman, J.H., 1995. Seasonal appearance of grass pollen allergen in natural pauci-micronic aerosol of various size fractions. Relationship with airborne grass pollen concentrations. Clinical and Experimental Allergy 25, 234–239.

Stock, T.H., Kotchmar, D.J., Contant, C.F., Buffler, P.A., Holguin, A.H., Gehan, B.M., Noel, L.M., 1985. The estimation of personal exposures to air pollutants for a community based study of health effects in asthmatics: design and results of air monitoring. Journal of Air Pollution Control Association 35, 1266–1273.

Suh, H.H., Koutrakis, P., Spengler, J.D., 1993. Validation of personal exposure models for sulfate and aerosol strong acidity. Journal of Air and Waste Management Association 43, 845–850.

Suh, H.H., Nishioka, Y., Allen, G.A., Koutrakis, P., Burton, R.M., 1997. The metropolitan acid aerosol characterization study: results from the summer 1994 Washington, D.C. field study. Environmental Health Perspectives 105 (8), 826–834.

Suphioglu, C., Singh, M.B., Knox, B., 1992. Mechanism of grass-pollen-induced asthma. Lancet 339, 569–572.

Swoboda, I., Scheiner, O., Kraft, D., Breitenbach, M., Heberle-Bors, E., Vicente, O., 1994. A birch gene family encoding pollen allergens and pathogenesis-related proteins. Biochimica et Biophysica Act 1219, 457–464.

Tager, I.B., 1993. Introduction to working group on tropospherical ozone, Health Effects Institute Environmental Epidemiology Planning Project. Environmental Health Perspectives 101 (Suppl. 4), 205–207.

Tamura, K.A.M., Matsumoto, Y., 1996. Estimation of levels of personal exposure to suspended particulate matter and nitrogen dioxide in Tokyo. US-EPA Report EPA/600/P-95/001aF, Vol. I, p. 7-99–7-100.

Thatcher, T.L., Layton, D.W., 1995. Deposition, resuspension and penetration of particles within a residence. Atmospheric Environment 29 (13), 1487–1497.

US-EPA, 1996a. Air quality criteria for particulate matter. EPA/600/P-93/004aF, Vol. I, US-Environmental Protection Agency, Washington, DC.

US-EPA, 1996b. Air Quality Criteria for Ozone and related photochemical oxidants. EPA/600/P-93/004aF, Vol. I, US-Environmental Protection Agency, Washington, DC.

Vandenplas, O.D.J., Evrared, G., Aimont, P., Van Der Brempt, S., Jamart, J., Delaunois, L., 1995. Prevalence of occupational asthma due to latex among hospital personnel. American Journal of Respiratory and Critical Care Medicine 151, 54–60.

Waldman, J.M., Lioy, P.J., Thurston, G.D., Lippmann, M., 1990. Spatial and temporal patterns in summertime sulfate aerosol and neutralization within a metropolitan area. Atmospheric Environment 24 (1), 115–126.

Wallace, L.A., 1996. Indoor particles: a review. Journal of Air and Waste Management Association 46, 98–127.

Wallace, L.A., Pellizzari, E.D., Hartwell, T.D., Whitmore, R., Zelon, H., Perritt, R., Sheldon, L., 1988. The California TEAM study: breath concentrations and personal exposures to 26 volatile organic compounds in air and drinking water of 188 residents in Los Angeles, Antioch and Pittsburg, CA. Atmospheric Environment 22 (10), 2141–2163.

Wallace, L.A., Pellizzari, E.D., Hartwell, T.D., Davis, V., Michael, L.C., Whitmore, R.W., 1989. The influence of personal activities on exposure to volatile organic compounds. Environmental Research 50, 37–55.

Weschler, C.J., Schields, H.C., Naik, D.V., 1989. Indoor ozone exposure. Journal of Air Pollution Control Association 39, 1562–1568.

Weschler, C.J., Shields, H.C., Rainer, D., 1990. Concentrations of volatile organic compounds at a building with health and comfort complaints. American Industrial Hygiene Association Journal 51 (5), 261–268.

Weschler, Ch.J., Brauer, M., Koutrakis, P., 1991. Indoor ozone and nitrogen dioxide: a potential pathway to the generation of nitrate radicals, dinitrogen pentaoxide, and nitric acid indoors. Environmental Science and Technology 26 (1), 179–184.

Whitby, K.T., Sverdrup, G.M., 1980. California Aerosols: their physical and chemical characteristics. Advances in Environmental Science and Technology 10, 477–483.

Wijnand, E., 1996. Measurements methods and strategies for non-infectious components in bioaerosols at the workplace. Analyst 121, 1197–1201.

Willeke, K., Baron, P.A., 1993. Aerosol Measurements. Van Nostrand Reinhold, New York.

Williams, M.L., 1995. Monitoring of exposure to air pollution. Science of the Total Environment 168, 169–174.

Wilson, A.F., Novey, H.S., Berke, R.A., Surprenant, E.L., 1973. Deposition of inhaled pollen and pollen extract in human airways. New England Journal of Medicine 288, 1056–1058.

Wilson, R., Spengler, J., 1996. Particles in Our Air. Harvard University Press, Boston.

Wilson, W.E., Suh, H.H., 1997. Fine particles and coarse particles: concentration relationships relevant to epidemiologic studies. Journal of Air and Waste Management Association 47, 1238–1249.

Wüthrich, B., Schindler, Ch., Leuenberger, Ph., Ackermann-Liebrich, U., SAPALDIA Team, 1995. Prevalence of atopy and pollinosis in the adult population of Switzerland (SAPALDIA study). International Archives of Allergy and Immunology 106, 149–156.

Yanagisawa, Y., Nishimura, J., 1982. A badge-type personal sampler for measurement of personal exposure to NO_2 and NO in ambient air. Environmental International 8, 235–242.

Yanagisawa, Y., Nishimura, H., Matsuki, H., Osaka, F., Ksuga, H., 1988. Urinary hydroxyproline to creatinine ratio as a biological effect marker of exposure to NO_2 and tobacco smoke. Atmospheric Environment 22 (10), 2195–2203.

Yeh, H.C., Phalen, R.F., Raabe, O.G., 1976. Factors influencing the deposition of inhaled particles. Environmental Health Perspectives 15, 147–156.

Yocom, J.E., 1982. Indoor–outdoor air quality relationships. A critical review. Journal of Air Pollution Control Association 32, 500–520.

Appendix

1. Introduction and methods

The purpose of this Appendix is to present some aspects of the results of the recently completed BRISKA- (Basle Risk Assessment Study of Ambient Air Pollutants) and EXPOLIS-study (Air Pollution Exposure in European Cities) (Braun-Fahrländer et al., 1999; Jantunen et al., 1998). The first study focussed on spatial variations of air pollutants in an urban environment and the latter on the relationships between indoor, outdoor, workplace and personal exposures to $PM_{2.5}$, CO and VOCs. In the BRISKA-study, suspended particulate matter (TSP, PM_{10}, $PM_{2.5}$) and gaseous air pollutants were measured at six sites within the urban area of Basle. In the EXPOLIS-Study, 50 (at one site 250) study persons participated in six cities in a 48-hour measurement of indoor, outdoor, workplace and personal exposures.

The objective of the BRISKA-project was to obtain small-scale, spatially resolved air pollution data within the urban area of Basle. In a second step, cancer risks of the population in Basle were estimated based on the spatial gradients of the pollutants in a unit-risk model. Measurements of air pollutants were performed during a one-year period in 1997. A mobile container, equipped with measuring devices for air pollutants was installed at six sites reflecting differences in pollutant patterns (traffic, residence, etc.). As the measurements did not cover a full year at each site, a model calculation was made in order to consider the temporal variation of the pollutants over the year and the meteorological conditions. The model incorporated air pollution data from an additional stationary or fixed site survey and meteorological data (Braun-Fahrländer et al., 1999; Röösli et al., 2000a,b). Of the meteorological variables, daily average temperature, relative humidity, sum of the precipitation, global radiation, wind speed, temperature gradient between 250 and 493 m (at day and night) and wind direction were included in the model. Based on the resulting multiple regression model, annual average levels were estimated for each site. The coefficients of determination (R^2) in the models were very good and ranged between 0.945–0.989. After gravimetric analyses, the filters were further analysed for their content of elements. Estimates for annual averages at each site for the chemicals were calculated based upon a model described in Braun-Fahrländer et al. (1999) and Röösli et al. (2000a). In the EXPOLIS-study, each participant measured personal exposure levels during 48 hours (study period: autumn 1996 winter 97/98). In addition, pollutant levels were measured in the bedroom, outdoors of the home and at the workplace (Koistinen et al., 1999).

2. Results and discussion

The estimates of the annual averages of the particle levels in BRISKA are shown in Fig. 10a. Within the class of suspended particulate matter, TSP had the largest spatial variation of the concentration, followed by $PM_{2.5}$ and PM_{10}. The difference in the ranking between $PM_{2.5}$ and PM_{10} was not significant. For the elements, largest spatial variation was found for Pb, elemental carbon (EC) and nitrate and least spatial variation for ammonium and sulphate. Secondary pollutants (sulphate, ammonium, organic matter) were more homogeneously distributed. The big coefficient of variation for NO_3^- was due some uncertainties in the model. For the gases (Fig. 10b), the spatial variation was largest for NO, followed by SO_2 and NO_2. Smallest spatial variations were observed

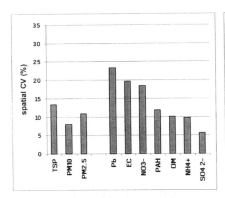 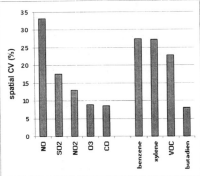

Figure 10. Spatial coefficient of variation (CV: standard deviation divided by average value) for suspended particulate matter (TSP: total suspended particulate matter, PM_{10}, $PM_{2.5}$: particles < 10, 2.5 µm) and elements (lead (Pb), elemental carbon (EC), nitrate (NO_3^-), polyaromatic hydrocarbons (PAH), ammonium (NH_4^+), organic matter (OM) and sulphate (SO_4^{2-})) (10a) and for gases (10b). (Data from Braun-Fahlränder et al., 1999.)

for CO and O_3. In the group of the VOCs, the aromatic hydrocarbons, benzene and xylene had largest variations, almost in the range of that of NO. The spatial variance of butadiene was about one third of that of the aromatics.

In conclusion, these results confirmed the spatial heterogeneity of traffic markers of primary emissions (Pb, EC) and the spatial homogeneity of secondary pollutants SO_4^{2-} and O_3. Some other results of the BRISKA-study (Junker et al., 2000) showed that for the ultrafine and fine particle mass and number (concentrations) the spatial non-homogeneity increased with decreasing particle size. This is due the direct emission of the ultrafine particles (e.g. from diesel sources) which undergo coagulation resulting in a more homogeneous distribution over an area than the primary emitted particles.

While the BRISKA-project focussed on health risks of ambient air pollutants, the EXPOLIS study aimed to assess the relationships between personal exposures and stationary measurements (home outdoor, indoor, workplace). In the data analysis from Oglesby et al. (2000a) the relationship between ambient levels of $PM_{2.5}$, sulphur and potassium, lead and calcium was determined. These elements were chosen, as they are distinct markers of certain emission sources. Lead and bromine stem for traffic related particulates, calcium is a soil derived crustal element, sulphur and potassium reflect regional air pollution. Table 11 shows the Spearman rank correlations between the personal measurements and the home outdoor levels of the $PM_{2.5}$ mass and some elements. Data was only available for the city of Basel. The strongest correlations were observed between personal and outdoor sulphur levels. Note that in the calculations smokers and homes with indoor activity such as indoor combustion or grilling were excluded. Weak to moderate correlations were found for the traffic-related elements Pb and Br. No correlation was observed for the ubiquitous element Ca. This analysis showed that the correlation between personal exposures to particles and outdoor measurements differs strongly between different particle parameters. For secondary pollutants (e.g. sulphur), reflecting particles in the accumulation mode and regional air pollution, the correlation

Table 11. Spearman rank correlation for $PM_{2.5}$ parameters between personal and home outdoor levels. (**: $p < 0.001$, *: $p < 0.05$). (Data from Oglesby et al., 2000)

	Spearman rank correlation (r)
mass	0.21
sulphur	0.85**
potassium	0.79**
lead	0.53*
bromine	0.49*
calcium	0.14

is good. For markers of primary traffic emissions (Pb), the correlation becomes weaker. Differences in time activity patterns (e.g. spending time near sources) and the strong spatial inhomogeneity of the ultrafine (traffic) primary particles are two important factors which bias the relationship.

In studies on long-term exposure and effects using the so-called semi-individual study design, where health parameters exist for all individuals and where exposure data is assigned to a group of people (living in a well-defined area) spatially aggregated data are used. Fig. 11 compares aggregated (non-weighted) geometric mean values of indoor and outdoor data from the cities Athens, Basle, Helsinki and Prague. Despite some potential interference of indoor smoking, a large R^2 ($= 0.9137$) was obtained. The personal $PM_{2.5}$ exposures were measured as night-time (private) and daytime (workday) samples. The personal night-time values correlated strongly with the home indoor values $R^2 = 0.92$ (figure not shown), indicating that the outdoor sample is a good surrogate for personal (private) exposure. In Fig. 12, the personal workday values are plotted against the workplace and the outdoor home values. For Basle, and Helsinki, the geometric means of the workplace levels corresponded with the home outdoor levels. For Milan and Athens, the workplace levels were much higher than the home outdoor levels. For both relationships the coefficients of variation were above 0.9, indicating good agreement. In conclusion, all these analyses based on aggregated city data showed that, on average, personal exposures are quite well reflected by outdoor (or workplace) measurements. A look

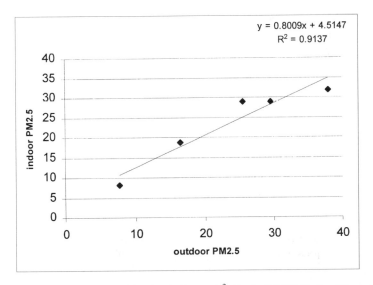

Figure 11. Outdoor vs. indoor $PM_{2.5}$ levels (in $\mu g\,m^{-3}$) for the EXPOLIS sites Athens, Basle, Helsinki, Milan and Prague. (Data from Jantunen, 1999).

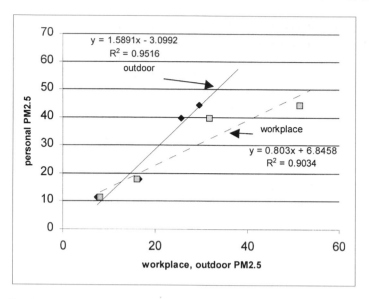

Figure 12. Workplace and outdoor home $PM_{2.5}$ levels vs. personal workday levels (in $\mu g\,m^{-3}$) for the EXPOLIS sites Athens, Basle, Helsinki and Prague. (Data from Jantunen, 1999).

at the absolute average levels of personal exposures and outdoor levels shows that the personal (workday) levels were about 40% higher than the outdoor levels. This finding corresponds to previous findings (Wallace, 1996). This difference also shows that careful attention has to be paid when calculating health effect-estimates based on outdoor levels.

3. Conclusion

In studies using aggregated data of air pollutants, ambient outdoor measurements are good surrogates of personal exposures. The difference in absolute terms between outdoor and personal levels has to be considered in order to reduce bias in the effect-estimate. Moreover, the "aggregate" has to reflect average population-exposure levels. This is not always guaranteed when exposure assignments are based on a single fixed site monitor. For pollutants with large spatial variation such as markers of primary traffic or NO_2 this can be critical. As it is not practical to perform measurements at each home some alternatives have to be evaluated. As an example, Oglesby et al. (2000b) evaluated a questionnaire based on an annoyance score. Annoyance due to traffic correlated well with measured PM_{10} and NO_2 levels when regressing average annoyance levels against measured values (between-area analysis). The validity of the method, however, is restricted.

In studies on long-term exposures, the loss of study persons due to moving is a problem. Moreover, data from ambient monitors are often restricted to short time periods (e.g. a few years). This shows that there is a need for spatially resolved long-term data. In order to solve the first problem, multiple regression models were calculated for the estimation of annual mean PM_{10} values in grids of 1 km^2 over the three countries Switzerland, Austria and France (Filliger et al., 1999). These values can be assigned to the population living in distinct square grids. Such an approach of multiple regression modelling (e.g. for NO_2, PM_{10} and $PM_{2.5}$) for spatially resolved pollution data helps to assign the exposure of people who moved away from the original study site. Furthermore, emission trends can be considered and data can be calculated in a retrospective manner for an estimation of long-term exposures. It is clear that these approaches need strong validations as the static multiple regression model may be restricted to a limited geographical area. Moreover, people are exposed to a mixture of pollutants and the single pollutant approach can be misleading. The exposure-effect relationship will show if the new approach reduces bias.

References

Braun-Fahrländer, Ch., Theis, G., Künzli, N., Camenzind, M., Röösli, M., Monn, Ch., 1999. Gesundheitsrisiken durch Luftverunreinigungen in der Stadt Basel. Analyse der Immissionsmessungen. 1. Zwischenbericht (1st Intermediate Report of the BRISKA Project).

Filliger, P., Puybonnieux-Texier, V., Schneider, J., 1999. Health costs due to road traffic-related air pollution. PM_{10} population exposure. Technical report on air pollution. (WHO-ADEME, FEA.) Published by the Federal Department of Environment, Transport, Energy and Communications, Berne, Switzerland.

Jantunen, M.J., Hänninen, O., Katsouyanni, K., Knöppel, H., Künzli, N., Lebret, E., Maroni, M., Saarela, K., Sram, R., Zmirou, D., 1998. The "EXPOLIS study". J Exp Analysis and Env Epidemiol 8 (4) 495–518.

Jantunen, M.J., (Coordinator) 1999. Final report: air pollution exposure in European cities: the EXPOLIS study. (EU contracts ENV4-CT96-0202.)

Junker, M., Kasper, M., Röösli, M., Camenzind, M., Künzli, N., Monn, Ch., Theis, G., Braun, Ch., 2000. Airborne particle number profiles, particle mass distribution and particle bound PAH concentrations within the city environment of Basle: an assessment of the BRISKA project. Atmospheric Environment 34 (19) 3171–3181.

Koistinen, K., Kousa, A., Tenhola, V., Hänninen, O., Jantunen, M., Oglesby, L., Künzli, N., Georgoulis, L., 1999. Fine particle (PM2.5) measurement methodology, quality assurance prosedures and pilot results of the EXPOLIS study. J Air & Waste Manage Assoc 49, 1212–1220.

Oglesby, L., Künzli, N., Röösli, M., Braun-Fahrländer, Ch., Mathys, P., Stern, W., Jantunen, M., Kousa, A., 2000a. Validity of ambient levels of fine particles as surrogate for personal exposure to outdoor air pollution-results of the European EXPOLIS-EAS study (Swiss Centre Basle). J Air & Waste Manage Assoc 50, 1251–1261.

Oglesby, L., Künzli, N., Monn Ch., Schindler, C., Ackermann, U., Leuenberger, Ph. and SAPAL-DIA Team 2000b. Validity of annoyance scores to estimate long-term air pollution exposure in epidemiological studies. Am J Epidemiology 152, 75–83.

Röösli, M., Braun-Fahrländer, Ch., Künzli, N., Oglesby, L., Theis, G., Camenzind, M., Mathys, P., Staehelin, J., 2000a. Spatial variability of different fractions of particulate matter within the urban environment and between urban and rural sites. J Air & Waste Manage Assoc 50, 1115–1124.

Röösli, M., Theis, G., Künzli, N., Staehelin, J., Mathys, P., Oglesby, L., Camenzind, M., Braun-Fahrländer, Ch., 2000b. Temporal and spatial variation of the chemical composition of PM10 at urban and rural sites in the Basle area, Switzerland. Atmospheric Environment, in press.

Wallace, L.A., 1996. Indoor particles: A review. J Air & Waste Manage Assoc 46, 98–127.

Air Pollution Science for the 21st Century
J. Austin, P. Brimblecombe and W. Sturges, editors
© 2002 Elsevier Science Ltd. All rights reserved.

Chapter 5

New Directions: Reducing the toxicity of vehicle exhaust

R.L. Maynard

Department of Health, Skipton House, 80 London Road, London, SE1 6LW, UK

Many studies have demonstrated associations between daily average concentrations of particles and the number of deaths and hospital admissions occurring each day. It is likely that these associations are causal. Annual average concentrations of fine particles ($PM_{2.5}$) are associated with a reduction in life expectancy (Department of Health, *Non Biological Particles and Health,* London, HMSO, 1995).

Accepting, then, that exposure to current ambient aerosols has adverse effects on health, it is reasonable to ask what can be done to reduce these effects. A first answer might be to argue simply for a reduction in ambient particle concentrations. This has been the approach taken by countries, including the UK, that have set particle standards and objectives in terms of the metrics used to monitor particle concentrations; PM_{10} for example. If it is accepted that the relationship between, say, PM_{10} and daily deaths is known and is immutable, reductions in PM_{10} will necessarily be accompanied by a predictable reduction in deaths. (The term "deaths" is used here in the sense in which the phenomenon is studied in time-series studies, i.e. deaths brought forward by an (at present) unknown period.)

Whether it is actually inevitable that a reduction in levels of a pollutant will be accompanied by a predicted reduction in effects has been questioned. The reasons for this doubt include difficulties in identifying when the benefits of reducing levels may become apparent, and the difficulties of predicting the impact of competing risks when one is reduced. Should the toxicity of the ambient aerosol change, then we should expect the relationship between particle concentrations and effects on health to also change. That this may be the case is an excellent reason for conducting further studies and for *not* assuming that a reduction in particles will be associated with a reduction in effects.

It is clearly necessary to examine closely the relationships that have been reported between daily levels of particles and effects on health. During the coal smoke smogs of London in the 1950s and early 1960s the curve relating daily concentrations of particles to health effects was less steep than curves

First published in Atmospheric Environment 34 (2000) 2667–2668

describing current relationships in the UK, US and Europe. How can this be explained? It is possible that people took evasive action on high pollution days, and thus actual exposure on such days was less than predicted by concentrations recorded at monitoring sites. Also, the very high levels seen in episodes of pollution that lasted up to a week might have killed a proportion of vulnerable subjects in the first days of the episode, leaving fewer to respond on the following days. Thus, when all the data were collected there would be a surprisingly low number of deaths on many high pollution days. But it is also possible that the toxicity of the ambient aerosol was lower in the 1950s than it is today.

One possible explanation for such a change in toxicity might be that the size of particles in the ambient aerosol has decreased. This is difficult to prove. Early studies showed that the number median diameter of the London aerosol in the early 1960s was 90 nm and many small particles were present. Aggregation of very small particles may have been aided by high concentrations of larger particles during intense episodes of air pollution.

The slope of the lines relating daily concentrations of particles and daily deaths in the US tend to be steeper than those reported in European studies. Again the possibility exists that US ambient aerosol is more toxic than that in Europe. It is also possible that in the eastern US, at least, the high acidity of the aerosol plays an important role. It has been reported that the slope of the line relating $PM_{2.5}$ to effects on health is steeper than that for PM_{10}, and that the association between $PM_{10-2.5}$ and health effects is less strong than the association between $PM_{2.5}$ and health effects. This suggests that some of the larger particles making up PM_{10} are acting as a more-or-less inert diluent, and that we should be searching in the $PM_{2.5}$ fraction for the biologically active components of the ambient aerosol.

Despite the convincing epidemiological studies, there is the nagging suspicion that the mass of particles deposited in the lung per day is too small to explain the reported effects. This could mean that the mass concentration of particles in the air is acting as a surrogate for something else that is toxicologically active. An obvious suggestion is some pollutant gas that happens to vary with particle concentration. Several gases have been looked at, and for none is the evidence fully convincing. (Carbon monoxide is a possible exception with regard to effects on the heart.) An alternative explanation is that effects are related to the number of very small particles present (Seaton et al., Lancet 345, 176–178, 1995). Should this number vary linearly with mass measurements of aerosol, e.g., PM_{10}, then clearly PM_{10} would appear to be related to effects.

Evidence to support the idea that very small particles may be playing a role has recently appeared. Peters et al. (*American Journal of Respiratory and Critical Care Medicine* 155, 1376–1383, 1997) have shown that some indices of ill-health may be related to the number concentration of very small particles. Toxicological studies of compounds such as TiO_2, Al_2O_3, and carbon black show that such compounds are much more active on a "per unit mass" basis when presented to animals and in vitro (e.g., cellular) test systems as ultrafine particles, than as particles of 250–500 nm diameter. Ultrafine particles deposit efficiently in the deep (alveolar) region of the lung. Alveolar deposition peaks at about 60% of inhaled particles of about 20–50 nm diameter. Particles of <20 nm are mainly deposited in the upper airways and do not reach the alveoli. On a "per unit mass" basis the total surface area of ultrafine particles is much greater than that of an aerosol of larger particles, i.e. the specific surface area is greater.

Motor vehicles make a large contribution to the primary aerosol in urban areas. Let us accept for the moment that automobile emissions of particles are contributing significantly to that component of the ambient aerosol that is linked with effects on health. Reducing the mass of particles emitted is an obvious first step towards reducing effects. This approach rests on the assumption that the toxicologically active components of the emissions will fall in line with the mass. More importantly, we should ask how the toxicity of the contribution made by motor vehicles to the ambient aerosol could be reduced. We may argue that to exert an effect the particles must find their way deep into the lung and be deposited there. It is possible that comparatively large particles depositing in the upper airways have an effect, perhaps by a reflex producing secondary effects deeper in the lung, but this will not be pursued here. Reduction of deposition of particles in the deep lung could be achieved in two ways. Firstly, if the particles were exceedingly small (<20 nm), and did not agglomerate, they would exhibit peak deposition in the upper airways. Secondly, if the particles could be made larger this would reduce deposition in the deep lung.

The first option seems to me likely to be very difficult to achieve. The second, however, seems to offer some chance of success. The deposition efficiency of 700 nm particles in the alveolar part of the lung is only about 10%, as compared with 40% for 50 nm particles. Thus, shifting the size distribution of the particles could significantly reduce the mass of particles deposited in the lung. Packaging the same mass of particles as 700 nm particles as compared with 50 nm particles would also greatly reduce the total number of particles deposited. It is clearly also important to examine the specific toxicity of various size fractions of the particles emitted. No data on the differential toxicity of various size fractions of vehicle exhaust particles are currently available. If different size fractions of motor vehicle-generated particles could be prepared then this would enable such studies. Particle composition may also be an important factor.

In conclusion, a properly designed study of the toxicity of different size fractions of vehicle-generated particles, and of their chemical components, would aid in the rational planning of strategies to reduce the effects of such particles on health. Reducing the total mass emitted, and increasing the size of the particles, are obvious first steps, but should be supported by experimental studies.

The views expressed in this paper are those of the author and should not be taken as those of the UK Department of Health.

Air Pollution Science for the 21st Century
J. Austin, P. Brimblecombe and W. Sturges, editors
© 2002 Elsevier Science Ltd. All rights reserved.

Chapter 6

The transport sector as a source of air pollution

R.N. Colvile, E.J. Hutchinson, R.F. Warren

T.H. Huxley School, Imperial College, London, SW7 2BP, UK

J.S. Mindell

*Department of Epidemiology and Public Health, Imperial College,
London, W2 1PG, UK*

Abstract

Transport first became a significant source of air pollution after the problems of sooty smog from coal combustion had largely been solved in western European and North American cities. Since then, emissions from road, air, rail and water transport have been partly responsible for acid deposition, stratospheric ozone depletion and climate change. Most recently, road traffic exhaust emissions have been the cause of much concern about the effects of urban air quality on human health and tropospheric ozone production. This article considers the variety of transport impacts on the atmospheric environment by reviewing three examples: urban road traffic and human health, aircraft emissions and global atmospheric change, and the contribution of sulphur emissions from ships to acid deposition. Each example has associated with it a different level of uncertainty, such that a variety of policy responses to the problems are appropriate, from adaptation through precautionary emissions abatement to cost–benefit analysis and optimised abatement. There is some evidence that the current concern for road transport contribution to urban air quality is justified, but aircraft emissions should also give cause for concern given that air traffic is projected to continue to increase. Emissions from road traffic are being reduced substantially by the introduction of technology especially three-way catalysts and also, most recently, by local traffic reduction measures especially in western European cities. In developing countries and Eastern Europe, however, there remains the possibility of great increase in car ownership and use, and it remains to be seen whether these countries will adopt measures now to prevent transport-related air pollution problems becoming severe later in the 21st Century.

First published in Atmospheric Environment 35 (2001) 1537–1565

1. Introduction

Transport is widely recognised to be a significant and increasing source of air pollution world wide. Several previous reviews have focussed on individual modes of transport and/or single environmental impacts of transport. For example, OECD (1988) briefly considers regional and global impacts of transport emissions of air pollution, but is mostly concerned with the impact of emissions on local urban air quality, and considers only road transport. The Third International Symposium on Transport and Air Pollution (Joumard, 1995) also has an emphasis on road traffic and urban air quality, but the Special Edition of *Science of the Total Environment* presenting highlights of the symposium also includes a few papers covering air and sea transport. Joumard comments on the value of the contributions from developing countries including Africa and Latin America; a review of road transport emissions and their impact on the environment at all scales from local to global was also published a couple of years earlier by Faiz (1993). One of the most comprehensive recent reviews of the environmental impacts of transport is that of the Royal Commission on Environmental Pollution (Houghton, 1994). This report includes a section on air transport, and the treatment of surface transport includes freight as well as passenger, rail as well as road. Shipping is mentioned briefly by reference to other work, especially Donaldson (1994). Urban air quality and global climate change are identified as major issues, but regional air quality, acidification, noise and impacts other than air pollution emissions are also considered. There is an emphasis on assessing possible solutions to environmental problems caused by transport, concluding with an exhaustive list of recommendations; these are for the UK, but the perspective is international. An update three years later (Houghton, 1997) has a narrower scope, restricting itself to inland surface transport, motivated by a concern that there was still too little action to limit the environmental impact of road traffic despite much debate on the subject having been stimulated. Most recently, the Intergovernmental Panel on Climate Change (Penner et al., 1999) has published a major report focussing on air transport and the global environment, in contrast to the emphasis on road transport in much of the earlier literature.

In addition to these reviews, a number of papers attempting to quantify the environmental cost of transport necessarily include a concise review of the subject, but since preparation of a complete impact valuation is a huge interdisciplinary task it is more common to consider road transport alone even if an attempt is made to quantify all its impacts. We will not attempt a survey of this area of economics here, although one example (Eyre et al., 1997), will be cited later as an example where the authors include a greater than usual emphasis on atmospheric science.

In the late 1990s, as reflected in the selection of previous reviews summarised above, the impact of road transport on urban air quality has had a very high profile in many countries. In the space of a couple of decades around the turn of the millennium, three-way catalytic converters are being fitted to every new petrol (gasoline)-engined car in the world, soon to be followed by similar developments in diesel emissions control. This is arguably the biggest exercise ever carried out in the application of end-of-pipe technology for the abatement of air pollution emissions from any type of source, certainly if the scale of the exercise is measured in terms of the number of individual people affected world-wide. Nearly, every family in the industrialised countries is already involved and increasing numbers of people in developing countries and the former Soviet Bloc are following, as the motor car is one of the great icons of 20th Century capitalism. In this review, we will ask the question, "Is the impact of road traffic emissions on urban air quality really currently the biggest issue concerning transport emissions of air pollution, and is it likely to remain so beyond the first few years of the 21st Century?" We will see that the extent to which we understand the relevant atmospheric science is different for each single impact of individual modes of transport that we will consider. Since it is very difficult to assess the relative severity of disparate impacts of air pollution emissions, this variability in the completeness of our understanding of the science is also having an impact on how different modes of transport are becoming subject to legislation, economic incentives to control emissions, and voluntary action to protect the environment.

We thus aim to provide a new, distinctive account of transport as a source of air pollution. The emphasis will be on the science of air pollution and recent developments in aspects of the assessment of urban air quality, regional atmospheric chemistry and global atmospheric change that are of relevance to transport. It is impossible to carry out an exhaustive review of the whole of this subject within a single journal article, so individual case studies will be presented, each case considering the contribution of a single mode of transport to one aspect of air pollution. These are selected to illustrate the range of issues that are of concern as we enter the 21st Century. We will restrict ourselves to gaseous and particulate emissions, whilst not forgetting that noise is also a major pollutant emitted by transport into the atmosphere. However, some of our conclusions will be relevant to noise as well as chemical pollutants, as some types of emissions abatement measures will deliver additional benefit through noise reduction.

The review starts with a summary of air pollution emissions from transport, by presenting an overview of how emissions inventories are compiled and used (Section 2), with particular emphasis on road traffic emissions. This is followed by a brief survey of the impacts of these emissions on the environment and society, presented chronologically to indicate that concern for the environmental impact of transport has evolved over the past three decades (Section 3). The three selected examples of impacts of air pollution emissions from individual transport sectors are presented in Section 4, from which it is possible to see how some of the lessons learnt in trying to control emissions from road transport might be applied to other modes in future, and vice versa. The review concludes with a discussion of whether the current preoccupation with road transport and urban air quality is likely to be long lasting given the magnitude of the impact and the level of uncertainty in our ability to quantify it, from which recommendations for further work to support future sustainable integrated transport systems are drawn.

2. Overview of transport contribution to emissions

2.1. Air pollutants emitted by transport sources

With a few exceptions, all modes of transport emit air pollution from the combustion of liquid fossil fuel. Most transport sources today therefore emit similar pollutants, although the relative abundance of these varies depending on the exact composition of the fuel and details of the combustion conditions.

The most significant transport emissions to the atmosphere by mass are carbon dioxide (CO_2) and water vapour (H_2O) from the complete combustion of the fuel. Some transport power sources achieve almost complete combustion by ensuring there is plenty of excess air, as in a diesel engine or a lean-burn petrol engine. A feature that distinguishes other mobile combustion sources from almost all stationary sources, however, is that combustion is incomplete, and a small fraction of the fuel is oxidised only to carbon monoxide (CO) with some volatile hydrocarbons also emitted as vapour in the exhaust and carbonaceous particles from incompletely burnt fuel droplets. The particles from a modern diesel engine, after modification by coagulation and other processes that occur in the first few seconds after emission, have a bimodal size spectrum with a large number of particles below 20 nm in size and another mode between about 30 and 100 nm (Shi et al., 1999), with approximately equal total mass in each mode.

In addition to the mixture of hydrocarbons, all fuels contain some impurities (with the possible exception of hydrogen obtained from a fuel cell, and the lightest hydrocarbon fuels such as methane which are available with very low levels of impurities). Sulphur is oxidised mostly to sulphur dioxide (SO_2) on combustion, and sometimes to sulphate which can assist in the nucleation of particles in the exhaust. Several other impurities such as vanadium in oil do not burn or have combustion products that have a low vapour pressure and so contribute further to particle formation. The organic lead compounds that are still added to high octane petrol only in parts of Africa and Asia, to prevent premature combustion, also form particles in the exhaust. Finally, at the high combustion temperatures of most transport sources of air pollution, atmospheric nitrogen (N_2) is oxidised to nitric oxide (NO) and small quantities of nitrogen dioxide (NO_2), in addition to smaller quantities from nitrogen-containing impurities in the fuel. Nitrous oxide (N_2O) is emitted only in small quantities from the combustion process, but is somewhat more abundant in the exhaust of cars fitted with catalytic converters.

2.2. Life-cycle assessment of emissions from transport

The air pollution emissions generated during use of any form of transport are only a part of the total amount of air pollution generated by transport-related activity. The techniques of Life-cycle Assessment (LCA) (ISO, 1997) can be used to identify which stage in the production, use and disposal of a given transport technology is responsible for the most significant atmospheric emissions. For the majority of examples, most of the emissions occur at the time and place of transport use. For example, 60–65% of life-cycle greenhouse gases from a petrol-engined car are CO_2 exhaust emissions during use with a further 10% being non-CO_2 exhaust emissions during use. The remainder is 10% associated with the car's manufacture (mostly energy use), and a further 15–20% emitted during extraction, refinery and transport of its fuel (OECD, 1993). It should be noted, however, that this calculation excludes significant quantities of CO_2 that are emitted in the production of materials to construct transport infrastructure such as roads and bridges, especially concrete. For hydrocarbon emissions, the pre-use part of the fuel life-cycle is even more important, as shown in Fig. 1 (Gover et al., 1996) (neglecting, for this example, the fact that different hydrocarbons emitted at different locations can have very different impacts), and other volatile organic compounds are emitted on evaporation of solvents during painting of bodywork as well as evaporation from the fuel tank and parts of some older engines when the vehicle is not in use. For airbags and air conditioning units, the major emission of the gases contained within them is on disposal at the end of their life.

Air pollution from the operation of electric railways and the small but growing number of road vehicles that are powered by electricity is all emitted some distance away from the place of use, which is a major attraction of electric power for urban transport. Coal-fired generation of electricity tends to produce a larger amount of SO_2 per unit mass of fuel than combustion of oil by stationary or mobile sources, because the amount of sulphur in coal is often higher (1–6% by mass) than in oil, and there is no refining process where sulphur can be removed; natural gas for electricity generation or used in a mobile source has negligible levels of sulphur, the same as the most recent clean automotive

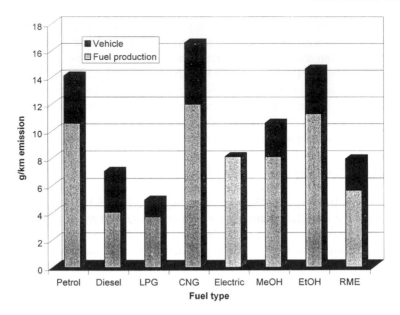

Figure 1. Life-cycle hydrocarbon emissions including methane for light goods vehicles as a function of fuel type.

fuels. Nuclear generation of electricity has the potential to emit zero levels of air pollution, although the Chernobyl accident illustrated that this is not always achieved in practice. Hydroelectric power's only emissions are during construction and demolition of the plant, and wind power is similar with the addition of noise emissions during use.

2.3. Quantification of emissions from transport

Atmospheric emissions can be quantified by adopting the so-called "top-down" or "bottom-up" methodology to generate an emissions inventory.

The top-down approach starts with data describing total polluting activity throughout the whole geographical area of interest, such as total national petrol sales for calculation of road transport emissions. This is related to the magnitude of the associated air pollution source by means of an emissions factor that can be obtained by laboratory measurement of a representative sample of engines or vehicles under simulated typical operating conditions, for example average NO_x emission per litre of petrol consumed. It is important to allow for the fact that engines in use are typically not maintained to the manufacturers' specified standards for emissions minimisation, and this is now taken into account for determination of road transport emissions though still not for some aircraft emissions. Spatial disaggregation of a top-down emissions inventory is then performed, if required, by assuming local emissions are proportional to some other variable that can reasonably be assumed to have a similar geographical distribution to that of the polluting activity, for example population density or length of road per unit area of land.

The bottom-up approach is different in that it starts with geographically resolved data, for example traffic flow on an individual length of road. For some sources (usually the larger stationary ones) emissions data are determined directly by measurement of each individual source. More usually however, especially for transport emissions where a large number of small individual sources are involved, emissions factors again need to be used, for example average emissions of NO_x per vehicle per kilometre driven. Total emissions for a geographical area of interest can then be obtained by summing all the individual contributions.

The top-down and bottom-up methods invariably give different total emissions, as each is subject to different sources of error (for example, Samaras et al., 1995). For road traffic, annual emissions for a typical whole city where activities are rather well characterised can be determined to within a factor of two or better using either method, while emissions from a smaller part of the city or a shorter averaging time, such as a single road during a specific hour, are generally known rather less accurately, with a more than a factor of 10 overestimation or underestimation being quite common. A "bottom-up" emissions inventory inherently suffers from requiring a very large amount of data, such that there is a tendency to make several assumptions and approximations. For example, traffic surveys quantifying the number and type of vehicles on every road in a town or city are usually taken manually, so each road will be sampled no more than a few days per year and average factors applied to relate this to weekends, nights and other seasons. Automatic traffic counts rarely give any information about vehicle type, although video cameras can now determine vehicle type by reading the registration plate. Computational models of traveller behaviour can be used to fill in data gaps on major roads, but are typically designed to study peak flow not daily or total emissions, and also might consider all vehicles as multiples of the number of passenger cars, still therefore needing factors to be applied to get hour-by-hour flows 365 $d\,yr^{-1}$ broken down into vehicle types for the quantification of air pollution emissions.

A convenient and regularly updated review of emissions factors for Western cities and inventory construction methodology is maintained by the London Research Centre (LRC, 1999), including data from the European Community DRIVE programme (Jost et al., 1992) and information from USEPA (1999). Fig. 2 shows data from two examples of emissions inventories: an urban inventory for fine particles (Buckingham et al., 1997), apparently showing a large contribution from diesel-engined road transport which will be discussed further below, and an older national inventory for carbon dioxide (ERR (1990) cited in Whitelegg (1993)) showing a significant contribution from transport, including air transport for which emissions are expected to increase while others decrease (Penner et al., 1999).

Emissions inventories can be valuable in providing a first estimate of the contribution of transport to air pollution emissions compared with other activity, or the relative contribution of alternative modes of transport when designing sustainable integrated transport systems. There are two respects, however, in which doing this can result in erroneous analysis. Firstly, different types of source of a given pollutant might have very different source–receptor relationships. This is unimportant for well-mixed pollutants such as CO_2, for which the total global concentration is of interest, but in an urban area emissions from vehicle exhausts are much closer to human receptors than tall chimneys on industrial point sources. For example, the contribution to pedestrian exposure per unit emission from vehicles in a city street is 300 times that of a 200 m high chimney in average dispersion conditions, even at the point of maximum ground-level concentration from the chimney.

The second factor that is ignored when an emissions inventory is used uncritically to assess transport contribution to local air quality is the possibility that sources outside the area of the inventory could make a significant contribution. The best example of this is fine particles, for which Fig. 2(a) gave the impression that abatement of diesel engine sources could have a large impact on air quality in a large non-industrial city. In fact, atmospheric dispersion modelling (Carruthers et al., 1999) has shown that future large cuts in particle emissions may have almost no impact on PM_{10} concentrations except immediately adjacent to the busiest roads and in the most severe winter stagnation air pollution episodes, because a significant contribution to annual average PM_{10} concentrations in the city as a whole is imported, in the case of the UK from as far away as Eastern Europe. This imported contribution is predicted to become a large fraction of the total in future when local sources are reduced. The importance of long-range transport is greatest for fine particles as a consequence of their long life in the atmosphere (APEG, 1999).

2.4. Summary

The factors that need to be taken into account when quantifying a given impact of transport emissions of air pollution are therefore as follows:

- emissions during complete life-cycle of vehicle, fuel and associated infrastructure;
- significance of transport emissions compared with other sources of the same pollutant(s) within a given geographical area, as shown by emissions inventory data;

a)

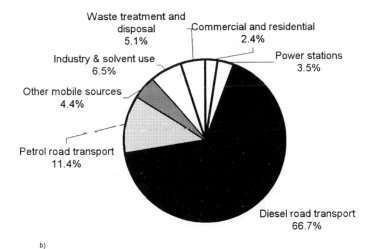

b)

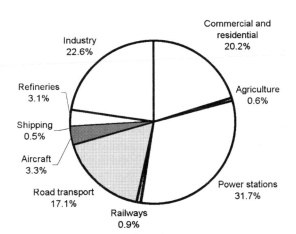

Figure 2. (a) Emissions of fine particles smaller than 10 μm aerodynamic diameter in London in 1995/1996; (b) UK emissions of carbon dioxide in 1988.

- contribution of sources outside the geographical area covered by the emissions inventory;
- source–receptor relationships;
- other pollutants contributing to or exacerbating the impact of interest;
- other impacts of the pollutant(s) of interest.

Frequent changes in public opinion and policy to control emissions from transport can occur when one or more of these factors is not considered, either through error of omission or through lack of the necessary understanding or information. Such changes that have occurred during the last four decades of the 20th Century will be outlined in the next section, as part of a general review of all the major impacts of transport emissions of air pollution.

3. Overview of transport air pollution emissions impacts

3.1. *"Clean air" in the 1960s and early 1970s*

At the beginning of the 1970s, widespread availability of electricity and clean fossil fuels coupled with the introduction of clean air legislation had resulted in the severe urban air quality problems of preceding decades being solved in most Western cities. The major emissions abatement measures then were not transport related. They were directed towards the formerly much more significant source of pollution: coal, burnt in inefficient boilers or in a separate grate for each individual room to heat offices and homes. This was replaced by cleaner central heating systems, especially in areas where natural gas became available at about the same time, on grounds of improved comfort and convenience as well as economic incentives and legislation. The major transport emissions abatement measure of this era was the replacement of steam traction with diesel and electric on the railways.

Ellison and Waller (1978) reviewed the evidence on the health effects of urban air pollution (principally sulphur dioxide and suspended particulates), with particular reference to the UK. They concluded that urban air pollution until around 1968 caused increased mortality and morbidity, with exacerbation of pre-existing chronic respiratory disease, but felt these effects were no longer occurring. The sooty smogs of the 1950s had also been highly visible and tangible, so the improvement in air quality could be seen, smelt and even tasted by the general public as well as monitored by scientists, adding to the general impression that the problem had been solved. Plentiful oil permitted the development of ever larger automobiles, especially in the United States. Comfort, status, mobility and vehicle performance were higher priorities for vehicle design than exhaust emissions or fuel economy. Aircraft design, similarly, focussed on speed and size, with the Anglo-French Concorde setting standards for supersonic passenger transport that have not been surpassed since but at the expense of emissions many times higher than those of more modern aircraft. The increase in prosperity in Western Europe and North America after the end of the Second World War also led to a rapid increase in the ability of ordinary people to travel using these more polluting modes of transport.

3.2. The return of smog

The first major automotive emissions control measures were stimulated by the infamous Los Angeles smog at a time when urban air quality had become much less of a problem in other parts of the world. This smog was (and is) of a different type to the sooty fog that had been tackled in cities with cooler, less sunny climates. The photochemical smog was produced by the action of sunlight on oxides of nitrogen and hydrocarbons, the very pollutants that were emitted in large quantities by the rapidly increasing numbers of automobiles in the 1950s and 1960s. In the 1960s, the first oxidation catalysts were fitted to convert vehicle emissions of carbon monoxide and hydrocarbons to carbon dioxide and water (Heck and Farrauto, 1995). Steadily increasing standards were then introduced at federal level throughout the 1970s, with upwards of 80% of new cars being fitted with a catalytic converter since 1975. The first car to be equipped with a three-way catalytic converter in the United States was imported by Volvo in 1977 (OECD, 1988), and the US 1981 emissions standards required every new car to be fitted with one.

3.3. The emergence of acidification

In the 1970s, it was suddenly noticed that trees were apparently dying en masse in the highly polluted "Black Triangle" of East Germany, the Czech Republic and Poland (Ulrich, 1990; Kandler and Innes, 1995; Bach, 1985) and numbers of dead fish floated to the surface of Swedish rivers and lakes (Borg, 1986) as well as in similar North American environments (Driscoll et al., 1980). Initially, it was largely the use of coal in large combustion plant that was to blame, with unabated emissions of sulphur dioxide converted to sulphuric acid by oxidation in the atmosphere (Hewitt, 2000). Following international agreements (Geneva in 1979 and Helsinki in 1985) to cut emissions (McCormick, 1997), the next step to reduce air pollution emissions might have been replacement of fossil fuels with nuclear power. The growth of green politics especially in nuclear cold-war front-line Germany, however, opposed this, and attention turned to private cars as a source of oxides of nitrogen precursors to the increasingly significant atmospheric concentrations of nitric acid. Three-way catalytic converters to tackle emissions of oxides of nitrogen, hydrocarbons and carbon monoxide have been in use in Germany since 1984, nine years ahead of the European legislation to make such emissions control mandatory (CONCAWE, 1997). Sweden and Switzerland also introduced vehicle emissions standards ahead of the rest of Europe, in 1976 and 1982 respectively. Europe thus started to catch up with the United States in control of emissions from road transport, but the environmental impact driving the change was different on the two sides of the Atlantic.

3.4. Climate change and stratospheric ozone depletion

The environmental pressure from acid rain in Europe and photochemical smog in California was combined with the oil price rises of 1973 and 1978, leading to fuel consumption by transport coming under scrutiny, especially larger automobiles. As the acidification issue became old news and efforts to solve the problem got under way (Stanners and Bordeau, 1995), the environmental agenda shifted and the 1980s became the decade of the global atmosphere. Predictions of widespread flooding (Carter, 1987) as thermal expansion of the oceans was predicted to cause sea-level rises up to 1 m (Houghton et al., 1996) focussed minds on global warming. Despite a fall in the price of oil (Hampton, 1991), this led to increasing popularity of the diesel engine over petrol especially in parts of Europe, on the grounds that emissions of greenhouse gas carbon dioxide from inherently more fuel efficient diesel engines are lower than those from equivalent three-way catalyst equipped petrol cars. The discovery of an annually occurring ozone hole over Antarctica, which deepened rapidly during the second half of the decade (Farman and Gardiner, 1987; Farman, 1987), as the first observation of a major catastrophic failure in natural regulation of the functioning of the global atmosphere, caused the spotlight to fall on emissions of long-lived, stable but catalytically active molecules such as chlorofluorocarbons, of which transport is far from being the largest source. In due course, however, concern began to grow over aircraft being possible contributors to ozone depletion through emissions of sulphur dioxide, soot and oxides of nitrogen. On the ground, mobile air conditioning units, which had been commonplace in North American cars since the 1970s and would become rapidly less unusual in European cars later in the 1990s (as a result of global warming perhaps, but more likely just a couple of hot summers), alongside other refrigeration systems came under the regulation of the Montreal Protocol (1987) to phase out the use of the most powerful ozone-depleting chemicals during the 1990s. Emissions of greenhouse gases came under global control somewhat less rapidly as a result of genuine scientific uncertainty concerning the magnitude of the problem combined with powerful lobbying by the fossil fuel industry. The Rio Summit of 1992 (Quarrie, 1992; Grubb et al., 1993) concluded that climate change is a serious problem such that action cannot wait for scientific uncertainty to be reduced, with developed countries being identified as having a responsibility to take the lead and compensate developing countries for the cost of controlling emissions of greenhouse gases, complete with proposals for far-reaching institutional change to integrate environmental protection with development. This was followed by the Kyoto summit of 1997 (Grubb, 1999), where the first international agreement was reached to make some small reductions in greenhouse gas emissions. These are, however, nowhere near the drastic global cuts that are required to bring about a return to pre-industrial or even current atmospheric levels of greenhouse gases before the end of the next century if ever, but are the first step towards stabilising atmospheric CO_2 during the later years of the 21st Century at around double its current concentration.

3.5. Urban air quality revisited

In the final decade of the century, the European and North American air pollution agenda has come back full circle and the issue of urban air quality that had last been at the top of the European agenda in the late 1950s rose again to the fore world wide. Diesel engines rather rapidly ceased to be cited as the environmentally friendly option as epidemiologists (Pope et al., 1992; Dockery et al., 1992), laboratory-based scientists (Diaz-Sanchez, 1997) and expert groups (QUARG, 1993) found evidence that the particles emitted might be responsible for measurable increases in the manifestations of cardiovascular and respiratory diseases even at the comparatively low levels of air pollution in modern Western cities. These had not been seen before because older statistical methods were not powerful enough to detect the very low signal-to-noise ratio of the effect of air pollution against other causes of health inequality and variability, and because computers to handle the large amounts of data required were not widely available. A large number of epidemiological studies followed on the effect of various road traffic emissions on a range of health end-points. Public concern over air quality is enhanced by its effects on children (Brunekreef et al., 1997) and has focussed in lay minds on associations with asthma, the incidence and prevalence of which have increased dramatically during the second half of the 20th Century (Holgate et al., 1995; Jarvis and Burney, 1998) in many countries (Miyamoto, 1997; Ninan and Russell, 1992). Current evidence suggests that air pollution exacerbates or provokes symptoms in those with pre-existing asthma (Krishna and Chauhan, 1996) but there is no good evidence that asthma is caused by air pollution (Holgate et al., 1995). There are also fears of cancer, as specific hydrocarbon components of vehicle exhaust, especially polycyclic aromatic hydrocarbons bound to diesel exhaust particulates, plus benzene and 1,3-butadiene (Perera, 1981; US EPA, 1990, 1993), are known carcinogens. CO is present in the cities of developing countries at levels high enough to exacerbate cardiovascular disease by impairment of the oxygen-carrying capacity of the blood, but the introduction of catalytic converters has meant that levels this high are a thing of the past elsewhere (DETR, 1998) unless, as has been the case with fine particles, improved statistical techniques allow detection of effects at much lower levels than had previously been found. The same is true of lead (Delves, 1998; SMEPB, 1994; Olaiz et al., 1996; Yang et al., 1996) which has been shown at levels in previous years to cause neurotoxicological damage and lower Intelligence Quotient scores in children (Smith, 1998; WHO, 1995; EPAQS, 1998). It has long been known from laboratory studies that SO_2 causes coughing on short-term exposure to high concentrations, particularly among people with asthma (Sheppard et al., 1980), although the older field measurements of effects on populations need to be applied with some care to modern traffic-dominated cities since the earlier high SO_2 levels from coal combustion were accompanied by particulate air pollution concentrations several times higher than today's.

The traffic-related pollutant most recently implicated in causing ill health in the cities of developed countries today is NO_2. Its effects are summarised in Table 1. Some studies have suggested that NO_2 is acting wholly or partly as a surrogate for another pollutant that has similar properties and source distribution (Poloniecki et al., 1997; Touloumi et al., 1997; Morgan et al., 1998). However, others have shown an effect of NO_2 after allowing for the effects of other pollutants (Castellsagué et al., 1995; Pantazopoulou et al., 1995; Linn et al., 1996). Another study revealed increased effects of NO_2 when other pollutants were included in the models (Sunyer et al., 1997). The debate continues, although recent studies have again found effects of NO_2 (Atkinson et al., 1999; Hajat et al., 1999; Garcia et al., 2000).

NO_2, along with volatile organic compounds (VOCs), is also a precursor of ground-level ozone (O_3) and other photochemical pollutants (Sillman, 1999). Not only has O_3 been shown to worsen asthma symptoms (Romieu et al., 1996) and be associated with an increase in emergency hospital respiratory admissions (Schwartz, 1996; Spix et al., 1998) but it also damages crops (Ashmore et al., 1980). A major difference between O_3 and primary emissions from transport sources is that the time taken to form O_3 is sufficiently long for the highest concentrations to be found typically 100 km from the source so it is a regional pollutant. Except in the most severe urban photochemical smog conditions (Apling et al., 1977) levels of O_3 at street-level in city centres tend to

Table 1.   Summary of associations between NO_2 and human health

Effect of daily rise in NO_2	Reference
Increase in total mortality	Touloumi et al. (1997)
Cardiovascular deaths	Zmirou et al. (1996)
Infant mortality	Bobák and Leon (1999)
Intrauterine deaths	Pereira et al. (1998)
Asthma emergency hospital admissions	Sunyer et al. (1997)
Chronic obstructive pulmonary disease hospital admissions	Anderson et al. (1997)
Cardiovascular disease hospital admissions, especially heart attack and angina	Poloniecki et al. (1997)
Hospital visits for asthma	Castellsagué et al. (1995)
Croup in pre-school children	Schwartz et al. (1991)
All emergency hospital admissions especially for the elderly	Ponce de Leon et al. (1996)

be lower than elsewhere or even zero because of the proximity of road traffic sources of nitric oxide (NO), which scavenges the O_3 to form NO_2. Some authors are now beginning to describe O_3 as a global pollutant as background levels rise across the whole of the North Atlantic area due to North American and Western European road traffic emissions combined (Johnson et al., 1999), heralding a return to increased concern about regional and global atmospheric problems as we enter the 21st Century. What remains to be seen is the extent to which transport emissions of air pollution are responsible for this, and which modes of transport cause the most or the least generation of ground-level O_3.

The widespread impression that visibly clean air is genuinely clean thus seems to have disappeared in the last two decades of the 20th Century, and unlike in the 1950s, transport is receiving the most attention as a source of air pollution. The fact that modern transport-related air pollution is largely invisible seems to be resulting in it not being ignored but instead in it being more frightening, rather as invisible ionising radiation has become a subject of much fear and suspicion in most societies. Added to this is the visible congestion, noise, stress and other inconvenience and annoyance that is the result of unrestrained growth of transport systems in nearly all cities (Forsberg et al., 1997; Lercher et al., 1995; Williams and McCrae, 1995), resulting in pressure for change that is probably irresistible. The remainder of this review will look in detail at three examples of environmental impact of air pollution emissions from individual modes of transport, to investigate whether current priorities for change are supported by scientific evidence.

4. Case studies

In this section, three contrasting examples will be examined in depth to illustrate the issues involved in quantifying the impacts of air pollution emissions from transport by land, air and sea. The currently highest profile example of road traffic contribution to the effects of urban air quality on human health is considered first, with an emphasis on particulate matter as the pollutant currently causing at least as much concern over health effects as any other. This is then compared with the impact of aircraft emissions on the global atmosphere. Finally, sulphur pollution from ships in Europe will be used as an example of emissions abatement policy to reduce acidification being applied to the transport sector. The aim is not to identify all the most significant impacts of transport on air quality, as some impacts that are not considered may be more important than those we focus on. Notably, rail transport is omitted almost completely from this review. The reason for this is not that its impacts on air quality are slight (indeed, its net impact is benefit if one takes into account road traffic reduction achievable by increased rail use), but the major issues concerning emissions, source–receptor relationships and multi-pollutant multi-effect analysis are illustrated adequately by the examples that are discussed in depth. Our discussion of urban road transport has been introduced with particular reference to the private car, although light goods, heavy goods and public service vehicles also contribute to air pollution. Our detailed discussion of goods transport will be limited to marine shipping and our discussion of commercial passenger transport limited to air traffic. For each example that we do consider in depth, the main issues that determine the nature and magnitude of the impact are reviewed, and a conclusion is reached concerning the extent to which we are currently capable of quantifying the impact. The aim is that these examples can then stimulate similar future analysis of impacts of other transport sub-sectors on other receptors as and when required.

4.1. Road traffic and effects of urban air quality (especially particulate matter) on human health

4.1.1. Factors determining magnitude of transport impact

Road transport is distinguished from other sources of air pollution, as mentioned already above, in that the emissions are released in very close proximity to human receptors. This reduces the opportunity for the atmosphere to dilute the emissions which would render them less likely to damage human health. Furthermore, in most city centre atmospheres, concentrations of vehicle exhaust are significantly enhanced by the fact that many roads have buildings alongside. The effect of such buildings is to shelter the road, reducing the wind speed at the source of emissions by as much as an order of magnitude relative to that on an open road. The contribution of emissions from traffic on that road to kerbside pollutant concentrations is increased by approximately the same factor. Such enhancement of transport emissions often has little impact on total daily population exposure to a given pollutant, largely because people spend much larger amounts of time indoors (Jantunen et al., 1998). Nevertheless, much air pollution work has focussed on the outdoor environment as individual citizens have less control over the air they breathe outdoors than they do in their own homes, and the high levels of air pollution in city street canyons coincide with noise, smell, dust and traffic congestion that people find unpleasant leading to further enhanced concern about possible health effects. Furthermore, the major impact of road traffic emissions on human health can occur inside the buildings that line city streets, where concentrations of pollutants from road traffic are determined largely by the outdoor concentration adjacent to windows and doors (for example, Kukadia and Palmer, 1998).

4.1.2. Current ability to quantify impact

Flow and dispersion patterns in two-dimensional city streets have been studied in the field and using computational modelling by Johnson et al. (1973) and Dabberdt et al. (1973), and in the wind tunnel by Yamartino and Wiegand (1986) and others. A semi-empirical operational model for a long street bounded by equal height buildings on either side has been developed by Berkovicz et al. (1997), and is now being increasingly used in air quality management especially in Europe (McHugh et al., 1997). Such modelling indicates that time-averaged concentrations vary by as much as a factor of two to three over distances as short as a few metres on the road, introducing the potential for different road-users (for example, cyclists versus car drivers) to be exposed to rather different levels of air pollution. Instantaneous concentrations exhibit greater variability associated with emissions from individual vehicles coupled with fluctuations in atmospheric turbulence, giving rise to further enhanced exposure of road users who preferentially occupy the most polluted parts of the road such as a cyclist in the slip-stream of a bus, but these transient phenomena are very difficult to model computationally. Even for time-averaged concentrations, extension of the simple idealised two-dimensional street canyon case to the simplest three-dimensional situation of an intersection of two building-lined streets (Hoydysh and Dabberdt, 1994; Scaperdas and Colvile, 1999) or unequal building heights (Hoydysh and Dabberdt, 1988) increases the complexity considerably. *CAR–International* (den Boeft et al., 1996) is an empirical model that does attempt to take some two-dimensional building shape factors into account when calculating annual average roadside pollutant concentrations. An alternative approach is to model real urban geometry computationally (for example Hunter et al., 1992; Lee and Park, 1994). In theory, such fluid dynamics models are capable of reproducing any urban geometry at any spatial resolution over any area, but in practice finite computational resources limit them to single street canyons or small groups of buildings, with buildings often represented as simple regular cuboids. To cover a larger area of a city, building-resolving computational fluid dynamics models will soon be nested within overlying meteorological boundary layer models.

In view of the complications and uncertainties that remain in high-resolution urban air quality modelling, most assessments of human exposure to date have used measurement, not modelling. The simplest approach is to use data from a single city-centre or suburban background air quality monitoring station as a surrogate for the daily level of air pollution to which the whole population of a city is exposed. This will be much more accurate for a pollutant such as PM_{10} that has major distant sources (as discussed in Section 2) for a pollutant such as CO or NO_x that is predominantly emitted by local road transport. For people such as children or the elderly who often spend all day in the urban or suburban back-street environment, using background air quality monitoring data will be a good approximation for exposure even to these traffic-related pollutants, but is less accurate for working populations who can spend as much as $3 \, h \, d^{-1}$ commuting. A roadside monitoring station gives a first indication of the extent to which such roaduser exposure is higher than the urban average, and will also provide a measure of the exposure of people who live or work alongside busy roads. Each individual roadside location is unique, though, so that it is impossible to obtain any sort of concentration map (as is provided by a dispersion model) without using a very dense network of measurements indeed. This has been attempted in a few studies (for example, Briggs et al., 1997), but several have gone one step further and measured the exposure of roadusers themselves, using air pollution monitoring equipment small and lightweight enough to be carried by a person as they go about their daily life or as they travel by car, bicycle or public transport (Monn, 2001). For example, Sitzmann et al. (1996) found that cyclists in London are exposed to concentrations of particulate air pollution significantly higher than those measured by fixed roadside air pollution monitors; Chan et al. (1991) found that commuters in Massachusetts were exposed to much higher levels of non-formaldehyde VOCs inside cars than when in subway electric trains, walking or cycling, and similar results may be found in a review for the Institute for European Environmental Policy (DETR, 1997). The Europe-wide EX-POLIS study has recently completed measurements of total daily exposure of 451 volunteers in six cities, with application of statistical methods to attribute total exposure to the sum of the different microenvironments through which the volunteers move (Jantunen et al., 1998), including transport microenvironments.

The most accurate method of assessing human exposure to air pollution is biological measurement. For example, exposure to 20 ppm of CO (such as might still be encountered in the most confined and heavily trafficked areas of European Cities, such as road tunnels, and which still commonly occurs in many cities in developing countries) will cause blood levels of carboxy-haemoglobin to rise to an equilibrium level of 3.2% in about 8 h if a person is carrying out light activity, or 4 h during more strenuous exercise (Forbes et al., 1945, cited in EPAQS, 1994). For lead, a blood sample reveals the level of exposure over a longer time period, and a rise from 10 to 20 $\mu g \, dl^{-1}$ has been found to be associated with a loss of up to two Intelligence Quotient points (EPAQS, 1998).

Using biological sampling or personal exposure monitoring, however, it is only possible to measure the exposure of a small number of people. To assess accurately the variability of exposure of entire populations, either a very large number of exposure measurements are required (as in EXPOLIS) followed by a statistical analysis of how exposure relates to daily lifestyle, or high-resolution mapping of the spatial and temporal variability of air pollution concentration must be used. There are now a few examples of high-resolution mapping techniques being applied to the assessment of exposure from road traffic, either empirically (Briggs et al., 1997) or more theoretically (Khandelwal, 1999; Grossinho et al., 1999). Similar methodology has been used somewhat more widely at lower spatial resolution for larger sources, for example McGavran et al. (1999) and Ihrig et al. (1998). The most sophisticated operational urban air quality models are probably now capable of starting to assess the exposure of moving roadusers as a function of the amount of time they spend in more or less polluted streets.

Quantification of the effect of urban road traffic pollution on human health can be attempted using any measure of individual or population exposure and correlating that with records or observations of the incidence or severity of, or mortality from, disease. A variety of designs of epidemiological study exist, looking at whole populations (ecological study) or closely monitored small groups of subjects (cohort study), and examining the effects of variations in air pollution concentrations in time or space. Time-series analysis can detect only short-term effects of air pollution, while geographical methods can also pick up chronic effects (Elliott et al., 1992). Usually, the statistical power of a large ecological study is required to detect the very small air pollution signal against the noise of other variability in health and the factors that influence it, such as weather and virus epidemics. Some of the exposure assessment methodologies outlined above for road transport pollution are more suitable for certain designs of epidemiological study than others, for example urban background monitoring for a time-series study, personal monitoring or biological sampling of a cohort, or high-resolution dispersion modelling for an ecological small-area geographical study.

For a pollutant such as CO, where most spatial and temporal variability in outdoor concentrations is due to road transport emissions, an observed relationship between air pollution levels and health can more easily be equated to a relationship between road transport emissions and health. For other pollutants, however, most studies look at the impact of a pollutant that has several sources of which road transport is only one. PM_{10} is an extreme example of this, where road traffic exhaust can be responsible for a rather small fraction of the total concentration, as discussed in Section 2. Similarly, for lead, even though road traffic exhaust particulate matter was formerly the main source in most urban atmospheres, there are many pathways of exposure in addition to inhalation of vehicle exhaust, including ingestion from old lead paint, in drinking water from lead pipes, and from dust deposited in carpets ingested during hand-to-mouth activity (especially for children). Not only other sources of air pollution but also other causes of variations in health need to be taken into account before the impact of road traffic emissions can be isolated. Where there is a high degree of correlation between these and the pollutant of interest, correction for confounding requires sophisticated statistical techniques. A major confounder in time-series studies is the effect on health of temperature changes associated with air pollution episodes. In a geographical study, it is necessary to correct for how low income rather than poor air quality is often a cause or consequence of ill health close to a pollution source such as a major road (Dockery, 1993; Schwartz et al., 1996). To circumvent all the problems of source apportionment and exposure pathway (but still leaving socio-economic confounding to be corrected for), there are a few small-area studies of geographical variations in health that look at road transport emissions in general instead of a single specific pollutant, or even a parameter such as distance of place of residence from a major road, to obtain a more direct measure of the association between road traffic and health (for example, Briggs et al., 1997).

The results of epidemiological studies can be applied to current air quality statistics to estimate the magnitude of the impact of air pollution on health. The World Health Organisation (WHO) has produced meta-analyses for the effects on mortality and morbidity of a number of pollutants (for example, WHO, 1996). Their effect estimates have been used by others to calculate aspects of the burden of poor health attributable to pollution. For example, in the UK, COMEAP (the UK Department of Health's Committee on the Medical Effects of Air Pollutants) calculated that PM_{10} was associated with 8100 deaths brought forward and with 10,500 emergency hospital respiratory admissions (brought forward and additional) in urban areas of Great Britain. The corresponding figures for SO_2 were 3500 deaths brought forward and 3500 early and extra hospital admissions. The effects of ozone were 700 deaths and 500 admissions if there is no health effect below 50 ppb, but 12,500 and 9900 if there is no threshold (COMEAP, 1998). The risk due to ozone is higher for residents of rural areas because urban road traffic emissions of NO_x scavenge ozone in cities. Various attempts have been made to quantify the economic value of such impacts on individuals, despite the very large uncertainties involved. Maddison and Pearce (1999), Ostro et al. (1999), DoH (1999), Spadaro et al. (1998) and Bickel et al. (1998) used exposure response functions derived from epidemiological studies to estimate the proportion of health endpoints, such as hospital admissions, attributable to air pollution, and then used inferred prices based on contingent valuation studies to calculate the value people attach to these health endpoints. Reports prepared for the World Health Organisation Ministerial conference on Environment and Health in London in June 1999 (Künzli et al., 2000) considered the chronic effects of air pollution (Künzli et al., 1999), population exposure to PM_{10} (Filliger et al., 1999) and an economic evaluation of the health effects (Sommer et al., 1999). These found that Austria, France and Switzerland bear almost €50 billion of air pollution related health costs, of which a little under €30 billion is related to road traffic. In the USA, Ostro and Chestnut (1998) calculated that the annual health benefits of achieving new standards for $PM_{2.5}$ relative to 1994–1996 ambient concentrations in the USA are likely to be between $14 billion and $55 billion annually, with a mean estimate of $32 billion. A major difficulty in quantifying the health impact of air pollution is that a very large number of people are exposed to relatively low levels over long periods of time, resulting in slight or rare health problems that are difficult to value or difficult to attribute to a given source of pollution, as illustrated in Fig. 3.

The examples cited above are estimates of the cost of health effects of current levels of certain pollutants for all sources, and for SO_2 and PM_{10} road transport is far from being the largest contributor to concentrations in most cities. Eyre et al. (1997) used emissions-based dispersion modelling to estimate the exposure of the population of London specifically to road transport emissions and compared this with other impacts. Their results reproduced in Table 2 suggest that urban diesel particulate emissions have by far the most significant impact of all road transport emissions. Interestingly, the next most significant impact is secondary nitrate particles formed from emissions of NO_x. The current European trend towards larger fractions of PM_{10} being composed of nitrate formed from NO_x, much of which is of road transport origin, is a trend towards the health effects of PM_{10} becoming increasingly an impact of road transport emissions.

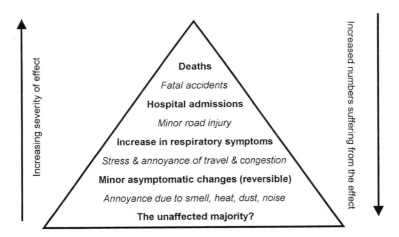

Figure 3.   The range and scale of health impacts associated with ambient air pollution (adapted from Walters and Ayres (1996), bold upright type), compared with range and scale of similar examples of impacts of road transport other than those attributable to exhaust emissions (light italic type).

Table 2. Damage costs of transport emissions (from Eyre et al., 1997)

| Emission | Impact | Damage costs in £0.01 km^{-1} (£0.6 ≈ $1 or €1) | | | | | |
| | | Rural emissions | | | Urban emissions | | |
		Petrol	Gas	Diesel	Petrol	Gas	Diesel
Carbon dioxide	Global warming	0.09	0.07	0.07	0.1	0.09	0.1
Methane	Global warming	< 0.001	0.005	< 0.001	< 0.001	0.006	< 0.001
Nitrous oxide	Global warming	0.003	0.003	0.001	0.006	0.006	0.001
Carbon monoxide	Global warming	0.001	0.001	< 0.001	0.003	0.001	0.001
Particulates	Health	0.003	< 0.001	0.2	0.003	< 0.001	1.7
Particulates	Buildings	< 0.001	< 0.001	0.003	< 0.001	< 0.001	0.04
Sulphur dioxide	Health	0.02	0.001	0.01	0.2	0.001	0.2
Sulphur dioxide	Crops	< 0.001	< 0.001	< 0.001	< 0.001	< 0.001	< 0.001
Sulphur dioxide	Timber	0.02	0.001	0.01	0.02	0.001	0.02
Sulphur dioxide	Buildings	0.005	< 0.001	0.003	0.04	< 0.001	0.04
Sulphate aerosol	Health	0.03	0.001	0.02	0.04	0.001	0.03
Oxides of nitrogen	Health	0.01	0.007	0.03	0.08	0.05	0.1
Oxides of nitrogen	Timber	0.02	0.01	0.05	0.04	0.02	0.05
Oxides of nitrogen	Buildings	0.006	0.003	0.013	0.034	0.024	0.051
Nitrate aerosol	Health	0.1	0.06	0.2	0.2	0.1	0.2
Ozone from NO$_x$	Health	0.05	0.03	0.1	0.07	0.05	0.1
Ozone from NO$_x$	Crops	0.003	0.001	0.006	0.004	0.003	0.006
Benzene	Health	0.01	< 0.001	0.004	0.1	0.001	0.05
Ozone from VOC	Health	0.1	0.02	0.02	0.1	0.02	0.04
Ozone from VOC	Crops	0.006	0.001	0.001	0.008	0.001	0.002
Non-methane VOC	Global warming	0.003	< 0.001	< 0.001	0.003	< 0.001	0.001
Totals		**0.5**	**0.2**	**0.7**	**1.1**	**0.4**	**2.7**

4.1.3. Sources of uncertainty and implications for transport

Despite the very large uncertainties in all valuations of impacts of air quality, the larger sums estimated in such studies have led to particulate air pollution causing at least as much concern over its effect on human health than any other ambient air pollutant world wide. The choice of PM$_{10}$ as the measure of particulate air pollution to be controlled is based on a biological plausibility argument given the aerodynamic characteristics of the human respiratory tract. The strength of the evidence for health effects of PM$_{10}$ in general may be assessed against the criteria proposed by Bradford Hill (1965), which are listed in Box 1. Much less certain is the extent to which primary PM$_{10}$ from road transport exhaust is responsible for the health effects of PM$_{10}$. The uncertainty here is so great that current action to control diesel particulate emissions in Europe (the focus in the US is much more on regional secondary PM$_{10}$ from large combustion sources) must be described as precautionary.

If an estimate of population exposure used in an epidemiological study is subject to error, this will cause the observed relationship between air quality

Box 1: Use of epidemiological evidence to infer the existence of a cause-and-effect relationship

Even a perfect epidemiological study cannot prove the existence of a cause-and-effect relationship between vehicle emissions and ill health, but it can contribute to the decision that an observed relationship is more likely than not to be causal. Other aspects (Bradford Hill, 1965), with examples for PM_{10}, are

- the cause must precede the effect (Abbey et al., 1995; Thurston et al., 1994);
- experimental evidence exists (such as Smith and Aust's (1997) laboratory measurement of free radicals generated in the lung by PM_{10});
- a physiological explanation of why damage might be expected to occur (Gilmour et al., 1996);
- coherence – demonstration of effects across the range of severity (for example, effects of PM_{10} on healthy lung function (Scarlett et al., 1996) and in individuals with asthma (Timonen and Pekkanen, 1997), symptoms (Braun–Fahrländer et al., 1992), medication use (Pope et al., 1991), emergency attendance at hospital for asthma (Castellsagué et al., 1995), hospital admission (Brunekreef et al., 1995) and mortality (Dockery et al., 1992) especially cardiovascular (Schwartz, 1993) and respiratory (Zmirou et al., 1996);
- consistency of results from different epidemiological study designs (for example time-series studies (Schwartz et al., 1996; Wordley et al., 1997) and examination of the causes of death in people dying on high pollution days (Schwartz, 1994));
- specificity of the effects (Bremner et al., 1999; Schwartz et al., 1993);
- demonstration of a biological gradient (Schwartz et al., 1993; Wordley et al., 1997).

and health to be an underestimate of the true effect (Elwood, 1988). Critically, concentrations of pollutants from non-road traffic sources tend to exhibit much less spatial variability than those from road networks, for example PM_{10} tends to be fairly constant over wide areas while NO_2 can vary by an order of magnitude over a hundred metres or so in a residential area close to a few major roads, such that it is much easier to assess exposure to PM_{10} accurately. This raises the prospect of PM_{10} appearing wrongly to be more strongly associated with health effects than NO_2, as discussed by Fairley (1990), on account of the effect of the road traffic related pollution being diluted by exposure misclassification in certain designs of epidemiological study (Lipfert and Wyzga, 1995).

The smallest diesel exhaust particles do not enter the human lung very easily because they undergo Brownian diffusion to the nose and throat, but the larger ones are close to the size that can penetrate the deepest into the alveolar regions of the lung where gas exchange with the blood occurs. Particles several micrometres in size from mechanical sources such as resuspension of road dust can dominate total PM_{10} mass but most of these are likely to be intercepted by impaction in the nose. In recent years, toxicological laboratory studies predominantly carried out on rats have been driving interest towards smaller particles (for example, Peters et al., 1997; Ferin et al., 1991; Li et al., 1999) and future legislation is currently expected to focus either on smaller particle mass fractions $PM_{2.5}$, PM_1, $PM_{0.1}$ or on particle number. This potentially has a very great impact on urban road traffic, especially if epidemiological studies can be designed to differentiate between the effects of diesel exhaust particles and the secondary acid sulphate and nitrate particles that are imported to an urban area from distant sources. The laboratory studies already show that particles from vehicle exhaust may not be the most toxic fine particles in the urban atmosphere, for example, quartz present in resuspended road dust appears to be much more toxic than diesel exhaust (Murphy et al., 1998), but if primary particulates from road traffic exhaust continue to be blamed for the observed health effects of PM_{10}, precise details of which particle sizes and composition are most important will determine which combination of fuel type, engine type and end-of-pipe emissions abatement technology is likely to be most effective at reducing impacts on human health. Such detailed information is currently unknown.

Finally, it must be noted that the chronic effect of particulate air pollution is potentially much larger and less socially acceptable than the acute, but is often omitted from attempts to quantify benefits of pollution control because estimates of the chronic effect are the most uncertain of all.

4.2. Impact of aircraft emissions in the upper troposphere and lower stratosphere on global atmospheric change

4.2.1. Factors determining magnitude of transport impact

The considerable visible and noise impact of large jet aircraft has resulted in their being considered frequently as potential significant sources of ground-level air pollution in the vicinity of major airports. Several studies (for example, ERL, 1993) have shown, however, that the emissions of the aircraft themselves contribute rather little compared with the great volumes of road traffic that large airports generate, plus other airport-related surface sources of air pollution. Even though an airport itself, typically located outside urban areas, can be the largest source of emissions in the vicinity, those from the aircraft themselves are efficiently dispersed before they reach the ground in the same way as the emissions from tall chimneys that were discussed in Section 2. Future reductions in road traffic emissions and growth of the air transport industry may mean the aircraft contribution to ground-level air quality will become more significant relative to other sources, especially as pressure on land for expansion produces a tendency for increased airport development in cleaner air further away from the major cities that airports serve. Meanwhile, however, to find the most significant contribution of aviation to air pollution, we must look to higher altitude, where the lower atmospheric pressure and lack of other nearby anthropogenic sources of trace gases and particles means that a given volume of emissions can have a much greater impact.

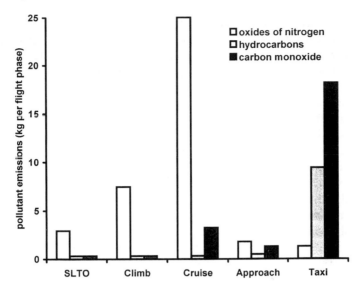

Figure 4. Pollutant emissions from aircraft during phases of flight (after Houghton (1994), citing Gried and Simon (1990)) (Estimates were made of emissions of NO_x, hydrocarbons and CO during the landing and take-off (LTO) cycle and cruise phase of a General Electric CF6-50C engine with a 30 min period of cruising at Mach 0.85 and an altitude of 10.7 km. SLTO = ICAO Standard Landing and Take-Off cycle.)

Fig. 4 shows that hydrocarbons and carbon monoxide are emitted from jet aircraft engines predominantly on the ground. The major trace gas emission during flight is oxides of nitrogen (NO_x). In addition, much larger quantities of carbon dioxide and water vapour are emitted, which illustrate some important issues of temporal evolution of emissions and impact for this rapidly expanding mode of transport. Furthermore, soot particles in the exhaust need to be considered for their roles as ice nuclei and in heterogeneous atmospheric chemistry.

The impact of these depends on how high the aircraft is flying. About half the NO_x emissions from sub-sonic aircraft occur at the main cruising altitude of 10–12 km. Since the top of the free troposphere varies from about 8 km in polar regions to 16 km in the tropics, subsonic flight is in the lower stratosphere at high latitudes and in the troposphere elsewhere. On busy North Atlantic routes, as much as 75% of the total fuel per flight may be used in the stratosphere (RNMI, 1994, cited in Houghton, 1994). Supersonic aircraft cruise higher, always in the stratosphere, but at the time of writing the tragic crash of the Air France Concorde seems to be indicating that civilian supersonic flight is unlikely to return to our skies for several years at least.

In the free troposphere, emissions of NO_x lead to the formation of ozone (Jenkin and Clemitshaw, 2000). This O_3 can be mixed down to ground level and contribute to regionally poor air quality during photochemical air pollution episodes in what is gradually becoming a global air pollution problem, and it is also a greenhouse gas. Formation of O_3 by NO_x is one or two orders of magnitude more efficient in the free troposphere than at the surface, and about 50% more efficient in the Southern Hemisphere than in the North, because of the cleaner air there. This latitudinal variation compensates in part for the fact that there is more air traffic in the Northern Hemisphere. The result of all these nonlinearities combined is that free troposphere aircraft emissions of about 3% of total NO_x emissions probably account for an approximately equal amount of global warming as total surface NO_x emissions (Johnson et al., 1992). This effect is reduced, however, by the way NO_x leads to increased levels of photochemical oxidants and hence shorter atmospheric life-time of methane.

In the stratosphere, the chemistry of ozone is totally different. At middle and low latitudes NO_x is involved in catalytic cycles that destroy the protective ozone layer and can allow dangerous ultra violet radiation from the sun to reach the surface of the Earth. Even though emissions of sulphur dioxide and soot particles from aircraft are negligible compared with total global emissions, their potential to damage the atmosphere is even more enhanced in the stratosphere relative to at the surface than is the case for NO_x. Sulphur dioxide in the stratosphere becomes oxidised to form droplets of sulphuric acid, and these with soot particles from aircraft exhaust promote heterogeneous chemical reaction cycles that destroy O_3. These stratospheric clouds also promote the conversion of NO_x to nitric acid (HNO_3), thus lessening the potential of the NO_x to destroy O_3 by gas-phase chemical reactions, but at very low temperatures the HNO_3 itself can form droplets with then add to the heterogeneous chemistry that destroys O_3. Depletion of stratospheric ozone has a cooling effect on climate, partially offsetting the warming effect of NO_x from aircraft in the troposphere.

In comparison to these two indirect impacts of aircraft emissions on global atmospheric chemistry, the direct effect of CO_2 from aircraft causing climate warming due to the ability of CO_2 to absorb outgoing infra-red radiation is conceptually simple. When comparing different impacts of aircraft upon the global atmosphere with each other, and with the effect of emissions from other transport sectors and non-transport related activity, the most challenging aspect of CO_2 is perhaps the time scale over which it has an effect. CO_2 is chemically sufficiently unreactive for its dominant removal process to be physical. Solution in the water of the upper ocean and exchange of carbon between the atmosphere and terrestrial biomass are relatively rapid, with the combined annual flux amounting to 20% of the atmospheric carbon reservoir mass of 750 Gt (Houghton et al., 1996), but these fluxes are bi-directional. The rate determining step for net removal of carbon is mixing from the surface and intermediate ocean to the much larger carbon reservoir of the deep oceans. At the turn of the 21st Century, anthropogenic carbon emissions of 7–8 Gt yr^{-1} (including deforestation) are greater than the equilibrium rate of removal at current atmospheric and surface ocean concentrations, such that an amount of carbon equal to around half the emissions each year is removed and the imbalance results in a steady increase in atmospheric carbon dioxide levels. Were emissions to remain constant at today's rate, the atmospheric concentration would reach an equilibrium level about one-third higher than today's value towards the end of the 21st Century. The global total emissions of CO_2 from aviation in 1990 was about 450 million tonnes of carbon (Barrett, 1991), which was less than 20% of global road transport emissions and about 3% of total anthropogenic emissions. Furthermore, historical emissions of CO_2 from aviation are almost zero going back just a few decades into the mid-20th Century, while around half the carbon dioxide from all anthropogenic sources currently in the atmosphere was emitted before 1980, so the overwhelming majority of the total is from non-aviation sources. The small contribution of aviation is, however, increasing, and the small amounts of CO_2 being emitted by aircraft now will remain in the air for many decades.

Finally, water vapour from jet engines can also form line-shaped clouds in the free troposphere. The temperature of these clouds is lower than that of Earth's surface, so their black body radiation is less than what would be emitted from Earth's surface were the clouds not there, resulting in warming. This is more significant than the cooling effect of the clouds reflection of incoming solar radiation, so that overall the contrails have a warming effect on climate at the surface. Usually, contrails evaporate again within minutes or even seconds such that their impact is negligible, but under certain meteorological conditions they can be sufficiently persistent for a large part of the sky to become obscured continually along a major flight path until weather conditions change many hours or days later. In the stratosphere, contrails are never persistent because of the low ambient relative humidity there, although the water vapour from aircraft is not removed rapidly by precipitation as it is in the troposphere so has a small warming effect on climate because of its greenhouse gas properties.

4.2.2. Current ability to quantify impact and major sources of uncertainty

In theory, the impact of aircraft emissions on upper troposphere and lower stratosphere chemistry can be quantified using global models of circulation and chemistry (such as Johnson et al., 1999). In practice, however, despite the fact that the reaction mechanisms are now qualitatively understood, quantifying the impact of aircraft emissions remains elusive. There are two main reasons for this.

Firstly, the chemical reaction cycles are complex, as different gas-phase and heterogeneous pathways become more important at different temperatures. Small errors in the predicted mix of different pollutants can propagate via resulting errors in the relative rates of two or more competing reactions to end up with quite unrealistic simulated O_3 concentrations. Not only must the chemical composition of the upper troposphere and stratosphere be simulated accurately, but rates of mixing between layers as well as chemistry determine the composition, the temperature needs to be known to determine where heterogeneous processes occur, and the temperature has a large influence on the mixing. The whole process of stratospheric O_3 destruction in particular is a highly nonlinear catastrophic process.

Secondly, emissions of aircraft in the upper troposphere and stratosphere occur along highly localised flight paths that vary in time and space. The physical size of these is much less than the resolution of the global-scale models that are required to simulate chemistry in the upper troposphere and stratosphere. This problem of scale is added to the fact that the total emissions from aircraft are at least as difficult to quantify as emissions for road traffic are on the ground. It is exacerbated by the fact that other sources of the same pollutants in the upper troposphere and lower stratosphere, such as lightning and mixing from the lower troposphere, are also very difficult to quantify accurately.

Any one of these difficulties would make calculations of the total atmospheric impact of aircraft emissions liable to error. Combined, they present a very formidable challenge indeed for the science of atmospheric chemistry modelling. The most recent calculations indicate that the effect of aircraft NO_x emissions on producing O_3 in the upper troposphere/lower stratosphere is greater than the effect of sulphur and soot emissions on destroying O_3, except at high latitudes (Penner et al., 1999). The greatest overall expected change in O_3 due to aircraft emissions is thus an increase of about 6% in the region 30–60° N at 9–13 km altitude. Observational evidence of this is very difficult to find, because O_3 variability is high near the tropopause and the expected 20% increase in NO_x due to aircraft emissions is substantially smaller than the observed variability. In presenting these results, the IPCC Working Groups stress that the models include some notable deficiencies in the physics and the gas-phase and heterogeneous chemistry of the problem. To reduce the uncertainties, a concerted effort is therefore required, combining model development with detailed field observation campaigns that recent instrument development and improvement have made possible.

Quantification of the direct climate change impacts of aircraft through their CO_2 emissions is arguably not so fraught with difficulty. The global climate models that are now in an advanced state of development are, nevertheless, extremely complex. A discussion of these and uncertainties therein is beyond the scope of this review. The sub-grid size of contrails however presents some problems similar to those discussed for NO_x chemistry, and the sensitivity of contrail persistence to meteorological parameters presents a test of model accuracy in regions where such performance requirements have not been demanded before and where validation data are sparse, on top of the effects of large uncertainty surrounding ice nucleation processes.

A significant area of debate is concerned not so much with the accuracy of our predictions about the impact of aviation on climate, but with how to respond to the implications. In terms of the size of the perturbation to the radiation balance, the latest calculations are sufficiently accurate to indicate that contrails have at least as large an impact on climate as CO_2 emissions from aviation, and that the radiative forcing due to contrails could be several times larger if less conservative estimates of the more uncertain contrails impact are used. If the formation of persistent contrails could be prevented, however, the radiation balance of the atmosphere would respond immediately, in contrast to CO_2 from aviation in the 20th Century which will remain in the atmosphere for several decades. Integrated over a long future time horizon, CO_2 emissions thus may be considered the most significant impact of aviation on the global atmosphere. Faced with a hypothetical choice between preventing contrails and reducing fuel consumption, the decision must therefore be whether to make a small contribution to the solution of a long term but possibly severe problem or if a larger, instantaneous benefit is sufficiently desirable to be worth paying for in the latter part of the century.

In contrast to the human health effects of urban air pollution, where economic valuation was attempted at least for the acute effects, assessments of the implications of global atmospheric change are mostly qualitative. A most comprehensive example of this is Watson et al. (1996), which catalogues a wide range of impacts of climate change, but does not attempt to judge whether or not these impacts constitute "dangerous anthropogenic interference with the climate system" on the grounds that definition of what is "dangerous" is a political not a scientific judgement. For the impact of stratospheric ozone depletion on human health, which is expected to result in increased incidence of skin cancer over several decades, the magnitude of the effect is very difficult to quantify, for similar reasons to the difficulty in quantifying chronic effects of urban air pollution on human health.

4.3. Controlling acidification: sulphur emissions from ships

4.3.1. Factors determining magnitude of transport impact

The acidifying potential of the polluted atmosphere is enhanced relative to the clean atmosphere by the oxidation of oxides of sulphur and nitrogen. Wet and dry deposition of sulphuric and nitric acid, together with ammonia, contribute to the acidification of a wide range of ecosystem types. As has already been discussed briefly in Section 3, catastrophic damage to upland forests attributable to acid deposition led to land-based European emissions of sulphur being reduced by 40% between 1980 and 1993 (Barrett and Seland, 1995). These emissions are principally from large coal-fired combustion plant, and reductions have been achieved by changing to a fuel with a lower sulphur content (including natural gas) or fitting flue-gas desulphurization abatement technology. At the time of writing, the Second Sulphur Protocol (UNECE, 1994) has been ratified by 22 of the 28 parties, whilst the Gothenburg Protocol to abate acidification, eutrophication and ground-level ozone has recently been signed by 31 parties. The Gothenburg protocol (UNECE, 1999) requires, by 2010, overall European emission cuts of 41% in oxides of nitrogen (from 23,330 to 13,846 kt yr^{-1}), and 18% in ammonia (from 7653 to 6280 kt yr^{-1}), and 43% in VOCs (from 23,964 to 13,590 kt yr^{-1}) relative to 1990 (adopting the geographical scope of RAINS, 2000). This protocol thus addresses the fact that emissions of NO_x have remained almost constant, as have emissions of ammonia (although the estimates of ammonia emissions are subject to large uncertainties). This means that nitrogen is now more significant than it has been in the past, and the major ground-level transport source of oxides of nitrogen is road traffic. The need to comply with the Gothenburg protocol, subsequent agreements such as the EU Daughter Directive on Tropospheric Ozone, and also urban air pollution problems, focusses the emphasis on road traffic as a major source of air pollution, as discussed at length with respect to particulates and health in Section 4.1. The Gothenburg Protocol will reduce sulphur emissions on land to 63% of their 1990 levels, from 38,040 to 13,990 kt SO_2 yr^{-1}. This includes some large reductions from some countries e.g. the UK (from 3805 to 625 kt SO_2). Attention on shipping as a potential contributor to acidification was increased following the publication of an updated inventory of shipping emissions of sulphur dioxide which demonstrated that these were greater than had previously been thought (Lloyd's Register of Shipping, 1995). For example, emissions from shipping in the North Sea were estimated at 439 kt SO_2 in 1990, which is comparable to the UK emission ceiling under the Gothenburg protocol. Although the relative contribution made by shipping to deposition of sulphur is thus much higher than in the past, the control of shipping emissions lies outside the remit of the UN ECE protocols under the Convention on Long-range Transboundary Air Pollution. Therefore, shipping remains a significant contributor to acid deposition, although arguably less so than the oxides of nitrogen emitted by road traffic. The argument that these emissions from marine transport should also be reduced will therefore be based on calculations of the costs and benefits associated with such measures.

There are several reasons why further reductions of acidification have been recommended for Europe. It has been noted that wet deposition fails to respond as rapidly to emission reductions than dry deposition (Downing et al., 1995). This is especially acute in upland areas where wet deposition is enhanced by the presence of orographic cloud, as small, highly acidic droplets in low-level cloud formed by forced uplift of polluted air over hills are washed out by cleaner rain falling from aloft (Weston and Fowler, 1991; Inglis et al., 1995; Dore et al., 1999). Areas of alkaline geology (for example, limestone) are able to withstand anthropogenic acid deposition many times greater than the natural background. Areas of already acid geology however (for example, granite), start to show signs of disruption of plant and animal life with much smaller changes in the rate of input of acidity. The amount of acid deposition that an area can tolerate is described in terms of a critical load, specifically the rate of acid deposition below which significant harmful effects on specified sensitive elements of the environment do not occur according to present knowledge (Posch et al., 1997). Deposition measurements and model calculations indicate that acid deposition will not be reduced to below the critical load everywhere in Europe even if the best available emissions abatement technology is applied to every sulphur source on land. Furthermore, if the necessary reductions in deposition could be made, it would take decades for the ecosystems to recover, particularly where base cations have been leached out (Stoddard et al., 1999).

4.3.2. Current ability to quantify impact

Early attempts to reduce atmospheric sulphur emissions were based upon equal abatement of industrial emissions everywhere in Europe. It was soon realised, however, that greater environmental benefit could be achieved at less cost if application of emissions abatement measures was concentrated in parts of Europe that are responsible for causing acidification damage to the most sensitive ecosystems. Initially, it is possible to apply relatively inexpensive abatement measures, but as tighter emissions limits are demanded, application of further technology becomes increasingly expensive and brings less additional benefit. The decision as to where these increasingly costly measures should most effectively be applied is informed by the results of integrated assessment modelling. Two examples of integrated assessment models (IAMs) are RAINS (Alcamo et al., 1990; Amman et al., 1999) and ASAM (ApSimon et al., 1994; ApSimon and Warren, 1996; Warren and ApSimon, 1999). These models use source–receptor relationships representing the long-range transport of pollutants originating from each individual country to each of several hundred 150 km grid cells. The matrices are calculated using a detailed meteorological model with a simplified chemical scheme, namely the EMEP Lagrangian model (Barrett and Seland, 1995). Modelled acid deposition is then compared with the critical load in each grid cell, which varies greatly geographically. Since these critical loads are not attainable even if Best Available Technology were to be applied across Europe as a whole, it is necessary to seek ways of partially closing the gap between deposition and critical load. In doing this it is important to maintain a degree of equity by seeking to attain the same percentage of gap closure in each 150 km grid cell (Warren and ApSimon, 2000a). IAMs find the most cost-effective means of doing this by linking the source–receptor relationships with information concerning the cost of reducing emissions in each country. It is these IAMs that were used to guide policy makers towards cost-effective strategies for emission abatement in their preparations for the Gothenburg protocol, (the RAINS model being the official model, whilst the ASAM model has been used for sensitivity analyses and model comparison studies).

In 1998 a major study was commissioned by the UK Department of Environment, Transport and the Regions to investigate the cost-effectiveness of control of shipping emissions in the North Sea and other areas (ApSimon, 1998). The main control option considered was switching to low-sulphur bunker fuel. Within this study the ASAM model was used to derive a cost-effective abatement strategy to reduce the area of ecosystems whose critical loads were exceeded as of 1990 by 50% by the year 2010. This environmental target (commonly referred to as the "50% gap closure") was being studied at the time (1997) within the context of the then proposed European Union acidification strategy. As a starting point, the model uses a reference scenario that assumes each country emits the maximum allowed while complying with all current legislation. It then selects the cheapest combination of abatement options for SO_2, NO_x and NH_3 which achieve 50% gap closure. The optimised cost of achieving this is 7.1 billion e.c.u. yr^{-1}.

Including abatement of SO_2 emissions from ships leads to large savings, reducing the overall cost to 4.6 million e.c.u. yr^{-1} for a reduction in the sulphur content of bunker fuel to 1.5%, or further to 3.6 billion e.c.u. yr^{-1} if the sulphur content is reduced to 0.5%. This can be understood in terms of the relative contributions of shipping and land-based sources to sensitive costal areas: 1 Mt emission of SO_2 from shipping in the North Sea contributes 202 eq H^+ ha^{-1} yr^{-1} to a typical sensitive grid cell located in the Netherlands, whereas 1 Mt of SO_2 emitted from the land-based sources in the Netherlands, UK, Eire and France contributes 374 eq H^+ ha^{-1} yr^{-1} to deposition in that cell. The contribution from shipping is thus seen to be highly significant and is almost as great as the 249 eq H^+ ha^{-1} yr^{-1} contributed by sources in the Netherlands. If a certain fraction abatement of emissions on land costs more than about double a comparable percentage abatement of emissions from ships, it will therefore be more cost effective to abate the emissions from ships first. Similar calculations have also been carried out using the RAINS model, allowing for reductions in the NO_x emissions from ships as well as in SO_2 emissions, and these also showed that significant cost savings could be made by taking into account emission reductions from shipping in addition to land-based sources.

4.3.3. The effect of scientific uncertainty

A detailed study of uncertainty in the ASAM model (Warren and ApSimon, 1999) showed a surprising degree of robustness to both random and systematic uncertainties in meteorological data, critical loads and cost information, provided that the model was used to examine the pattern of country expenditures at a given overall emission abatement cost to Europe. However, there is a specific problem in using integrated assessment modelling to assess the role of shipping because of the limited spatial resolution. This is particularly important for shipping since some occurs close to the shore whilst the rest may be far out to sea; it is also concentrated in shipping lanes, and close to ports, where emissions are likely to impact coastal regions. The spatial resolution of an IAM is limited by that of the EMEP model which provides the meteorological information at 150 km resolution. It therefore fails to resolve very localised dispersion and deposition of emissions close to a source, and does not include the orographic enhancement of wet deposition. Both these factors are highly relevant when considering the influence of emissions from shipping in the vicinity of coastal mountains or upland areas. To study the effect of finer spatial resolution, a further study was therefore carried out in which the OPCD model (Lowles and ApSimon, 1996) was used in place of the EMEP model. This shows that shipping alone is not able to cause exceedences of critical loads in the absence of land-based sources, but that there are some sensitive ecosystem areas close to ports where critical loads are exceeded by marine- and land-based sources combined, with shipping locally responsible for a highly significant fraction of the total. A rapid decline in cost-effectiveness of placing control on ships was observed with distance from the shore.

Clearly, the reductions of NO_x suggested by both ASAM and RAINS also lead to a reduction in tropospheric ozone concentrations (Amman et al., 1999) and the reductions in SO_2, NO_x and NH_3 also lead to a reduction in human exposure to secondary particulate matter (Warren and ApSimon, 2000b).

In summary, the study showed that control of shipping emissions might play a useful role in the cost-effective achievement of environmental targets for acidification in Europe, whilst control of shipping in port was shown to greatly benefit the immediate local environment. It may be more beneficial to both the environment and for ease of implementation to apply controls to ships in port or to categories of shipping that tend to remain close to the shore, rather than applying a blanket abatement control across the whole of the North Sea. However, the degree of significance of emission controls for shipping in controlling acid deposition in Europe is clearly smaller than that of control of road transport in controlling air pollution in general.

5. Discussion

5.1. Implications of scientific uncertainty

In Section 4, we examined three specific impacts of transport on the atmospheric environment. In each case, factors exist to magnify the impact of transport emissions relative to other source types, and attempts have been made to quantify the impact to inform policy to control the emissions. This analysis reveals three distinct approaches to uncertainty depending on how well we are able to quantify each impact of the pollution of which transport is the source:

- For acidification, cause and effect is well established and source–receptor relationships are relatively well known, such that detailed cost–benefit studies are already being used to inform policy. These have indicated there would be significant benefits from reducing emissions from marine transport, but that it is important to disaggregate the shipping emissions in ports and close to coasts from others out to sea, and this is not covered by the Integrated Assessment Models that are currently used for the analysis. In recent years, these models have begun to include some quantitative analysis of uncertainty, and it has been demonstrated that their main conclusions are very robust.
- For global atmospheric change due to aircraft emissions, the uncertainties are great and it is difficult to say whether or not a large impact could occur. Action to protect the stratospheric ozone layer has therefore focussed more on terrestrial sources of ozone-depleting chemicals than on aviation. For climate change, some attempts at reducing greenhouse gas emissions have been made, but it is now almost certainly inevitable that climate change will occur as a result of past polluting activity. The emphasis here is therefore on improving predictions of the likelihood and nature of catastrophic change so that we can plan strategies to adapt if this does occur (Watson et al., 1996), and on assessing the extent to which future damage can be reduced by early action to abate emissions.
- For urban air pollutants from road traffic, and their effects on human health, there remain some uncertainties concerning both the existence and the mechanisms of cause and effect, especially for particulate air pollution and NO_2, which are currently two of the pollutants causing most concern. However, if we apply a precautionary principle in assuming that the damage is occurring and is attributable to transport emissions, we do have sufficient evidence to attempt a quantification of the magnitude of the effects including the economic cost of the impact of road transport emissions on health.

5.2. The current emphasis on abatement of road traffic emissions

During the 20th Century, the major response of transport to the general level of certainty discussed above, that environmental damage can be attributed to air pollution emissions, has been to reduce the emissions per vehicle from road transport. Table 3 (CONCAWE, 1997; EC, 1996) shows how the emissions of CO, hydrocarbons, NO_x and particulate matter have been reduced in Europe (note that CO_2 is omitted, for which similar reductions in emissions have not been attempted), reflecting the ability of technology to deliver reductions in emissions. The data show how the largest reductions in emissions have already taken place, with projections that further reductions will be possible by the introduction of on-board diagnostic systems, in-service emissions testing, recall programmes and fuel quality improvements (CONCAWE, 1997). These reductions in petrol- and diesel-engined vehicle emissions are sufficient to leave little room for improvement by switching to alternative hydrocarbon fuels such as natural gas or vegetable oil. The major advantage of non-fossil fuel hydrocarbon energy sources is that their contribution to carbon emissions to the atmosphere is offset by the return of carbon from the atmosphere to whichever crop is grown to provide the oil. The only cleaner option, as far as local emissions are concerned, is for a zero-emissions vehicle powered by electricity or hydrogen fuel cells. For such vehicles, it is important to consider, however, the total environmental impact of their use, as the air pollution emissions from remote generation of electricity or production of hydrogen fuel could possibly exceed the exhaust emissions that a conventional vehicle would produce. The main advantage of zero-emission vehicles is that the emissions can be relocated to where they are further from human receptors, so benefits to human health can be obtained although other environmental impacts are not reduced (see Fig. 1).

CEC (1996) studied seven European cities and investigated whether or not the then proposed improvements in petrol- and diesel-engined vehicle and fuel technology would be sufficient to meet health-related air quality standards in Europe. With the exception of Athens, where the geographical situation and climate make air quality an especially difficult problem, the study concluded that technological improvements would eliminate widespread exceedences of current health-related air quality standards by the end of the first decade of the 21st Century, but that limited local air quality problems would remain. It is proposed that these should be dealt with by means of local traffic management initiatives (Houghton, 1994). For example, altering road design can reduce traffic speed or acceleration, or can reallocate road space taking it away from cars and reserving it for buses and bicycles, hence physically restricting the volume of traffic. Where rapid intervention is required, without the delay associated with planning and financing physical changes in road design, speed

Table 3. European emissions standards for cars as a function of year when each standard comes into force

EU passenger car emission limits (g km^{-1})					
Petrol engines	CO	HC	NO$_x$	HC + NO$_x$	PM
1991[a]	14.3–27.1	1.5–2.4	2.1–3.4	4.7–6.9	
1993	3.2	–	–	1.1	
1996	2.2	–	–	0.5	
1997[b]	2.7	0.34	0.25	–	
2001[b]	2.3	0.20	0.15	–	
2006[b]	1.0	0.10	0.08	–	
Diesel engines	CO	HC	NO$_x$	HC + NO$_x$	PM
1991[a]	14.3–27.1	1.5–2.4	2.1–3.4	4.7–6.9	
1993[c]	3.2	–	–	1.1	0.18
1996[c]	1.0	–	–	0.70	0.08
1997[b]	1.0	0.71	0.63	–	0.08
2001[b]	0.64	–	0.50	0.56	0.05
2006[b]	0.50	–	0.25	0.30	0.025

[a] Limits in g test^{-1} converted to approximate average g km^{-1} over 4.052 km test distance for comparison with later data.
[b] Proposed modified test cycle, starting with cold engine.
[c] Indirect injection diesels only; limits apply in later years to direct injection engines.

limits as low as 30 km h^{-1} can be used, and physical traffic calming only sub-sequently introduced where speeds remain too high. (Such a combination of approaches has been used in Hamburg, for example.) Economic disincentives for drivers to enter or remain in polluted areas can be unpopular, but have been applied in several cities, such as area licensing in Singapore and cordon pricing in Bergen and Oslo, although air pollution control is often not the main reason for applying such measures. Reduction of road traffic can in theory be achieved by encouraging lone drivers to share their cars with others who make similar journeys, for example by high vehicle-occupancy lanes that are well-established on motorways in Los Angeles and Washington and were more recently introduced in Amsterdam. In Athens, a more integrated approach to chronic transport-related air pollution problems is being used, where the relocation of the airport is supposed to result in improved air quality. It is noteworthy that this development is also justified on the grounds that it allows for increase in air traffic that would not have been possible with the airport in its former, highly polluted urban location. The trend in many cities is thus away from taxation of vehicle ownership, purchase and fuel which are rather ineffective at controlling emissions, to measures designed to have an impact on traffic more specifically on certain congested roads. The public protests in recent months in many European Community member states may accelerate the trend away from fuel taxation, but whether widespread use of road pricing and congestion charging will be any more acceptable to our democratic societies remains to be seen.

5.3. Potential for further increased pressure on road transport

As discussed in Section 3, the emphasis on urban air quality that we find in many cities at the end of the 20th Century is not new, but this is the first time transport has been primarily blamed for this. In many countries, attempting to reduce road traffic instead of building infrastructure for its growth is a major change in policy. It is the first time traffic reduction has been considered in London since Hackney coaches, the 17th Century precursors of the taxi, caused such problems of congestion in the 1650s that Cromwell brought in regulations for their control (Hudson, 1998). The extent to which air pollution control is likely to continue to exert pressure on road transport into the 21st Century will differ from city to city. Three main categories of urban area can be identified (examples of which are discussed in Fenger, 1999):

1. already highly motorised cities in countries with low population growth (or slow population decline) where existing air quality standards are likely to be met by a combination of reduction in emissions per vehicle and reduction of traffic volumes;
2. cities where there remains huge potential for growth of private car ownership, often also in countries experiencing rapid population growth and urbanisation;
3. cities where local topography, meteorology and climate give rise to especially difficult air pollution climatology, usually photochemical smog.

For the first category only, there is a possibility that pressure for further control of road transport will reduce and be balanced or partially reversed by the very great demand for freedom of movement, especially in North America. Alternatively, however, one of two developments could cause the pressure to be maintained, at least for one or two decades.

The most probable is that air quality standards will be progressively tightened as technology and integrated transport system development steadily becomes able to deliver ever lower emissions. Our review of the major issues in Section 4.1 included an emphasis on particulate air pollution, which is already providing a good example of this trend as it becomes apparent that current air quality standards for PM_{10} will be met in most European cities by application of emissions abatement technology alone and no additional restraint on road traffic volumes. Given the lack of evidence of any threshold below which there is no effect of PM_{10} on health, it can be argued that tighter standards should be applied as long as costs to meet them are not considered to be excessive. However, proposed future tighter standards are close to background levels in parts of Europe close to the sea and the Sahara Desert. These will therefore be increasingly difficult to meet, especially as exhaust emissions are abated to the level where they are hardly a significant contribution to the total particulate mass even at busy roadside locations. Resuspended dust from roads includes sufficient particles below $10\,\mu m$ in size for it to begin to dominate PM_{10} emissions from road transport as exhaust emissions are reduced. Unless toxicological and epidemiological evidence can specifically exclude this aerosol fraction from being responsible for adverse effects of particulate air pollution on human health, even zero (exhaust) emissions vehicles will therefore require control from being capable of causing the most stringent proposed PM_{10} air quality standards to be exceeded. Conversely, if current toxicological studies confirmed by future epidemiological investigation can demonstrate that ultrafine particles are the fraction of PM_{10} responsible for most of the observed health effects, then increased attention in future is likely to be focussed on rather low concentrations of exhaust particulates from petrol- as well as diesel-engined vehicles.

Even in the absence of tighter air quality standards, there are arguments for environmental improvement by road traffic reduction. This is in recognition that the effects of air pollution emissions from road transport are far from being the largest road transport related impact on population health. In addition to road injury risk, large but very poorly quantified health benefits of road traffic reduction include increased physical exercise associated with most modes of transport other than the private car (BMA, 1998).

In Eastern European, Asian and South American cities, car ownership is still much lower than the western norm, and economic growth brings with it the expectation that more people will drive cars. Air quality is already much worse in cities such as Mexico City (Borja-Aburto et al., 1997) and Beijing (Xu et al., 1995) than would be considered acceptable by Western European or North American populations today. At the time of writing, alarmingly high levels of local urban air pollution in Dhaka following rapid conversion of the city's transport system of bicycle rickshaws to polluting two-stroke engines (Hussain, 2000) is likely soon to become the focus of rapid action to control transport emissions of air pollution. Many such rapidly growing cities are also in climate regions prone to photochemical smog, such that items 2 and 3 in the list at the beginning of this section will coincide where large numbers of the 21st Century global population will live. Not only are the effects of tropospheric ozone on human health becoming clearer, its impact on crops is a source of worry as we try to feed a population that is still growing rapidly.

There is some advantage to be gained by developing economies adopting new technology more rapidly than was the case in Europe and North America, but this is unlikely to be sufficient to deliver acceptable air quality due to rapid growth and very high population density over large urban areas. In addition, due to the inconsistencies in fuel quality and less advanced engine designs, very good quality catalysts would have to be fitted to achieve the same level of reduction in emissions as in the West. The longevity of the heavily engineered cars of the 1950s and 1960s in some third-world cities where a dry climate prevents rapid corrosion is remarkable.

It remains to be seen how the nations where these cities are found will respond. Developing countries could introduce schemes to prevent the growth of dependence on the private car that has occurred elsewhere, or alternatively will simply follow a number of years behind North America and Western Europe in allowing such a culture to develop in the name of progress before the problems become so severe that action cannot be delayed any further. It is possible, irrespective of the need to limit transport emissions to atmosphere, that 21st Century cities in what today are rapidly industrialising areas of the world will simply be too large for their transport systems to bear any resemblance at all to those in the 20th Century's so-called modern cities. Instead, the sheer scale of the demand for mobility may result in the development of alternative systems that deliver environmental sustainability as a side effect of greatly improved efficiency in moving people and goods compared with existing methods.

5.4. Future dominant transport and air pollution issues beyond urban air quality and road transport

The human health benefits of large reductions in emissions of NO_x, VOCs and CO resulting from the use of a three-way catalyst, plus the indirect elimination of lead from vehicle exhaust as vehicles equipped with catalysts are required to run on unleaded fuel, are widely believed to outweigh any undesirable side effects through increases in other impacts. Disadvantages from use of catalytic converters include increased emissions of CO_2, N_2O and NH_3 contributing to climate change and acid deposition. It is difficult to assess the extent to which CO_2 emissions have increased as a result of fitting catalytic converters, because improvements in fuel economy have been made at the same time as development of the engine management systems that are required to minimise NO_x and VOC emissions. While emissions of N_2O have been suggested by some authors to increase by as much as a factor of 10 (Wade et al., 1994; de Soete and Sharp, 1991; Dasch, 1992), N_2O is responsible for only a few per cent of the total global warming potential of road transport emissions (Wade et al., 1994; OECD, 1993; Gwilliam, 1993), so only a small increase in CO_2 emissions would have a greater impact. The contrast is stark between reductions of more than 90% in emissions of pollutants of concern to urban air quality at the same time as pollutants responsible for global warming stay approximately constant or, in cases where larger cars become fashionable, are allowed to increase. Even in a country such as Bangladesh, where changes in sea level and monsoon rainfall due to climate change have a very great impact, it is the local air quality in the capital city that has caused transport emissions of air pollution to come under scrutiny (see above). In global carbon emissions negotiations so far, there has been an emphasis on the industrialised world reducing its fossil fuel consumption first, and the majority global population who currently consume much less energy per capita not being expected to pay for the damage caused by the mobility and prosperity that the wealthy minority have enjoyed. If these attitudes continue to prevail, it is likely that urban air quality will continue to enjoy more prominence than climate change in exerting pressure on urban road transport world-wide. However, there have in the recent past, been decades when acidification or global warming has competed with urban air quality for highest prioritisation, and it is likely that this will occur again. As the problem of urban air quality is being solved, others are once again rising to the fore.

If the rapidly forced climate system suddenly exhibits a marked nonlinear or chaotic mode-switching response however, this is likely to return suddenly to being the biggest air pollution problem world wide, as is already happening anyway rather more slowly. Transport will thus in future be examined more closely than hitherto for the magnitude of its contribution to greenhouse gas emissions, although by then it will probably be too late to bring about reversal of the damage to the atmosphere other than very slowly. Even if future climate change assessments provide more certain predictions of the impact of further emissions, the moral and political issues of who should pay will not go away.

In our analysis of three case studies of impacts of transport emissions of air pollution, the impact of aviation on the global atmosphere was found to be subject to the greatest scientific uncertainty. At the same time, aviation is projected to grow rapidly, for example Archer (1993) quoted global growth of 6.5% yr^{-1} in the first 10 yr of the 21st Century with as rapid as 12% for international and 20% for domestic air travel in China. In a highly competitive industry, a recent development is the entrance of low-cost airlines who intend to make money by stimulating growth of the demand for air travel. Furthermore, a new fleet of supersonic aircraft, if widely adopted, will be likely to alter radically the impact of aircraft emissions on the atmosphere, on account of the higher altitude at which they fly compared with current aircraft. The only factor preventing the number of aircraft movements from expanding at the same rate as the growth in traffic is the trend towards larger aircraft. This shifts the burden from increasingly congested air space to limited capacity of airports on the ground. Pressure to build more and bigger airport facilities is intense as countries and regions compete to attract air traffic, and protection of the upper troposphere and lower stratosphere is not high on the agenda when airport developments are proposed. Ground-level air quality as well as noise, however, is currently an issue in determining whether or not airport expansion is allowed, for example in the recent enquiry over a fifth terminal at London Heathrow Airport. The consultants' report submitted by the airport operators to the enquiry (ERL, 1993) specifically states that greenhouse gas emissions and the ozone layer are not considered relevant to the question of whether or not Heathrow should be allowed to expand, and even points out that the impact of airport-related emissions of NO_x and VOCs on tropospheric ozone will be felt too far away to be of relevance to the planning application process. Unless the air pollution agenda changes, we are thus faced with the prospect of the major constraint on upper troposphere and lower stratosphere emissions being imposed indirectly by way of restrictions on airport expansion because of concern over ground-level air quality due to primary emissions in the immediate vicinity of the airport, a far from coherent approach to the application of atmospheric science to transport development.

Of our three case studies, the most mature branch of atmospheric science considered was acidification. Here, cause and effect is proven at least as far as the link between precursor emissions and deposition to the ground is concerned. Unlike the other two examples, it has therefore been possible to carry out rather detailed attempts to evaluate the emissions abatement strategies economically using integrated assessment modelling. We found, however, that scientific uncertainty still needs to be taken into account, but that it is playing a rather different role in the assessment of the environmental impact of air pollution emissions from marine transport. Instead of resulting in debate as to whether or not action needs to be taken, the uncertainty is at the level of how much action is justified, at what price, who should pay and what is the most efficient way of protecting the environment. Even without the application of technology to reduce emissions, the contribution of transport by sea and inland waterway to air pollution is much less per unit mass carried per unit distance travelled than other modes, especially transport by air. At present, however, the global economy is often willing to pay the environmental price of air travel in return for the benefit of the journey times that can be achieved. It is also difficult to make direct comparisons between short-term, local impacts and much longer-term impacts that may be greatest thousands of kilometres away from the beneficiaries of the transport responsible for the emissions. Integrated assessment modelling has not yet been widely applied to comparison of such markedly different impacts as on urban air quality and upper troposphere chemistry, but its application to future integrated transport systems, especially where there are conflicts between local and global environmental priorities, has the potential to be extremely valuable and would therefore be an intellectually challenging and worthwhile development to pursue.

6. Conclusions and recommendations

Our comparison of the distinct impacts of air pollution emissions from three different modes of transport illustrates how current transport management is influenced by the availability of good scientific understanding and ability to make quantitative estimates of the magnitude of impacts (see summary in Section 5.1). Looking to the future, the global potential for growth of the transport sector is immense, and greatly reduced air pollution emissions per person-km could be a most welcome side effect of more efficient integrated transport systems that will be required to meet demand for mobility of people and goods over short and long distances.

There remains a need to continue research to improve our understanding of the mechanisms leading to impacts of air pollution emissions from transport, to reduce uncertainty in our ability to quantify relationships between all emissions and all impacts.

The scale of the current preoccupation with the effects of local air pollution emissions from road transport on the health of urban human populations does appear to be temporary despite its continuing importance for the coming one to two decades, especially in more recently industrialising countries of the world. As road transport for the first time becomes subject to widespread demand management to meet environmental objectives, unrestrained growth of aviation starts to appear unfashionable. Some of the largest gaps in our understanding of transport impacts of air pollution emissions are also concerned with aviation emissions and their impact on global atmospheric chemistry. Future atmospheric science research should therefore be integrated more effectively with transport research, including urban air quality and human health impacts research directed to the needs of large developing cities, in order to contribute to the stimulation of more imaginative and sustainable development of future integrated local and global transport systems on our increasingly crowded planet. It is hoped that this Millennial Review can provide some stimulus for discussion of what the priorities for such research should be.

Acknowledgements

Emma Hutchinson gratefully acknowledges receipt of a CASE studentship from the UK Economics and Social Research Council and the British Geological Survey. Jennifer Mindell is supported by the North Thames Regional Training Budget for Public Health and Rachel Warren by the UK Department of the Environment, Transport and the Regions. Roy Colvile acknowledges receipt of support for research on air pollution and transport from the UK Engineering and Physical Science Research Council through contracts GR/L91085 and GR/M70773, and from the Rees Jeffreys Road Fund, which also form a contribution to the *EUROTRAC* urban air pollution sub-project *SATURN*, and on particulate air pollution and health from the Higher Education Funding Council for England and the City of Westminster (NERC award GR3/E004).

References

Abbey, D.E., Hwang, B.L., Burchette, R.J., Vancuren, T., Mills, P.K., 1995. Estimated long-term ambient concentrations of PM_{10} and development of respiratory symptoms in a nonsmoking population. Archives of Environmental Health 50, 139–152.

Alcamo, J., Shaw, R., Hordijk, L. (Eds.), 1990. The RAINS Model of Acidification: Science and Strategies in Europe. Kluwer Academic Publishers, Dordrecht, The Netherlands. Also partly available at http://www.iiasa.ac.at/~rains/cgi-bin/rains_web.pl.

Amman, M., Bertok, I., Cofala, J., Gyarfas, F., Heyes, C., Klimont, Z., Makowski, M., Schopp, W., Syri, S., 1999. Seventh Interim Report: Cost-effective Control of Acidification and Ground-level Ozone. International Institute for Applied Systems Analysis, Laxenburg, Austria.

Anderson, H.R., Spix, C., Medina, S., Schouten, J.P., Castellsagué, J., Rossi, G., Zmirou, D., Touloumi, G., Wojtyniak, B., Pönkä, A., Bachárová, L., Schwartz, J., Katsouyanni, K., 1997. Air pollution and daily admissions for chronic obstructive pulmonary disease in 6 European cities: results from the APHEA project. European Respiratory Journal 10, 1064–1071.

APEG (Atmospheric Particles Expert Group), 1999. Source apportionment of airborne atmospheric fine particulate matter in the United Kingdom. Report to the UK Department of Environment, Transport and the Regions.

Apling, A.J., Sullivan, E.J., Williams, M.L., Ball, D.J., Bernard, R.E., Derwent, R.G., 1977. Ozone concentrations in South East England during the summer heatwave of 1976. Nature 269, 569–573.

ApSimon, H.M. (Ed.), 1998. Costs and benefits of controlling sulphur dioxide emissions from ships in the North Sea and the seas to the west of Britain. Report to UK DETR.

ApSimon, H.M., Warren, R.F., 1996. Transboundary air pollution in Europe. Energy Policy 24, 631–640.

ApSimon, H.M., Warren, R.F., Wilson, J.J.N., 1994. The abatement strategies assessment model – ASAM: applications to reductions of sulphur dioxide emissions across Europe. Atmospheric Environment 28, 649–663.

Archer, L., 1993. Aircraft emissions and the environment. Oxford Institute for Energy Studies. Papers on Energy and the Environment EV17. ISBN 0948061-79-0.

Ashmore, M.R., Bell, J.N.B., Dalpra, C., Runeckles, V.C., 1980. Visible injury of crop species by ozone in the United Kingdom. Environmental Pollution 21, 209–215.

Atkinson, R.W., Anderson, H.R., Strachan, D.P., Bland, J.M., Bremner, S.A., Ponce de Leon, A., 1999. Short-term associations between outdoor air pollution and visits to accident and emergency departments in London for respiratory complaints. European Respiratory Journal 13 (2), 257–265.

Bach, W., 1985. Waldsterben: our dying forests – part III. Forest dieback: extent of damage and control strategies. Experientia 41 (9), 1095–1104.

Barrett, K., Seland, O., 1995. European transboundary air pollution: 10 years calculated fields and budgets to the end of the first sulphur protocol. EMEP/MSC-W Report 1/95. EMEP MSC-W, Norwegian Meteorological Institute, P.O. Box 43-Blindern, N-0313 Oslo 3, Norway.

Barrett, M., 1991. Aircraft pollution – environmental impacts and future solutions. World Wide Fund for Nature. Research paper.

Berkovicz, R., Hertel, O., Sørensen, N.N., Micelsen, J.A., 1997. Modelling air pollution from traffic in urban areas. In: Perkins, R.J., Belcher, S.E. (Eds.), Flow and Dispersion through Groups of Obstacles. Proceedings of Conference on Flow and Dispersion through Groups of Obstacles. Institute of Mathematics and its Applications, University of Cambridge, March 1994. Clarendon Press, Oxford.

Bickel, P., Schmid, S., Krewitt, W., Friedrich, R., Watkiss, P., Collings, S., Holland, M.R., Pilkington, A., Hurley, F., Donnan, P., Landrieu, G., Eyre, N., Navrud, S., Rabl, A., Spadaro, J., Assimacopoulos, D., Vossiniotis, G., Fontana, M., Frigerio, M., Dorland, K., 1998. In: Friedrich, R., Bickel, P., Krewitt, W. (Eds.), External Costs of Transport. IER, Universität Stuttgart.

BMA (British Medical Association), 1998. Health and Environmental Impact Assessment. Earthscan Publications, London.

Bobák, M., Leon, D.A., 1999. The effect of air pollution on infant mortality appears specific for respiratory causes in the postneonatal period. Epidemiology 10, 666–670.

Borg, H., 1986. Metal speciation in acidified mountain streams in Central Sweden. Water, Air and Soil Pollution 30, 1007–1014.

Borja-Aburto, V.H., Loomis, D.P., Bangdiwala, S.I., Shy, C.M., Rascon-Pacheco, R.A., 1997. Ozone, suspended particulates, and daily mortality in Mexico City. American Journal of Epidemiology 145, 258–268.

Bradford Hill, A., 1965. The environment and disease: association or causation? Proceedings of the Royal Society of Medicine 58, 295–300.

Braun-Fahrländer, C., Ackermann-Liebrich, U., Schwartz, J., Gnehm, H.P., Rutishauser, M., Wanner, H.K., 1992. Air pollution and respiratory symptoms in preschool children. American Review of Respiratory Disease 145, 42–47.

Bremner, S.A., Anderson, H.R., Atkinson, R., McMichael, A.J., Strachan, D.P., Bland, J.M., Bower, J.S., 1999. Short term associations between outdoor air pollution and mortality in London 1992–4. Occupational and Environmental Medicine 56, 237–244.

Briggs, D.J., Collins, S., Elliott, P., Fischer, P., Kingham, S., Lebret, E., Pryl, K., van Reeuwijk, H., Smallbone, K., van der Veen, A., 1997. Mapping urban air pollution using GIS: a regression-based approach. International Journal of Geographical Information Science 11 (7), 699–718.

Brunekreef, B., Dockery, D.W., Krzyzanowski, M., 1995. Epidemiologic studies on short-term effects of low levels of major ambient air pollution components. Environmental Health Perspectives 103 (S2), 3–13.

Brunekreef, B., Janssen, N.A.H., de Hartog, J., Harssema, H., Knape, M., Van, V.P., 1997. Air pollution from truck traffic and lung function in children living near motorways. Epidemiology 8, 298–303.

Buckingham, C., Clewley, L., Hutchinson, D., Sadler, L., Shah, S., 1997. London atmospheric emissions inventory. London Research Centre. ISBN 1-85261-267-3.

Carruthers, D.J., Singles, R.J., Nixon, S.G., Ellis, K.L., Pendrey, M., Harwood, J., 1999. Modelling air quality in Central London. Report FM 327 to the Central London Boroughs and DETR, Cambridge Environmental Research Consultants.

Carter, W., 1987. Headline: Huge tides threaten to engulf Britain. Quoted in The Times.

Castellsagué, J., Sunyer, J., Saez, M., Antó, J.M., 1995. Short-term association between air pollution and emergency room visits for asthma in Barcelona. Thorax 50, 1051–1056.

CEC (Commission of the European Communities), 1996. Communication from the Commission to the European Parliament and the Council on a future strategy for the control of atmospheric emissions from road transport taking into account the results from the Auto/Oil programme. COM (96) 248. Catalogue Number CB-CO-96-310-EN-C, ISSN 0254-1475, ISBN 92-78-05787-8. Office for Official Publications of the European Communities, Luxembourg.

Chan, C.C., Spengler, J.D., Özkaynak, H., Lefkopoulou, M., 1991. Commuter exposures to VOCs in Boston, Massachusetts. Journal of the Air and Waste Management Association 41, 1594–1600.

COMEAP (Department of Health Committee on the Medical Effects of Air Pollutants), 1998. Quantification of the effects of air pollution on health in the United Kingdom. HMSO, London.

CONCAWE, 1997. Motor vehicle emission regulations and fuel specifications. Part 2: detailed information and historic review (1970–1996). Report No. 6/97, CONCAWE, Brussels.

Dabbeidt, W.F., Ludwig, F.L., Johnson, W.B., 1973. Validation and applications of urban diffusion model for vehicular applications. Atmospheric Environment 7, 603–618.

Dasch, J.M., 1992. Nitrous oxide emissions from vehicles. Journal of the Air and Waste Management Association 42 (1), 63–67.

De Soete, G., Sharp, 1991. Nitrous oxide emissions: modifications as a consequence of current trends in industrial fossil fuel combustion and in land use. European Commission Report EUR 13473 EN, Luxembourg.

Delves, H.T., 1998. Overview of UK and international studies on trends in blood lead and the use of lead isotope ratios to identify environmental sources. In: Gompertz, D. (Ed.), IEH Report on Recent UK Blood Lead Surveys. Institute for Environmental Health, Leicester, UK.

Den Boeft, J., Eerens, H.C., den Tonkelaar, W.A.M., Zandfeld, P.Y.J., 1996. CAR International: a simple model to determine city street air quality. Science of the Total Environment 190, 321–326.

DETR (UK Department of the Environment, Transport and the Regions), 1997. Environmental Transport Association. Road User Exposure to Air Pollution. A Literature Review. DETR, London.

DETR (UK Department of the Environment, Transport and the Regions), 1998. Transport Statistics for Great Britain 1998. The Stationery Office, London.

Diaz-Sanchez, D., 1997. The role of diesel exhaust particles and their associated polyaromatic hydrocarbons in the induction of allergic airway disease. Allergy 52, 52–56.

Dockery, D.W., 1993. Epidemiologic study design for investigating respiratory health effects of complex air pollution mixtures. Environmental Health Perspectives 101, 187–191.

Dockery, D.W., Schwartz, J., Spengler, J.D., 1992. Air pollution and daily mortality: associations with particulates and acid aerosols. Environmental Research 62, 362–373.

DoH (UK Department of Health Ad-Hoc Group on the Economic Appraisal of the Health Effects of Air Pollution), 1999. Economic Appraisal of the Health Effects of Air Pollution. The Stationery Office, London.

Donaldson, Lord of Lymington, 1994. Safer ships, cleaner seas. Inquiry into the prevention of pollution from merchant shipping. Cm 2560. HMSO, London.

Dore, A.J., Sobik, M., Migala, K., 1999. Patterns of precipitation and pollutant deposition in the western Sudete mountains, Poland. Atmospheric Environment 33 (20), 3301–3312.

Downing, C.E.H., Vincent, K.J., Campbell, G.W., Fowler, D., Smith, R.I., 1995. Trends in wet and dry deposition of sulphur in the United Kingdom. Water, Air and Soil Pollution 85 (2), 659–664.

Driscoll, C.T., Jr., Baker, J.P., Bisogni, J.J., Jr., Schofield, C.L., 1980. Effect of aluminium speciation on fish in dilute acidified waters. Nature 284, 161–164.

EC (European Commission), 1996. On a Future Strategy for the Control of Atmospheric Emissions from Road Transport Taking into Account the Results from the Auto/Oil programme. COM (96) 248 Final, Commission of the European Communities.

Elliott, P., Cuzick, J., English, D., Stern, R. (Eds.), 1992. Geographical and Environmental Epidemiology: Methods for Small-Area Studies. Oxford University Press, Oxford.

Ellison, J.M., Waller, R.E., 1978. A review of sulphur oxides and particulate matter as air pollutants with particular reference to effects on health in the United Kingdom. Environmental Research 16, 302–325.

Elwood, J.A., 1988. Causal Relationships in Medicine. Oxford University Press, Oxford.

EPAQS (Expert Panel on Air Quality Standards), 1994. Carbon Monoxide. HMSO, London.

EPAQS (Expert Panel on Air Quality Standards), 1998. Lead. HMSO, London.

ERL (Environmental Resources Limited), 1993. Terminal 5 Heathrow, Environmental Statement. BAA plc, Crawley, West Sussex.

ERR (Earth Resources Research), 1990. Atmospheric Emissions from the Use of Transport in the UK, Vol. 2. The Effect of Alternate Policies. World Wide Fund for Nature, London. Cited in: Whitelegg, J., 1993. Transport for a Sustainable Future. The Case for Europe. Belhaven Press, London. ISBN 1852931469.

Eyre, N.J., Ozdemiroglu, E., Pearce, D.W., Steele, P., 1997. Fuel and location effects on the damage costs of transport emissions. Journal of Transport Economics and Policy 31, 5–24.

Fairley, D., 1990. The relationship of daily mortality to suspended particulates in Santa Clara County, 1980–1986. Environmental Health Perspectives 89, 159–168.

Faiz, A., 1993. Automotive emissions in developing countries – relative implications for global warming, acidification and urban air quality. Transportation Research 27 (3), 167–186.

Farman, J.C., 1987. Recent measurements of total ozone at British Antarctic Survey stations. Philosophical Transactions of the Royal Society of London, Series A 323, 629–644.

Farman, J.C., Gardiner, B.G., 1987. Ozone depletion over Antarctica. Nature 329, 574.

Fenger, J., 1999. Urban air quality. Atmospheric Environment 33, 4877–4900.

Ferin, J., Oberdörster, G., Soderholm, S.C., Gebein, R., 1991. Pulmonary tissue access of ultrafine particles. Journal of Aerosol Medicine 4, 57–58.

Filliger P., Puybonnieux-Texier, V., Schneider, J., 1999. Health costs due to road traffic-related air pollution. An impact assessment project of Austria, France and Switzerland. PM_{10} population exposure. Technical Report on Air Pollution. GVF-Report No. 326-TEH05. Federal Department for Environment, Transport, Energy and Communications Bureau for Transport Studies, Bern.

Forbes, W.H., Sargent, F., Roughton, F.J.W., 1945. The rate of carbon monoxide uptake by normal men. American Journal of Physiology 143, 594–608.

Forsberg, B., Stjernberg, N., Wall, S., 1997. People can detect poor air quality well below guideline concentrations: a prevalence study of annoyance reactions and air pollution from traffic. Occupational and Environmental Medicine 54, 44–48.

Garcia-Aymerich, J., Tobias, A., Antó, J.M., Sunyer, J., 2000. Air pollution and mortality in a cohort of patients with chronic obstructive pulmonary disease: a time series analysis. Journal of Epidemiology and Community Health 54 (1), 73–74.

Gilmour, P.S., Brown, D.M., Lindsay, T.G., Beswick, P.H., MacNee, W., Donaldson, K., 1996. Adverse health effects of PM_{10} particles: involvement of iron in generation of hydroxyl radical. Occupational and Environmental Medicine 53, 817–822.

Gover, M.P., Collings, S.A., Hitchcock, G.S., Moon, D.P., Wilkins, G.T., 1996. Alternative Road Transport Fuels – a Preliminary Life-Cycle Study for the UK, Vol. 1. HMSO, London.

Gried, H., Simon, B., 1990. Pollutant emissions of existing and future engines for commercial aircraft, pp. 43–83. In: Schumann, U. (Ed.), Air Traffic and the Environment – Background Tendencies and Potential Global Atmospheric Effects. Proceedings of a DLR International Colloquium, 15/16 November. Lecture Notes in Engineering, Vol. 60. Springer, Berlin.

Grossinho, A., Gulliver, J., Briggs, D., Machin, F., Tate, J., Ashmore, M., Catena, B., Elliott, P., 1999. Air pollution assessment of Northampton school children using a network model. Epidemiology 10, S104.

Grubb, M., 1999. The Kyoto Protocol, a guide and assessment. Energy and Environment Programme, Royal Institute of International Affairs. Earthscan Publications, London. ISBN 1853835811.

Grubb, M., Koch, M., Munson, A., Sullivan, F., Thomson, K., 1993. The Earth Summit Agreement, a guide and assessment. Energy and Environment Programme, Royal Institute of International Affairs. Earthscan Publications, London. ISBN 1853831778.

Gwilliam, K.M., 1993. On reducing transport's contribution to global warming. In: ECMT (European Conference of Ministers of Transport). Transport Policy and Global Warming. ECMT, Paris.

Hajat, S., Haines, A., Goubet, S.A., Atkinson, R.W., Anderson, H.R., 1999. Association of air pollution with daily GP consultations for asthma and other lower respiratory conditions in London. Thorax 54, 597–605.

Hampton, M., 1991. Cycling towards low prices. Petroleum Economist 58 (4).

Heck, R.M., Farrauto, R.J., 1995. Catalytic Air Pollution Control: Commercial Technology. Van Nostrand Reinhold. International Thomson Publishing, New York.

Hewitt, C.N., 2000. Sulphur and nitrogen chemistry in power plant plumes. Atmospheric Environment 35, 1155–1170.

Holgate, S.T., Commins, B.T., Anderson, H.R., 1995. Asthma and outdoor air pollution. Department of Health Committee on the Medical Effects of Air Pollutants. HMSO, London.

Houghton, J. (Chairman), 1994. Transport and the environment. Eighteenth Report of the Royal Commission on Environmental Pollution, Cm 2674. The Stationery Office, London. ISBN 0-10-137522-0.

Houghton, J. (Chairman), 1997. Transport and the environment – developments since 1994. Twentieth Report of the Royal Commission on Environmental Pollution, Cm 3752. The Stationery Office, London. ISBN 0-10-137522-0.

Houghton, J.T., Meira Filho, L.G., Callander, B.A., Harris, N., Kattenberg, A., Maskell, K. (Eds.), 1996. Climate Change 1995, the Science of Climate Change. Oxford University Press, Oxford.

Hoydysh, W.G., Dabberdt, W.F., 1988. Kinematics and dispersion characteristics of flows in asymmetric street canyons. Atmospheric Environment 22 (12), 2677–2689.

Hoydysh, W.G., Dabberdt, W.F., 1994. Concentration fields and urban intersections: fluid modelling studies. Atmospheric Environment 28 (11), 1849–1860.

Hudson, R. (Ed.), 1998. London, Portrait of a City. The Folio Society, London.

Hunter, L.J., Johnson, G.T., Watson, I.D., 1992. An investigation of 3-dimensional characteristics of flow regimes within the urban canyon. Atmospheric Environment 26 (4), 425–432.

Hussain, M., 2000. Dreadful effect of air pollution on childhealth: worsening situation in Dhaka city. Bangladesh Environmental Newsletter 11 (1), 1–2.

Ihrig, M.M., Shalat, S.L., Baynes, C., 1998. A hospital-based case–control study of still births and environmental exposure to arsenic using an atmospheric dispersion model linked to a geographical information system. Epidemiology 9, 290–294.

Inglis, D.W.F., Choularton, T.W., Wicks, A.J., Fowler, D., Leith, I.D., Werkman, B., Binnie, J., 1995. Orographic enhancement of wet deposition in the United Kingdom: case studies and modelling. Water, Air and Soil Pollution 85 (4), 2119–2124.

ISO (International Standards Organisation), 1997. ISO 14040. Environmental management – life cycle assessment – principles and framework. ISO, Geneva.

Jantunen, M.J., Hänninen, O., Katsouyanni, K., Knöppel, H., Kuenzlii, N., Lebret, E., Maroni, M., Saarela, K., Sram, R., Zmirou, D., 1998. Air pollution exposure in European cities:

the 'EXPOLIS'-study. Journal of Exposure Analysis and Environmental Epidemiology 8 (4), 495–518.

Jarvis, D., Burney, P., 1998. The epidemiology of allergic disease. British Medical Journal 316, 607–610.

Jenkin, M.E., Clemitshaw, K.C., 2000. Ozone and other photochemical pollutants: chemical processes governing their formation in the planetary boundary layer. Atmospheric Environment 34 (16), 2499–2527.

Johnson, C., Henshaw, H., McInnes, G., 1992. Impact of aircraft and surface emissions of nitrogen oxides on tropospheric ozone and global warming. Nature 355, 69–71.

Johnson, C.E., Collins, W.J., Stevenson, D.S., Derwent, R.G., 1999. Relative roles of climate and emissions changes on future tropospheric oxidant concentrations. Journal of Geophysical Research 104 (D15), 18631–18645.

Johnson, W.B., Ludwig, F.L., Dabberdt, W.F., Allen, R.J., 1973. An urban diffusion simulation model for carbon monoxide. Journal of the Air Pollution Control Association 23, 490–498.

Jost, P., Hassel, D., Weber, F.J., Sonnborn, K.-S., 1992. Emission and fuel consumption modelling based on continuous measurements. Deliverable Nr. 7 of the DRIVE PROJECT V 1053: Modelling of emissions and consumption in urban areas – MODEM. TÜV Rheinland Institut für Umweltschutz und Energietechnik, Stabsabteilung Verkehr und Umwelt, Postfach 101750, D-5000, Köln 1, Germany.

Joumard, R., 1995. Transport and air pollution – some conclusions. Science of the Total Environment 169 (1–3), 1–5.

Kandler, O., Innes, J.L., 1995. Air pollution and forest decline in Central Europe. Environmental Pollution 90 (2), 171–180.

Khandelwal, P., 1999. An assessment of personal exposure to $PM_{2.5}$ for bicycle commuters in Central London using personal measurements and applications of dispersion modelling. M.Sc. Dissertation, Imperial College Centre for Environmental Technology.

Krishna, M.T., Chauhan, A.J., 1996. Air pollution and health. Journal of the Royal College of Physicians, London 30, 448–452.

Kukadia, V., Palmer, J., 1998. The effect of external atmospheric pollution on indoor air quality: a pilot study. Energy in Buildings 27, 223.

Künzli, N., Kaiser, R., Medina, S., Studnicka, M., Chanel, O., Filliger, P., Herry, M., Horak, F., Jr., Puybonnieux-Texier, V., Quenel, P., Schneider, J., Seethaler, R., Vergnaud, J.-C., Sommer, H., 2000. Public-health impact of outdoor and traffic-related air pollution: a European assessment. Lancet 356, 795–800.

Künzli, N., Kaiser, R., Medina, S., Studnicka, M., Oberfeld, G., Horak, F., 1999. Health costs due to road traffic-related air pollution. An impact assessment project of Austria, France and Switzerland. Air pollution attributable cases. Technical Report on Epidemiology. GVF-Report No. 326-TEH06. Federal Department for Environment, Transport, Energy and Communications Bureau for Transport Studies, Bern.

Lee, I.Y., Park, H.M., 1994. Parametrisation of pollutant transport and dispersion in urban street canyons. Atmospheric Environment 28 (14), 2343–2349.

Lercher, P., Schmitzberger, R., Kofler, W., 1995. Perceived traffic air pollution, associated behavior and health in an alpine area. Science of the Total Environment 169, 71–74.

Li, X.Y., Brown, D., Smith, S., MacNee, W., Donaldson, K., 1999. Short-term inflammatory responses following intratracheal instillation of fine and ultrafine carbon clack in rats. Inhalation Toxicology 11 (8), 709–731.

Linn, W.S., Shamoo, D.A., Anderson, K.R., Peng, R.-C., Avol, E.L., Hackney, J.D., Gong, H.J., 1996. Short-term air pollution exposures and responses in Los Angeles area schoolchildren. Journal of Exposure Analysis & Environmental Epidemiology 6, 449–472.

Lipfert, F.W., Wyzga, R.E., 1995. Air pollution and mortality: issues and uncertainties. Journal of the Air and Waste Management Association 45, 949–966.

Lloyd's Register of Shipping, 1995. Marine Exhaust Emissions Research Programme, London, UK.

Lowles, I., ApSimon, H.M., 1996. The contribution of sulphur dioxide emissions from ships to coastal acidification. International Journal of Environmental Studies 51, 21–34.

LRC (London Research Centre), 1999. Atmospheric emissions inventories. Web pages at http://www.london-research.gov.uk/et/etemiss.htm.

Maddison, D., Pearce, D., 1999. Costing the health effects of poor air quality. In: Holgate, S.T., Samet, J.M., Koren, J.M., Maynard, R.L. (Eds.), Air Pollution and Health. Academic Press, London.

McCormick, J., 1997. Acid Earth, the Politics of Acid Pollution, 3rd ed. Earthscan publications Ltd., London.

McGavran, P.D., Rood, A.S., Till, J.E., 1999. Chronic beryllium disease and cancer risk estimates with uncertainty for beryllium released to the air from the rocky flats plant. Environmental Health Perspectives 107, 731–744.

McHugh, C.A., Carruthers, D.J., Edmunds, H.A., 1997. ADMS – Urban: an air quality management system for traffic, domestic and industrial pollution. International Journal of Environmental Pollution 8 (3–6), 666–674.

Miyamoto, T., 1997. Epidemiology of pollution-induced airway disease in Japan. Allergy 52, 30–34.

Monn, C., 2001. Exposure assessment of air pollutants: a review on spatial heterogeneity and indoor/outdoor/personal exposure to suspended particulate matter, nitrogen dioxide and ozone. Atmospheric Environment 35, 1–32.

Montreal Protocol, 1987. Montreal protocol on substances that deplete the ozone layer, Montreal, 16 September 1987. HMSO, London. Treaty Series No. 19(1990). ISBN 0101097727.

Morgan, G., Corbett, S., Wlodarcyzk, J., Lewis, P., 1998. Air pollution and daily mortality in Sydney, Australia, 1989 through 1993. American Journal of Public Health 88, 759–764.

Murphy, S.A., BeruBe, K.A., Pooley, F.D., Richards, R.J., 1998. Response of lung epithelium to well characterised fine particles. Life Sciences 62 (19), 1789–1799.

Ninan, T.K., Russell, G., 1992. Respiratory symptoms and atopy in Aberdeen schoolchildren: evidence from two surveys 25 years apart. British Medical Journal 304, 873–875.

OECD (Organisation for Economic Co-operation and Development), 1988. Transport and the Environment. OECD, Paris.

OECD (Organisation for Economic Co-operation and Development), 1993. Cars and Climate Change. Energy and Environment Series. OECD/IEA, Paris.

Olaiz, G., Fortoul, T.I., Rojas, R., Doyer, M., Palazuelos, E., Tapia, C.R., 1996. Risk factors for high levels of lead in blood of school children in Mexico City. Archives of Environmental Health 51, 122–125.

Ostro, B., Chestnut, L., 1998. Assessing the health benefits of reducing particulate matter air pollution in the United States. Environmental Research 76, 94–106.

Ostro, B.D., Chestnut, L.G., Mills, D.M., Watkins, A.M., 1999. Estimating the effects of air pollutants on the population: human health benefits of sulphate aerosol reductions under Title IV of the 1990 Clean Air Act Amendments. In: Holgate, S.T., Samet, J.M., Koren, J.M., Maynard, R.L. (Eds.), Air Pollution and Health. Academic Press, London.

Pantazopoulou, A., Katsouyanni, K., Kourea-Kremastinou, J., Trichopoulos, D., 1995. Short-term effects of air pollution on hospital emergency outpatient visits and admissions in the greater Athens, Greece area. Environmental Research 69, 31–36.

Penner, J.E., Lister, D.H., Griggs, D.J., Dokken, D.J., McFarland, M. (Eds.), 1999. Aviation and the Global Atmosphere. Special Report of IPCC Working Groups I and III in collaboration with the Scientific Assessment Panel to the Montreal Protocol on Substances that Deplete the Ozone Layer. Cambridge University Press, Cambridge.

Pereira, L.A.A., Loomis, D., Conceicao, G.M.S., Braga, A.L.F., Arcas, R.M., Kishi, H.S., Singer, J.M., Bohm, G.M., Saldiva, P.H.N., 1998. Association between air pollution and intrauterine mortality in Sao Paulo, Brazil. Environmental Health Perspectives 106, 325–329.

Perera, F., 1981. Carcinogenicity of airborne fine particulate benzo(a)pyrene: an appraisal of the evidence and the need for control. Environmental Health Perspectives 42, 163–185.

Peters, A., Wichmann, H.E., Tuch, T., Heinrich, J., Heyder, J., 1997. Respiratory effects are associated with the number of ultrafine particles. American Journal of Respiratory and Critical Care Medicine 155, 1376–1383.

Poloniecki, J.D., Atkinson, R.W., Ponce de Leon, A., Anderson, H.R., 1997. Daily time series for cardiovascular hospital admissions and previous day's air pollution in London, UK. Occupational and Environmental Medicine 54, 535–540.

Ponce de Leon, A., Andersen, H.R., Bland, J.M., Strachan, D.P., Bower, J., 1996. Effects of air pollution on daily hospital admissions for respiratory disease in London between 1987–88 and 1991–92. Journal of Epidemiology and Community Health 50, S63–S70.

Pope, C.A.I., Dockery, D.W., Spengler, J.D., Raizenne, M.E., 1991. Respiratory health and PM_{10} pollution: a daily time series analysis. American Review of Respiratory Disease 144, 668–674.

Pope, C.A.I., Schwartz, J., Ransom, M.R., 1992. Daily mortality and PM_{10} pollution in Utah Valley. Archives of Environmental Health 47, 211–217.

Posch, M., Hettelingh, J.-P., de Smet, P.A.M., Downing, R.J. (Eds.), 1997. Calculation and Mapping of Critical Thresholds in Europe: Status Report 1997. Coordinating Centre for Effects, Rijksinstitut Voor Volksgezondheid en Milieu, Netherlands.

QUARG (Quality of Urban Air Review Group), 1993. Diesel vehicle emissions and urban air quality: second report of the Quality of Urban Air Review Group. Institute of Public and Environmental Health, University of Birmingham.

Quarrie, J. (Ed.), 1992. Earth Summit 1992. Regency Press, London. ISBN 09520469-0-3.

RAINS, 2000. RAINS-Europe Web Edition On-line Model, Scenario Analysis Functions and emissions data, http://www.iiasa.ac.at/~rains/cgi-bin/rains_web.pl, accessed November 2000.

RNMI, 1994. Report by Royal Netherlands Meteorological Institute to the Ministry of Transport, June 1994.

Romieu, I., Meneses, F., Ruiz, S., Sienra, J.J., Huerta, J., White, M.C., Etzel, R.A., 1996. Effects of air pollution on the respiratory health of asthmatic children living in Mexico City. American Journal of Respiratory and Critical Care Medicine 154, 300–307.

Samaras, Z., Kyriakis, N., Zachariadis, Th., 1995. Reconciliation of macroscale and microscale motor vehicle emission estimates. The Science of the Total Environment 169, 231–239.

Scaperdas, A., Colvile, R., 1999. Assessing the representativeness of monitoring data from an urban intersection site in Central London, UK. Atmospheric Environment 33, 661–674.

Scarlett, J.F., Abbott, K.J., Peacock, J.L., Strachan, D.P., Anderson, H.R., 1996. Acute effects of summer air pollution on respiratory function in primary school children in southern England. Thorax 51, 1109–1114.

Schwartz, J., 1993. Air pollution and daily mortality in Birmingham, Alabama. American Journal of Epidemiology 137, 1136–1147.

Schwartz, J., 1994. What are people dying of on high air pollution days? Environmental Research 64, 26–35.

Schwartz, J., 1996. Air pollution and hospital admissions for respiratory disease. Epidemiology 7, 20–28.

Schwartz, J., Slater, D., Larson, T.V., Pierson, W.E., Koenig, J.Q., 1993. Particulate air pollution and hospital emergency room visits for asthma in Seattle. American Review of Respiratory Disease 147, 826–831.

Schwartz, J., Spix, C., Touloumi, G., Bachárová, L., Barumamdzadeh, T., Le, T.A., Piekarski, T., Ponce de Leon, A., Pönkä, A., Saez, M., Schouten, J.P., Rossi, G., 1996. Methodological issues in studies of air pollution and daily counts of deaths or hospital admissions. Journal of Epidemiology and Community Health 50, S3–S11.

Schwartz, J., Spix, C., Wichmann, H.-E., Malin, E., 1991. Air pollution and acute respiratory illness in five German communities. Environmental Research 56, 1–14.

Sheppard, D., Wong, W.S., Uehara, C.F., Nadel, J.A., Boushey, H.A., 1980. Lower threshold and greater bronchomotor responsiveness of asthmatic subjects to sulfur dioxide. American Review of Respiratory Disease 122, 873–878.

Shi, J.P., Harrison, R.M., Brear, F., 1999. Particle size distribution emitted from a modern diesel engine. The Science of the Total Environment 235, 305–317.

Sillman, S., 1999. The relation between ozone, NO_x and hydrocarbons in urban and polluted rural environment. Atmospheric Environment 33, 1821–1845.

Sitzmann, B., Kendall, M., Watt, J., Williams, I., 1996. Personal exposure study of cyclists to airborne particulate matter in London. Journal of Aerosol Science 27, S499–S500.

SMEPB (Shanghai Municipality Environmental Protection Bureau), 1994. Shanghai Environmental Quality Bulletin. SMEPB, Shanghai.

Smith, K.R., Aust, A.E., 1997. Mobilization of iron from urban particulates leads to generation of reactive oxygen species in vitro and induction of ferritin synthesis in human lung epithelial cells. Chemical Research and Toxicology 10, 828–834.

Smith, M., 1998. Studies of lead and children's IQ with respect to blood lead levels of the populations studied. In: Gompertz, D. (Ed.), Report on Recent UK Blood Lead Surveys. Institute for Environmental Health, Leicester, UK.

Sommer, H., Seethaler, R., Chanel, O., Herry, M., Masson, S., Vergnaud, J.-C., 1999. Health costs due to road traffic-related air pollution. An impact assessment project of Austria, France and Switzerland. Economic evaluation. Technical Report on Economy. GVF-Report No. 326-TEH07. Federal Department for Environment, Transport, Energy and Communications Bureau for Transport Studies, Bern.

Spadaro, J.V., Rabl, A., Jourdain, E., Coussy, P., 1998. External costs of air pollution: case study and results for transport between Paris and Lyon. International Journal of Vehicle Design 20 (1–4), 274–282.

Spix, C., Anderson, H.R., Schwartz, J., Vigotti, M.A., LeTerte, A., Vonk, J.M., Touloumi, G., Balducci, F., Piekarski, T., Bachárová, L., Tobias, A., Pönkä, A., Katsouyanni, K., 1998. Short-term effects of air pollution on hospital admissions of respiratory diseases in Europe: a quantitative summary of APHEA study results. Archives of Environmental Health 53, 54–64.

Stanners, D., Bordeau, P., 1995. The Dobris Assessment. European Environment Agency. ISBN 92 826 5409 5.

Stoddard, J.L., Jeffries, D.S., Lukewille, A., Clair, T.A., Dillon, P.J., Driscoll, C.T., Forsius, M., Johannessen, M., Kahl, J.S., Kellogg, J.H., Kemp, A., Mannio, J., Monteith, D.T., Murdoch, P.S., Patrick, S., Rebsdorf, A., Skjelkvale, B.L., Stainton, M.P., Traeen, T., van Dam, H., Webster, K.E., Wieting, J., Wilander, A., 1999. Regional trends in aquatic recovery from acidification in North America and Europe. Nature 401, 575–578.

Sunyer, J., Spix, C., Quenel, P., Ponce de Leon, A., Barumandzadeh, T., Touloumi, G., Bachárová, L., Wojtyniak, B., Vonk, J., Bisanti, L., Schwartz, J., Katsouyanni, K., 1997. Urban air pollution and emergency admissions for asthma in four European cities: the APHEA project. Thorax 52, 760–765.

Thurston, G.D., Ito, K., Hayes, C.G., Bates, D.V., Lippmann, M., 1994. Respiratory hospital admissions and summertime haze air pollution in Toronto, Ontario: consideration of the role of acid aerosols. Environmental Research 65, 271–290.

Timonen, K.L., Pekkanen, J., 1997. Air pollution and respiratory health among children with asthmatic or cough symptoms. American Journal of Respiratory and Critical Care Medicine 156, 546–552.

Touloumi, G., Katsouyanni, K., Zmirou, D., Schwartz, J., Spix, C., Ponce de Leon, A., Tobias, A., Quennel, P., Rabczenko, D., Bachárová, L., Bisanti, L., Vonk, J.M., Ponka, A., 1997. Short-term effects of ambient oxidant exposure on mortality: a combined analysis within the APHEA project. Air pollution and health: a European approach. American Journal of Epidemiology 146, 177–185.

Ulrich, B., 1990. Waldsterben: forest decline in West Germany. Environmental Science and Technology 24 (4), 436–441.

UNECE (United Nations Economic Commission for Europe), 1994. Protocol to the 1979 Convention on Long Range Transboundary Air Pollution on Further Reduction of Sulphur Emissions, Oslo. Also available at http://www.unece.org.

UNECE (United Nations Economic Commission for Europe), 1999. Protocol to Abate Acidification, Eutrophication, and Ground-Level Ozone, Gothenburg. Also available at http://www. unece.org.

US EPA (United States Environmental Protection Agency), 1990. Cancer Risks from Outdoor Exposure to Air Toxics, Vol. 1. Final Report. Washington, DC, USA.

US EPA (United States Environmental Protection Agency), 1993. Motor vehicle related air toxic study. Public Review Draft. Washington, DC, USA.

US EPA (United States Environmental Protection Agency), 1999. Current EPA emission factor and inventory guidance and resource material. Clearinghouse for Inventories and Emissions Factors, Info CHIEF, Emission Factor and Inventory Group (MD–14), Office of Air Quality Planning and Standards, US EPA, Research Triangle Park, NC 27711. Also available at http:// www.epa.gov/ttn/chief/doclist.pdf.

Wade, J., Holman, C., Fergusson, M., 1994. Passenger car global warming potential: current and projected levels in the UK. Energy Policy 22 (6), 509–522.

Walters, S., Ayres, J., 1996. The health effects of air pollution. In: Harrison, R. (Ed.), Pollution Causes, Effects and Control, 3rd ed. Royal Society of Chemistry (Chapter 11, London).

Warren, R.F., ApSimon, H.M., 1999. Uncertainties in integrated assessment modelling of abatement strategies: illustrations with the ASAM model. Journal of Environmental Science & Policy 2, 439–456.

Warren, R.F., ApSimon, H.M., 2000a. Selection of target loads for acidification in emissions abatement policy: the use of gap closure approaches. Water, Air and Soil Pollution 121, 229–258.

Warren, R.F., ApSimon, H.M., 2000b. The role of secondary particulates in European emission abatement strategies. Integrated Assessment 1, 63–68.

Watson, R.T., Zinyowera, M.C., Moss, R.H. (Eds.), 1996. Climate Change 1995 – Impacts, Adaptions and Mitigation of Climate Change: Scientific – Technical Analyses – Contribution of Working Group II to the Second Assessment Report of the Intergovernmental Panel on Climate Change. Oxford University Press, Oxford.

Weston, K., Fowler, D., 1991. The importance of orography in spatial patterns of rainfall acidity in Scotland. Atmospheric Environment 25 (8), 1517–1522.

Whitelegg, J., 1993. Transport for a sustainable future: the case for Europe. Belhaven Press, London. ISBN 1-85293-146-9.

WHO (World Health Organisation, Regional Office for Europe), 1995. Lead and Health. WHO, Copenhagen.

WHO (World Health Organisation, Regional Office for Europe), 1996. Update and revision of the air quality guidelines for Europe. Meeting of the Working Group on Volatile Organic Compounds, Brussels, Belgium, 2–6 October 1995. WHO, Copenhagen.

Williams, I.D., McCrae, I.S., 1995. Road traffic nuisance in residential and commercial areas. Science of the Total Environment 169, 75–82.

Wordley, J., Walters, S., Ayres, J.G., 1997. Short term variations in hospital admissions and mortality and particulate air pollution. Occupational and Environmental Medicine 54, 108–116.

Xu, X., Ding, H., Wang, X., 1995. Acute effects of total suspended particles and sulfur dioxides on preterm delivery: a community-based cohort study. Archives of Environmental Health 50, 407–415.

Yamartino, R.J., Wiegand, G., 1986. Development and evaluation of simple models for the flow, turbulence and pollutant concentration fields within an urban street canyon. Atmospheric Environment 20 (11), 2137–2156.

Yang, J.S., Kang, S.K., Park, I.J., Rhee, K.Y., Moon, Y.H., Sohn, D.H., 1996. Lead concentration in blood among the general population of Korea. International Archives of Occupational and Environmental Health 68, 199–202.

Zmirou, D., Barumamdzadeh, T., Balducci, F., Ritter, P., Laham, G., Ghilhardi, J.-P., 1996. Short term effects of air pollution on mortality in the city of Lyon, France, 1985–90. Journal of Epidemiology and Community Health 50, S30–S35.

Air Pollution Science for the 21st Century
J. Austin, P. Brimblecombe and W. Sturges, editors
© 2002 Elsevier Science Ltd. All rights reserved.

Chapter 7

New Directions: Air pollution and road traffic in developing countries

A. Faiz

The World Bank, Washington, DC 20433, USA
E-mail: afaiz@worldbank.org

P.J. Sturm

Institute for Internal Combustion Engines and Thermodynamics,
Technical University of Graz, A-8010, Austria
E-mail: sturm@vkmb.tu-graz.ac.at

Motorized road vehicles are the primary means of transporting passengers and freight throughout the developing world because of their versatility, flexibility, and low initial cost, as compared to other transport modes. In all but the poorest developing countries, economic growth, rising incomes and urbanization are contributing to a rapid increase in vehicle ownership and use. Over the last two decades, motor vehicles have emerged as a critical source of urban air pollution in much of the developing world. For example, motor vehicles are the largest source of PM_{10} emissions in most Asian cities, exceeding the contributions from resuspended road dust, heavy fuel oil and coal combustion, and refuse burning. The incidence of other transport-related pollutants (e.g. CO, NO_x, SO_2, O_3) in developing countries also exceeds international and national norms. The associated human health and welfare costs run into hundreds of millions of dollars and far exceed the prevention costs in terms of the control measures.

Air pollution is a serious public health problem in most major metropolitan areas in the developing world. Pollutant levels in megacities such as Bangkok, Mexico City, and Cairo far exceed those in any city in the industrialized countries. Epidemiological studies show that air pollution in developing countries annually accounts for tens of thousands of excess deaths and billions of dollars in medical costs and lost productivity. These losses and the associated degradation in quality of life, impose a major burden on people in all sectors of society, but especially the poor. The air pollution problem is particularly serious in the rapidly urbanizing cities of South and East Asia (especially in mega cities such as Bombay, Calcutta, Delhi, Dhaka, and Karachi in South

First published in Atmospheric Environment 34 (2000) 4745–4746

Asia, and Bangkok, Beijing, Shanghai, Jakarta, and Manila in East Asia). Here a growing number of urban dwellers are exposed to unacceptable levels of atmospheric pollutants from a variety of sources. In addition, the problem is rapidly spreading to other urban centers (such as Colombo, Dhaka, Lahore, Jaipur, Kathmandu, Kuala Lumpur, Surabaya and numerous regional cities in China). Air quality in several Asian megacities is approaching the dangerous levels recorded in London in the 1950s. In these cities, pollution levels often exceed World Health Organization (WHO) air quality guidelines by a factor of three or more (World Resources: A Guide to the Global Environment 1998–99, Oxford University Press, New York, 1998, and Urban Air Pollution in Megacities of the World, Blackwell, Oxford, 1992). In China's major urban centers, particulate levels are as much as six times the WHO guidelines. The rapid and continuing increase in motorized and polluting forms of transport combined with inadequate transport infrastructure, lax environmental legislation and enforcement, weak institutions, and lack of skilled manpower, have resulted in poorly planned urban growth with severe air pollution problems.

With economic prosperity and urbanization, there has been an unabated increase in motor vehicles bringing unprecedented mobility to the burgeoning middle class in many Asian countries. Although per capita vehicle ownership in most Asian countries is low compared to OECD countries (for example, in China there are about 8 vehicles per 1000 persons and in India only 7 vehicles per 1000 persons compared to 750 vehicles per 1000 persons in the US), vehicle growth in the region has been phenomenal. The growth of motor vehicles in China has averaged about 11% annually in the last 30 years doubling every 5 years, and in India the growth has been around 7% per annum for the past 10 years. The motor vehicle fleet is predominantly concentrated in urban areas and the growth of the urban fleet tends to be much faster. This concentration of the motor vehicle fleet, particularly cars, in major urban centers has led to severe traffic congestion as well as concomitant problems of air pollution and road accidents. It is estimated that over 90% of the urban population in Asia is exposed to particulate matter (PM) concentrations above the WHO guidelines.

In terms of human exposure to harmful emissions, motor vehicle traffic has become a major problem in large cities. As reported cases of lung disease and breathing disorders increase, many governments are beginning to search for solutions to the problem of air pollution. Although it is difficult to isolate air pollution-related health impacts, estimates have been made of the health and related economic toll of air pollution in Asian cities. For example, it has been estimated that about 10,000 people may die prematurely due to air pollution in Delhi and many hundreds of thousands of cases of respiratory illnesses have been attributed to poor air quality (Anand, Proceedings of the Workshop on Integrated Approach to Vehicular Pollution Control in Delhi, Central Pollution Control Board, New Delhi, pp. 110–117, 1998). Recent surveys in Thailand have found that Bangkok residents have a much higher occurrence rate of pollution-related respiratory diseases (19%) as compared to rural cities with lower levels of air pollution (8%). Blood lead levels remain high in urban residents in Asia, especially in children, although unleaded gasoline is becoming widely available in parts of Asia (e.g. Thailand, South Korea) and has been mandated for use by governments in China, and India. Diesel in most developing Asian economies continues to have high levels of sulfur, with high costs being the primary hurdle to switching to low-sulfur diesel.

Experimentation with alternative fuels (Faiz et al., Air Pollution from Motor Vehicles: Standards and Technologies for Controlling Emissions, World Bank, Washington, DC, 1996) such as electricity, LPG, CNG, etc., has met with varying degrees of success. Factors such as higher initial costs and subsidies, relative fuel prices, public inertia and acceptance, and lack of a policy framework, impede the widespread implementation of alternative fuel technologies in transport vehicles. As the smaller and medium Asian cities continue to grow economically and physically, many of them will experience the same problems that have so far plagued the large metropolitan centers. Planning and management, as well as regular monitoring and evaluation of air quality, vehicle emissions, and health impacts are essential if human health and productivity in these countries are to be protected. These actions need to be supported by technical measures involving the vehicle–fuel complex to reduce automotive emissions. In addition, transport demand management and market incentives, and urban transport and road infrastructure improvements are also required.

Air pollution is, of course, a problem that has yet to be solved even in industrialized countries. Unabated demand for transportation, as reflected in increasing personal mobility and globalization of industry and services, has counterbalanced the technical improvements in vehicle and fuel technologies to reduce vehicular emissions. Although impressive improvements in vehicle emission controls have been achieved in industrialized countries for CO, SO_2, and lead, ambient concentrations of NO_x and tropospheric ozone still remain at high levels, and the problem of fine particulate matter is yet to be resolved.

Air Pollution Science for the 21st Century
J. Austin, P. Brimblecombe and W. Sturges, editors
© 2002 Elsevier Science Ltd. All rights reserved.

Chapter 8

New Directions: Assessing the real impact of CO_2 emissions trading by the aviation industry

David S. Lee

Defence Evaluation and Research Agency, Propulsion and Performance Department,
Pyestock, GU14 0LS, UK
E-mail: dslee@dera.gov.uk

Robert Sausen

Deutsches Zentrum für Luft und Raumfahrt e.V., Institut für Physik der Atmosphäre,
Oberpfaffenhofen, D-82234, Wessling, Germany
E-mail: robert.sausen@dlr.de

The recent report of the Intergovernmental Panel on Climate Change (IPCC, Aviation and the Global Atmosphere 1999) assessed the impacts of subsonic aviation on the radiative forcing of climate for 1992 and 2050. Radiative forcing (RF) effects arise from CO_2 from the fuel burned, plus other emissions that result in aerosol, contrails, ozone (O_3) formation, methane (CH_4) destruction (the latter two from NO_x emissions), and possibly enhanced cloudiness. These effects are shown in Fig. 1.

The CH_4 loss results in a negative RF. The effect of positive RF from O_3 formation and negative RF from CH_4 removal do not, however, cancel each other out, nor do they imply a null climatic effect. The negative RF from CH_4 is rather uniform across latitudes (like the positive RF from CO_2), whereas the RF from O_3 is concentrated in the Northern Hemisphere. The overall climate effect from inhomogeneous forcing (remembering that RF is a proxy for climate change) is unknown. There are indications, however, that it might be larger than that expected from homogeneous forcing (Ponater et al., Climate Dynamics, Vol. 15 (1999) pp. 631–642). Contrails and aviation-induced cloudiness also result in a large RF, but there are large uncertainties associated with the estimated effects.

The RF from aviation CO_2 amounts to 37% of the total aviation RF (excluding the possibility of enhanced cirrus formation); the rest being attributed to other effects associated with aviation emissions. The RF effect of tropospheric O_3 formation and contrails is dependent upon factors such as altitude, latitude and longitude, which define chemical and physical conditions. Thus, the RF

First published in Atmospheric Environment 34 (2000) 5337–5338

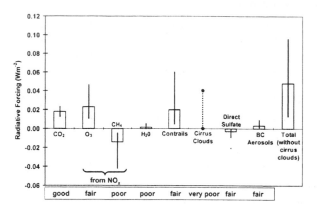

Figure 1.   Globally averaged radiative forcing (RF) from 1992 subsonic aviation (IPCC, 1999). The evaluations ('good', 'fair' etc.) are a relative appraisal associated with each component, and indicate the level of scientific understanding.

effects from O_3 and contrails are aviation specific, whereas CO_2 RF (from burning aviation fuel) is not.

Currently, emissions of CO_2 from international aviation are not covered by the Kyoto Protocol. Following the publication of the IPCC report, the International Civil Aviation Organisation (ICAO) Assembly adopted a resolution that its Committee on Aviation Environmental Protection (CAEP) would study options to limit or reduce 'greenhouse'gases (GHGs) from aviation. CAEP is looking at possibilities from technology and standards, operational measures, and market-based options (ICAO Journal, Vol. 54 (1999) pp. 5–8). Market-based options might include emission charges, fuel taxes and emissions trading. Gander and Helme suggest (ICAO Journal, Vol. 54 (1999) pp. 12–14) that emissions trading of CO_2 may offer a workable solution to emissions reductions from aviation. One of CAEP's Working Groups is considering market-based options and focussing its efforts on CO_2.

Here we consider emissions trading from an atmospheric impacts perspective. Emissions trading can be instituted in a number of ways but for simplicity we will define 'open trading' as being inter-sector, and 'closed trading' as being intra-sector (i.e. within the aviation industry). We consider the potential consequences assuming that aviation is, overall, a purchaser rather than a seller of CO_2 emissions permits, which we believe to be a reasonable assumption.

Considering *'open trading'* first, two major issues need to be discussed: firstly, NO_x emissions associated with fossil fuel combustion (and therefore CO_2 emission) at the Earth's surface are involved in O_3 production, but the overall production efficiency is higher at cruise altitudes in the upper troposphere. Moreover, the temperature response (and thus RF) of O_3 is very altitudinally dependent. Secondly, fossil fuel combustion at the Earth's surface results in water vapour, but for aviation the resultant water vapour is involved in contrail formation resulting in an RF that would not occur in the case of surface emissions.

Next, we consider '*closed trading*': within the aviation sector the magnitude of the RF from other emitted species is also variable for the same global fuel burn and thus, CO_2 emissions. For example, the O_3 production efficiency of the upper troposphere depends upon the background NO_x concentration. An approximately linear relationship (Grewe et al., Geophysics Research Letters, Vol. 26 (1999) pp. 47–50) has been found between increases in global aviation NO_x and increases in global tropospheric O_3 in modelling studies. However, when these changes are examined by latitude, different sensitivities are found. For example, incremental increases in O_3 are much greater at southerly latitudes than for the whole atmosphere for the same increase in NO_x emissions. This implies that for capped global CO_2 emissions from aviation, if emissions were traded such that more flights occurred in the tropics or the Southern Hemisphere, then increases in O_3 (and its associated RF) would be much stronger than if emissions occurred over areas with relatively high background NO_x (e.g. northerly latitudes).

A similar phenomenon will occur for contrails. The tropics are more susceptible to contrail formation because of the higher humidity (Sausen et al., Theoretical and Applied Climatology, Vol. 61 (1998) pp. 127–141) such that a shift of growth to lower latitudes enabled by CO_2 emissions trading would increase RF. This would be exacerbated if the fleet growth occurred in these regions from newer aircraft, as they have a higher propulsive efficiency than older aircraft and trigger contrail formation more easily (Schumann, Aerospace Science and Technology (2000), in press).

The differential sensitivity of the atmosphere to aircraft emissions of NO_x, particles and water vapour dictates the amount of O_3 and persistent contrails formed, and therefore the overall RF effect. In addition, these spatially variable effects may amplify responses under the open trading scenario.

We conclude the following:

- If aviation participates in an open regime of inter-sector CO_2 emissions trading and aviation is an overall purchaser, then for capped global CO_2 emissions, any purchase of additional CO_2 emission permits by aviation from other sectors will result in a larger RF from associated aviation emissions than if the CO_2 had been emitted at the Earth's surface.
- If a closed intra-sector emissions trading scheme for aviation CO_2 is envisaged, the total RF effects from associated emissions could be greater or lesser, depending upon the latitude, longitude and altitude at which they are emitted for the same global capped CO_2 emissions.

It is evidently important to consider RF effects from other aviation emissions in addition to CO_2, as aviation is a unique source sector in terms of RF effects. It is possible that emissions trading – which has the ambition of reducing RF – could actually increase the RF from aviation for the same global capped CO_2 emission. Aviation's unique emission characteristics and potential RF effects should, therefore, be considered in formulating policies to mitigate climate change.

The obvious way forward in an open sector emissions trading regime is to weight any CO_2 permits purchased by aviation such that the additional RF effects are accounted for. However, such weighting functions would be spatially and temporally variable, and the supporting science to define these require further development.

Air Pollution Science for the 21st Century
J. Austin, P. Brimblecombe and W. Sturges, editors
© 2002 Elsevier Science Ltd. All rights reserved.

Chapter 9

The atmospheric chemistry of sulphur and nitrogen in power station plumes

C.N. Hewitt

*Department of Environmental Science, Institute of Environmental and Natural Sciences,
Lancaster University, Lancaster, LA1 4YQ, UK
E-mail: n.hewitt@lancaster.ac.uk*

Abstract

Emissions of sulphur and nitrogen compounds from power stations represent a significant fraction of the total emissions of these elements to the atmosphere. Understanding their subsequent chemical reactions in the atmosphere is of fundamental importance as without it, a quantitative assessment of their contribution to local and regional scale air pollution is not possible. Here the atmospheric chemistry of sulphur dioxide and the oxides of nitrogen, and their resultant likely behaviour in the plumes of power stations are reviewed.

1. Introduction

At least 90% of the sulphur present in fossil fuel enters the gas phase in the form of sulphur dioxide (SO_2) during combustion, and, unless it is deliberately removed from the flue gas, is emitted to the atmosphere, leading to the global anthropogenic emission of about 65 Tg SO_2 (as S) yr^{-1} (Benkovitz et al., 1996; Spiro et al., 1992). Globally, an increasing proportion of this reactive sulphur is emitted from power plants, as power generation becomes more and more concentrated on large units. High-temperature combustion processes also inevitably produce oxides of nitrogen ($NO_x = NO + NO_2$), leading to the global emission of about 21–25 Tg NO_x (expressed as N) yr^{-1} (Benkovitz et al., 1996; Dignon, 1992). SO_2 and NO_x are important air pollutants with both environmental and health effects attributed to them. Both compounds may lead to the deposition of acidity ("acid deposition" or "acid rain") on the regional scale and NO_x takes part in chemical reactions that lead to the formation of ozone (O_3) in the troposphere, also on the regional scale. However, here the

First published in Atmospheric Environment 35 (2001) 1155–1170

focus is on the processes that affect SO_2 and NO_x oxidation in power station plumes, particularly with respect to how these processes influence the concentrations of the primary pollutants and their products in the relatively near field, for example at receptor sites located tens of kilometres from the source.

2. Review of the chemistry of sulphur dioxide and the oxides of nitrogen in the atmosphere

A very considerable body of literature exists on the atmospheric chemistry of sulphur and nitrogen occurring inside power station plumes. This is derived from theoretical and experimental studies of the chemistry of S and N compounds in the gas and aerosol phases, supplemented by observations of concentration changes in the atmosphere, both inside and outside of power station plumes.

2.1. Gas-phase oxidation of sulphur dioxide

In the gas phase, SO_2 can potentially be subject to a large number of reactions involving reactive transient oxidants (Calvert et al., 1978). These reactions, shown in Table 1, are all thermodynamically possible in the atmosphere under ambient conditions of temperature and pressure, but have widely differing elementary rate constants, ranging from about 10^{-12} to $\sim 10^{-20}$ cm^3 molec^{-1} s^{-1}. Coupled with the wide range of concentrations of these various oxidants in the atmosphere, this variation in rate constants ensures that only very few of these possible reactions play any appreciable role in the oxidation of SO_2 in the atmosphere. In fact, the reactions of SO_2 with oxygen atoms and ozone are unlikely to be important in the gas phase, with the possible exception of the reaction with the $O(^3P)$ atom in the very early stages of stack gas dispersion. $O(^3P)$ is formed by the photodissociation of NO_2, and under conditions of very high NO_2 concentrations, such as might exist in a plume close to the point of emission, it is possible that significant $O(^3P)$ concentrations may result in appreciable SO_2 oxidation rates. However, as plume dilution with background air occurs, the instantaneous rate of this reaction will quickly fall.

Oxidation of SO_2 by ozone (O_3) is highly exothermic $(\Delta H = -242\,\text{kJ mol}^{-1})$. However with a gas-phase rate constant of $k = \sim 8 \times 10^{-24}\,\text{m}^3\,\text{molec}^{-1}\,\text{s}^{-1}$ and a maximum likely ozone concentration of $\sim 5 \times 10^{12}\,\text{molec cm}^{-3}$ there will normally be insignificant SO_2 oxidation by this route $(< 1.4 \times 10^{-5}\%\,\text{h}^{-1})$. Although the rate of the gas-phase reaction of SO_2 with ozone is extremely slow, the addition of an alkene to a dilute O_3–SO_2 mixture in air results in the significant oxidation of the SO_2 (Cox and Penkett, 1971, 1972). The mechanisms behind this enhancement in oxidation rate are not fully elucidated but

Table 1. Enthalpy changes and rate constants for potentially important reactions of ground state SO_2 and SO_3 molecules in the lower atmosphere (adapted from Calvert, 1984)

Reaction	$-\Delta H°$ (kJ mol^{-1}) 25°C	k (cm^3 molec^{-1} s^{-1})
$O_2(^1\Delta_g) + SO_2 \rightarrow SO_4$ (biradical)	105	
$O_2(^1\Delta_g) + SO_2 \rightarrow SO_4$ (cyclic)	~ 117	
$O_2(^1\Delta_g) + SO_2 \rightarrow SO_3 + O(^3P)$	−56.5	3.9×10^{-20}
$O_2(^1\Delta_g) + SO_2 \rightarrow SO_2(^3\Sigma_g^-) + SO_2$	94	
$O_2(^1\Sigma_g^+) + SO_2 \rightarrow SO_4$ (biradical)	167	
$O_2(^1\Sigma_g^+) + SO_2 \rightarrow SO_4$ (cyclic)	180	
$O_2(^1\Sigma_g^+) + SO_2 \rightarrow SO_3 + O(^3P)$	6.3	6.6×10^{-16}
$O_2(^1\Sigma_g^+) + SO_2 \rightarrow SO_2 + O_2(^1\Delta_g)$	62.8	
$O(^3P) + SO_2(+M) \rightarrow SO_3(+M)$	347	5.7×10^{-14}
$SO_3 + SO_2 \rightarrow O_2 + SO_3$	241	$< 8 \times 10^{-24}$
$NO_2 + SO_2 \rightarrow NO + SO_3$	41.4	8.8×10^{-30}
$NO_3 + SO_2 \rightarrow NO_2 + SO_3$	136	$< 7 \times 10^{-21}$
$ONOO + SO_2 \rightarrow NO_2 + SO_3$	~ 126	$< 7 \times 10^{-21}$
$N_2O_5 + SO_2 \rightarrow N_2O_4 + SO_3$	100	$< 4 \times 10^{-23}$
$HO_2 + SO_2 \rightarrow HO + SO_3$	69.9	
$HO_2 + SO_2(+M) \rightarrow HO_2SO_2(+M)$	29.3	$< 1 \times 10^{-18}$
$CHO_3O_2 + SO_2 \rightarrow CHO_3O + SO_3$	~ 113	$< 1 \times 10^{-18}$
$CHO_3O_2 + SO_2(+M) \rightarrow CHO_3O_2SO_2(+M)$	~ 130	$~ 1.4 \times 10^{-14}$
$(CH_3)_3CO_2 + SO_2 \rightarrow (CH_3)_3CO + SO_3$	~ 109	
$(CH_3)_3CO_2 + SO_2 \rightarrow (CH_3)_3CO_2SO_2$	~ 126	$< 7.3 \times 10^{-19}$
$CHO_3COO_2 + SO_2 \rightarrow CH_3CO_2 + SO_3$	~ 138	
$CHO_3COO_2 + SO_2 \rightarrow CH_3COO_2SO_2$	~ 155	$< 7 \times 10^{-19}$
$HO + SO_2(+M) \rightarrow HOSO_2(+M)$	~ 155	1.1×10^{-12}
$CHO_3O + SO_2(+M) \rightarrow CH_3OSO_2(+M)$	~ 100	$~ 5.5 \times 10^{-13}$
$SO_3 + H_2O \rightarrow H_2SO_4$	24.8	9.1×10^{-13}

are most likely due to the formation of the Criegee intermediate (CH_2O_2) and its reaction with SO_2. However, the significance of this reaction in the real atmosphere is difficult to assess, since the effect of addition of NO to the reaction mixture, and competition between NO and SO_2 for reaction with CH_2O_2, is not fully known.

By far the most important route of oxidation of SO_2 in the gas phase is by reaction with the hydroxyl radical, OH (e.g. McAndrew and Wheeler, 1962; Harris and Wayne, 1975; Castleman and Tang, 1976; Atkinson et al., 1976; Davis et al., 1979a, b; Cox and Sheppard, 1980; Harris et al., 1980; Leu, 1982).

In the unpolluted atmosphere, OH is produced by the photolysis of O_3 and the subsequent reaction of oxygen atoms with water vapour:

$$O_3 + h\nu \rightarrow O(^3P) + O_2 \quad \lambda > 315 \, nm,$$

$$O(^3P) + O_2 + M \rightarrow O_3 + M$$

or

$$O_3 + h\nu \rightarrow O(^1D) + O_2 \quad \lambda < 315\,\text{nm},$$

$$O(^1D) + M \rightarrow O(^3P) + M \quad 96\%$$

but

$$O(^1D) + H_2O \rightarrow 2OH \quad 4\%.$$

In polluted air, the photolysis of nitrous acid (HONO) and hydrogen peroxide (H_2O_2) produce OH directly

$$HONO + h\nu \rightarrow OH + NO \quad \lambda < 400\,\text{nm},$$

$$H_2O_2 + h\nu \rightarrow 2OH \quad \lambda < 360\,\text{nm}.$$

Under more polluted conditions, OH radicals are produced by the photolysis of the products of the incomplete combustion of fossil fuels, especially aldehydes, ketones and other oxygenated organic compounds. For example, formaldehyde photolysis leads to

$$HCHO + h\nu \rightarrow H + CHO \quad \lambda < 335\,\text{nm}$$

$$CHO + O_2 \rightarrow HO_2 + CO$$

$$H + O_2 + M \rightarrow HO_2 + M$$

$$HO_2 + HO_2 \rightarrow H_2O_2 + O_2$$

then

$$H_2O_2 + h\nu \rightarrow 2OH \quad \lambda < 360\,\text{nm}.$$

Hydroxyl then rapidly reacts with SO_2 by addition:

$$OH + SO_2 + M \rightarrow HOSO_2 + M$$

where M is another molecule (usually N_2) that is required to absorb excess kinetic energy from the reactants. At the pressures pertaining in the lower atmosphere the rate of this reaction is independent of the concentration of M and hence it is effectively a second-order reaction.

Many attempts have been made to measure the rate constant for the OH–SO_2 reaction in the laboratory (e.g. Cox, 1975; Castleman and Tang, 1976;

Atkinson et al., 1976; Cox and Sheppard, 1980; Harris et al., 1980) and rather a wide range of values have been obtained. However, Atkinson and Lloyd (1984) recommend an effective bimolecular rate constant at 1 atm and 298 K of $9 \times 10^{-13} \, cm^3 \, molec^{-1} \, s^{-1}$, with an uncertainty of around $\pm 50\%$. For an average ambient OH concentration of 1×10^6 radicals cm^{-3} (Hewitt and Harrison, 1985) the lifetime of SO_2 with respect to this one gas phase reaction will be ~13 d in the "typical" troposphere.

Once formed, the free radical $HOSO_2$ will rapidly react with oxygen to form SO_3, which in turn will rapidly react with water vapour to form sulphuric acid, H_2SO_4 (Goodeve et al., 1934; Cox and Penkett, 1971; Calvert et al., 1978; Stockwell and Calvert, 1983):

$$HOSO_2 + O_2 \rightarrow HO_2 + SO_3,$$

$$SO_3 + H_2O \rightarrow H_2SO_4.$$

Eatough et al. (1994) point out that the rate of conversion of SO_2 to sulphate by OH will increase with both increasing temperature and relative humidity. In practise, the rate of the homogeneous gas-phase conversion of SO_2 to sulphate can vary from much less than $1\% \, h^{-1}$ up to a probable maximum of $10\% \, h^{-1}$ under optimum atmospheric conditions.

2.2. Aqueous phase oxidation of sulphur dioxide

The presence of aqueous droplets in the atmosphere offers another phase in which oxidation of SO_2 can occur. After transport of the gas to the surface of the droplet and transfer of the gas across the air–liquid interface, SO_2 can dissolve in water and establish an equilibrium with its ionic products, with the aqueous-phase concentrations described at equilibrium by the Henry's law constant. As a result, dissolved SO_2 actually consists of three species, hydrated SO_2 ($SO_2 \cdot H_2O$), the bisulphite ion (HSO_3^-) and the sulphite ion (SO_3^{2-}):

$$SO_2 + H_2O \rightleftharpoons SO_2 \cdot H_2O,$$

$$SO_2 \cdot H_2O \rightleftharpoons HSO_3^- + H^+,$$

$$HSO_3^- \rightleftharpoons SO_3^{2-} + H^+.$$

The predominant form depends upon the acidity of the solution in which the SO_2 dissolves, since the hydrogen ion concentration will drive the equilibrium either to the left or right (Martin and Damschen, 1981). The oxidation state of the sulphur in the various products is $+4$, and hence S(IV) is used to denote all these forms of sulphur taken together. In contrast, the oxidised form of sulphur (i.e. H_2SO_4 and SO_4^{2-}) is in the $+6$ oxidation state and hence is referred to as S(VI). There are several possible routes of oxidation of S(IV) to S(VI) and these are now briefly considered.

Molecular oxygen may oxidise S(IV) to S(VI) but in the absence of a catalyst the reaction is insignificant. Although both iron (III) and manganese (II) appear to catalyse the reaction it is still relatively slow and unlikely to be of significance (e.g. Huss et al., 1982; Martin and Good, 1991; Martin et al., 1991).

As noted above, O_3 reacts very slowly with SO_2 in the gas phase, but in the liquid phase the reaction is rapid (e.g. Penkett et al., 1979):

$$S(IV) + O_3 \rightarrow S(VI) + O_2.$$

At ambient gas-phase concentrations of 30–60 ppb and a Henry's law constant of $0.01 \, mol \, l^{-1} \, atm^{-1}$, the expected aqueous-phase equilibrium concentration of O_3 will be around $(3–6) \times 10^{-10} \, mol \, l^{-1}$. Although this concentration is around 6 orders of magnitude lower than that for dissolved O_2, oxidation of S(IV) by O_3 is much more important under most conditions because of the much higher rate constant for the reaction, which is pH dependent (e.g. Lagrange et al., 1994; Botha et al., 1994). At a pH greater that around 4 this is likely to be an important route of oxidation of S(IV) in water droplets.

Hydrogen peroxide has been shown to oxidise S(IV) relatively rapidly in solution and in fact is one of the most effective oxidants of S(IV) in solution (Penkett et al., 1979). Because of its high solubility, with a Henry's law constant of $1 \times 10^5 \, M \, atm^{-1}$, a 1 ppb gas-phase concentration of H_2O_2 would produce an equilibrium aqueous-phase concentration at 298 K of around $1 \times 10^{-4} \, mol \, l^{-1}$, or 6 orders of magnitude greater than the solution-phase equilibrium O_3 concentration. The rate constant for the reaction of H_2O_2 and S(IV) is largely pH independent, and at a gas-phase H_2O_2 concentration of 1 ppb is around $300 \times 10^{-6} \, M \, h^{-1} \, (ppb \, SO_2)^{-1}$. Organic peroxides have also been proposed as potential aqueous phase S(IV) oxidants but have lower Henry's law constants and lower gas-phase concentrations than does H_2O_2 and are considered to normally be of minor importance in this regard (Lind et al., 1987).

The different routes of aqueous-phase oxidation of S(IV) can be compared as a function of pH and temperature (Seinfeld and Pandis, 1998), using starting conditions of, for example, 5 ppb SO_2, 1 ppb NO_2, 1 ppb H_2O_2, 50 ppb O_3, 0.3 μM Fe(III) and 0.03 μM Mn(II). Under practically all conditions, oxidation by dissolved H_2O_2 is the predominant pathway for sulphate formation. Only at pH values greater than 5 does oxidation by O_3 start to dominate, and at pH 6 this route is around 10 times faster than that by H_2O_2. Assuming a liquid cloud water content of $1 \, g \, m^{-3}$, the rate of oxidation by H_2O_2 can exceed $500\% \, h^{-1}$. In contrast to the homogeneous gas-phase reactions of SO_2 the aqueous-phase process is therefore largely controlled by mixing and reactant limitations, rather than kinetic considerations (Eatough et al., 1994).

2.3. Gas-phase oxidation of the oxides of nitrogen

Homogeneous formation of nitric acid by the oxidation of NO and NO_2 appears to be a much less complicated, and better understood process than the formation of sulphuric acid from sulphur dioxide. Most work on NO_x oxidation has been carried out in polluted air and this may hinder extrapolation of predicted oxidation rates to the case of power station plumes. NO and smaller amounts of NO_2 are formed during high temperature combustion processes, with power stations emitting NO_x (the sum of NO and NO_2) as $\sim 95\%$ NO and $\sim 5\%$ NO_2.

The termolecular reaction of NO with O_2:

$$2NO + O_2 \rightarrow 2NO_2$$

is very slow at atmospheric temperatures and, except at extremely high NO concentrations, is insignificant compared with the reactions of NO with O_3, OH, HO_2 (hydroperoxy) and RO_2 (alkylperoxy):

$$NO + O_3 \rightarrow NO_2 + O_2,$$

$$NO + HO_2 \rightarrow NO_2 + OH,$$

$$NO + RO_2 \rightarrow NO_2 + RH.$$

The peroxyradicals, HO_2 and RO_2, are formed in chain reactions in the atmosphere, initiated by the attack of hydroxyl on reactive hydrocarbons. For example, in the simplest (but slowest) case involving methane

$$CH_4 + OH \rightarrow CH_3 + H_2O,$$

$$CH_3 + O_2 + M \rightarrow CH_3O_2 + M,$$

$$CH_3O_2 + NO \rightarrow NO_2 + CH_3O,$$

$$CH_3O + O_2 \rightarrow HCHO + HO_2,$$

$$H_2O + NO \rightarrow NO_2 + OH.$$

There is some observational evidence that the termolecular reaction with O_2 may play a role in the rapid oxidation of NO to NO_2 in highly polluted urban air in the wintertime, when the concentrations of the photochemically produced oxidants O_3, OH, HO_2 and RO_2 are likely to be very low. However the role of free radical chemistry is also now thought to be significant, even in wintertime smog episodes (Harrison and Shi, 1996; Shi and Harrison, 1997; Harrison et al., 1998).

Overall, then, the predominant fate of emitted NO is oxidation to NO_2 which can then react with OH during the daytime:

$$NO_2 + OH + M \rightarrow HNO_3 + M.$$

A great variety of experimental studies support the view that this reaction is relatively rapid, with an effective 2nd order rate constant at 1 atm in air of around 1.1×10^{-11} cm^3 $molec^{-1}$ s^{-1} at 298 K (Atkinson and Lloyd, 1984), approximately ten times the rate of the SO_2–OH reaction.

A second potentially important source of nitric acid involves the reaction of the nitrate radical, NO_3, with some organic compounds. The nitrate radical is formed in the presence of NO_2 and O_3 and is in equilibrium with N_2O_5 in the atmosphere:

$$O_3 + NO_2 \rightarrow NO_3 + O_2$$

$k = 3.2 \times 10^{-17}$ cm^3 $molec^{-1}$ s^{-1} (Atkinson and Lloyd, 1984)

$$NO_3 + NO_2 \rightleftharpoons N_2O_5.$$

NO_3 reacts relatively rapidly with a variety of organics, and in the case of alkanes and aldehydes, the reaction is believed to proceed by hydrogen abstraction to form nitric acid:

$$NO_3 + RH \rightarrow HNO_3,$$

$$NO_3 + RCHO \rightarrow RCO + HNO_3.$$

In polluted urban air, alkane concentrations (excluding CH_4) are typically around 100 ppb (2.5×10^{12} $molec$ cm^{-3}) and total aldehyde concentrations are of the order of tens of ppb. Using an estimated NO_3 concentration of 100 ppt, the calculated rate of formation of HNO_3 formation by this route would be around 0.3 ppb h^{-1}. This can be compared to the daytime rate of formation of HNO_3 by reaction of NO_2 with the OH radical of around 2 ppb h^{-1}, at an NO_2 concentration of 50 ppb and an average OH concentration of 1×10^6 cm^{-3} (Hewitt and Harrison, 1985). Thus the rate of nighttime formation of HNO_3 is likely to be rather insignificant relative to the rate of daytime formation.

Another potential source of nitric acid is the hydrolysis of N_2O_5 (Platt and Perner, 1980):

$$N_2O_5 + H_2O \rightarrow 2HNO_3$$

where $k < 1.3 \times 10^{-11} \, cm^3 \, molec^{-1} \, s^{-1}$ (Tuazon et al., 1983). This rate constant is sufficiently large to ensure that this reaction may contribute significantly to HNO_3 formation in the polluted atmosphere. In fact, Tuazon et al. (1983) estimated a formation rate of $0.3 \, ppb \, h^{-1}$ at NO_2 and NO_3 concentrations of 3 ppb and 100 ppt, respectively, at 50% relative humidity. This is similar to the rate of formation due to the reactions of the nitrate radical with alkanes and aldehydes. Other studies, including Jones and Seinfeld (1983) and Richards et al. (1981, 1983) confirm this.

2.4. Aqueous phase oxidation of the oxides of nitrogen

Some field studies (e.g. Lazrus et al., 1983) have indicated appreciable formation of nitric acid in the aqueous phase in the atmosphere (NO_2 conversion rates of $\sim 8\% \, h^{-1}$, Gertler et al., 1984), but laboratory studies are less conclusive. In fact, aqueous phase oxidation of the oxides of nitrogen probably proceeds far too slowly under ambient conditions to contribute to either the significant removal of these gas-phase compounds or to cloud-water acidification (rate constants of 2×10^{10} and $1.3 \times 10^9 \, M^{-1} \, s^{-1}$, for the reactions of NO and NO_2 with OH, respectively).

2.5. Sulphuric and nitric acids compared

Once formed in the atmosphere sulphuric and nitric acids show very different behaviours, both physically and chemically. Nitric acid is more volatile and hence exists in significant concentrations in the gas phase, while sulphuric acid has a very low vapour pressure under ambient conditions and hence exists in the form of aerosol phase particles.

Both nitric acid and sulphuric acid will react with alkaline substances in the atmosphere to produce salts, the most important of these under ambient conditions being the ammonium compounds resulting from reactions with ammonia (APEG, 1999; Lee and Atkins, 1994; Langford et al., 1992).

When HNO_3 reacts with NH_3 an equilibrium is established (Hidleman et al., 1984):

$$HNO_3 + NH_3 \rightleftharpoons NH_4NO_3.$$

The ammonium nitrate exists as a solid if the relative humidity is less than that of deliquescence. At higher relative humidities it will exist as aerosol droplets. However at elevated temperatures, above $\sim 310 \, K$, little NH_4NO_3 will exist in the atmosphere, since the dissociation constant is temperature dependent and the above reaction will move to the left (Mozurkewich, 1993).

In polluted air, where high concentrations of both NO_x and SO_2 are present, the oxidation chemistries of these compounds become intertwined. As NO_x concentrations increase, so the gas-phase concentration of OH will decrease, hence lowering the rate of oxidation of SO_2. Similarly, high concentrations of NO_x will lead to a decrease in the rate of formation of H_2O_2 as the NO molecules will compete for reaction with HO_2, so reducing the number of HO_2 radicals available for the self-reaction needed to produce H_2O_2.

2.6. Removal of sulphur and nitrogen by wet and dry deposition processes

Both the primary pollutants, SO_2 and NO_x, and their secondary products, H_2SO_4, HNO_3, SO_4^{2-} and NO_3^-, as well as their further reaction products, e.g. $(NH_4)_2SO_4$ and NH_4NO_3, are subject to continuous removal from the boundary layer by the process of dry deposition to the Earth's surface. This process may be parameterized by the use of a deposition velocity, the magnitude of which varies from compound to compound and with environmental conditions and the nature of the surface. A typical range of deposition velocities reported for various pollutants and surfaces is $0.1–2\,\mathrm{cm\,s^{-1}}$. Additionally, intermittent deposition may occur for soluble gases and particles by the process of wet deposition. Although wet deposition may be extremely efficient, it is, by its nature, sporadic.

The current understanding of dry deposition has recently been authoritatively reviewed by Wesely and Hicks (2000) while recent measurements of the deposition velocity of SO_2 include those of Horvath et al. (1998), who found values of $0.19–0.20\,\mathrm{cm\,s^{-1}}$ over short vegetation. However, it is recognized that more direct measurements of deposition fluxes and deposition velocities are required for the improved parametization of deposition models. Such models (e.g. those contained within the Acid Deposition and Oxidant Model (ADOM: Pleim et al., 1984), the Regional Acid Deposition Model (RADM: Walmsley and Wesely, 1996) and the Routine Deposition Model (RDM: Brook et al., 1999) appear to give a reasonable description of pollutant deposition integrated over time and space (e.g. monthly deposition fluxes on a $10\,\mathrm{km} \times 10\,\mathrm{km}$ grid) but their utility in describing removal rates from a plume as it rapidly moves over complex and varying terrain are very limited (Park, 1998). Similarly, wet deposition, resulting from cloud and precipitation scavenging, can be very effective in removing pollutants (Martin, 1984; Pruppacher and Klett, 1978; Bidleman, 1988) but is extremely difficult to model for a plume, due to its intermittent nature.

3. Power station plume chemistry

3.1. Observations

A large number of field studies have been carried out with the aim of validating the understanding of sulphur and nitrogen pollution chemistry obtained theoretically or in the laboratory and/or with the aim of obtaining observational estimates of the rate of oxidation of sulphur and nitrogen (and hence the rate of formation of sulphate and nitrate) in power station plumes. The oxidation rates observed in plumes may well be expected to differ from those predicted or observed in ambient or background air for a number of reasons, including the possibility of enhanced concentrations of catalysts and the rapid depletion of oxidants (e.g. removal of O_3 by reaction with NO) within the plume. However, very few studies have procured sufficient simultaneous measurements of SO_2 and NO_x concentrations to allow a full understanding of the relative rates of removal of these gases from discrete plumes. This poses a major limitation on the conclusions that can be drawn.

The SO_2 oxidation rates seen in plumes from point sources vary from nearly zero to more that $16\% \, h^{-1}$. In urban plumes even higher oxidation rates, up to $30\% \, h^{-1}$, have been reported. Part of this variation may be attributable to uncertainties in data collection and interpretation, but the wide variations in plume compositions and background conditions, both chemical and meteorological, are also important.

As already noted above, heterogeneous oxidation of SO_2 may be very rapid under certain conditions and will depend on the amount of water vapour available for the formation of droplets. Additionally, experimental studies (e.g. Haury et al., 1977) and modelling of the H_2SO_4–H_2O gas-phase binary system have shown that the rates of nucleation and condensation and hence of the rate of formation of sulphate aerosol is largely controlled by the concentration of water vapour present. However, most field studies of plume chemistry, particularly those involving aircraft, have been conducted in conditions where the aqueous-phase reactions were not favoured. The SO_2 oxidation rates discussed below are summarised in Table 2. Although the focus of most early field studies was the formation of sulphate and nitrate (in the frame of a general concern about acid deposition and environmental acidification), more recently the focus has shifted somewhat to the role of power plant plumes to the formation of ozone and other oxidants in the lower atmosphere. For example, Gillani et al. (1998a, b), Ryerson et al. (1998), Luria et al. (1999, 2000), Sillman (2000), St. John and Chameides (2000) and Nunnermacker et al. (2000) all studied the formation of O_3 in power plant plumes, rather than, explicitly, sulphate and nitrate formation.

Table 2. Oxidation rates of S(IV) in plumes (adapted from Harter, 1985)

Location of source	Plume age (h)	Stack distance (km)	Travel time (h) of sampling	Month and year rate	Oxidation data (% h^{-1})	No of rate points	Average (% h^{-1})	Rate derivation, method comments	Reference
Coal fired power plants									
Cumberland, TN, USA	1.5–10	8–160	1.67–12	Aug. 1978	< 0.1–7.5	19	3	Particulate S/total S ratio	Gillani et al. (1977, 1978, 1981, 1983)
Johnsonville, TN, USA	6–8	56–160	6–8	Aug. 1978	0.8–8.5	6	3	Dry conditions only	
Cumberland, TN, USA	1.3–10.7		2.15–4.77	Jul./Aug. 1976	1.1–8.5	5	5.5	Average daytime rate	Zak (1981)
	0–7		1.35–7.03	Aug. 1978	0.4–16.7	5	6.1	Stack to first measurement Particulate S/gaseous S ratio	
Keystone, PA, USA			0.5–2.67	Apr./May/Sep./Oct. 1978	0.01–5.92	13	0.05	At r.h 42–64%	Dittenhoeffer and De Pena (1980)
							0.78	At r.h 65–90%	
							3.31	At r.h 91–100% Particulate S/total S ratio	
Paradise, KY, USA			0.2–3.8	Jun.	0–1.3	5		Particulate S/total S ratio	Meagher et al. (1981)
Colbert, AL, USA		7–50	0.28–5.20	May/Jun.	0.78–2.79	12	1.3	Morning rate	Meagher and Luria (1982)

Table 2. (Continued)

Location of source	Plume age (h)	Stack distance (km)	Travel time (h) of sampling	Month and year rate	Oxidation data (% h^{-1})	No of rate points	Average (% h^{-1})	Rate derivation, method comments	Reference
Widows Creek, AL, USA		2–49.3	0.43–6.01	Aug.	0.34–3.41	17	1	Morning rate	
							2.4	Afternoon rate Particulate S/total S ratio	Forrest et al. (1981)
Cumberland, TN, USA	0.76–9	11–200		Aug. 1978	0.1–7	21	3	Late morning & afternoon	
							0.5	Night & early morning	
							2	Average diurnal rate Particulate S/total S ratio	
Bowen, GA, USA			0.12–3.33	Dec. 1979	0–2.3	7	< 0.2	Particulae S/total S ratio	Liebsch and De Pena (1982)
Breed, IN, USA	1.3–6.8			Jun./Nov. 1977	0–3.7	7	1	Particulate S/total S ratio	Easter et al. (1980)
Cobb, MI, USA	1.1–3			May/Nov. 1977	0.2–8.4	6	2.6	Particulate S/total S ratio	
Labadie, MO, USA	0.8–12	12–320	0.83–12	Jul. 1976	0.3–3.2	17	1–3	Particulate S/total S ratio	Gillani and Wilson (1980)
Labadie, MO, USA	0.7–12.5	14–360		Aug. 1974/Jul. 1976	0.1–4.8	50	1.6	Particulate S/total S ratio Rate 1–4% h^{-1} d^{-1}, < 0.5 night	Husar et al. (1978)
Leland-Olds, ND, USA			0.38–1.08	Jun. 1978	0–0.06	4		Change in total particle volume & particulate S/total S ratio	Hegg and Hobbs (1980)

Table 2. (Continued)

Location of source	Plume age (h)	Stack distance (km)	Travel time (h) of sampling	Month and year rate	Oxidation data (% h^{-1})	No of rate points	Average (% h^{-1})	Rate derivation, method comments	Reference
Big Brown, TX, USA			0.63–1.32	Jun. 1978	0.15–5.7	4			
Sherburne Co. MN, USA			0.17–2.7	Jun. 1978	0–2.2	4			
Centralia, WA, USA			0.03–1.42	Mar./Oct. 1976 Sep./Oct. 1977	0.03–0.56	5		Change in total particle volume	Hobbs et al. (1979)
Four Corners, NM, USA			0.78–0.87	Jun. 1977	0.34–6.6	3			
Navajo AZ, USA	3–6.1			Jun./Jul. 1979	0.7–13	22	1.9	Rate at noon	Wilson and McMurry (1981)
							0.9	Diurnal average Particle volume/ Total S ratio	
Great Basin, NV, USA	0.36–10.5			Jul./Aug. 1979	1–7	16		Particulate S/total S ratio	Eatough et al. (1981)
Navajo, AZ, USA	2.5–11	25–115	2.67–10.92	Jul./Dec. 1979	0–0.8	13		Particulate S/total S ratio	Richards et al. (1981)
Four Corners, NM, USA		2–90	0.3–12.5	Jun. 1978	0.15–0.5	3		Varies with stack distance CN production/ SO$_2$ ratios	Mamane and Pueschel (1980)
Nanticoke Ont., Canada	0.15–1.93	3–43		Jun. 1978	0–8.7	7	4	Plume age < 2 h with fumigation Particulate S/total S ratio	Anlauf et al. (1982)

Table 2. (Continued)

Location of source	Plume age (h)	Stack distance (km)	Travel time (h) of sampling	Month and year rate	Oxidation data (% h^{-1})	No of rate points	Average (% h^{-1})	Rate derivation, method comments	Reference
Nanticoke Ont., Canada	0.07–2.9	2–93		Nov. 1975	0.32–12.6	19		Varies with stack distance Particulate S/total S ratio	Melo (1977)
Oil fired power plants									
Anclote, FL, USA		0.50	0–1.67	Aug. 1976/Feb. 1977			≤ 0.25	Steady state rate Particulate S/total S ratio	Forrest et al. (1979b)
Andrus, MS, USA	1.2–6.6			May/Oct. 1977	0–5.1	9	2.2	Particulate S/total S ratio	Easter et al. (1980)
Northport, NY, USA	0–3.3		0–2	Various, over 3 yr		60	< 1	Rate essentially incalculable Particulate S/total S ratio	Garber et al. (1981)
Metal smelters									
Mt Isa, Qld, Australia	0.08–14.83	2–256		Jun. 1977	0.06–0.45	65	0.25	Diurnally averaged rate Gaseous S/total S ratio	Roberts and Williams (1979)
Mt Isa, Qld, Australia	2.2–42.5	60–1001		Jul. 1979			0.15	Diurnally averaged rate	Williams et al. (1981)

Table 2. (Continued)

Location of source	Plume age (h)	Stack distance (km)	Travel time (h) of sampling	Month and year rate	Oxidation data (% h^{-1})	No of rate points	Average (% h^{-1})	Rate derivation, method comments	Reference
Urban plumes									
St Louis, MO, USA				Aug. 1975	10–14	18		Gaseous S/total S ratio	Alkezweeny and Powell (1977)
Milwaukee, WI, USA			0–3	Aug. 1976/Jul. 1977	1–9		4	Sulphate and light scattering	Miller and Alkezweeny (1980)
St Louis, MO, USA				Jun. 1976	0–4			Measurements	Forrest et al. (1979a)
								Particulate S/total S ratio	
Budapest Hungary		50	~3h	Jul./Aug. 1978	3–31	8	10		Horvath and Bonis (1980)
Long-range transport trajectories									
Sweden		900–1900	23–61	Summer (Apr.–Sep.)	0.3–5.2	12	1.4	Particulate S/gaseous S ratio	Traegaardh (1980)
		1300–2500	27–55	Winter (Oct.–Mar.)	0.4–1.3	4	0.8		

One of the first airborne investigations of SO_2 oxidation in the plume of a power station was that of Flyger et al. (1978), who used an SF_6 tracer to monitor dispersion and deposition from the plume emitted from a 122 m high stack. They estimated that half the SO_2 was lost from the plume within 45 km of the source, corresponding to a travel time (and half-life) of about 90 min.

Gillani et al. (1978) studied the gas-to-particle conversion rate of sulphur emitted from a coal fired power station in Kansas. They observed that the ratio of particle phase to total sulphur was related linearly to the total solar radiation dose experienced by the plume. This resulted in the amount of sulphur lost by deposition to the ground surface being about 25% in the first 200 km of travel, comparable to the amount of particulate sulphate formed. The maximum rate of particulate sulphur formation was less than $3\% \, h^{-1}$ and only occurred during the daytime. Further analysis of the same data set was carried out by Husar et al. (1978) who concluded that SO_2 oxidation rates were $1-4\% \, h^{-1}$ during the daytime and $< 0.5\% \, h^{-1}$ during the night. In a further re-examination of these data, Gillani and Wilson (1980) concluded that the condition of the background airmass which receives the plume, the extent of plume – background air interactions and photochemical processes are the most important factors in determining the SO_2 oxidation rate. This is probably because of the dominant role of O_3, both directly and as a source of OH. Additionally, the entertainment of non-methane hydrocarbons from polluted background air can provide a source of peroxy radicals HO_2 and RO_2), allowing replenishment of the O_3 removed by reaction with NO in the plume.

Put together, the Kansas experiment suggests that sulphate aerosol forms relatively rapidly during the daytime when sufficient oxidants (OH, HO_2 and RO_2), are available in the plume. Initially, close to the point of emission, oxidant depletion occurs rapidly within the plume by consumption of O_3 by emitted NO, thus limiting the concentrations of OH (which is formed by the photolysis of O_3). If sufficient mixing with background air occurs, O_3 concentrations will recover, either by mixing in of O_3 or of non-methane hydrocarbons, which can in turn form peroxy radicals. These can both contribute to O_3 formation, by rapidly converting NO to NO_2 without consumption of O_3, and can directly offer a route of oxidation of SO_2.

Clearly, SO_2 oxidation rates in the bulk plume are likely to be lower than in background air due to oxidant limitation. However, it is possible that at the plume top and edges, where plume dilution is greatest, conversion rates may be significantly enhanced. Zak (1981) observed daytime conversion rates up to $5.5\% \, h^{-1}$ in a plume edge. Similarly, when the plume top is in contact with clouds, fast liquid-phase reactions will enhance the SO_2 oxidation rate, up to $10\% \, h^{-1}$ (Gillani et al., 1981).

Another important consideration is the SO_2/NO_x ratio. This is a critical parameter in determining sulphate formation rates because SO_2 and NO_x compete for the same oxidising radicals, and oxidation of NO_2 by OH is much faster ($\times 10$) than that of SO_2. Hence the presence of NO_x in a plume will inhibit SO_2 oxidation rates (and sulphate formation rates) relative to the rates in background air.

Mamane and Pueschel (1980) used the concentrations of particles in the plume of the Four Corners power station to estimate an SO_2 oxidation rate of about $0.15\% \, h^{-1}$ in the first $0.3 \, h$, increasing to $0.5\% \, h^{-1}$ in the next two hours of travel time, decreasing again to $\sim 0.3\% \, h^{-1}$. Outside the Four Corners plume, Davis et al. (1979a, b) estimated SO_2 oxidation rates due to reaction with OH of $0.2\% \, h^{-1}$ in the early morning and $2.3\% \, h^{-1}$ around noon, with an average conversion rate of $0.7\% \, h^{-1}$.

Dittenhoeffer and De Pena (1978) found that gas-phase photochemical reactions were the dominant sulphate formation mechanisms during the daytime in a power station-plume at Keystone, Pennsylvania, at low relative humidities. However, when the plume merged with a cooling tower plume, liquid-phase oxidation predominated. In a later study of the Keystone plume they found an average SO_2 oxidation rate of about $1\% \, h^{-1}$ in the first $2 \, h$ of plume travel, rising to $6\% \, h^{-1}$ when the plume encountered clouds (Dittenhoeffer and De Pena, 1982).

Meagher et al. (1977) used an instrumented aircraft to determine the rate of SO_2 oxidation within the plume of the Cumberland Valley coal-fired power station in Tennessee. Most of the oxidation of SO_2 that was observed appeared to occur in the immediate vicinity of the power plant. Beyond $10 \, km$, an average oxidation rate of $0.2\% \, h^{-1}$ was found. In a later modelling study, Meagher and Luria (1982) simulated the chemistry of a plume from a power station of similar size and location to the Cumberland Valley plant. Hydroxyl was found to be the most important oxidant of SO_2, except when high background concentrations of hydrocarbons were used, in which case oxidation by HO_2 and RO_2 became significant. Rapid attenuation of oxidation rates were observed, as oxidant limitation became significant, following the removal of O_3 by NO, and plume dilution by incorporation of background air was required before radical chemistry became established. Forrest et al. (1981) found a daytime average conversion rate of $3\% \, h^{-1}$ in summer, varying with time of day, and an average of $0.5\% \, h^{-1}$ at night. The Cumberland Valley plant was studied more recently, during the 1995 Southern Oxidant Study, when Gillani et al. (1998a, b) found nitrate formation rates of $10-15\% \, h^{-1}$ and a high differential loss rate of nitrogen species (relative to SO_2) of about $0.12 \, h^{-1}$.

Summer and wintertime oxidation rates of SO_2 were obtained in the plume from a northern Alberta power station by Lusis et al. (1978). In February the oxidation rate was found to be low (less than $0.5\% \, h^{-1}$), increasing to $1-3\% \, h^{-1}$ in June, again suggesting the role played by photochemical processes. Another wintertime study (Liebsch and De Pena, 1982), this time of the

Bowen, Georgia, power station plume, found that the highest conversion rate was $0.2\%\,h^{-1}$, except when high relative humidity was observed, when the conversion rate increased to $2.3\%\,h^{-1}$. Meagher and Luria (1981) also observed increased conversion rates, up to $2.8\%\,h^{-1}$, in periods of high relative humidity.

Data obtained from five different coal-fired power stations in the western USA at times favourable to gas-phase photochemical reactions gave SO_2 to SO_4^{2-} conversion rates of $0–5.7\%\,h^{-1}$ for travel times of $10–162\,min$ (Hegg and Hobbs, 1980). In a later study carried out in Arizona, they failed to find measurable conversion rates in four flights out of five (Hegg and Hobbs, 1983), while in another study on the plume from the coal-fired Mohave station, an SO_2 to particle conversion rate of $0.6\%\,h^{-1}$ was estimated (Hegg et al., 1985).

Anlauf et al. (1982) found an average summertime SO_2 to SO_4^{2-} oxidation rate of $4\%\,h^{-1}$ in the plume of the Nanticoke coal-fired power station in Ontario at distances downwind from the stack of $3–43\,km$ at relative humidities of $30–50\%$. Eatough et al. (1982) studied sulphur chemistry in the plume of the Kennecott copper smelter in Utah and found SO_2 to sulphate conversion rates as high as $6\%\,h^{-1}$ in hot, dry and sunny summer conditions.

Luria et al. (1983) attempted to obtain SO_2 oxidation rates by airborne sampling of the plume from the coal-fired Colbert Steam Plant, Alabama, and by studying stack gases from the plant in a reaction chamber. Quite good agreement was obtained between the two methods of estimation, with an SO_2 to SO_4^{2-} oxidation rate of $2.2\%\,h^{-1}$ found on the first day of study. On the second day the power station plume merged and mixed with an urban plume and an enhanced oxidation rate of $4.1\%\,h^{-1}$ was seen, presumably because of enhanced free radical chemistry occurring as a result of the presence of reactive hydrocarbons. This study also provided one of the very few estimates of NO_x oxidation rates obtained from plume observations and a first-order rate coefficient for NO_x removal of $27\%\,h^{-1}$ was found.

Cheng et al. (1987) studied the plume from an oil sand extraction plant in Alberta. This contained relatively high aerosol concentrations close to the source, and the aerosol surface was believed to be wet. SO_2 to SO_4^{2-} conversion rates of $0–2.8\%\,h^{-1}$ in winter, and $0–6\%\,h^{-1}$ in summer, were found. Heterogeneous processes were believed to be appreciable on the wet aerosol surfaces close to the point of emission.

The heterogeneous oxidation of SO_2 in power station plumes has also been studied in field experiments. Gillani and Wilson (1983) used measurements made under wet conditions from three plumes in Missouri and Tennessee when the power station plumes interacted with clouds and during light rain. Variable oxidation rates were observed, but during light rain the liquid-phase rate was around $8\%\,h^{-1}$. In a modelling study of plume–cloud interactions, Gillani et al. (1983) estimated a daytime liquid-phase sulphate formation rate of about $(12 \pm 6)\% \, h^{-1}$.

In one of the few European field studies of plume sulphur chemistry, Clark et al. (1984) followed the chemical composition of cloud water in a power station plume over the North Sea in winter. The plume was trapped in a shallow layer filled with stratocumulus cloud, restricting dilution and allowing total depletion of O_3 in the boundary layer. An SO_2 conversion rate of about $1\% \, h^{-1}$ was estimated at 5 h travel time, comparable to an estimated rate of $4.3\% \, h^{-1}$ in background air over the same period.

3.2. Other plume observations

The occurrence and chemistry of sulphur, and to a lesser extent, nitrogen, compounds has been studied in plumes other than those from coal-fired power stations. Oil-fired stations, metal smelters and urban plumes have all been studied, and have yielded some information relevant to understanding plume chemistry. However, since the chemical composition of emissions from these types of plumes will be different to those from coal-fired power stations, so the dominant routes of oxidation and their rates will be different. For example, the concentrations of trace metals will be different, giving rise to different rates of catalysis (Garber et al., 1981). Newman (1981) observed sulphur oxidation rates five times greater in an oil-fired power station plume than in a coal-fired station plume, but this has not been borne out by more recent work.

Forrest et al. (1979a) studied the Anclote oil-fired power station in Florida and measured an SO_2 oxidation rate of $< 0.25\% \, h^{-1}$. Garber et al. (1981) measured a similarly low rate in a variety of meteorological conditions in the Northport, NY, oil-fired plume. However, Eatough et al. (1981) found no significant differences in SO_2 oxidation rate in an oil-fired plume to that from a coal-fired station. Somewhat higher rates $(3.1 \pm 0.8\% \, h^{-1})$ were observed by Eatough (1984) in the plume from a Pacific coast oil-fired power station, with a rate of oxidation of $30 \pm 4\% \, h^{-1}$ when the plume passed through a fog bank. Enger and Hoegstoem (1979) qualitatively describe a similarly enhanced SO_2 oxidation rate under conditions of very high relative humidity in Sweden.

The rate of conversion of SO_2 to sulphate aerosol particles was determined by a comprehensive measurement programme in the plume emitted from an oil sands extraction plant at Fort McMurray in Alberta (Cheng et al., 1987). A single parcel of air was tracked and sampled by an aircraft and conversion rates of $0-2.8\% \, h^{-1}$ were observed in winter and $0-6\% \, h^{-1}$ in summer. It was suggested that heterogeneous processes were responsible for the oxidation in winter and close to the point of emission when the aerosol particles in the plume were wet.

Plumes from metal smelters might be expected to contain relatively high concentrations of trace metals and hence have elevated SO_2 oxidation rates due to catalytic effects. However, Roberts and Williams (1979) found an SO_2 conversion rate of only $0.15\%\,h^{-1}$ in a smelter plume at Mt Isa, Australia, and Eatough et al. (1981, 1982) found similarly low rates.

Urban plumes differ from point source plumes in having relatively higher NO_x and hydrocarbon concentrations, derived from vehicle exhausts. In general it might be expected that higher oxidation rates of SO_2 and NO_x will be experienced in an urban plume than in a point source plume, due to the more reactive nature of a hydrocarbon-rich plume (Winchester, 1980; Ellestad, 1980). The St Louis, Missouri, plume was extensively studied in the 1970 s. Alkezweeny and Powell (1977) found S(IV) conversion rates of $10-14\%\,h^{-1}$ in the summertime while Zak (1981) estimated a daytime rate of $8.5 \pm 4\%\,h^{-1}$ and a nighttime rate of $1.1 \pm 0.5\%\,h^{-1}$. In a study of the Milwaukee urban plume a rate of $6-8\%\,h^{-1}$ was seen on one day and $< 1\%\,h^{-1}$ on the following day, in similar meteorological and precursor conditions (Miller and Alkezweeny, 1980; Alkezweeny et al., 1982). Recently, the urban plumes of Atlanta (e.g. St. John and Chameides, 2000) and Nashville (e.g. Nunnermacker et al., 2000) and other major conurbations have been extensively studied, but with a clear focus on O_3 formation.

3.3. Plume fringe activity

It is quite clear from the above review of sulphur and nitrogen chemistry and of power station plume studies that the oxidation of SO_2 and NO_2 is likely to be rather slower in a power station plume than in ambient air, since the supply of oxidants will quickly become limiting, at least in clear air where homogeneous photochemically driven gas-phase reactions predominate. Primary emissions of NO will soon eliminate O_3 from the plume, so cutting off the production of OH. However, the situation is likely to be somewhat different at the fringes of the plume where the plume mixes with ambient air. Increased chemical reactivity on the fringes of a power station plume was first observed by Davis and Klauber (1975) who saw an increase in the O_3 concentration at the edges of a power station plume, which they erroneously attributed to photochemical reactions involving SO_2. In fact O_3 formation in the plume fringe is almost certainly due to the photochemical reactions of NO, NO_2 and hydrocarbons, the latter derived largely from the background air, as described by Calvert et al. (1978). Such O_3 enhancements in plume fringes have now been repeatedly observed (e.g. Gillani et al., 1978; Lusis et al., 1978; Miller et al., 1978; Forrest et al., 1979b).

3.4. Model calculations

As well as the experimental studies summarised above, several attempts have been made to estimate sulphur and nitrogen oxidation rates in power station plumes using models of the gas and aerosol phase chemistry believed to be occurring in such plumes. Eltgroth and Hobbs (1979) developed an early model, supported by data, that suggested that homogeneous gas to particle conversion of SO_2 to SO_4^{2-} is greatest at the edges of a plume, due to the mixing in of ambient air. Hov and Isaksen (1981) developed a comprehensive plume model where the chemistry and meteorology of the boundary layer interact with a power station plume, which was given spatial resolution in the cross wind direction. The model predicted the formation of O_3 in the plume fringes, with chemical activity enhanced in the fringes, but becoming more pronounced towards the centreline of the plume with time. SO_2 to SO_4^{2-} oxidation rates varied $(1–5\% \, h^{-1})$, with nitric acid formation proceeding more rapidly. Depletion of NO and NO_2 occurred very rapidly, with about 80% of the NO_x being converted to HNO_3 and NO_3 within three hours.

Seigneur (1982) modelled sulphate aerosol formation in power station plumes, using a gas and aerosol phase chemistry scheme and a particle growth model. Rather slow SO_2 oxidation rates were predicted: $0.5\% \, h^{-1}$ by homogeneous processes. Importantly, the rate was found to be sensitive to the background hydrocarbon and NO_x concentrations, as well as to humidity and to the photolysis rate of NO_2. A similar dependence on the background hydrocarbon and NO_x concentrations was found in the plume chemistry model of Meagher and Luria (1982). SO_2 oxidation rates of $1–2\% \, h^{-1}$ were predicted during the spring and summer, reducing in winter.

Joos and Mendonca (1986) integrated a chemistry model with a comprehensive model of secondary aerosol formation processes, with validation of the output using observational data from a plume study. The sulphate formation rate was found to be: very sensitive to relative humidity (increasing with relative humidity); sensitive to temperature only at high relative humidities, where SO_2 solubility decreases as the temperature increases; very sensitive to solar intensity; and not very sensitive to background O_3 concentration. The formation rate of nitrate aerosol in the plume was found to be: very sensitive to temperature, because the $HNO_3/NH_3/NH_4NO_3$ equilibrium is temperature dependent; not very sensitive to relative humidity, although model limitations may make this conclusion invalid; very sensitive to solar irradiation; and slightly more sensitive to the background ozone concentration than is the SO_2 oxidation rate. The relative insensitivity of this model to background O_3 is important since it implies that the OH radical concentrations in the plume depend more upon NO_x/reactive hydrocarbon photochemical reactions in the plume than on the amount of O_3 mixed into the plume. In addition, the model showed that the net contribution of the plume to sulphate aerosol concentrations was always significant, but that the plume contribution to nitrate aerosol was only significant when there was a relatively low background NO_x concentration and a reactive hydrocarbon/NO_x ratio greater than 10.

Recently, a reactive plume model incorporating transport, chemistry and aerosol dynamics has been used by Karamchandani and Seigneur (1999) to simulate sulphate and nitrate chemistry in power station plumes. They made simulations for winter and summer emissions with travel times of 8 and 10 h, respectively, and with different start times. NO_x and SO_2 emission rates were held constant at 88 and 161 ton d^{-1} respectively. The effects of changes in the chemical composition of the background air were investigated by varying ambient O_3, hydrocarbon and PAN concentrations, the hydrolysis rate of N_2O_5 and the horizontal and vertical dispersion coefficients.

The model results very largely confirmed the many earlier predictions of plume chemistry: oxidation rates in the plume are lower than in background air; excess nitrate concentrations in the plume were 4–7 times greater than the excess sulphate concentrations (i.e. NO_x removal was 4–7 times greater than SO_2 removal in the same travel time); conversion of NO_x and SO_2 was at a maximum in the daytime in summer and at a minimum during the night and in winter; NO_x conversion proceeded at night in clear air by the hydrolysis of N_2O_5 whereas SO_2 conversion did not; and in the presence of clouds SO_2 is rapidly oxidised during both the day and night. Background O_3, reactive hydrocarbon and PAN concentrations had significant effects on the SO_2 and NO_x conversion rates.

Duncan et al. (1995) demonstrated the possible use of emission inventories coupled with chemistry and transport modelling as a tool for apportionment of the sources of NO_x and SO_2 measured at particular locations. Since the sulphur content of gasoline differs from that of coal and of heavy fuel oil, plumes from mobile and point sources should be identifiable by their characteristic SO_2 to NO_x ratios. However, since these ratios will vary over time, due to the different oxidation rates of these species in the atmosphere, and will vary with ambient conditions, as one or other of the multitude of possible NO_x and/or SO_2 oxidation routes becomes dominant, this is not a trivial task. In this study, these changes over time were ignored and an average SO_2/NO_x ratio of 0.05 was used for mobile sources and 2.67–4.56 for power station plumes. This was justified on the grounds that the study was confined to a relatively small area with plume travel times of 3–10 h at maximum. They concluded that the four major power stations in the study area accounted for about 15% of the ambient NO_x at the surface measurement sites on average, but that in the short term this value was extremely variable, due to the intermittent nature of plume fumigation.

3.5. *Summary of plume chemistry*

As can be seen from the foregoing review of SO_2 and NO_x chemistry, the field observations of concentration changes in plumes and the theoretical modelling studies, there is now significant convergence in our understanding and prediction of processes affecting the removal of SO_2 and NO_x and the formation of sulfate and nitrate in power station plumes. This understanding can be summarised as:

(a) In non-cloudy conditions, SO_2 removal in power station plumes occurs primarily during the daytime by reaction with the OH radical, whereas NO_x removal occurs both during the daytime, by fast reaction with OH, and at nighttime by the NO_3/N_2O_5 pathway.

(b) In non-cloudy conditions, NO_x removal will occur much more rapidly ($\sim$10 times faster) than SO_2 removal.

(c) In cloudy conditions, SO_2 will be removed rapidly by reactions in the aqueous phase, but NO_x will not.

(d) The dry deposition velocity of SO_2 is greater than that of NO_x leading to more rapid removal of SO_2 by this process. Conversely, nitrate aerosol is likely to be removed more rapidly by dry deposition than is sulphate aerosol.

(e) These differences in removal rates will cause changes in the ratios of S and N concentrations with time of travel from the point of emission. In clear air, as the plume travels downwind, the SO_2/NO_x ratio will increase. Conversely, the ratio of sulphate aerosol to nitrate aerosol concentration will decrease downwind. In cloudy conditions, where aqueous-phase reactions become important and photochemical processes become less important, SO_2 oxidation will proceed faster than NO_x oxidation and hence the SO_2/NO_x ratio may decrease.

(f) SO_2 and NO_x removal rates will normally be lower in a plume than in background air, due to oxidant limitations, in both the gas and aerosol phases, with plume fringes offering an intermediate oxidation environment.

(g) Absolute oxidation rates of SO_2 and NO_x will vary with plume and background air composition and ambient conditions. In sunny conditions a maximum SO_2 conversion rate of around $3\% \, h^{-1}$ and a maximum NO_x conversion rate of around $30\% \, h^{-1}$ might be expected. However, lower rates may be expected in a "normal" power station plume as oxidant supply becomes diminished by consumption of O_3, although the rate of oxidation of NO_x will remain $\sim$10 times that of SO_2 in photochemically active conditions.

Unfortunately, however, there do remain significant restrictions in our ability to translate these points into a complete description of SO_2 and NO_x behaviour in a power station plume. These arise because of the lack of field data of NO_x oxidation rates inside discrete plumes, a poor understanding of the effects of oxidant limitation on NO_x oxidation and a poor understanding of the rapidity with which oxidant limitation can be reversed once a plume encounters polluted urban air or mixes with background air. Despite these limitations, it is clear that ground level SO_2/NO_x ratios may vary with travel time and ambient conditions. In conditions of simple meteorology and plume dynamics, with clear air and adequate sunlight, the photochemical removal of NO_x will proceed up to 10 times faster than that of SO_2. At night, SO_2 conversion will be effectively zero in clear air while NO_x conversion proceeds. In cloudy conditions the opposite effect will pertain, as the SO_2 conversion rate will be increased, but the NO_x conversion rate will be reduced. Whether or not the resultant ratio changes are observable in the ground-level concentrations of the plume will depend upon the travel time, the ambient conditions and on the nature and magnitude of other sources of SO_2 and NO_x.

3.6. Source apportionment using ratios of S/N

The discussion above of the atmospheric chemistry of SO_2 and NO_x clearly suggests that when elevated concentrations of these pollutants are observed at ground level as a result of emissions from multiple sources, the unambiguous identification of these sources, from consideration of the ratios of their concentrations, is not straightforward. Indeed, such unambiguous source apportionment is not likely without use of a sophisticated chemical and transport model of the atmosphere and then only under ideal conditions. However, in an idealised dry atmosphere with simple photochemistry occurring and in the absence of other sources of the pollutants, an observable increase in the SO_2/NO_x ratio in a power station plume as it travels downwind may be expected. In fact, such elevated ratios of SO_2/NO_x measured downwind could be taken as evidence that power station emissions were responsible for the elevated concentrations of SO_2. However, if other sources of pollutants are present then this idealised pattern of changing ratios could be distorted. In particular, vehicular or other sources of NO_x could substantially reduce the SO_2/NO_x ratio below that expected for a power station plume. Thus high SO_2/NO_x ratios ($> \sim 5$) are almost certainly indicative of pollution from major combustion processes, but low SO_2/NO_x ratios ($< \sim 5$) do not unambiguously rule out power stations as being the source of the SO_2. In wet weather, when liquid phase chemical conversion processes will occur, the situation is much more complex. In both wet and dry conditions it is necessary to correct for background concentrations of the pollutants before consideration of the ratios.

4. Conclusions

Pollution by the products of fossil fuel combustion remains a major environmental issue, with ecological, health and material damage effects well documented. As well as the direct effects of SO_2 and NO_x and their indirect effects on acid deposition and visibility reduction, consideration is now being given to the role of power plant plumes to O_3 formation. Current understanding of the chemistries of SO_2 and NO_x in plumes is sufficiently advanced to allow the broad description and prediction of their behaviour under ideal or near-ideal conditions. However, the effects of oxidant limitation and of more complicated atmospheric conditions cause major problems to the prediction of plume behaviour in the ambient atmosphere. Chemical models of plume behaviour require integration with models of the physical dispersion and transport of the plume if a full description of concentration changes is to be effected. The need for large-scale integrated field experiments for the validation of chemistry-transport models remains.

Acknowledgements

This work was funded by TXU Europe Power.

References

Alkezweeny, A.J., Laulainen, N.S., Thorp, J.M., 1982. Physical, chemical and optical characteristics of a clean air mass over Northern Michigan. Atmospheric Environment 16, 2421–2430.

Alkezweeny, A.J., Powell, D.C., 1977. Estimation of transformation rate of SO_2 to SO_4 from atmospheric concentration data. Atmospheric Environment 11, 179–182.

Anlauf, K.G., Fellin, P., Wiebe, H.A., 1982. The Nanticoke shoreline diffusion experiment, June 1978–IV. A. Oxidation of sulphur dioxide in a power plant plume. B. Ambient concentrations and transport of sulphur dioxide, particulate sulphate and nitrate, and ozone. Atmospheric Environment 16, 455–466.

APEG, 1999. Source apportionment of airborne particulate matter in the United Kingdom. Report of the Air Particles Expert Group, DETR.

Atkinson, R., Lloyd, A.C., 1984. Evaluation of kinetic and mechanistic data for modelling of photochemical smog. Journal of Physical and Chemical Reference Data 13, 315–444.

Atkinson, R., Perry, R.A., Pitts, Jr., J.N., 1976. Rate constants for the reactions of the OH radical with NO_2 (M = Ar and N_2) and SO_2 (M = Ar). Journal of Chemical Physics 65, 306–310.

Benkovitz, C.M., Scholtz, M.T., Pacyna, J., Tarrason, L., Dignon, J., Voldner, E.C., Spiro, P.A., Logan, J.A., Graedel, T.E., 1996. Global gridded inventories of anthropogenic emissions of sulphur and nitrogen. Journal of Geophysical Research 101, 29239–29253.

Bidleman, T.F., 1988. Atmospheric processes. Environmental Science and Technology 22, 361–367.

Botha, C.F., Hahn, J., Pienaar, J.J., Vaneldik, R., 1994. Kinetics and mechanism of the oxidation of sulphur (IV) by ozone in aqueous solutions. Atmospheric Environment 28, 3207–3212.

Brook, J.R., Zhang, L.M., Li, Y.F., Johnson, D., 1999. Description and evaluation of a model of deposition velocities for routine estimates of dry deposition over North America. Part II: review of past measurements and model results. Atmospheric Environment 33, 5053–5070.

Calvert, J.G., Su, F., Bottenheim, J.W., Strausz, O.P., 1978. Mechanism of the homogeneous oxidation of sulphur dioxide in the troposphere. Atmospheric Environment 12, 197–226.

Castleman, A.W., Tang I.N., 1976/7. Kinetics of the association reaction of SO_2 with hydroxyl radical. Journal of Photochemistry 6, 349–354.

Cheng, L., Peake, E., Davis, A., 1987. The rate of SO_2 to sulphate particle formation in an air parcel from an oil sands extraction plant plume. Journal of the Air Pollution Control Association 37, 163–167.

Clark, P.A., Fletcher, I.S., Kallend, A.S., McElroy, W.J., Marsh, A.R.W., Webb, A.H., 1984. Observations of cloud chemistry during long-range transport of power plant plumes. Atmospheric Environment 18, 1849–1858.

Cox, R.A., 1975. The photolysis of gaseous nitrous acid – a technique for obtaining kinetic data on atmospheric photooxidation reactions. International Journal of Chemical Kinetics Symposium, Vol. 1, pp. 379–398.

Cox, R.A., Penkett, S.A., 1971. Photooxidation of atmospheric SO_2. Nature 229, 486–488.

Cox, R.A., Penkett, S.A., 1972. Aerosol formation from sulphur dioxide in the presence of ozone and olefinic hydrocarbons. Journal of Chemical Society Faraday Transactions 68, 1735–1753.

Cox, R.A., Sheppard, D., 1980. Reactions of OH radicals with gaseous sulphur compounds. Nature 284, 330–331.

Davis, D.D., Klauber, G., 1975. Atmospheric gas phase oxidation mechanisms for the molecule SO_2. International Journal of Chemical Kinetics Symposium, Vol. 1, pp. 543–556.

Davis, D.D., Heaps, W., Philen, D., McGee, T., 1979a. Boundary layer measurements of the OH radical in the vicinity of an isolated power plant plume: SO_2 and NO_2 chemical conversion times. Atmospheric Environment 13, 1197–1203.

Davis, D.D., Ravishankara, A.R., Fischer, S., 1979b. SO_2 oxidation via the hydroxyl radical: atmospheric fate of HSO_x radicals. Geophysical Research Letters 6, 113–116.

Dignon, J., 1992. NO_x and SO_x emissions from fossil fuels: a global distribution. Atmospheric Environment 26A, 1157–1163.

Dittenhoefer, A.C., De Pena, R.G., 1978. A study of production and growth of sulphate particles in plumes from a coal-fired power plant. Atmospheric Environment 12, 297–306.

Dittenhoefer, A.C., De Pena, R.G., 1980. Sulphate aerosol production and growth in coal-operated power plant plumes. Journal of Geophysical Research 85, 4499–4506.

Duncan, B.N., Stelson, A.W., Kiang, C.S., 1995. Estimated contribution of power plants to ambient nitrogen oxides measured in Atlanta, Georgia in August 1992. Atmospheric Environment 29, 3043–3054.

Eatough, D.J., 1984. Rapid conversion of $SO_2(g)$ to sulphate in a fog bank. Environmental Science and Technology 18, 855–859.

Eatough, D.J., Caka, F.M., Farber, R.J., 1994. The conversion of SO_2 to sulphate in the atmosphere. Israel Journal of Chemistry 34, 301–314.

Eatough, D.J., Christensen, J.J., Eatough, N.L., Hill, M.W., Major, T.D., Mangelson, N.F., Post, M.E., Ryder, J.F., Hansen, L.D., 1982. Sulphur chemistry in a copper smelter plume. Atmospheric Environment 16, 1001–1015.

Eatough, D.J., Richter, B.E., Eatough, N.L., Hansen, L.D., 1981. Sulphur chemistry in smelter and power plant plumes in the Western U.S. Atmospheric Environment 15, 2241–2253.

Ellestad, T.G., 1980. Aerosol composition of urban plumes passing over a rural monitoring site. Annals of the New York Academy of Science 338, 202–218.

Eltgroth, M.W., Hobbs, P.V., 1979. Evolution of particles in the plumes of coal fired power plants II. A numerical model and comparisons with field measurements. Atmospheric Environment 12, 953–975.

Enger, L., Hoegstroem, U., 1979. Dispersion and wet deposition of sulphur from a power plant plume. Atmospheric Environment 13, 797–810.

Flyger, H., Lewin, E., Lund Thomsen, E., Fenger, J., Lyck, E., Gryning, S.E., 1978. Airborne investigations of SO_2 oxidation in the plumes from power stations. Atmospheric Environment 12, 295–296.

Forrest, J., Garber, R., Newman, L., 1979a. Formation of sulphate, ammonium and nitrate in an oil-fired power plant plume. Atmospheric Environment 13, 1287–1297.

Forrest, J., Schwartz, S.E., Newman, L., 1979b. Conversion of sulphur dioxide to sulphate during the Da Vinci flights. Atmospheric Environment 13, 157–167.

Forrest, J., Garber, R., Newman, I., 1981. Conversion rates in power plant plumes based on filter pack data: the coal-fires Cumberland plume. Atmospheric Environment 13, 1287–1297.

Garber, R., Forrest, J., Newman, L., 1981. Conversion rates in power plant plumes based on filter pack data: the oil-fired Northport plume. Atmospheric Environment 15, 2283–2292.

Gertler, A.W., Miller, D.F., Lamb, D., Katz, U., 1984. Studies of sulphur dioxide and nitrogen dioxide reactions in haze and cloud. In: Durham, J.L. (Ed.), Teasley, J.I. (Series, Ed.), Chemistry of Particles, Fogs and Rain, Acid Precipitation Series, Vol. 2. Butterworth, Boston, pp. 131–160.

Gillani, N.V., Colby, J.A., Wilson, W.E., 1983. Gas to particle conversion of sulphur in power plant plumes – II. Parameterization of plume-cloud interactions. Atmospheric Environment 17, 1753–1763.

Gillani, N.V., Husar, R.B., Husar, J.D., Patterson, D.E., Wilson, W.E., 1978. Project MISTT: kinetics of particulate sulphur formation in a power plant plume out to 300 km. Atmospheric Environment 12, 589–598.

Gillani, N.V., Kohli, S., Wilson, W.E., 1981. Gas-to-particle conversion of sulphur in power plant plumes – I. Parametrization of the conversion rate for dry, moderately polluted ambient conditions. Atmospheric Environment 15, 2293–2313.

Gillani, N.V., Luria, M., Valente, R.J., Tanner, R.L., Imhoff, R.E., Meagher, J.F., 1998a. Loss rate of NO_y from a power plant plume based on aircraft measurements. Journal of Geophysical Research 103, 22585–22592.

Gillani, N.V., Meagher, J.F., Valente, R.J., Imhoff, R.E., Tanner, R.L., Luria, M., 1998b. Relative production of ozone and nitrates in urban and rural power plant plumes 1. Composite results based on data from 10 field measurement days. Journal of Geophysical Research 103, 22593–22615.

Gillani, N.V., Wilson, W.E., 1980. Formation and transport of ozone and aerosols in power plant plumes. Annals of the New York Academy of Sciences 338, 276–296.

Gillani, N.V., Wilson, W.E., 1983. Gas to particle conversion of sulphur in power plant plumes – II. Observations of liquid-phase conversions. Atmospheric Environment 17, 1739–1752.

Goodeve, C.F., Eastman, A.S., Dooley, A., 1934. The reaction between sulphur trioxide and water vapors and a new periodic phenomenon. Transactions of Faraday Society 30, 1127–1133.

Harris, G.W., Atkinson, R., Pitts, Jr., J.N., 1980. Temperature dependence of the reaction OH + SO_2 + M → HSO_3 + M for Ar and SF_6. Chemical and Physical Letters 69, 378–382.

Harris, G.W., Wayne, R.P., 1975. Reaction of hydroxyl radicals with NO, NO_2 and SO_2. Journal of the Chemical Society of Faraday Transactions 71, 610–617.

Harrison, R.M., Shi, J.P., 1996. Sources of nitrogen dioxide in winter smog episodes. Science of the Total Environment 190, 391–399.

Harrison, R.M., Shi, J.P., Grenfell, J.L., 1998. Novel night-time free radical chemistry in severe nitrogen dioxide pollution episodes. Atmospheric Environment 32, 2769–2774.

Harter, P., 1985. Sulphates in the Atmosphere. IEA Coal Research, London, pp. 155.

Haury, G., Jordan, S., Hofmann, C., 1977. Experimental investigation of the aerosol-catalyzed oxidation of SO_2 under atmospheric conditions. Atmospheric Environment 12, 281–287.

Hegg, D.A., Hobbs, P.V., 1980. Measurements of gas-to-particle conversion in the plumes from five coal-fired electric power plants. Atmospheric Environment 14, 99–116.

Hegg, D.A., Hobbs, P.V., Lyon, J.H., 1985. Field studies of a power plant plume in the arid southwestern United States. Atmospheric Environment 19, 1147–1167.

Hewitt, C.N., Harrison, R.M., 1985. Tropospheric concentrations of the hydroxyl radical – a review. Atmospheric Environment 19, 545–554.

Hidleman, L.M., Russell, A.G., Cass, G.R., 1984. Ammonia and nitric acid concentrations in equilibrium with atmospheric aerosols: experiment vs. theory. Atmospheric Environment 18, 1737–1750.

Horvath, L., Nagy, Z., Weidinger, T., 1998. Estimation of dry deposition velocities of nitric oxide, sulphur dioxide, and ozone by the gradient method above short vegetation during the tract campaign. Atmospheric Environment 32, 1317–1322.

Hov, O., Isaksen, I.S.A., 1981. Generation of secondary pollutants in a power plant plume: a model study. Atmospheric Environment 15, 2367–2376.

Husar, R.B., Patterson, D.E., Husar, J.D., Gillani, N.V., Wilson, W.E., 1978. Sulphur budget of a power plant plume. Atmospheric Environment 12, 549–568.

Huss, Jr., A., Lim, P.K., Eckert, C.A., 1982. Oxidation of aqueous SO_2. 1. Homogeneous manganese(II) and iron(II) catalysis at low pH. Journal of Physical Chemistry 86, 4224–4228.

Jones, C.L., Seinfeld, J.H., 1983. The oxidation of NO_2 to nitrate – day and night. Atmospheric Environment 17, 2370.

Joos, E., Mendonca, A., 1986. Evaluation of a reactive plume model with power plant plume data – Application to the sensitivity analysis of sulfate and nitrate formation. Atmospheric Environment 21, 1331–1343.

Karamchandani, P., Seigneur, C., 1999. Simulation of sulphate and nitrate chemistry in power plant plumes. Journal of Air & Waste Management Association 49, 175–181.

Lagrange, J., Pallares, C., Lagrange, P., 1994. Electrolyte effects on aqueous atmospheric oxidation of sulphur dioxide by ozone. Journal of Geophysical Research 99, 14595–14600.

Langford, A.O., Fehsenfeld, F.C., Zachariassen, J., Schimel, D.S., 1992. Gaseous ammonia fluxes and background concentrations in terrestrial ecosystems in the United States. Global Biogeochemical Cycles 6, 459–483.

Lazrus, A.L., Haagenson, P.L., Kok, G.L., Huebert, B.J., Kreitzberg, C.W., Likens, G.E., Mohnen, V.A., Wilson, W.E., Winchester, J.W., 1983. Acidity in air and water in a case of warm frontal precipitation. Atmospheric Environment 17, 581.

Lee, D.S., Atkins, D.H.F., 1994. Atmospheric ammonia emissions from agricultural waste combustion. Geophysical Research Letters 21, 281–284.

Leu, M.-T., 1982. Rate constants for the reaction of OH with SO_2 at low pressure. Journal of Physical Chemistry 86, 4558–4562.

Liebsch, E.J., De Pena, R.G., 1982. Sulphate aerosol production in coal-fired power plant plumes. Atmospheric Environment 16, 1323–1331.

Lind, J.A., Lazrus, A.L., Kok, G.L., 1987. Aqueous phase oxidation of sulphur (IV) by hydrogen peroxide, methylhydroperoxide, and peroxyacetic acid. Journal of Geophysical Research 92, 4171–4177.

Luria, M., Tanner, R.L., Imhoff, R.E., Valente, R.J., Bailey, E.M., Mueller, S.F., 2000. Influence of natural hydrocarbons on ozone formation in an isolated power plant plume. Journal of Geophysical Research 105, 9177–9188.

Luria, M., Olszyna, K.J., Meagher, J.F., 1983. The atmospheric oxidation of flue gases from a coal-fired power plant: a comparison between smog chamber and airborne plume sampling. Journal of the Air Pollution Control Association 483–487.

Luria, M., Valente, R.J., Tanner, R.L., Gillani, N.V., Imhoff, R.E., Mueller, S.F., Olszyna, K.J., Meagher, J.F., 1999. The evolution of photochemical smog in a power plant plume. Atmospheric Environment 33, 3023–3036.

Lusis, M.A., Anlauf, K.G., Barrie, L.A., Wiebe, H.A., 1978. Plume chemistry studies at a Northern Alberta power plant. Atmospheric Environment 12, 2429–2437.

Mamane, Y., Pueschel, R.F., 1980. Formation of sulphate particles in the plume of the Four Corners power plant. Journal of Applied Meteorology 19, 779–790.

Martin, A., 1984. Estimating washout coefficients for sulphur dioxide, nitric oxide, nitrogen dioxide and ozone. Atmospheric Environment 19, 1955–1961.

Martin, L.R., Damschen, D.E., 1981. Aqueous oxidation of suphur dioxide by hydrogen peroxide at low pH. Atmospheric Environment 15, 1615.

Martin, L.R., Good, T.W., 1991. Catalyzed oxidation of sulphur dioxide in solution: the iron–manganese synergism. Atmospheric Environment 25A, 2395–2399.

Martin, L.R., Hill, M.W., Tai, A.F., Good, T.W., 1991. The iron catalyzed oxidation of sulphur(IV) in aqueous solution: differing effects of organics at high and low pH. Journal of Geophysical Research 96, 3085–3097.

McAndrew, R., Wheeler, R., 1962. The recombination of atomic hydrogen in propane flame gases. Journal of Physical Chemistry 66, 229–232.

Meagher, J.F., Luria, M., 1982. Model calculations of the chemical processes occurring in the plume of a coal-fired power plant. Atmospheric Environment 16, 183–195.

Meagher, J.F., Stockburger, L., Bailey, E.M., Huff, O., 1977. The oxidation of sulphur dioxide to sulphate aerosols in the plume of a coal-fired power plant. Atmospheric Environment 12, 2197–2203.

Meagher, J.F., Stockburger, L., Bonanno, R.J., Bailey, E.M., Luria, M., 1981. Atmospheric oxidation of flue gases from coal fired power plants – a comparison between conventional and scrubbed plumes. Atmospheric Environment 15, 749–762.

Miller, D.F., Alkezweeny, A.J., 1980. Aerosol formation in urban plumes over Lake Michigan. Annals of the New York Academy of Sciences 338, 219–232.

Miller, D.F., Alkezweeny, A.J., Hales, J.M., Lee, R.N., 1978. Ozone formation related to power plant emissions. Science 202, 1186–1190.

Mozurkewich, M., 1993. The dissociation constant of ammonium nitrate and its dependence on temperature, relative humidity and particle size. Atmospheric Environment 27, 261–270.

Newman, L., 1981. Atmospheric oxidation of sulphur dioxide: a review as viewed from power plant and smelter plume studies. Atmospheric Environment 15, 2231–2239.

Nunnermacker, L.J., Kleinman, L.I., Imre, D., Daum, P.H., Lee, Y.N., Lee, J.H., Springston, S.R., Newman, L., Gillani, N., 2000. NO_y lifetimes and O_3 production efficiencies in urban and power plant plumes: analysis of field data. Journal of Geophysical Research 105, 9165–9176.

Park, S.U., 1998. Effects of dry deposition on near-surface concentrations of SO_2 during medium-range transport. Journal of Applied Meteorology 37, 486–496.

Penkett, S.A., Jones, B.M.R., Brice, K.A., Eggleton, A.E.J., 1979. The importance of atmospheric ozone and hydrogen peroxide in oxidizing sulphur dioxide in cloud and rainwater. Atmospheric Environment 13, 123–137.

Platt, U., Perner, D., 1980. Direct measurement of atmospheric CH_2O, HNO_2, O_3 and SO_2 by differential optical absorption in the near UV. Journal of Geophysical Research 85, 7453.

Pleim, J.E., Venkatram, A., Yamartino, R., 1984. ADOM/TADAP Model Development Program. The Dry Deposition Module. Vol. 4. Ontario Ministry of the Environment, Rexdale, Canada.

Pruppacher, H., Klett, J.D., 1978. Microphysics of Clouds and Precipitation. Dreidel, Dordecht, The Netherlands.

Richards, L.W., Anderson, J.A., Blumenthal, D.L., Brandt, A.A., McDonald, J.A., Watus, N., Macias, E.S., Bhardwaja, P.S., 1981. The chemistry, aerosol physics, and optical properties of a western coal-Fred power plant plume. Atmospheric Environment 15, 2111.

Richards, L.W., Anderson, J.A., Blumenthal, D.L., McDonald, J.A., Kok, G.L., Lazrus, A.L., 1983. Hydrogen peroxide and sulphur (IV) in Los Angeles cloud water. Atmospheric Environment 17, 911.

Roberts, D.B., Williams, D.J., 1979. The kinetics of oxidation of sulphur dioxide within the plume from a sulphide smelter in a remote region. Atmospheric Environment 13, 1485–1499.

Ryerson, T.B., Buhr, M.P., Frost, G.J., Goldan, P.D., Holloway, J.S., Hubler, G., Jobson, B.T., Kuster, W.C., McKeen, S.A., Parrish, D.D., Roberts, J.M., Sueper, D.T., Trainer, M., Williams, J., Fehsenfeld, F.C., 1998. Emissions lifetimes and ozone formation in power plant plumes. Journal of Geophysical Research 103, 22569–22583.

Seigneur, C., 1982. A model of sulphate aerosol dynamics in atmospheric plumes. Atmospheric Environment 16, 2207–2228.

Seinfeld, J.H., Pandis, S.N., 1998. Atmospheric Chemistry and Physics. Wiley, New York, pp. 1326.

Shi, J.P., Harrison, R.M., 1997. Rapid NO$_2$ formation in diluted petrol fuelled engine exhaust: a source of NO$_2$ in winter smog episodes. Atmospheric Environment 31, 3857–3866.

Sillman, S., 2000. Ozone production efficiency and loss of NO$_x$ in power plant plumes: photochemical model and interpretation of measurements in Tennessee. Journal of Geophysical Research 105, 9189–9202.

Spiro, P.A., Jacob, D.J., Logan, J.A., 1992. Global inventory of sulphur emissions with a $1° \times 1°$ resolution. Journal of Geophysical Research 97, 6023–6036.

St. John, J.C., Chameides, W.L., 2000. Possible role of power plant plume emissions in fostering O-3 exceedence events in Atlanta, Georgia. Journal of Geophysical Research 105, 9203–9211.

Stockwell, W.R., Calvert, J.G., 1983. The mechanism of the HO–SO$_2$ reaction. Atmospheric Environment 17, 2231–2235.

Tuazon, E.C., Atkinson, C.R., Plum, C.N., Winer, A.M., Pitts, Jr., J.N., 1983. The reaction of gas phase N$_2$O$_5$ with water vapor. Geophysical Research Letters 10, 953.

Walmsley, J.L., Wesely, M.L., 1996. Modification of coded parameterizations of surface resistances to gaseous dry deposition. Atmospheric Environment 30, 1181–1188.

Wesely, M.L., Hicks, B.B., 2000. A review of the current status of knowledge on dry deposition. Atmospheric Environment 34, 2261–2282.

Winchester, J.W., 1980. Sulphate formation in urban plumes. Annals of the New York Academy of Sciences 338, 297–308.

Zak, B.D., 1981. Lagrangian measurements of sulphur dioxide to sulphate conversion rates. Atmospheric Environment 15, 2583–2591.

Air Pollution Science for the 21st Century
J. Austin, P. Brimblecombe and W. Sturges, editors
© 2002 Elsevier Science Ltd. All rights reserved.

Chapter 10

New Directions: Fugitive emissions identified by chemical fingerprinting

Jon Peters and Ag Stephens

School of Environmental Sciences, University of East Anglia, Norwich, NR4 7TJ, UK
E-mail: j.peters@uea.ac.uk

Air pollution monitoring techniques of increased resolution and specificity will become incorporated into pollution control strategies as they become available and economically viable. We speculate on how future technology may be used to identify fugitive emissions from industrial sources, which currently represent a deficiency in emission inventories and pollution legislation. Although we focus here on the situation in the United Kingdom (UK), similar arguments apply internationally.

In the UK an operator must gain Integrated Pollution Control (IPC) authorisation from the Environment Agency (EA) in order to operate a particular industrial process at a specific site. The focus of IPC in the UK has always been on emissions that result from the normal operation of industrial processes. Normal 'process' emissions are released from well-defined points on industrial sites such as stacks or vents and are relatively easy to monitor and quantify. In most cases authorisations cover explicitly only those emissions released in this way and take little account of non-process or 'fugitive' emissions. These unexpected or unplanned emissions are difficult to quantify in any meaningful way because they are released at non-specific points in space and time. They may result from ageing equipment, bad process management or the occurrence of accidents. In the case of industrial processes using, producing or involving volatile organic compounds (VOCs), fugitive emissions can account for a significant percentage of the total emission that will in reality be released from a site. That these emissions are not included in IPC authorisations for industrial processes is based more upon practicality than sound scientific rationale. The lack of authoritative information on the magnitude of fugitive emissions reflects both the difficulty and cost of monitoring at individual industrial sites, and the diversity of industrial processes that may produce them. Clearly, a technique that could be used to identify, quantify and attribute multiple fugitive emissions from a single monitoring point would be beneficial.

First published in Atmospheric Environment 35 (2001) 1347–1348

The industrial component of emission inventories such as the National Atmospheric Emissions Inventory (NAEI) in the UK is compiled using data based directly upon the IPC authorisations issued by the EA. Clearly, there are major implications for operators, policymakers and scientists if a substantial component of industrial emissions is not properly quantified and included in emission inventories. A comparison of the total known emission of VOCs in the NAEI with monitored data shows the inventory to be underestimating emissions by around 40% (*Ozone in the United Kingdom: Fourth Rep. Photochemical Oxidants Rev. Group*, Institute of Terrestrial Ecology, Edinburgh, 1997). Significant progress towards validation of the inventory could be made by including industrial non-process and fugitive emissions.

Currently in the UK, 25 species of VOCs are monitored by the Automated Hydrocarbon Network across one rural and 11 urban sites. Gas chromatography (GC) methods provide high-resolution hourly VOC concentration data. Samples are analysed and the data automatically sent to the Department of the Environment, Transport and the Regions (DETR) database. Recent developments in the separation of VOCs suggest that in the future it will be possible to separate, identify and quantify a much greater number of compounds. A.C. Lewis, for example, has illustrated that whilst a portion of a conventional GC analysis of urban air revealed a dozen or so identifiable peaks, a two-dimensional "comprehensive" chromatography separation run in parallel revealed the presence of literally hundreds of different VOCs (New Directions: Novel separation techniques in VOC analysis pose new challenges to atmospheric chemistry, Atmospheric Environment, vol. 34 (2000) pp. 1155–1156). Once such techniques have been fully validated they will provide a far superior monitoring tool for the environmental manager.

The use of *chemical fingerprints* has been applied to the study of various atmospheric pollutants. Khallili et al. (Atmospheric Environment, vol. 29 (1995) pp. 533–542) used them to examine polycyclic aromatic hydrocarbons (PAHs) whilst Alcock et al. (Chemosphere, vol. 38 (1999) pp. 759–770) recently applied a fingerprinting technique to the identification of industrial sources of polychlorinated dibenzo-*p*-dioxins (PCDDs) and polychlorinated biphenyls (PCBs). A fingerprint can be obtained by monitoring and quantifying emissions from a particular industrial location. The emission of each compound will lead to an expected concentration in the air above the site. This air mass will have an identifiable composition of various VOCs at certain concentrations, representing the fingerprint of the source. As the air mass moves away from the site, it can be assumed that the initial changes in concentrations will relate to dispersion, and that this effect will be equal for all species. In the case of reactive species a model of atmospheric decomposition could be incorporated. Hence the concentration of components relative to one another can be predicted and the fingerprint will be identifiable at monitoring sites.

The focus of this article is to suggest how the use of industrial fingerprints combined with advanced analytical techniques could be used to identify fugitive emissions. The IPC legislation provides the regulating body (the EA in England and Wales) with emissions data for each industrial site. The formulation of a chemical fingerprint from each source is a straightforward process, requiring a dilution of the stack concentrations into the receiving air volume. We envisage that this approach will primarily be used in highly industrialised areas. Monitoring stations would be sited in locations where a number of point sources influence the air quality. At the monitoring station automated monitoring systems would be coupled to an atmospheric computer model. Inputs of source fingerprints and real-time meteorological data such as prevailing winds would assist in the identification of any sources influencing that area. Once key emission sources are located, we suggest that by matching the air quality data with chemical fingerprints, each fingerprint can be subtracted from the data to reveal any residual VOC emissions. It is these residuals that may indicate the presence of fugitive emissions. If a residual is always detected when a certain fingerprint dominates the atmospheric composition, it can be assumed with a reasonable degree of confidence to be linked to the fingerprint source. Hence this method can potentially be used to identify fugitive emissions from industrial locations. After a residual emission has been identified it should be possible to use a reverse-modelling approach to estimate the size of the emission at the source.

One problem may be the concurrence of other VOC emitters in the same area as the industrial sources of interest. In particular, vehicles are known to account for around 30% of the total VOC emission in the UK (*Fourth Rep. Photochemical Oxidants Rev. Group,* see above). Fortunately, in many cases the fingerprint of industrial sites may be of esoteric compounds that do not present in vehicle exhaust fumes. If this is not the case it should be possible to create fingerprints for petrol and diesel emissions and thus deduct their influence in the same way as for normal industrial process emissions. If necessary we envisage that this technique could be extended to other non-industrial VOC sources.

Identification of fugitive emissions would increase the accuracy of knowledge of both regulators and operators regarding the actual emission from industrial sites. The EA could assess the environmental impact of newly identified fugitive emissions in accordance with the current classifications of VOCs in terms of their harmfulness. If the chemical components of such emissions are relatively innocuous then the emission fingerprint may simply be added to the standard chemical fingerprint of the identified site. However, in the case of harmful compounds the regulator may require the improvement of environmental management systems. This may for example involve modifications to storage techniques, site operations and training of personnel. Economic benefits to the operators may become apparent with increasing efficiency in the use of raw materials.

A primary application of fingerprint identification of fugitive industrial emissions would be the validation and improvement of emission inventories, thereby reducing their tendency to underestimate concentrations of VOCs in comparison with monitoring data. As emission inventory data is improved in terms of speciation and accuracy of VOC emissions, identification of new compounds of interest in the atmosphere may also follow. It is likely that, as in the past, the focus of atmospheric chemistry will change as new species are highlighted due to their atmospheric concentrations or their effect upon human health or the environment.

Concepts such as photochemical ozone creation potential (POCP), global warming potential (GWP) and ozone depletion potential (ODP) are often calculated empirically using both monitored and emissions data coupled with a knowledge of atmospheric chemistry. For these concepts to be useful in a framework of international environmental policy making it is essential that the data upon which they are based is accurate and validated. If fugitive emissions are indeed as important as suggested here, then they play a significant and currently unquantified role in industrial air pollution. Improvements to POCP, GWP, ODP and emission inventories are likely to occur as a result of future implementation of the methodology proposed here. We therefore hope that the technology needed to monitor VOCs to the required specificity becomes available sooner rather than later.

Air Pollution Science for the 21st Century
J. Austin, P. Brimblecombe and W. Sturges, editors
© 2002 Elsevier Science Ltd. All rights reserved.

Chapter 11

Ozone and other secondary photochemical pollutants: chemical processes governing their formation in the planetary boundary layer

Michael E. Jenkin

National Environmental Technology Centre, AEA Technology plc, Culham, Abingdon, Oxfordshire OX14 3ED, UK
E-mail: michael.e.jenkin@aeat.co.uk

Kevin C. Clemitshaw

TH Huxley School of Environment, Earth Science and Engineering, Centre for Environmental Technology, Imperial College of Science, Technology and Medicine, Silwood Park, Ascot, Berkshire SL5 7PY, UK

Abstract

The chemical processing of pollutants emitted into the atmosphere leads to a variety of oxidised products, which are commonly referred to as secondary pollutants. Such pollutants are often formed on local or regional scales in the planetary boundary layer, and may have direct health impacts and/or play wider roles in global atmospheric chemistry. In the present review, a comparatively detailed description of our current understanding of the chemical mechanisms leading to the generation of secondary pollutants in the troposphere is provided, with particular emphasis on chemical processes occurring in the planetary boundary layer. Much of the review is devoted to a discussion of the gas-phase photochemical transformations of nitrogen oxides (NO_x) and volatile organic compounds (VOCs), and their role in the formation of ozone (O_3). The chemistry producing a variety of other oxidants and secondary pollutants (e.g., organic oxygenates; oxidised organic and inorganic nitrogen compounds), which are often formed in conjunction with O_3, is also described. Some discussion of nighttime chemistry and the formation of secondary organic aerosols (SOA) in tropospheric chemistry is also given, since these are closely linked to the gas-phase photochemical processes. In many cases, the discussion of the relative importance of the various processes is illustrated by observational data, with emphasis generally placed on conditions appropriate to the UK and northwest continental Europe.

First published in Atmospheric Environment 34 (2000) 2499–2527

1. Introduction

The emission of a variety of pollutant gases (e.g., nitrogen oxides, NO_x, and volatile organic compounds, VOCs) into the troposphere may present a health risk either directly, or as a result of their oxidation. This can lead to a variety of secondary oxidised products, many of which are potentially more harmful than their precursors. Because much of the chemistry is driven by the presence of sunlight, the oxidised products are commonly referred to as *secondary photochemical pollutants*, and include photochemical oxidants such as ozone (O_3). The production of elevated levels of O_3 at ground level is of particular concern, since it is known to have adverse effects on human health, vegetation (e.g., crops) and materials (PORG, 1997). Established air quality standards for O_3 are frequently exceeded, and the formulation of control policies is therefore a major objective of environmental policy (UNECE, 1992,1993,1994). Nevertheless, other pollutants that are formed on local or regional scales in the planetary boundary layer may also have direct health impacts (e.g. peroxy acetyl nitrate, PAN), and/or play wider roles in global atmospheric chemistry.

Photochemical air pollution, first identified in Los Angeles in the 1940s, is now a widespread phenomenon in many of the world's population centres (e.g., see NRC, 1991; PORG, 1997). Consequently, considerable attention has been given to identifying and quantifying chemical processes leading to the generation of O_3 and other secondary photochemical pollutants in the planetary boundary layer. This has involved the laboratory study of many hundreds of chemical reactions, and a significant body of evaluated chemical kinetics and photochemical data has accumulated for elementary atmospheric reactions (e.g., Atkinson et al., 1997a,b; DeMore et al., 1997). Computer models have provided a useful means of assembling these data, and of describing the likely behaviour and interconversion of various atmospheric pollutants, and such models play a central role in policy development and implementation. This work has been driven, of course, by the need to interpret the results of field studies of atmospheric chemical processes. In recent years, an enormous variety of observational data has become available for molecular and free radical species involved in atmospheric chemical processes in both polluted and clean environments.

The aim of this review is to provide a comparatively detailed description of our current understanding of the chemical mechanisms leading to the generation of secondary photochemical pollutants in the troposphere, with particular emphasis on chemical processes occurring in the planetary boundary layer. Much of the review is devoted to a discussion of the gas-phase photochemical transformations of nitrogen oxides and volatile organic compounds, and their role in the formation of O_3. The chemistry producing a variety of other oxidants and secondary pollutants, which are often formed in conjunction with O_3, is also described. Some discussion of nighttime chemistry and the formation and role of secondary organic aerosols (SOA) in tropospheric chemistry is also given, since these are closely linked to the gas-phase photochemical processes. Where possible, the relative importance of the various processes is discussed and illustrated by observational data, with emphasis generally placed on conditions appropriate to the UK and northwest continental Europe. Although some reference to heterogeneous reactions and aqueous uptake is made, multiphase chemical processes are not considered in detail, and the reader is referred to other texts (e.g., Jonson and Isaksen, 1993; Ravishankara, 1997) for further information on this important area of tropospheric chemistry.

2. Photochemical transformations of oxidised nitrogen species

2.1. Daytime interconversion of NO and NO_2

Nitrogen oxides are released into the troposphere from a variety of biogenic and anthropogenic sources (Logan, 1983; IPCC, 1995; Lee et al., 1997). Approximately 40% of the global emissions, and the largest single source, results from the combustion of fossil fuels, which almost exclusively leads to emission directly into the planetary boundary layer, mainly in the form of NO. A small fraction (generally $\leqslant 10\%$) may be released as NO_2 (PORG, 1997), or is produced close to the point of emission from the termolecular reaction of NO with O_2:

$$2NO + O_2 \rightarrow 2NO_2. \tag{1}$$

The rate of this reaction is strongly dependent on the NO concentration. Thus, at high levels typical of those close to points of emission, the rate of conversion of NO to NO_2 is rapid (e.g. ca. $0.5\% \, s^{-1}$ at 1000 ppmv NO), but the significance of reaction (1) decreases dramatically as NO is diluted, with a fractional conversion rate of only $5 \times 10^{-6} \, s^{-1}$ at 1 ppmv NO. Under most tropospheric conditions, reaction (1) is insignificant, and the dominant pathway by which NO is converted to NO_2 is via the reaction with O_3 (see Fig. 1):

$$NO + O_3 \rightarrow NO_2 + O_2. \tag{2}$$

At a typical boundary layer concentration of 30 ppbv O_3, this reaction occurs on a timescale of ca. 1 min. During daylight hours, however, NO_2 is converted back to NO as a result of photolysis, which also leads to the regeneration of

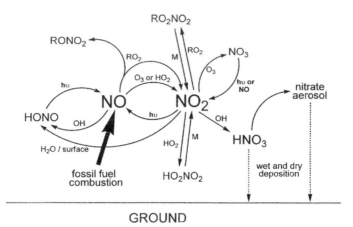

GROUND

Figure 1.   Daytime interconversions of oxidised nitrogen compounds in the troposphere.

O_3 as follows,

$$NO_2 + h\nu(\lambda < 420\,nm) \rightarrow NO + O(^3P) \tag{3}$$

$$O(^3P) + O_2(+M) \rightarrow O_3(+M) \tag{4}$$

where M is a third body, most commonly N_2. Thus, reactions (2)–(4) constitute a cycle with no net chemistry. In the absence of competing interconversion reactions, this cycle leads to a photostationary state in which the concentrations of NO and NO_2 are related to the O_3 concentration by the following expression (Leighton, 1961):

$$[O_3] = (J_3[NO_2]/k_2[NO]) \tag{i}$$

where J_3 is the rate of NO_2 photolysis, and k_2 is the rate coefficient for the reaction of NO with O_3. As a result of this rapid interconversion, the behaviour of NO and NO_2 is highly coupled, and they are usually collectively referred to as NO_x. The lifetime of NO_2 with respect to photolysis in the boundary layer depends on latitude, season and time of day. Reported midsummer measurements of J_3 indicate that the minimum lifetime of NO_2 under conditions typical of the UK is of the order of 1.5 min, with a mean daylight lifetime of ca. 3 min (e.g., Carpenter et al., 1998). In the wintertime, this is typically a factor of two or three longer.

Other daytime chemical processes interconverting the NO_x species generally involve free radicals. Of particular importance are the hydroperoxy radical (HO_2) and organic peroxy radicals (RO_2), which are mainly produced in the troposphere as intermediates in the photochemical oxidation of carbon monoxide (CO) and volatile organic compounds, as described in Section 3. Both HO_2

and RO_2 provide additional NO to NO_2 conversion routes to supplement reaction (2):

$$HO_2 + NO \rightarrow OH + NO_2, \tag{5}$$

$$RO_2 + NO \rightarrow RO + NO_2. \tag{6a}$$

However, since conversion of NO to NO_2 as a result of these reactions does not consume O_3, the subsequent photolysis of NO_2 (reaction (3)), followed by reaction (4), represents a net source of O_3. These reactions form the core of photochemical O_3 production in the troposphere for which NO_x, organic compounds and sunlight are essential ingredients, as discussed further in Section 3. The net production rate of O_3 resulting from this *fast photochemistry* is given by the expression:

$$d[O_3]/dt = (k_5[HO_2] + \Sigma k_{6a}[RO_2])[NO]. \tag{ii}$$

The production rate is very variable, because the peroxy radical concentration is a strong function of factors such as location, season, time of day and cloud cover. However, for an ambient NO concentration of 1–2 ppbv (typical of a UK rural site) and a total peroxy radical concentration of ca. 10 pptv, which has been observed in the UK under summertime conditions (Clemitshaw et al., 1997a), O_3 production rates of the order of 5–10 ppbv h^{-1} may be calculated.

Reactions (5) and (6a) also perturb the photostationary state, and observed deviations from equation (i) are therefore often interpreted in terms of the presence of significant concentrations of the peroxy radicals (Ridley et al., 1992; Cantrell et al., 1993a,1997a; Kleinman et al., 1995; Hauglustaine et al., 1996; Carpenter et al., 1998). A modified equation allowing for deviations can be defined as follows:

$$[O_3] + \Psi_{ox} = (J_3[NO_2]/k_2[NO]) \tag{iii}$$

where "Ψ_{ox}" represents the observed deviation, which is usually termed the *missing oxidant* (Cantrell et al., 1993b) and is effectively the ambient level of oxidising free radicals expressed as an O_3-*equivalent concentration*. If the observed deviation is a consequence of reactions (5) and (6a), then

$$\Psi_{ox} = \{(k_5[HO_2]/k_2) + (\Sigma k_{6a}[RO_2]/k_2)\} \tag{iv}$$

and the net O_3 production rate is $k_2[NO]\cdot\Psi_{ox}$. When significant deviations from the photostationary state are observed, Ψ_{ox} tends to show a diurnal variation consistent with that expected for photochemically derived free radicals (e.g., Cantrell et al., 1993c; Cape et al., 1994), and their capacity for oxidising

NO to NO_2 can readily be compared with that due to O_3. Where simultaneous measurements are available, Ψ_{ox} usually shows a positive correlation with the total ambient concentration of peroxy radicals, $[HO_2] + \Sigma[RO_2]$, measured using the chemical amplification technique (e.g., Cantrell et al., 1993b). However, the precise calculation of the total peroxy radical concentration from measurements of Ψ_{ox} using equation (iv) is often not possible, since the value of k_{6a} depends on the structure of the RO_2 radical, with currently measured values for a variety of radicals covering a range of more than a factor of three (e.g., Atkinson et al., 1997a). It is also possible that deviations from Eq. (i) result from reactions other than (5) and (6a) (e.g., see Calvert and Stockwell, 1983), and peroxy radical concentrations calculated by this method are typically higher than measured values by as much as factors of 2 to 4 (Cantrell et al., 1992,1993a,1996a,b,1997a; Hauglustaine et al., 1996; Volz-Thomas et al., 1998). These large discrepancies either imply significant errors in the measured data for NO_x, O_3, J_3 and peroxy radicals, or infer the presence of unidentified species which oxidise NO to NO_2, or both (Ridley et al., 1992; Cantrell et al., 1996a,1997a).

Other chemical transformations of NO_x lead to the generation of a variety of inorganic and organic oxidised nitrogen compounds. Oxidised nitrogen species of atmospheric significance are usually collectively referred to as "NO_y", which is taken to consist of NO, NO_2, higher oxides (NO_3 and N_2O_5), oxyacids (HNO_3, HO_2NO_2 and HONO), organic peroxy nitrates (RO_2NO_2), organic nitrates ($RONO_2$) and aerosol nitrate. The component of NO_y excluding NO and NO_2 is sometimes defined as "NO_z" (Volz-Thomas et al., 1995; Colvile et al., 1996). The majority of these species are generated during daylight, and their formation and removal chemistry is described in the subsections which follow. The significance of the formation of the higher oxides (NO_3 and N_2O_5), is greatest for nighttime tropospheric chemistry, and is therefore discussed in Section 4.

2.2. The formation and removal of oxyacids of nitrogen

The addition reactions of the NO_x species with HO_x radicals (OH and HO_2) lead to the generation of the oxyacids, nitric acid (HNO_3), peroxynitric acid (HO_2NO_2) and nitrous acid (HONO). The reaction of OH with NO_2 to form HNO_3, is of particular importance, since it provides the predominant chemical removal route for NO_x during daylight, and therefore plays a major role in controlling its concentration:

$$OH + NO_2(+M) \rightarrow HNO_3(+M). \tag{7}$$

The boundary layer lifetime of NO_2 with respect to this reaction is ca. 1 d for a typical background OH concentration of 0.04 pptv (10^6 molecule cm^{-3}), which decreases to ca. 2 h for an elevated OH concentration of 0.4 pptv (10^7 molecule cm^{-3}) consistent with a photochemical episode. HNO_3 is therefore usually a significant component of NO_z (e.g., Fahey et al., 1986; Nielsen et al., 1995; Singh et al., 1996; Aneja et al., 1996; Harrison et al., 1999). It is removed comparatively efficiently from the troposphere by both wet and dry deposition (Huebert and Robert, 1985; Hov et al., 1987; Derwent et al., 1988; Dentener, 1993), and also by adsorption on, or reaction with, the tropospheric aerosol (Cox, 1988; Fenter et al., 1995).

The addition reaction of HO_2 with NO_2 leads to the formation of HO_2NO_2:

$$HO_2 + NO_2(+M) \rightleftharpoons HO_2NO_2(+M). \tag{8}$$

The lifetime of NO_2 with respect to this reaction in the sunlit boundary layer is typically of the order of 2 h ($[HO_2] = 10$ pptv; 2.5×10^8 molecule cm^{-3}). However, HO_2NO_2 is thermally unstable and only has a lifetime of the order of 30 s with respect to decomposition by the reverse reaction at 288 K and atmospheric pressure. Consequently, NO_2 is readily regenerated and the contribution of HO_2NO_2 to NO_z is believed to be limited, although this has not been confirmed by direct observations.

The addition reaction of OH with NO leads to the production of HONO, with the boundary layer lifetime of NO with respect to this reaction being ca. 2 d for $[OH] = 0.04$ pptv (10^6 molecule cm^{-3}), and ca. 5 h for $[OH] = 0.4$ pptv (10^7 molecule cm^{-3}):

$$OH + NO(+M) \rightarrow HONO(+M). \tag{9}$$

Once again, HONO only acts as a temporary reservoir for NO_x, because it is readily photolysed by near ultraviolet radiation, such that its photolysis lifetime is typically less than 1 h (see Section 3):

$$HONO + h\nu \rightarrow OH + NO. \tag{10}$$

As a result, HONO is unable to accumulate significantly during daylight hours, and is often undetectable or close to detection limits of current HONO instrumentation (30–100 pptv). Where daytime measurements of HONO are possible, it provides a potential tracer for OH radicals, provided its photolysis rate (J_{10}) and the concentration of NO are also measured:

$$[OH] = J_{10}[HONO]/k_9[NO]. \tag{v}$$

This relation only holds, however, if reaction (9) is the sole (or dominant) daytime source of HONO.

Observational data are consistent with the existence of additional thermal sources of HONO, which may operate throughout the diurnal cycle, leading to an accumulation of HONO during the night, followed by photolysis at sunrise (e.g. Harris et al., 1982; Kessler and Platt, 1984; Harrison et al., 1996). The available information is consistent with HONO production from heterogeneous reactions involving the NO_x species and H_2O, for which the following have been postulated:

$$2NO_2(g) + H_2O(ads) \rightarrow HNO_3(ads) + HONO(g), \qquad (11)$$

$$NO(g) + NO_2(g) + H_2O(ads) \rightarrow 2HONO(g). \qquad (12)$$

There have been numerous laboratory kinetic investigations of these reactions (England and Corcoran, 1974; Sakamaki et al., 1983; Pitts et al., 1984; Svensson et al., 1987; Jenkin et al., 1988). These studies indicate that reaction (11) is probably more important than reaction (12), but that the measured rates in laboratory reactors are typically too slow to explain the HONO concentrations observed in the atmosphere (Jenkin et al., 1988; Lammel and Cape, 1996). More recent studies have demonstrated, however, that carbonaceous surfaces (e.g. soot aerosols) are particularly reactive, and may represent an important substrate for HONO formation (Ammann et al., 1998; Kalberer et al., 1999).

Atmospheric observations are also mainly consistent with HONO formation by reaction (11) on land or aerosol surfaces. Kessler and Platt (1984) derived a conversion rate of NO_2 to HONO of ca. 0.6% h^{-1} in the urban boundary layer, from nighttime measurements in Juelich, Germany. Kitto and Harrison (1992) have confirmed that there is a surface source of nitrous acid by reaction of NO_2, and Harrison et al. (1996) have derived an effective rate coefficient of $5.6 \times 10^{-6} \times 100/h \, s^{-1}$ (where h is the mixing height in m) for reaction (11) in the suburban boundary layer.

2.3. The formation and removal of organic nitrates

The reactions of the organic peroxy radicals (RO_2) with NO, in addition to converting NO to NO_2 as described above (reaction (6a)), also have minor channels leading to the production of organic nitrates ($RONO_2$):

$$RO_2 + NO(+M) \rightarrow RONO_2(+M). \qquad (6b)$$

The majority of information available for these reactions is for simple alkyl peroxy radicals derived from the oxidation of alkanes which have been studied in some detail, and for the corresponding alkyl nitrate products (Carter and

Atkinson, 1989; Roberts, 1990; Atkinson, 1990; Lightfoot et al., 1992). Reaction channel (6b) is extremely minor for small peroxy radicals such as CH_3O_2 ($< 0.5\%$) and $C_2H_5O_2$ ($< 1.4\%$), but increases with radical size up to ca. 35% of the overall reaction for the secondary $C_8H_{17}O_2$ isomers under boundary layer conditions (Carter and Atkinson, 1989; Lightfoot et al., 1992). Reaction (6b) has been shown to be more important for secondary alkyl peroxy radicals than for isomeric primary radicals, but there is currently insufficient information to draw any firm conclusions for tertiary radicals. The limited information available for peroxy radicals containing oxygenated functional groups also suggests that the formation of $RONO_2$ species is less efficient than for the corresponding unsubstituted alkyl peroxy radicals (Lightfoot et al., 1992; Muthuramu et al., 1993; Atkinson, 1994; Eberhard et al., 1995).

Alkyl nitrates have been routinely detected in the atmosphere in numerous field studies over the past decade (e.g., Buhr et al., 1990; Flocke et al., 1991, 1998; Atlas et al., 1992; Shepson et al., 1993; Williams, 1994; O'Brien et al., 1995, 1997). For example, a series of nine C_1–C_5 alkyl nitrates has been detected, both in the boundary layer over southern England, and at various altitudes at locations over the North Atlantic (Williams, 1994). Although these measurements have shown the alkyl nitrates to be present at comparatively low levels in the springtime boundary layer (typically in the range 10–20 pptv for each of methyl, ethyl, 1 and 2 propyl and 2-butyl nitrate), the data also indicate that the compounds are comparatively stable under tropospheric conditions, persisting at a similar level in aged boundary layer air masses and being transported into the background troposphere. Laboratory studies have established that alkyl nitrates are insoluble in water, and do not readily transfer to the particulate phase (Roberts, 1990; Kames and Schurath, 1992), which indicates that they are ultimately degraded either by reaction with OH or by photolysis. For small alkyl nitrates, the major loss process is photolysis (Yang et al., 1993; Talukdar et al., 1997a; Zhu and Ding, 1997; Zhu and Kellis, 1997), which occurs on the timescale of about 2–4 weeks during spring and summer at northern mid-latitudes (Luke et al., 1989; Roberts and Fajer, 1989; Turberg et al., 1990; Clemitshaw et al., 1997b). For the larger compounds, however, attack of OH on the carbon skeleton becomes important (Atkinson, 1990, 1994; Roberts, 1990). Consequently, for butyl and pentyl nitrates, removal by reaction with OH is competitive with photolysis, and the tropospheric lifetimes of the compounds become progressively shorter as the size of the carbon skeleton increases $> C_5$, owing to more rapid removal by reaction with OH. Although photolysis is known to release NO_x in the form of NO_2 (Roberts, 1990; Yang et al., 1993; Talukdar et al., 1997a; Zhu and Ding, 1997; Zhu and Kellis, 1997), and measurements of rate coefficients for OH radical reactions have been carried out (Roberts, 1990; Shallcross et al., 1997; Talukdar et al., 1997b), there are currently no reported studies of the products of the reactions of OH with alkyl nitrates.

Ambient measurements of a series of organic nitrates containing hydroxy groups in the position β to the nitrate group (i.e., β-nitrooxy alcohols) have also been reported (O'Brien et al., 1995, 1997). Under the photochemical episodic conditions of those field studies, the nitrates were believed to be formed exclusively from the reactions of β-hydroxy peroxy radicals with NO (reaction (6b)). Such peroxy radicals are generated predominantly from the OH-initiated oxidation of alkenes, and the observations confirmed their significant participation in regional scale O_3 formation downwind of major population centres (Toronto and Vancouver). It should be noted, however, that β-nitrooxy alcohols and other bifunctional organic nitrates may also be generated from the NO_3 radical initiated oxidation of alkenes, which occurs predominantly at night-time (as discussed further in Section 4.2).

There is only limited information available on the further oxidation of bifunctional oxidised organic nitrogen compounds in general, and this relates solely to the rate of removal (e.g. Roberts and Fajer, 1989; Zhu et al., 1991; Barnes et al., 1993). It is probable that β-nitrooxy alcohols will be removed predominantly by reaction with OH. This may ultimately re-release NO_x, if the oxidation occurs in the gas phase. However, as the further oxidation is likely to generate intermediate products containing a large number of polar substituent groups (e.g., $-ONO_2$, $-OH$, $-CHO$), uptake into aqueous droplets may also be particularly important (Kames and Schurath, 1992).

2.4. The formation and removal of organic peroxy nitrates

The reactions of NO_2 with organic peroxy radicals (RO_2) lead to the production of compounds most commonly referred to as *organic peroxy nitrates* (RO_2NO_2), although such compounds are also (more correctly) called *peroxycarboxylic nitric anhydrides* by some authors (see discussion in Roberts, 1990):

$$RO_2 + NO_2(+M) \rightleftharpoons RO_2NO_2(+M). \tag{13}$$

Conversion of NO_2 to organic peroxy nitrates by reaction (13) occurs only about an order of magnitude more slowly than photolysis (reaction (3)). The thermal stability of peroxy nitrates is, however, strongly dependent on the structure of the organic group "R", as shown in Table 1. Simple alkyl peroxy derivatives, such as $CH_3O_2NO_2$, are unstable and decompose on the timescale of about 1 s under typical boundary layer conditions. Consequently, the equilibrium (13) is rapidly established, with only a very small amount of NO_x sequestered in the form of the peroxy nitrate. The presence of electron-withdrawing substituents in the organic group, however, tends to increase the

Table 1. Thermal decomposition rates (k_{-13}) and lifetimes (τ) for a series of peroxy nitrates (RO_2NO_2) at 298 K and 760 Torr

RO_2NO_2 (+M) $\rightarrow$ RO_2 + NO_2 (+M)		
R	$k_{-13}\,s^{-1}$	τ
H	0.076^a	13 s
CH_3	1.6^b	0.61 s
C_2H_5	4.0^b	0.25 s
CH_3CO	0.00033^b	50 min
$CH_2{=}CH(CH_3)CO$	0.00035^c	48 min

[a]Based on parameters recommended by Atkinson et al. (1997b).
[b]Based on parameters recommended by Atkinson et al. (1997a).
[c]Roberts and Bertman (1992).

thermal stability of the compounds (Lightfoot et al., 1992). Thus, a carbonyl (C=O) group adjacent to the peroxy radical centre has a particularly marked effect on the strength of the OO–NO$_2$ bond, and the resultant *peroxy acyl nitrates* are significantly more stable than the alkyl peroxy derivatives. The simplest example of this class of compound, peroxy acetyl nitrate or PAN (CH$_3$C(O)OONO$_2$), was first detected in the atmosphere about thirty years ago (Stevens, 1969), and subsequent field measurements have established that PAN often makes a significant contribution to NO$_z$ in the boundary layer, particularly at high northern latitudes (e.g., Singh et al., 1992; Bottenheim et al., 1993; Solberg et al., 1997). Furthermore, PAN is invariably more abundant in ambient air than its higher homologues (Singh and Salas, 1989; Walega et al., 1992; Singh et al., 1993; Altschuller, 1993). This is mainly because the precursor peroxy radical, CH$_3$C(O)O$_2$, is potentially produced from the degradation of a large number of organic compounds $\geqslant$ C$_2$, whereas the abundance of potential source compounds systematically diminishes for the larger RC(O)O$_2$ radicals.

Whereas PAN itself is potentially generated from many organic compounds (e.g., Derwent et al., 1998; Jenkin et al., 1999), other peroxy acyl nitrates may be generated from only a limited number of organic precursors. This is particularly true for those peroxy acyl nitrates which contain other specific functional groups. One such compound which has received particular interest (e.g., Bertman and Roberts, 1991; Williams et al., 1997), is peroxy methacroyl nitrate or MPAN (CH$_2$=C(CH$_3$)C(O)OONO$_2$). The only known source of MPAN is the degradation of methacrolein which, in turn, is only believed to be generated significantly from the abundant natural hydrocarbon isoprene. Consequently, MPAN represents a unique marker for isoprene degradation, and is generally present at about 10% of the PAN concentration under conditions where isoprene represents the dominant local hydrocarbon emis-

Table 2. Thermal decompostion lifetimes (τ) of PAN as a function altitude at mid latitudes[a,b]

$CH_3C(O)OONO_2$ (+M) $\rightarrow$ $CH_3C(O)O_2$ + NO_2 (+M)			
Approx. altitude/km	T (K)	P (Torr)	τ
0	288	760	4.2 hr
2.5	273	540	2.4 day
5.0	258	390	46 day
7.5	243	280	3.4 yr
10.0	228	200	144 yr

[a]Based on parameters recommended by Atkinson et al. (1997a).
[b]The increase of τ with altitude is almost entirely due to the decrease in temperature.

sion (Williams et al., 1997). Interestingly, however, current knowledge of isoprene degradation is also consistent with significant generation of other peroxy acyl nitrates containing additional hydroxy functionalities (Jenkin et al., 1999), namely $HOCH_2C(O)OONO_2$, $HOCH_2C(CH_3)=CHC(O)OONO_2$ and $HOCH_2CH=C(CH_3)C(O)OONO_2$, and such compounds are represented in some detailed chemical mechanisms (e.g., Jenkin et al., 1997a; Saunders et al., 1997). Indeed there are almost certainly many unidentified peroxy acyl nitrates containing additional oxygenated functionalities present in the atmosphere which are difficult to detect with established techniques.

Laboratory studies have established that PAN (and other peroxy acyl nitrates) typically has a lifetime of the order of hours with respect to thermal decomposition at temperatures characteristic of the planetary boundary layer (e.g., Lightfoot et al., 1992; Wallington et al., 1992; Roberts and Bertman, 1992; Grosjean et al., 1994a,b). As shown in Table 2, however, the stability of PAN is strongly dependent on temperature, its lifetime increasing dramatically as the temperature is lowered. Consequently, under meteorological conditions characterised by rapid vertical transport accompanied by rapid decrease in temperature, PAN and other peroxy acyl nitrates become stable molecules which may only be degraded to release NO_x on much longer timescales, either by reaction with OH radicals, or by photolysis (e.g., Talukdar et al., 1997a,b). At higher altitudes, concentrations of PAN have been shown to exceed those of the NO_x species (Singh et al., 1992, 1993). Consequently, the long-range transport of peroxy acyl nitrates (and possibly the organic nitrates referred to in Section 2.3) is believed to provide a major source of NO_x in the background troposphere (Singh et al., 1992; Horowitz et al., 1998; Thakur et al., 1999).

3. The role of volatile organic compounds in photochemical ozone formation

3.1. General description

It has been established for some decades (Haagen-Smit and Fox, 1954, 1956; Leighton, 1961) that the formation of O_3 in the troposphere is promoted by the presence of volatile organic compounds (VOCs), NO_x and sunlight, and the mechanism by which this occurs is now well understood (e.g., Atkinson, 1990, 1994, 1998a,b). The sunlight initiates the process by providing near ultra-violet radiation which dissociates certain stable molecules, leading to the formation of hydrogen-containing free radicals (HO_x). In the presence of NO_x, these free radicals catalyse the oxidation of VOCs, ultimately to carbon dioxide and water vapour. Partially oxidised organic species such as aldehydes, ketones and carbon monoxide are produced as intermediate oxidation products, with O_3 formed as a by-product. An enormous variety of VOC classes may be emitted from numerous anthropogenic and biogenic sources (e.g. Rudd, 1995; Guenther et al., 1995) and, depending on location, either or both categories can make a major contribution to photochemical ozone formation (e.g., Sillman, 1999, and references therein).

The details of the chemistry are shown schematically in Fig. 2 for the oxidation of a generic saturated hydrocarbon, RH (i.e., an alkane), into its first generation oxidised products. In common with the tropospheric oxidation of most organic compounds, the oxidation is initiated by reaction with the hydroxyl radical (OH), leading to the following rapid sequence of reactions:

$$OH + RH \rightarrow R + H_2O, \tag{14}$$

$$R + O_2(+M) \rightarrow RO_2(+M), \tag{15}$$

$$RO_2 + NO \rightarrow RO + NO_2, \tag{6a}$$

$$RO \rightarrow\rightarrow \text{carbonyl product(s)} + HO_2, \tag{16}$$

$$HO_2 + NO \rightarrow OH + NO_2. \tag{5}$$

Since OH is regenerated, this mechanism is a catalytic cycle with OH, R (alkyl radical), RO_2, RO (alkoxy radical) and HO_2 acting as chain propagating radicals. Reactions (5) and (6a), involving the peroxy radicals, play a key role in O_3 formation by oxidising NO to NO_2. As discussed in Section 2.1, NO_2 is efficiently photodissociated by near ultra-violet and visible radiation to generate O_3 by reactions (3) and (4).

The abbreviated reaction (16) shows that the RO radical is converted into HO_2 as part of the chain propagating process leading to the regeneration of

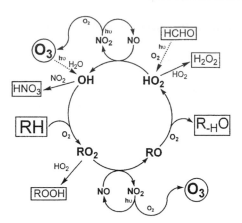

Figure 2.   Schematic representation of the free radical-catalysed oxidation of a generic saturated hydrocarbon, RH, to its first generation oxidised product, $R_{-H}O$. The key role played by the NO_x species in the chain-propagating process is also illustrated, which leads to the generation of O_3 as a by-product. The major sources and sinks of the free radicals are also shown (detailed discussion is given in the text).

OH. The mechanism of reaction (16) is strongly dependent on the structure of RO, and therefore on the structure of the parent organic compound. For small alkoxy radicals (e.g., CH_3O), conversion to HO_2 is achieved in a single step by reaction with O_2, also yielding an aldehyde or ketone oxidation product $R_{-H}O$ (e.g., in the case of CH_3O, the product is formaldehyde, HCHO):

$$RO + O_2 \rightarrow R_{-H}O + HO_2. \qquad (16)$$

Larger or more complex RO radicals may also thermally decompose or isomerise (if they contain a carbon chain $\geqslant C_4$). In these cases, the mechanism is frequently multi-step, possibly involving further peroxy and oxy radical intermediates but, in the majority of cases, still ultimately yielding HO_2. Assuming reaction (16′) is the fate of RO, the overall chemistry of the oxidation of RH into its first generation product $R_{-H}O$ (i.e., one cycle in Fig. 2) is given by the following overall equation:

$$RH + 4O_2 \rightarrow R_{-H}O + H_2O + 2O_3.$$

Thus, the oxidation of one molecule of RH to $R_{-H}O$ (catalysed by HO_x and NO_x) is accompanied by the generation of two molecules of O_3. As discussed in the next section and shown in the figure, O_3 photolysis is a major source of HO_x radicals. It may therefore be regarded as an autocatalyst, since it stimulates its own production via the chemistry described above.

The further oxidation of the first generation product, $R_{-H}O$, also follows the same general pattern (i.e. generating O_3), and subsequent organic products are, in turn, oxidised until CO_2 is eventually produced. The penultimate oxidised product is commonly CO, for which the atmospheric oxidation of organic compounds is therefore the major global source. Its oxidation to CO_2 (achieved by reaction with OH radicals) also generates HO_2, and therefore O_3 by the chemistry described above:

$$OH + CO \rightarrow H + CO_2, \tag{17a}$$

$$H + O_2(+M) \rightarrow HO_2(+M), \tag{18}$$

$$OH + CO(+M) \rightarrow HOCO(+M), \tag{17b}$$

$$HOCO + O_2 \rightarrow HO_2 + CO_2. \tag{19}$$

In the polluted troposphere, direct emissions of CO as a result of combustion processes are often the major local source, and therefore make some contribution to O_3 production in the boundary layer.

The chemistry outlined above specifically describes the oxidation of an alkane. The OH radical initiated oxidation of other VOCs (and of the oxygenated products of alkane degradation referred to above) generally occurs by very similar mechanisms, the details of which are reasonably well understood. For many small VOCs (e.g., methane, ethane, ethene, methanol, *t*-butanol, acetaldehyde, acetone and dimethyl ether), the rates and products of the elementary reactions involved in their oxidation have received a great deal of attention, and the mechanisms are well established (see, for example, Atkinson, 1994, 1997; Atkinson et al., 1997a; Jenkin et al., 1997a). For many other VOCs, oxidation mechanisms defined by analogy are fully consistent with available experimental data, and it is also possible to predict with reasonable confidence oxidation mechanisms for some compounds for which little or no experimental data exist, by use of structure–reactivity correlations (e.g., Atkinson, 1987; Dagaut et al., 1989; Kwok and Atkinson, 1995; Jenkin et al., 1997a; Porter et al., 1997). In view of the very large number of VOCs emitted into the atmosphere, the use of such correlations is essential in mechanism construction. There are, however, still significant uncertainties in the oxidation mechanisms of some common complex VOCs, such as aromatic hydrocarbons and terpenes, although new data are constantly emerging (Kwok et al., 1997; Atkinson and Arey, 1998; Barnes et al., 1998; BIOVOC, 1998; Calogirou et al., 1999).

The rate of oxidation of VOCs (and therefore production of O_3) is governed by the ambient concentration of the catalytic HO_x radicals, which is controlled by a balance between production and removal routes. In the subsections which follow, the major sources and sinks of HO_x radicals are described.

3.2. Sources of HO_x radicals

HO_x radicals are generated by the photolysis of certain atmospheric trace species. A strict definition of HO_x radicals would include only OH and HO_2. For the purposes of the present discussion, however, organic radicals are also classed as HO_x since (as described above and shown in Fig. 2), they are readily converted to OH and HO_2 under atmospheric conditions.

The incident sunlight which penetrates to the lower layers of the atmosphere is almost entirely at wavelengths longer than ca. 290 nm, and potential radical precursors are therefore those species which absorb light at these wavelengths. A further requirement is that the energy of the absorbed radiation is sufficient to break the weakest bond in the molecule. Major photolytic sources of HO_x radicals, listed in Table 3, are the photolysis of O_3 and aldehydes. In global terms, the most important tropospheric source of free radicals results from the photolysis of O_3 in its near ultra-violet band (e.g., Meier et al., 1997):

$$O_3 + h\nu \rightarrow O(^1D) + O_2, \tag{20}$$

$$O(^1D) + H_2O \rightarrow 2OH. \tag{21}$$

However, the required wavelengths are near the atmospheric cut-off at ca. 290 nm, where both the O_3 absorption cross-section and the quantum yield for $O(^1D)$ formation are wavelength dependent, decreasing with increasing wavelength (e.g. see discussion in Ravishankara et al., 1998). Consequently, the photolysis rate J_{20} varies strongly with the changes in the atmospheric pathlength which accompany variations in altitude, latitude and season, and other sources of free radicals may be more important under specific conditions. A further influence on the production rate of OH results from competing "quenching" reactions for the electronically excited $O(^1D)$ atom:

$$O(^1D)(+M) \rightarrow O(^3P)(+M), \tag{22}$$

$$O(^3P) + O_2(+M) \rightarrow O_3(+M). \tag{4}$$

Available kinetic data for reactions (21) and (22) (Atkinson et al., 1997b) suggest that the fractional conversion, f, of $O(^1D)$ into OH is given approximately by the expression

$$f = P_{H_2O}/(P_{H_2O} + 0.13(P - P_{H_2O})) \tag{vi}$$

where P_{H_2O} is the partial pressure of water vapour and P is the total pressure. Thus, this factor has to be taken into consideration when comparing the relative importance of photolysis reactions as free radical sources.

Table 3. Comparison of boundary layer photolytic radical sources for various scenarios under midsummer and midwinter conditions

Reaction	Photolysis rate[a] $(10^{-6}\,s^{-1})$	Precursor concentration $(10^9\,molecule\,cm^{-3})$	Radical production rate $(10^4\,molecule\,cm^{-3}s^{-1})$
Midsummer			
(a) Rural average			
$O_3 \rightarrow O(^1D) + O_2$	12.8	981 (40.0 ppbv)[b]	252[c]
$HCHO \rightarrow H + HCO$	15.7	41.0 (1.67 ppbv)[d]	129
$RCHO \rightarrow R + HCO$	9.53[e]	20.4 (0.83 ppbv)[d]	38.9
(b) Rural episode			
$O_3 \rightarrow O(^1D) + O_2$	12.8	2450 (100 ppbv)[f]	651[c]
$HCHO \rightarrow H + HCO$	15.7	94.0 (3.83 ppbv)[g]	295
$RCHO \rightarrow R + HCO$	9.53[e]	39.5 (1.61 ppbv)[g]	75.3
(c) Urban average			
$O_3 \rightarrow O(^1D) + O_2$	12.8	613 (25 ppbv)[h]	157[c]
$HCHO \rightarrow H + HCO$	15.7	245 (10 ppbv)[i]	769
$RCHO \rightarrow R + HCO$	9.53[e]	123 (5 ppbv)[i]	234
Midwinter			
(a) Rural average			
$O_3 \rightarrow O(^1D) + O_2$	0.486	794 (30 ppbv)[b]	7.73[c]
$HCHO \rightarrow H + HCO$	2.42	22.8 (0.86 ppbv)[k]	11.0
$RCHO \rightarrow R + HCO$	1.46[e]	17.5 (0.66 ppbv)[k]	5.11
(b) Urban average			
$O_3 \rightarrow O(^1D) + O_2$	0.486	397 (15 ppbv)[h]	3.87[c]
$HCHO \rightarrow H + HCO$	2.42	397 (15 ppbv)[i]	192[j]
$RCHO \rightarrow R + HCO$	1.46[e]	199 (7.5 ppbv)[i]	58.1

[a] Daylight average, 50° Lat., calculated using parameters reported by Jenkin et al. (1997b).
[b] Typical daytime concentrations at rural sites in the southern UK (PORG, 1997).
[c] Calculated assuming 11 Torr water vapour.
[d] Mean of data measured at Harwell, UK April–September 1993–1997 (Solberg, 1999; Dollard and Jenkin, 1999).
[e] Using propanal (propionaldehyde) as a representative.
[f] Typical level during photochemical pollution episode in the southern UK (PORG, 1997).
[g] Simulated concentration for southern UK under episodic conditions (Dollard and Jenkin, 1999).
[h] Typical daytime concentrations at urban centre sites in the UK (PORG, 1997).
[i] Typical seasonal average for European cities (Dollard and Jenkin, 1999).
[j] Harrison et al., 1996.
[k] Mean of data measured at Harwell, UK October–March 1993–1997 (Solberg, 1999; Dollard and Jenkin, 1999).

As shown in Table 3, the photolysis of HCHO (reaction (23)) makes an important contribution to free radical production at the concentrations observed in the rural UK, and dominates radical production under urban conditions, when the concentration of O_3 tends to be suppressed to a certain extent by reaction with NO (reaction (2)), and the concentration of HCHO is elevated due to high local emissions or production rates from hydrocarbon oxidation (see above):

$$HCHO + h\nu \rightarrow H + HCO, \tag{23}$$

$$RCHO + h\nu \rightarrow R + HCO, \tag{24}$$

$$H + O_2(+M) \rightarrow HO_2(+M), \tag{18}$$

$$R + O_2(+M) \rightarrow RO_2(+M), \tag{15}$$

$$HCO + O_2 \rightarrow HO_2 + CO. \tag{25}$$

Similarly, the photolysis of other aldehydes (reaction (24)) can collectively make a significant contribution to radical production, particularly in polluted regions. Clearly, the photolysis of HCHO, and to a lesser extent RCHO, can play an important role in the initiation of summertime pollution at urban locations, when O_3 concentrations may initially be significantly suppressed.

Also worthy of note is the photolysis of HONO (reaction 10), which is very efficient under both summertime and wintertime conditions, and is a potentially important free radical source, despite being present at very low concentrations during daylight hours. On the basis of a daytime concentration of 100 pptv, typically observed in urban locations (Kitto and Harrison, 1992; Harrison et al., 1996; Andres Hernandez et al., 1996), and typical mid-latitude photolysis rates of ca. $10^{-3}\,s^{-1}$, HONO photolysis potentially makes a major contribution to free radical production compared with the reactions shown in Table 3. However, since a large proportion of the observed daytime HONO is derived from the reaction of OH with NO, particularly under summertime conditions,

$$OH + NO(+M) \rightarrow HONO(+M) \tag{9}$$

the net radical source is invariably substantially lower than the actual flux through the photolysis reaction. Thus, only HONO generated from alternative sources (e.g., thermal heterogeneous reactions of NO_x and water vapour as described in Section 2.2, or direct emission in vehicle exhaust) leads to net radical production upon photolysis. This is particularly significant immediately after sunrise, when observed concentrations of HONO are generally at their highest owing to its nighttime generation from thermal reactions, and accumulation in the absence of sunlight (Harris et al., 1982; Kessler and Platt, 1984; Kitto and Harrison, 1992; Harrison et al., 1996). Consequently, HONO photolysis potentially provides a pulse of free radical production in the early morning when the photolysis rate of O_3 and aldehydes is very slow owing to the long atmospheric pathlength. This has been confirmed by boundary layer model calculations (Harris et al., 1982; Jenkin et al., 1988) which have demonstrated a substantial enhancement of early morning OH radical production, following the photolysis of HONO, and increased O_3 generation.

As discussed further in Section 3.5, another significant source of HO_x radicals throughout the diurnal cycle results from the reactions of O_3 with alkenes. Paulson and Orlando (1996) have demonstrated that these reactions are possibly major daytime sources of HO_x under some urban settings, even exceeding radical production rates from the photolysis reactions discussed above.

3.3. Sinks and reservoirs for HO_x radicals

HO_x radicals are removed from the atmosphere by a variety of termination reactions. The molecular products of these reactions are usually termed reservoirs, since the possibility of HO_x regeneration through thermal decomposition or photolysis often exists. The lifetimes of reservoirs with respect to these processes are very variable, and even for a given reservoir, may be a strong function of time and location (i.e., temperature, pressure, solar intensity). If a reservoir is short-lived, it has only a minor effect on HO_x (e.g., the formation and thermal decomposition of HO_2NO_2 in the boundary layer, reaction (8)). If the reservoir is comparatively stable with respect to thermal decomposition or photolysis, however, it is likely that its removal by either physical processes (e.g., dry deposition) or chemical ones (e.g., reaction with OH radicals) becomes competitive or even dominant. Since this precludes the quantitative regeneration of HO_x radicals, the formation of the reservoir represents a sink for HO_x radicals. In addition, the formation and transport of reservoirs with intermediate thermal lifetimes in the boundary layer (in particular PAN), can lead to significant net radical removal locally (e.g., Sillman and Samson, 1995), with the possibility of subsequent radical regeneration.

On the basis of measured rate coefficients (Atkinson et al., 1997a,b; DeMore et al., 1997), the most significant tropospheric sinks for HO_x radicals involve either the mutual termination of two HO_x species,

$$HO_2 + HO_2 \rightarrow H_2O_2 + O_2, \tag{26}$$

$$RO_2 + HO_2 \rightarrow ROOH + O_2, \tag{27}$$

$$OH + HO_2 \rightarrow H_2O + O_2, \tag{28}$$

or the removal of HO_x by reaction with the NO_x species by the following reactions:

$$OH + NO_2(+M) \rightarrow HNO_3(+M), \tag{7}$$

$$RO_2 + NO(+M) \rightarrow RONO_2(+M). \tag{6b}$$

Table 4. Illustrative comparison of boundary layer HO_x radical removal rates from major sink reactions in the southern UK[a]

Midsummer, 50° Lat		Urban[b]	Rural[c]	Rural episode[d]
$HO_x + HO_x$ reactions				
$HO_2 + HO_2 \rightarrow H_2O_2 + O_2$	(26)	42	42	260
$RO_2 + HO_2 \rightarrow ROOH + O_2$	(27)	28	28	320
$OH + HO_2 \rightarrow H_2O + O_2$	(28)	7.0	7.0	94
$HO_x + NO_x$ reactions				
$OH + NO_2 \ (+M) \rightarrow HNO_3 \ (+M)$	(7)	1300	280	1100
$RO_2 + NO \ (+M) \rightarrow RONO_2 \ (+M)$	(6b)	470	23	48

[a] Units 10^4 molecule cm^{-3} s^{-1}. The figures given are based on illustrative midday reactant concentrations, inferred from ambient measurements and modelling studies appropriate to the southern UK, as given in the following notes.
[b] Urban conditions, $[OH] = 1.6 \times 10^6$ molecule cm^{-3}; $[HO_2] = 2.0 \times 10^8$ molecule cm^{-3}; $[RO_2] = 1.0 \times 10^8$ molecule cm^{-3}; $[NO] = [NO_2] = 6.2 \times 10^{11}$ molecule cm^{-3} (ca. 25 ppbv).
[c] Rural conditions, $[OH] = 1.6 \times 10^6$ molecule cm^{-3}; $[HO_2] = 2.0 \times 10^8$ molecule cm^{-3}; $[RO_2] = 1.0 \times 10^8$ molecule cm^{-3}; $[NO] = 3.0 \times 10^{10}$ molecule cm^{-3} (ca. 1.2 ppbv) $[NO_2] = 1.3 \times 10^{11}$ molecule cm^{-3} (ca. 5 ppbv).
[d] Rural episode conditions, $[OH] = 8.5 \times 10^6$ molecule cm^{-3}; $[HO_2] = 5.0 \times 10^8$ molecule cm^{-3}; $[RO_2] = 4.5 \times 10^8$ molecule cm^{-3}; $[NO] = 1.4 \times 10^{10}$ molecule cm^{-3} (ca. 0.6 ppbv) $[NO_2] = 9.5 \times 10^{10}$ molecule cm^{-3} (ca. 3.9 ppbv).

As illustrated in Table 4, the relative importance of these reactions depends strongly on ambient conditions. Reactions (26)–(28) show a second order (quadratic) dependence on the concentration of HO_x and therefore vary greatly with time of day, location and season. Reactions (6b) and (7) involving NO_x are clearly much more important in the boundary layer over populated areas than in remote tropospheric environments, owing to the higher levels of NO_x. As discussed in Section 2.3, the efficiency of reaction (6b) relative to the alternative channel (6a) is a strong function of the size and structure of the RO_2 radical. Consequently the importance of reaction (6b) as a radical sink is influenced by the precise composition of the peroxy radical population. The data presented in Table 4 were calculated assuming reaction channel (6b) accounts for 1% of the overall reaction, which is consistent with the major contribution to the RO_2 radical population being made by small peroxy radicals such as CH_3O_2 and $C_2H_5O_2$, and oxygenated peroxy radicals such as $HOCH_2CH_2O_2$ and $CH_3C(O)O_2$.

The data presented in Table 4 provide an illustrative comparison, and clearly only consider selected boundary layer conditions. Nevertheless, the data demonstrate that reaction (7) tends to be the major radical sink for conditions appropriate to the boundary layer in the UK and other populated regions.

This reaction totally dominates radical removal at urban levels of NO_2, and also makes the main contribution under representative rural conditions. At the elevated radical concentrations consistent with a photochemical episode in the southern UK, however, reactions (26) and (27) compete with reaction (7), owing to their second-order dependence on $[HO_x]$. At other locations with lower levels of NO_x, or high local emissions of natural hydrocarbons (in particular, isoprene), reactions (26) and (27) are often the dominant radical removal processes.

Since reactions (7), (26) and (27) are collectively the major radical sinks, the overall removal rate of HO_x (φ) is given by the following equation, to a first approximation:

$$\varphi = k_7[OH][NO_2] + 2k_{26}[HO_2]^2 + 2k_{27}[HO_2][RO_2]. \qquad \text{(vii)}$$

An additional contribution to radical removal is made by reactions (6b) and (28). However, reaction (6b), involving NO_x, is generally minor in comparison with reaction (7), and the second order reaction (28) is minor in comparison with reactions (26) and (27), although it does gain in significance in the middle and upper troposphere.

In addition to being an important radical sink, the self-reaction of HO_2 (reaction (26)) is the major source of hydrogen peroxide (H_2O_2) in the atmosphere. Consequently, H_2O_2 is a further indicator of free-radical driven photochemical processes (e.g., Kleinman, 1986; Ayers et al., 1992), and it is routinely measured in the polluted and remote planetary boundary layer (e.g., Dollard et al., 1991; Slemr and Tremmel, 1994). Similarly, reaction (27) for a variety of organic peroxy radicals leads to the generation of analogous organic hydroperoxides (ROOH), which have been detected in numerous field studies (e.g., Hellpointer and Gab, 1989; Hewitt and Kok, 1991; Slemr and Tremmel, 1994; Jackson and Hewitt, 1996; Staffelbach et al., 1996; Ayers et al., 1996). H_2O_2 readily transfers to the aqueous phase, and is particularly important as an oxidant for sulphur dioxide (e.g., Penkett et al., 1979; Martin and Damschen, 1981; Chandler et al., 1988; Gervat et al., 1988).

3.4. Sensitivity of O_3 formation chemistry to changes in NO_x and VOC concentration

The production of O_3 from the ultra-violet irradiation of mixtures of VOCs and NO_x in air was first demonstrated in the smog chamber studies of Haagen-Smit and coworkers (e.g., Haagen-Smit and Fox, 1954,1956). The peak O_3 concentrations generated from various initial concentrations of NO_x and VOCs were usually presented as an O_3 isopleth diagram of the form shown in Fig. 3, in

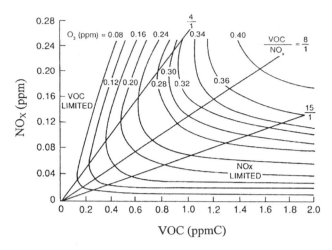

Figure 3. Example of an isopleth diagram illustrating calculated peak O_3 concentrations generated from various initial concentrations of NO_x and a specified VOC mixture using the US EPA
empirical kinetic modelling approach (diagram adapted from Dodge, 1977). Although the methodology was originally developed for highly polluted scenarios, and the reagent concentrations on
the axes are thus significantly greater than typically observed in the boundary layer, the characteristic shape of the O_3 isopleths as a function of VOC/NO_x ratio also applies to lower reagent
concentrations (detailed discussion is given in the text).

which initial mixture compositions giving rise to the same peak O_3 concentration are connected by the appropriate isopleth. Although such diagrams were
originally defined by experiment, the detailed chemistry leading to O_3 formation is now well understood, and isopleth diagrams can be generated from
modelling studies using validated chemical mechanisms. In this way, diagrams
can be generated for different VOCs (or VOC mixtures) and for different levels of solar intensity. Such diagrams are sometimes used to assess the effect on
O_3 which would result from NO_x and VOC emissions control strategies (e.g.,
Dodge, 1977; NRC, 1991). It is clear from Fig. 3 that the influence of changing
the concentration of NO_x or VOCs on the production of O_3 is strongly dependent on the ambient conditions (in particular on the relative concentrations of
NO_x and VOCs). In this subsection, some features of the isopleth diagram are
explained in terms of the chemistry described in the preceding subsections,
with the species RH as a representative VOC.

   The propensity for O_3 formation in a given air mass is essentially proportional to the number of free radical-propagated cycles (in Fig. 2) which can
occur before radical removal. This is usually referred to as the chain-length
for O_3 formation. Thus, when considering the influence of changing the concentration of NO_x or VOC on the production of O_3, it is the effect of such a
change on the chain-length which is important. This is determined primarily

by the effect the change has on the rate of the chain terminating reactions (i.e., the radical sinks) compared with the competing chain propagating reactions.

Let us first consider the situation when the relative concentration $[NO_x]/[VOC]$ is high, the condition usually referred to as VOC limited. As indicated in the previous subsection, the dominant chain terminating reaction under these circumstances is reaction (7). As shown in Fig. 2, the competing chain propagating reaction is reaction (14), which leads to O_3 formation:

$$OH + RH \rightarrow R + H_2O, \tag{14}$$

$$OH + NO_2(+M) \rightarrow HNO_3(+M). \tag{7}$$

Clearly, a decrease in the concentration of RH, which might result from VOC emission controls, would decrease the chain length and hence the rate of O_3 formation. In contrast, a decrease in the concentration of NO_2, which might result from NO_x emission controls, would increase the chain length and hence the rate of O_3 formation unless a simultaneous reduction in the concentration of RH occurs. Thus, under VOC-limited conditions, O_3 formation correlates positively with the ratio $[VOC]/[NO_x]$, and the O_3 isopleths tend towards lines which pass though the origin (i.e., lines of constant $[VOC]/[NO_x]$) in this region of the diagram.

At high $[VOC]/[NO_x]$ ratios, or NO_x-limited conditions, the dominant chain-terminating reactions are (26) and (27). As discussed in the previous subsection, this circumstance is more readily achieved at higher total radical concentrations, so that the $[VOC]/[NO_x]$ ratio required for NO_x-limited conditions to prevail is also influenced by the solar intensity. As presented schematically in Fig. 2, the ozone formation chain length under these conditions is determined by a competition between reactions (5) and (26),

$$HO_2 + NO \rightarrow OH + NO_2, \tag{5}$$

$$HO_2 + HO_2 \rightarrow H_2O_2 + O_2, \tag{26}$$

and between reactions (6a) and (27):

$$RO_2 + NO \rightarrow RO + NO_2, \tag{6a}$$

$$RO_2 + HO_2 \rightarrow ROOH + O_2. \tag{27}$$

In contrast to the VOC-limited conditions, the key chain propagating reactions (5) and (6a) involve NO_x and, consequently any reduction in NO_x decreases the O_3 formation chain length. Since RH is not itself directly involved in these competitions, the O_3 formation chain length is insensitive to changes in the level of RH which might result from VOC emission control. Thus, under NO_x-limited conditions, O_3 formation correlates positively with $[NO_x]$, and the O_3 isopleths tend towards lines which run parallel with the VOC axis (i.e., lines of constant $[NO_x]$) in this region of the diagram. It should be noted, however, that variation of the concentration of some VOCs has an effect on O_3 formation under NO_x-limited conditions, resulting from an indirect influence on the ambient concentration of NO_x. This arises because the degradation of some VOCs can lead to the significant removal of NO_x as organic nitrates and peroxy nitrates (see Sections 2.3 and 2.4).

Fig. 3 clearly demonstrates that reduction of ozone formation is best achieved by a decrease in the VOC concentration under VOC-limited conditions, and by a reduction of NO_x concentration under NO_x-limited conditions. It is apparent from the above discussion, however, that VOC-limited conditions correspond to when the dominant radical sink is reaction (7), and that NO_x-limited conditions correspond to when reactions (26) and (27) are the major sinks. Indeed, the use of the ratio of the products of reactions (26) and (7), $[H_2O_2]/[HNO_3]$, as an indicator for VOC or NO_x limitation is well established and documented (e.g., Sillman, 1995,1999; Sillman et al., 1998). As implied by the illustration in Table 4, the conditions encountered in the boundary layer (particularly over Europe) tend to be intermediate to these extremes, or possibly varying between them for a given air mass trajectory. Consequently, a thorough appraisal of the influence of reductions of VOC and NO_x emissions on O_3 production is often achieved by use of suitable boundary layer airshed or trajectory models incorporating the precursor emissions and appropriate chemical mechanisms (e.g., Derwent and Davies, 1994). Such models are also used to consider the relative contributions to ozone formation made by many different emitted VOCs, recognising that the contribution (per unit mass emission) can vary from one compound to another by virtue of differences in reactivity and structure (e.g., Carter, 1991,1994; Derwent and Jenkin, 1991; Andersson-Skold et al., 1992; Simpson, 1995; Derwent et al., 1998). This has given rise to the definition of scales of so-called *reactivity* or *ozone formation potential* of which the most widely publicised and applied are the Maximum Incremental Reactivity (MIR) scale, developed by Carter and co-workers to assess ozone formation over periods of up to a day in urban scenarios in the USA (e.g., Carter, 1994,1995; Carter et al., 1995), and the Photochemical Ozone Creation Potential (POCP) scale, developed by Derwent and co-workers to investigate regional scale ozone formation over periods of up to five days in northwest Europe (e.g., Derwent and Jenkin, 1991; Derwent et al., 1996,1998; Jenkin and Hayman, 1999).

A further parameter of interest is the number of molecules of O_3 generated for every molecule of NO_x oxidised (e.g., to HNO_3), which is sometimes referred to as the *ozone production efficiency*, OPE. On the basis of the above discussion, the OPE would be expected to vary with ambient conditions. Under VOC-limited conditions, when the cyclic mechanism is terminated by reaction (7) (i.e., converting NO_2 to HNO_3), the OPE is essentially equal to the chain length (i.e., the number of O_3 molecules generated by the free radical-propagated mechanism before termination occurs). Thus, as described above, the precise value depends on the VOC/NO_x ratio. Values of the OPE inferred from ambient measurements, or calculated using boundary layer models, are consistent with values in the region of 1–5 under comparatively polluted conditions in the boundary layer over the US (Ryerson et al., 1998; Sillman et al., 1998; Nunnermacker et al., 1998) and Europe (Derwent et al., 1994; Derwent and Davies, 1994). Under NO_x-limited conditions, when reactions (24) and (25) are the dominant termination reactions, the O_3 formation mechanism is less efficient at oxidising NO_x and the OPE is potentially much greater. However, the minor participation of reaction (7), and the additional removal of NO_x as organic nitrates and peroxy nitrates tends to place an upper limit on the values which may be achieved. Available ambient measurements at rural sites in the US and Canada are consistent with OPE values of approximately 10 (Liu et al., 1987; Trainer et al., 1993; Olszyna et al., 1994), although calculations indicate that this is likely to increase dramatically as NO_x is further reduced (Lin et al., 1988), and values approaching 100 may be possible in the remote boundary layer and the free troposphere (e.g., Collins et al., 1995). It should also be noted that the observed yield of O_3 relative to the oxidation of NO_x is also influenced by the night-time oxidation of NO_x by thermal reactions, which leads to the removal of O_3 (i.e., the OPE is effectively negative at night). The chemistry involved is discussed further below in Section 4.

The HO_X species, OH and HO_2, also react directly with O_3. Consequently, VOC oxidation can lead to the removal of O_3 at very low levels of NO_x, mainly as a result of the reaction of HO_2 with O_3 (reaction (29)):

$$HO_2 + O_3 \rightarrow OH + 2O_2, \tag{29}$$

$$HO_2 + NO \rightarrow OH + NO_2. \tag{5}$$

The competition between reactions (5) and (29) therefore effectively determines whether net O_3 production or removal occurs. Assuming a background concentration of O_3 of ca. 30 ppbv, the available rate coefficients for reactions (5) and (29) (Atkinson et al., 1997b) indicate that only ca. 7 pptv NO is required for O_3 production from reaction (5) to balance the destruction from reaction (29). At higher levels of NO, the photochemical oxidation of VOCs leads to net O_3 production by the mechanism described above.

3.5. VOC oxidation initiated by reaction with O_3

Unsaturated VOCs such as alkenes dienes and monoterpenes, in addition to reacting rapidly with OH, may also be oxidised by reaction with O_3 (e.g., Atkinson and Carter, 1984; Atkinson, 1997). The importance of reaction with OH and O_3 for a series of VOCs detected at rural locations in the UK is compared in Table 5. It is clear that O_3-initiated oxidation makes an important contribution, in particular for the more alkyl-substituted alkenes such as the 2-butene isomers and 2-methyl 2-butene.

Although the details of the oxidation mechanisms are less well known than for those following OH attack, the main features are reasonably well established (Atkinson, 1997). The mechanism proceeds *via* addition of O_3 to the double bond, leading initially to formation of an energy rich ozonide. This ozonide decomposes rapidly by two possible channels, each forming a carbonyl compound and a Criegee biradical which also possesses excess energy (denoted by $\ddagger$). For example, in the case of 2-methyl-2-butene, this may be represented as follows:

$$O_3 + CH_3C{=}C(CH_3)_2 \rightarrow CH_3CHO + [(CH_3)_2COO]^{\ddagger} \qquad (30a)$$

$$\rightarrow CH_3C(O)CH_3 + [CH_3CHOO]^{\ddagger}. \qquad (30b)$$

The energy rich Criegee biradicals are either collisionally stabilised, or decompose to yield a series of radical and molecular products. For $[CH_3CHOO]^{\ddagger}$, the following have been postulated as being the probable major reaction channels (Atkinson, 1997):

$$[CH_3CHOO]^{\ddagger}(+M) \rightarrow CH_3CHOO(+M), \qquad (31)$$

$$[CH_3CHOO]^{\ddagger} \rightarrow CH_2CHO + OH, \quad \text{or } CH_3 + CO + OH, \qquad (32a)$$

$$\rightarrow CH_3 + CO_2 + H, \qquad (32b)$$

$$\rightarrow CH_4 + CO_2. \qquad (32c)$$

The predominant reaction for the stabilised biradicals (such as CH_3CHOO formed in reaction (31)) under tropospheric conditions is believed to be the reaction with water vapour, which can lead to the formation of carboxylic acids, hydroxyalkyl hydroperoxides and H_2O_2 (see Atkinson, 1997, and references therein), e.g.:

$$CH_3CHOO + H_2O \rightarrow CH_3C(O)OH + H_2O \qquad (33a)$$

$$\rightarrow CH_3CH(OH)OOH \qquad (33b)$$

$$\rightarrow CH_3CHO + H_2O_2. \qquad (33c)$$

Table 5. Comparison of chemical lifetimes of selected VOC detected at UK rural sites with respect to reaction with OH, O_3 and NO_3 at assumed ambient levels

VOC	OH^a	$O_3{}^a$	$NO_3{}^a$
Alkanes			
Ethane	29 d		91 yr
Propane	6.3 d		7.8 yr
Butane	2.9 d		2.7 yr
2-Methyl propane	3.1 d		1.5 yr
Pentane	1.8 d		1.5 yr
2-Methyl butane	1.9 d		1.3 yr
Alkenes			
Ethene	20 h	9.7 d	7.3 month
Propene	6.6 h	1.5 d	4.9 d
1-Butene	5.5 h	1.6 d	3.5 d
2-Butene	2.9 h	2.4 h	2.9 h
2-Methyl propene	3.4 h	1.4 d	3.4 h
1-Pentene	5.5 h	1.5 d	$3.5 d^b$
2-Pentene	2.6 h	$2.4 h^b$	$2.9 h^b$
2-Methyl 1-butene	2.8 h	$1.4 d^b$	$3.4 h^b$
3-Methyl 1-butene	5.5 h	$1.6 d^b$	$3.5 d^b$
2-Methyl 2-butene	2.0 h	55 min	7.1 min
1,3-Butadiene	2.6 h	2.4 d	11 h
Isoprene	1.7 h	1.2 d	1.7 h
Aldehydes			
Formaldehyde	18 h		2.7 month
Acetaldehyde	11 h		17 d
Aromatics			
Benzene	5.7 d		
Toluene	1.2 d		1.8 yr
Ethyl benzene	23 h		
o-Xylene	12 h		4.1 month
m-Xylene	7.1 h		6.6 month
p-Xylene	12 h		3.4 month
Sulpur-containing organics			
Dimethyl sulphide	1.5 d		1.0 h
Dimethyl disulphide	46 min		1.5 h

[a]Concentrations used in calculations: $[OH] = 1.6 \times 10^6$ molecule cm^{-3} (ca. 0.06 pptv), $[O_3] = 7.5 \times 10^{11}$ molecule cm^{-3} (ca. 30 ppbv), $[NO_3] = 2.5 \times 10^8$ molecule cm^{-3} (ca. 10 pptv); 1/e lifetimes calculated using rate coefficients taken from the evaluations of Atkinson (1991,1994) except where indicated.

[b]Rate coefficient estimated by analogy.

However, they may also play a minor role in the oxidation of trace atmospheric species, for example SO_2 (Cox and Penkett, 1971,1972):

$$CH_3CHOO + SO_2 \rightarrow CH_3CHO + SO_3. \tag{34}$$

Reactions (32a) and (32b) show that the production of free radicals from the O_3-initiated oxidation of unsaturated VOCs can occur. Since free radicals catalyse the formation of O_3, as described above, it is apparent that the reactions of O_3 with unsaturated VOCs do not necessarily constitute a sink for O_3 in the troposphere. Most attention has been given to the generation of OH, which has been observed in significant yield from the reactions of O_3 with more than 20 alkenes, dienes and monoterpenes (e.g., Niki et al., 1987; Atkinson and Aschmann, 1993; Paulson et al., 1997; Paulson et al., 1998; Marston et al., 1998; Pfeiffer et al., 1998; Donahue et al., 1998). These laboratory studies indicate that the OH yield increases with alkyl substitution of the double bond, for example from ca. 30% for propene to ca. 60% and 90% for trans-2-butene and 2-methyl-2-butene respectively. The more substituted alkenes also tend to have an increased reactivity towards O_3 and (as shown in Table 5) are therefore those most likely to react with O_3 under tropospheric conditions. Consequently the tropospheric reactions of O_3 with alkenes potentially lead to significant radical formation (e.g., see Paulson and Orlando, 1996) and, depending on the ambient conditions, possible net O_3 production. It is also clear that these reactions can generate radicals at night, as will be discussed further below in Section 4.3.

4. Nighttime chemistry

Although the major oxidation processes in the troposphere are initiated by the presence of sunlight, there are potentially significant chemical processes which can occur during the night. These processes cannot generate O_3 (indeed, they lead to O_3 removal), but potentially do produce a series of secondary pollutants, including H_2O_2. The chemistry also oxidises NO_x and VOCs which, as described above, are precursors to the formation of O_3 and other secondary photochemical pollutants during daylight. In this section, current understanding of nighttime chemistry is summarised, with particular emphasis placed on the role of the nitrate radical, NO_3.

4.1. The formation of NO_3 and N_2O_5

Throughout the diurnal cycle, NO_2 is slowly converted into NO_3 by reaction with O_3, which occurs on the timescale of ca. 12 h at a typical boundary layer

O_3 concentration of 30 ppbv:

$$NO_2 + O_3 \rightarrow NO_3 + O_2. \tag{35}$$

During daylight, however, NO_3 is photolysed extremely efficiently (on the timescale of a few seconds), leading mainly to the regeneration of both NO_2 and O_3:

$$NO_3 + h\nu \rightarrow NO_2 + O(^3P), \tag{36}$$

$$O(^3P) + O_2(+M) \rightarrow O_3(+M). \tag{4}$$

NO_3 also reacts rapidly with NO, leading to the regeneration of NO_2:

$$NO_3 + NO \rightarrow 2NO_2. \tag{37}$$

Consequently, the importance of NO_3 in daytime chemistry is severely limited. At night, however, the chemistry of both NO_3 and NO_x differs from the daytime behaviour. In addition to the absence of sunlight itself, the concentration of the OH radical is significantly suppressed, since it is produced mainly from the photolysis of stable molecules. Thus, once formed from reaction (2), NO_2 cannot be photolysed to regenerate NO, or removed at a significant rate by reaction with OH (reaction (7)). Provided the ambient concentration of O_3 is sufficiently high, therefore, NO is rapidly converted to NO_2 (reaction (2)), which in turn is slowly converted to NO_3 by reaction (35), as shown schematically in Fig. 4. Reaction (37) is therefore generally unimportant at night, because NO is only present in significant concentrations close to points of emission where O_3 has been completely titrated, and NO_3 cannot be formed. The principle reaction of NO_3 at night is often with NO_2:

$$NO_3 + NO_2(+M) \rightleftharpoons N_2O_5(+M). \tag{38}$$

The lifetime of NO_3 with respect to this reaction is about 2 s for an ambient [NO_2] of ca. 10 ppbv (Atkinson et al., 1997b). However, since the product N_2O_5 is thermally unstable and decomposes on a similar timescale (ca. 15 s at 298 K), equilibrium (38) is readily established with NO_3 and N_2O_5 present in comparable concentrations. As a result, their behaviour is strongly coupled and any process removing one of the species is also a sink for the other. Thus, at sunrise both species rapidly fall to very low concentrations due to the efficient photolysis of NO_3.

A major removal process for N_2O_5 (and therefore NO_3) at night is the reaction of N_2O_5 with water:

$$N_2O_5 + H_2O \rightarrow 2HNO_3. \tag{39}$$

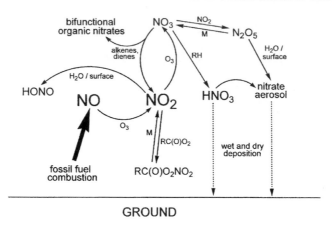

Figure 4.   Nighttime interconversions of oxidised nitrogen compounds in the troposphere.

Laboratory studies of this reaction have shown it to occur extremely slowly in the gas phase (Mentel et al., 1996; Atkinson et al., 1997b). In the troposphere it is believed to occur predominantly in cloud water and on the surface of particulate, presumably involving the charge transfer, or ionic, intermediate $NO_2^+NO_3^-$, with the product being nitrate aerosol rather than gaseous HNO_3. Laboratory studies have also established that N_2O_5 uptake is efficient on water droplets, and on aerosols of sulphuric acid or ammonium sulphate (Mozurkewichz and Calvert, 1988; Van Doren et al., 1990; Hanson and Ravishankara, 1991; Lovejoy and Hanson, 1995), and it is concluded that the heterogeneous hydrolysis of N_2O_5 is an important loss process for tropospheric NO_x. Owing to its heterogeneous nature, however, the precise rate of this reaction is variable and difficult to define under tropospheric conditions. The timescale for removal of N_2O_5 on the tropospheric aerosol is believed to vary from the order of minutes in highly polluted air characterised by high particle densities and surface areas, to several hours in more remote, continental regions (e.g., Dentener and Crutzen, 1993). The heterogenous removal of NO_3 on water droplets may also occur, but it is believed to be much less important than hydrolysis of N_2O_5 (Dentener, 1994).

4.2. The reactions of NO₃ with volatile organic compounds

The reactions of NO_3 with trace organic compounds also potentially contribute to its removal at night. On the basis of available rate coefficients (e.g., see Wayne et al., 1991; Atkinson, 1991,1994), and concentrations of a series of organic compounds observed at rural sites in the UK (PORG, 1997), it is clear that alkenes and sulphur-containing organics are particularly significant for scavenging NO_3, with minor contributions made by aldehydes, aromatic hydrocarbons and alkanes. From these data it is clear that the sulphur-containing organics are very important for scavenging NO_3 at coastal locations. For example, extensive measurements made on the north Norfolk coast have demonstrated that reaction with dimethyl sulphide (DMS) is the dominant loss process for NO_3 in clean air during spring (Carslaw et al., 1997) and summer (Allan et al., 1997). At other times, and generally at inland rural locations, the most important class of organic compound is the alkenes. In particular, 2-methyl propene (*i*-butene) appears to make a notable contribution to NO_3 removal. Whereas abstraction reactions with other organic compounds convert NO_3 to HNO_3,

$$NO_3 + RH \rightarrow HNO_3 + R \tag{40}$$

the reactions with the alkenes occur by an addition mechanism, initiating a complex chemistry involving nitro-oxy subsitituted organic radicals, which can either regenerate NO_2 or produce comparatively stable bifunctional organic nitrate products (Wayne et al., 1991). For 2-methyl propene, the initial reaction has two channels, as follows:

$$NO_3 + CH_2=C(CH_3)_2(+M) \rightarrow CH_2(ONO_2)C(CH_3)_2(+M) \tag{41a}$$

$$\rightarrow CH_2C(ONO_2)(CH_3)_2(+M). \tag{41b}$$

The subsequent reaction mechanism (shown schematically in Fig. 5) is propagated by reactions of nitro-oxy substituted peroxy and oxy radicals, and potentially leads to the production of the bifunctional organic nitrate product 2-nitro-oxy 2-methyl propanal, via the minor reaction channel (41b). As shown in Fig. 5, this reaction sequence also yields HO_2, and is therefore potentially a nighttime source of OH radicals by reactions (29) and (42), and H_2O_2 by reaction (26):

$$HO_2 + NO_3 \rightarrow OH + NO_2 + O_2. \tag{42}$$

However, the major reaction pathway (via channel (41a)) leads to the production of the unsubstituted carbonyl compounds formaldehyde and acetone, and regenerates NO_2. Consequently, the NO_3-initiated oxidation of 2-methyl propene is believed to lead to significant regeneration of NO_x, but is a comparatively minor source of HO_x. Since the same is true for branched alkenes in general, which tend to be the most reactive towards NO_3 (Wayne et al., 1991; Atkinson, 1991,1994), it is probable that, for the most part, the reactions of NO_3 with alkenes lead to significant regeneration of NO_x, thereby inhibiting the conversion of NO_x to nitrate aerosol or HNO_3 at night. The NO_3-initiated oxidation of the less alkyl substituted alkenes (e.g., the 2-butene

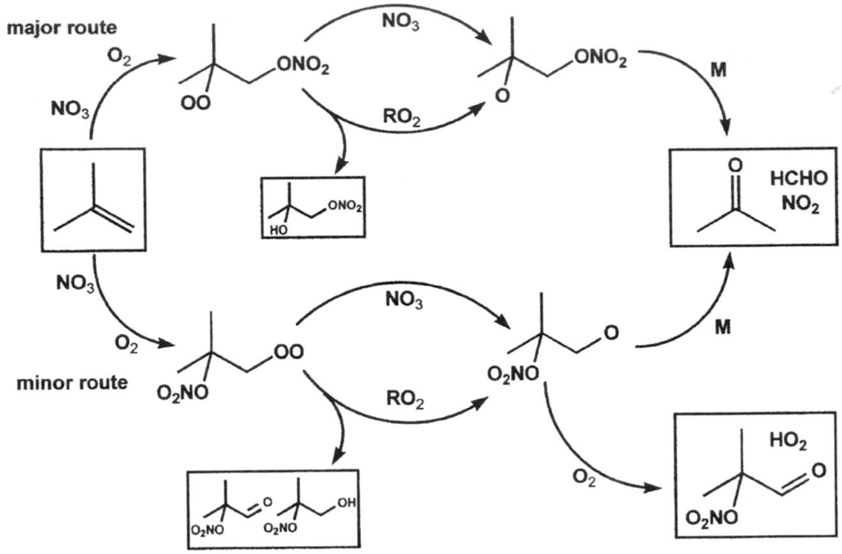

Figure 5. Schematic diagram of the NO_3 radical initiated oxidation of 2-methyl propene (*i*-butene).

isomers) should result in a greater yield of HO_x and bifunctional organic nitrate products, but is also likely to lead to substantial regeneration of NO_x. It should also be noted that terminating reactions of RO_2 and HO_2 with the intermediate nitro-oxy peroxy radicals produced from alkenes in general, can lead to the generation of bifunctional products containing carbonyl, hydroxy or hydroperoxy groups in addition to the nitro-oxy substitution (see Fig. 5), and measurements of such products in ambient air have been reported (e.g., Kastler and Ballschmiter, 1998).

On the basis of the reported concentrations of alkenes and sulphur-containing organics (PORG, 1997), and the likely removal rate of NO_3 via the heterogeneous hydrolysis of N_2O_5, a mean removal rate of NO_3 at night is calculated to be ca. 2×10^{-2} s^{-1}, under UK conditions. This corresponds to a lifetime of ca. 1 min. Assuming ambient concentrations of NO_2 and O_3 of 10 ppbv and 30 ppbv, typical of rural southern UK (PORG, 1997), this implies a mean rural nighttime NO_3 concentration of ca. 10 pptv. Although this is consistent with the limited available UK data (e.g. see Fig. 6), the NO_3 concentration is expected to be strongly dependent on the prevailing ambient conditions (e.g., concentration of NO_x concentration of alkenes, particle density, humidity, temperature).

The above discussion has been concerned primarily with the role of the reactions of NO_3 with organic compounds in sequestering and recycling NO_x. The

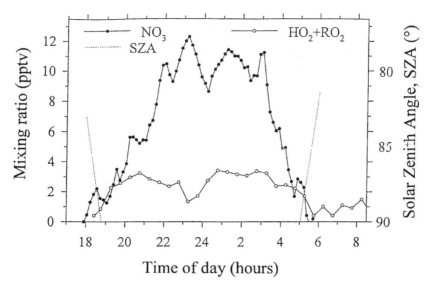

Figure 6. Overnight mixing ratios of NO3 and peroxy radicals measured at the Weybourne Atmospheric Observatory, Norfolk, UK (15–16th April 1994). The variation of solar zenith angle indicates the times of sunset and sunrise (data taken from Carslaw et al., 1997).

significance of these reactions as removal routes for the organic compounds, in comparison with removal by reactions with OH or O_3 (see Section 3), is also worthy of note. Using the mean nighttime NO_3 concentration calculated above, the lifetimes of a series of organic compounds with respect to reaction with NO_3 can be calculated, as shown in Table 5. Clearly, certain compounds are potentially removed by reaction with NO_3 at a rate which is comparable with, or greater than, the rate of removal by OH and O_3, and these reactions are therefore often included in chemical mechanisms for use in tropospheric models (e.g. Jenkin et al., 1997a; Saunders et al., 1997). However, it should be emphasised that NO_3 initiates, but does not catalyse, the removal of organic compounds. Consequently, its concentration can be substantially suppressed (i.e., to very much lower than the 10 pptv average) by the presence of an organic compound with which it reacts rapidly. Under such circumstances, the lifetimes presented in Table 5 may be significantly underestimated.

4.3. Night-time sources of HOx radicals

Although major sources of HO_x radicals result from photolytic processes (see Section 3.2), generation can also occur from a number of thermal reactions discussed above, namely the reactions of NO_3 with organic compounds (Section 4.2), the reactions of O_3 with alkenes (Section 3.5) and the thermal decomposition of peroxy acyl nitrates (Section 2.4). These reactions therefore potentially generate HO_x at night, and in the winter months when production by photolysis processes is slow. The reactions of O_3 with alkenes have received particular attention, and can clearly represent significant sources of OH (and other radicals) within the nocturnal boundary layer in rural and semi-polluted environments (Paulson and Orlando, 1996; Bey et al., 1997).

Substantial evidence for nighttime radical generation has been provided by observations of significant peroxy radical concentrations at a variety of locations including north Norfolk, England (Carslaw et al., 1997), Brittany, France (Behmann et al., 1993; Cantrell et al., 1995,1996c), Mace Head, Eire (Carpenter et al., 1997), Hawaii, USA (Cantrell et al., 1997b), Schauinsland, Germany (Mihelcic et al., 1993; Volz-Thomas et al., 1997) and northern Portugal (Kluepfel et al., 1996). In some of these studies, a strong correlation between NO_3 and peroxy radicals has been observed. The data shown in Fig. 6, for example, demonstrate a sharp rise in peroxy radical concentrations from the low values observed at sunset, with a corresponding drop at sunrise (Carslaw et al., 1997). Consequently, this suggests a major role for NO_3 in the chemistry leading to measurable HO_x, which may result both from its reactions with VOCs (as described above), and from its involvement in radical propagation reactions. An example of the latter is the reactions of NO_3 with acyl peroxy radicals, which has been found to be rapid in the case of $CH_3C(O)O_2$ (Canosa-Mas et al., 1996). This can promote the decomposition of peroxy acyl nitrates by competing with the recombination reaction of the acyl peroxy radical with NO_2, e.g., for PAN:

$$CH_3C(O)O_2NO_2(+M) \rightleftharpoons CH_3C(O)O_2 + NO_2(+M), \qquad (43)$$

$$CH_3C(O)O_2 + NO_3 \rightarrow CH_3 + CO_2 + NO_2 + O_2, \qquad (44)$$

$$CH_3 + O_2(+M) \rightarrow CH_3O_2(+M). \qquad (45)$$

Thus, the presence of NO_3 facilitates the conversion of the peroxy radical reservoir, PAN, into methyl peroxy radicals (CH_3O_2) and, subsequently, HO_2 at night. The importance of the reactions of NO_3 with peroxy radicals in general has also been demonstrated in modelling studies (Kirchner and Stockwell, 1996).

A further potential source of free radicals results from the reactions of NO_2 with some reactive conjugated dienes found in vehicle exhaust (Shi and Harrison, 1997; Harrison et al., 1998). Although these reactions may occur throughout the diurnal cycle, they have been postulated as playing an important role in the generation of free radicals under conditions when photochemical sources are inoperative, and concentrations of O_3 and NO_3 are suppressed by high levels of NO (Shi and Harrison, 1997). They have particular significance, therefore, to winter-time urban pollution episodes when boundary layer levels of NO_x and hydrocarbons from vehicle exhaust are elevated (Bower et al., 1994). The probable subsequent generation of nitro-substituted peroxy radicals may play a key role in the observed enhanced conversion of NO to NO_2 under such conditions (Harrison et al., 1998).

5. Chemical processes leading to secondary organic aerosol (SOA) formation

The formation of aerosols in the atmosphere has an important influence on visibility, climate and chemical processes, and is of concern since fine particulate matter is inhalable. The reduction of visibility observed in power station plumes and during photochemical episodes is mainly due to the formation and growth of large numbers of particles or droplets, which are able to absorb and scatter radiation. Similarly, the scattering and absorption of incoming solar radiation by aerosols throughout the atmosphere has a direct effect on the Earth's radiative balance (and therefore climate), by influencing the energy reaching ground level. A further indirect effect results from the role of aerosols in cloud formation (i.e., hygroscopic aerosols act as cloud condensation nuclei), since clouds reflect incoming radiation. Both these effects lead to atmospheric cooling (i.e., negative radiative forcing) which offsets the warming influence of radiatively active trace gases such as CO_2 (IPCC, 1995).

Aerosols in the tropospheric boundary layer may be emitted directly, or formed *in situ* as a result of chemical processes. Similarly to gaseous pollutants, therefore, they are usually classified as either primary or secondary respectively. There are numerous sources within these categories, which may either be natural phenomena or as a result of anthropogenic pollution (e.g., see Finlayson-Pitts and Pitts, 1986; Jaenicke, 1993). The sizes (i.e., aerodynamic diameters) of both primary and secondary aerosols may also vary over many orders of magnitude. The complete range of sizes which is usually of interest varies from about 2 nm (the smallest size detectable with a condensation nuclei counter, and effectively a molecular cluster) to about 10 μm. At greater diameters, aerosols are not readily inhaled, and are removed comparatively efficiently from the air by sedimentation.

Secondary aerosols are generated by gas-to-particle conversion, following the formation of products of particularly low-volatility, or high solubility, from gas-phase oxidation processes. Since these processes are often photochemically driven, the resultant aerosol usually falls into the category of secondary photochemical pollutant. A discussion of the chemistry leading to the formation of such products is therefore of relevance to the present review. An essential prerequisite for new particle formation to occur is the presence of a species in the gas phase at a concentration in excess of its saturation vapour concentration with respect to the condensed phase (i.e., condensable material). This can be achieved by emission of hot gas mixtures into a cool environment (e.g., resulting from combustion processes), with the rapidly formed particulate matter falling into the category of primary aerosol. However, a major contribution to the formation of condensable material in the troposphere results from the significant emission of comparatively volatile trace gases, which are oxidised in the gas phase to yield products of much lower volatility. The subsequently-formed secondary aerosol is generated either by condensation onto existing aerosol, or by nucleation to form new particles or droplets (e.g., see Seinfeld, 1986; Clement and Ford, 1996). The most significant condensable molecule formed in the atmosphere is sulphuric acid (generated from the oxidation of SO_2 and reduced sulphur-containing trace gases, in particular dimethyl sulphide), which has also been long recognised as the most important from the point of view of the nucleation of new particles (e.g., Jaenicke, 1993; Clement and Ford, 1999a,b). However, increasing attention in recent years has been given to the contribution to secondary aerosol formation made by organic material. The main aim of the present section is to give an overview of the gas-phase chemical processes leading to low-volatility organic oxygenates, which can contribute to the formation of secondary organic aerosol (SOA).

5.1. The formation of organic oxygenates

Numerous studies in recent years have established that the production of condensable material from the tropospheric oxidation of emitted VOCs leads to the formation of SOA (e.g., Simoneit, 1986; Pandis et al., 1993; Saxena et al., 1995; Kavouras et al., 1998). This can occur, for example, during photochemical smog episodes (i.e., driven by anthropogenic pollution), but also as a natural phenomenon resulting from the rapid oxidation of some biogenic hydrocarbons (e.g. monoterpenes).

As described in Section 3, the gas-phase oxidation of VOCs proceeds by complex mechanisms, leading initially to the production of a variety of first generation oxidised organic products. These products are either of the same carbon number as the parent VOC, or of a lower carbon number if a fragmentation process has occurred. Those of the same carbon number are invariably less volatile than the parent VOC, since they are of higher molecular weight and contain one or more polar functional groups. Key functional groups which tend to reduce the volatility of a product are, in particular, carboxylic acid (–C(=O)OH), but also aldehyde (–C(=O)H), ketone (–C(=O)–), alcohol (–OH) and nitrate (–ONO_2). The possible transfer of oxidised products to the condensed phase occurs in competition with further oxidation in the gas phase which, once again, can either generate even less volatile multi-functional products of the same carbon number, or products of lower carbon number following fragmentation steps.

For all classes of VOC, the propensity for aerosol formation increases as the size of the VOC is increased. This arises partially because the reactivity of larger organic molecules in a given class is generally greater than that of smaller ones (i.e., the rate of accumulation of oxidised products increases with the size of the VOC), and partially because the volatility of the oxidation products of larger VOCs is lower. Consequently, the oxidation of larger VOCs is more likely to generate oxidised products at concentrations in excess of the saturation vapour concentration.

For many VOCs known to be emitted, the fractional conversion to aerosol under typical tropospheric conditions is estimated to be very low indeed (Grosjean and Seinfeld, 1989). However, it is also clear that certain classes of VOC are more likely to lead to aerosol formation by virtue of their general high reactivity and types of oxidation product formed. Of particular significance are cyclic compounds, since the products of fragmentation (i.e., ring opening) processes are often of the same (or similar) carbon number as the parent compound. Furthermore, in the cases of cycloalkenes, aromatic compounds and terpenes, oxidation occurs predominantly by an addition mechanism, so that the first generation products contain two polar functional groups. Consequently the oxidation of these classes of compound is more likely to lead to the generation of aerosol than the oxidation of similar sized compounds in other classes. For cycloalkenes and terpenes, oxidation by reaction with O_3 is potentially very important, with the possibility of the generation of low volatility bifunctional carboxylic acids as first generation products (e.g., Hoffmann et al., 1997; Christoffersen et al., 1998; Glasius et al., 1999).

The ability of monoterpenes to generate condensable material, and therefore SOA, has received particular attention in recent years, owing to the magnitude of their global emissions (Guenther et al., 1995), and their is a growing body of information on their formation and composition, both in enviromental chambers and the troposphere. The available chamber studies (e.g. Pandis et al., 1991; Odum et al., 1996; Hoffmann et al., 1997; Hallquist et al., 1999; Jang and Kamens, 1999) have demonstrated that aerosol yields are very variable, depending on the identity and starting concentration of the terpene, and on the experimental conditions (e.g. whether oxidation is initiated predominantly by O_3, OH or NO_3). In particular, these studies have confirmed that dark ozonolysis experiments tend to lead to the largest aerosol yields (Hoffmann et al., 1997), and that the fractional yields in all experiments generally depend on the organic aerosol mass concentration, commonly denoted M_0 (Odum et al., 1996; Hoffmann et al., 1997).

The available results have been simulated reasonably successfully using gas-aerosol partitioning models (e.g. Pankow, 1994a,b; Odum et al., 1996; Hoffmann et al., 1997; Kamens et al., 1999), with the methodology gaining further support from measurements of aerosol composition. These studies have shown that, as M_0 increases, a progressively larger proportion of the aerosol mass is made up of comparatively volatile compounds which are present in the gas phase at concentrations significantly below their saturation vapour concentration. Despite this, these compounds are able to partition a proportion of their mass (i.e., "dissolve") into the condensed organic phase (Pankow, 1994a,b). Consequently, at high values of M_0, the major condensed and gas-phase products are often the same species. For example, in the case of the most abundant and well-studied monoterpene, α-pinene, pinonaldehyde (a C_{10} ketoaldehyde) often represents the majority of the aerosol mass in chamber experiments, even though its gas-phase concentration is typically three orders of magnitude below its saturation vapour concentration (i.e., 3.8×10^{-2} Torr at 298 K, corresponding to ca. 50 ppmv at 1 bar; Hallquist et al., 1997). Even at the lower values of M_0 typical of ambient air, pinonaldehyde and nopinone (the major first generation product of β-pinene degradation) have been observed as significant components of secondary organic aerosol (Kavouras et al., 1998).

As indicated above, however, the formation of new aerosol particles requires the production of involatile degradation products which are likely to exceed their saturation vapour concentrations under atmospheric conditions. Recent studies have identified pinic acid (a C_9 dicarboxylic acid) as a promptly formed condensed product of the ozonolysis of both α- and β-pinene (Christoffersen et al., 1998; Kamens et al., 1999; Glasius et al., 1999). On the basis of observed structure–volatility trends for carboxylic acids and dicarboxylic acids (Tao and McMurry, 1989), it is probable that pinic acid is very involatile (its saturation vapour concentration possibly being below 10^{-9} Torr, i.e. ca. 1 pptv at 1 bar), and it is therefore currently believed to be the most likely degradation product of both α- and β-pinene leading to new aerosol formation by nucleation (Christoffersen et al., 1998). At present, however, only speculative mechanisms have been proposed to explain the prompt formation of pinic acid as a first generation product of α- and β-pinene ozonolysis, and further mechanistic information on the formation of condensable products from the oxidation of organics in general is required.

5.2. The hygroscopic properties of SOA

The growth of aerosols by absorption of water, and their ability to act as cloud condensation nuclei, not only has an influence on their interaction with light, but also on their role in chemical processes. The hygroscopic properties of the inorganic acids and salts commonly observed in tropospheric aerosols have been well studied (e.g., Orr et al., 1958; Potukuchi and Wexler, 1995). H_2SO_4, for example, is completely miscible with water, and the quantity absorbed is proportional to the water vapour pressure. Above a critical relative humidity (known as the deliquescence point), the same is true for soluble inorganic salts such as ammonium sulphate ($(NH_4)_2SO_4$), which deliquesces at ca. 80% relative humidity, and sodium chloride (NaCl) for which the deliquescence point is ca. 75% relative humidity. Consequently, aerosols composed of such inorganic compounds take up water efficiently, and are good cloud condensation nuclei.

The nature of organic aerosols, or aerosols containing organic material, is also of considerable interest, particularly with regard to their hygroscopic properties. A common assumption has been that water absorption by aerosols is due solely to the presence of water soluble inorganic compounds. This would also suggest that the condensation of organic material on to an existing inorganic (e.g., sulphate) core might inhibit its growth by adsorption of water. Although this may be true for condensable hydrocarbons, the same may not be the case for organic oxygenates, since they contain both polar functional groups (which are hydrophilic), and hydrocarbon chains (which are hydrophobic). Indeed, organic compounds have a wide range of aqueous solubilities, and clearly have various possible effects on the water absorption properties of aerosols, and their ability to act as cloud condensation nuclei (CCN).

Observations of particle chemical composition and water content at urban and rural locations in the USA (Saxena et al., 1995), have identified interesting effects of organics on the hygroscopic properties of the particles. The rural data were consistent with an enhanced effect of the organic content on water absorption, whereas the urban data suggested an inhibiting influence. This can be explained in terms of the organic component being dominated by primary organic material (e.g., emitted polycyclic aromatic hydrocarbons) in urban aerosol, but by secondary organic material in rural aerosol. The organic material in primary organic aerosol is likely to be less oxygenated, and therefore more hydrophobic. In contrast, the secondary organic material, formed from hydrocarbon oxidation, contains polar functional groups and is therefore more hydrophilic.

Cruz and Pandis (1997) have recently considered the ability of pure dicarboxylic acid aerosols to act as CCN. As already discussed, such compounds are likely to be important components of secondary organic aerosols. The results indicated that submicron aerosols of glutaric acid (C_5) and adipic acid (C_6) were activated at low supersaturations (1.003 and 1.01 respectively). Furthermore, it was shown that a coating of glutaric acid increased the CCN activation of ammonium sulphate particles. In contrast, aerosols formed from the oxidation of α-pinene, β-pinene and limonene in the EUPHORE smog chamber have been found to be only slightly hygroscopic, with growth factors significantly lower than those of ammonium sulphate (Virkkula et al., 1999). As discussed above, these aerosols are likely to be multi-component, with a significant mass contribution from aldehyde and ketone products, which are likely to be less hygroscopic than the condensable dicarboxylic acids. Consequently, it is possible that organic aerosols formed under ambient conditions at lower mass concentrations (i.e., predominantly as a result of nucleation/condensation processes) are hygrosopic, but become more hydrophobic at higher mass concentrations when the dicarboxylic acids represent a smaller fraction of the aerosol mass. It is probable, therefore, that CCN activity of secondary organic aerosols is very variable, and further study is clearly required.

6. Conclusions

Considerable progress has been made in identifying chemical processes responsible for the generation of O_3 and other secondary photochemical pollutants in the planetary boundary layer. This has been achieved by a combination of field observations, laboratory investigations and numerical modelling studies. However, further research in all three areas is necessary to improve our quantitative understanding of the impact of the chemical processing of pollutants emitted into the atmosphere.

The general features of O_3 formation from the sunlight-initiated oxidation of VOCs and NO_x are well established, with the rates and mechanisms of the oxidation of numerous VOCs reasonably well characterised and quantified by laboratory study. Clearly it is not feasible to study the complete degradation of all VOCs emitted into the troposphere, and the present database provides a framework for defining oxidation mechanisms for many unstudied VOCs by analogy and with the aid of stucture–reactivity correlations. However, there are certain classes of VOCs for which the detailed oxidation mechanisms are still uncertain (most notably aromatic hydrocarbons, terpenes and sulphur-containing organics), and the degradation of some classes of VOC has received little or no attention. The latter case includes emitted organic compounds containing nitrogen and silicon.

For the aromatic hydrocarbons and terpenes in particular, the remaining uncertainties have repercussions for understanding their role in the generation of both O_3 and SOA. Identifying the detailed chemical processes leading to the generation of products which contribute to the atmospheric burden of SOA (whether by nucleation, condensation or gas-particle partitioning), and establishing the hygroscopic properties of such aerosols, is almost certainly one of the major current challenges in tropospheric chemistry research.

More generally, further information is required on the atmospheric chemistry of specific classes of product formed from VOC oxidation, such as carbonyl compounds, organic hydroperoxides and organic nitrates and peroxy nitrates. For carbonyl compounds, there are comparatively few studies of the UV absorption cross-sections and quantum yields to allow atmospheric photolysis rates and products to be determined. This is particularly the case for multifunctional compounds containing more that one carbonyl group, or carbonyl groups in addition to other functionalities (e.g., $-OH$, $-OOH$, $-ONO_2$). For hydroperoxides, the photolysis and OH reaction has only been studied for CH_3OOH: clearly, additional data for other hydroperoxides would help the assignment of structure–reactivity relationships. For the oxidised organic nitrogen compounds, various aspects of their atmospheric chemistry are incompletely understood, even though their formation, transport and degradation potenially has an impact on the environment and climate in a number of ways, ranging from the inhibition of ozone formation on local/regional scales to influencing the global budget and distribution of NO_x and O_3. For example, there are no reported product studies of the OH initiated degradation of any organic nitrate or peroxy nitrate. Given that the yields of these products can be substantial (e.g., up to 25% from monoterpene oxidation), it is clearly important to establish whether NO_x is ultimately released when they are degraded in the atmosphere. The further development and application of methods of detecting all these classes of oxygenated product in the field, in addition to the HO_x radical intermediates which drive the oxidation mechanisms, is also essential for the validation of our understanding of atmospheric chemical processes, and for pointing out where knowledge is lacking.

In other respects, the chemistry interconverting oxidised nitrogen compounds is reasonably well understood, and explains the formation of a series of inorganic and organic species which can act as reservoirs for NO_x and HO_x. The chemical and photochemical reactions of inorganic oxidised nitrogen species in the gas phase have generally received considerable attention, leading to a large body of evaluated data for these species (Atkinson et al., 1997b; DeMore et al., 1997). Such evaluated data are of clearly invaluable for atmospheric modellers, although there are instances where the evaluations of the two data review panels differ considerably (albeit very few). Most notably, the current recommendations for the reaction of OH with NO_2 (reaction (7)) differ by a factor of ca. 5 at the high pressure limit, such that the recommended values at 298 K and 760 Torr differ by a factor of ca. 1.7. These two different interpretations of the same database is clearly a cause for great concern for a reaction which is a major sink for HO_x and NO_x on local, regional and global scales, and this has stimulated further recent study of the reaction (Dransfield et al., 1999; Brown et al., 1999).

Although not covered in detail in the present review, the role of heterogeneous and multi-phase chemical processes potentially has an important impact on the gas-phase processes leading to the formation of O_3 and other secondary pollutants in the boundary layer. This results not only from the provision of a substrate which allows alternative reactions to occur, but also from an indirect influence on the gas-phase mechanisms through the removal of key species such as HO_x, NO_x and even O_3 itself. There is still a great deal to be learnt about the role of heterogeneous and multi-phase chemistry in the atmosphere, and this is necessarily an active area of research.

Acknowledgements

MEJ gratefully acknowledges the support of the Department of the Environment, Transport and the Regions, both in the preparation of this review (under contract EPG 1/3/70), and for the some of the work described. KCC gratefully acknowledges financial support from the European Union (under contract ENV4-CT97-0404). Thanks are also due to members of the Photochemical Oxidants Review Group (PORG), in particular Tony Cox (University of Cambridge) and Dick Derwent (UK Meteorological Office) for comments on some of the material presented.

References

Allan, B.J., Carslaw, N., Coe, H., Burgess, R.A., Plane, J.M.C., 1997. Observations of the nitrate radical in the marine boundary layer. Journal of Atmospheric Chemistry 33 (2), 129–155.

Altschuller, A.P., 1993. PANs in the atmosphere. Journal of Air and Waste Pollution Association 43 (9), 1221–1230.

Ammann, M., Kalberer, M., Jost, D.T., Tobler, L., Rossler, E., Piguet, D., Gaggeler, H.W., Baltensperger, U., 1998. Heterogeneous production of nitrous acid on soot in polluted air masses. Nature 395 (6698), 157–160.

Andersson-Skold, Y., Grennfelt, P., Pleijel, K., 1992. Photochemical ozone creation potentials: a study of different concepts. Journal of the Air and Waste Management Association 42, 1152–1158.

Andres Hernandez, M.D., Notholt, J., Hjorth, J., Schrems, O., 1996. A DOAS study on the origin of nitrous acid at urban and non-urban sites. Atmospheric Environment 30 (2), 175–180.

Aneja, V.P., Kim, D.S., Hartsell, B.E., 1996. Measurements and analysis of reactive nitrogen species in the rural troposphere of southeast United States. Atmospheric Environment 23 (4), 1591–1599.

Atkinson, R., 1987. A structure-activity relationship for the estimation of rate constants for the gas-phase reactions of OH radicals with organic compounds. International Journal of Chemical Kinetics 19, 799–828.

Atkinson, R., 1990. Gas-phase tropospheric chemistry of organic compounds: a review. Atmospheric Environment 24A, 1–41.

Atkinson, R., 1991. Kinetics and mechanisms of the gas-phase reactions of the nitrate radical with organic compounds. Journal of Physical and Chemical Reference Data 20 (3), 459–507.

Atkinson, R., 1994. Gas-phase tropospheric chemistry of organic compounds. Journal of Physical Chemistry Reference Data, Monograph 2, 1–216.

Atkinson, R., 1997. Gas-phase tropospheric chemistry of volatile organic compounds: 1. Alkanes and alkenes. Journal of Physical and Chemical Reference Data 26 (2), 215.

Atkinson, R., 1998a. Gas-phase degradation of organic compounds in the troposphere. Pure and Applied Chemistry 70 (7), 1327–1334.

Atkinson, R., 1998b. Product studies of gas-phase reactions of organic compounds. Pure and Applied Chemistry 70 (7), 1335–1343.

Atkinson, R., Carter, W.P.L., 1984. Kinetics and mechanisms of the gas-phase reactions of ozone with organic compounds under atmospheric conditions. Chemical Reviews 84 (5), 437–470.

Atkinson, R., Aschmann, S.M., 1993. OH radical production from the gas-phase reactions of O_3 with a series of alkenes under atmospheric conditions. Environmental Science and Technology 27 (7), 1357–1363.

Atkinson, R., Arey, J., 1998. Atmospheric chemistry of biogenic organic compounds. Accounts of Chemical Research 31 (9), 574–583.

Atkinson, R., Baulch, D.L., Cox, R.A., Hampson, R.F., Kerr, J.A., Jr., Rossi, M.J., Troe, J., 1997a. Evaluated kinetic and photochemical data for atmospheric chemistry; supplement V-IUPAC sub-committee on gas kinetic data evaluation for atmospheric chemistry. Journal of Physical Chemistry Reference Data 26 (3), 521–1011.

Atkinson, R., Baulch, D.L., Cox, R.A., Hampson, R.F., Kerr, J.A., Jr., Rossi, M.J., Troe, J., 1997b. Evaluated kinetic and photochemical data for atmospheric chemistry; supplement VI-IUPAC sub-committee on gas kinetic data evaluation for atmospheric chemistry. Journal of Physical Chemistry Reference Data 26 (6), 1329–1499.

Atlas, E., Ridley, B.A., Hubler, G., Walega, J.G., Carroll, M.A., Montzka, D.D., Huebert, B.J., Norton, R.B., Grahek, F.E., Shaufler, S., 1992. Partitioning and budget of NO_y species during the Mauna Loa Observatory Photochemistry Experiment. Journal of Geophysical Research 97, 10449–10462.

Ayers, G.P., Penkett, S.A., Gillett, R.W., Bandy, B., Galbally, I.E., Meyer, C.P., Elsworth, C.M., Bentley, S.T., Forgan, B.W., 1992. Evidence for photochemical control of ozone concentrations in unpolluted marine air. Nature 360 (6403), 446–449.

Ayers, G.P., Penkett, S.A., Gillett, R.W., Bandy, B.J., Galbally, I.E., Meyer, C.P., Elsworth, C.M., Bentley, S.T., Forgan, B.W., 1996. Annual cycle of peroxides and ozone in marine air at Cape Grim, Tasmania. Journal of Atmospheric Chemistry 23 (3), 221–252.

Barnes, I., Becker, K.H., Zhu, T., 1993. Near UV absorption spectra and photolysis products of difunctional organic nitrates: possible importance as NO_x reservoirs. Journal of Atmospheric Chemistry 17, 353–373.

Barnes I., Becker, K.H., Klotz, B., Sorenssen, S., 1998. Atmospheric oxidation of arene oxides. Proceedings of the second workshop of the EUROTRAC-2 subproject Chemical Mechanism Development (CMD), Karlsruhe, Germany, September 1998, p. GPP-2.

Behmann, T., Weissenmayer, M., Burrows, J.P., 1993. Peroxy radicals in the night-time oxidation chemistry. In: Restelli, G., Angeletti, G. (Eds.), Proceedings of the Sixth European Symposium on the Physico-Chemical Behaviour of Atmospheric Pollutants, European Commission, Brussels, pp. 259–264.

Bertman, S.B., Roberts, J.M., 1991. A PAN analog from isoprene photooxidation. Geophysical Research Letters 18 (8), 1461–1464.

Bey, I., Aumont, B., Toupance, G., 1997. The nighttime production of OH radicals in the continental troposphere. Geophysical Research Letters 24 (9), 1067–1070.

BIOVOC, 1998. Final report on the EU project on Degradation Mechanisms of Biogenic VOC 'BIOVOC'. Contract reference: ENV-CT95-0059: Co-ordinator Dr. J. Hjorth, JRC Ispra Italy.

Bottenheim, J.W., Barrie, L.A., Atlas, E., 1993. The partitioning of nitrogen oxides in the lower arctic troposphere during spring 1988. Journal of Atmospheric Chemistry 17 (1), 15–27.

Bower, J.S., Broughton, G.F.J., Stedman, J.R., Williams, M.L., 1994. A winter NO_2 smog episode in the UK. Atmospheric Environment 28 (3), 461–475.

Brown, S.S., Talukdar, R.K., Ravishankara, A.R., 1999. Rate constants for the reaction OH + NO_2 + M $\rightarrow$ HNO_3 + M under atmospheric conditions. Chemical Physics Letters 299 (3–4), 277–284.

Buhr, M.P., Parrish, D.D., Norton, R.B., Fehsenfeld, F.C., Sievers, R.E., Roberts, J.M., 1990. Contribution of organic nitrates to the total reactive nitrogen budget at a rural eastern U.S. site. Journal of Geophysical Research 95, 9809–9816.

Calogirou, A., Larsen, B.R., Kotzias, D., 1999. Gas-phase terpene oxidation products: a review. Atmospheric Environment 33, 1423–1439.

Calvert, J.G., Stockwell, W.R., 1983. Deviations from the O_3–NO–NO_2 photostationary state in tropospheric chemistry. Canadian Journal of Chemistry 61 (5), 983–991.

Canosa-Mas, C.E., King, M.D., Lopez, R., Percival, C.J., Wayne, R.P., Shallcross, D.E., Pyle, J.A., Daele, V., 1996. Is the reaction between $CH_3C(O)O_2$ and NO_3 important in the night-time troposphere? Journal of the Chemistry Society Faraday Transactions 92 (12), 2211–2222.

Cantrell, C.A., Lind, J.A., Shetter, R.E., Calvert, J.G., Goldan, P.D., Kuster, W., Fehsenfeld, F.C., Montzka, S.A., Parrish, D.D., Williams, E.J., Buhr, M.P., Westberg, H.H., Allwine, G., Martin, R., 1992. Peroxy radicals in the ROSE experiment: measurement and theory. Journal of Geophysical Research 97 (D18), 20671–20686.

Cantrell, C.A., Shetter, R.E., Calvert, J.G., Parrish, D.D., Fehsenfeld, F.C., Goldan, P.D., Kuster, W., Williams, E.J., Westberg, H.H., Allwine, G., Martin, R., 1993a. Peroxy radicals as measured in ROSE and estimated from photostationary state deviations. Journal of Geophysical Research 98 (D10), 18355–18366.

Cantrell, C.A., Shetter, R.E., McDaniel, A.H., Calvert, J.G., 1993b. Measurement methods for peroxy radicals in the atmosphere. In: Newman, L. (Ed.), Measurement Challenges in Atmospheric Chemistry, ACS Advances in Chemistry Series No. 232. American Chemical Society, pp. 290–322.

Cantrell, C.A., Shetter, R.E., Lind, J.A., McDaniel, A.H., Calvert, J.G., Parrish, D.D., Fehsenfeld, F.C., Buhr, M.P., Trainer, M., 1993c. An improved chemical amplifier technique for peroxy radical measurements. Journal of Geophysical Research 98 (D2), 2897–2909.

Cantrell, C.A., Shetter, R.E., Calvert, J.G., 1995. Comparison of peroxy radical concentrations at several contrasting sites. Journal of Atmospheric Chemistry 52 (19), 3408–3412.

Cantrell, C.A., Shetter, R.E., Gilpin, T.M., Calvert, J.G., 1996a. Peroxy radicals measured during Mauna Loa Photochemistry Experiment 2: The data and first analysis. Journal of Geophysical Research 101 (D9), 14643–14652.

Cantrell, C.A., Shetter, R.E., Gilpin, T.M., Calvert, J.G., Eisele, F.L., Tanner, D.J., 1996b. Peroxy radical concentrations measured and calculated from trace gas measurements in the Mauna Loa Photochemistry Experiment 2. Journal of Geophysical Research 101 (D9), 14653–14664.

Cantrell, C.A., Shetter, R.E., Calvert, J.G., 1996c. Peroxy radical chemistry during FIELDVOC 1993 in Brittany. France. Atmospheric Environment 30 (23), 3947–3957.

Cantrell, C.A., Shetter, R.E., Calvert, J.G., Williams, E., Baumann, K., Brune, W.H., Stevens, P.S., Mather, J.H., 1997a. Peroxy radicals from photostationary state deviations and steady state calculations during the Tropospheric OH Photochemistry Experiment at Idaho Hill, Colorado, 1993. Journal of Geophysical Research 102 (D5), 6369–6378.

Cantrell, C.A., Shetter, R.E., Calvert, J.G., Eisele, F.L., Tanner, D.J., 1997b. Some considerations of the origin of night-time peroxy radicals observed in MLOPEX-2c. Journal of Geophysical Research 102 (D13), 15899–15914.

Cape, J.N., Smith, R.I., Fowler, D., 1994. The influence of ozone chemistry and meteorology on plant exposure to photo-oxidants. Proceedings of the Royal Society of Edinburgh 102B, 11–31.

Carpenter, L.J., Monks, P.S., Galbally, I.E., Meyer, C.P., Bandy, B.J., Penkett, S.A., 1997. A study of peroxy radicals and ozone photochemistry at coastal sites in the Northern and Southern hemispheres. Journal of Geophysical Research 102 (21), 25417–25428.

Carpenter, L.J., Clemitshaw, K.C., Burgess, R.A., Penkett, S.A., McFadyen, G.G., Cape, J.N., 1998. Investigation and evaluation of the O_3/NO_x photochemical stationary state. Atmospheric Environment 32 (19), 3353–3365.

Carslaw, N., Carpenter, L.J., Plane, J.M.C., Allan, B.J., Burgess, R.A., Clemitshaw, K.C., Coe, H., Penkett, S.A., 1997. Simultaneous measurements of nitrate and peroxy radicals in the marine boundary layer. Journal of Geophysical Research 102 (D15), 18917–18933.

Carter, W.P.L., 1991. Development of ozone reactivity scales for volatile organic compounds. U.S. Environmental Protection Agency Report, EPA–600/3–91/050.

Carter, W.P.L., 1994. Development of ozone reactivity scales for volatile organic compounds. Journal of the Air and Waste Management Association 44, 881–899.

Carter, W.P.L., 1995. Computer modelling of environmental chamber measurements of maximum incremental reactivities of volatile organic compounds. Atmospheric Environment 29, 2513–2527.

Carter, W.P.L., Atkinson, R., 1989. Alkyl nitrate formation from the atmospheric photo-oxidation of alkanes – a revised estimation method. Journal of Atmospheric Chemistry 8 (2), 165–173.

Carter, W.P.L., Pierce, J.A., Luo, D., Malkina, I.L., 1995. Environmental chamber study of maximum incremental reactivities of organic compounds. Atmospheric Environment 29, 2499–2511.

Chandler, A.S., Choularton, T.W., Dollard, G.J., Eggleton, A.E.J., Gay, M.J., Hill, T.A., Jones, B.M.R., Tyler, B.J., Bandy, B.J., Penkett, S.A., 1988. Measurements of H_2O_2 and SO_2 in clouds and estimates of their reaction rate. Nature 336 (6199), 562–565.

Christoffersen, T.S., Hjorth, J., Horie, O., Jensen, H.R., Kotzias, D., Molander, L.L., Neeb, P., Ruppert, L., Winterhalter, R., Virkkula, A., Wirtz, K., Larsen, B.R., 1998. Cis-pinic acid, a possible precursor from organic aerosol formation from ozonolysis of α-pinene. Atmospheric Environment 32, 1657–1661.

Clement, C.F., Ford, I.J., 1996. The competition between aerosol growth and nucleation in the atmosphere. Journal of Aerosol Science 27, S39–S40.

Clement, C.F., Ford, I.J., 1999a. Gas-to-particle conversion in the atmosphere: I. Evidence from empirical atmospheric aerosols. Atmospheric Environment 33 (3), 475–487.

Clement, C.F., Ford, I.J., 1999b. Gas-to-particle conversion in the atmosphere: II. Analytical models of nucleation bursts. Atmospheric Environment 33 (3), 489–499.

Clemitshaw, K.C., Carpenter, L.J., Penkett, S.A., Jenkin, M.E., 1997a. A calibrated peroxy radical chemical amplifier (PERCA) instrument for ground-based measurements in the troposphere. Journal of Geophysical Research 102 (D21), 25405–25416.

Clemitshaw, K.C., Williams, J., Rattigan, O.V., Shallcross, D.E., Law, K.S., Cox, R.A., 1997b. Gas-phase ultra-violet absorption cross-sections and atmospheric lifetimes of several C_2-C_5 alkyl nitrates. Journal of Photochemistry and Photobiology A: Chemistry 102 (2–3), 117–126.

Collins, W.J., Stevenson, D.S., Johnson, C.E., Derwent, R.G., 1995. Tropospheric ozone modelled with a global-scale Lagrangian model of diurnal atmospheric chemistry. Met O(APR) Turbulence and Diffusion Note No. 224, Meteorological Office. Bracknell, England.

Colvile, R.N., Choularton, T.W., Cape, J.N., Bandy, B.J., Bower, K.N., Burgess, R.A., Davies, T.J., Dollard, G.J., Gallagher, M.W., Hargreaves, K.J., Jones, B.M.R., Penkett, S.A., Storeton-West, R.L., 1996. Processing of oxidised nitrogen compounds by passage through winter-time orographic cloud. Journal of Atmospheric Chemistry 24 (3), 211–239.

Cox, R.A., 1988. Atmospheric chemistry of NO_x and hydrocarbons influencing tropospheric ozone. In: Isaksen, I.S.A. (Ed.), Tropospheric Ozone. D. Reidel, Dordrecht, pp. 263–292.

Cox, R.A., Penkett, S.A., 1971. Oxidation of SO_2 by oxidants formed in the ozone-olefin reaction. Nature 230, 321.

Cox, R.A., Penkett, S.A., 1972. Aerosol formation from sulphur dioxide in the presence of ozone and olefinic hydrocarbons. Journal of the Chemistry Society Faraday Transactions I 68, 1735.

Cruz, C.N., Pandis, S.N., 1997. A study of the ability of pure secondary organic aerosol to act as cloud condensation nuclei. Atmospheric Environment 31, 2205–2214.

Dagaut, P., Liu, R., Wallington, T.J., Kurylo, M.J., 1989. Kinetic measurements of the gas-phase reactions of OH radicals with hydroxy ethers, hydroxy ketones and ketoethers. Journal of Physical Chemistry 93, 7838–7840.

DeMore, W.B., Sander, S.P., Golden, D.M., Hampson, R.F., Kurylo, M.J., Howard, C.J., Ravishankara, A.R., Kolb, C.E., Molina, M.J., 1997. Chemical kinetics and photochemical data for use in stratospheric modelling. Evaluation number 12. NASA panel for data evaluation. Jet Propulsion Laboratory Report, 97–104.

Dentener, F.J., 1993. Heterogeneous chemistry in the troposphere. Ph.D. Thesis, Universiteit Utrecht.

Dentener, F.J., Crutzen, P.J., 1993. Reaction of N_2O_5 on tropospheric aerosols: impact on the global distributions of NO_x, O_3 and OH. Journal of Geophysical Research 98 (D4), 7149–7163.

Dentener, F.J., Crutzen, P.J., 1994. A three dimensional model of the global ammonia cycle. Journal of Atmospheric Chemistry 19 (4), 331–369.

Derwent, R.G., Jenkin, M.E., 1991. Hydrocarbons and the long-range transport of ozone and PAN across Europe. Atmospheric Environment 25A, 1661–1678.

Derwent, R.G., Davies, T.J., 1994. Modelling the impact of NO_x and hydrocarbon control on photochemical ozone formation in Europe. Atmospheric Environment 28 (12), 2039–2052.

Derwent, R.G., Dollard, G.J., Metcalfe, S.E., 1988. On the nitrogen budget for the United Kingdom and north-west Europe. Quarterly Journal of the Royal Meteorological Society 114, 1127–1152.

Derwent, R.G., Simmonds, P.G., Collins, W.J., 1994. Ozone and carbon monoxide measurements at a remote maritime location, Mace Head, Ireland, from 1990 to 1992. Atmospheric Environment 28 (16), 2623–2637.

Derwent, R.G., Jenkin, M.E., Saunders, S.M., 1996. Photochemical ozone creation potentials for a large number of reactive hydrocarbons under European conditions. Atmospheric Environment 30 (2), 181–199.

Derwent, R.G., Jenkin, M.E., Saunders, S.M., Pilling, M.J., 1998. Photochemical ozone creation potentials for organic compounds in north west Europe calculated with a master chemical mechanism. Atmospheric Environment 32, 2419–2441.

Dodge, M.C., 1977. Combined use of modeling techniques and smog chamber data to derive ozone-precursor relationships. International Conference on Photochemical Oxidant Pollution and its Control: Proceedings, Vol. II B, EPA/600/3-77-001b. U.S. Environmental Protection Agency, Research Triangle Park, NC, pp. 881–889.

Dollard, G.J., Jones, B.M.R., Davies, T.J., 1991. Measurements of gaseous hydrogen peroxide and PAN in rural southern England. Atmospheric Environment 25A (9), 2039–2053.

Dollard, G.J., Jenkin, M.E., 1999. Baseline review of aldehydes in ambient air in the UK. Draft report prepared for the UK Department of the Environment, Transport and the Regions, London, March 1999.

Donahue, N.M., Kroll, J.H., Anderson, J.G., Demerjian, K.L., 1998. Direct observation of OH production from the ozonolysis of olefins. Geophysical Research Letters 25 (1), 59–62.

Dransfield, T.J., Perkins, K.K., Donahue, N.M., Anderson, J.G., 1999. Temperature and pressure dependent kinetics of the gas-phase reaction of the hydroxyl radical with nitrogen dioxide. Geophysical Research Letters 26 (6), 687–690.

Eberhard, J., Müller, C., Stocker, D.W., Kerr, J.A., 1995. Isomerizations of alkoxy radicals under atmospheric conditions. Environmental Science and Technology 29 (1), 232–241.

England, C., Corcoran, W.H., 1974. Kinetics and mechanism of the gas phase reaction of water vapour with nitrogen dioxide. Indian Engineering Chemical Fundamentals 13, 373–384.

Fahey, D.W., Hubler, G., Parrish, D.D., Williams, E.J., Norton, R.B., Ridley, B.A., Singh, H.B., Liu, S.C., Fehsenfeld, F.C., 1986. Reactive nitrogen species in the troposphere: measurements of NO, NO_2, HNO_3, particulate nitrate, peroxyacetyl nitrate (PAN), O_3, and total reactive odd nitrogen (NO_y) at Niwot Ridge. Colorado. Journal of Geophysical Research 91 (D9), 9781–9793.

Fenter, F.F., Caloz, F., Rossi, M.J., 1995. Experimental evidence for the efficient deposition of nitric acid on calcite. Atmospheric Environment 29 (22), 3365–3372.

Finlayson-Pitts, B.J., Pitts, J.N., 1986. Atmospheric Chemistry: Fundamentals and Experimental Techniques. Wiley Interscience, New York.

Flocke, F., Volz-Thomas, A., Kley, D., 1991. Measurements of alkyl nitrates in rural and polluted air masses. Atmospheric Environment 25A, 1951–1960.

Flocke, F., Volz-Thomas, A., Buers, H.J., Patz, W., Garthe, H.J., Kley, D., 1998. Long-term measurements of alkyl nitrates in southern Germany 1. General behavior and seasonal and diurnal variation. Journal of Geophysical Research 103 (D5), 5729–5746.

Gervat, G.P., Clark, P.A., Marsh, A.R.W., Teasdale, I., Chandler, A.S., Choularton, T.W., Gay, M.J., Hill, M.K., Hill, T.A., 1988. Nature 333 (6170), 241–243.

Glasius, M., Lahaniati, M., Calogirou, A., Di Bella, D., Jensen, N.R., Hjorth, J., Kotzias, D., Larsen, B.R., 1999. Carboxylic acids in secondary aerosols from the oxidation of cyclic monoterpenes by ozone. Environmental Science and Technology, in press.

Grosjean, D., Seinfeld, J.H., 1989. Parameterisation of the formation potential of secondary organic aerosols. Atmospheric Environment 23 (8), 1733–1747.

Grosjean, D., Williams, E.L., Grosjean, E., 1994a. Gas-phase thermal decomposition of peroxy-n-butyryl nitrate. International Journal of Chemical Kinetics 26 (3), 281–387.

Grosjean, D., Grosjean, E., Williams, E.L., 1994b. Formation and thermal decomposition of butyl-substituted peroxy acyl nitrates: $n\text{-}C_4H_9C(O)OONO_2$ and $i\text{-}C_4H_9C(O)OONO_2''$. Environmental Science and Technology 28 (6), 1099–1105.

Guenther, A., Hewitt, C.N., Erickson, D., Fall, R., Geron, C., Graedel, T., Harley, P., Klinger, L., Lerdau, M., McKay, W.A., Pierce, T., Scholes, B., Steinbrecher, R., Tallamraju, R., Taylor, J., Zimmerman, P., 1995. A global model of natural volatile organic compound emissions. Journal of Geophysical Research 100, 8873–8892.

Haagen-Smit, A.J., Fox, M.M., 1954. Photochemical ozone formation with hydrocarbons and automobile exhaust. Journal of Air Pollution Control Association 4, 105–109.

Haagen-Smit, A.J., Fox, M.M., 1956. Ozone formation in photochemical oxidation of organic substances. Indian Enginering Chemistry 48, 1484.

Hallquist, M., Wangberg, I., Ljungstrom, E., 1997. Atmospheric fate of dicarbonyl products originating from α-pinene, and Δ^3-Carene: determination of rate of reaction with OH and NO_3 radicals, UV absorption cross sections and vapor pressures. Environmental Science and Technology 31, 3166–3172.

Hallquist, M., Wangberg, I., Ljungstrom, E., Barnes, I., Becker, K.H., 1999. Aerosol and product yields from NO_3 radical-initiated oxidation of selected monoterpenes. Environmental Science and Technology 33 (4), 553–559.

Harris, G.W., Carter, W.P.L., Winer, A.M., Pitts, J.N., Platt, U., Perner, D., 1982. Observations of nitrous acid in the Los Angeles atmosphere and implications for predictions of ozone precursor relationships. Environmental Science and Technology 16 (7), 414–419.

Harrison, R.M., Peak, J.D., Collins, G.M., 1996. Tropospheric cycle of nitrous acid. Journal of Geophysical Research 101 (D9), 14429–14439.

Harrison, R.M., Shi, J.P., Grenfell, J.L., 1998. Novel nighttime free radical chemistry in severe nitrogen dioxide pollution episodes. Atmospheric Environment 32 (16), 2769–2774.

Harrison, R.M., Grenfell, L.J., Yamulki, S., Clemitshaw, K.C., Penkett, S.A., Cape, J.N., McFadyen, G.G., 1999. Budget of NO_y species measured at a coastal site in the United Kingdom. Atmospheric Environment 33 (26), 4255–4272.

Hauglustaine, D.A., Madronich, S., Ridley, B.A., Walega, J.G., Cantrell, C.A., Shetter, R.E., Hubler, G., 1996. Observed and model calculated photostationary state at Mauna Loa Observatory during MLOPEX 2. Journal of Geophysical Research 101 (D9), 14681–14696.

Hellpointer, E., Gab, S., 1989. Detection of methyl, hydroxymethyl and hydroxyethyl hydroperoxides in air and precipitation. Nature 337 (6208), 631–634.

Hewitt, C.N., Kok, G.L., 1991. Occurrence of organic hydroperoxides in the troposphere: laboratory and field observations. Journal of Atmospheric Chemistry 12 (2), 181–194.

Hoffmann, T., Odum, J.R., Bowman, F., Collins, D., Klockow, D., Flagan, R.C., Seinfeld, J.H., 1997. Formation of organic aerosols from the oxidation of biogenic hydrocarbons. Journal of Atmospheric Chemistry 26, 189–222.

Horowitz, L.W., Liang, J., Gardner, G.M., Jacob, D.J., 1998. Export of reactive nitrogen from North America during summertime: sensitivity to hydrocarbon chemistry. Journal of Geophysical Research 103 (D11), 13451–13476.

Hov, O., Allegrini, I., Beilke, S., Cox, R.A., Eliassen, A., Elshout, A.J., Gravenhorst, G., Penkett, S.A., Stern, R., 1987. Evaluation of atmospheric processes leading to acid deposition in Europe. Commission of the European Communities Air Pollution Research Report 10, EUR 11441, CEC.

Huebert, B.J., Robert, C.H., 1985. The dry deposition of nitric acid to grass. Journal of Geophysical Research 90 (D1), 2085–2090.

IPCC, 1995. Climate Change 1994 – Radiative Forcing and Climate Change and an Evaluation of the IPCC IS92 Emission Scenarios. Cambridge University Press, Cambridge, UK.

Jackson, A.V., Hewitt, C.N., 1996. Hydrogen peroxide and organic hydroperoxide concentrations in air in a eucalyptus forest in central Portugal. Atmospheric Environment 30 (6), 819–830.

Jaenicke, R., 1993. Tropospheric Aerosols – Aerosol-Cloud-Climate Interactions. Academic Press, New York, 1–31.

Jang, M., Kamens, R.M., 1999. Newly characterised products and composition of secondary aerosols from the reaction of α-pinene with ozone. Atmospheric Environment 33 (3), 459–464.

Jenkin, M.E., Hayman, G.D., 1999. Photochemical ozone creation potentials for oxygenated volatile organic compounds: sensitivity to variations in kinetic and mechanistic parameters. Atmospheric Environment 33, 1275–1293.

Jenkin, M.E., Cox, R.A., Williams, D.J., 1988. Laboratory studies of the kinetics of formation of nitrous acid from the thermal reaction of nitrogen dioxide and water vapour. Atmospheric Environment 22 (3), 487–498.

Jenkin, M.E., Saunders, S.M., Pilling, M.J., 1997a. The tropospheric degradation of volatile organic compounds: a protocol for mechanism development. Atmospheric Environment 31 (1), 81–104.

Jenkin, M.E., Hayman, G.D., Derwent, R.G., Saunders, S.M., Pilling, M.J., 1997b. Tropospheric chemistry modelling: improvements to current models and application to policy issues. First Annual Report (Reference AEA/RAMP/20150/R001 Issue 1) prepared for the UK Department of the Environment, London, on Contract EPG 1/3/70, May 1997.

Jenkin, M.E., Hayman, G.D., Derwent, R.G., Saunders, S.M., Carslaw, N., Pascoe, S., Pilling, M.J., 1999. Tropospheric chemistry modelling: improvements to current models and application to policy issues. Final Report (Reference AEAT-4867/20150/R004) prepared for the UK Department of the Environment, Transport and the Regions, London, on Contract PECD 1/3/70, March 1999.

Jonson, J.E., Isaksen, I.S.A., 1993. Tropospheric ozone chemistry: impact of cloud chemistry. Journal of Atmospheric Chemistry 16 (2), 99–122.

Kalberer, M., Ammann, M., Gaggeler, H.W., Baltensperger, U., 1999. Adsorption of NO_2 on carbon aerosol particles in the low ppb range. Atmospheric Environment 33 (17), 2815–2822.

Kamens, R., Jang, M., Chien, C.J., Leach, K., 1999. Aerosol formation from the reaction of α-pinene and ozone using a gas phase kinetics-aerosol partitioning model. Environmental Science and Technology 33 (9), 1430–1438.

Kames, J., Schurath, U., 1992. Alkyl nitrates and bifunctional nitrates of atmospheric interest: Henry's law constants and their temperature dependencies. Journal of Atmospheric Chemistry 15 (1), 79–95.

Kastler, J., Ballschmiter, K., 1998. Bifunctional alkyl nitrates – trace constituents of the atmosphere. Fresenius Journal of Analytical Chemistry 360 (7–8), 812–816.

Kavouras, I.G., Mihalopoulos, N., Stephanou, E.G., 1998. Formation of atmospheric particles from organic acids produced by forests. Nature 395, 683–686.

Kessler, C., Platt, U., 1984. Nitrous acid in polluted air masses – sources and formation pathways. CEC, proceedings of the third European symposium on the physico-chemical behaviour of atmospheric pollutants. Varese, Italy, 10–12 April, pp. 412–422.

Kirchner, F., Stockwell, W.R., 1996. Effect of peroxy radical reactions on the predicted concentrations of ozone, nitrogenous compounds, and radicals. Journal of Geophysical Research 101 (D15), 21007–21022.

Kitto, A.-M.N., Harrison, R.M., 1992. Nitrous and nitric acid measurements at sites in south-east England. Atmospheric Environment 26A (2), 235–241.

Kleinman, L.I., 1986. Photochemical formation of peroxides in the boundary layer. Journal of Geophysical Research 91 (D10), 889–904.

Kleinman, L., Lee, Y.N., Springston, S.R., Lee, J.H., Nunnermacker, L., Weinsteinlloyd, J., Zhou, X.L., Newman, L., 1995. Peroxy radical concentration and ozone formation rate at rural site in the Southeastern United States. Journal of Geophysical Research 100 (D4), 7263–7273.

Kluepfel, T., Arnold, T., Perner, D., Harder, H., Volz-Thomas, A., 1996. FIELDVOC 1994: Measurements of peroxy radicals and comparison with the photostationary state in an eucalyptus forest at Tabua, Portugal. In: Borrell, P.M., Borrell, P., Cvitas, T., Kelly, K., Seiler, W. (Eds.), Transport and Transformation of Pollutants in the Troposphere, Proceedings of EUROTRAC Symposium '96, Garmisch-Partenkirchen. Computational Mechanics Publications, Southampton, pp. 207–211.

Kwok, E.S.C., Atkinson, R., 1995. Estimation of hydroxyl radical reaction rate constants for gas-phase organic compounds using a structure-reactivity relationship: an update. Atmospheric Environment 29 (14), 1685–1695.

Kwok, E.S.C., Aschmann, S.M., Akinson, R., Arey, J., 1997. Products of the gas-phase reactions of o-, m- and p-xylene with the OH radical in the presence and absence of NO_x. Journal of the Chemical Society, Faraday Transactions 93 (16), 2847–2854.

Lammel, G., Cape, N., 1996. Nitrous acid and nitrite in the atmosphere. Chemical Society Reviews, 361–369.

Lee, D.S., Kohler, I., Grobler, E., Rohrer, F., Sausen, R., Gallardo-Klenner, L., Olivier, J.J.G., Dentener, F.J., 1997. Estimation of global NO_x emissions and their uncertainties. Atmospheric Environment 31 (12), 1735–1749.

Leighton, P.A., 1961. Photochemistry of Air Pollution. Academic Press, New York.

Lightfoot, P.D., Cox, R.A., Crowley, J.N., Destriau, M., Hayman, G.D., Jenkin, M.E., Moortgat, G.K., Zabel, F., 1992. Organic peroxy radicals: kinetics, spectroscopy and tropospheric chemistry. Atmospheric Environment 26A (10), 1805–1964.

Lin, X., Trainer, M., Liu, S.C., 1988. On the non-linearity of the tropospheric ozone production. Journal of Geophysical Research 93 (D12), 15879–15888.

Liu, S.C., Trainer, M., Fehsenfeld, F.C., Parrish, D.D., Williams, E.J., Fahey, D.W., Hubler, G., Murphy, P.C., 1987. Ozone production in the rural troposphere and the implications for regional and global ozone distributions. Journal of Geophysical Research 92 (D4), 4191–4207.

Logan, J.A., 1983. Nitrogen oxides in the troposphere: global and regional budgets. Journal of Geophysical Research 88 (NC15), 785–807.

Lovejoy, E.R., Hanson, D.R., 1995. Measurement of kinetics of reactive uptake on submicron sulphuric acid aerosols. Journal of Physical Chemistry 99 (7), 2080–2087.

Luke, W.T., Dickerson, R.R., Nunnermacker, L.J., 1989. Direct measurements of the photolysis rate coefficients and Henry's Law constants of several alkyl nitrates. Journal of Geophysical Research 94 (D12), 14905–14921.

Marston, G., McGill, C.D., Rickard, A.R., 1998. Hydroxyl-radical formation in the gas-phase ozonolysis of 2-methylbut-2-ene. Geophysical Research Letters 25 (12), 2177–2180.

Martin, L.R., Damschen, D.E., 1981. Aqueous oxidation of sulphur dioxide by hydrogen peroxide at low pH. Atmospheric Environment 15, 1615–1622.

Meier, R.R., Anderson, G.P., Cantrell, C.A., Hall, L.A., Lean, J., Minschwaner, K., Shetter, R.E., Shettle, E.P., Stamnes, K., 1997. Actinic radiation in the terrestrial atmosphere. Journal of Atmospheric Solar-Terrestrial Physics 59 (17), 2111–2157.

Mentel, T.F., Bleilebens, D., Wahner, A., 1996. A study of nighttime nitrogen oxide oxidation in a large reaction chamber – The fate of NO_2 N_2O_5, HNO_3 and O_3 at different humidities. Atmospheric Environment 30 (23), 4007–4020.

Mihelcic, D., Klemp, D., Muesgen, P., Paetz, H.W., Volz-Thomas, A., 1993. Simultaneous measurements of peroxy and nitrate radicals at Schauinsland. Journal of Atmospheric Chemistry 16 (4), 313–335.

Mozurkewicz, M., Calvert, J.G., 1988. Reaction probability of N_2O_5 on aqueous aerosols. Journal of Geophysical Research 93 (D12), 15889–15896.

Muthuramu, K., Shepson, P.B., O'Brien, J.M., 1993. Preparation, analysis and atmospheric production of multifunctional organic nitrates. Environmental Science and Technology 27 (6), 1117–1124.

NRC, 1991. Report of the US National Research Council. Rethinking the ozone problem in urban and regional air pollution. National Academy Press, Washington, D.C.

Nielsen, T., Egelov, A.H., Granby, K., Skov, H., 1995. Observations of particulate organic nitrates and unidentified components of NO_y. Atmospheric Environment 29 (15), 1757–1769.

Niki, H., Maker, P.D., Savage, C.M., Breitenbach, L.P., Hurley, M.D., 1987. FTRI spectroscopic study of the mechanism for the gas-phase reaction between ozone and tetramethylethylene. Journal of Physical Chemistry 91 (4), 941–946.

Nunnermacker, L.J., Imre, D., Daum, P.H., Kleinman, L., Lee, Y.N., Lee, J.H., Springston, S.R., Newman, L., Weinsteinlloyd, J., Luke, W.T., Banta, R., Alvarez, R., Senff, C., Sillman, S., Holdren, M., Keigley, G.W., Zhou, X., 1998. Characterisation of the Nashville urban plume on July 3 and July 18, 1995. Journal of Geophysical Research 103 (D21), 28129–28148.

O'Brien, J.M., Shepson, P.B., Muthuramu, K., Hao, C., Niki, H., Hastie, D.R., 1995. Measurements of alkyl and multifunctional alkyl nitrates at a rural site in Ontario. Journal of Geophysical Research 100 (D11), 22795–22804.

O'Brien, J.M., Shepson, P.B., Wu, Q., Biesenthal, T., Bottenheim, J.W., Wiebe, H.A., Anlauf, K.G., Brickell, P., 1997. Production and distribution of organic nitrates, and their relationship to carbonyl compounds in an urban environment. Atmospheric Environment 31 (14), 2059–2069.

Odum, J.R., Hoffmann, T., Bowman, F., Collins, D., Flagan, R.C., Seinfeld, J.H., 1996. Gas/particle partioning and secondary organic aerosol yields. Environmental Science and Technology 30, 2580–2585.

Olszyna, K.J., Bailey, E.M., Simonaitis, R., Meagher, J.F., 1994. Ozone and NO_y relationships at a rural site. Journal of Geophysical Research 99 (D7), 14557–14563.

Orr, C., Hurd, F.K., Corbett, W.J., 1958. Aerosol size and relative humidity. Journal of Colloid Science 13, 472–482.

Pandis, S.N., Paulson, S.E., Seinfeld, J.H., Flagan, R.C., 1991. Aerosol formation in the photo-oxidation of isoprene and β-pinene. Atmospheric Environment 25A, 997–1008.

Pandis, S.N., Wexler, A.S., Seinfeld, J.H., 1993. Secondary organic aerosol formation and transport II. Predicting the ambient secondary organic aerosol size distribution. Atmospheric Environment 27A (15), 2403–2416.

Pankow, J.F., 1994a. An absorption model of gas/particle partitioning of organic compounds in the atmosphere. Atmospheric Environment 28A, 185–188.

Pankow, J.F., 1994b. An absorption model of gas/particle partitioning involved in the formation of secondary organic aerosol. Atmospheric Environment 28A, 189–193.

Paulson, S.E., Orlando, J.J, 1996. The reactions of ozone with alkenes: an important source of HO_x in the boundary layer. Geophysical Research Letters 23 (25), 3727–3730.

Paulson, S.E., Sen, A.D., Liu, P., Fenske, J.D., Fox, M.J., 1997. Evidence for formation of OH radicals from the reaction of O_3 with alkenes in the gas phase. Geophysical Research Letters 24 (24), 3193–3196.

Paulson, S.E., Chung, M., Sen, A.D., Orzechowska, G., 1998. Measurement of OH radical formation from the reaction of ozone with several biogenic alkenes. Journal of Geophysical Research 103 (D19), 25533–25539.

Penkett, S.A., Jones, B.M.R., Brice, K.A., Eggleton, A.E.J., 1979. The importance of atmospheric ozone and hydrogen peroxide in oxidising sulphur dioxide in cloud and rainwater. Atmospheric Environment 13, 123–137.

Pfeiffer, T., Forberich, O., Comes, F.J., 1998. Tropospheric OH formation by ozonolysis of terpenes. Chemical Physics Letters 298 (4–6), 351–358.

Pitts, J.N., Sanhueza, E., Atkinson, R., Carter, W.P.L., Winer, A.M., Harris, G.W., Plum, C.N., 1984. An investigation of the dark formation of nitrous acid in environmental chambers. International Journal of Chemical Kinetics 16 (7), 919–939.

PORG, 1997. Ozone in the United Kingdom. Fourth Report of the UK Photochemical Oxidants Review Group, Department of the Environment, Transport and the Regions, London.

Porter, E., Wenger, J., Treacy, J., Sidebottom, H., Mellouki, A., Teton, S., Le Bras, G., 1997. Kinetic studies on the reactions of hydroxyl radicals with diethers and hydroxyethers. Journal of Physical Chemistry A 101 (32), 5770–5775.

Potukuchi, S., Wexler, A.S., 1995. Identifying solid-aqueous phase transitions in atmospheric aerosols-I. Neutral acidity solutions. Atmospheric Environment 29 (22), 1663–1676.

Ravishankara, A.R., 1997. Heterogeneous and multiphase chemistry in the troposphere. Science 276 (5315), 1058–1065.

Ravishankara, A.R., Hancock, G., Kawasaki, M., Matsumi, Y., 1998. Photochemistry of ozone: surprises and recent lessons. Science 280 (5360), 60–61.

Ridley, B.A., Madronich, S., Chatfield, R.B., Walega, J.G., Shetter, R.E., Carroll, M.A., Montzka, D.D., 1992. Measurements and model simulations of the photostationary state during the Mauna Loa Observatory Experiment: implications for radical concentrations and ozone production and loss rates. Journal of Geophysical Research 97 (D10), 10375–10388.

Roberts, J.M., 1990. The atmospheric chemistry of organic nitrates. Atmospheric Environment 24A (2), 243–287.

Roberts, J.M., Fajer, R.W., 1989. UV absorption cross-sections of organic nitrates of potential importance and estimation of atmospheric lifetimes. Environmental Science and Technology 23 (8), 945–951.

Roberts, J.M., Bertman, S.B., 1992. The thermal decomposition of peroxyacetic nitric anhydride (PAN) and peroxymethacrylic nitric anhydride (MPAN). International Journal of Chemical Kinetics 24 (3), 297–307.

Rudd, H.J., 1995. Emissions of volatile organic compounds from stationary sources in the United Kingdom: speciation. AEA Technology Report, AEA/CS/REMA-029. AEA Technology. Oxfordshire, UK.

Ryerson, T.B., Buhr, M.P., Frost, G.J., Goldan, P.D., Holloway, J.S., Hubler, G., Jobson, B.T., Kuster, W.C., McKeen, S.A., Parrish, D.D., Roberts, J.M., Sueper, D.T., Trainer, M., Williams, J., Fehsenfeld, F.C., 1998. Emissions lifetimes and ozone formation in power plant plumes. Journal of Geophysical Research 103 (D17), 22569–22583.

Sakamaki, F., Hatakeyama, S., Akimoto, H., 1983. Formation of nitrous acid and nitrous oxide in the heterogeneous reaction of nitrogen dioxide and water vapour in a smog chamber. International Journal of Chemical Kinetics 15 (10), 1013–1029.

Saunders, S.M., Jenkin, M.E., Derwent, R.G., Pilling, M.J., 1997. Report summary: World Wide Web site of a Master Chemical Mechanism (MCM) for use in tropospheric chemistry models. Atmospheric Environment 31, 1249–1250.

Saxena, P., Hildemann, L.M., McMurry, P.H., Seinfeld, J.H., 1995. Organics alter hygroscopic behaviour of atmospheric particles. Journal of Geophysical Research 100 (D9), 18755–18770.

Seinfeld, J.H., 1986. Atmospheric chemistry and physics of air pollution. Wiley Interscience, New York.

Shallcross, D.E., Biggs, P., Canosa-Mas, C.E., Clemitshaw, K.C., Harrison, M.G., Reyes Lopez Alanon, M., Pyle, J.A., Vipond, A., Wayne, R.P., 1997. Rate coefficients for the reaction between OH and CH_3ONO_2 and $C_2H_5ONO_2$ over a range of pressure and temperature. Journal of the Chemistry Society Faraday Transactions 93 (16), 2807–2811.

Shepson, P.B., Anlauf, K.G., Bottenheim, J.W., Wiebe, H.A., Gao, N., Muthuramu, K., Mackay, G.I., 1993. Alkyl nitrates and their contribution to reactive nitrogen at a rural site in Ontario. Atmospheric Environment. 27A (5), 749–757.

Shi, J.P., Harrison, R.M., 1997. Rapid NO_2 formation in diluted petrol-fuelled engine exhaust – a source of NO_2 in winter smog episodes. Atmospheric Environment 31 (23), 3857–3866.

Sillman, S., 1995. The use of NO_y, H_2O_2 and HNO_3 as indicators for O_3–NO_x–VOC sensitivity in urban locations. Journal of Geophysical Research 100, 14175–14188.

Sillman, S., 1999. The relationship between ozone, NO_x and hydrocarbons in urban and polluted rural environments. Atmospheric Environment 33 (12), 1821–1845.

Sillman, S., Samson, F.J., 1995. Impact of temperature on oxidant photochemistry in urban, polluted rural and remote environments. Journal of Geophysical Research 100 (D6), 11497–11508.

Sillman, S., He, D.Y., Pippin, M.R., Daum, P.H., Imre, D.G., Kleinman, L.I., Lee, J.H., Weinstein-lloyd, J., 1998. Model correlations for ozone, reactive nitrogen and peroxides for Nashville in comparison with measurements: implications for O_3-NO_x-hydrocarbon chemistry. Journal of Geophysical Research 103 (D17), 22629–22644.

Simoneit, B.R.T., 1986. Characterisation of organic constituents in aerosols in relation to their origin and transport – a review. International Journal of Environmental Analytical Chemistry 23 (3), 207–237.

Simpson, D., 1995. Hydrocarbon reactivity and ozone formation in Europe. Journal of Atmospheric Chemistry 20, 163–177.

Singh, H.B., Salas, L.J., 1989. Measurements of peroxyacetylnitrate (PAN) and peroxypropionyl nitrate (PPN) at selected urban, rural and remote sites. Atmospheric Environment 23A (1), 231–328.

Singh, H.B., O'Hara, D., Herlth, D., Bradshaw, J.D., Sandholm, S.T., Gregory, G.L., Sachse, G.W., Blake, D.R., Crutzen, P.J., Kanakidou, M., 1992. Atmospheric measurements of peroxy acetyl nitrate and other organic nitrates at high latitudes: possible sources and sinks. Journal of Geophysical Research 97 (D15), 16511–16522.

Singh, H.B., Herlth, D., O'Hara, D., Zahnle, K., Bradshaw, J.D., Sandholm, R., Talbot, R., Crutzen, P.J., Kanakidou, M., 1993. Relationship of PAN to active and total odd nitrogen at northern high latitudes: influence of reservoir species on NO_x and O_3. Journal of Geophysical Research 97 (D15), 16523–16530.

Singh, H.B., Herlth, D., Kolyer, R., Salas, L., Bradshaw, J.D., Sandholm, S.T., Davis, D.D., Crawford, J., Kondo, Y., Koike, M., Talbot, R., Gregory, G.L., Sachse, G.W., Browell, E., Blake, D.R., Rowland, F.S., Newell, R., Merill, J., Heines, B., Liu, S.C., Crutzen, P.J., Kanakidou, M., 1996. Reactive nitrogen and ozone over the western Pacific: distribution, partitioning and sources. Journal of Geophysical Research 101 (D1), 1793–1808.

Slemr, F., Tremmel, H.G., 1994. Hydroperoxides in the remote troposphere over the Atlantic Ocean. Journal of Atmospheric Chemistry 19 (4), 371–404.

Solberg, S., Krognes, T., Stordal, F., Hov, O., Beine, H.J., Jaffe, D.A., Clemitshaw, K.C., Penkett, S.A., 1997. Reactive nitrogen compounds at Spitzbergen in the Norwegian Arctic. Journal of Atmospheric Chemistry 28 (1–3), 209–225.

Solberg, S., 1999. NILU, Kjeller, Norway. Private communication.

Staffelbach, T.A., Kok, G.L., Heikes, B.G., McCully, B., Mackay, G.I., Karecki, D.R., Schiff, H.I., 1996. Comparison of hydroperoxide measurements made during the Mauna Loa observatory photochemistry experiment 2. Journal of Geophysical Research 101 (D9), 14729–14739.

Stevens, E.R., 1969. The formation, reactions, and properties of peroxyacetyl nitrates (PANs) in photochemical air pollution. Advances in Environmental Science and Technology 1, 119–147.

Svensson, R., Ljungstrom, E., Lindqvist, O., 1987. Kinetics of the reaction between nitrogen dioxide and water vapour. Atmospheric Environment. 21 (7), 1529–1539.

Talukdar, R.K., Burkholder, J.B., Hunter, M., Gilles, M.K., Roberts, J.M., Ravishankara, A.R., 1997a. Atmospheric fate of several alkyl nitrates. 2. UV absorption cross-sections and photodissociation quantum yields. Journal of the Chemistry Society Faraday Transactions 93 (16), 2787–2796.

Talukdar, R.K., Burkholder, J.B., Herndon, S.C., Roberts, J.M., Ravishankara, A.R., 1997b. Atmospheric fate of several alkyl nitrates. 1. Rate coefficients of the reactions of alkyl nitrates with isotopically labelled hydroxyl radicals. Journal of the Chemistry Society Faraday Transactions 93 (16), 2797–2805.

Tao, Y., McMurry, P.H., 1989. Vapour pressures and surface free energies of C_{14}–C_{18} monocarboxylic acids and C_5–C_6 dicarboxylic acids. Environmental Science and Technology 25, 1519–1523.

Thakur, A.N., Singh, H.B., Mariani, P., Chen, Y., Wang, Y., Jacob, D.J., Brasseur, G., Muller, J.-F., Lawrence, M., 1999. Distribution of reactive nitrogen species in the remote free troposphere: data and model comparisons. Atmospheric Environment 33, 1403–1422.

Trainer, M., Parrish, D.D., Buhr, M.P., Norton, R.B., Fehsenfeld, F.C., Anlauf, K.G., Bottenheim, J.W., Tang, Y.Z., Wiebe, H.A., Roberts, J.M., Tanner, R.L., Newman, L., Bowersox, V.C., Meagher, J.F., Olszyna, K.J., Rodgers, M.O., Wang, T., Berresheim, H., Demerjian, K.L., Roychowdhury, U.K., 1993. Correlation of ozone with NO_y in photochemically aged air. Journal of Geophysical Research 98 (D2), 2917–2925.

Turberg, M.P., Giolando, D.M., Tilt, C., Soper, T., Mason, S., Davies, M., Klingensmith, P., Takacs, G.A., 1990. Atmospheric photochemistry of alkyl nitrates. Journal of Photochemistry and Photobiology A 51 (3), 281–292.

UNECE, 1992. Air Pollution Studies 8: Impacts of Long-range Transboundary Air Pollution. Published by United Nations.

UNECE, 1993. Air Pollution Studies 9: The State of Transboundary Air Pollution 1992 Update. Published by United Nations.

UNECE, 1994. Environmental Conventions. Published by United Nations.

Van Doren, J.M., Watson, L.R., Davidovits, P., Worsnop, D.R., Zahniser, M.S., Kolb, C.E., 1990. Temperature dependence of the uptake coefficients of HNO_3, HCl and N_2O_5 by water droplets. Journal of Physical Chemistry 94 (8), 3265–3269.

Volz-Thomas, A., Ridley, B.A., Andrae, M.O., Chameides, W.L., Derwent, R.G., Galbally, I.E., Lelieveld, J., Penkett, S.A., Rodgers, M.O., Trainer, M., Vaughan, G., Zhou, X.J., 1995. In: Ennis, C.A. (Ed.), Tropospheric Ozone. World Meteorological Organisation Global Ozone Research and Monitoring Project-Report No. 37, Scientific Assessment of Ozone Depletion: 1994. WMO, Geneva (Chapter 5).

Volz-Thomas, A., Arnold, T., Behmann, T., Borrell, P., Borrell, P.M., Burrows, J.P., Cantrell, C.A., Carpenter, L.J., Clemitshaw, K.C., Gilge, S., Heitlinger, M., Kluepfel, T., Kramp, F., Mihelcic, D., Muesgen, P., Paetz, H.W., Penkett, S.A., Perner, D., Schultz, M., Shetter, R.E., Slemr, J., Weissenmayer, M., 1998. Peroxy Radical InterComparison Exercise. A formal comparison of methods for ambient measurements of peroxy radicals. Berichte des Forshungszentrums Juelich; 3597; ISSN 0944-2952; Institute fuer Chemie und Dynamik der Geosphaere-2.

Virkkula, A., Van Dingenen, R., Raes, F., Hjorth, J., 1999. Hygroscopic properties of aerosol formed by oxidation of limonene, α-pinene and β-pinene. Journal of Geophysical Research 104, 3569–3579.

Walega, J.G., Ridley, B.A., Madronich, S., Grahek, F.E., Shetter, J.D., Sauvain, T.D., Hahn, C.J., Merill, J.T., Bodhaine, B.A., Robinson, E., 1992. Observations of peroxyacetyl nitrate, peroxypropionyl nitrate, methyl nitrate and ozone during the Mauna-Loa-Observatory photochemistry experiment. Journal of Geophysical Research 97 (D10), 10311–10330.

Wallington, T.J., Dagaut, P., Kurylo, M.J., 1992. Ultra-violet absorption cross sections and reaction kinetics and mechanisms for peroxy radicals in the gas phase. Chemical Reviews 92 (4), 667–710.

Wayne, R.P., Barnes, I., Biggs, P., Burrows, J.P., Canosa-Mas, C.E., Hjorth, J., Le Bras, G., Moortgat, G.K., Perner, D., Poulet, G., Restelli, G., Sidebottom, H., 1991. The nitrate radical: physics, chemistry and the atmosphere. Atmospheric Environment 25A (1), 1–206.

Williams, J., 1994. A study of the atmospheric chemistry of alkyl nitrates. Ph.D. Thesis, University of East Anglia, Norwich.

Williams, J., Roberts, J.M., Fehsenfeld, F.C., Bertman, S.B., Buhr, M.P., Goldan, P.D., Hubler, G., Kuster, W.C., Ryerson, T.B., Trainer, M., Young, V., 1997. Regional ozone from biogenic hydrocarbons deduced from airborne measurements. Geophysical Research Letters 24, 1099–1102.

Yang, X., Felder, P., Huber, J.R., 1993. Photo-dissociation of methyl nitrate in a molecular beam. Journal of Physical Chemistry 97 (42), 10903–10910.

Zhu, L., Ding, C.F., 1997. Temperature dependence of the near UV absorption spectra and photolysis products of ethyl nitrate. Chemical Physics Letters 265 (1–2), 177–184.

Zhu, L., Kellis, D., 1997. Temperature dependence of the UV absorption cross sections and photodissociation products of C_3–C_5 alkyl nitrates. Chemical Physics Letters 278 (1–3), 41–48.

Zhu, T., Barnes, I., Becker, K.H., 1991. Relative rate study of the gas-phase reaction of hydroxyl radicals with difunctional organic nitrates at 298 K and atmospheric pressure. Journal of Atmospheric Chemistry 13, 301–311.

Chapter 12

The relation between ozone, NO$_x$ and hydrocarbons in urban and polluted rural environments

Sanford Sillman

*Department of Atmospheric, Oceanic and Space Sciences, University of Michigan,
Ann Arbor, MI 48109-2143, USA
E-mail: sillman@umich.edu*

Abstract

Research over the past ten years has created a more detailed and coherent view of the relation between O$_3$ and its major anthropogenic precursors, volatile organic compounds (VOC) and oxides of nitrogen (NO$_x$). This article presents a review of insights derived from photochemical models and field measurements. The ozone–precursor relationship can be understood in terms of a fundamental split into a NO$_x$-sensitive and VOC-sensitive (or NO$_x$-saturated) chemical regimes. These regimes are associated with the chemistry of odd hydrogen radicals and appear in different forms in studies of urbanized regions, power plant plumes and the remote troposphere. Factors that affect the split into NO$_x$-sensitive and VOC-sensitive chemistry include: VOC/NO$_x$ ratios, VOC reactivity, biogenic hydrocarbons, photochemical aging, and rates of meteorological dispersion. Analyses of ozone–NO$_x$–VOC sensitivity from 3D photochemical models show a consistent pattern, but predictions for the impact of reduced NO$_x$ and VOC in indivdual locations are often very uncertain. This uncertainty can be identified by comparing predictions from different model scenarios that reflect uncertainties in meteorology, anthropogenic and biogenic emissions. Several observation-based approaches have been proposed that seek to evaluate ozone–NO$_x$–VOC sensitivity directly from ambient measurements (including ambient VOC, reactive nitrogen, and peroxides). Observation-based approaches have also been used to evaluate emission rates, ozone production efficiency, and removal rates of chemically active species. Use of these methods in combination with models can significantly reduce the uncertainty associated with model predictions.

First published in Atmospheric Environment 33 (1999) 1821–1845

1. Introduction

The relation between ozone and its two main precursors, NO_x ($= NO + NO_2$) and volatile organic compounds (VOC), represents one of the major scientific challenges associated with urban air pollution. It is generally known that for some conditions the process of ozone formation is controlled almost entirely by NO_x and is largely independent of VOC, while for other conditions ozone production increases with increasing VOC and does not increase (or sometimes even decreases) with increasing NO_x. However it has been difficult to determine whether ozone production during specific events is associated with NO_x-sensitive chemistry or VOC-sensitive chemistry. Particulates and other secondary air pollutants also show a complex dependence on NO_x and VOC (Meng et al., 1997). There is also an analogous split into NO_x-sensitive and NO_x-saturated chemistry in the remote troposphere. The relation between ozone, NO_x and VOC is especially important as a basis for environmental policy. Ozone is a major environmental concern because of its adverse impacts on human health (Lippman, 1993; Bascomb et al., 1996) and also because of its impact on crops and forest ecosystems (NRC, 1991). Most major metropolitan areas in the US have continually violated government health standards for ozone continually since the passage of the first Clean Air Act in 1970. Throughout this period policy plans to have been developed for lowering ambient ozone and bringing cities into compliance with the law. These plans have had only modest success (Fiore et al., 1998). Elevated ozone has also been observed in Europe since the 1970s (e.g. Guicherit and van Dop, 1977) and is especially bad in Athens and other cities of southern Europe (e.g. Moussiopoulos et al., 1995; Giovannoni and Russell, 1995; Prevot et al., 1997). Elevated ozone has also been observed in urban areas in Canada, Japan, China, India and, most notably, Mexico City (MARI, 1994). Because ozone forms most rapidly in conditions with warm temperatures and sunshine, cities with warm climates (including cities in developing nations) are especially likely to experience high ozone. In each of these locations it is necessary to understand how ozone depends on NO_x and VOC in order to develop an effective policy response.

In addition to its importance for policy, the relation between ozone, NO_x and VOC is worthy of attention as a purely scientific problem. The process of ozone formation provides a case study of the interaction between nonlinear chemistry and dynamics in the earth sciences, and frequently calls for sophisticated mathematical and analytical treatment. The nonlinear ozone chemistry extends into rural and remote areas as well, where it has important implications for global photochemical equilibria in the atmosphere. Interpretation of ozone chemistry also raises questions about the use of models and the nature of scientific proof in the environmental sciences.

In recent years there have been major advances in understanding the process of ozone formation, based in part on the development and use of 3D models for atmospheric processes and in part on interpretation of field measurement campaigns. This has resulted in a more sophisticated understanding of ozone–NO_x–VOC sensitivity relative to the last major review (NRC, 1991). Sections 2 and 3 below summarize the current understanding. Section 2 presents an overview of ozone–NO_x–VOC sensitivity. Section 3 describes the pattern that has emerged from recent investigations of ozone–NO_x–VOC sensitivity, including the factors that tend to produce NO_x-sensitive and VOC-sensitive chemistry in models, geographical variations and sources of model uncertainty. These sections are also intended to provide an overview for readers whose primary expertise lies in other fields or in policy. Section 4 presents the chemistry that drives the relation between ozone, NO_x and VOC in greater detail, including the role of odd hydrogen radicals (OH, HO_2, etc.), ozone production efficiency and comparisons with the chemistry of the remote troposphere. Section 5 describes some of the recent innovative attempts to evaluate ozone–NO_x–VOC chemistry based on interpretation of field measurements rather than the more traditional model-based evaluations. This section also includes a brief overview of methods to evaluate other features of ozone chemistry (ozone production efficiency, NO_x removal rates, and emission rates) based on measurements. Although these methods have been developed and applied primarily in the US it is likely that they will be applicable to events in other locations as well.

A central theme throughout this paper is the uncertain nature of predictions concerning the response of ozone to reductions in NO_x and VOC emissions, and the difficulty of obtaining scientifically valid evidence. Most evaluations of ozone–NO_x–VOC sensitivity are based on predictions from 3D Eulerian models which contain representations of emission rates, dynamics and photochemistry. The predicted response to reduced emissions is derived by repeating the model base case with reduced NO_x or VOC. These models are useful for identifying general features of ozone–NO_x–VOC chemistry, but predictions for specific events and specific urban areas are uncertain. There is also no direct way to test whether these NO_x–VOC predictions are accurate. In recent years considerable skepticism has been expressed about the use of models as the basis for environmental policy (Oreskes et al., 1994). The viewpoint in this paper is that model predictions for ozone–NO_x–VOC sensitivity should be accepted as scientifically valid only when there is extensive measurement-based evidence to show that the specific model prediction is true. The description of observation-based methods and interpretations of field campaigns illustrate some recent approaches to this difficult task.

2. Overview of ozone–NO$_x$–VOC sensitivity

The central features of the relation between ozone, NO$_x$ and VOC can be illustrated by ozone isopleth plots, a form of which is shown in Fig. 1. This plot shows the rate of ozone production (ppb h^{-1}, where ppb is parts per billion) as a function of NO$_x$ and VOC concentrations. A more familiar form of this plot (e.g. NRC, 1991, see also Fig. 4 below) shows ozone concentrations as a function of NO$_x$ and VOC emission rates or initial concentrations. The isopleth plot with ozone production rates is used here because it provides a representation of instantaneous ozone chemistry that would apply to a broad range of atmospheric conditions and is less dependent on (though not totally independent of) assumptions of the individual calculation.

The isopleth plot shows that ozone formation is a highly nonlinear process in relation to NO$_x$ and VOC. When NO$_x$ is low the rate of ozone formation increases with increasing NO$_x$ in a near-linear fashion. As NO$_x$ increases the rate of increase in ozone formation slows and eventually reaches a local maximum. At higher NO$_x$ concentrations the rate of ozone formation would decrease with increasing NO$_x$. The line representing the local maxima for the rate of ozone formation (the "ridge line") can be thought of as a dividing line separating two

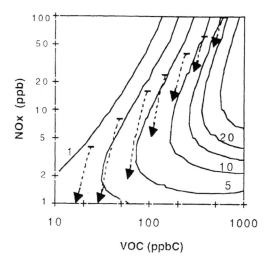

Figure 1. Isopleths giving net rate of ozone production (ppb/h, solid lines) as a function of VOC (ppbC) and NO$_x$ (ppb) for mean summer daytime meteorology and clear skies. The solid lines represent production rates of 1, 2.5, 5, 10, 15, 20 and 30 ppb/h. The dashed lines and arrows show the calculated evolution of VOC and NO$_x$ concentrations in a series of air parcels over an 8 h period (9 am–5 pm), each with initial VOC/NO$_x$ = 6 and speciation typical of urban centers in the US, based on calculations shown in Milford et al. (1994).

different photochemical regimes. In the *NO$_x$-sensitive* regime below the ridge line, ozone increases with increasing NO$_x$ and shows relatively little change in response to increased VOC. In the *VOC-sensitive* (or *NO$_x$-saturated*) regime, ozone increases with increasing VOC and decreases with increasing NO$_x$. The contrast between NO$_x$-sensitive and VOC-sensitive regimes in Fig. 1 (see also Fig. 13) illustrates the difficulties involved in developing policies to reduce ozone in polluted regions. Ambient ozone can be reduced only by reducing emissions of its precursors, NO$_x$ and VOC (and CO). Reductions in VOC will only be effective in reducing ozone if VOC-sensitive chemistry predominates. Reductions in NO$_x$ will be effective only if NO$_x$-sensitive chemistry predominates and may actually increase ozone in VOC-sensitive regions.

The "ridge line" that divides NO$_x$-sensitive and VOC-sensitive regimes generally follows a line of constant VOC/NO$_x$ ratio, with high VOC/NO$_x$ ratios corresponding to NO$_x$-sensitive chemistry and low VOC/NO$_x$ ratios corresponding to VOC-sensitive chemistry. It should be noted that the actual split between NO$_x$-sensitive and VOC-sensitive chemistry includes a broad transitional region rather than a sharp dividing line. There are also ambiguities associated with the definition of terms such as "NO$_x$-sensitive", "VOC-sensitive" and "NO$_x$-saturated". The divide between the two regimes is sometimes defined relative to the maxima for ozone formation as a function of NO$_x$ and VOC, so that the NO$_x$-saturated regime refers specifically to conditions in which increased NO$_x$ would result in lower O$_3$. This is the most broadly useful definition because it applies equally to regions in the remote troposphere and to power plants, where concepts from urban chemistry such as "VOC-sensitive" are not applicable. In the context of urban chemistry the divide is sometimes defined based on the relative impact of a given percent reduction in NO$_x$ relative to VOC. In this definition the VOC-sensitive regime refers to situations in which a percent reduction in anthropogenic VOC would result in a significantly greater decrease in O$_3$ relative to the same percent reduction in NO$_x$. The NO$_x$-sensitive regime refers to situations in which a percent reduction in NO$_x$ results in a significantly greater decrease in O$_3$ relative to the same percent reduction in anthropogenic VOC. This definition contains a number of ambiguities (e.g. in relation to the role of biogenic hydrocarbons or to power plants with little anthropogenic VOC) but it has a number of advantages for describing urban chemistry. In this paper the latter definition for the terms "NO$_x$-sensitive" and "VOC-sensitive" will be used in general, although it is recognized that the former definition and use of the term "NO$_x$-saturated" (rather than VOC-sensitive) is more broadly applicable to the troposphere as a whole. The term "NO$_x$-saturated" will be used in reference to the remote troposphere and power plant plumes where the term "VOC-sensitive" does not apply.

The isopleth plot (Fig. 1) illustrates many important features of ozone–NO_x–VOC sensitivity but it does not provide a complete understanding. The most important feature in addition to the isopleth plot is the pattern of evolution of an air mass as it moves downwind from emission sources. Typically (though not always) freshly emitted pollutants are characterized by VOC-sensitive chemistry and evolve towards NO_x-sensitive chemistry as the air mass ages. This process is illustrated by the air mass trajectories superimposed on the isopleth plot in Fig. 1 (from Milford et al., 1994). These trajectories illustrate the change in VOC and NO_x concentrations calculated for air parcels that were initialized with a fixed concentration of VOC and NO_x (with a 6 : 1 ratio) and allowed to react chemically for an 8 h period. As the air parcels age the VOC/NO_x ratios increase and the chemistry changes from the VOC-sensitive to the NO_x-sensitive regime. The speed of conversion from VOC-sensitive to NO_x-sensitive depends on how rapidly the NO_x in the air parcel reacts away. For real-world conditions this upwind-downwind pattern would be modified by the complex geography of emissions. However the split between VOC-sensitive conditions near urban centers and NO_x-sensitive conditions further downwind also appears in 3D models (Milford et al., 1989, 1994; Sillman et al., 1990). In general, NO_x emissions within an urban area determine the total amount of ozone that is formed after the air moves downwind and chemistry has run to completion, while VOC emissions control the rate of the initial buildup of O_3.

The pattern of downwind evolution is also related to a fundamental feature of ozone chemistry: the NO_x-saturated regime is associated with lack of sunlight to fuel the ozone formation process. The split between the NO_x-sensitive and NO_x-saturated regimes is related to the relative supply of NO_x (from emissions) in comparison with the supply of radicals generated by sunlight (Kleinman, 1991, 1994). In a freshly emitted plume of polluted air the initial NO_x supply greatly exceeds the supply of radicals. As the air mass ages the total amount of radicals created during the process of photochemical evolution catches up with and eventually surpasses the initial NO_x source, causing a switch from NO_x-saturated to NO_x-sensitive conditions. The chemistry of odd hydrogen radicals in connection with the ozone–NO_x–VOC system is described in detail in Section 4.

The above description of O_3–NO_x–VOC sensitivity refers only to the process of ozone production in association with NO_x and VOC emissions. There is also an important process of ozone removal associated with directly emitted NO. This process, referred to as *NO_x titration*, occurs because freshly emitted NO (typically, 90% or more of total NO_x emitted) reacts rapidly with O_3 to produce NO_2. In situations with significant ozone production (including most urban and polluted rural areas during meteorological conditions favorable to ozone formation) this removal of O_3 is small compared to the rate of ozone production. The process of NO$_x$ titration can only remove at most one O$_3$ per emitted NO (up to 1.5 O$_3$ per NO$_x$ at night), wheras the process of ozone formation typically produces four or more O$_3$ per emitted NO$_x$ (Lin et al., 1988; Liu et al., 1987; Trainer et al., 1993; Sillman et al, 1998; see Section 4). However, the process of NO$_x$ titration has a large impact in three situations: night time, winter and large power plants. At night there is no ozone formation and loss through NO$_x$ titration becomes the dominant process. O$_3$ at night in urban centers is often lower than in the surrounding rural area for this reason. Similarly, in cold-weather climates during winter the process of ozone formation is very slow and polluted plumes can be characterized by net loss of ozone for a long distance downwind (Parrish et al., 1991).

Power plants are characterized by very large emissions of NO$_x$ (with emissions from a single plant often exceeding total NO$_x$ emissions in many urban centers) and very low emissions of VOC and CO. NO$_x$ concentrations in power plant plumes are often high enough to prevent any ozone production near the plume source and to cause significant loss of ozone through NO$_x$ titration. As shown in Fig. 2, power plants are often associated with decreased O$_3$ within 80 km of the plume source. As the plume moves further downwind NO$_x$ concentrations are reduced by dilution and chemistry and production of ozone in the plume replaces the initial loss of O$_3$. During meteorological conditions

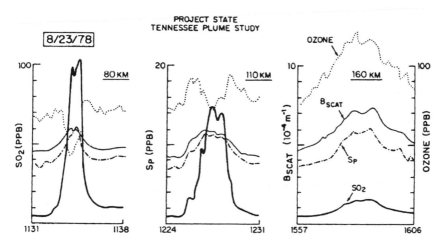

Figure 2. Stages in the chemical development of a power plant plume. The three sets of profiles show measurements of SO$_2$ (surrogate for NO$_x$, heavy solid line), ozone (dotted line), particulate sulfur (S$_p$, line-dot-line), all in ppb; and the light scattering coefficient (B_{scat}, 10^{-4} m, light solid line) made during crosswind aircraft traverses through the plume of the Cumberland power plant in NW Tennessee on 23 August 1978. The traverses at 80, 110 and 160 km downwind distances illustrate the "early", the "intermediate" and the "mature" stages of chemical development of the plume, respectively. From Gillani et al. (1996).

that favor ozone formation (high sunlight and warm temperatures) ozone in the downwind plume can be higher than in the surrounding rural area by as much as 50 ppb (Miller et al., 1978; White et al., 1983; Gillani and Pleim, 1996). The early stage of a power plant plume represents an extreme case of NO_x-saturated chemistry, but the loss of ozone through NO_x titration should not be confused with the NO_x-saturated or VOC-sensitive chemistry discussed above. VOC-sensitive chemistry occurs in regions characterized by net production of ozone, and the tendency for ozone to decrease with increasing NO_x is due to the chemistry of ozone production (see Section 4) rather than to NO_x titration.

3. The relation between ozone, NO_x and VOC in photochemical models

The development and expansion of 3D Eulerian models for ozone photochemistry and transport has lead to extensive use of these models to predict the response of ozone to reductions in NO_x and VOC. These types of predictions, if accurate, are very useful for evaluating the benefits of pollution control policies. It must be emphasized that NO_x–VOC predictions for specific events are subject to large uncertainties and that all models include assumptions that can cause predictions for the relative impact of NO_x vs. VOC to be biased in one direction or another. However, model NO_x–VOC results provide an excellent basis for identifying the causal factors that distinguish NO_x-sensitive and VOC-sensitive cases. Model-based studies are especially useful for identifying the way NO_x–VOC predictions depend on model assumptions.

This section contains a summary of: (a) factors that affect model predictions for NO_x–vs.–VOC sensitivity; (b) geographical variations in NO_x–VOC sensitivity; and (c) sources of uncertainty in model NO_x–VOC predictions. Throughout this section an attempt is made to distinguish between speculative or uncertain NO_x–VOC predictions as opposed to predictions that have some level of support through interpretation of field measurements or model-measurement comparisons. Section 5 below describes methods for evaluating the accuracy of NO_x–VOC predictions.

3.1. Factors that affect NO_x–VOC sensitivity

NO_x–VOC chemistry in models is affected by five major factors: the VOC/NO_x ratio; the reactivity of the VOC mix; the role of biogenic hydrocarbons; the extent of photochemical aging; and the severity of the air pollution event. These factors are described here.

3.1.1. VOC/NO_x ratio

The impact of VOC/NO_x ratios on ozone–NO_x–VOC chemistry was first identified by Haagen-Smit (1954) as part of the earliest investigations into the ozone formation process. Since then, the impact of VOC–NO_x ratios has been demonstrated in model calculations and in smog chamber experiments (see summary in NRC, 1991). The isopleth plot (Fig. 1) discussed in the previous section illustrates this impact.

In the 1980s analyses of NO_x–VOC chemistry were frequently made with a Lagrangian model (EKMA) that calculated the evolution of a specific VOC/NO_x mixture (representing morning VOC and NO_x concentrations in an urban center) through the course of a day (see summary in NRC, 1991). Based on these analyses a simple rule was developed in which morning VOC/NO_x ratios lower than 10 were equated with VOC-sensitive peak ozone and morning VOC/NO_x ratios greater than 20 correspond to NO_x-sensitive peak ozone. This rule does not account for the impact of VOC reactivity, biogenic hydrocarbons, geographic variations or the severity of the event, all discussed later. It has also been repeatedly shown to fail in more sophisticated photochemical models (Chameides et al., 1988; Milford et al., 1989, 1994). Despite these failings, the morning VOC/NO_x rule is still used to justify NO_x–VOC predictions and policies (e.g. Hanna et al., 1996).

3.1.2. VOC reactivity

Although the impact of VOC on ozone chemistry is frequently expressed in terms of VOC/NO_x ratios, the true impact of VOC is related more closely to the reactivity of the VOC species with respect to OH rather than to the total amount of VOC. Locations with highly reactive VOC, e.g. xylenes or isoprene, are more likely to have NO_x-sensitive chemistry than locations with similar total VOC but lower reactivity. The impact of VOC reactivity is especially important with regard to biogenic hydrocarbons, which typically have relatively low ambient concentrations but high reactivity. The importance of VOC reactivity has also been highlighted by studies that show uncertainties in the emission rates of some of the more reactive anthropogenic hydrocarbons (Fujita et al., 1992). In general, model evaluations based on measured total VOC have the potential to be misleading because total VOC concentrations are dominated by relatively less reactive alkanes.

Chameides et al. (1992) developed the concept of "propylene-equivalent carbon" as a simple method for quantifying the impact of VOC reactivity. This represents a weighted sum of VOC in which each VOC species is weighted by its reaction rate with OH divided by the reaction rate of propene (i.e. so that the weighting factor for propene is one); and then multiplied by the number of carbon atoms it contains (with 0.7 for CO). The number of carbon atoms is used to approximate the impact of relatively short-lived intermediate hydrocarbons (e.g. aldehydes) that are generated following reaction of the primary VOC with OH. Carter (1994, 1995) and earlier works by Carter and Atkinson developed a more sophisticated method for estimating the contribution of individual VOC species. His method is based on incremental reactivity, defined as the relative impact of changes in individual VOC concentrations on calculated rates of ozone formation. Either of these formulations can be used to generate a reactivity-weighted sum of VOC species, which in theory should represent the impact of VOC on ozone formation more accurately than the total concentration of VOC. However, it should be emphasized that no simple rule has been established that would relate reactivity-weighted VOC/NO$_x$ ratios with predicted NO$_x$–VOC sensitivity. Section 5 describes recent attempts to derive a rule of this type.

3.1.3. Biogenic hydrocarbons

Biogenic hydrocarbons, emitted primarily by deciduous trees, have a major impact on ozone formation (Trainer et al., 1987; Chameides et al., 1988). Within the US emission of biogenic hydrocarbons during summer is estimated to equal or exceed the total emission of anthropogenic hydrocarbons (Lamb et al., 1985; Geron et al., 1994). Since the initial study by Chameides et al. (1988) it has been recognized that biogenic hydrocarbons, especially isoprene (C$_5$H$_8$) have a large impact on ozone formation in urban areas as well.

Two properties of biogenic hydrocarbons tend to lead to errors in interpreting their role in urban ozone chemistry. First, emission rates for isoprene shows a strong diurnal signal with zero emissions at night and maximum emissions between noon and 4 pm local time. Evaluations of ozone chemistry based on measured VOC concentrations during the morning hours do not account for the impact of biogenics (Chameides et al., 1992). As illustrated in Fig. 3 for a suburban site near Los Angeles, biogenic VOC form a negligible component of measured reactivity-weighted VOC at 8–12 am but account for 25% of reactivity-weighted VOC at 12–4 pm. Second, biogenic VOC species are extremely reactive relative to most anthropogenic VOC. Consequently the impact of biogenic VOC is large relative to their ambient concentrations. For the 12–4 pm measurements shown in Fig. 3 the average isoprene concentration was just 1.5 ppb, representing less than 1% of total VOC, but it accounts for 25% of VOC reactivity (Chameides et al., 1992).

The role of biogenic VOC deserves special emphasis because it has often been ignored or underestimated. The most recent inventories in the US (BEIS2, Geron et al., 1994) (see also Geron et al., 1995; Guenther et al., 1995) have emission rates for isoprene, the most important biogenic hydrocarbon,

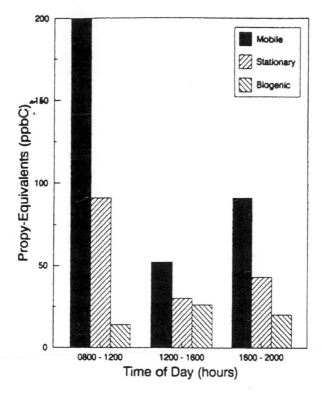

Figure 3. Reactivity weighted as propylene equivalents (ppbC) observed at Glendora, CA (near Los Angeles) as a function of time of day and apportioned by source category. Sampling period was from 8–20 August 1986, a period of extremely high temperatures. From Chameides et al. (1992).

that are 3–5 times larger than previous estimates. Many published analyses of NO$_x$–VOC sensitivity are based on older inventories with underestimated biogenic emissions (e.g. Roselle and Schere, 1995) or ignore biogenic emissions entirely (Hanna et al., 1996). An underestimate of the magnitude identified by Geron et al. (1994) is likely to have a large impact on model NO$_x$–VOC predictions in the US. Unpublished model applications developed for the Ozone Transport Assessment Group (1996) showed that the predicted chemistry for peak ozone in most major urban areas in the eastern US would shift from VOC-sensitive to NO$_x$-sensitive if the older emission estimates were replaced by BEIS2. Similar results were reported by Sillman et al. (1995a) for Atlanta. Thus, the choice of biogenic emission inventories is possibly the most important science issue associated with NO$_x$–VOC policy in the US.

It is unclear whether biogenic hydrocarbons have a similar impact on NO_x–VOC chemistry in Europe. Simpson et al. (1996) reported that biogenic emissions had little impact on model NO_x–VOC predictions throughout Europe. Vogel et al. (1995) found that biogenic hydrocarbons affect ozone formation rates in Karlsruhe, Germany, although they did not analyze the impact on NO_x–VOC sensitivity. Biogenic emissions are generally smaller in Europe than in the eastern US, due in part to smaller forest cover and in part to the types of trees. The same considerations may limit the role of isoprene elsewhere. Terpenes (emitted by conifers) also tend to be relatively more important in Europe.

3.1.4. Photochemical aging

The impact of photochemical aging on VOC–NO_x chemistry is approximately equal in importance to the impact of VOC/NO_x ratios. A polluted air mass is most likely to have VOC-sensitive chemistry when it is close to its emission sources. As the air mass ages, chemistry tends to shift to NO_x-sensitive chemistry. The impact of photochemical aging was first demonstrated in peer-reviewed literature by Milford et al. (1989). Milford et al. showed that models for Los Angeles generate VOC-sensitive chemistry close to downtown and NO_x-sensitive chemistry in downwind locations. Johnson (1984, 1990) and Hess et al. (1992) showed a similar effect in smog chamber experiments. They described a situation in which the initial rate of ozone formation associated with a given VOC–NO_x mixture is controlled by VOC, but the total amount of ozone produced at the end of the aging process is controlled by NO_x. The shift between a VOC-sensitive downtown and NO_x-sensitive downwind is often accelerated by the higher rate of biogenic emissions in downwind locations.

In addition to the difference between downtown and downwind, the process of photochemical aging can create differences in predicted NO_x–VOC sensitivity between events dominated by local photochemistry in a single city and events characterized by multiday transport. Transport events are common in regions that are characterized by a high overall emissions density and numerous cities in close proximity to each other. Multiday transport can also be associated with "recirculation" of air within a single metropolitan area, as has been hypothesized for Los Angeles (Jacobson et al., 1996; Lu and Turco, 1996) and in Spain (Millan et al., 1996) and Israel (Tov et al., 1997). Multiday transport involves photochemically aged air which is more likely to have NO_x-sensitive chemistry. Models that evaluate NO_x–VOC sensitivity relative to region-wide VOC and NO_x reductions during multiday events with significant transport or recirculation tend to predict greater sensitivity to NO_x then models for a single city without recirculation (Sillman et al., 1990; Winner et al., 1995).

3.1.5. Severity of the event

A combination of theoretical analyses and results from photochemical models have suggested that events with higher overall concentrations of ozone precursors are more likely to have peak ozone sensitive to VOC-sensitive chemistry, while events with lower precursors are more likely to have peak ozone sensitive to NO$_x$. Higher precursor concentrations are associated with either higher emissions densities (i.e. larger cities) or events with more stagnant meteorology (i.e. light winds and lower daytime vertical mixing).

This effect is demonstrated by the air parcel trajectories in Fig. 1, as analyzed by Milford et al. (1994). The series of air parcels in Fig. 1 were all initiated with the same VOC/NO$_x$ ratio and VOC speciation but with different VOC and NO$_x$ concentrations. At the end of an 8 h calculation the air parcels with lower initial VOC and NO$_x$ had NO$_x$-sensitive chemistry while the air parcels with higher initial VOC and NO$_x$ still had VOC-sensitive chemistry. The effect can be explained using the theoretical description of NO$_x$-sensitive and NO$_x$-saturated chemistry from Kleinman (1991, 1994), discussed in the previous section. NO$_x$-saturated chemistry occurs so long as the NO$_x$ source exceeds the accumulated source of radicals. When the NO$_x$ source per unit volume is high (after accounting for dilution through daytime vertical mixing) it takes more time for the accumulated source of radicals to catch up with and surpass the NO$_x$ source, so that NO$_x$-saturated conditions persist for a longer time. When the NO$_x$ source is low then the accumulated radical source exceeds the NO$_x$ source after a short period of photochemical aging and the situation shifts to NO$_x$-sensitive conditions.

Milford et al. (1994) found that a shift occurs in NO$_x$–VOC chemistry in box model calculations with low and high precursor concentrations even with identical VOC/NO$_x$ emission ratios. Milford et al. also found that VOC-sensitive chemistry in 3D models can be correlated with high total reactive nitrogen (NO$_y$), which is indicative of high precursor concentrations. A similar finding was reported by Simpson et al. (1995). Roselle and Schere (1995) found in a 3D simulation for the eastern US that VOC-sensitive chemistry was associated both with the largest metropolitan areas (New York and Chicago) and with the most severe events. Their evaluation ranked events in each geographic region based on peak O$_3$ and found that events with the highest O$_3$ were more likely to have VOC-sensitive chemistry (Fig. 4). The studies by Milford et al. and Roselle both used older emission estimates for biogenic hydrocarbons and are therefore likely to overestimate the extent of VOC-sensitive chemistry, but the tendency towards greater sensitivity to VOC in large cities and during more severe events is likely to remain in models with corrected biogenic emissions. More recently Sistla et al. (1996) showed that wind speeds and mixing heights had a comparable effect on NO$_x$–VOC predictions in simulations for

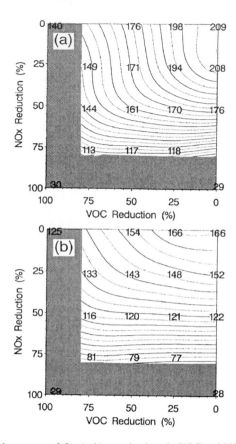

Figure 4. Simulated response of O_3 (ppb) to reductions in VOC and NO_x emissions ranging from 0% (base case) to 100% (zero emissions) in the northeast corridor of the US on 9 July 1988. Results are shown for (a) peak O_3 and (b) 99th percentile O_3 in the region. Results for (a) show greater sensitivity to VOC for emission reductions lower than 50% and greater sensitivity to NO_x for higher emission reductions. Results for (b) show greater sensitivity to NO_x for all levels of emission reductions. From Roselle and Scbere (1995).

New York. Their scenarios with lighter winds and lower mixing heights generated higher peak O_3 and VOC-sensitive chemistry while the scenarios with more vigorous mixing generated NO_x-sensitive chemistry.

There is one important caveat for this effect. The results described here all refer to NO_x–VOC sensitivity associated with peak ozone during an event. They may not apply to individual locations within a metropolitan area. A. Russell (Georgia Institute of Technology, private communication, 1998) pointed out that light winds would allow air to undergo photochemical aging for a longer period of time before being advected out of a metropolitan area while strong winds would advect emissions away from a metropolitan area before there has been significant photochemical aging. This would counteract the tendency towards VOC-sensitive chemistry during stagnant events and NO$_x$-sensitive chemistry during events with more rapid dispersion. Future research may clarify this issue.

3.1.6. Other factors

Sunlight, cloud cover and water vapor concentration are all expected to impact NO$_x$–VOC chemistry. Decreasing sunlight, increased cloud cover and decreased water vapor all cause a reduction in the source of radicals Following Kleinman (1991), these can all be expected to cause a shift towards VOC-sensitive chemistry. These factors are also expected to reduce O$_3$. None of these factors have been studied explicitly in 3D models, but Jacob et al. (1995) (see also Hirsch et al., 1996) identified a seasonal transition between NO$_x$-sensitive chemistry during summer and VOC-sensitive chemistry during autumn in rural Virginia, partly associated with reduced sunlight during the autumn. This finding may be especially relevant for studies in Europe, where models have shown a greater tendency towards VOC-sensitive chemistry in northern locations (Simpson, 1995; Simpson et al., 1996). Walcek et al. (1997) and Matthijsen et al. (1997) found that aqueous chemistry significantly reduced rates of ozone formation in urban areas, but did not discuss implications for NO$_x$–VOC chemistry.

Temperature is expected to have no direct effect on NO$_x$–VOC chemistry. Lower temperatures are associated with lower O$_3$. Cardelino et al. (1990) and Sillman et al. (1995b) both attributed the reduced rates of ozone formation to the increased photochemical lifetime of peroxyacetyl nitrate (PAN) at lower temperatures, and PAN acts as a sink for both NO$_x$ and radicals. Since NO$_x$–VOC chemistry is associated with the relative amounts of NO$_x$ and radicals the increased rate of formation of PAN should have little impact. However, lower temperature has several indirect effects on NO$_x$–VOC chemistry. Lower temperature usually coincides with decreased sunlight, decreased water vapor and sharply lower biogenic emissions, all of which favor VOC-sensitive chemistry.

A final NO$_x$–VOC factor relates to the definitions used to identify NO$_x$- and VOC-sensitive conditions. The above discussion presumes that NO$_x$- and VOC-sensitive chemistry is defined based on model responses to moderate reductions (25–50%) in VOC and NO$_x$ concentrations; so that VOC-sensitive chemistry occurs when a 25 or 50% reduction in VOC is more effective in reducing O$_3$ than a corresponding percent reduction in NO$_x$. If VOC–NO$_x$ chemistry were defined based on larger percentage reductions in VOC and NO$_x$, reduced NO$_x$ is likely to be more effective relative to reduced VOC. Simulations with 100% reductions in anthropogenic VOC in the eastern US predict 80 ppb O_3, but simulations with 100% reductions in NO_x predict that ozone concentrations reduce to background values (30–40 ppb) (Roselle and Schere, 1995). In other words, ozone chemistry is always "NO_x-sensitive" when analyzed based on 100% reductions in NO_x or VOC. The impact of 100% reductions predicted by Roselle and Schere are also consistent with observed O_3 in power plants (with high NO_x but little anthropogenic VOC) in contrast to sites in Brazil (with high biogenic VOC but little NO_x, e.g. Chameides et al., 1992). Alternately, VOC–NO_x analyses based on very small percent decreases or on percent increases in NO_x and VOC emissions are more likely to show VOC-sensitive chemistry then the definition used here (see Fig. 4).

3.2. Geographical variation in NO_x–VOC chemistry

Results from Roselle and Schere (1995) and other 3D models (Sillman et al., 1990, 1993; Milford et al., 1989, 1994; McKeen et al., 1991, Simpson et al., 1996) have generated the following picture of the geography of NO_x–VOC chemistry.

Rural areas appear to be predominantly NO_x-sensitive. Ozone formation in rural areas of the eastern US has been studied in great detail, including both model and measurement-based studies (Trainer et al., 1987, 1993; Sillman et al., 1990, 1993; McKeen et al., 1991; Jacob et al., 1993, 1995; Kleinman et al., 1994; Olszyna et al., 1994; Buhr et al., 1995; Roselle and Schere, 1995). It has long been recognized that elevated ozone in eastern North America and western Europe frequently extends over 500 km or more (e.g. Vukovich et al., 1977; Samson and Ragland, 1977; Guicherit and van Dop, 1977; Logan, 1989). These regional events are characterized by nearuniform high ozone (80–100 ppb) in rural locations and intermittent higher ozone associated with urban plumes. The uniform nature of elevated ozone concentrations during these events and their role in regional transport was demonstrated in aircraft measurements by Clark and Ching (1983). This type of regional ozone results from a mixture of widely distributed small emission sources and urban and power plant plumes that have been aged for more than 24 h (Sillman et al., 1990). Multiday transport during these events also contributes significantly to elevated ozone in cities within the region. Similar events have been studied in Europe (e.g. Guicherit and van Dop, 1977; Vogel et al., 1995; Simpson et al., 1996).

Studies of this type of background rural ozone have almost always found evidence of NO_x-sensitive chemistry, but there are some exceptions. Jacob et al. (1995) reported a seasonal transition between NO_x-sensitive conditions during summer and VOC-sensitive conditions in autumn. VOC-sensitive chemistry is also possible in rural locations that are directly impacted by large urban plumes (e.g. Sillman et al., 1993; Hanna et al., 1996), and NO$_x$-saturated chemistry occurs in the early stages of power plant plumes. Results from Simpson et al. (1996) suggest that rural areas in densely populated parts of northern Europe may have VOC-sensitive chemistry, although Simpson et al. (1995) and others (e.g. Dommen et al., 1996, 1998; Prevot et al., 1997; Kuebler et al., 1996) found evidence of NO$_x$-sensitive rural chemistry elsewhere in Europe.

VOC-sensitive chemistry is most likely to occur in central locations in large cities. Urban chemistry has been studied most extensively in the city of Los Angeles, including model-based studies (Milford et al., 1989; Harley et al., 1993; Kumar et al., 1994, 1996; Winner et al., 1995; Reynolds et al., 1996; Jacobson et al., 1996; Lu and Turco, 1996), field measurements (Lawson, 1990; Drummond et al., 1989; Williams and Grosjean, 1990; Sakugawa and Kaplan, 1989; Sillman et al., 1997) and evaluation of emission inventories (Fujita et al., 1992). These studies have found strong evidence for VOC-sensitive chemistry in downtown Los Angeles and possibly in much of the metropolitan region. NO$_x$-sensitive chemistry is possible in outlying regions (Milford et al., 1989). Even in Los Angeles there is significant uncertainty based on the possibility of "recirculation" of air exported from Los Angeles back into the city center on subsequent days, as a result of the complex ocean and mountain circulation in the region (Jacobson et al., 1996; Lu and Turco, 1996). The impact of recirculation on NO$_x$–VOC chemistry has not been reported, but transported air is more likely to have NO$_x$-sensitive chemistry (Winner et al., 1995). Nonetheless, most evidence suggests that the urban center in Los Angeles has VOC-sensitive chemistry. VOC-sensitive chemistry has also been reported in urban centers in New York (Sistla et al., 1996), Chicago (Hanna et al., 1996) and Milan (Prevot et al., 1996), although the evidence is less clear than in Los Angeles. In each of these cases it may also be possible to find NO$_x$-sensitive chemistry at downwind locations (e.g. Sillman et al., 1993; Prevot et al., 1996).

This split between NO$_x$-sensitive rural areas and VOC-sensitive urban centers has inspired an unofficial debate on policy. Advocates of VOC controls emphasize the impact of VOC in locations with the highest population density. Advocates of NO$_x$ controls, including environmental groups, emphasize the fact that VOC controls have little impact on the eventual total ozone produced, but merely delay the process of ozone formation until the air has moved further downwind. Evaluations of ozone policy that weigh impacts based on population exposure are more likely to favor VOC controls. By contrast, evaluations that include the impact of ozone concentrations at levels below the current US ambient standard (125 ppb), possibly including the recently proposed switch from a 1 h to an 8 h standard, are more likely to favor NO$_x$ controls.

The highest ozone concentrations are typically found in urban plumes as they move downwind of the city center. Peak O$_3$ usually occurs 50–100 km from the city center but in some instances the peak occurs much further downwind, especially in coastal environments (e.g. Maine, Lake Michigan). Peak O_3 often represents an intermediate point, both geographically and chemically, between VOC-sensitive urban centers and NO_x-sensitive rural areas. Peak O_3 is also associated with the greatest uncertainties in NO_x–VOC predictions. It is frequently possible to generate both NO_x-sensitive and VOC-sensitive model scenarios for these locations by making reasonable variations in model assumptions (e.g. Sillman et al., 1995a; Reynolds et al., 1996) (see Section 3.3). Model predictions for NO_x–VOC chemistry associated with peak ozone are the most important results in terms of policy, but they are also the results that should be viewed with the greatest skepticism.

3.3. Uncertainties in photochemical models

This paper has repeatedly emphasized the uncertain nature of model VOC–NO_x predictions. Sources of uncertainty will be summarized here.

3.3.1. Emission rates

Emission rates are probably the largest source of uncertainty in NO_x–VOC predictions. The uncertainty associated with emission of biogenic hydrocarbons and its impact on NO_x–VOC chemistry has already been discussed. Similar uncertainties are associated with emission inventories for anthropogenic VOC. Fujita et al. (1992) identified possible underestimates in anthropogenic VOC inventories of a factor of two or more, associated with auto emissions, in southern California. Evaluations of photochemical models tend to confirm this (Harley et al., 1993; Jacobson et al., 1996; Lu and Turco, 1996; Sillman et al., 1997). Fujita et al. (1992) also found underestimates in the reactivity of the anthropogenic VOC mix. By contrast, recent studies in Atlanta (Cardelino et al., 1994) and Baltimore (Pierson et al., 1996) found no evidence of underestimated VOC emissions. NO_x emissions are generally regarded as more accurate than VOC emissions. These estimates of uncertainty associated with emissions apply only to the US.

3.3.2. Meteorology

Meteorology especially wind speed is frequently the largest source of uncertainty in individual model applications. Most modelers are familiar with model applications with large underestimates or overestimates in ozone concentrations, with errors in the location of the ozone peak, both of which are attributed to errors in wind speed. High-ozone events are frequently associated with very low wind speeds (<2 m/s). In these situations the uncertainty in wind speeds is frequently has the same magnitude as the wind speed itself (Kumar and Russell, 1996; Sistla et al., 1996; Al-Wali et al., 1996). Uncertainties are generated by imprecision in measurements, the stochastic nature of wind and the need for interpolation based on measurements separated by 200 km or more, especially in prognostic models. Uncertainties in the height of the convective mixed layer (Marsik et al., 1995) also contribute to errors in model ozone.

It is important to recognize that uncertainties in emission rates and in wind speed have very different impacts on model performance. Uncertain emission rates (especially VOC/NO$_x$ ratios, VOC reactivity and biogenics) have a direct impact on model NO$_x$–VOC chemistry but may have less impact on model ozone concentrations. By contrast, uncertainties in wind speeds or mixing heights have a direct impact on ozone formation but only a secondary impact on NO$_x$–VOC chemistry. Thus, model performance evaluations based on measured ozone do not provide evidence for the accuracy of model VOC–NO$_x$ predictions.

3.3.3. Chemistry

Chemistry as a source of uncertainty was analyzed by Gao et al. (1996). Gao found that known uncertainties in reaction rates and stoichiometries caused a 20% uncertainty in simulated concentrations of ozone and most other species (40% for H$_2$O$_2$). Gao et al. (1996) did not report impacts of uncertain chemistry on NO$_x$–VOC predictions.

3.3.4. Evaluating the uncertainties

Few studies have attempted to derive a quantitative estimate for the uncertainty associated with model predictions for the impact of reduced NO$_x$ and VOC on ozone. Sillman et al. (1995a) examined the impact of changed model assumptions on NO$_x$–VOC predictions in Atlanta. They found that the size of the reduction in peak O$_3$ resulting from reduced VOC varied by a factor of two or more in model scenarios with 25% changes in anthropogenic emissions, wind speeds and mixing heights (see Fig. 5, results for 10 August 1990). Even larger changes were found if uncertainties associated with biogenic VOC were included. The size of the reduction in peak O$_3$ resulting from reduced NO$_x$ varied by up to a factor of ten. Uncertainties in model NO$_x$–VOC predictions were also reported for Los Angeles, based on uncertain emissions and transport (Winner et al., 1995), and for New York, based on model representation of vertical mixing (Sistla et al., 1996).

To some extent the format of Fig. 5 exaggerates the uncertainty associated with model NO$_x$–VOC predictions. The level of uncertainty is associated specifically with the split between NO$_x$-sensitive and VOC-sensitive chem-

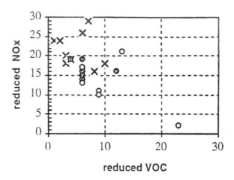

Figure 5.   Predicted reduction in peak O_3 (ppb) resulting from either a 35% reduction in anthropogenic VOC or a 35% reduction in NO_x, from different model scenarios for Atlanta. The closed and open circles represent scenarios for 8 October 1992. Closed circles represented scenarios that were in agreement with measured O_3/NO_y; open circles represented scenarios that differed from measured O_3/NO_y. The X's represent scenarios for 8 November 92.

istry. If models were evaluated relative to the predicted response to simultaneous reductions in both VOC and NO_x, the variation among different scenarios would have been much smaller. Uncertainties are also much smaller in models for events with strongly NO_x-sensitive or strongly VOC-sensitive chemistry. During a second event in Atlanta with strongly NO_x-sensitive chemistry (11 August 1998 in Fig. 5) the predicted reduction resulting from reduced NO_x varied by less than 30% and variations in the predicted response to reduced VOC were all small in magnitude. These results demonstrate central importance of the NO_x–VOC split as a source of uncertainty. The biggest concern associated with model performance is the possibility of bias in predictions for the impact of NO_x vs. VOC.

The uncertainty in model predictions can also be greatly reduced if model predictions are evaluated based on comparisons with ambient measurements for species other than ozone. In the Atlanta event shown in Fig. 5, it was possible to reject several model scenarios based on discrepancies between model and measured reactive nitrogen. If the rejected scenarios were excluded, then the uncertainty in predicted reductions in peak O_3 among the remaining model scenarios would be ±30%. Observation-based methods for determining O_3–NO_x–VOC sensitivity and evaluating model predictions will be discussed further in Section 5.

4. Chemistry of ozone, NO_x and VOC

This section describes the chemical factors that create the split between NO_x-sensitive and VOC-sensitive regimes for ozone, which were presented in general terms in Section 2. It also presents urban NO$_x$–VOC chemistry in a broader context, linked to photochemical processes in the remote troposphere. Several common analytical terms (odd hydrogen radicals, odd oxygen, and ozone production efficiency) are also presented and defined here.

Ozone is produced directly by photolysis of NO$_2$

$$NO_2 + h\nu \rightarrow NO + O \qquad (R1)$$

where the oxygen atom (O) rapidly recombines with molecular oxygen (O$_2$) to produce ozone (O$_3$). Normally, this reaction is counterbalanced by the reaction of NO with ozone:

$$NO + O_3 \rightarrow NO_2. \qquad (R2)$$

Taken together, reactions (R1) and (R2) produce no net change in ozone. Each of these reactions occurs rapidly, on a time scale of 200 s or less. Typically, the two major components of NO$_x$ (NO and NO$_2$) adjust to establish a near-steady state between reactions (R1) and (R2). However, there are two distinct situations in which these reactions result in a net change in ozone concentration: removal of ozone via reaction (R2) at nighttime or in the vicinity of large NO$_x$ sources (sometimes referred to as NO$_x$ titration) and ozone production associated with daytime NO$_x$–VOC–CO chemistry. Removal of ozone occurs when O$_3$ + NO $\rightarrow$ NO$_2$ (i.e. reaction (R2)) dominates over NO$_2$ + $h\nu$ $\rightarrow$ NO + O (i.e. reaction (R1)). There is always net removal of ozone at nighttime since photolysis rates are zero. Surface O$_3$ is normally low at night ($<$30 ppb) and high NO emissions are associated with lowest nighttime O$_3$. During the daytime significant removal of ozone via reaction (R2) occurs in the vicinity of large NO emission sources, especially large point sources. In these situations ambient NO$_x$ concentrations reach 50 ppb or higher, equal to or greater than ambient O$_3$. Since ambient NO$_x$ originates mostly from emission of NO, the rapid interconversion of O$_3$, NO and NO$_2$ via reactions (R1) and (R2) results in a photochemical equilibrium with significant loss of O$_3$. This process, sometimes referred to as NO$_x$ titration, results in reduced O$_3$ in the vicinity of large emission sources of NO, especially in power plant plumes. Analyses of ozone chemistry often use the concept of odd oxygen, O$_x$ = O$_3$ + O + NO$_2$ (Logan et al., 1981) as a way to separate the process of NO$_x$ titration from the processes of ozone formation and removal that occur on longer time scales. Odd oxygen is unaffected by reactions (R1) and (R2) and remains constant in situations dominated by NO$_x$ titration, such as the early states of a power plant plume. Production of odd oxygen occurs only through NO$_x$–VOC–CO chemistry, and loss of odd oxygen occurs through conversion of NO$_2$ to PAN and HNO$_3$ or through slower ozone loss reactions (e.g. reaction (R7) below), rather than through the more rapid back-and-forth reactions (R1) and (R2).

The chemical lifetime of odd oxygen relative to these losses is typically 2–3 d in the lower troposphere. This lifetime is often more useful for describing atmospheric processes associated with ozone then the chemical lifetime of ozone relative to reaction (R2).

The chemical process of ozone formation occurs through reaction sequences involving VOC, CO and NO_x, which result in the conversion of NO to NO_2 through processes other than reaction (R2). The NO-to-NO_2 conversion is followed by $NO_2 + h\nu \rightarrow NO + O$ (i.e. reaction (R1)) and results in additional O_3. These reaction sequences are almost always initiated by reactions of hydrocarbons (RH) or CO with OH:

$$RH + OH \xrightarrow{[O_2]} RO_2 + H_2O, \tag{R3}$$

$$CO + OH \xrightarrow{[O_2]} HO_2 + CO_2, \tag{R4}$$

followed by reactions of RO_2 and HO_2 radicals with NO

$$RO_2 + NO \xrightarrow{[O_2]} R'CHO + HO_2 + NO_2, \tag{R5}$$

$$HO_2 + NO \xrightarrow{[O_2]} OH + NO_2. \tag{R6}$$

Reactions (R5) and (R6) convert NO to NO_2 and result in the formation of ozone when followed by reaction (R2). $R'CHO$ represents intermediate organic species, typically including aldehydes and ketones. The directly emitted hydrocarbons and intermediate organics are collectively referred to as volatile organic compounds (VOC). Since these reactions also affect the ratio NO_2/NO, measured values of this ratio can be used (especially in the remote troposphere) to identify the process of ozone formation. When the ratio NO_2/NO is higher than it would be if determined solely by reactions (R1) and (R2), it provides evidence for ozone formation (e.g. Ridley et al., 1992). Reactions (R1), (R2), (R5) and (R6) can be combined to derive the summed concentration of HO_2 and RO_2 radicals from measured O_3, NO, NO_2 and solar radiation (e.g. Duderstadt et al., 1998).

For $NO_x > 0.5$ ppb (typical of urban and polluted rural sites in the eastern US and Europe) reactions (R5) and (R6) represent the dominant reaction pathways for HO_2 and RO_2 radicals. In this case the rate of ozone production is controlled by the availability of odd hydrogen radicals (defined by Kleinman (1986) as the sum of OH, HO_2 and RO_2) and in particular by the OH radical in connection with the rate-limiting reactions with CO and hydrocarbons reactions (R3) and (R4). The split into NO_x-sensitive and VOC-sensitive regimes is closely associated with sources and sinks of radicals.

Odd hydrogen radicals are produced by photolysis of ozone, formaldehyde and other intermediate organics:

$$O_3 + h\nu \xrightarrow{[H_2O]} 2OH, \tag{R7}$$

$$HCHO_2 + h\nu \xrightarrow{[O_2]} HO_2 + CO. \tag{R8}$$

They are removed by reactions that produce peroxides and nitric acid:

$$HO_2 + HO_2 \rightarrow H_2O_2 + O_2, \tag{R9}$$

$$RO_2 + HO_2 \rightarrow ROOH + O_2, \tag{R10}$$

$$OH + NO_2 \rightarrow HNO_3. \tag{R11}$$

Formation of peroxy acetyl nitrate (PAN) is also a significant sink for odd hydrogen. The split into NO$_x$-sensitive and VOC-sensitive regimes is determined by the size of the peroxide- and nitric-acid-forming reactions (Sillman et al., 1990; Kleinman, 1991). When nitric acid represents the dominant sink for odd hydrogen, then the concentration of OH is determined by the equilibrium between reactions (R7) and (R11). In this case OH decreases with increasing NO$_x$ and either remains constant or increases (due to the impact of reaction (R8) with increasing VOC. The rate of ozone formation is controlled by the hydrocarbon–OH reactions (R5) and increases with increasing VOC and decreases with increasing NO$_x$. This is the VOC-sensitive regime. When peroxides represent the dominant sink for odd hydrogen, then the sum HO$_2$ + RO$_2$ is relatively insensitive to changes in NO$_x$ or VOC. The rate of ozone formation, approximately equal to the rate of reactions (R5) and (R6), increases with increasing NO$_x$ and is largely unaffected by VOC. This is the NO$_x$-sensitive regime. These patterns can be seen in Figs. 6 and 7, which show OH and HO$_2$ + RO$_2$ as a function of NO$_x$ and VOC for conditions corresponding to the isopleths in Fig. 1. The "ridge line" in Fig. 1 that separates NO$_x$-sensitive and VOC-sensitive chemistry corresponds to high OH, while HO$_2$ + RO$_2$ is highest in the region corresponding to NO$_x$-sensitive chemistry. OH is lowest for conditions with either very high NO$_x$ (due to removal of OH through formation of nitric acid, reaction R5) or very low NO$_x$ (due to the slow rate of conversion from HO$_2$ to OH through reaction (R6)).

Fig. 8 shows the ratio of the rate of formation of peroxides (reactions (R5) and (R6)) divided by the rate of formation of HNO$_3$ (based on Kleinman et al., 1997; Sillman et al., 1990). A comparison with Fig. 1 shows that this ratio is closely associated with the split between NO$_x$- and VOC-sensitive regimes. The ratio is typically 0.9 or higher for NO$_x$-sensitive conditions, where peroxides dominate over HNO$_3$ as a sink for odd hydrogen, and 0.1 or less for

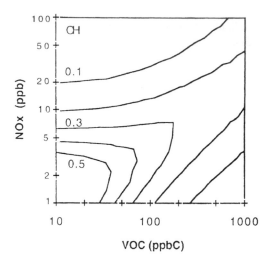

Figure 6. Isopleths showing the concentration of OH (ppt) as a function of VOC (ppbC) and NO$_x$ (ppb) for mean summer daytime meteorology and clear skies, based on 0D calculations shown in Milford et al. (1994) and in Fig. 1. The isopleths represent 0.1, 0.2, 0.3, 0.4 and 0.5 ppt.

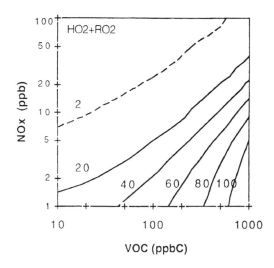

Figure 7. Isopleths showing the concentration of HO$_2$ + RO$_2$ (ppt) as a function of VOC (ppbC) and NO$_x$ (ppb) for mean summer daytime meteorology and clear skies, based on 0D calculations shown in Milford et al. (1994) and in Fig. 1. The isopleths represent 2 ppt (dashed line) and 20, 40, 60, 80 and 100 ppt (solid lines).

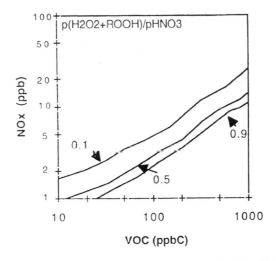

Figure 8. Isopleths showing the rate of production of peroxides (including H_2O_2 and organics) divided by the rate of production of HNO_3 as a function of VOC (ppbC) and NO_x (ppb) for mean summer daytime meteorology and clear skies, based on 0D calculations shown in Milford et al. (1994) and in Fig. 1. The isopleths represent ratios of 0.1, 0.5 and 0.9.

VOC-sensitive conditions, where HNO_3. The "ridge line" that separates NO_x-sensitive and VOC-sensitive chemistry corresponds to a ratio of 0.5 (Sillman, 1995; Kleinman et al., 1997). This result provides the basis for using peroxides and nitric acid as "indicators" for NO_x–VOC chemistry, described in Section 5.

Kleinman (1991, 1994a) found that NO_x–VOC chemistry is related to the relative size of the sources of odd hydrogen radicals (from reactions (R7) and (R8)) and NO_x (determined by emissions and/or transport). If the source of radicals exceeds the source of NO_x, then peroxides become the dominant sink for odd hydrogen and NO_x-sensitive conditions apply. If the source of NO_x exceeds the source of radicals, then the supply of OH to initiate the ozone-forming reaction sequence is limited by NO_x. Nitric acid becomes the dominant sink for radicals and NO_x-saturated conditions apply. This analysis has been used in Sections 2 and 3 as a basis for understanding NO_x–VOC chemistry.

Another central concept for NO_x–VOC chemistry is the ozone production efficiency (Liu et al., 1987; Lin et al., 1988; Trainer et al., 1993). Ozone production efficiency represents the ratio of production of odd oxygen to removal of NO_x ($= P(O_3 + NO_2)/L(NO_x)$). Liu et al. (1987) and Lin et al. (1988) found that production efficiencies are highest at low NO_x concentrations, even when VOC is assumed to increase with increasing NO_x. Lin et al. (1988) also found that production efficiencies increase with VOC. In theory, ozone production efficiencies are given by the ratio between reactions (R3 + R4) and

(R11), i.e. by the ratio of the sum of reactivity-weighted VOC and CO to NO_x, although they are also influenced by the rate of formation of organic nitrates. An updated analysis (Fig. 9, adapted from unpublished work by Greg Frost, NOAA Aeronomy lab) showed the same pattern but with lower values than initially reported by Liu and Lin. Ozone production efficiencies in polluted regions are likely to be even lower than shown in Fig. 9 because these calculations typically do not include removal of ozone (even though removal of NO_x is directly linked to removal of ozone through the reaction sequence (R2) followed by (R11), and also do not count net formation of PAN or nighttime formation of HNO_3 in the sum of NO_x losses. Recent studies (e.g. Sillman et al., 1998; Ryerson et al., 1998; Nunnermacker et al., 1998; Trainer et al., 1995; NARSTO review, in preparation) estimated an ozone production efficiency of 3–5 during pollution events.

The characteristics of the NO_x-saturated regime can be explained in part by the chemistry of odd hydrogen radicals and in part by the ozone production efficiency. In the NO_x-saturated regime the rate of removal of NO_x is limited by the availability of radicals, so that the rate of chemical processing

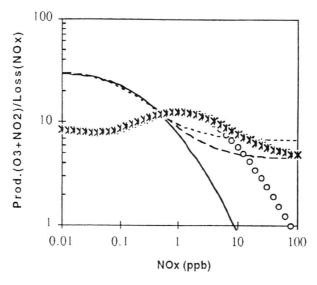

Figure 9. Ozone production efficiency, expressed as the rate of production of odd oxygen ($O_3 + NO_2$) divided by the loss of NO_x, from steady state calculations. The calculations assume: (i) CO and CH_4 only (solid line); (ii) anthropogenic VOC with $VOC/NO_x = 10$ (dashed line); (iii) anthropogenic VOC with $VOC/NO_x = 20$ (short dashed line); (iv) CH, CH_4 and 1 ppb isoprene (circles); and (v) anthropogenic $VOC/NO_x = 10$ and 1 ppb isoprene (asterisks). Calculations use chemistry described in Sillman et al. (1998), with PAN at steady state, and are based on similar unpublished analyses by Greg Frost (NOAA Aeronomy lab).

of NO$_x$ does not increase with increasing NO$_x$. At the same time, increased NO$_x$ is associated with lower ozone production efficiency. These two factors in combination result in a lower rate of ozone production as NO$_x$ increases. In the NO$_x$-sensitive regime an increase in NO$_x$ concentrations is always associated with a higher absolute rate of removal of NO$_x$, and consequently with increased ozone production. The rate of ozone production is determined by the rate of NO$_x$ removal and the ozone production efficiency.

Jaegle et al. (1998) has described a similar split between NO$_x$-sensitive and NO$_x$-saturated photochemical regimes in the remote troposphere. This split is associated with the relative rates of formation of peroxides and nitric acid and the relative source strength for radicals vs. NO$_x$, as described by Sillman and Kleinman for the polluted troposphere. The source of radicals greatly exceeds the source of NO$_x$ for the troposphere as a whole, so that the oxidizing capacity of the troposphere is more than sufficient to process the burden of NO$_x$ imposed by human activities. For this reason most of the troposphere is in a NO$_x$-sensitive rather than a NO$_x$-saturated state. However, model calculations suggest that ozone increases with increasing VOC even in the NO$_x$-sensitive remote troposphere (e.g. Jaegle et al., 1998; Kanakidou et al., 1991), apparently because the reaction sequence initiated by hydrocarbons (R1) leads to greater ozone formation per OH then the reaction sequence initiated by CO (R2) which otherwise dominates in the remote troposphere. Ozone increases with increasing VOC in both the NO$_x$-sensitive and NO$_x$-saturated regimes in the remote troposphere. NO$_x$-saturated chemistry occurs at lower NO$_x$ concentrations in the remote troposphere than in polluted regions (1 ppb or higher in the remote troposphere, 5–10 ppb or higher in polluted regions) because the radical source (driven by lower H$_2$O in reaction (R7)) is lower.

5. Observation-based methods for evaluating ozone–NO$_x$–VOC chemistry

Observation-based methods refer to a number of techniques that may be used to derive features of ozone–NO$_x$–VOC chemistry directly from ambient measurements. In recent years there has been interest in developing methods for determining the sensitivity of ozone to NO$_x$ and VOC from measurements (e.g. Chameides et al., 1992). This interest in observation-based methods has been motivated by the high level of uncertainty associated with O$_3$–NO$_x$–VOC predictions from models, discussed in Section 3. Observation-based methods have been used more broadly in the field of atmospheric chemistry to identify important features of ozone chemistry other than O$_3$–NO$_x$–VOC sensitivity. These include observation-based estimates of emission rates, ozone production efficiency and removal rates for NO$_x$. This section presents a brief description of the recent attempts to develop observation-based methods.

It has occasionally been suggested that NO_x–VOC predictions from observation-based methods might be used by themselves as a replacement for model-based predictions (Chameides et al., 1992; Cardelino et al., 1995; Sillman, 1995). This use of observation-based NO_x–VOC predictions would require enormous confidence in the method and an extensive measurement network. A more realistic goal for the observation-based methods would be to provide a basis for evaluating the accuracy of model NO_x–VOC predictions. As discussed in Section 3, model NO_x–VOC predictions are often critically dependent on assumptions in the individual model scenario, and would change significantly if different assumptions were used. Observation-based methods can be used to establish limits on the uncertainty associated with model assumptions, or to evaluate the accuracy of NO_x–VOC predictions from individual model scenarios. This could significantly reduce the level of uncertainty associated with model NO_x–VOC predictions.

Models for urban ozone have always been subject to evaluation based on ambient measurements, chiefly O_3. The US EPA has recommended an extensive set of criteria for model performance vs. ambient O_3, which must be passed before the model can be used for policy-making purposes (NRC, 1991). However, measured O_3 provides little basis for confidence in model NO_x–VOC predictions. Sillman et al. (1995a) and Reynolds et al. (1996) have both shown that alternative model base cases can generate similar O_3 along with very different predictions for NO_x–VOC sensitivity. In recent years, model results have been evaluated against a more complete set of ambient measurements (e.g. Harley et al. 1993, 1995; Giovanoni and Russell, 1995; Jacobson et al., 1996), although these evaluations have been limited to a small number of cities (mostly, Los Angeles) during specific events. The observation-based approaches provide a basis for a different type of model evaluation which would be targeted specifically at the accuracy of model NO_x–VOC predictions.

Observation-based methods can be divided into two broad categories: methods based on ambient VOC, NO_x and CO; and methods based on secondary reaction products, usually involving reactive nitrogen and peroxides. These methods will be briefly reviewed here. More detailed descriptions will be included in articles associated with the current NARSTO critical review series (Sillman, 1998; Kleinman, 1998; Trainer, 1998; Cardelino, 1998).

5.1. Evaluations based on ambient VOC, NO_x and CO

Ambient VOC and NO_x probably represent the most important single factor that determines the predicted response of O_3 to reduced emissions in models. For this reason attempts to evaluate NO_x–VOC sensitivity have always emphasized the need to include measured VOC and NO_x as part of the analysis. Comparisons between model and measured VOC and NO_x have been used as part of general model evaluations and to evaluate the accuracy of emission inventories (e.g. Harley et al., 1993, 1995; Giovannoni and Russell, 1995; Jacobson et al., 1996). Chameides et al. (1992) proposed that the ratio of reactivity-weighted VOC/NO$_x$ can be used directly to obtain information about NO$_x$–VOC sensitivity. Rappengluck et al. (1998) also used reactivity-weighted VOC/NO$_x$ ratios to evaluate VOC–NO$_x$ sensitivity in Athens.

Cardelino et al. (1995, 1998) developed a more detailed observation-based model that would calculate the dependence of ozone on NO$_x$ and VOC based on a network of measured ambient O$_3$, NO$_x$ and VOC. In this method, emission inventories are regarded as the largest source of uncertainty in VOC–NO$_x$ predictions from Eulerian models ("emission-based models"). The observation-based model concept seeks to use ambient measurements for NO$_x$ and VOC as a replacement for the emission inventory in a calculation of ozone chemistry. A series of 0-dimensional photochemical calculations are performed at each measurement site in which measured NO$_x$ and VOC are used to calculate concentrations of unmeasured secondary species (intermediate VOC, PAN, etc.) and production rates for ozone. The impact of reduced VOC or NO$_x$ emissions on ozone production is identified by repeating the calculation with assumed lower ambient concentrations of VOC or NO$_x$.

There are two types of problems associated with this and other attempts to calculate O$_3$–NO$_x$–VOC sensitivity directly from measured VOC and NO$_x$. First, measured VOC and NO$_x$ are associated with the instantaneous rate of production of O$_3$ (see Fig. 1), but do not provide a basis for evaluating long-term ozone chemistry and transport. Ozone concentrations at individual locations are the result of photochemical production that has occurred over several hours (or sometimes 2–3 d) in a moving air mass. NO$_x$ has a photochemical lifetime of 2–4 h, and some important hydrocarbons (e.g. isoprene) have an even shorter lifetime. Ambient VOC and NO$_x$ represent immediate local conditions rather than the history of ozone production in the air mass. Chameides et al. (1992), Tonnesen and Dennis (1998) and Sillman (1998) found that the ratio of reactivity-weighted VOC to NO$_x$ was strongly correlated with the NO$_x$–VOC dependence associated with instantaneous rates of ozone production. However, Sillman (1998) failed to find a similar correlation between VOC/NO$_x$ ratios and NO$_x$–VOC sensitivity associated with ozone concentrations in 3D models.

Because ambient VOC and NO$_x$ are correlated with instantaneous rather than long-term ozone chemistry, attempts to use ambient VOC and NO$_x$ to evaluate the sensitivity of ozone concentrations to NO$_x$ and VOC often implicitly assume a specific pattern of emissions and transport history. For example, the old rule that used morning VOC/NO$_x$ ratios to identify NO$_x$-sensitive vs. VOC-sensitive conditions, described in Section 3, was based on an assumed pattern of transport and diffusion of air as it left an urban center, including zero

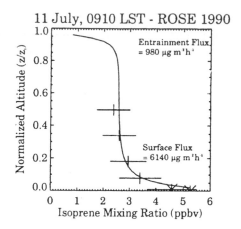

Figure 10. Observed isoprene mixing ratio (ppb) vs. altitude (crosses and asterisks) and fitted theoretical vertical profile based on flux-gradient calculations. Error bars represent theoretical variances. Altitude (vertical axis) is expressed relative to z_i, the daytime convective mixing height, which is typically 500–2000 m. From Davis et al. (1994).

downwind emissions and zero biogenics. The method developed by Cardelino et al. is more sophisticated, but the method of aggregating the total response to reduced NO_x and VOC may be dependent on assumed patterns of transport and geography. The method has been tested against results from a 3D Eulerian model only for a single event in Atlanta.

A second problem associated with NO_x–VOC sensitivity evaluations based on ambient NO_x and VOC concerns the impact of vertical mixing and surface emissions. Measured VOC and NO_x are typically available only at sites near the ground, and measured concentrations are influenced by near-surface emissions and the rate of vertical diffusion away from the surface. By contrast, the process of ozone formation typically takes place in a convective mixed layer which extends 500 m above the surface in Los Angeles and 1000–2000 m above the surface in most continental cities in the US and Europe. During conditions associated with elevated O_3 (that is, sunny afternoons) there is typically little variation in the concentration of ozone vs. height, but primary NO_x and VOC may have significantly higher concentrations near the surface than throughout the convective mixed layer.

The variation in species concentrations with height within the convective mixed layer has been studied most intensely for isoprene at rural locations (Davis et al., 1994; Andronache et al., 1994; Guenther et al., 1996a, b). Davis et al. (1994) and Guenther et al. (1996a) have both found that the rate of decrease of isoprene with height during convective conditions can be approximated by using a mixed layer gradient and other mesoscale models to represent the rate

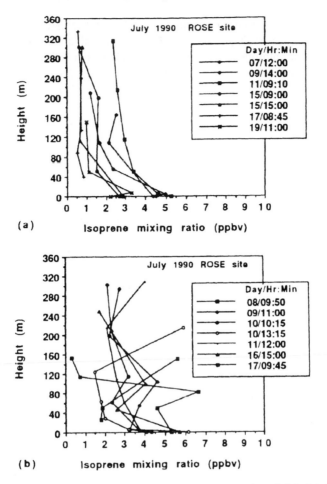

Figure 11. Variations in the observed vertical profile of isoprene (ppb) vs. height (m): (a) profiles with relatively simple patterns, (b) more complex patterns. The times denote the hour and minute of the start of each air sampling segment. Measurements were made at Rose, AL, in July 1990. From Andronache et al. (1994).

of near-surface mixing (see Fig. 10). However, Andronache et al. (1994) found considerable day-to-day variation in observed vertical profiles for isoprene and also found some cases with maximum isoprene at 100 m rather than near the surface (see Fig. 11). Guenther et al. (1996a, b) report that mixed-layer average concentration for isoprene is 38–58% lower than the near-surface (0–150 m) concentration. These ratios, along with the pattern shown in Fig. 11, might be used as a basis for estimating mixed-layer average concentrations of isoprene when only surface measurements are available. Typically, mixed-layer average values must be used when ambient VOC and NO_x are used to evaluate NO_x–VOC chemistry. There have been few studies that would identify vertical profiles for anthropogenic VOC in urban areas, but Sillman et al. (1995a) reported that the xylenes vary with height in a way similar to isoprene.

Apart from the attempts to evaluate NO_x–VOC chemistry, ambient NO_x and VOC are widely used as a basis for evaluating the accuracy of emission inventories and for evaluating 3D ozone models. The most direct way to evaluate emission inventories is to compare measured NO_x and VOC with model results (e.g. Harley et al., 1993, 1995; Jacobson et al., 1996). Chang et al. (1997) also developed a series of calculations in which the emission inventory used in a 3D model were modified to obtain the closest possible agreement with ambient measurements. The method described by Chang et al. (1997) implicitly assumes that emission inventories are uncertain and derives emission rates from model analyses of ambient measurements. In all the model-measurement comparisons for VOC and NO_x, it is often uncertain whether differences between models and measurements are due to errors in the emission inventory or other factors (rates of vertical mixing, horizontal transport, impact of direct emission sources near the measurement site, etc.).

A variety of methods have been developed for evaluating emission inventories directly from measured VOC and NO_x, often based on measured ratios between species. The morning VOC/NO_x ratio in urban centers has frequently been used to evaluate emissions ratios (e.g. Fujita et al., 1992). Early morning measurements are used for this purpose because chemical losses are relatively small at that time and ambient concentrations reflect the ratio of directly emitted species. Profiles of speciated VOC have also been used along with the general techniques of chemical mass balance and receptor modeling to identify emission sources of VOC (Henry et al., 1984). Parrish et al. (1991) used ambient CO/NO_y, ratios at rural sites (where NO_y represents the sum of reactive nitrogen species, including NO_x, HNO_3, PAN and other organic nitrates) as a basis for evaluating the CO/NO_x emissions ratio. Goldan et al. (1995) used CO/NO_y, speciated VOC/NO_y and VOC/VOC ratios for individual VOC species to evaluate urban emissions ratios. McKeen et al. (1996) extended the techniques for interpreting ambient VOC/VOC ratios in remote locations, accounting for photochemical losses and dilution as emission sources move downwind. Buhr et al. (1995) was able to derive information about emission inventories and NO_x–VOC sensitivity by performing a principal component analysis on a set of measured VOC, NO_x, NO_y and CO. These evaluations based on ambient species ratios (especially the comprehensive evaluation in Goldan et al., 1995) provide an excellent basis for evaluating urban emission inventories (see Fig. 12). In the case of isoprene and other biogenic hydrocarbons, Guenther et al. (1996b) also developed methods for deriving surface fluxes based on measured vertical profiles.

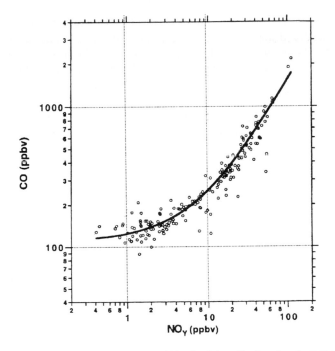

Figure 12. Observed mixing ratio of CO vs. NO$_x$ (both in ppb) plotted on logarithmic scales. The line represents a linear fit. [CO] = 110 + 14.4 NO$_y$. Measurements are from February 1991 in Boulder, CO. From Goldan et al. (1995).

5.2. Evaluations based on secondary reaction products: reactive nitrogen and peroxides

Methods for evaluating ozone chemistry based on secondary reaction products fall into two broad categories: methods to evaluate the ozone production efficiency (e.g. Trainer et al., 1993) and methods to evaluate O$_3$–NO$_x$–VOC sensitivity directly (e.g. Sillman, 1995).

Trainer et al. (1993) found that in rural areas during periods of photochemical activity, a strong correlation is found between measured ozone and the sum of NO$_x$ reaction products (NO$_y$–NO$_x$, or NO$_z$). Sillman et al. (1990) and Trainer et al. (1993) tried to use the positive correlation between O$_3$ and NO$_y$, and between O$_3$ and NO$_z$ as evidence for NO$_x$-sensitive chemistry. However, the main use of the O$_3$–NO$_z$ correlation, proposed by Trainer et al. (1993), has been to estimate the ozone production efficiency. Because NO$_z$ represents the sum of species produced by the removal of NO$_x$, the notion that ozone production efficiency is associated with the O$_3$–NO$_z$ slope follows directly from the definition of ozone production efficiency (Section 4).

Since the initial work by Trainer et al. (1993) there have been extensive measurement and analysis of the correlation between O_3 and NO_z at rural sites in the US and in a few urban areas in the US and Europe (Olszyna et al., 1994; Kleinman et al., 1994; Jacob et al., 1995; Trainer et al., 1995; Daum et al., 1996; Hirsch et al., 1996; Prevot et al., 1997; Ridley et al., 1998; Staffel-bach et al., 1998; Dommen et al., 1998). Slopes between O_3 and NO_z during periods of photochemical activity typically range from 5 to 10. Several authors (e.g. Chin et al., 1994; Jacob et al., 1995) expressed concern that the true ozone production efficiency is lower than the observed O_3–NO_z slope because NO_z species (chiefly HNO_3) are removed from the atmosphere more rapidly than ozone. Ridley et al. (1994, 1998), Atherton et al. (1996) and Roberts et al. (1996) also found that the O_3–NO_z slope is higher in air that is several days downwind from emission sources, presumably due to removal of NO_z as the air travels downwind. Recently Nunnermacker et al. (1998), Ryerson et al. (1998) and Sillman et al. (1998) estimated that ozone production efficiencies in the eastern US were approximately 3, significantly lower than the observed O_3–NO_z slope. Chin et al. (1994) and Hirsch et al. (1996) also derived ozone production efficiencies based on the observed slope between O_3 and CO ($\Delta O_3 / \Delta CO = 0.3$), using an assumed emissions ratio for CO/NO_x. A more extensive review of this subject is found in Trainer et al. (1998).

The concept of evaluating ozone–NO_x–VOC sensitivity directly from measured reactive nitrogen and other secondary reaction products was developed by Milford et al. (1994), Sillman (1995, 1998) and Sillman et al. (1997, 1998). They found that NO_x-sensitive conditions in 3D models for ozone nearly always coincided with predicted high values for certain species ratios, and that VOC-sensitive conditions in models nearly always coincided with low values for the same ratios. They proposed that ratios with this type of behavior might be regarded as "indicators" for ozone–NO_x–VOC sensitivity. Ambient measurements of the indicator ratios would be interpreted as evidence for NO_x-sensitive or VOC-sensitive conditions if the measurements corresponded to values that were associated with NO_x-sensitive or VOC-sensitive conditions in models. The indicator ratios identified by Sillman et al. (1997) were O_3/NO_y, O_3/NO_z, H_2O_2/HNO_3, and other similar ratios involving ozone, reactive nitrogen and peroxides. The ratio H_2O_2/HNO_3 in particular is closely related to the chemical factors that create the split between NO_x-sensitive and VOC-sensitive conditions (Section 4). In contrast with VOC–NO_x ratios (Fig. 1) or with production rates of nitric acid and peroxides (Fig. 8), the proposed indicators represent relatively long-lived species (12 h or more during daytime) and are associated with the long-term process of ozone formation rather than instantaneous ozone chemistry.

The indicator concept is illustrated in Fig. 13, which shows the predicted reduction in O_3 in response to reduced NO_x and reduced anthropogenic VOC in

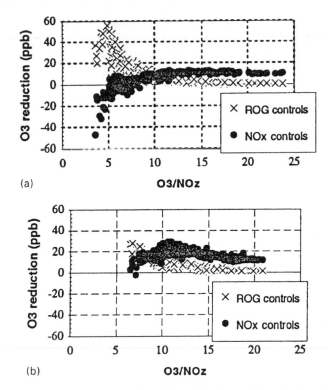

Figure 13.   Predicted reduction in peak O$_3$ (ppb) resulting from a 35% reduction in the emission rate for anthropogenic VOC (crosses) and from a 35% reduction in the emission rate for NO$_x$ (circles), plotted against predicted O$_3$/NO$_z$ concurrent with the ozone peak, in simulations for (a) Chicago/Lake Michigan and (b) New York/northeast corridor. From Sillman (1995) based on models described in Sillman et al. (1993).

3D Eulerian models for the Chicago/Lake Michigan region and for the northeast corridor (New York to Boston) in the US. This figure also serves to illustrate the sharply divergent responses to reduced NO$_x$ and VOC associated with the NO$_x$-sensitive and VOC-sensitive regimes. The model for Lake Michigan includes a large VOC-sensitive region in which reduced VOC would cause a large reduction in O$_3$, while reduced NO$_x$ would cause either little change or an increase in O$_3$. The model for the northeast corridor includes a large NO$_x$-sensitive region in which reduced NO$_x$ would cause a large reduction in O$_3$ and reduced VOC would cause little change in O$_3$. In both models the locations with predicted VOC-sensitive conditions also have low values (< 8) for the ratio O$_3$/NO$_z$, while the locations with predicted NO$_x$-sensitive conditions have high values (> 10) for O$_3$/NO$_z$. This result suggests that the model NO$_x$–VOC predictions could be confirmed if measured indicator ratios showed low values in predicted VOC-sensitive locations and high values in predicted NO_x-sensitive locations. Conversely, measured indicator ratios that contrast with model predictions (NO_x-sensitive indicator values vs. VOC-sensitive model prediction, or vice versa) would provide evidence for erroneous NO_x–VOC predictions. In contrast with Trainer et al. (1993), this interpretation is based on the ratio O_3/NO_z rather than the slope of the correlation between O_3 and NO_z.

Ambient measurements show that the proposed indicator ratios vary in a way that is consistent with the hypothesized difference between NO_x-sensitive and VOC-sensitive locations. High values for several indicator ratios were observed at rural sites in Colorado (Watkins et al., 1995) and in the eastern US (Jacob et al., 1995; Daum et al., 1996; Sillman et al., 1998) and in Atlanta (Sillman et al., 1995a, 1997), consistent with expected NO_x-sensitive conditions in those locations. Low values for the indicator ratios were observed in Los Angeles (Sillman, 1995; Sillman et al., 1997), consistent with expected VOC-sensitive conditions there. Measurements in the urban plume from Milan, Italy, showed values consistent with VOC-sensitive chemistry 1–3 h downwind of Milan, transitional or NO_x-sensitive chemistry further downwind in the Milan plume, NO_x-sensitive chemistry in the surrounding rural region (Prevot et al., 1997; Staffelbach et al., 1997). Dommen et al. (1995, 1998), (see also Kuebler et al., 1996) also found indicator values suggestive of NO_x-sensitive chemistry over most of the Swiss plateau. Jacob et al. (1995) also reported a large shift in measured values of the ratios O_3/NO_z and H_2O_2/NO_z between summer and autumn at a rural site in Virginia, consistent with a hypothesized seasonal transition from NO_x-sensitive chemistry in summer to VOC-sensitive chemistry in autumn.

In a case study for Atlanta, measured O_3 and NO_y was used as a basis for evaluating NO_x–VOC predictions from a series of model scenarios. As illustrated in Fig. 5 and discussed in Section 3, predictions for the impact of reduced NO_x and VOC in Atlanta varied greatly among model scenarios with different assumptions about emissions and meteorology. As reported in Sillman et al. (1995a, 1997), the model scenarios with predicted NO_x-sensitive chemistry showed good agreement with measured O_3 and NO_y. The scenario with predicted VOC-sensitive chemistry showed good agreement with measured O_3 but seriously underestimated the indicator ratio O_3/NO_y (measured $O_3/NO_y = 14$, model $O_3/NO_y = 6$). If the model scenarios with erroneous O_3/NO_y were rejected, the remaining model scenarios would show much less variation in their NO_x–VOC predictions. Thus, the use of measured indicator ratios has the potential to reduce the uncertainty associated with model NO_x–VOC predictions.

Deposition and other removal processes represent a major problem for the proposed NO_x–VOC indicators. The indicator ratios all involve species

(HNO_3, H_2O_2) that are rapidly removed by wet deposition. They also have relatively rapid dry deposition rates and may be subject to removal through interaction with aerosols. It is possible that the indicator–NO_x–VOC correlations may change with time of day or with aging as an air mass moves downwind. The indicator ratios may also be affected by uncertain peroxide chemistry. Lu and Chang (1998) reported a case in which the correlation between NO_x–VOC predictions and indicator ratios was significantly different from the results reported by Sillman. A more complete description of these issues is presented in Sillman (1998).

Because removal of NO_y is a critical uncertainty for both the ozone production efficiency and for the NO_x–VOC indicators, several efforts have been made to estimate this removal rate. Munger et al. (1996, 1998) used eddy covariance methods to derive removal rates for NO_y from field measurements. Hall and Claiborn (1997) also measured deposition of H_2O_2. Ryerson et al. (1998) and Nunnermacker et al. (1998) both derived removal rates for NO_x by evaluating changes in the ratios CO/NO_x and SO_2/NO_x in urban and power plant plumes. In each case, increases in the ratio as the plume moved downwind was interpreted as evidence for the removal of NO_x. These methods are expected to be used and refined in future research.

The ratio $O_3/(NO_z + 2H_2O_2)$ has been proposed as a critical test for the proposed NO_x–VOC indicators. Sillman (1995) and Sillman et al. (1998) found that this ratio assumed a near-constant value ($= 6$–7) in photochemical models for conditions with high ozone (that is, sunny, warm afternoons). This ratio is related to radical chemistry, since O_3 represents the major source or radicals and NO_z and H_2O_2 represent major sinks. Unlike the proposed indicator ratios, the ratio $O_3/(NO_z + 2H_2O_2)$ does not appear to change when model conditions vary from NO_x-sensitive to VOC-sensitive. However, the ratio $O_3/(NO_z + 2H_2O_2)$ is sensitive to removal rates and peroxide chemistry, both of which represent major uncertainties for the indicator ratios. If measured values for $O_3/(NO_z + 2H_2O_2)$ differ from the model values in Sillman et al. (1998), it would suggest that the indicator ratios O_3/NO_z and H_2O_2/NO_z are affected by processes that were not accurately represented in the models used to derive the indicator interpretation, and their use would be suspect.

The relation between ozone, reactive nitrogen and peroxides is conveniently summed up in Fig. 14, which shows correlations in a polluted air mass from the eastern US as it travels over the nearby Atlantic Ocean (Daum et al., 1996). Ozone increases with NO_z throughout the air mass, but the O_3–NO_z slope decreases as O_3 and NO_z get higher. This pattern of decreasing O_3–NO_z slope appears in some but not all sets of measurements. Following Trainer et al. (1993), the O_3–NO_z slope would be interpreted as evidence for the ozone production efficiency in the air mass. The decrease in slope might suggest that ozone production efficiency is lower in highly polluted regions. However the

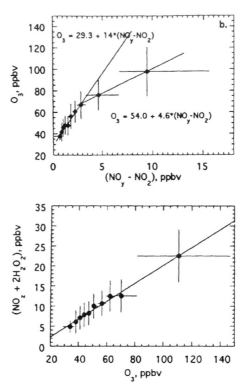

Figure 14. Measured O_3 as a function of $(NO_y–NO_2)$ (a, upper panel), and the sum $(NO_z + 2H_2O_2)$ as a function of O_3 (b, lower panel), all in ppb. Measurements were made the North Atlantic Ocean downwind from the northeast corridor of the US. Measurements are presented as bins in which the original data set was ordered based on the abcissa and then divided into ten intervals. Horizontal bars represent the range of abcissa values; vertical bars represent the standard deviation of the variable on the vertical axis for each interval. Linear fits with slopes and intercepts were calculated from the individual data points. From Daum et al. (1996).

decrease in slope might also be caused by a higher removal rate of NO_z in the regions with lower O_3 and NO_z (presumably because these regions are further from emission sources) rather than by differences in ozone production efficiency. Following Sillman (1995), the ratios O_3/NO_z and H_2O_2/NO_z (not the $O_3–NO_z$ slope) would be interpreted as indicators for NO_x–VOC sensitivity. In this view the split between NO_x-sensitive and VOC-sensitive regimes is caused by differences in radical chemistry (as evidenced by changes in the indicator ratios) and not by differences in the ozone production efficiency. The high O_3/NO_z and H_2O_2/NO_z in the region with lower O_3 would be indicative of NO_x-sensitive chemistry while the lower values ($O_3/NO_z = 10$, $H_2O_2/NO_z = 0.3$) in the more polluted region would suggest conditions that are closer to the transition between NO$_x$-sensitive and VOC-sensitive chemistry. This interpretation is also dependent on assumptions about removal rates for NO$_z$ and H$_2$O$_2$. The linear correlation between O$_3$ and the sum NO$_z$ + 2H$_2$O$_2$ with near-constant ratio (= 5) would be interpreted as evidence that the NO$_x$–VOC interpretation of the indicator ratios is valid in this case.

6. Conclusions

This review of ozone–NO$_x$–VOC sensitivity has emphasized three themes.

(1) The relation between ozone, NO$_x$ and VOC can be understood in terms of a few theoretical concepts. These include: the split into VOC-sensitive (or NO$_x$-saturated) and NO$_x$-sensitive photochemical regimes; the evolution from VOC-sensitive to NO$_x$-sensitive chemistry as a plume moves downwind; the role of odd hydrogen radicals and the supply of radicals relative to NO$_x$; and the ozone production efficiency.

(2) Predictions for the impact of reduced NO$_x$ and VOC on ozone derived from 3D Eulerian photochemical models have large uncertainties. The uncertainty in model predictions is associated specifically with the difference between the NO$_x$-sensitive and VOC-sensitive regimes, and can be quantified by comparing NO$_x$–VOC predictions from different model scenarios that reflect uncertainties in emission rates and meteorology.

(3) The uncertainties associated with NO$_x$–VOC predictions can be significantly reduced if investigations place greater emphasis on ambient measurements rather than just models. Observation-based techniques have been developed that seek to identify NO$_x$-sensitive or VOC-sensitive chemistry and also to evaluate emission inventories, ozone production efficiency and removal rates for chemically active species.

Measurement-based studies have traditionally received heavy emphasis as part of investigations into the chemistry of the remote troposphere and at rural sites. Rural and remote sites have been the subject of frequent measurement intensives (e.g. Hoell et al., 1996; Fehsenfeld et al., 1996) or long-term monitoring of chemically active species (e.g. Trainer et al., 1993). By contrast, investigations of urban chemistry, especially in the US, has emphasized the development and use of photochemical models. These model-based studies are often closely associated with specific issues of regulatory policy (e.g. Hanna et al., 1996). Consequently, the question of ozone–NO$_x$–VOC sensitivity is sometimes viewed as primarily a question of policy rather than a subject for scientific investigation. The view presented here is that model predictions for ozone–NO$_x$–VOC sensitivity should be regarded as scientific hypotheses, and their validity must be established by comparison with ambient measurements. These model-measurement comparisons should be designed specifically to evaluate the accuracy of model NO_x–VOC predictions or to evaluate critical model assumptions and results (e.g. emission inventories, ozone production efficiencies), and should involve species other than just O_3. They should also be combined with measurement-based investigations that are comparable to recent efforts at rural and remote sites.

Investigation of ozone–NO_x–VOC sensitivity has been given impetus by its close connection to regulatory policy, but this can also be a disadvantage. Research in this field has been influenced by political considerations to a much greater extent than normally occurs in the geophysical sciences. This review article has sought to demonstrate that the issue of ozone–NO_x–VOC chemistry can and should be addressed as a purely scientific issue, which is separate from questions of policy.

Acknowledgements

This research is part of the Southern Oxidants Study (SOS) – a collaborative university, government, and private industry study to improve scientific understanding of the accumulation and effects of photochemical oxidants. Financial and in-kind support for SOS research and assessment activities is provided by the Environmental Protection Agency, National Oceanic and Atmospheric Administration, Department of Energy, Tennessee Valley Authority, Electric Power Research Institute, The Southern Company, Coordinating Research Council, and the States of Alabama, Florida, Georgia, Kentucky, Louisiana, Mississippi, North Carolina, South Carolina, Tennessee, and Texas. Although the research described in this article has been funded wholly or in part by the Environmental Protection Agency under Assistance Agreement No. CR818336 to University of Alabama in Huntsville, it has not been subjected to the Agency's peer and administrative review, and therefore may not necessarily reflect the views of the Agency, and no official endorsement should be inferred. Support was also provided by the National Science Foundation under grant #ATM-9713567.

References

Al-Wali, K.I., Samson, P.J., 1996. Preliminary sensitivity analysis of Urban Airshed Model simulations to temporal and spatial availability of boundary layer wind measurements. Atmospheric Environment 30 (12), 2027–2042.

Andronache, C., Chameides, W.L., Rodgers, M.O., Martinez, J.E., Zimmerman, P., Greenberg, J., 1994. Vertical distribution of isoprene in the lower boundary layer of the rural and urban southern United States. Journal of Geophysical Research 99, 16,989–17,000.

Atherton, C.S., Sillman, S., Walton, J., 1996. Three dimensional global modeling studies of transport and photochemistry over the North Atlantic Ocean. Journal of Geophysical Research 101, 29,289–29,304.

Bascomb, R., Bromberg, P.A., Costa, D.L., Devlin, R., Dockery, D.W., Frampton, M.W., Lambert, W., Samet, J.M., Speizer, F.E., Utell, M., 1996. Health effects of outdoor air pollution. American Journal of Respiration and Critical Care in Medicine 153, 477–498.

Buhr, M., Parrish, D., Elliot, J., Holloway, J., Carpenter, J., Goldan, P., Kuster, W., Trainer, M., Montzka, S., McKeen, S., Fehsenfeld, F.C., 1995. Evaluation of ozone precursor source types using principal component analysis of ambient air measurements in rural Alabama. Journal of Geophysical Research 100, 22,853–22,860.

Cardelino, C.A., Chameides, W.L., 1998. The application of the observation based model to atmospheric measurement datasets. Atmospheric Environment in association with the NARSTO critical review series (submitted).

Cardelino, C., Chameides, W.L., 1995. An observation-based model for analyzing ozone-precursor relationships in the urban atmosphere. Journal of Air and Waste Management Association 45, 161–180.

Cardelino, C., Chang, W.L., Chang, M.E., 1994. Comparison of emissions inventory estimates and ambient concentrations of ozone precursors in Atlanta, Georgia. Presented at the Air and Waste Management Association International Conference on the Emission Inventory: Applications and Improvement, Raleigh, NC. 1–3 November 1994.

Cardellino, C.A., Chameides, W.L., 1990. Natural hydrocarbons, urbanization, and urban ozone. Journal of Geophysical Research 95, 13,971–13,979.

Carter, W.P.L., 1994. Development of ozone reactivity scales for volatile organic compounds. Journal of Air and Waste Management Association 44, 881–899.

Carter, W.P.L., 1995. Computer modeling of environmental chamber studies of maximum incremental reactivities of volatile organic compounds. Atmospheric Environment 29–18, p. 2513.

Chameides, W.L., Fehsenfeld, F., Rodgers, M.O., Cardellino, C., Martinez, J., Parrish, D., Lonneman, W., Lawson, D.R., Rasmussen, R.A., Zimmerman, P., Greenberg, J., Middleton, P., Wang, T., 1992. Ozone precursor relationships in the ambient atmosphere. Journal of Geophysical Research 97, 6037–6056.

Chameides, W.L., Lindsay, R.W., Richardson, J., Kiang, C.S., 1988. The role of biogenic hydrocarbons in urban photochemical smog: Atlanta as a case study. Science 241, 1473–1474.

Chang, M.E., Hartley, D., Cardelino, C., Chang, W.-L., 1997. Temporal and spatial distribution of biogenic emissions of isoprene based on an inverse method using ambient isoprene observations from the 1992 Southern Oxidants Study, Atlanta Intensive Atmospheric Environment (in press).

Chin, M., Jacob, D.J., Munger, J.W., Parrish, D.D., Doddridge, B.G., 1994. Relationship of ozone and carbon monoxide over North America. Journal of Geophysical Research 99, 14,565–14,573.

Clarke, J.F., Ching, J.K.S., 1983. Aircraft observations of regional transport of ozone in the northeastern United States. Atmospheric Environment 17, 1703–1712.

Cleveland, W.S., Kleiner, B., McRae, J.E., Warner, J.L., 1976. Photochemical air pollution: transport from the New York City area into Connecticut and Massachusetts. Science 191, 179–181.

Daum, P.H., Kleinman, L.I., Newman, L., Luke, W.T., Weinstein-Lloyd, J., Berkowitz, C.M., Busness, K.M., 1996. Chemical and physical properties of anthropogenic pollutants transported over the North Atlantic during NARE. Journal of Geophysical Research 101, 29,029–29,042.

Davis, K.J., Lenschow, D.H., Zimmerman, P.R., 1994. Biogenic nonmethane hydrocarbon emissions estimated from tethered balloon observations. Journal of Geophysical Research 99, 25,587–25,598.

Dommen, J., Neftel, A., Sigg, A., Jacob, D.J., 1995. Ozone and hydrogen peroxide during summer smog episodes over the Swiss Plateau: measurements and model simulations. Journal of Geophysical Research 100, 8953–8966.

Dommen, J., Prevot, A.S.H., Hering, A.M., Staffelbach, T., Kok, G.L., Schillawski, R.D., 1998. Photochemical production and the aging of an urban air mass. Journal of Geophysical Research (submitted).

Drummond, J., Schiff, H., Karecki, D., Mackay, G., 1989. Measurements of NO_2, O_3, PAN, HNO_3, H_2O_2 and H_2CO during the Southern California Air Quality Study. Presented at the 82nd annual meeting of the Air and Waste Management Association, Anaheim, CA, 1989; Paper 89–139.4.

Duderstadt, K.A., Carroll, M.A., Sillman, S., Wang, T., Albercook, G.M., Feng, L., Parrish, D.D., Holloway, J.S., Fehsenfeld, F., Blake, D.R., Blake, N.J., Forbes, G., 1998. Photochemical production and loss rates at Sable Island, Nova Scotia during the North Atlantic Regional Experiment 1993 Summer Intensive. Journal of Geophysical Research 103, 13,531–13,555.

Fehsenfeld, F.C., Daum, P., Leaitch, W.R., Trainer, M., Parrish, D.D., Hubler, G., 1996. Transport and processing of O_3 and O_3 precursors over the North Atlantic: an overview of the 1993 North Atlantic Regional Experiment (NARE) summer intensive. Journal of Geophysical Research 101, 28,877–28,891.

Fiore, A.M., Jacob, D.J., Logan, J.A., Yin, J.H., 1998. Long-term trends in ground level ozone over the contiguous United States, 1980–1995. Journal of Geophysical Research 103, 1471–1480.

Fujita, E.M., Croes, B.E., Bennett, C.L., Lawson, D.R., Lurmann, F.W., Main, H.H., 1992. Comparison of emission and ambient concentration ratios of CO, NO_x, and NMOG in California's south coast air basin. Journal of Air and Waste Management Association 42, 264–276.

Gao, D., Stockwell, W.R., Milford, J.B., 1996. Global uncertainty analysis of a regional-scale gas phase chemical mechanism. Journal of Geophysical Research 101, 9071–9078.

Geron, C.D., Guenther, A.B., Pierce, T.E., 1994. An improved model for estimating emissions of volatile organic compounds from forests in the eastern United States. Journal of Geophysical Research 99, 12,773–12,791.

Geron, C.D., Pierce, T.E., Guenther, A.B., 1995. Reassessment of biogenic volatile organic compound emissions in the Atlanta area. Atmospheric Environment 29, 1569–1578.

Gillani, N.V., Pleim, J.E., 1996. Sub-grid-scale features of anthropogenic emissions of NO_x and VOC in the context of regional Eulerian models. Atmospheric Environment 30, 2043–2059.

Giovannoni, J.-M., Russell, A., 1997. Impact of using prognostic and objective wind fields on the photochemical modeling of Athens, Greece. Atmospheric Environment 29, 3633–3654.

Goldan, P.D., Trainer, M., Kuster, W.C., Parrish, D.D., Carpenter, J., Roberts, J.M., Yee, J.E., Fehsenfeld, F.C., 1995. Measurements of hydrocarbons, oxygenated hydrocarbons, carbon monoxide, and nitrogen oxides in an urban basin in Colorado: implications for emission inventories. Journal of Geophysical Research 100, 22,771–22,783.

Guenther, A.B. et al., 1995. A global model of natural volatile organic compound emissions. Journal of Geophysical Research 100, 8873–8892.

Guenther, A., Zimmerman, P., Klinger, L., Greenberg, J., Ennis, C., Davis, K., Pollock, W., Westberg, H., Allwine, G., Geron, C., 1996a. Estimates of regional natural volatile organic compound fluxes from enclosure and ambient measurements. Journal of Geophysical Research 101, 1345–1359.

Guenther, A., Baugh, W., Davis, K., Hampton, G., Harley, P., Klinger, L., Vierling, L., Zimmerman, P., Allwine, E., Dilts, S., Lamb, B., Westberg, H., Baldocchi, D., Geron, C., Pierce, T., 1996b. Isoprene fluxes measured by enclosure, relaxed eddy accumulation, surface layer gradient, mixed layer gradient, and mixed layer mass balance techniques. Journal of Geophysical Research 101, 18,555–18,567.

Guicherit, R., Van Dop, H., 1977. Photochemical production of zone in Western Europe (1971–1975) and its relation to meteorology. Atmospheric Environment 11, 145–155.

Haagen-Smit, A.J., Fox, M.M., 1954. Photochemical ozone formation with hydrocarbons and automobile exhaust. Journal of Air Pollution Control Association 4, 105–109.

Hall, B.D., Claiborn, C.S., 1997. Measurements of the dry deposition of peroxides to a Canadian boreal forest. Journal of Geophysical Research 102, 29,343–29,353.

Hanna, S.R., Moore, G.E., Fernau, M.E., 1996. Evaluation of photochemical grid models (UAM-IV, UAM-V, and the ROM/UAM-IV couple) using data from the Lake Michigan Ozone Study (LMOS). Atmospheric Environment 30, 3265–3279.

Harley, R.A., Russell, A.G., McRae, G.J., Cass, G.R., Seinfeld, J.H., 1993. Photochemical modeling of the Southern California Air Quality Study. Environmental Science and Technology 27, 378–388.

Harley, R.A., Cass, G.R., 1995. Modeling the atmospheric concentrations of individual volatile organic compounds. Atmospheric Environment 29, 905–922.

Henry, R.C., Lewis, C.W., Hopke, P.K., Williamson, H.W., 1984. Review of receptor modeling fundamentals. Atmospheric Environment 18, 1507–1515.

Hess, G.D., Carnovale, F., Cope, M.E., Johnson, G.M., 1992. The evaluation of some photochemical smog reaction mechanisms-I. Temperature and initial composition effects. Atmospheric Environment 26A, 625–641.

Hirsch, A.I., Munger, J.W., Jacob, D.J., Horowitz, L.W., Goldstein, A.H., 1996. Seasonal variation of the ozone production efficiency per unit NO$_x$ at Harvard Forest, Massachusetts. Journal of Geophysical Research 101, 12,659–12,666.

Hoell, J.M., Davis, D.D., Liu, S.C., Newell, R., Shipham, M., Akimoto, H., McNeal, R.J., Bendura, R.J., Drewry, J.W., 1996. Pacific Exploratory Mission-West A (PEM-West A): September–October 1991. Journal of Geophysical Research 101, 1641–1655.

Jacob, D.J., Heikes, B.G., Dickerson, R.R., Artz, R.S., Keene, W.C., 1995. Evidence for a seasonal transition from NO$_x$- to hydrocarbon-limited ozone production at Shenandoah National Park, Virginia. Journal of Geophysical Research 100, 9315–9324.

Jacob, D.J., Logan, J.A., Gardner, G.M., Yevich, R.M., Spivakowsky, C.M., Wofsy, S.C., Sillman, S., Prather, M.J., 1993. Factors regulating ozone over the United States and its export to the global atmosphere. Journal of Geophysical Research 98, 14,817–14,827.

Jacobson, M.Z., Lu, R., Turco, R.P., Toon, O.P., 1996. Development and application of a new air pollution modeling system – Part I: gas-phase simulations. Atmospheric Environment 30, 1939–1963.

Jaegle, L., Jacob, D.J., Brune, W.H., Tan, D., Faloona, I., Weinheimer, A.J., Ridley, B.A., Campos, T.L., Sachse, G.W., 1998. Sources of HO$_x$ and production of ozone in the upper troposphere over the United States. Geophysical Research Letters 25, 1705–1708.

Johnson, G.M., 1984. A simple model for predicting the ozone concentration of ambient air. Proc. 8th Int. Clean Air Conf. Melbourne, Australia, May 2, pp. 715–731.

Johnson, G.M., Quigley, S.M., Smith, G.J., 1990. Management of photochemical smog using the AIRTRAK approach. 10th International Conference of the Clean Air Society of Australia and New Zealand, Auckland, New Zealand, March, pp. 209–214.

Kanakidou, M., Singh, H.B., Valentin, K.M., Crutzen, P.J., 1991. A two-dimensional study of ethane and propane oxidation in the troposphere. Journal of Geophysical Research 96, 15,395–15,414.

Kleinman, L.I., 1986. Photochemical formation of peroxides in the boundary layer. Journal of Geophysical Research 91, 10,889–10,904.

Kleinman, L.I., 1991. Seasonal dependence of boundary layer peroxide concentration: the low and high NO$_x$ regimes. Journal of Geophysical Research 96, 20,721–20,734.

Kleinman, L.I., 1994. Low and high-NO$_x$ tropospheric photochemistry. Journal of Geophysical Research 99, 16,831–16,838.

Kleinman, L.I., Lee, Y.-N., Springston, S.R., Nunnermacker, L., Zhou, X., Brown, R., Hallock, K., Klotz, P., Leahy, D., Lee, J.H., Newman, L., 1994. Ozone formation at a rural site in the southeastern United States. Journal of Geophysical Research 99, 3469–3482.

Kleinman, L.I., Daum, P.H., Lee, J.H., Lee, Y.-N., Nunnermacker, L.J., Springston, S.R., Newman, L., Weinstein-Lloyd, J., Sillman, S., 1997. Dependence of ozone production on NO and hydrocarbons in the troposphere. Geophysical Research Letters 24, 2299–2302.

Kleinman, L.I., 1998. Observation based analysis for ozone production. Atmospheric Environment as part of the NARSTO critical review series (submitted).

Kuebler, J., Giovannoni, J.-M., Russell, A.G., 1996. Eulerian modeling of photochemical pollutants over the Swiss Plateau and control strategy analyses. Atmospheric Environment 30, 951–966.

Kumar, N., Odman, M.T., Russell, A.G., 1994. Multiscale air quality modeling: application to southern California. Journal of Geophysical Research 99, 5385–5397.

Kumar, N., Russell, A.G., 1996. Comparing prognostic and diagnostic meteorological fields and their impacts on photochemical air quality modeling. Atmospheric Environment 30, 1989–2010.

Lamb, B., Westberg, H., Allwine, G., Quarles, T., 1985. Biogenic hydrocarbon emissions from deciduous and coniferous trees in the United States. Journal of Geophysical Research 90, 2380.

Lawson, D.R., 1990. The Southern California air quality study. Journal of Air and Waste Management Association 40,156–165.

Lippman, M., 1993. Health effects of tropospheric ozone: review of recent research findings and their implications to ambient air quality standards. Journal of Exposure Anal. Environment Epidemiol. 3, 103–128.

Lin, X., Trainer, M., Liu, S.C., 1988. On the nonlinearity of tropospheric ozone. Journal of Geophysical Research 93, 15,879–15,888.

Liu, S.C., Trainer, M., Fehsenfeld, F.C., Parrish, D.D., Williams, E.J., Fahey, D.W., Hubler, G., Murphy, P.C., 1987. Ozone production in the rural troposphere and the implications for regional and global ozone distributions. Journal of Geophysical Research 92, 4191–4207.

Logan, J.A., 1989. Ozone in rural areas of the United States. Journal of Geophysical Research 94, 8511–8532.

Lu, R., Turco, R.P., 1995. Air pollution transport in a coastal environment: Part II: three-dimensional simulations over the Los Angeles basin. Atmospheric Environment 29, 1499–1518.

Lu, R., Turco, R.P., 1996. Ozone distributions over the Los Angeles basin: three-dimensional simulations with the SMOG model. Atmospheric Environment 30, 4155–4176.

MARI, The Mexico City Air Quality Research Initiative. Los Alamos Nat. Lab. Rep., LA-12699, 1994.

Marsik, F.J., Fischer, K.W., McDonald, T.D., Samson, P.J., 1995. Comparison of methods for estimating mixing heights used during the 1992 Atlanta field intensive. Journal of Applied Meteorology 34, 1802–1814.

Matthijsen, J., Builtjes, P.J.H., Meijer, E.W., Boersen, G., 1997. Modeling cloud effects on ozone on a regional scale: a case study. Atmospheric Environment 31, 3227–3238.

McKeen, S.A., Hsie, E.-Y., Liu, S.C., 1991. A study of the dependence of rural ozone on ozone precursors in the eastern United States. Journal of Geophysical Research 96, 15,377–15,394.

McKeen, S.A., Liu, S.C., Hsie, E.-Y., Lin, X., Bradahaw, J.D., Smyth, S., Gregory, G.L., Blake, D.R., 1996. Hydrocarbon ratios during PEM-WEST: a model perspective. Journal of Geophysical Research 101, 2087–2109.

Meng, Z., Dabdub, D., Seinfeld, J.H., 1997. Chemical coupling between atmospheric ozone and particulate matter. Science 277, 116–119.

Millan, M., Salvador, R., Mantilla, E., Artinano, B., 1996. Meteorology and photochemical air pollution in southern Europe: Experimental results from EC research projects. Atmospheric Environment 30, 1909–1924.

Milford, J., Gao, D., Sillman, S., Blossey, P., Russell, A.G., 1994. Total reactive nitrogen (NO$_y$) as an indicator for the sensitivity of ozone to NO$_x$ and hydrocarbons. Journal of Geophysical Research 99, 3533–3542.

Milford, J., Russell, A.G., McRae, G.J., 1989. A new approach to photochemical pollution control: implications of spatial patterns in pollutant responses to reductions in nitrogen oxides and reactive organic gas emissions. Environmental Science and Technology 23, 1290–1301.

Miller, D.F., Alkezweeny, A.J., Hales, J.M., Lee, R.N., 1978. Ozone formation related to power plant emissions, Science 202, 1186–1188.

Moussiopoulos, N., Sahm, P., Kessler, Ch., 1997. Numerical simulation of photochemical smog formation in Athens, Greece – a case study. Atmospheric Environment 29, 3619–3632.

Munger, J.W., Fan, S.-M., Bakwin, P.S., Goulden, M.L., Goldstein, A.H., Colman, A.S., Wofsy, S.C., 1998. Regional budgets for nitrogen oxides from continental sources: Variations of rates for oxidation and deposition with season and distance from source regions. Journal of Geophysical Research 103, 8355–8368.

National Research Council (NRC), 1991. Committee on Tropospheric Ozone Formation and Measurement. Rethinking the Ozone Problem in Urban and Regional Air Pollution, National Academy Press.

Nunnermacker, L.J., Imre, D., Daum, P.H., Kleinman, L., Lee, Y.N., Lee, J.H., Springston, S.R., Newman, L., Weinstein-Lloyd, J., Luke, W.T., Banta, R., Alvarez, R., Senff, C., Sillman, S., Holdren, M., Keigley, G.W., Zhou, X., 1998. Characterization of the Nashville urban plume on July 3 and July 18, 1995. Journal of Geophysical Research 103, 28,129–28,148.

Olszyna, K.J., Bailey, E.M., Simonaitis, R., Meagher, J.F., 1994. O$_3$ and NO$_y$ relationships at a rural site. Journal of Geophysical Research 99, 14,557–14,563.

Oreskes, N., Shrader-Frechette, K., Beliz, K., 1994. Verification, validation, and confirmation of numerical models in the earth sciences. Science 263, 641–646.

Parrish, D.D., Trainer, M., Buhr, M.P., Watkins, B.A., Fehsenfeld, F.C., 1991. Carbon monoxide concentrations and their relation to concentrations of total reactive nitrogen at two rural U.S. sites. Journal of Geophysical Research 96, 9309–9320.

Pierson, W.R., Gertler, A.W., Robinson, N.F., Sagebiel, J.C., Zielinska, B., Bishop, O.A., Stedman, D.H., Zweidinger, R.B., Ray, W.D., 1996. Real-world automotive emissions – summary of studies in the Fort McHenry and Tuscarora Mountain Tunnels. Atmospheric Environment 30, 2233–2256.

Prevot, A.S.H., Staehelin, J., Kok, G.L., Schillawski, R.D., Neininger, B., Staffelbach, T., Neftel, A., Wernli, H., Dommen, J., 1997. The Milan photooxidant plume. Journal of Geophysical Research 102, 23,375–23,388.

Rappengluck, B., Fabian, P., Kalabokas, P., Viras, L.G., Ziomas, I.C., 1998. Quasi-continuous measurements of non-methane hydrocarbons (NMHC) in the greater Athens area during MEDCAPHOT-TRACE. Atmospheric Environment 32, 2103–2121.

Reynolds, S., Michaels, H., Roth, P., Tesche, T.W., McNally, D., Gardner, L., Yarwood, G., 1996. Alternative base cases in photochemical modeling: their construction, role, and value. Atmospheric Environment 30 (12), 1977–1988.

Ridley, B.A., Walega, J.G., Lamarque, J.-F., Grahek, F.E., Trainer, M., Hubler, G., Lin, X., Fehsenfeld, F.C., 1998. Measurements of reactive nitrogen and ozone to 5-km altitude in June 1990 over the southeastern United States. Journal of Geophysical Research 103, 8369–8388.

Ridley, B.A., Walega, J.G., Dye, J.E., Grahek, F.E., 1994. Distributions of NO, NO$_x$, NO$_y$, and O$_3$ to 12 km altitude during the summer monsoon season over New Mexico. Journal of Geophysical Research 99, 25,519–25,534.

Ridley, B.A., Madronich, S., Chatfield, R.B., Walega, J.G., Shelter, R.E., Carroll, M.A., Montzka, D.D., 1992. Measurements and model simulations of the photostationary state during the Mauna Loa Observatory Photochemistry Experiment: implications for radical concentrations and ozone production and loss rates. Journal of Geophysical Research 97, 10,375–10,388.

Roberts, J.M., Parrish, D.D., Norton, R.B., Bertman, S.B., Holloway, J.S., Trainer, M., Fehsenfeld, F.C., Carroll, M.A., Albercook, G.M., Wang, T., Forbes, G., 1996. Episodic removal of NO$_y$ species from the marine boundary layer of the North Atlantic. Journal of Geophysical Research 101, 28,947–28,960.

Roselle, S.J., Schere, K.L., 1995. Modeled response of photochemical oxidants to systematic reductions in anthropogenic volatile organic compound and NO$_x$ emissions. Journal of Geophysical Research 100, 22929–22941.

Ryerson, T.B., Buhr, M.B., Frost, G., Goldan, P.D., Holloway, J.S., Hubler, G., Jobson, B.T., Kuster, W.C., McKeen, S.A., Parrish, D.D., Roberts, J.M., Sueper, D.T., Trainer, M., Williams, J., Fehsenfeld, F.C., 1998. Emissions lifetimes and ozone formation in power plant plumes. Journal of Geophysical Research 103, 22,569–22,584.

Sakugawa, H., Kaplan, I.R., 1989. H$_2$O$_2$ and O$_3$ in the atmosphere of Los Angeles and its vicinity: Factors controlling their formation and their role as oxidants of SO$_2$. Journal of Geophysical Research 94, 12,957–12,974.

Samson, P.J., Ragland, K., Ozone and visibility reduction in the midwest: evidence for large-scale transport. Journal of Applied Meteorology 16, 1101–1106, 1077.

Sillman, S., 1998. The method of photochemical indicators as a basis for analyzing O$_3$–NO$_x$-hydrocarbon sensitivity. Atmospheric Environment as part of the NARSTO critical review series (submitted).

Sillman, S., He, D., Pippin, M., Daum, P., Kleinman, L., Lee, J.H., Weinstein-Lloyd, J., 1998. Model correlations for ozone, reactive nitrogen and peroxides for Nashville in comparsion with measurements: implications for NO$_x$-hydrocarbon sensitivity. Journal of Geophysical Research 103, 22,629–22,644.

Sillman, S., He, D., Cardelino, C., Imhoff, R.E., 1997. The use of photochemical indicators to evaluate ozone–NO$_x$-hydrocarbon sensitivity: Case studies from Atlanta, New York and Los Angeles. Journal of Air and Waste Management Association 47, 1030–1040.

Sillman, S., 1995. The use of NO$_y$, H$_2$O$_2$ and HNO$_3$ as indicators for O$_3$–NO$_x$–VOC sensitivity in urban locations. Journal of Geophysical Research 100, 14,175–14,188.

Sillman, S., Al-Wali, K., Marsik, F.J., Nowatski, P., Samson, P.J., Rodgers, M.O., Garland, L.J., Martinez, J.E., Stoneking, C., Imhoff, R.E., Lee, J.H., Weinstein-Lloyd, J.B., Newman, L., Aneja, V., 1995a. Photochemistry of ozone formation in Atlanta, GA: models and measurements. Atmospheric Environment 29, 3055–3066.

Sillman, S., Samson, P.J., 1995b. The impact of temperature on oxidant formation in urban, polluted rural and remote environments. Journal of Geophysical Research 100, 11,497–11,508.

Sillman, S., Samson, P.J., Masters, J.M., 1993. Ozone production in urban plumes transported over water: photochemical model and case studies in the northeastern and midwestern U.S. Journal of Geophysical Research 98, 12,687–12,699.

Sillman, S., Logan, J.A., Wofsy, S.C., 1990. The sensitivity of ozone to nitrogen oxides and hydrocarbons in regional ozone episodes. Journal of Geophysical Research 95, 1837–1851.

Simpson, D., 1995. Biogenic emissions in Europe, 2, Implications for ozone control strategies. Journal of Geophysical Research 100, 22,891–22,906.

Simpson, D., Guenther, A., Hewitt, C.N., Steinbrecher, R., 1995. Biogenic emissions in Europe 1. Estimates and uncertainties. Journal of Geophysical Research 100, 22,875–22,890.

Sistla, G., Zhou, N., Hou, W., Ku, J.-Y., Rao, S.T., Bornstein, R., Freedman, F., Thuns, P., 1996. Effects of uncertainties in meteorological inputs on Urban Airshed Model predictions and ozone control strategies. Atmospheric Environment 30, 2011–2025.

Staffelbach, T., Neftel, A., Blatter, A., Gut, A., Fahrni, M., Stahelin J., Prevot, A., Hering, A., Lehning, M., Neininger, B., Baumie, M., Ko, G.L., Dommen, J., Hutterli, M., Anklin, M., 1997. Photochemical oxidant formation over southern Switzerland, part I: Results from summer, 1994. Journal of Geophysical Research 102, 23,345–23,362.

Staffelbach, T., Neftel, A., Horowitz, L.W., 1997. Photochemical oxidant formation over southern Switzerland, part II: Model results. Journal of Geophysical Research 102, 23,363–23,374.

Tov, D.A.-S., Peleg, M., Matveev, V., Mahrer, Y., Seter, L, Luria, M., 1997. Recirculation of polluted air masses over the east Mediterranean coast Atmospheric Environment 31, 1441–1448.

Trainer, M., Parrish, D.D., Goldan, P.D., Roberts, J., Fehsenfeld, F.C., 1998. Regional factors influencing ozone concentrations. Atmospheric Environment as part of the NARSTO critical reviews (submitted).

Trainer, M., Ridley, B.A., Buhr, M.P., Kok, G., Walega, J., Hubler, G., Parrish, D.D., Fehsenfeld, F.C., 1995. Regional ozone and urban plumes in the southeastern United States: Birmingham, a case study. Journal of Geophysical Research 100, 18,823–18,834.

Trainer, M., Parrish, D.D., Buhr, M.P., Norton, R.B., Fehsenfeld, F.C., Anlauf, K.G., Bottenheim, J.W., Tang, Y.Z., Wiebe, H.A., Roberts, J.M., Tanner, R.L., Newman, L., Bowersox, V.C., Maugher, J.M., Olszyna, K.J., Rodgers, M.O., Wang, T., Berresheim, H., Demerjian, K., 1993. Correlation of ozone with NO$_y$ in photochemically aged air. Journal of Geophysical Research 98, 2917–2926.

Trainer, M., Williams, E.J., Parrish, D.D., Buhr, M.P., Allwine, E.J., Westberg, H.H., Fehsenfeld, F.C., Liu, S.C., 1987. Models and observations of the impact of natural hydrocarbons on rural ozone. Nature 329,6141, 705–707.

Vukovich, P.M., Bach, W.D., Crisman, B.W., King, W.J., 1977. On the relationship between high ozone in the rural surface layer and high pressure systems. Atmospheric Environment 11, 967–983.

Walcek, C.J., Yuan, H.-H., Stockwell, W.R., 1997. The influence of aqueous-phase chemical reactions on ozone formation in polluted and nonpolluted clouds. Atmospheric Environment 31, 1221–1237.

Watkins, B.A., Parrish, D.D., Trainer, M., Norton, R.B., Yee, J.E., Fehsenfeld, F.C., Heikes, B.G., 1995. Factors influencing the concentration of gas phase hydrogen peroxide during the summer at Niwot Ridge, Colorado. Journal of Geophysical Research 100, 22,831–22,840.

White, W.H., Patterson, D.E., Wilson Jr., W.E., 1983. Urban exports to the nonurban troposphere: results from project MISTT. Journal of Geophysical Research 88, 10,745–10,752.

Williams, E.L., Grosjean, D., 1990. Southern California Air Quality Study: Peroxyacetyl nitrate. Atmospheric Environment 24, 2369–2377.

Winner, D.A., Cass, G.R., Harley, R.A., 1995. Effect of alternative boundary conditions on predicted ozone control strategy performance: a case study in the Los Angeles area. Atmospheric Environment 29, 3451–3464.

Air Pollution Science for the 21st Century
J. Austin, P. Brimblecombe and W. Sturges, editors
© 2002 Elsevier Science Ltd. All rights reserved.

Chapter 13

New Directions: VOCs and biosphere–atmosphere feedbacks

J.D. Fuentes, B.P. Hayden, M. Garstang

Department of Environmental Sciences, University of Virginia, Charlottesville, VA 22903, USA
E-mail: jf6s@virginia.edu

M. Lerdau

Ecology and Evolution Department, State University of New York, Stony Brook, NY 11794-5245,
USA

D. Fitzjarrald

Jungle Research Group, ASRC, SUNY – Albany,
NY 12203, USA

D.D. Baldocchi

Department of Environmental Science, Policy and Management, University of California,
Berkeley, CA 94720-3110, USA

R. Monson

Department of EPO Biology, University of Colorado, Boulder, CO 80309-0334, USA

B. Lamb

Dept. of Civil & Environmental Engineering, Washington State University, Pullman,
WA 99164-2910, USA

C. Geron

National Risk Management Research Laboratory, U.S. Environmental Protection Agency,
Research Triangle Park, NC 27711, USA

Shallcross and Monks (New Directions: A role for isoprene in biosphere-climate-chemistry feedbacks, Atmospheric Environment, Vol. 34 (2000) pp. 1659–1660) recently summarised the importance of biogenic isoprene in a biosphere-atmosphere system under constant change. In this article, we expand this synthesis to include the biophysical feedbacks between plants that produce volatile organic compounds (VOCs) and the overlying atmosphere.

Plants produce a wide range of hydrocarbons including isoprene, terpenes, hemiterpenes, oxygenated species, and cuticular waxes. At the global scale it is estimated that vegetation emits 1.2×10^{15} g C per year, an amount equivalent to global methane (CH_4) emissions.

First published in Atmospheric Environment 35 (2001) 189–191

There has been little direct experimental research on how global environmental change can affect emissions of phytogenic hydrocarbons. A broad-based research programme is needed because VOCs rapidly react with hydroxyl radical (OH), ozone (O_3), and nitrate (NO_3). Such reactions lead to the formation of secondary chemical species (e.g., formaldehyde, peroxy radicals, carbonyl compounds, etc.) that can enhance O_3 and other oxidant levels in locales rich in nitrogen oxides. The reaction of VOCs and OH can lead to enhanced CH_4 levels, as OH is the major atmospheric sink for CH_4. Poisson and co-workers (Journal of Atmospheric Chemistry, Vol. 36, (2000) pp. 157–230) estimated that VOC emissions have the global effect of increasing the lifetime of CH_4 by 15% and enhancing background O_3 levels, by 18%. Once reacted, terpene compounds can also generate carbonaceous aerosols whose impacts may include feedback mechanisms in the Earth radiation balance. Finally, VOCs constitute a source of atmospheric carbon and can thus play key functions in the global carbon budget and cycling.

The potential for feedbacks among global environmental change, phytogenic hydrocarbons, and associated atmospheric chemistry, requires coordinated research efforts. In particular, new research is required to ascertain how environmental changes occurring at the global scale might affect VOC emissions from ecosystems to the atmosphere. This research should take into account: (1) warming from increased greenhouse gas concentrations and from other perturbations, (2) elevated carbon dioxide (CO_2) levels, (3) increased deposition of nitrogen, and (4) conversion of land from unmanaged to managed ecosystems.

Warming can augment emissions of phytogenic hydrocarbons. As isoprene and terpene production strongly depends on temperature, the temperature increases predicted by global climate models suggest that VOC emissions may increase exponentially. This change could, with the increase in reactive nitrogen oxides that result from fertilisation and fossil fuel combustion, lead to increasing levels of tropospheric carbon monoxide and O_3: both important pollutants and greenhouse gases.

Similarly, as elevated CO_2 increases productivity in terrestrial systems, the standing biomass available to produce VOCs will also rise, and emissions may increase proportionally. Elevated CO_2 may also increase the relative availability of carbon to nitrogen in ecosystems and this will lead to a proportional increase in the production of carbon-based compounds such as isoprenoids, but experiments are only beginning to examine this possibility.

An environmental change that is occurring locally and across the globe, and that may be of great relevance to biogenic VOC production, is the increased deposition of reactive nitrogen. Nitrogen could serve to 'fertilise' ecosystems and lead to large increases in productivity. Experiments show that plant monoterpene and isoprene emissions respond strongly to nitrogen fertilisation (see, for example, Litvak and co-workers writing in Plant, Cell and Environment, Vol. 19 (1996) pp. 549–559). Increases appear to occur as a result of both enhanced carbon fixation and increased activity of enzymes responsible for isoprenoid production. Nitrogen deposition may also alter VOC emissions through its impact on biological diversity. It is not yet known whether nitrogen deposition will tend to favour hydrocarbon-producing species or not, but effects could be profound.

Perhaps, the single most important change in the global environment that is occurring with respect to VOCs is the expansion of agriculture, especially in tropical regions, and the resulting conversion of forest to cropland, pasture, and then secondary forest. In addition, some of the most important species in modern agroforestry, poplars and eucalyptus, emit large quantities of VOCs. These land-use changes have the potential to dramatically shift both regional and global budgets of biogenic VOC emissions.

Although most studies of VOC emission impacts have focused on changes in atmospheric redox potential, there is emerging evidence that these VOCs can also affect local radiative balance. In areas with active photosynthesis and relatively dry air masses, VOCs contribute to elevated atmospheric emissivities and thus retard nocturnal cooling (Hayden, Philosophical Transactions of the Royal Society of London, B, Vol. 353 (1998) pp. 5–18). In such regions, minimum temperatures are significantly higher than the dewpoint, but in regions without vegetation, the average daily minimum temperatures are close to the dewpoint.

Field observations by Garstang and co-workers (Journal of Experimental Biology, Vol. 200 (1997) pp. 421–431) show that extreme nocturnal inversions can exist over surfaces covered by high terpene emitters (e.g., mopane – *Colophospermum mopane*). Temperatures at 50 m can be $\sim 10°C$ warmer than at the surface. This extreme positive lapse rate may be the result of two concomitant processes.

Firstly, hydrocarbons such as α-pinene and β-pinene effectively ($\sim 90\%$) absorb thermal energy within the 'atmospheric window' (8–13 μm) of the electromagnetic spectrum. These hydrocarbons, in place of water vapour, act as greenhouse gases thereby elevating minimum temperatures. Note that for this greenhouse effect to be detectable, the ambient abundance of VOCs must be high, with mixing ratios perhaps reaching parts per million (ppm) levels. Utilising the meteorology data reported by Garstang and co-workers together with vegetation characteristics and reported emissions for mopane, we estimate ambient terpene mixing ratios ranging from 10 to 50 ppm. Haze layers near the surface, observed near sunset as the nocturnal inversion forms, suggest trapping of emissions and formation of visible aerosols close to the surface. These aerosols, ranging in size from 0.01 to 0.4 μm (Hoffmann and co-workers, Journal of Atmospheric Chemistry, Vol. 26 (1997) pp. 189–222), can form at relative humidities above 60% on particulate hydrocarbons, resulting in "wet" haze particles.

Secondly, the release of latent heat of condensation can increase the heat content of a thermally insulated atmospheric column $1 \text{ m}^2 \times 100$ m by as much as 30 K at relative humidities greater than 60% (> 10 g (water) per kg of air). Although such extreme heating is never realised in the open atmosphere, owing to the presence of turbulent diffusion and radiative flux divergence, this process requires further investigation.

The regional implications of this VOC heating within the atmospheric surface layer is that there can be a tendency to change the atmosphere from the dry adiabatic state to more stable conditions, and hence statically stable regimes can be augmented at night. As a consequence, the lapse rate of the atmospheric boundary layer becomes less negative, and thus the static stability is increased. We propose as a working hypothesis that the biogenic hydrocarbon 'greenhouse effect', combined with latent heat of condensation released to the environment due to water condensation onto hydrocarbon-derived aerosols, can retard heat losses and add heat to the lower atmosphere and alter the thermal structure of the atmospheric boundary layer.

We therefore propose here a framework of vegetation controlled atmosphere dynamics consisting of three linked feedback loops, operating on different time scales. First, Loop I represents the vegetation-nocturnal temperature feedback. Vegetation produces isoprene (day), as well as terpenes and other chemical species (day and night). Under low humidity levels, these compounds strongly absorb terrestrial radiation and elevate nocturnal minimum temperatures above the dewpoint. Feedback Loop II comprises VOC conversion to aerosols on which water vapour condensation occurs. Such condensation produces latent heat, which is converted to sensible heat throughout the atmospheric boundary layer. The resulting vertical change in temperature with height gives rise to increased atmospheric static stability. With this stability comes increased residence time of gaseous VOCs, aerosols, increased haze, and hence additional heating. We believe that this feedback operates at time scales ranging from days to weeks, and at spatial scales from the ecosystem to synoptic levels. All these processes lead to the long-term feedback Loop III, which operates at time scales of years to decades.

We propose that vegetation adjusts to these atmospheric dynamics in a synergistic manner. For Loop III, we further propose that with these elevated nocturnal temperatures, neo-tropical plant species become more successful at higher latitudes with photosynthetic and VOC production capabilities over a longer portion of the year. These systematic changes give rise to altered plant species assemblages, biodiversity changes, and shifted ecotones. If land-use changes increase the proportion of VOC plant emitters then this positive feedback will be amplified.

To address the processes described here co-ordinated and integrated research efforts are required involving atmospheric scientists, ecologists, and atmospheric chemists.

Air Pollution Science for the 21st Century
J. Austin, P. Brimblecombe and W. Sturges, editors
© 2002 Elsevier Science Ltd. All rights reserved.

Chapter 14

Chemistry of HO$_x$ radicals in the upper troposphere

Lyatt Jaeglé, Daniel J. Jacob

Harvard University, Division of Engineering and Applied Sciences and Department of Earth and Planetary Sciences, 29 Oxford St., Cambridge, MA 02138, USA
E-mail: jaegle@atmos.washington.edu

William H. Brune

Department of Meteorology, Pennsylvania State University, University Park, PA 16802, USA

Paul O. Wennberg

Division of Geological and Planetary Sciences, 1200 E. California Blvd., California Institute of Technology, Pasadena, CA 91125, USA

Abstract

Aircraft observations from three recent missions (STRAT, SUCCESS, SO-NEX) are synthesized into a theoretical analysis of the factors controlling the concentrations of HO$_x$ radicals (HO$_x$ = OH + peroxy) and the larger reservoir family HO$_y$ (HO$_y$ = HO$_x$ + 2H$_2$O$_2$ + 2CH$_3$OOH + HNO$_2$ + HNO$_4$) in the upper troposphere. Photochemical model calculations capture 66% of the variance of observed HO$_x$ concentrations. Two master variables are found to determine the variance of the 24 h average HO$_x$ concentrations: the primary HO$_x$ production rate, P(HO$_x$), and the concentration of nitrogen oxide radicals (NO$_x$ = NO + NO$_2$). We use these two variables as a coordinate system to diagnose the photochemistry of the upper troposphere and map the different chemical regimes. Primary HO$_x$ production is dominated by the O(^{1}D) + H$_2$O reaction when [H$_2$O] > 100 ppmv, and by photolysis of acetone (and possibly other convected HO$_x$ precursors) under drier conditions. For the principally northern midlatitude conditions sampled by the aircraft missions, the HO$_x$ yield from acetone photolysis ranges from 2 to 3. Methane oxidation amplifies the primary HO$_x$ source by factors of 1.1–1.9. Chemical cycling within the HO$_x$ family has a chain length of 2.5–7, while cycling between the HO$_x$ family and its HO$_y$ reservoirs has a chain length of 1.6–2.2. The number of ozone molecules produced per HO$_y$ molecule consumed ranges from 4 to 12, such that ozone production rates vary between 0.3 and 5 ppbv d^{-1} in the upper troposphere. Three chemical regimes (NO$_x$-limited, transition, NO$_x$-saturated) are identified to describe the

First published in Atmospheric Environment 35 (2001) 469–489

dependence of HO_x concentrations and ozone production rates on the two master variables $P(HO_x)$ and $[NO_x]$. Simplified analytical expressions are derived to express these dependences as power laws for each regime. By applying an eigen-lifetime analysis to the HO_x–NO_x–O_3 chemical system, we find that the decay of a perturbation to HO_y in the upper troposphere (as from deep convection) is represented by four dominant modes with the longest time scale being factors of 2–3 times longer than the steady-state lifetime of HO_y.

1. Introduction

The first measurements of HO_x radicals (HO_x = OH + peroxy) in the upper troposphere were obtained over the last five years in four aircraft missions: ASHOE/MAESA (Airborne Southern Hemisphere Ozone Experiment/Measurements of Atmospheric Effects of Supersonic Aircraft), STRAT (Stratospheric Tracers for Atmospheric Transport), SUCCESS (Subsonic Aircraft: Contrail and Cloud Effects Special Study), and SONEX (Subsonic Assessment: Ozone and Nitrogen Oxide Experiment) (Folkins et al., 1997; Wennberg et al., 1998; Brune et al., 1998, 1999). These measurements revealed HO_x levels frequently 2–4 times higher than expected from the commonly assumed primary source (Levy, 1971):

$$O_3 + h\upsilon \rightarrow O(^1D) + O_2, \tag{R1}$$

$$O(^1D) + H_2O \rightarrow OH + OH. \tag{R2}$$

Such elevated concentrations of HO_x imply a more photochemically active upper troposphere than previously thought, with enhanced rates of ozone formation through reactions (R3)–(R5), and thus a potentially more important role in ozone greenhouse forcing from anthropogenic emissions of nitrogen oxides (NO_x = NO + NO_2) (Roelofs et al., 1997; Berntsen et al., 1997; Haywood et al., 1998; Mickley et al., 1999):

$$CO + OH\,(+O_2) \rightarrow CO_2 + HO_2, \tag{R3}$$

$$HO_2 + NO \rightarrow OH + NO_2, \tag{R4}$$

$$NO_2 + h\upsilon\,(+O_2) \rightarrow NO + O_3. \tag{R5}$$

Another implication is faster oxidation of SO_2 by OH, producing $H_2SO_4(g)$ which promotes the formation of new aerosol particles in the upper troposphere (Clarke, 1993; Raes, 1995; Arnold et al., 1997).

Convective injection to the upper troposphere of HO_x precursors other than H_2O, including acetone (Singh et al., 1995; Arnold et al., 1997), peroxides

(Chatfield and Crutzen, 1984; Prather and Jacob, 1997; Jaeglé et al., 1997; Cohan et al., 1999) and aldehydes (Müller and Brasseur, 1999), may provide the missing primary sources of HO$_x$ in the upper troposphere. Inclusion of these sources in photochemical models improves the simulation of observed HO$_x$ levels (McKeen et al., 1997; Folkins et al., 1997; Jaeglé et al., 1997, 2000; Wennberg et al., 1998; Brune et al., 1998).

Fig. 1 schematically illustrates the chemistry of HO$_x$ radicals in the upper troposphere as described in current models. Because of the cycling taking place between HO$_x$ radicals and their non-radical reservoirs, we define the larger HO$_y$ family as HO$_y$ = HO$_x$ + 2H$_2$O$_2$ + 2CH$_3$OOH + HNO$_2$ + HNO$_4$. We do not include HNO$_3$ in the HO$_y$ family because its lifetime (a few weeks) is much longer than that of peroxides (a few days), and under most conditions in the upper troposphere its photolysis is a small source of HO$_x$. Four factors control the HO$_x$ and HO$_y$ concentrations: the primary HO$_x$ sources, the amplification of these primary sources through methane oxidation (Logan et al., 1981), the chemical cycling between HO$_x$ and its reservoirs, and the sinks of HO$_y$. *Primary* sources in the upper troposphere are sources which are independent of local HO$_x$ concentrations, such as (R2) and photolysis of short-lived HO$_x$ precursors transported from the lower troposphere by convection. *Secondary* sources depend on the presence of a preexisting pool of HO$_x$. They include photolysis of locally produced peroxides and of CH$_2$O produced from methane oxidation by OH.

In this paper we synthesize the knowledge gained from the aircraft observations into a theoretical analysis of HO$_x$ chemistry in the upper troposphere. We use the concentration of NO$_x$ and the primary HO$_x$ production rate, P(HO$_x$), as master variables to map out the chemical regimes, generalizing from our previous work which showed that HO$_x$ concentrations and ozone production in the upper troposphere can be largely described on the basis of these two variables alone (Jaeglé et al., 1998a, 1999, 2000). We show how [NO$_x$] and P(HO$_x$) provide a unified framework to interpret HO$_x$ observations not only for individual aircraft missions, but across missions spanning several seasons and locations. We examine the following questions: (1) What is the relative importance of different HO$_x$ precursors as primary HO$_x$ sources? (2) To what extent does methane oxidation provide an autocatalytic source of HO$_x$? (3) What is the role of cycling within the HO$_x$ and HO$_y$ families? (4) What are the chemical regimes for HO$_x$ and ozone production? (5) What are the time scales for the relaxation of HO$_y$ following episodic perturbations from convection?

Our analysis focuses on daytime conditions outside of clouds. The chemistry of HO$_x$ at night and close to sunrise and sunset is still poorly understood (Brune et al., 1999; Wennberg et al., 1999; Jaeglé et al., 1999). Cirrus clouds appear to provide an efficient sink of HO$_x$ (Faloona et al., 1998; Jaeglé et al., 2000), but their overall importance is limited because of the small volume they

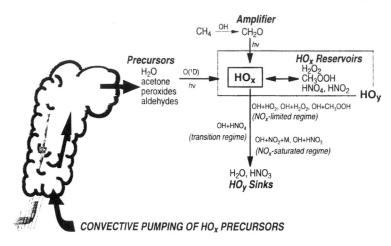

Figure 1. Schematic of the chemistry of HO_x radicals ($= OH + $ peroxy radicals) and the HO_y reservoir family ($= HO_x + 2\ H_2O_2 + 2\ CH_3OOH + HNO_2 + HNO_4$) in the upper troposphere. The schematic illustrates the factors controlling HO_x concentrations: primary sources of HO_x, amplification of the HO_x primary source by CH_4 oxidation, cycling between HO_x and HO_y, and loss of HO_y. Depending on the levels of NO_x, three distinct chemical regimes exist for HO_y loss: the NO_x-limited regime (dominant loss pathways through $OH + HO_2$, $OH + H_2O_2$, and $OH + CH_3OOH$), a transition regime ($OH + HNO_4$) and the NO_x-saturated regime ($OH + NO_2 + M$, $OH + HNO_3$). These regimes in turn determine the dependence of HO_x concentrations and ozone production on NO_x concentrations. For a detailed discussion of the factors controlling NO_x radicals, see the review by Bradshaw et al. (2000).

occupy in the upper troposphere and the short lifetime of HO_x. It has been proposed that halogen radicals might play a role in the photochemistry of the upper troposphere (Borrmann et al., 1996; Crawford et al., 1996; Solomon et al., 1997), but there is no observational evidence for such a role.

Faloona et al. (2000) report that models systematically underestimate the concentrations of HO_x observed under very high NO_x conditions ($[NO_x] > 300$ pptv). They suggest that this could reflect flaws in our understanding of the chemistry coupling HO_x and NO_x radicals. Very high concentrations of NO_x in the observations can usually be linked to lightning or convection (Thompson et al., 1999; Liu et al., 1999). Another possible explanation for the model underestimates of HO_x under such circumstances is the presence of unmeasured HO_x precursors transported by convection or produced by lightning (Müller and Brasseur, 1999; Zuo and Deng, 1999; Jaeglé et al., 2000), combined with the increased efficiency of these precursors at producing HO_x under high NO_x conditions (Folkins and Chatfield, 2000). It remains unclear whether the results presented in Faloona et al. (2000) imply major flaws in our understanding of chemistry in the high-NO_x regime or simply an insufficient accounting of primary HO$_x$ sources under these unusual conditions (Brune et al., 1999). Given this remaining uncertainty, we assume the latter here and use standard tropospheric chemistry.

Section 2 begins by a presentation of the aircraft data sets providing the basis for our analysis and by an assessment of the ability of current photochemical models to simulate observed HO$_x$ concentrations. In Section 3 we discuss the factors controlling the chemistry of HO$_x$ and HO$_y$ in the upper troposphere. The chemical regimes defining the dependence of HO$_x$ concentrations and ozone production on NO$_x$ and the primary HO$_x$ source are analyzed in Section 4. In Section 5, the relaxation of HO$_y$ following episodic perturbations by convection is examined with an eigenlifetime analysis (Prather, 1994, 1996). Conclusions are in Section 6.

2. Observations and models

2.1. Aircraft data sets

Extensive measurements of OH and HO$_2$ concentrations by laser-induced fluorescence (Stevens et al., 1994; Wennberg et al., 1995) were made in the upper troposphere (between 8 km and the local tropopause) during STRAT, SUCCESS, and SONEX. Measurements of OH and HO$_2$ concentrations were also obtained in the lower stratosphere during these missions, especially during STRAT where altitudes up to 21 km were sampled by the ER-2 aircraft, but few observations were made in the lower troposphere. Tropospheric HO$_x$ measurements during ASHOE/MAESA were restricted to two profiles (Folkins et al., 1997) and we do not consider them in our analysis.

The observations extend from 15°N to 65°N latitude and cover different seasons. Median values (as well as 20th and 80th percentiles) observed between 8 km and the local tropopause are summarized in Table 1. In addition to HO$_x$, many other chemical species and meteorological parameters were measured. Observations at night and near the terminator (SZA > 80°), in stratospheric air ([O$_3$] > 100 ppbv with [CH$_4$] < 1760 ppbv), inside clouds, and in fresh aircraft exhaust plumes were excluded from Table 1.

STRAT took place over the North Pacific between California and Hawaii during 1995–1996 (Wennberg et al., 1998). The aircraft sampled the upper troposphere up to 17 km in tropical air, for a total of ten flights in three seasonal deployments (October–November 1995, January–February 1996, and August 1996). Frequent influence of deep marine convection transporting clean surface air to the upper troposphere can be seen in the observed 20th percentiles in Table 1 (27 ppbv O$_3$, 63 ppbv CO, and 24 pptv NO$_x$) (Jaeglé et al., 1997).

SUCCESS took place over the central United States from April 15 to May 15 1996 (Toon and Miake-Lye, 1998). While OH measurements were available

Table 1. Median observations in the upper troposphere (8 km altitude to tropopause)[a]

Observations	Aircraft mission			Overall medians used in standard model calculation
	STRAT	SUCCESS	SONEX	
Time period	1995–1996[b]	15 Apr.–15 May 1996	13 Oct.–12 Nov. 1997	March 1
Number of observations[c]	560	468	1741	2769
Latitude, °N	22 (21–38)	37 (36–40)	47 (39–54)	41
Pressure, hPa	198 (163–240)	240 (209–302)	290 (242–331)	263
Temperature, K	217 (209–224)	219 (213–231)	229 (222–235)	225
Ozone column, DU	288 (257–317)	333 (329–339)	283 (251–301)	292
H_2O, ppmv	38 (9–72)	80 (37–204)	148 (74–294)	114
Relative humidity (ice), %	40 (17–66)	82 (48–104)	69 (35–105)	
OH, pptv	0.23 (0.11–0.39)	0.28 (0.17–0.51)	0.092 (0.051–0.15)	
HO_2, pptv	3.6 (2.5–5.3)	10.3 (6.3–13.4)	2.5 (1.6–3.5)	
$(OH)_{24h}$[d], pptv	0.12 (0.095–0.15)	0.13 (0.072–0.26)	0.036 (0.022–0.054)	
$(OH)_{24h}$[d], pptv	1.6 (0.93–3.7)	5.2 (4.1–6.9)	0.95 (0.63–1.5)	
H_2O_2, pptv	N/A	N/A	78 (49–147)	
CH_3OOH, pptv	N/A	N/A	25 (< 25–66)	
CH_2O, pptv	N/A	N/A	< 50 (< 50–74)	
NO, pptv	84 (20–158)	42 (19–89)	56 (20–137)	59
NO_x[e], pptv	99 (24–179)	54 (27–118)	87 (38–202)	84
HNO_4[e], pptv	21 (11–33)	31 (19–53)	56 (37–81)	
HNO_3, pptv	N/A	N/A	106 (59–191)	
PAN, pptv	N/A	N/A	60 (39–87)	
NO_y, pptv	217 (90–484)	317 (212–461)	307 (200–598)	
O_3, ppbv	48 (27–75)	64 (50–80)	53 (38–66)	54
CH_4, ppbv	1768 (1763–1774)	1774 (1766–1786)	1771 (1764–1780)	1772
CO, ppbv	72 (63–83)	105 (95–131)	86 (75–99)	87
Ethane, pptv	444 (384–677)	N/A	677 (590–810)	620
Propane, pptv	31 (24–95)	N/A	84 (55–144)	71
C_{4-5} alkanes, pptv	9 (5–21)	N/A	26 (12–63)	22

Table 1. (Continued)

Observations	Aircraft mission			Overall medians used in standard model calculation
	STRAT	SUCCESS	SONEX	
Acetone, pptv	N/A	N/A	515 (427–607)	4.7
Aerosol surface area, μm^2 cm^{-3}	1.8 (0.7–2.8)	2.8 (0.56–13)	6.1 (3.6–12)	
$J(NO_2)$[f], 10^{-3} s^{-1}	4.6 (4.0–5.7)	6.4 (5.8–8.0)	2.9 (2.4–3.6)	
$J(O_3 \rightarrow O(^1D))$[f], 10^{-6} s^{-1}	11 (5.9–23)	16 (14–20)	3.0 (2.1–6.1)	
$P(HO_x)$[g], pptv d^{-1}	65 (48–103)	212 (147–328)	54 (34–105)	83
HO$_x$ yield from acetone + $h\nu$[h]	2.9 (2.1–3.0)	2.5 (1.9–2.9)	2.8 (2.4–3.0)	
HO$_x$ yield from CH$_4$ + OH[i]	0.52 (0.36–0.55)	0.41 (0.24–0.53)	0.47 (0.37–0.51)	
Amplification factor A_{CH4}[j]	1.4 (1.3–1.9)	1.2 (1.1–1.5)	1.4 (1.2–1.7)	
HO$_x$ chain length[k]	5.5 (2.8–7)	3.6 (2.5–5.2)	5.0 (3.7–6.3)	
HO$_y$ chain length[l]	1.7 (1.6–2.2)	1.8 (1.6–2.1)	1.8 (1.7–2.1)	
HO$_x$ lifetime, min	37 (19–50)	25 (13–33)	36 (20–51)	
HO$_y$ lifetime, d	1.9 (1.2–3.0)	2.3 (1.1–3.4)	4.6 (2.7–8.1)	
$P(O_3)$[m], ppbv d^{-1}	0.85 (0.6–1.1)	2.3 (0.97–5)	0.54 (0.35–0.8)	

Table 1. (Continued)

[a]The values are medians. The 20th and 80th percentiles are listed in parentheses. Observations below the limit of detection are included in the statistics. Observations in the lower stratosphere ($[O_3] > 100$ ppbv, $[CH_4] < 1760$ ppbv), in clouds, in fresh aircraft exhaust, and outside daytime (SZA $> 80°$) are excluded.

[b]Upper tropospheric flights during the STRAT mission took place on three deployments: October–November 1995, January–February 1996, and August 1996.

[c]Number of 1 min HO_2 observations in the upper troposphere.

[d]24 h average concentrations of OH and HO_2 are obtained by scaling observed HO_x with a model derived diel factor (Appendix A).

[e]Model calculated values for NO_x ($NO_x = $ observed NO + modeled NO_2) and HNO_4.

[f]Photolysis frequencies calculated from observed actinic fluxes and then averaged over 24 h using a model scaling factor.

[g]The primary HO_x production rate, $P(HO_x)$, is a 24 h average value calculated using observed H_2O and acetone concentrations. During STRAT and SUCCESS acetone was not measured, we use instead a correlation between acetone and CO derived from previous aircraft missions (McKeen et al., 1997). The HO_x yield from acetone photolysis is calculated for the observed conditions and ranges from 2 to 3 (see entry below and Section 3.1).

[h]Net number of HO_x molecules produced in the complete oxidation of acetone to CO, including contributions from acetone photolysis and from CH_2O photolysis (Section 3.1).

[i]Net number of HO_x molecules produced in the oxidation of methane to CO (Section 3.2).

[j]The amplification factor for methane oxidation A_{CH_4} is defined as the relative increase in the primary source of HO_x due to oxidation of CH_4 by OH (Eq. (4)).

[k]Ratio of HO_2 to HO_x loss rates, measuring the efficacy of cycling within the HO_x family (Eq. (5) in Section 3.3).

[l]Ratio of HO_x to HO_y loss rates, measuring the efficacy of cycling between HO_x and its HO_y reservoirs (Eq. (6) in Section 3.3).

[m]The ozone production rate, $P(O_3)$, is calculated using observed NO and HO_2 concentrations and averaged over 24 h using a model-derived diel factor (Appendix A).

throughout the mission, HO$_2$ was only measured for the flights in May. The composition of the upper troposphere was strongly impacted by convective injection of continental air from the United States as indicated by measurements of high concentrations of CO and NO$_y$ (Table 1) (Jaeglé et al., 1998b), and of aerosols rich in crustal material (Talbot et al., 1998) and sulfate (Dibb et al., 1998).

SONEX focused on the North Atlantic region between 13 October and 12 November 1997 (Singh et al., 1999). It included measurements of acetone, peroxides, and CH$_2$O, which were missing in STRAT and SUCCESS. The atmosphere sampled by SONEX was influenced by an ensemble of NO$_x$ sources: lightning, convection and aircraft emissions (Liu et al., 1999; Thompson et al., 1999; Koike et al., 2000). Photochemistry was less active than in STRAT or SUCCESS because of higher latitudes and the fall season.

The composition of the upper troposphere has several features that distinguish it from the rest of the troposphere. Levels of H$_2$O are low (10–300 ppmv) because of the cold temperatures (210–235 K), and result in a primary source of HO$_x$ radicals from O(^{1}D) + H$_2$O which is 1–2 orders of magnitude smaller than in the lower troposphere. Other primary sources of HO$_x$ such as convective injection of acetone, peroxides and aldehydes from the lower troposphere can play an important role in the upper troposphere. Concentrations of NO$_x$ are relatively high in the upper troposphere (typically 50–100 pptv, Table 1), due to high-altitude sources (lightning, convective transport of surface pollution, transport from the stratosphere, aircraft emissions) combined with a long lifetime against oxidation to HNO$_3$. The lifetime of NO$_x$ is proportional to the NO$_x$/NO$_2$ ratio as NO$_2$ is the reactant species for the conversion to HNO$_3$:

$$NO + O_3 \rightarrow NO_2 + O_2, \tag{R6}$$

$$NO_2 + OH + M \rightarrow HNO_3 + M, \tag{R7}$$

$$NO_2 + NO_3 + M \rightarrow N_2O_5 + M, \tag{R8}$$

$$N_2O_5 + H_2O \text{ (aerosols)} \rightarrow 2\,HNO_3. \tag{R9}$$

The partitioning of NO$_x$ in the daytime upper troposphere favors NO relative to NO$_2$ ([NO]/[NO$_x$] $\approx$ 0.6–0.9, Table 1) because of the temperature dependence of reaction (R6) and the low O$_3$ concentrations. The lifetime of NO$_x$ in the upper troposphere is 5–10 d compared to $\sim$1 d in the lower troposphere (Jacob et al., 1996; Jaeglé et al., 1998b).

The last column in Table 1 summarizes median upper tropospheric conditions across all three missions. These median conditions will be used as constraints for our standard model calculations presented in Sections 2.2, 3, 4 and 5.

2.2. Comparison between models and observations

Fig. 2 and Table 2 show summary comparisons between observations and model calculations of HO_2, OH, and the HO_2/OH ratio for STRAT, SUC-CESS, and SONEX. Each point is a 1 min average. The calculations were done with the Harvard photochemical zero-dimensional (0-D) model (see Appendix A) constrained with local observations of NO, O_3, H_2O, acetone, CO, CH_4, C_2H_6, C_3H_8, C_{4-5} alkanes, aerosol surface area, pressure, temperature, and actinic flux. Concentrations of acetone in STRAT and SUCCESS are estimated on the basis of correlations with CO (McKeen et al., 1997); since acetone varies only over a narrow range this assumption is of little consequence. Peroxides and aldehydes are assumed to be in chemical steady state. We restrict the comparison to daytime, upper tropospheric observations (> 8 km altitude) in clear air and outside of aircraft exhaust plumes, following the criteria of Table 1. Comparisons for individual missions have been discussed previously (Brune et al., 1998, 1999; Faloona et al., 2000; Jaeglé et al., 1997, 1998a, 2000; McKeen et al., 1997; Wennberg et al., 1998).

As seen in Fig. 2c, the HO_2/OH concentration ratio is generally reproduced by model calculations to well within the combined uncertainties of observations (reported accuracy of $\pm 20\%$) and rate coefficients ($+120$ to -70%, Wennberg et al. (1998)). The median model-to-observed ratio for HO_2/OH is 1.08 with a correlation coefficient $r^2 = 0.76$ (Table 2). The cycling between OH and HO_2 in the model takes place on a time scale of a few seconds and is mainly controlled by

$$OH + CO\,(+O_2) \rightarrow HO_2 + CO_2, \tag{R3}$$

$$OH + O_3 \rightarrow HO_2 + O_2, \tag{R10}$$

$$HO_2 + NO \rightarrow OH + NO_2, \tag{R4}$$

$$HO_2 + O_3 \rightarrow OH + 2\,O_2. \tag{R11}$$

In the upper troposphere, loss of OH through reaction with CO (R3) largely dominates over its reactions with O_3 (R10), methane, and non-methane hydrocarbons. Assuming pseudo-steady state for OH and HO_2, and neglecting (R10), we have the concentration ratio:

$$\frac{[HO_2]}{[OH]} \approx \frac{k_3\,[CO]}{k_4\,[NO] + k_{11}\,[O_3]}, \tag{1}$$

where k_i represents the rate constant for reaction (Ri). Eq. (1) is generally accurate to within 25%. Reaction (R11) dominates over (R4) only for very low NO concentrations ($[NO_x] < 10\text{--}20\,$pptv). Under most conditions in the upper

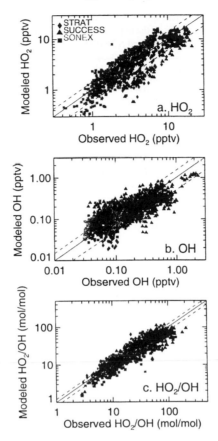

Figure 2. Comparison between observations and model calculations for (a) HO$_2$ concentrations, (b) OH concentrations, and (c) the HO$_2$/OH concentration ratio in the upper troposphere (8 km altitude to the local tropopause). One-minute averages corresponding to STRAT, SUCCESS, and SONEX are shown respectively by diamonds, triangles, and squares. Observations inside cirrus clouds, influenced by stratospheric air, in fresh aircraft exhaust, or at high solar zenith angle (SZA) (SZA > 80°) are excluded from this figure. The calculations were obtained with a diel steady-state photochemical model constrained with local observations of HO$_x$ precursors (see Appendix A). The 1 : 1 line is shown as solid. The dashed lines represent the instrumental accuracies ($\pm 40\%$ for OH and HO$_2$, $\pm 20\%$ for HO$_2$/OH).

troposphere, k_4[NO] is the dominant term of the denominator. The observed [HO$_2$]/[OH] ratio varies between ~ 5 and 100 (Fig. 2); changes in [NO] and [CO] account respectively for 60 and 15% of this variability. As shown in Table 1, [NO] varies over almost one order of magnitude in the upper troposphere while [CO] varies by only a factor of 2. When the sources or sinks of HO$_x$ com-

Table 2. Comparison between observed and model-calculated HO_x concentrations in the upper troposphere (> 8 km altitude) during STRAT, SUCCESS, and SONEX[a]

Variable		STRAT	SUCCESS	SONEX	All
HO_2	R	0.97	0.70	1.20	1.09
	r^2	0.53	0.61	0.84	0.66
	N	330	232	1163	1725
OH	R	0.80	0.78	1.20	0.98
	r^2	0.84	0.60	0.64	0.63
	N	357	635	1081	2073
HO_2/OH	R	1.15	0.78	0.96	1.08
	r^2	0.87	0.61	0.76	0.76
	N	313	229	1072	1614

[a] R is the median model-to-observed ratio, r^2 is the correlation coefficient between the model and observations, and N is the number of points for which the comparison is conducted.

pete with HO_x cycling reactions (for example, for $[NO_x] < 10$ pptv and large HO_x precursor concentrations, see Section 3.3) Eq. (1) no longer applies.

The median ratios of model-to-observed concentrations ($R_{HO_x} = [HO_x]_{model}/[HO_x]_{obs}$) are 0.98 for OH and 1.09 for HO_2 (Table 2). In addition to uncertainties in model calculations, the accuracy of observations ($\pm40\%$) needs to be considered in evaluating R_{HO_2} and R_{OH}. We find that 75% of the HO_2 points and 60% of the OH points fall within the reported instrumental accuracies. The measurement precision is better for HO_2 than for OH because of higher mixing ratios, which likely explains the smaller scatter between model and observations. In some cases (10 and 14% of HO_2 and OH observations, respectively), the calculations underestimate the observations by factors of 1.6–4. Many of these cases have been identified as fresh convective outflows where high levels of peroxides and aldehydes might have provided additional sources of HO_x (Brune et al., 1998; Jaeglé et al., 1997, 1998a; Wennberg et al., 1998). As discussed in the Introduction, missing HO_x precursors could explain the systematic model underestimates apparent at very high NO_x. For the remaining observations (15% for HO_2, and 26% for OH), the model overestimates the concentrations of HO_x by factors of 1.4–2. Such model overestimates were observed mostly during SONEX and could be due in part to the influence of evaporating cirrus clouds (Jaeglé et al., 2000).

The model accounts for 66% of the observed variance in HO_2 concentrations (Table 2), meaning that most of this variance can be explained on the basis of current understanding. The concentration of HO_x depends strongly on

SZA (Ehhalt and Rohrer, 2000 and references therein). Once this dependence is taken into account by examining diurnal averages, we find that the variance in the model is mostly driven by changes in $[NO_x]$ and in the HO_x primary production rate, $P(HO_x)$. As we will demonstrate in Section 4, the dependence of $[HO_x]$ on $[NO_x]$ and $P(HO_x)$ is non-linear and can be approximated by power laws of the form $[HO_x] \propto [NO_x]^a \, P(HO_x)^b$. For each point shown in Fig. 2, we calculate 24 h averages of the observed $[HO_2]$ and $P(HO_x)$ following the methodology discussed in the Appendix A. $P(HO_x)$ includes the contributions from (R2) and acetone photolysis (see Section 3.1). We find that a linear regression with $\ln([NO_x])$ and $\ln(P(HO_x))$ accounts for 57% of the variance of the observed $\ln([HO_2]_{24h})$ concentrations in Fig. 2. Individual regressions with $\ln([NO_x])$ and $\ln(P(HO_x))$ account for 30 and 38% of the variance, respectively.

Fig. 3 shows the ensembles of NO_x concentrations ($=$ observed $NO +$ modeled NO_2) and 24 h average $P(HO_x)$ for all three missions. The top panel of Fig. 3 shows the $P(HO_x)$ calculated from (R2) and acetone only, while the bottom panel also includes a hypothesized source from convection of additional HO_x precursors as needed to match the HO_2 concentrations (about 10% of observations are affected). Both $[NO_x]$ and $P(HO_x)$ vary across two orders of magnitude. As can be seen in Table 1 and Fig. 3, $P(HO_x)$ during SUCCESS was three times higher than during STRAT or SONEX. This reflects the low amounts of UV radiation during SONEX (fall, northern midlatitudes) and the low levels of water vapor at the higher altitudes sampled in the tropical upper troposphere during STRAT. The NO_x concentrations in STRAT, SUCCESS, and SONEX varied over fairly similar ranges (Table 1 and Fig. 3).

Fig. 4 plots observed values of HO_2, OH and the HO_2/OH concentration ratio, scaled to 24 h averages, in the coordinate system $\{[NO_x], P(HO_x)\}$. Values of $P(HO_x)$ include the hypothesized contribution from convected precursors (bottom panel of Fig. 3). As noted above, we use 24 h averages to remove the obvious influence of the diurnal cycle of solar radiation. Photochemical model calculations for varying $[NO_x]$ and $P(HO_x)$, and otherwise median conditions (last column in Table 1), are shown as contour lines. In these calculations, $P(HO_x)$ is varied by specifying a generic HO_x source with a solar zenith angle dependence assumed to be the same as photolysis of CH_3OOH (see Appendix A for further details). Model [OH] increases with $[NO_x]$ up to ~ 100–500 pptv NO_x due to shift in the $[HO_2]/[OH]$ ratio towards OH (Eq. (1)), and decreases with further increases of NO_x concentrations due to HO_y loss by reactions involving NO_2, HNO_4, and HNO_3 (Fig. 1). Model $[HO_2]$ decreases with increasing $[NO_x]$ due to shift in the $[HO_2]/[OH]$ ratio towards OH and thus faster HO_y loss, since HO_y is lost by reactions involving OH (Fig. 1). The $[HO_2]/[OH]$ ratio has a strong dependence on $[NO_x]$ and little dependence on

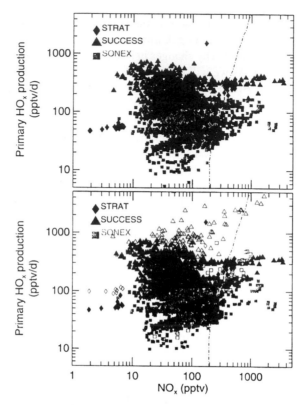

Figure 3. Ranges of [NO$_x$] and the primary HO$_x$ production rate $P(\text{HO}_x)$ in the upper troposphere during STRAT (diamonds), SUCCESS (triangles), and SONEX (squares). Values of
$P(\text{HO}_x)$ are 24 h averages. In the top panel, $P(\text{HO}_x)$ is calculated using observed concentrations
of H$_2$O and acetone only. In the bottom panel, an additional source from convective injection
of other HO$_x$ precursors is also added where necessary to match observed HO$_2$ concentrations
(see Appendix A). The cases where this additional source represents more than 50% of $P(\text{HO}_x)$
are flagged by open symbols. The threshold NO$_x$ value for the transition from the NO$_x$-limited
regime to the NO$_x$-saturated regime for ozone production $(\partial P(\text{O}_3)/\partial[\text{NO}_x] = 0)$ is indicated as
a function of $P(\text{HO}_x)$ for otherwise median conditions in the upper troposphere (see Table 1 and
Section 4).

$P(\text{HO}_x)$ as expected from Eq. (1). These model dependences are also found in
the observations, as can be seen from inspection of Fig. 4.

The rate of ozone production can be expressed as

$$P(\text{O}_3) \approx k_4[\text{NO}][\text{HO}_2],\tag{2}$$

where we neglect the contributions of organic peroxy radical which account for
less than 15% of $P(\text{O}_3)$ in the upper troposphere (Jacob et al., 1996; Crawford

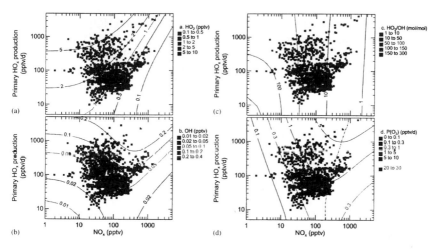

Figure 4. Dependence of (a) HO$_2$ concentrations, (b) OH concentrations, (c) the HO$_2$/OH concentration ratio, and (d) the ozone production rate $P(O_3)$ on the concentrations of NO$_x$ and $P(HO_x)$ in the upper troposphere for STRAT, SUCCESS, and SONEX. The contour lines correspond to model calculations for median conditions in the upper troposphere (Table 1) with varying [NO$_x$] and $P(HO_x)$. Observations are shown with color-coded squares corresponding to concentration ranges. Both model and observations for OH and HO$_2$ are 24 h averages (see Appendix A). Values of $P(O_3)$ are computed from observed NO and HO$_2$ and averaged over 24 h (see Appendix A). The dashed line in panel (d) corresponds to $\partial P(O_3)/\partial[NO_x] = 0$.

et al., 1997). Chemical loss of O$_3$ (reactions (R10), (R11), (R7) and (R2)) generally accounts for less than 20% of $P(O_3)$ in the upper troposphere so that Eq. (2) effectively represents the ozone tendency (Davis et al., 1996; Folkins and Chatfield, 2000). We use Eq. (2) to calculate instantaneous values of $P(O_3)$ from observed [NO] and [HO$_2$], and then average $P(O_3)$ over 24 h using a model-derived scaling factor (see Appendix A). For most observations in Fig. 4d, $P(O_3)$ increases with increasing [NO$_x$], representing the NO$_x$-limited regime. However, a significant fraction of observations during SONEX lie in the transition regime where $P(O_3)$ is independent of [NO$_x$] (Section 4.2). By sampling $P(O_3)$ at constant $P(HO_x)$ in SONEX, Jaeglé et al. (1999) observed the transition from NO$_x$-limited to NO$_x$-saturated regime in a manner consistent with model calculations. Very few observations are in the NO$_x$-saturated regime where $P(O_3)$ decreases with increasing [NO$_x$].

Fig. 4 demonstrates that the two master variables {[NO$_x$], $P(HO_x)$} can describe the overall behavior of observed HO$_x$ and ozone production rates across different seasons and regions. In the model calculations of Fig. 4 (contour lines) we vary only [NO$_x$] and $P(HO_x)$ while fixing all other variables to their median observed values. Where does this analysis fail in describing the observations? Some discrepancies can be noted in Fig. 4. For $P(HO_x) < 20$ pptv d^{-1}

the model overestimates the concentrations of both OH and HO_2. These observations were obtained at high latitudes during SONEX (> 50°N), where recycling of HO_x radicals by photolysis of peroxides is much slower than for the median conditions of Table 1. They are found to be well simulated by a model constrained with the appropriate local radiative environment.

3. Chemistry of HO_x and HO_y

In this section we examine in a broader context the different chemical processes that affect HO_x and the reservoir HO_y family in the upper troposphere.

3.1. Primary sources of HO_x

As discussed above, the primary sources of HO_x in the upper troposphere are the photooxidation of water vapor (R2), the photolysis of acetone, and the photolysis of other HO_x precursors such as peroxides and aldehydes transported to the upper troposphere by convection. The primary HO_x production rate can be expressed as:

$$P(HO_x) = 2k_2[O(^1D)][H_2O] + \sum_i y_i J_i [X_i]. \qquad (3)$$

The sum on the right-hand side is over all HO_x primary precursors X_i with photolysis frequency J_i and corresponding HO_x yield y_i.

An abbreviated version of the acetone oxidation chain following initial photolysis is given in (R12)–(R21). The HO_x yield from photolysis of acetone includes contributions from the initial photolysis step (R12), and the subsequent photolysis of CH_2O by (R20a):

$$CH_3C(O)CH_3 + h\upsilon\,(+2\,O_2) \rightarrow CH_3CO_3 + CH_3O_2, \qquad (R12)$$

$$CH_3CO_3 + NO\,(+O_2) \rightarrow CH_3O_2 + CO_2 + NO_2, \qquad (R13)$$

$$CH_3CO_3 + NO_2 \rightleftharpoons CH_3C(O)O_2NO_2, \qquad (R14)$$

$$CH_3CO_3 + HO_2 \rightarrow CH_3C(O)OOH + O_2 \qquad (R15a)$$

$$\rightarrow CH_3COOH + O_3, \qquad (R15b)$$

$$CH_3O_2 + NO\,(+O_2) \rightarrow CH_2O + HO_2 + NO_2, \qquad (R16)$$

$$CH_3O_2 + HO_2 \rightarrow CH_3OOH + O_2, \qquad (R17)$$

$$CH_3OOH + h\upsilon\,(+O_2) \rightarrow OH + HO_2 + CH_2O, \qquad (R18)$$

$$CH_3OOH + OH \rightarrow OH + CH_2O + H_2O \qquad \text{(R19a)}$$

$$\rightarrow CH_3O_2 + H_2O, \qquad \text{(R19b)}$$

$$CH_2O + h\nu \,(+2O_2) \rightarrow 2\,HO_2 + CO, \qquad \text{(R20a)}$$

$$CH_2O + h\nu \rightarrow CO + H_2, \qquad \text{(R20b)}$$

$$CH_2O + OH \,(+O_2) \rightarrow HO_2 + CO + H_2O. \qquad \text{(R21)}$$

In addition to photolysis, acetone can be lost through reaction with OH. In the upper troposphere, this pathway represents on average less than 10% of acetone loss. It becomes important at lower altitudes, as OH concentrations increase, the temperature-dependent rate constant for OH + CH$_3$C(O)CH$_3$ increases, and the acetone photolysis quantum yield decreases (Gierczak et al., 1998).

We derive the HO$_x$ yield from acetone photolysis ($y_{acetone}$) using fractional reaction rates, in a manner similar to Arnold et al. (1997) and Folkins and Chatfield (2000). The yield increases with increasing [NO$_x$] and decreasing P(HO$_x$) (Folkins and Chatfield, 2000) (Fig. 5), and lies generally between 2 and 3. The theoretical maximum is 3.3. For very low [NO$_x$] (< 10 pptv) and high P(HO$_x$) (> 200 pptv d^{-1}), $y_{acetone}$ can be negative: the dominant loss pathways for CH$_3$CO$_3$ and CH$_3$O$_2$ are then (R15) and (R17) followed by (R19) which lead to net loss of HO$_x$. The median yields for the conditions of the aircraft missions ranged from 2.5 (SUCCESS) to 2.9 (STRAT). While concentrations of acetone are fairly uniform in the upper troposphere with a typical range from 300 to 1000 pptv (Singh et al., 1995; Wohlfrom et al., 1999), the levels of water vapor show much more variability. A linear regression of 24 h average P(HO$_x$) with [H$_2$O], [CH$_3$C(O)CH$_3$], latitude, ozone column, yields r^2 correlation coefficients of 0.71, 0.03, 0.25, and 0.004, respectively. The importance of acetone relative to water vapor as a primary source of HO$_x$ thus largely depends on the abundance of H$_2$O. Fig. 6 shows the relative contributions of H$_2$O and acetone to P(HO$_x$) for the ensemble of observations in SONEX. Acetone becomes a significant HO$_x$ source when H$_2$O drops below 200 ppmv, and becomes dominant when H$_2$O drops below 100 ppmv. For air masses in the lowermost stratosphere ([O$_3$] > 100 ppbv, and [CH$_4$] < 1760 ppbv), (R2) dominates again due to increasing ozone levels and decreasing acetone levels. During SONEX, acetone photolysis and H$_2$O + O(^{1}D) were on average of comparable magnitude as sources of HO$_x$. A more dominant role for acetone was inferred for SUCCESS and especially for STRAT due to lower H$_2$O concentrations (Brune et al., 1998; McKeen et al., 1997; Jaeglé et al., 1997) (Table 1).

The contribution to P(HO$_x$) from convective transport of peroxides and aldehydes from the boundary layer was examined by Müller and Brasseur (1999) using a 3-D global model. They find a contribution of more than 60%

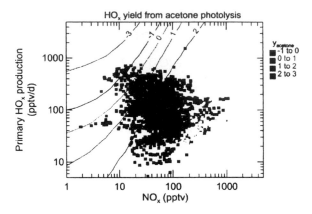

Figure 5. 24 h average HO$_x$ yields from acetone photolysis in the upper troposphere. Contour lines show results for the median conditions in Table 1 with varying [NO$_x$] and P(HO$_x$). The colored symbols represent the values calculated for individual points during STRAT, SUCCESS, and SONEX.

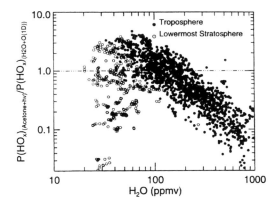

Figure 6. Ratio of the contributions from acetone and water vapor to 24 h average P(HO$_x$), as a function of H$_2$O concentrations for the SONEX mission between 8 and 12 km altitude. Observations in the lowermost stratosphere ([O$_3$] > 100 ppbv and [CH$_4$] < 1760 ppbv) are shown by open circles. The HO$_x$ yield from acetone photolysis varies mainly between 2 and 3 (Section 3.1). The HO$_x$ yield from H$_2$O photooxidation is 2.

in the tropics at 10–14 km altitude, in agreement with the study of Prather and Jacob (1997), with peroxides dominating over oceans and aldehydes over continents. In their model, this source increases HO$_x$ concentrations in the upper troposphere by 10–20% on average.

Enhancements of HO$_x$ concentrations by factors of 2–5 were observed in convective outflows during STRAT and SUCCESS (Brune et al., 1998; Jaeglé et al., 1997, 1998a). The additional source of HO_x necessary to account for these observations is 2–10 times higher than the local sources from H_2O and acetone. During SONEX, less than 15% of upper tropospheric observations appeared to have been recently affected by convection (Jaeglé et al., 2000). The lower altitudes and warmer temperatures during SONEX (10 K warmer compared to STRAT and SONEX, see Table 1) resulted in larger amounts of water vapor in convective outflows. A 10 K increase in temperature from 220 to 230 K results in ice saturation water vapor concentrations increasing from 90 to 300 ppmv. The high levels of H_2O in convective outflows reduced the influence of other transported HO_x precursors during SONEX.

3.2. Autocatalytic production through methane oxidation

The atmospheric lifetime of CH_4 is about 10 yr, with oxidation by OH providing the main sink. Because of its long lifetime, CH_4 has a uniform concentration of 1.7–1.8 ppmv in the upper troposphere (Table 1). The mechanism for oxidation of CH_4 to CO is initiated by

$$OH + CH_4(+O_2) \rightarrow CH_3O_2 + H_2O \tag{R22}$$

followed by reactions (R16)–(R21). Methane oxidation by OH may either be a source of HO_x (through (R20a)) or a sink of HO_x (through (R15) and (R19)) (Logan et al., 1981). We view the source of HO_x from CH_4 as secondary rather than primary because it depends on the local supply of OH. Because of its autocatalytic character, it represents an amplifying factor for the primary HO_x sources. Similar to the case of acetone photolysis, the overall HO_x yield (y_{CH_4}) from oxidation of methane increases with increasing [NO_x] and decreasing $P(HO_x)$ (Fig. 7a). The theoretical maximum of y_{CH_4} (0.62) is determined by the pressure- and temperature-dependent quantum yield of (R20a). Under most conditions observed in the upper troposphere, y_{CH_4} varies between 0.3 and 0.5.

Formaldehyde is the key intermediate species for HO_x production from methane oxidation. Model concentrations of CH_2O, from methane oxidation only, are shown by contour lines in Fig. 7b as a function of [NO_x] and $P(HO_x)$ for the median upper tropospheric conditions of Table 1. The model dependence of CH_2O on both parameters is similar to that of OH (Fig. 4b) as the source of CH_2O is proportional to OH (through (R22)). Reaction of CH_2O with OH (R21) represents typically 30–50% of the CH_2O sink, buffering the dependence of CH_2O on OH. The calculated concentration of CH_2O varies between 10 and 100 pptv. Observations of CH_2O, available from SONEX, are reported as the square symbols in Fig. 7b. A large fraction of the observations (55%) are below the limit of detection (LOD) of the instrument (50 pptv), consistent with model results. However, the remaining 45% of observations reach

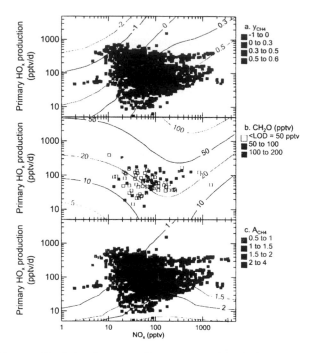

Figure 7. Same as Fig. 5, but the contour lines are for model calculated (a) HO_x yield from methane oxidation; (b) concentrations of CH_2O due to methane oxidation only; (c) amplification factor associated with methane oxidation (see text for definition). Observations of CH_2O obtained during SONEX are shown by the square symbols in panel b. Open symbols correspond to observations below the 50 pptv LOD of the instrument.

up to 300 pptv and do not exhibit any clear dependence on $[NO_x]$ or $P(HO_x)$. Sources of CH_2O from organic species other than CH_4 (including in particular acetone and methanol) are calculated to be only 30% of the source from CH_4 oxidation during SONEX and cannot resolve the discrepancy between model and observations in Fig. 7b. Convective transport from the lower troposphere is not a satisfactory explanation either, as high CH_2O is not correlated with tracers of convection. The only correlation of high CH_2O in the observations is with methanol (Jaeglé et al., 2000), suggesting a possible source from reaction of methanol on aerosols (Singh et al., 2000). In the case of SONEX, using the observed CH_2O concentrations instead of steady-state model values leads to an increase in the calculated HO_x by 30% (Jaeglé et al., 2000).

We define the amplification factor of $P(HO_x)$ associated with methane oxidation as

$$A_{CH_4} = \frac{P(HO_x) + y_{CH_4} k_{22}[OH][CH_4]}{P(HO_x)}. \tag{4}$$

Values of A_{CH_4} smaller than unity ($y_{CH_4} < 0$) correspond to quenching of $P(HO_x)$, while values of A_{CH_4} larger than unity ($y_{CH_4} < 0$) correspond to amplification of $P(HO_x)$. A_{CH_4} is shown in Fig. 7c as a function of [NO$_x$] and $P(HO_x)$. The amplification is largest when $P(HO_x)$ is low. The dependence of A_{CH_4} on $P(HO_x)$ acts as a weak buffer for HO$_x$ concentrations (Crawford et al., 1999): as $P(HO_x)$ decreases, A_{CH_4} increases. For median SONEX conditions a doubling of $P(HO_x)$ at constant [NO$_x$] increases the actual HO$_x$ source $A_{CH_4} P(HO_x)$ by $\sim 70\%$. Similarly, for constant $P(HO_x)$ and increasing [NO$_x$] up to 500 pptv, the increase in A_{CH_4} weakly offsets the effect of increasing HO$_x$ loss.

3.3. Loss of HO$_x$ and recycling from HO$_y$ reservoirs

Loss of HO$_x$ occurs through self-reactions and reactions with NO$_y$ species:

$$HO_2 + OH \rightarrow H_2O + O_2, \tag{R23}$$

$$HO_2 + HO_2 + M \rightarrow H_2O_2 + M, \tag{R24}$$

$$CH_3O_2 + HO_2 \rightarrow CH_3OOH + O_2, \tag{R17}$$

$$NO_2 + HO_2 + M \rightarrow HNO_4 + M, \tag{R25}$$

$$NO_2 + OH + M \rightarrow HNO_3 + M, \tag{R7}$$

$$NO + OH + M \rightarrow HNO_2 + M, \tag{R26}$$

$$HNO_3 + OH \rightarrow H_2O + NO_3. \tag{R27}$$

The resulting lifetime of HO$_x$ in the upper troposphere is typically 30 min (Table 1). The HO$_x$ reservoirs H$_2$O$_2$, CH$_3$OOH, HNO$_2$ and HNO$_4$ regenerate HO$_x$ through photolysis and (in the case of HNO$_4$) thermal decomposition:

$$H_2O_2 + h\upsilon \rightarrow 2\,OH, \tag{R28}$$

$$CH_3OOH + h\upsilon \rightarrow CH_3O + OH, \tag{R18}$$

$$HNO_4 + h\upsilon \rightarrow OH + NO_3, \tag{R29}$$

$$HNO_4 + M \rightarrow NO_2 + HO_2 + M, \tag{R30}$$

$$HNO_2 + h\upsilon \rightarrow OH + NO. \tag{R31}$$

However, reaction of these reservoir species with OH represents a permanent sink of HO$_x$:

$$H_2O_2 + OH \rightarrow H_2O + HO_2, \tag{R32}$$

$$CH_3OOH + OH \rightarrow OH + CH_2O + H_2O, \tag{R19a}$$

$$CH_3OOH + OH \rightarrow CH_3O_2 + H_2O, \tag{R19b}$$

$$HNO_4 + OH \rightarrow H_2O + NO_2 + O_2. \tag{R33}$$

Typical upper tropospheric lifetimes of H_2O_2, CH_3OOH, HNO_4 and HNO_2 are 2–5 d, 1–4 d, 1–4 d, and 10–30 min, respectively. The lifetime of HNO_2 with respect to photolysis is short and it is thus not an important player in HO_x chemistry except as a nighttime reservoir, and for very high NO_x concentrations. In addition, nighttime heterogeneous production of HNO_2 on aerosols might be an important source of HO_x at sunrise (Jaeglé et al., 2000).

The cycling within HO_x and between HO_x and its HO_y reservoirs play critical roles in controlling the concentration of HO_x. To illustrate their effect, we define two chain lengths. The HO_x chain length, $n(HO_x)$, is the average number of times that a HO_2 radical cycles within the HO_x family before being lost, while the HO_y chain length, $n(HO_y)$, is the average number of times that a HO_x radical cycles within the HO_y reservoirs before being lost through one of the HO_y sinks:

$$n(HO_x) = \frac{\text{Loss}(HO_2)}{\text{Loss}(HO_x)} \tag{5}$$

$$n(HO_y) = \frac{\text{Loss}(HO_x)}{\text{Loss}(HO_y)} \tag{6}$$

where $\text{Loss}(HO_2)$, $\text{Loss}(HO_x)$, and $\text{Loss}(HO_y)$ represent the ensemble of reaction rates leading to the loss of HO_2 (these reactions include (R4), (R11), (R17), (R24) and (R25)), HO_x ((R7), (R17), (R23) and (R27)) and HO_y ((R23), (R27), (R32), (R19) and (R33)), respectively. Fig. 8 shows the model-calculated dependences of HO_x and HO_y chain lengths on $[NO_x]$ and $P(HO_x)$. For $[NO_x] < 100$ pptv, $n(HO_x)$ increases with increasing $[NO_x]$ due to cycling of HO_x via $HO_2 + NO$ (R4), while at higher NO_x concentrations $n(HO_x)$ decreases because of loss of HO_x to HNO_3 and HNO_4 ((R7), (R25) and (R27)). At low $[NO_x]$, $n(HO_x)$ decreases with increasing $P(HO_x)$ because the HO_x sinks are quadratic in $[HO_x]$. The highest values of $n(HO_x)$, exceeding 6, are for low $P(HO_x)$ and intermediate NO_x (50–200 pptv). For the conditions of the aircraft missions, $n(HO_x)$ ranges from 2.5 to 7 (Table 1).

The dependence of $n(HO_y)$ on $[NO_x]$ is opposite to that of $n(HO_x)$: $n(HO_y)$ decreases with increasing $[NO_x]$ for $[NO_x] < 100$ pptv, and then it increases with further increases in $[NO_x]$. At low NO_x concentrations, HO_x cycling with its HO_y reservoirs occurs mostly through $HO_2 + HO_2$, while HO_y loss is through $HO_2 + OH$. As a result, and following Eq. (1), $n(HO_y)$ is inversely

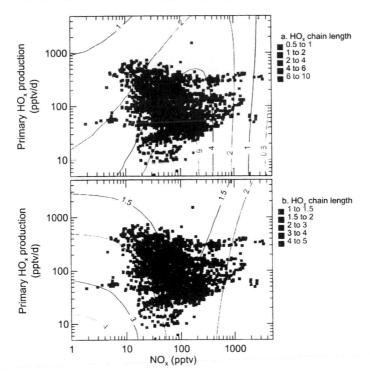

Figure 8. Model-calculated chain lengths $n(HO_x)$ and $n(HO_y)$ for HO$_x$ and HO$_y$ cycling (Eqs. (5) and (6)) as a function of [NO$_x$] and $P(HO_x)$. The contour lines show results for the median conditions described in Table 1. The colored symbols show values calculated over the ensemble of conditions during STRAT, SUCCESS, and SONEX.

proportional to NO. For the range of conditions encountered in the upper troposphere, $n(HO_y)$ varies between 1.6 and 2.2 (Table 1).

The product $n(HO_x) \times n(HO_y)$ represents approximately the ozone production efficiency per unit of primary HO$_x$ produced, as for [NO$_x$] > 10 pptv the cycling between HO$_2$ and OH is mostly through reaction with NO (R4). For most conditions sampled during the three missions, $n(HO_x) \times n(HO_y)$ ranges between 4 and 12.

Model-calculated steady-state concentrations of the peroxide reservoirs are shown in Fig. 9 as a function of [NO$_x$] and $P(HO_x)$ (contour lines) for the median upper tropospheric conditions in Table 1. The dependences of H$_2$O$_2$ and CH$_3$OOH on [NO$_x$] and $P(HO_x)$ follow that of HO$_2$ (Fig. 4a). The concentration of HNO$_4$ increases with increasing NO$_x$ to reach a maximum at 100 pptv NO$_x$, beyond which it decreases. This decrease is due to depletion of HO$_2$ by increasing NO$_x$ (Fig. 4a). For the median upper tropospheric conditions in Ta-

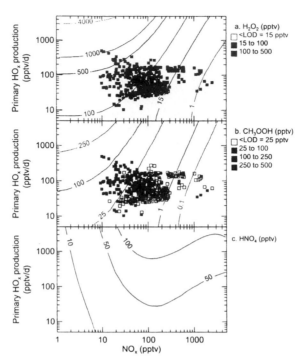

Figure 9. Steady-state concentrations of peroxides as a function of [NO$_x$] and $P(HO_x)$ in the upper troposphere: (a) H$_2$O$_2$, (b) CH$_3$OOH, and (c) HNO$_4$. The contour lines show model results for the median conditions of Table 1. Observations of H$_2$O$_2$ and CH$_3$OOH obtained during SONEX are shown as square symbols on panels a and b. Open symbols correspond to observations below the LOD of the instrument (15 pptv for H$_2$O$_2$ and 25 pptv for CH$_3$OOH).

ble 1, HNO$_4$ loss is 10% by thermolysis (R30), 30% by photolysis (R29) and 60% by reaction with OH (R33). At the warmer temperatures and low $P(HO_x)$ of SONEX, HNO$_4$ loss is equally divided between (R29), (R30) and (R33).

Observed H$_2$O$_2$ and CH$_3$OOH concentrations in SONEX (Snow et al., 2000) are shown in Figs. 9a and b. The concentrations of H$_2$O$_2$ decrease with increasing NO$_x$ as expected from the model calculations. The observations of H$_2$O$_2$ are reproduced by model calculations to within 50% if one allows for heterogeneous reaction of HO$_2$ in aerosols producing H$_2$O$_2$ (Jaeglé et al., 2000). A large fraction (47%) of the reported CH$_3$OOH values were below the 25 pptv LOD of the instrument. These observations generally fall below the 25 pptv contour line of the model (Fig. 9b). However, the remaining observations exhibit much higher values (up to 340 pptv) than calculated. The model underestimate of CH$_3$OOH by factors of 2–3 is systematic throughout SONEX and does not appear to be associated with recent convection (Jaeglé et al., 2000). It could reflect uncertainties in the rate constant for (R17) at low temperatures (Jacob et al., 1996; Schultz et al., 1999). Despite this lack of agreement on the absolute value of CH_3OOH, the observations do show the expected trend of increase with decreasing [NO$_x$].

The MarkIV balloon-borne instrument measured HNO_4 concentrations in the lowermost stratosphere which are a factor of 2 higher than expected. Inclusion of a speculated HNO_4 photodissociative transition in the near IR reconciles models and observations for HNO_4 and provides a missing source of HO$_x$ at high SZA (Wennberg et al., 2000).

3.4. HO$_y$ sinks

As we saw in the previous section, the sinks of HO$_y$ are through reactions of HO$_y$ and NO$_y$ species with OH ((R7), (R19), (R23), (R27), (R32) and (R33)). The lifetime of HO$_y$ varies between 1 and 6 days (Table 1). The relative importance of the different HO$_y$ sinks depends mostly on the levels of NO$_x$, as shown by the model calculations in Fig. 10. There is usually not one single dominant pathway. For STRAT and SUCCESS, the observed ranges in [NO$_x$] and $P(HO_x)$ (Table 1 and Fig. 3) place the upper troposphere predominantly in a regime where OH + HO$_2$ is the most important HO$_y$ loss pathway, accounting for 30–50% of total HO$_y$ loss (Wennberg et al., 1998; Jaeglé et al., 1998a). In contrast, for ASHOE/MAESA and SONEX, the low $P(HO_x)$ and relatively high [NO$_x$] led to the sampling of a second regime where OH + HNO$_4$ is the most important HO$_y$ loss pathway accounting for $\sim 35\%$ of total HO$_y$ loss (Folkins et al., 1997; Jaeglé et al., 2000). At very high [NO$_x$] (> 300 pptv), HO$_y$ is expected to be mostly lost through OH + NO$_2$ + M and OH + HNO$_3$. Only very few observations have been obtained under those conditions. Consideration of the relative importance of the different HO$_y$ loss pathways lays the foundation for our analysis of chemical regimes in the next section.

4. Chemical regimes for HO$_x$ and ozone production

Based on the dominant HO$_y$ loss pathways (Fig. 10), three main chemical regimes can be defined: (1) a low NO$_x$ regime where the dominant sinks of HO$_y$ are reactions of OH with HO$_2$, H$_2$O$_2$, and CH$_3$OOH ((R23), (R32) and (R19)); (2) an intermediate NO$_x$ regime where reaction of OH with HNO$_4$ becomes important (R33); and (3) a high-NO$_x$ regime where reactions of OH with NO$_2$ and HNO$_3$ dominate ((R7) and (R27)). We refer to these three regimes as the NO$_x$-limited regime, the transition regime, and the NO$_x$-saturated regime, respectively. The NO$_x$-saturated regime is commonly known

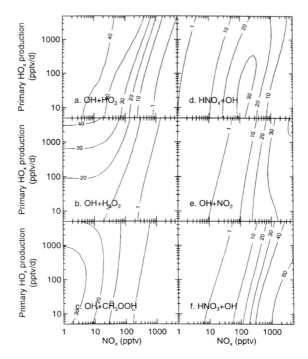

Figure 10. Percentage contributions (%) of individual reactions to the HO_y loss rate as a function of $[NO_x]$ and $P(HO_x)$. Values are model results for the median upper tropospheric conditions of Table 1.

as the hydrocarbon-limited regime in the lower troposphere, however "NO_x-saturated" is a better terminology in the upper troposphere where non-methane hydrocarbons do not have the capacity to influence HO_x chemistry. We define the relative sensitivity of variable Y to variable X as

$$S(Y, X) = \frac{\partial \ln(Y)}{\partial \ln(X)} = \frac{X}{Y} \frac{\partial Y}{\partial X}. \tag{7}$$

$S(Y, X)$ represents the exponent in the local power law relating Y to X, i.e., $Y = X^{S(Y, X)}$. Sensitivities of $P(O_3)$ and $[HO_x]$ relative to $[NO_x]$ and $P(HO_x)$ are shown in Fig. 11 as a function of $[NO_x]$ for the otherwise median conditions of Table 1. The values represent the partial derivatives for the data shown as contour lines in Fig. 4. In the NO_x-limited regime $S(P(O_3), [NO_x])$ is positive; at very low $[NO_x]$ it has a value close to 1 and then decreases with increasing $[NO_x]$. In the NO_x-saturated regime $S(P(O_3), [NO_x])$ becomes negative and it reaches a value close to -0.5 for very high $[NO_x]$. We define the transition regime as the region where $P(O_3)$ is the least sensitive to changes in

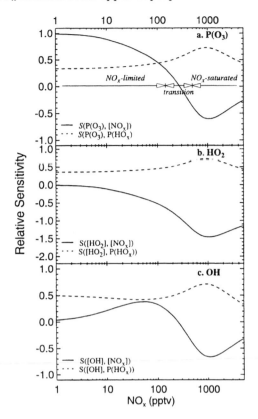

Figure 11.   Relative sensitivities $S(Y, X)$ of $P(O_3)$, [HO$_2$] and [OH] to [NO$_x$] and $P(HO_x)$, as a function of [NO$_x$] (Eq. (7)). Values are model results for the median upper tropospheric conditions of Table 1. The relative sensitivities were obtained by perturbing the standard model simulations by 10% increases in either [NO$_x$] or $P(HO_x)$.

NO$_x$: $|S(P(O_3), [NO_x])| < 0.25$. The relative sensitivity of $P(O_3)$ to $P(HO_x)$, $S(P(O_3), P(HO_x))$, is always positive and varies between about 0.3 and 1. In the NO$_x$-limited regime $P(O_3)$ is sensitive to changes in [NO$_x$], while in the NO$_x$-saturated regime $P(O_3)$ becomes more sensitive to changes in $P(HO_x)$: $|S(P(O_3), P(HO_x))| > |S(P(O_3), [NO_x])|$ (Fig. 11). As expected from equation (2) and shown on Fig. 11, the relative sensitivities of [HO$_2$] and $P(O_3)$ are such that $S([HO_2], [NO_x]) = S(P(O_3), [NO_x]) - 1$.

In the following, we derive simple analytical expressions approximately describing these sensitivities in each regime. Namely, we derive expressions describing the variables [OH], [HO$_2$], and $P(O_3)$ as power laws of the form $[NO_x]^a P(HO_x)^b$ where the exponents a and b are regime-dependent and correspond to the relative sensitivities defined in Eq. (7). We begin by assuming that the HO_y family is in chemical steady state:

$$P(HO_x) + y_{CH_4} k_{22}[CH_4][OH] = [OH](2k_{23}[HO_2] + 2k_{32}[H_2O_2]$$
$$+ 2k_{19}[CH_3OOH] + 2k_{33}[HNO_4] + k_7[NO_2][M] + k_{27}[HNO_3]). \quad (8)$$

The assumption of chemical steady state is not applicable in convective out-flows, where HO_y species can be enhanced or depleted relative to steady state. We will treat separately the effect of convection as a perturbation to the solution of Eq. (8) in Section 5.

Our approach for achieving simple analytical expressions describing the asymptotic behavior shown in Fig. 11, is to drastically simplify Eq. (8) in a manner that can characterize the different chemical regimes. Specifically we assume that each regime is characterized by a single dominant HO_y loss pathway and neglect all other terms on the right-hand side of (8). This is clearly an ovesimplification as there is always more than one significant pathway for HO_y loss (Fig. 10), but it allows an analysis of the key differences between the regimes. We also neglect the methane amplification term on the left-hand side of (8). As methane oxidation acts as a weak buffer to changes in NO_x and $P(HO_x)$ (see Section 3.2), this simplification will have for effect to slightly overestimate the dependence of HO_x on both variables.

In addition to $[NO_x]$ and $P(HO_x)$, the $[NO]/[NO_x]$ ratio is a third parameter controlling HO_x chemistry. We introduce it as a variable α, which we define by assuming $NO-NO_2-O_3$ photostationary steady state following (R5)–(R6):

$$\alpha = \frac{[NO]}{[NO_x]} = \frac{J_5}{J_5 + k_6[O_3]}. \quad (9)$$

Eq. (9) is accurate to within 10–30% in the upper troposphere, where the contribution of peroxy radicals to the conversion of NO to NO_2 is relatively small (Crawford et al., 1996).

4.1. NO_x-limited regime

At low NO_x concentrations (< 100 pptv), the loss of HO_y occurs mainly through (R23), (R32), and (R19) (Fig. 10). Under the median conditions corresponding to Fig. 11, (R23) is most important. Neglecting all other terms on the right-hand side of (8) and using the $[HO_2]/[OH]$ ratio from (1), we obtain

$$[OH] = \left[\frac{k_4\alpha[NO_x] + k_{11}[O_3]}{2k_3 k_{23}[CO]} P(HO_x) \right]^{1/2}, \quad (10)$$

$$[HO_2] = \left[\frac{k_3[CO]}{2k_{23}(k_4\alpha[NO_x] + k_{11}[O_3])} P(HO_x) \right]^{1/2}, \quad (11)$$

$$P(O_3) = k_4\alpha[NO_x]\left[\frac{k_3[CO]}{2k_{23}(k_4\alpha[NO_x]+k_{11}[O_3])}P(HO_x)\right]^{1/2}. \quad (12)$$

For very low NO$_x$ concentrations ($<$ 10 pptv), most of the cycling between OH and HO$_2$ takes place through (R11). As a result, both [OH] and [HO$_2$] are independent of [NO$_x$] and $P(O_3)$ is proportional to [NO$_x$] (Figs. 4 and 11). As [NO$_x$] increases above 10 pptv, (R4) dominates over (R11) and [OH] increases as [NO$_x$]^a, where a asymptotically increases towards 1/2; [HO$_2$] is then proportional to [NO$_x$]$^{a-1}$ and $P(O_3)$ increases as [NO$_x$]^a. The dependences of [HO$_x$] and $P(O_3)$ on $P(HO_x)$ can be expressed by the functionality $P(HO_x)^b$ with $b = 1/2$ over the extent of the NO$_x$-limited regime. The full model calculation shows a smaller value for b (Fig. 11), reflecting the importance of other HO$_y$ losses, (R32) and (R19), which are cubic in [HO$_x$].

4.2. Transition regime

To describe the transition regime we assume in (8) that (R33) is the only significant contributor to HO$_y$ loss. As noted in Section 3.3, for the median upper tropospheric conditions in Table 1, HNO$_4$ loss is dominated by reaction with OH (R33). Assuming steady state for HNO$_4$, we have

$$[OH] = \left[\frac{k_4(J_{29}+k_{30})}{2k_3k_{25}k_{33}[M][CO]}\frac{\alpha}{1-\alpha}P(HO_x)\right]^{1/2}, \quad (13)$$

$$[HO_2] = \left[\frac{1}{[NO_x]}\frac{k_3(J_{29}+k_{30})[CO]}{2k_4k_{25}k_{33}[M]}\frac{1}{\alpha(1-\alpha)}P(HO_x)\right]^{1/2}, \quad (14)$$

$$P(O_3) = \left[\frac{k_3(J_{29}+k_{30})[CO]}{2k_{25}k_{33}[M]}\frac{\alpha}{1-\alpha}P(HO_x)\right]^{1/2}. \quad (15)$$

In the transition regime, [OH] is thus independent of [NO$_x$] while [HO$_2$] is proportional to [NO$_x$]$^{-1}$. One important consequence is that $P(O_3)$ is independent of [NO$_x$] (Jaeglé et al., 1999). Both [HO$_x$] and $P(O_3)$ increase as $P(HO_x)^b$ with $b = 1/2$.

4.3. NO$_x$-saturated regime

When the concentration of NO$_x$ exceeds a few hundreds of pptv, loss of HO$_y$ proceeds mainly by conversion to HNO$_3$ (R7) and subsequent reaction of HNO$_3$ with OH (R27). Photolysis of HNO$_3$:

$$HNO_3 + h\upsilon \rightarrow OH + NO_2 \quad (R34)$$

becomes an important source of HO_x. In this regime, we assume that HNO_3 is in steady state and we treat it as another HO_y reservoir such that (8) is simplified as

$$P(HO_x) = 2 k_{27}[OH][HNO_3],\qquad(16)$$

where we neglect the role of heterogeneous production of HNO_3 on aerosols (R9) ((R9) represents 20–60% of HNO_3 production for the median conditions of Table 1). For the median conditions of Table 1, the dominant HNO_3 sink is photolysis (R34), such that

$$[OH] = \left[\frac{J_{34}}{2k_7k_{27}[M]} \frac{1}{1-\alpha} \frac{P(HO_x)}{[NO_x]} \right]^{1/2},\qquad(17)$$

$$[HO_2] = \frac{k_3\,[CO]}{k_4} \left[\frac{J_{34}}{2k_7k_{27}[M]} \frac{1}{\alpha^2(1-\alpha)} \frac{P(HO_x)}{[NO_x]^3} \right]^{1/2},\qquad(18)$$

$$P\,[O_3] = k_3[CO] \left[\frac{J_{34}}{2k_7k_{27}[M]} \frac{1}{1-\alpha} \frac{P(HO_x)}{[NO_x]} \right]^{1/2}.\qquad(19)$$

As seen in Figs. 4 and 11 and the above expressions in the NO_x-saturated regime, [OH] as well as $[HO_2]$ and $P(O_3)$ decrease with increasing $[NO_x]$. [OH] and $P(O_3)$ decrease with the dependence $[NO_x]^a$ and $[HO_2]$ as $[NO_x]^{a-1}$ with $a = -1/2$. The concentrations of HO_x increase as $P(HO_x)^b$, with $b = 1/2$. For very high $[NO_x]$ (> 1000 pptv), reaction (R9) can no longer be neglected as it becomes a large source of HO_y, causing the upturn (downturn) of the NO_x ($P(HO_x)$) relative sensitivities shown in Fig. 11.

As noted in the Introduction, observed $[HO_x]$ appears to increase with increasing NO_x for $[NO_x] > 300$ pptv (Brune et al., 1999; Faloona et al., 2000), contrary to the expected dependence of Eqs. (17) and (18). Enhanced $P(HO_x)$ associated with high $[NO_x]$ due to convection or lightning could explain this increase, but one cannot exclude the possibility of major shortcomings in our understanding of chemistry in NO_x-saturated regime.

5. Eigenlifetimes for relaxation of a HO_y perturbation

A large fraction of the turnover of the upper troposphere may take place by deep convective injection of air from the lower troposphere, carrying with it high concentrations of HO_x precursors (Fig. 1). Prather and Jacob (1997) have suggested that with a 10 d overturning rate of the tropical upper troposphere, deep convection could cause a persistent chemical imbalance in HO_y. Such a chemical imbalance has indeed been observed a few days downwind of convective events (Cohan et al., 1999).

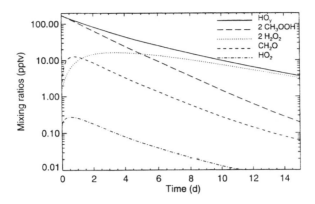

Figure 12. Temporal evolution of ΔH_2O_2, ΔCH_3OOH, ΔCH_2O, ΔHO_2, and ΔHO_y following a 100 pptv convective injection of CH_3OOH into the median atmosphere of Table 1. The species concentrations are initially at chemical steady state, and Δ represents the perturbations to the steady-state concentrations averaged over 24 h to remove the effect of diurnal variability.

We apply here our photochemical model to better understand the time scale for relaxation of the HO$_y$ family to chemical steady state following an episodic perturbation such as convection. The coupling of the chemical reactions in the O$_3$–NO$_x$–HO$_x$ system makes the evaluation of the chemical cycling within the HO$_y$ family a non-trivial problem. Let us consider for example a 100 pptv perturbation to CH$_3$OOH applied to the median conditions of Table 1 with HO$_y$ initially at chemical steady state. Fig. 12 shows the decay of this perturbation. Photolysis of CH$_3$OOH to HO$_x$ is followed by the formation of H$_2$O$_2$, CH$_2$O, and HNO$_4$. The decay of CH$_3$OOH after the perturbation initially proceeds with a 2.1 d e-folding decay time, which is close to the steady-state lifetime of CH$_3$OOH (1.9 d, defined as [CH$_3$OOH]/Loss(CH$_3$OOH) at steady state), but then approaches a 5.7 d decay time. After its initial production, H$_2$O$_2$ decays with a time constant of 5.7 d, longer than its steady state lifetime of 4.5 d. This 5.7 d time constant is more than twice as long as the steady-state lifetime of HO$_y$ (2.6 d). Thus, the steady-state lifetimes of HO$_y$ species do not describe accurately the decay time scales of a HO$_y$ perturbation. The decay time scales are longer because of cycling within the HO$_y$ family.

The true time scale for decay of a HO$_y$ perturbation in the upper troposphere can be obtained with an eigenlifetime analysis (Prather, 1994, 1996) which identifies the independent modes of variability of the chemical system. For a system of n species, and corresponding concentration vector $C\{C_1, C_2, \ldots, C_n\}$, let C be a steady-state solution of the set of mass balance equations,

$$\frac{dC_i}{dt} = P_i(C) - L_i(C) = 0 \quad \text{for } i = 1, n \tag{20}$$

where $P_i(C)$ and $L_i(C)$ are the production and loss rates of species i. Let J be the $n \times n$ Jacobian matrix of the functions dC_i/dt, with elements $J_{ij} = \partial(dC_i/dt)/\partial C_j$ calculated by perturbation to the steady-state solution. To first order, the decay of a perturbation ΔC about the steady-state solution C can be expressed by the first term of a Taylor series expansion (Prather, 1994):

$$\frac{d(\Delta C)}{dt} = J\Delta C. \tag{21}$$

Each of the n eigenvectors, A_k, of the Jacobian matrix J is associated with an eigenvalue, c_k. A perturbation to the concentration of a species can be expressed by a unique linear combination of these eigenvectors:

$$\Delta C(t) = \sum_{k=1}^{n} s_k A_k e^{c_k t} \tag{22}$$

where s_k are coefficients. The characteristic e-folding time constants with which the system responds to perturbations are thus the eigenlifetimes, $-1/c_k$.

We apply this analysis to our photochemical model for the median conditions of Table 1. We solve the steady-state Eq. (20) for $n = 30$ chemical species describing the O_3–NO_x–HO_x system, and using 24 h average values of photolysis frequencies. We calculate the corresponding value of the Jacobian and retrieve the ensemble of eigenvectors and eigenvalues. Seventeen eigenvectors, or modes, are relevant to the HO_y system; four have eigenlifetimes longer than one hour and are listed in Table 3 as dimensionless relative perturbations in the mixing ratios (expressed in percentage). Each mode (which corresponds to a column in the table) is labeled with its eigenlifetime. The longest time constant of the HO_y system is the 5.7 d mode (mode 1), which corresponds to a H_2O_2 perturbation, with a small CH_3OOH component. Other modes include two degenerate 2.09 d CH_3OOH–HNO_4 modes (modes 2 and 3) and a 0.4 d CH_2O mode (mode 4). On shorter time scales, there is a 28-min HNO_2 mode, a 13-min OH–HO_2–CH_3O_2 mode, and a 5 s OH mode.

Let us consider again our example of a 100 pptv perturbation to CH_3OOH from convective injection. We can use the eigenlifetime analysis to derive the exact time dependence of the response of HO_y species according to (22), as shown in Fig. 12 and expressed mathematically in Table 3 (we have neglected the modes with time scales shorter than 1 h). Modes 2 and 3 are degenerate and their terms have been combined. Each species responds as a superimposition of the same three exponential decay terms but the coefficients, $s_k A_k$, are species dependent. For CH_3OOH, of the initial 100 pptv perturbation, 0.6 pptv decay with the 5.7 d time scale, 96 pptv with the 2.09 d time scale, and 0.4 with the 0.4 d time scale. For H_2O_2, the perturbation of 100 pptv of CH_3OOH will result

Table 3. Eigenvectors (or modes) and eigenlifetimes for the decay of a HO$_y$ perturbation in the upper troposphere[a]

	Modes (A_k)				Steady-state lifetime, d[b]
	1	2	3	4	
Eigenlifetime ($-1/c_k$), d	5.70	2.09	2.09	0.40	
H$_2$O$_2$	100.00	0.0	-4.7	-2.0	4.55
HNO$_4$	9.9	47.8	-46.5	-2.0	2.13
CH$_3$OOH	28.2	100.00	100.00	-4.2	1.93
CH$_2$O	13.7	7.2	13.1	100.00	0.34
HNO$_2$	14.2	-2.1	4.3	9.9	1.86×10^{-2}
CH$_3$O$_2$	15.3	1.9	8.3	9.6	3.21×10^{-3}
HO$_2$	14.5	0.0	4.1	14.0	2.43×10^{-3}
OH	15.0	-2.1	4.5	10.2	6.35×10^{-5}

Decomposition of a perturbation $\Delta CH_3OOH[t=0] = 100$ pptv into modes ($\sum_k s_k A_k e^{c^k t}$)[c]

$\Delta H_2O_2[t] = 25e^{-t/5.70} - 26.8e^{-t/2.09} + 2.4e^{-t/0.4}$

$\Delta HNO_4[t] = 2.6e^{-t/5.70} - 2.1e^{-t/2.09} + 2.5e^{-t/0.4}$

$\Delta CH_3OOH[t] = 0.61e^{-t/5.70} + 95.9e^{-t/2.09} + 0.4e^{-t/0.4}$

$\Delta CH_2O[t] = 0.74e^{-t/5.70} + 24.5e^{-t/2.09} - 26.3e^{-t/0.4}$

$\Delta HNO_2[t] = 0.005e^{-t/5.70} + 0.019e^{-t/2.09} - 0.017e^{-t/0.4}$

$\Delta CH_3O_2[t] = 0.007e^{-t/5.70} + 0.111e^{-t/2.09} - 0.022e^{-t/0.4}$

$\Delta HO_2[t] = 0.068e^{-t/5.70} + 0.423e^{-t/2.09} - 0.318e^{-t/0.4}$

$\Delta OH[t] = 0.002e^{-t/5.70} + 0.006e^{-t/2.09} - +0.006e^{-t/0.4}$

$\Delta HO_x[t] = 0.08e^{-t/5.70} + 0.54e^{-t/2.09} - 0.35e^{-t/0.4}$

$\Delta HO_y[t] = 53.9e^{-t/5.70} + 136.6e^{-t/2.09} + 7.9e^{-t/0.4}$

[a]Eigenlifetime analysis applied to perturbation of the O$_3$–NO$_x$–HO$_x$ chemical system for the median upper tropospheric conditions of Table 1. The system of chemical steady-state equations is solved and the corresponding Jacobian matrix is calculated. Four principal modes with eigenlifetimes longer than 1 h are identified for the HO$_y$ system: Mode 1 (H$_2$O$_2$ mode), mode 2 (CH$_3$OOH–HNO$_4$ mode), mode 3 (CH$_3$OOH–HNO$_4$ mode) and mode 4 (CH$_2$O mode). These modes are listed in the table as relative perturbations in the mixing ratios of HO$_y$ species expressed as percentages.
[b]The steady state lifetime of a species is defined as the inverse of its 24 h average loss frequency at steady state.
[c]The time dependence of the decay of HO$_y$ species is obtained according to Eq. (22). Modes with time constants shorter than 1 h are neglected. Modes 2 and 3 are degenerate and have been combined.

in the production of H$_2$O$_2$ with a time scale of 2.09 d, which will then decay mostly with the 5.7 d time scale. HO$_y$ as a result will first decay with the 2.09 d time scale, with increasing contributions from the 5.7 d time scale after the first few days. For the conditions of Table 3 and Fig. 12, the decay time of HO$_y$ following a perturbation is thus twice as long as the steady-state lifetime of HO$_y$ (2.6 d). The 5.7 d time scale reflects the cycling between HO$_x$ and its reservoir species H$_2$O$_2$ and CH$_3$OOH, effectively increasing the persistence

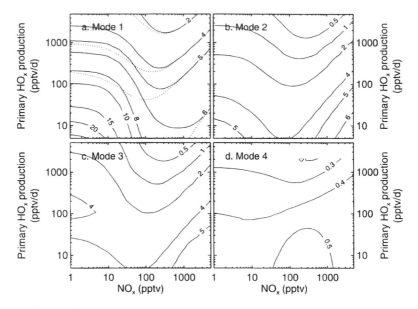

Figure 13. Eigenlifetimes, in days, for the four main HO_y modes (Table 3) as a function of $[NO_x]$ and $P(HO_x)$: (a) mode 1 (H_2O_2 mode), (b) mode 2 (CH_3OOH–HNO_4 mode), (c) mode 3 (CH_3OOH–HNO_4 mode), (d) mode 4 (CH_2O mode). These eigenlifetimes represent the e-folding decay times of a HO_y perturbation in the upper troposphere for the median conditions of Table 1. The steady-state lifetime of H_2O_2 (defined as the loss frequency of H_2O_2 under steady-state conditions) is shown by the dashed line in panel a.

of the perturbation. HO_2 in turn will respond to the perturbations in the HO_y reservoirs and decay with the same eigenlifetimes. When other perturbations are applied to the system in the form of any other HO_y species (H_2O_2, HNO_4, HO_2, OH), or of CH_2O, the same long-lived modes control the decay of the perturbation.

In this discussion, we have not taken into account the role of mixing with surrounding air which will also affect the evolution of HO_y concentrations following a perturbation. Mixing can be represented with its associated eigenlifetime, and will be an additional term in the decay equations of Table 3.

The compositions of the four main HO_y modes and the associated eigenlifetimes vary with $[NO_x]$ and $P(HO_x)$. Fig. 13 shows the eigenlifetimes of these four modes as a function of $[NO_x]$ and $P(HO_x)$ for the otherwise median conditions of Table 1. The longest eigenlifetime (mode 1), which is dominantly a H_2O_2 perturbation, is always longer than the steady-state lifetime of H_2O_2 (dashed line in Fig. 13a). At high $[NO_x]$, this eigenlifetime asymptotically approaches the H_2O_2 steady-state lifetime because the cycling between H_2O_2 and the other HO_y species becomes inefficient: OH produced from photolysis of H$_2$O$_2$ is lost by reaction with NO$_2$ or with HNO$_3$ instead of regenerating H$_2$O$_2$.

For a given perturbation, the fraction of HO$_y$ decaying with each eigenlifetime varies with [NO$_x$] and P(HO$_x$). For the median conditions of Table 1, a perturbation to HO$_y$ will decay principally with responses from the H$_2$O$_2$ mode (mode 1) and the coupled CH$_3$OOH–HNO$_4$ mode (modes 2 and 3). As [NO$_x$] increases, the role of H$_2$O$_2$ as a reservoir of HO$_y$ diminishes and the dominant decay time is associated with the CH$_3$OOH–HNO$_4$ mode. For STRAT (low [NO$_x$] and P(HO$_x$)) we find that the decay times of a HO$_y$ perturbation are 5–10 d (mode 1) and 2–3 d (modes 2 and 3), while for the higher P(HO$_x$) conditions of SUCCESS these decay times are 4–8 d and 1–2 d. For the tropical Pacific conditions at 8–12 km altitude examined by Cohan et al. (1999), P(HO$_x$) is very high and the dominant decay time of a HO$_y$ perturbation is 2–3 d, consistent with their observations in aged convective outflows. In that case the decay time is still longer than the local steady-state lifetime of HO$_y$ (1.4 d).

6. Conclusions

We have presented a theoretical analysis of HO$_x$ and HO$_y$ chemistry in the upper troposphere using photochemical model calculations anchored by recent observations of HO$_x$ from the STRAT, SUCCESS, and SONEX aircraft missions. Model calculations along the flight tracks generally agree with the observed daytime concentrations of OH and HO$_2$ to within the instrumental accuracy of $\pm$40%. Cycling between OH and HO$_2$ is understood to within the instrumental accuracy of the observed HO$_2$/OH concentration ratio ($\pm$20%). The model can account for 66% of the variance in the observed concentrations of OH and HO$_2$. After eliminating the diurnal dependence on solar UV radiation, we find that the variance is largely determined by two master variables: the concentration of NO$_x$ and the primary HO$_x$ production rate, P(HO$_x$). Consideration of these two variables allows a synthesis of the aircraft data sets taken in different regions and in different seasons. It also provides a framework for examining other aspects of the photochemistry of the upper troposphere.

The importance of acetone relative to water vapor as a primary HO$_x$ source is determined by the abundance of H$_2$O, which is far more variable than the abundance of acetone. Acetone photolysis dominates over H$_2$O + O(^{1}D) for [H$_2$O] < 100 ppmv. The HO$_x$ yield from acetone photolysis varies between 2 and 3 for observed conditions in the upper troposphere. Convective injection of peroxides and CH$_2$O from the boundary layer can also provide important primary HO$_x$ sources under low water vapor conditions, as often observed during STRAT and SUCCESS. Methane oxidation by OH followed by photolysis of amplifies the primary HO_x sources by a factor of 1.1–1.9.

The HO_x chain length, defined as the average number of times that HO_2 radicals cycle within the HO_x family before being lost to non-radical forms, varies between 2.5 and 7 for the conditions observed in the three aircraft missions. The HO_y chain length, defined as the average number of times that HO_x radicals cycle with their HO_y reservoirs before being lost to non-HO_y forms, varies between 1.6 and 2.2. The number of ozone molecules produced per HO_y molecule consumed (HO_y-based ozone production efficiency) varies from 4 to 12. Ozone production rates derived from observed HO_2 and NO concentrations range from 0.3 to 5 ppbv d^{-1}.

The pathways for HO_y loss determine the forms of the dependences of HO_x chemistry and the ozone production rate, $P(O_3)$, on [NO_x] and $P(HO_x)$; i.e. the chemical regime. In the NO_x-limited regime ($NO_x < 100$ pptv) reactions of OH with HO_2, H_2O_2 and CH_3OOH dominate the HO_y sink; $P(O_3)$ increases with increasing NO_x. In the transition regime ($100 < NO_x < 300$ pptv), reaction of OH with HNO_4 becomes important; $P(O_3)$ is independent of NO_x. Finally, in the NO_x-saturated regime ($NO_x > 300$ pptv), reactions of OH with NO_2 and HNO_3 are the strongest sinks of HO_y; $P(O_3)$ decreases with increasing NO_x. Qualitative analytic expressions for each regime provide distinct power law dependences of [OH], [HO_2], and $P(O_3)$ on [NO_x] and $P(HO_x)$ of the form [NO_x]^a $P(HO_x)^b$, where values of the exponents vary for the different regimes.

Using an eigenlifetime analysis, we find that the time scale for the relaxation of HO_y following a perturbation (such as from convection) can be factors of 2–3 longer than the steady-state lifetime of HO_y because of cycling within the HO_y family, in particular between HO_x and H_2O_2. The time scale of decay of such a perturbation is 2–15 d depending on [NO_x] and $P(HO_x)$; it is longest for low [NO_x] and low $P(HO_x)$, as was observed during the STRAT mission and which explains the high sensitivity of HO_y concentrations to convection more than a week upwind during that mission (Jaeglé et al., 1997). In contrast, during the PEM-Tropics (A) mission over the tropical South Pacific, $P(HO_x)$ was high and the perturbation to HO_y from convection decayed after a few days (Cohan et al., 1999).

It appears that HO_x chemistry and ozone production in the upper troposphere can be largely described on the basis of only two master variables, the NO_x concentration and $P(HO_x)$. A complete representation of upper tropospheric chemistry thus needs to accurately describe these two variables and the factors that control them. The sources of NO_x in the upper troposphere are difficult to characterize as illustrated by the differing interpretations of observations from the SONEX mission (Thompson et al., 1999; Liu et al., 1999; Koike et al., 2000). The factors controlling water vapor in the upper troposphere are not entirely elucidated. Our understanding of acetone sources is unsatisfactory (Singh et al., 2000). The importance of scavenging on deep convective transport of soluble HO$_x$ precursors such as H$_2$O$_2$, CH$_2$O and HNO$_3$ needs to be better quantified (Mari et al., 2000). Further work on these issues is critical for improving our understanding of the chemistry of the upper troposphere and how it may change in the future.

Acknowledgements

This work was supported by the National Science Foundation (NSF) and by the National Aeronautics and Space Administration (NASA).

Appendix A. Photochemical model

Our calculations use the Harvard 0-D photochemical model (Jaeglé et al., 2000; Schultz et al., 1999). The photochemical mechanism is based on the recommendations of DeMore et al. (1997) and Atkinson et al. (1997), with a few updates including: temperature-dependent cross-sections and pressure-dependent quantum yields for acetone photolysis (Gierczak et al., 1998), O(^{1}D) quantum yields from ozone photolysis (Talukdar et al., 1998), rate constants for the OH + NO$_2$ reaction (Dransfield et al., 1999), and for the OH + HNO$_3$ reaction (Brown et al., 1999). Hydrolysis of N$_2$O$_5$ in aerosols is included with a reaction probability $\gamma_{N_2O_5} = 0.1$ (DeMore et al., 1997). We also include reaction of HO$_2$ in aerosols (Hanson et al., 1992; Cooper and Abbatt, 1996). Following the recommendation of Jacob (2000), we assume that the uptake of HO$_2$ by the aerosols can be described by first-order kinetics with $\gamma_{HO_2} = 0.2$ and the stoichiometry HO$_2$ + aerosol $\rightarrow \frac{1}{2}$H$_2$O$_2$ + $\frac{1}{2}$O$_2$.

The model solves the system of kinetic equations using the diel steady-state assumption (i.e. by forcing the system to reach a periodic solution with a period of 24 h). We apply the model to calculate HO$_x$ concentrations along the flight tracks for the STRAT, SUCCESS, and SONEX missions, and compare the results to 1 min average observations (Jaeglé et al., 1997, 1998a, 2000). The model calculations are constrained by the ensemble of observations along the flight tracks for species other than HO$_x$, including NO, O$_3$, H$_2$O, CO, CH$_4$, acetone, propane, ethane, and C$_{4-5}$ alkanes, as well as temperature, pressure, actinic fluxes, and aerosol surface area. Peroxide and CH$_2$O concentrations are calculated from chemical steady state. The concentration of NO$_t$($=$ NO $+$ NO$_2$ $+$ NO$_3$ $+$ 2N$_2$O$_5$ $+$ HNO$_4$) is assumed constant and is calculated in the model such that the calculated NO matches the observed NO at the time of day of observations. Acetone was not observed during either STRAT or SUCCESS, and we infer its concentration from a relationship between CO and acetone obtained during PEM-West B (McKeen et al., 1997). Observations of H_2O_2, CH_3OOH and CH_2O are available from SONEX (Snow et al., 2000); comparisons to the steady-state model values are presented by Jaeglé et al. (2000) and are discussed in Section 3.3 of the present paper.

Comparisons of simulated and observed $[HO_2]$ and $[OH]$ in Fig. 2 use instantaneous values $[HO_2]_{inst}$ and $[OH]_{inst}$ sampled in the model at the time of day of observation. Scaling of observations to 24 h averages (e.g., in Fig. 4) is done using the 24 h average concentrations from the model, for example:

$$[HO_2]_{24\,h,obs} = [HO_2]_{inst,obs} \frac{[HO_2]_{24\,h,model}}{[HO_2]_{inst,model}}. \qquad (A.1)$$

We use a similar approach to scale instantaneous rates to 24 h averages (such as for ozone production in Fig. 4):

$$Rate_{24\,h,obs} = Rate_{inst,obs} \frac{Rate_{24\,h,model}}{Rate_{inst,model}}. \qquad (A.2)$$

The 24 h average values of the primary HO_x production rate $P(HO_x)$ along the aircraft flight tracks (e.g., Fig. 3) are derived from the local concentrations of water vapor and acetone (Section 3.1). In cases where the model underestimates the observed levels of HO_2, we increase $P(HO_x)$ in the model to match the observations (Fig. 3, bottom panel). This additional $P(HO_x)$ is intended to represent convective transport of other primary HO_x precursors to the upper troposphere such as peroxides and aldehydes (Fig. 1; Müller and Brasseur (1999)). It is implemented as a HO_x source with the same diel dependence as photolysis of CH_3OOH and with a stoichiometric $OH : HO_2$ ratio of unity.

The diel steady-state model is also used here to map HO_x concentrations and other parameters over the ranges of $[NO_x]$ and $P(HO_x)$ in the upper troposphere (e.g., Fig. 3). These calculations use the median conditions described in Table 1, but with NO_x and $P(HO_x)$ varying over their upper tropospheric ranges (1–5000 pptv for $[NO_x]$ and 5–5000 pptv d^{-1} for $P(HO_x)$). Similarly to above, the values of $P(HO_x)$ are specified by a generic source with the same diel dependence as photolysis of CH_3OOH and a stoichiometric $OH : HO_2$ ratio of unity.

References

Arnold, F., Burger, V., Droste-Franke, B., Grimm, F., Krieger, A., Schneider, J., Stilp, T., 1997. Acetone in the upper troposphere and lower stratosphere: impact on trace gases and aerosols. Geophysical Research Letters 24, 3017–3020.

Atkinson, R., et al., 1997. Evaluated kinetic and photochemical data for atmospheric chemistry – supplement VI. Journal of Physical and Chemical Reference Data 26, 1329–1499.

Berntsen, T.K., et al., 1997. Effects of anthropogenic emissions on tropospheric ozone and its radiative forcing. Journal of Geophysical Research 102, 28101–28126.

Borrmann, S., Solomon, S., Dye, J.E., Luo, B., 1996. The potential of cirrus clouds for heterogeneous chlorine activation. Geophysical Research Letters 23, 2133–2136.

Bradshaw, J., Davis, D., Grodzinsky, G., Smyth, S., Newell, R., Sandholm, S., Liu, S., 2000. Observed distributions of nitrogen oxides in the remote free troposphere from the NASA global tropospheric experiment programs. Reviews of Geophysics 38, 61–116.

Brown, S.S., Talukdar, R.K., Ravishankara, A.R., 1999. Reconsideration of the rate constant for the reaction of hydroxyl radicals with nitric acid. Journal of Physical Chemistry 103, 3031–3037.

Brune, W.H., et al., 1998. Airborne in-situ OH and HO$_2$ observations in the cloud-free troposphere and lower stratosphere during SUCCESS. Geophysical Research Letters 25, 1701–1704.

Brune, W.H., et al., 1999. OH and HO$_2$ chemistry in the North Atlantic free troposphere. Geophysical Research Letters 26, 3077–3080.

Chatfield, R.B., Crutzen, P.J., 1984. Sulfur dioxide in remote oceanic air: cloud transport of reactive precursors. Journal of Geophysical Research 89, 7111–7132.

Clarke, A.D., 1993. Atmospheric nuclei in the Pacific mid troposphere: their nature, concentration, and evolution. Journal of Geophysical Research 98, 20633–20647.

Cohan, D.S., Schultz, M.G., Jacob, D.J., Heikes, B.G., Blake, D.R., 1999. Convective injection and photochemical decay of peroxides in the upper troposphere: methyl iodide as a tracer of marine convection. Journal of Geophysical Research 104, 5717–5724.

Cooper, P.L., Abbatt, J.P.D., 1996. Heterogeneous interactions of OH and HO$_2$ radicals with surfaces characteristics of atmospheric particulate matter. Journal of Physical Chemistry 100, 2249–2254.

Crawford, J., et al., 1996. Photostationary state analysis of the NO$_2$–NO system based on airborne observations from the western and central North Pacific. Journal of Geophysical Research 101, 2053–2072.

Crawford, J., et al., 1997. An assessment of ozone photochemistry in the extratropical western North Pacific: impact of continental outflow during the late winter early spring. Journal of Geophysical Research 102, 28469–28487.

Crawford, J., et al., 1999. Assessment of upper tropospheric HO$_x$ sources over the tropical Pacific based on NASA GTE/PEM data: net effect on HO$_x$ and other photochemical parameters. Journal of Geophysical Research 104, 16255–16273.

Davis, D.D., et al., 1996. Assessment of ozone photochemistry in the western North Pacific as inferred from PEM-West A observations during the fall 1991. Journal of Geophysical Research 101, 2111–2134.

DeMore, W.B., et al., 1997. Chemical kinetics and photochemical data for use in stratospheric modeling. JPL Publications, 97-4.

Dibb, J.E., Talbot, R.W., Loomis, M.B., 1998. Tropospheric sulfate distribution during SUCCESS: contributions from jet exhaust and surface sources. Geophysical Research Letters 25, 1375–1378.

Dransfield, T.J., Perkins, K.K., Donahue, N.M., Anderson, J.G., Sprengnether, M.M., Demerjian, K.L., 1999. Temperature and pressure dependent kinetics of the gas-phase reaction of the hydroxyl radical with nitrogen dioxide. Geophysical Research Letters 26, 687–690.

Ehhalt, D.H., Rohrer, F., 2000. Dependence of the OH concentration on solar UV. Journal of Geophysical Research 105, 3565–3571.

Faloona, I., et al., 2000. Observations of HO$_x$ and its relationship with NO$_x$ in the upper troposphere during SONEX. Journal of Geophysical Research 105, 3771–3783.

Faloona, I., Tan, D., Brune, W.H., 1998. Observations of HO_x in and around cirrus clouds. Atmospheric Effects of Aviation Program meeting, NASA SASS, Virginia Beach, VA.

Folkins, I., et al., 1997. OH and HO_2 in two biomass burning plumes: source of HO_x and implications for ozone production. Geophysical Research Letters 24, 3185–3188.

Folkins, I., Chatfield, R., 2000. Impact of acetone on ozone production and OH in the upper troposphere at high NO_x. Journal of Geophysical Research 105, 11585–11599.

Gierczak, T., et al., 1998. Photochemistry of acetone under tropospheric conditions. Chemical Physics 231, 229–244.

Hanson, D.R., Burkholder, J.B., Howard, C.J., Ravishankara, A.R., 1992. Measurement of OH and HO_2 radical uptake coefficients on water and sulfuric acid surfaces. Journal of Physical Chemistry 96, 4979–4985.

Haywood, J.M., Schwarzkopf, M.D., Ramaswamy, V., 1998. Estimates of radiative forcing due to modeled increases in tropospheric ozone. Journal of Geophysical Research 103, 16999–17007.

Jacob, D.J., et al., 1996. Origin of ozone and NO_x in the tropical troposphere: a photochemical analysis of aircraft observations over the South Atlantic Basin. Journal of Geophysical Research 101, 24235–24250.

Jacob, D.J., 2000. Heterogeneous chemistry and tropospheric ozone. Atmospheric Environment 34, 2131–2159.

Jaeglé, L., et al., 1997. Observations of OH and HO_2 in the upper troposphere suggest a strong source from convective injection of peroxides. Geophysical Research Letters 24, 3181–3184.

Jaeglé, L., et al., 1998a. Sources of HO_x and production of ozone in the upper troposphere over the United States. Geophysical Research Letters 25, 1709–1712.

Jaeglé, L., et al., 1998b. Sources and chemistry of NO_x in the upper troposphere over the United States. Geophysical Research Letters 25, 1705–1708.

Jaeglé, L., et al., 1999. Ozone production in the upper troposphere and the influence of aircraft: approach of NO_x-saturated conditions. Geophysical Research Letters 26, 3081–3084.

Jaeglé, L., et al., 2000. Photochemistry of HO_x in the upper troposphere at northern midlatitudes. Journal of Geophysical Research 105, 3877–3892.

Koike, M., et al., 2000. Impact of aircraft emissions on reactive nitrogen over the North Atlantic Flight Corridor region. Journal of Geophysical Research 105, 3665–3677.

Levy II, H., 1971. Normal atmosphere: large radical and formaldehyde concentrations predicted. Science 173, 141–143.

Liu, S.C., et al., 1999. Sources of reactive nitrogen in the upper troposphere during SONEX. Geophysical Research Letters 26, 2441–2444.

Logan, J.A., et al., 1981. Tropospheric chemistry: a global perspective. Journal of Geophysical Research 86, 7210–7254.

Mari, C., Jacob, D.J., Bechtold, P., 2000. Transport and scavenging of soluble gases in a deep convective cloud. Journal of Geophysical Research, in press.

McKeen, S.A., Gierczak, T., Burkholder, J.B., Wennberg, P.O., Hanisco, T.F., Keim, E.R., Gao, R.S., Liu, S.C., Ravishankara, A.R., Fahey, D.W., 1997. The photochemistry of acetone in the upper troposphere: a source of odd-hydrogen radicals. Geophysical Research Letters 24, 3177–3180.

Mickley, L.J., Murti, P.P., Jacob, D.J., Logan, J.A., Rind, D., Koch, D., 1999. Radiative forcing from tropospheric ozone calculated with a unified chemistry-climate model. Journal of Geophysical Research 104, 30153–30172.

Müller, J.-F., Brasseur, G., 1999. Sources of upper tropospheric HO_x: a three-dimensional study. Journal of Geophysical Research 104, 1705–1715.

Prather, M.J., 1994. Lifetimes and eigenstates in atmospheric chemistry. Geophysical Research Letters 21, 801–804.

Prather, M.J., 1996. Time scales in atmospheric chemistry: theory, GWPs for CH$_4$ and CO, and runaway growth. Geophysical Research Letters 23, 2597–2600.

Prather, M.J., Jacob, D.J., 1997. A persistent imbalance in HO$_x$ and NO$_x$ photochemistry of the upper troposphere driven by deep tropical convection. Geophysical Research Letters 24, 3189–3192.

Raes, F., 1995. Entrainment of free tropospheric aerosol as a regulating mechanism for cloud condensation nuclei in the remote marine boundary layer. Journal of Geophysical Research 100, 2893–2903.

Roelofs, G.-J., Lelieveld, J., van Dorland, R., 1997. A three-dimensional chemistry/general circulation model simulation of anthropogenic derived ozone in the troposphere and its radiative climate forcing. Journal of Geophysical Research 102, 23389–23401.

Schultz, M., et al., 1999. On the origin of tropospheric ozone and NO$_x$ over the tropical South Pacific. Journal of Geophysical Research 104, 5829–5843.

Singh, H.B., Kanakidou, M., Crutzen, P.J., Jacob, D.J., 1995. High concentrations and photochemical fate of oxygenated hydrocarbons in the global troposphere. Nature 378, 50–54.

Singh, H.B., Thompson, A.M., Schlager, H., 1999. SONEX airborne mission and coordinated POLINAT-2 activity: overview and accomplishments. Geophysical Research Letters 26, 2053–3056.

Singh, H.B., et al., 2000. Distribution and fate of selected oxygenated organic species in the troposphere and lower stratosphere over the Atlantic. Journal of Geophysical Research 105, 3795–3805.

Snow, J., et al., 2000. Peroxides during SONEX. Journal of Geophysical Research, in preparation.

Solomon, S., Borrmann, S., Garcia, R.R., Portmann, R., Thomason, L., Poole, L.R., Winker, D., McCormick, M.P., 1997. Heterogeneous chlorine chemistry in the tropopause region. Journal of Geophysical Research 102, 21411–21429.

Stevens, P.S., Mather, J.H., Brune, W.H., 1994. Measurement of OH and HO$_2$ by laser induced fluorescence at low pressure. Journal of Geophysical Research 99, 3543–3557.

Talbot, R.W., Dibb, J.E., Loomis, M.B., 1998. Influence of vertical transport on free tropospheric aerosols over the central USA in springtime. Geophysical Research Letters 25, 1367–1370.

Talukdar, R.K., Longfellow, C.A., Gilles, M.K., Ravishankara, A.R., 1998. Quantum yields of O(^{1}D) in the photolysis of ozone between 289 and 329 nm as a function of temperature. Geophysical Research Letters 25, 143–146.

Thompson, A.M., Sparling, L.C., Kondo, Y., Anderson, B.E., Gregory, G.L., Sachse, G.W., 1999. Perspectives on NO, NO$_y$ and fine aerosol sources and variability during SONEX. Geophysical Research Letters 26, 3073–3076.

Toon, O.B., Miake-Lye, R.C., 1998. Subsonic aircraft: contrail and cloud effects special study (SUCCESS). Geophysical Research Letters 25, 1109–1112.

Wennberg, P.O., Hanisco, T.F., Cohen, R.C., Stimpfle, R.M., Lapson, L.B., Anderson, J.B., 1995. In-situ measurements of OH and HO$_2$ in the upper troposphere and stratosphere. Journal of Atmospheric Sciences 52, 3413–3420.

Wennberg, P.O., et al., 1998. Hydrogen radicals, nitrogen radicals and the production of ozone in the middle and upper troposphere. Science 279, 49–53.

Wennberg, P.O., et al., 1999. Twilight observations suggest unknown sources of HO$_x$. Geophysical Research Letters 26, 1373–1376.

Wohlfrom, K.-H., Hauler, T., Arnold, F., Singh, H.B., 1999. Acetone in the free troposphere and lower stratosphere: aircraft-based CIMS and GC measurements over the North Atlantic and a first comparison. Geophysical Research Letters 26, 2849–2852.

Zuo, Y.G., Deng, Y.W., 1999. Evidence for the production of hydrogen peroxide in rainwater by lightning during thunderstorms. Geochimica et Cosmochimica Acta 63, 3451–3455.

Air Pollution Science for the 21st Century
J. Austin, P. Brimblecombe and W. Sturges, editors
© 2002 Elsevier Science Ltd. All rights reserved.

435

Chapter 15

Future Directions: Satellite observations of tropospheric chemistry

Hanwant B. Singh

NASA Ames Research Center, Moffett Field, CA 94035, USA
E-mail: hsingh@mail.arc.nasa.gov

Daniel J. Jacob

Harvard University, Cambridge, MA 02138, USA
E-mail: djj@io.harvard.edu

The troposphere is an essential component of the earth's life support system as well as the gateway for the exchange of chemicals between different geochemical reservoirs of the earth. The chemistry of the troposphere is sensitive to perturbation from a wide range of natural phenomena and human activities. The societal concern has been greatly enhanced in recent decades due to ever increasing pressures of population growth and industrialization. Chemical changes within the troposphere control a vast array of processes that impact human health, the biosphere, and climate. A main goal of tropospheric chemistry research is to measure and understand the response of atmospheric composition to natural and anthropogenic perturbations, and to develop the capability to predict future change.

Atmospheric chemistry measurements are extremely challenging due to the low concentrations of critical species and the vast scales over which the observations must be made. Available tropospheric data are mainly from surface sites and aircraft missions. Because of the limited temporal extent of aircraft observations, we have very limited information on tropospheric composition above the surface. This situation can be contrasted to the stratosphere, where satellites have provided critical and detailed chemical data on the global distribution of key trace gases (e.g. UARS, http://uarsfot08.gsfc.nasa.gov/). (See explanation of acronyms at end of the article.)

Satellite observation of tropospheric composition is considerably more difficult than for the stratosphere because of the complexities involved in accounting for effects due to clouds, aerosols, water vapor, and the ozone layer overhead. To date, tropospheric chemistry observations from space are limited to a

First published in Atmospheric Environment 34 (2000) 4399–4401

few weeks of CO measurements from the MAPS instrument aboard the space shuttle, indirect inferences of tropical ozone from satellite measurements of the total ozone column, and preliminary data sets for the column content of a few species (O_3, NO_2, HCHO, SO_2, BrO) from the GOME solar backscatter instrument launched in 1995. These measurements provided no information on the vertical structure of these species within the troposphere. The MOPITT instrument, a gas correlation spectrometer launched in December 1999, is expected to provide global observations of CO starting in mid-2000 with reasonable vertical resolution. Also launched on the same satellite was MODIS, which has aerosol column measurement capabilities well beyond those of earlier sensors.

GOME and MOPITT herald the era of tropospheric chemistry measurements from space. Two major launches over the next few years are the ESA ENVISAT satellite (mid 2001; http://envisat.estec.esa.nl/) and the NASA AURA (previously known as CHEM) satellite (late 2002; http://eos-chem.gsfc.nasa.gov/). ENVISAT will include MIPAS, a high-resolution FTIR spectrometer observing in the limb, and SCIAMACHY, a solar backscatter instrument with both nadir and limb viewing capabilities. AURA will include HIRDLS, a limb-scanning IR radiometer with high-vertical resolution; MLS, an advanced version of the same instrument flown on UARS; OMI, a solar backscatter instrument that observes radiation in the visible and UV; and TES, a high-resolution FTIR spectrometer observing in both the limb and nadir. The instruments on board ENVISAT and AURA will provide the first space-based vertical profiles of a suite of gases in the troposphere. Not only ozone but also its key precursors (e.g. NO_x) may be globally mapped.

Table 1 compiles the principal satellite instruments expected to provide information on tropospheric composition over the next few years. Only those standard products for which retrieval capability has been carefully assessed are listed in the Table. The high-resolution spectroscopic measurements from MIPAS and TES have the potential to yield data on a large number of additional species. TES could provide global maps of species such as H_2O_2, acetone, methanol, HCN, HNO_4, SO_2, and PAN. There are plans for new launches within the next 3–5 years (e.g. GOME-2, IASI, PICASSO) which will provide a great deal more information on the distribution of gases and aerosols in the troposphere. PICASSO will use active remote sensing and provide extremely high resolution.

All satellite instruments in Table 1 are on polar orbiting platforms, which have the advantage of global coverage but the disadvantage of data sparsity (return time over a scene is typically a few days). Geostationary observation, which allow instruments to stare at a scene for an extended period, would be of considerable value for studying patterns of pollution outflow from large-scale source regions. The GIFTS instrument, an FTIR spectrometer recently selected by the NASA New Millenium Program for launch on a geostation-

Table 1. Major space-based tropospheric chemistry and aerosol data sets

Sensors	TOMS/TRIANA[a] 1979–	GOME 1995	MOPITT 1999	MODIS 1999	SCIAMACHY 2001	MIPAS 2001	TES[b] 2002	HRDLS 2002	OMI[a] 2002	MLS 2002	SAGE III 20003
O_3	Column	Column			Column +, $z = 3$–4 km limb	$z > 5$ km limb	$z = 2/4$ km limb/nadir	$z = 1$ km (UT)	Column	UT	$z = 1$ km (UT)
H_2O	Column			Column	Column +, $z = 3$–4 km limb	$z > 5$ km limb	$z = 2/4$ km limb/nadir	$z = 1$ km (UT)	Column	UT	$z = 1$ km (UT)
CO			3–4 levels		Column +, $z = 3$–4 km limb	$z > 5$ km limb	$z = 2/4$ km limb/nadir			UT	
NO							Tropical UT; $z = 2$ km				
NO_2		Column			Column +, $z = 3$–4 km limb				Column		
HNO_3						UT	UT; $z = 2$ km	$z = 1$ km (UT)			
CH_4			Column		Column +, $z = 3$–4 km limb		Column				
CH_2O		Column			Column +, $z = 3$–4 km limb				Column		
SO_2		Column			Column		Column		Column		
BrO		Column			Column +, $z = 3$–4 km limb						
Aerosol	Column			Column	Column/profiles				Column		$z = 1$ km (UT)

[a] TOMS has been in operation since 1979. Last launch was in 1996 and data continues to be collected at this time. The next TOMS launch is expected to be in 2000. TRIANA and OMI will take over TOMS functions in 2002. Much information on the AURA instruments (TES, HRDLS, OMI and MLS) is available from http://eos–chem.gsfc.nasa.gov/.

[b] A number of additional derived chemical products such as acetone, methanol, H_2O_2, HCN, NH_3, HNO_4, SO_2, and PAN are possible. UT = upper troposphere.

ary platform in 2003, will provide the first-geostationary data for tropospheric CO and possibly ozone. Instruments such as GIFTS may prove critical for the design and monitoring of international agreements controlling the export of environmentally important gases from geopolitical entities.

The growth of population and rapid industrialization in the developing world will lead to further globalization of air pollution concerns. Cumulative increases in global emission may tend to offset attempted improvements in local emission controls. Within the next 5 or so years, many studies will be performed to investigate the export of air pollution from major regions around the world. Satellite sensors will clearly be the platform of choice for such studies. These will allow the first detailed testing of global models of tropospheric chemistry.

In conclusion, the oncoming era of satellite observation could revolutionize tropospheric chemistry research. The limitations of satellite observations with regard to vertical resolution, precision, and the suite of observable species must however be kept in mind. These limitations can be overcome by in situ measurements from aircraft. Satellite and aircraft missions thus naturally complement each other, and a challenging task in the years ahead will be to design campaigns that take full advantage of this synergy to advance our scientific knowledge.

Acronyms. ESA: European Space Agency, FTIR: Fourier Transform Infrared Spectroscopy, GIFTS: Geostationary Imaging Fourier Transform Spectrometer; GOME: Global Ozone Monitoring Experiment, HIRDLS: High Resolution Dynamic Limb Sounder, IASI: Infrared Atmospheric Sounder Interferometer, MAPS: Measurement of Atmospheric Pollution from Satellites, MIPAS: Michelson Interferometer for Passive Atmospheric Sounding, MODIS: Moderate Resolution Imaging Spectroradiometer, MOPITT: Measurement of Pollution in the Troposphere, NASA: National Aeronautics and Space Administration, OMI: Ozone Monitoring Instrument, SAGE: Stratospheric Aerosol and Gas Experiment, SCIAMACHY: Scanning Imaging Absorption Spectrometer for Atmospheric Chartography, TES: Tropospheric Emission Spectrometer, TOMS: Total Ozone Mapping Spectrometer, PICASSO: Pathfinder Instruments for Cloud and Aerosol Spaceborne Observation, UARS: Upper Atmosphere Research Satellite.

Air Pollution Science for the 21st Century
J. Austin, P. Brimblecombe and W. Sturges, editors
© 2002 Elsevier Science Ltd. All rights reserved.

Chapter 16

New Directions: Rebuilding the climate change negotiations

Philippe Pernstich

Global Commons Institute, 42 Windsor Road, London, NW2 5DS, UK
E-mail: philippe@gci.org.uk

The Buenos Aires round of climate change negotiations have demonstrated that the Kyoto Protocol is a landmark on the road to nowhere. The continuing divisions over the details of the so-called Kyoto Mechanisms are an indication that the Protocol is not only inadequate in addressing the scientific facts of climate change, but it is also politically unworkable. The debate over voluntary targets, emissions trading and the Clean Development Mechanism (CDM) have distracted the attention away from the second review of the adequacy of commitments under the Convention. Consequently, the only conclusion to emerge from Buenos Aires on this point was an acceptance that the Kyoto Protocol was not sufficient to prevent harmful climatic change. Any talks about more meaningful measures, however, have been postponed.

The present impasse in the negotiations is the result of a failure to address the fundamental problem of distribution of a limited resource that is far outstripped by demand. By taking an historic perspective on the matter and trying to agree on cuts of emissions from present and past levels, we are bound to miss both the scientific goal of concentration stabilisation and the political requirement for equity. Instead, we should be looking at the resources safely available to us in the future and solve the question of distribution from that angle.

Equity has so far been the greatest stumbling block of negotiations since the very beginning of the process in 1990. The resulting division into Annex I and non-Annex I countries along the North–South divide seemed the obvious answer from the historic perspective. Taking this division into the future, however, will preserve the imbalance without slowing the climatic change. There is no question that developing countries will not be able to increase emissions indefinitely, so any delay in the shift towards more sustainable development paths ultimately represents a loss of opportunity for these countries. No one can deny the United States' claim that climate change is a global problem, and the conclusion that it therefore requires a global solution should be obvious.

The problem of distributing a scarce resource on a global scale can only be solved on an equitable basis. This is not for any ethical considerations, but

First published in Atmospheric Environment 33 (1999) 2297–2298

simply because it is the only chance of reaching an agreement that all major parties can accept. There are five criteria which will determine the success of any distribution model:

The basis of allocation must be known to each party and known to be known by other parties (Barret, 1992 in: Combatting Global Warming: Study on a global system of tradeable carbon emission entitlements, United Nations, New York).

Moral arbitrariness should be avoided (Kverndokk, Environmental Ethics, 4, 2, pp. 129–148, 1995).

The system should follow a simple allocation rule (Kverndokk, 1995 and Barret, 1992).

It should be consistent with other international policy goals, e.g. poverty alleviation in developing countries (Rose, 1992 in: Combatting Global Warming).

Any reallocation of emission permits should cause minimal disruption in the short term.

The targets set in the Kyoto Protocol clearly fail the first three of these criteria. The complete lack of any underlying structure to the Protocol means that it can only lead to a dead-end. The focus of negotiations needs to shift towards establishing a framework upon which to build a long-term, efficient and effective solution to global warming.

One proposal for such a framework that arises out of the consideration of the five criteria listed above is known as 'Contraction and Convergence'. Unlike the present approach, this takes the ultimate objective of emission stabilisation as its starting point to determine global emissions curve over a fixed period of 50–100 years or more. This global budget is then allocated to countries according to a convergence path to equal per capita entitlements by an agreed date. The entitlements are allocated in budget periods of up to five years and start out in the first period with the current distribution of per capita emissions. In each subsequent period the allocation is adjusted to narrow the present inequity in emissions until all countries receive equal per capita entitlements.

'Contraction and Convergence' is a political framework that can only work if all parties accept the need to compromise in order to achieve the Convention's ultimate objective. If this is achieved, then 'Contraction and Convergence' is the structure that can form the basis of negotiations regarding global budgets and target dates. Without it, the acceptance of compromise will never be turned into commitments if 160 countries each apply their own criteria.

In practical terms, for a stabilisation scenario of CO_2 at 450 ppmv, for example, this would mean that most developing countries would be allocated an increasing budget up to 2030 (see Fig. 1). In the case of the least developed countries, entitlements would grow well beyond any reasonably realistic growth of actual consumption, resulting in a surplus of entitlements. At the

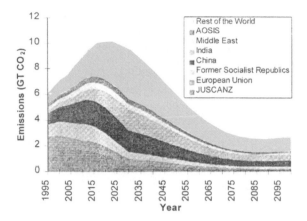

Figure 1. Stabilisation scenario of CO_2 at 450 ppmv under "Contraction and Convergence" including the Kyoto commitments. Convergence is completed by 2030 with a 70% reduction in CO_2 over 1990 levels by 2100. (AOSIS = Association of Small Island States; JUSCANZ = Japan, USA, Canada, Australia, New Zealand).

same time, industrialised countries would face quite rapid cuts in their entitlements reflecting the present gross over-consumption. Under a regime of convergence of emission entitlements, emissions trading is not only efficient but necessary. Reductions are achieved at least cost, a transfer of resources to developing countries occurs and even those countries without any real constraints on emissions in the near future have a real incentive to minimise their emissions.

Trading under this circumstance would be very different from the present proposals, where a weak trading regime including 'hot air' amongst industrialised countries only is further undermined by hypothetical savings achieved through the CDM and Joint Implementation. If credits from these mechanisms can be used to offset domestic action, the Kyoto commitment of a 5.2% reduction may well turn out to be stabilisation at best.

If the climate change negotiations are not to fail or become meaningless in the next few years, it will be necessary to take a big step back before progressing on a more principled basis. Politically, the challenge will be to achieve this without a seeming loss of face on any side. For this reason, the new approach would have to be initiated in parallel with the conclusion of the Kyoto Protocol.

Air Pollution Science for the 21st Century
J. Austin, P. Brimblecombe and W. Sturges, editors
© 2002 Elsevier Science Ltd. All rights reserved.

Chapter 17

A review of atmospheric aerosol measurements

Peter H. McMurry

*Particle Technology Laboratory, Department of Mechanical Engineering, University of
Minnesota, 111 Church St. SE, Minneapolis, MN 55455 0111, USA
E-mail: mcmurry@me.umn.edu*

Abstract

Recent developments in atmospheric aerosol measurements are reviewed. The topics included complement those covered in the recent review by Chow (JAWMA 45: 320–382, 1995) which focuses on regulatory compliance measurements and filter measurements of particulate composition. This review focuses on measurements of aerosol integral properties (total number concentration, CCN concentration, optical coefficients, etc.), aerosol physical chemical properties (density, refractive index, equilibrium water content, etc.), measurements of aerosol size distributions, and measurements of size-resolved aerosol composition. Such measurements play an essential role in studies of secondary aerosol formation by atmospheric chemical transformations and enable one to quantify the contributions of various species to effects including light scattering/absorption, health effects, dry deposition, etc. Aerosol measurement evolved from an art to a science in the 1970s following the development of instrumentation to generate monodisperse calibration aerosols of known size, composition, and concentration. While such calibration tools permit precise assessments of instrument responses to known laboratory-generated aerosols, unquantifiable uncertainties remain even when carefully calibrated instruments are used for atmospheric measurements. This is because instrument responses typically depend on aerosol properties including composition, shape, density, etc., which, for atmospheric aerosols, may vary from particle-to-particle and are often unknown. More effort needs to be made to quantify measurement accuracies that can be achieved for realistic atmospheric sampling scenarios. The measurement of organic species in atmospheric particles requires substantial development. Atmospheric aerosols typically include hundreds of organic compounds, and only a small fraction ($\sim$10%) of these can be identified by state-of-the-art analytical methodologies. Even the measurement of the total particulate organic carbon mass concentration is beset by difficulties including the unknown extent

---

First published in Atmospheric Environment 34 (2000) 1959–1999

of evaporative losses during sampling, adsorption of gas-phase organic compounds onto sampling substrates, and the unknown relationship between carbon mass and mass of the particulate organics. The development of improved methodologies for such measurements should be a high priority for the future. Mass spectrometers that measure the composition of individual particles have recently been developed. It is not clear that these instruments will provide quantitative information on species mass concentrations, and more work is needed to routinely interpret the vast quantities of data generated during field sampling. Nevertheless, these instruments substantially expand the range of atmospheric aerosol issues that can be explored experimentally. These instruments represent the most significant advance in aerosol instrumentation in recent years.

1. Introduction

Atmospheric aerosol particles range in size over more than four orders of magnitude, from freshly nucleated clusters containing a few molecules to cloud droplets and crustal dust particles up to tens of microns in size. Average particle compositions vary with size, time, and location, and the bulk compositions of individual particles of a given size also vary significantly, reflecting the particles' diverse origins and atmospheric processing. Particle surface composition is also an important characteristic since it affects interfacial mass transfer and surface reactions, which play a role in atmospheric chemical transformations. Such transformations can be significant both for their effects on gas-phase composition, as in stratospheric ozone depletion, and for their effects on particle composition. The production of fine (sub 2.5 μm) sulfates by liquid transformations in clouds is an example of a process that involves gas-to-particle mass transfer of species including water, sulfur dioxide, and oxidants.

An aerosol is defined as a suspension of liquid or solid particles in a gas. In reviewing aerosol measurement it is important to remember the gas. While atmospheric particles contain nonvolatile species such as salt, soot, metals, and crustal oxides, they also contain semivolatile compounds such as nitrates and many organic compounds. The distribution of such semivolatile compounds between the gas and particle phases varies with the amount of available particulate matter on which they can accumulate, the thermodynamic properties of the semivolatile compounds, and the gas and particle composition. Furthermore, fine (<2.5 μm) atmospheric particles are mostly hygroscopic and the water mass fraction in the condensed phase increases with relative humidity. Water typically constitutes more than half of the atmospheric fine particle mass at relative humidities exceeding roughly 80%. Thus, particle composition is inextricably linked with the composition of the gas phase, adding to the challenge of adequately characterizing the aerosol. Furthermore, sampling and/or measurement can change the thermodynamic environment or gas-phase composition thereby causing changes in particle composition before measurements are carried out.

In his visionary articles Friedlander (1970,1971) introduced a conceptual framework for characterizing instruments used for aerosol measurement. In these articles, he defined the aerosol size-composition probability density function $g(v, n_1, \ldots, n_{k-1})$ for an aerosol containing k chemical species. This function is defined such that the fraction of the total number concentration N_∞ having particle volume between v and $v + dv$, and molar composition of species i between n_i and $n_i + dn_i$ at time t is

$$\frac{\mathrm{d}N}{N_\infty} = g(t, v, n_1, \ldots n_{k-1}) \, \mathrm{d}v \, \mathrm{d}n_1 \ldots \mathrm{d}n_{k-1}. \tag{1}$$

Only $k - 1$ species are specified as independent variables because particle volume depends on the species' molar composition:

$$v = \sum_{i=1}^{k} n_i \bar{v}_i, \tag{2}$$

where $\bar{v}_i$ is the partial molar volume of species i. This formulation does not explicitly account for particle charge states, surface composition, morphologies, phase composition, etc., but it could in principle be generalized to include such information. Gas-phase compositions are implicitly coupled through the dependence of particle composition n_i on the gas phase.

Knowledge of $N_\infty g(t, v, n_1, \ldots, n_{k-1})$ would provide a comprehensive characterization of the size-resolved aerosol composition, including variations in composition among particles of a given size. Advances in single-particle mass spectrometry during the past several years have moved us closer to making such information a reality. Most aerosol measurements, however, provide integrals over time, size, and/or composition.

Fig. 1, adapted from Friedlander (1971), illustrates the type of information provided by various aerosol instruments in terms of $N_\infty g(t, v, n_1, \ldots, n_{k-1})$. The following notation is used to indicate integrations over size, time, and composition:

$$\frac{\int_{t_1}^{t_2} \int_{v_1}^{v_2} \int g \, \mathrm{d}n_1 \ldots \mathrm{d}n_{k-1} \, \mathrm{d}v \, \mathrm{d}t}{t_2 - t_1} \equiv \overline{\int_{v_1}^{v_2} \int g \, \mathrm{d}n_i \, \mathrm{d}v}. \tag{3}$$

The weighting factor, $W(v)$, for continuous integral measurements depends on the integral aerosol property being measured. Examples of weighting factors include:

$$W(v) = 1.0 \qquad \text{for CNCs,}$$

$$W(v) = \frac{\pi D_p^2}{4} \cdot K_{sp} \quad \text{for integrating nephelometers,}$$

$$W(v) = \rho_p \cdot v \qquad \text{for mass measurement,} \tag{4}$$

where D_p is the particle diameter, K_{sp} the single-particle scattering efficiency, and ρ_p the particle density. Additional information on integral measurements is available from various sources (e.g., Friedlander, 1977; Hinds, 1982; Seinfeld, 1986).

Because the available instruments use a variety of approaches to measure particle size, different sizes can be reported for the same particle. For example, the "aerodynamic size" obtained with impactors and aerodynamic particle sizers depends on particle shape, density, and size, while the "electrical mobility size" obtained by electrostatic classification depends on particle shape and size but not on density. "Optical sizes", which are determined from the amount of light scattered by individual particles, depend on particle refractive index, shape, and size. These sizes can be quite different from the "geometric" or "Stokes" sizes that would be observed in a microscope. Converting from one measure of size to another typically involves significant uncertainty. Such conversions, however, are often essential in utilizing aerosol measurements. These observations underline the importance of understanding the means used to measure sizes and of developing techniques to measure such properties including shape, density, and refractive index.

Laboratory calibrations can provide a misleading impression of accuracies that can be achieved when an instrument is used to measure atmospheric aerosols. Similar instruments that have been carefully calibrated in the laboratory may disagree when used for ambient aerosol measurements due to subtle difference in size cuts, or different sensitivities to aerosol hygroscopic properties, particle density or hygroscopicity. Therefore, rather than provide a misleading table of measurement precision and accuracy, I have discussed factors that affect measurement accuracy when discussing individual measurement techniques.

This review of aerosol instrumentation is organized according to the categories suggested by Friedlander with the order of presentation following Fig. 1. We first discuss measurements that provide a single piece of information integrated over size and composition and progress towards instruments that provide more detailed resolution with respect to size and time. We then follow a similar progression for instruments that measure aerosol chemical composition.

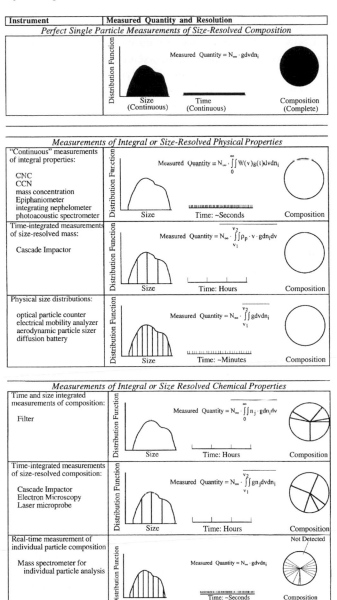

Figure 1. Classification of aerosol instruments according to their capacity to resolve size, time and composition.

A previous comprehensive review on ambient particulate measurements was written by Chow (1995). Chow's paper focuses on fixed-site sampling and includes comprehensive discussions of size-selective inlets, flow measurement, filter media, methods and sensitivities of analytical methodologies, etc. Much of the material that was discussed in Chow's article is pertinent to the NARSTO review, and the reader is referred to her paper for an in-depth critical review of measurements used for compliance monitoring. The present paper complements this earlier review.

2. Aerosol sampling inlets

The ideal aerosol sampling inlet would draw in 100% of the particles in a specified size range and would transport them all without modification to the detector or collector. Unfortunately, obtaining representative samples of aerosols can be difficult. The efficiency with which particles enter the inlet can be more or less than 100% and varies with particle size, wind speed, and direction. Particles can be lost en route from the inlet to the measurement device, and thermodynamic changes in the sampled air can lead to changes in particle size and/or chemical composition. Because problems that are encountered with inlets for fixed-point samplers are distinct from those encountered with aircraft inlets, they are discussed separately.

2.1. Fixed-point sampling

Inlets for fixed-point sampling must operate with minimal maintenance over extended periods in all weather conditions. Size-dependent sampling efficiencies depend on wind speed, and effective samplers deliver nearly 100% of the particles in the size range of interest for the usual range of wind speeds. Most commonly, inlets are designed to deliver all particles smaller than a specified size. Important design characteristics of such inlets include the particle size that is collected with 50% efficiency (i.e., the d_{50}), and the size range over which collection efficiencies rise from 0 to 100% (the "sharpness" of cut). Vincent (1989) presented a systematic discussion of practical and theoretical issues associated with sampler design and evaluation, with a particular emphasis on aspiration efficiencies. Hering (1995) discussed inertial classification techniques that are commonly used to remove particles above a specified "cut" size, and Chow (1995) critically reviewed the literature on size-selective samplers for fixed-point sampling of atmospheric aerosols. In particular, Chow summarized the characteristics of all inlets used in EPA-approved PM_{10} reference or equivalent samplers.

Size-selective inlets typically use inertial classification to remove particles larger than a specified aerodynamic size. The most commonly used inertial classifiers are impactors and cyclones. Conventional impactors accelerate the aerosol through a circular jet or a slit towards an impaction substrate located normal to the axis of the flow. Particles having sufficient inertia cross the flow streamlines to impact on this substrate. An advantage of impactors is that they can easily be designed to provide known size-dependent collection efficiencies (Marple and Liu, 1974,1975; Rader and Marple, 1985) and can be designed with sharper cutpoints than other size classifying devices that use inertial sep aration. A disadvantage of impactors is that a fraction of the dry, solid particles bounce upon impact (e.g., Dzubay et al., 1976; Wesolowski et al., 1977; Rao and Whitby, 1978; Cheng and Yeh, 1979; Wang and John, 1987a). Bounce can be avoided by coating substrates with oil or grease (Rao and Whitby, 1977; Turner and Hering, 1987; Pak et al., 1992), although such surfaces become ineffective at preventing bounce when heavily loaded (Reischl and John, 1978).

Virtual impactors are also used as size-selective inlets (Loo et al., 1976; Jaklevic et al., 1981a). With a virtual impactor, the impactor collection substrate is replaced by a receiving tube. Particles larger than the aerodynamic cut size are thrown due to their inertia into the receiving tube and delivered by the minor flow (typically 5 to 20% of the total flow) to the coarse particle filter. Particles smaller than the cut size are delivered by the major flow to the fine particle filter. Although laboratory measurements are often required to determine size-dependent losses on the receiving tube, size cuts can be calculated with reasonable confidence (Marple and Chien, 1980). Recent work has extended the flow rates and reduced the size cuts that can be achieved with virtual impactors (Solomon et al., 1983; Marple et al., 1990; Sioutas et al., 1994a–c). Virtual impactors are not affected by particle bounce and reentrainment, and they effectively collect both wet and dry particles without oil or grease-coated substrates, and therefore require less maintenance than conventional impactors when used as size-selective inlets for routine field monitoring.

Cyclones are cone-shaped or cylindrical devices in which the sampled aerosol enters tangentially, rotates several times about the axis, and exits vertically though an opening located on the axis at the top. Particles are transported to the wall by centrifugal force. Liquid particles adhere to the wall, while solid particles settle into a collecting cup located at the bottom of the cyclone. Cyclones are not affected by particle bounce or reentrainment, and they effectively collect both wet and dry particles without oil or grease-coated substrates. Cyclones can be inexpensive and are easy to maintain and operate. Unlike impactors, however, no theory provides reliable design criteria; cyclone performance must be determined empirically (Leith and Mehta, 1973; Chan and

Lippmann, 1977; Dirgo and Leith, 1985; Ramachandran et al., 1991). Also, cyclones tend to be somewhat more bulky than impactors.

Nuclepore filters with large cylindrical pores (5–12 µm) have occasionally been used to provide size-selective inlets (Flocchini et al., 1981; Cahill et al., 1990). Small particles pass through such filters with high efficiency, but large particles do not. John et al. (1983a,b) showed that the size-dependent collection efficiency and the collection mechanism depend on flow rate. At low flow rates interception is the dominant collection mechanism, while inertial collection is dominant at high flow rates. Interception depends on the particles geometric size, while inertial collection depends on aerodynamic size. Evidence for bounce of large, dry particles was observed, leading one to question the effectiveness of these filters as quantitative separators for fine and coarse particles.

2.2. Aircraft sampling

Aircraft used for atmospheric measurements typically fly at speeds ranging from 40 to 200 $m\,s^{-1}$ while the flow speed through filter samplers or aerosol counting/sizing instrumentation is typically less than 3 $m\,s^{-1}$ (Jonsson et al., 1995). Furthermore, flight maneuvers lead to variations of $\pm 10°$ in pitch and $\pm 5°$ in roll and yaw (Baumgardner and Huebert, 1993). These sampling conditions lead to significant uncertainty in aircraft aerosol measurements (Baumgardner and Huebert, 1993; Foltescu et al., 1995). Uncertainties arise from the effect of the flow around the airframe on particle size distributions near the probe inlet, unknown efficiencies with which particles of various sizes enter the sampling probe, unknown transport efficiencies through the probe to the particle measurement device, and the increase in the aerosol temperature caused by the rapid deceleration. These uncertainties can lead to measured concentrations that are either greater than or less than the true values.

Work by Huebert and coworkers (Huebert et al., 1990) demonstrated significant particulate species losses for marine sampling with isokinetic aircraft probes (i.e., probes for which the air flow speed through the inlet equals the speed of the aircraft). For an aircraft traveling at 100 $m\,s^{-1}$, they found that 50–90% of the sampled sodium deposited within the probe immediately downstream of the inlet where the flow was highly turbulent due to the small probe diameter and the high sampling speed. The fractional loss of sulfate within the probe was less than for sodium but was still significant. Although size distribution data were not available, previous work in a similar environment showed that most sulfate was submicron and most sodium was supermicron. Because losses by turbulent inertial deposition increase with size (Friedlander and Johnstone, 1957; Liu and Agarwal, 1974), they speculated that the higher losses for sodium likely resulted from its larger mean size. The fact that sulfate losses were significant led them to conclude that losses of submicron particles may have been significant and that "many of the existing literature values for aerosol concentrations above the surface [may possibly be] underestimates of the actual ambient values by factors of 2–10."

Daum and coworkers obtained data which led them to believe that isokinetic probes collect and transport submicron particles with high efficiency, as would be expected based on available theories for aerosol sampling and transport (Belyaev and Levin, 1974; Pui et al., 1987; Rader and Marple, 1988; Tsai and Pui, 1990). They found that concentrations of sulfates measured with an aircraft traveling at 50 m s^{-1} agreed to within experimental uncertainties with concentrations obtained at a fixed site sampler located on the ground (Peter Daum, personal communication, July 1997).

Fahey et al. (1989) took advantage of the non-ideal sampling characteristics of a sub-isokinetic inlet for NASAs ER-2 aircraft to obtain important information on heterogeneous chemical processes in the stratosphere. The air speed into a subisokinetic inlet is less than the speed of the aircraft. Therefore, air streamlines rapidly diverge as they approach the inlet. Particles with sufficient inertia cross streamlines and enter the inlet, leading to an enhancement in the concentration of large particles. They were able to confirm in this way that NO_y is a major constituent of polar stratospheric cloud particles.

Another factor that can play an important role in aerosol sampling is the temperature increase associated with "ram heating", which occurs when the flow speed is rapidly reduced in the probe. These temperature increases are in the range of 5–20°C, depending on the aircraft speed, and lead to evaporation of water and other volatile species. Losses of such species can affect size distributions and chemical composition of the sampled aerosol. Wilson et al. (1992) analyzed the effect of such heating on stratospheric sulfate particles as they traveled through the particle sampling inlet of NASA's ER-2 aircraft, which travels at 200 m s^{-1}, and concluded that sizes decrease by as much as 20%. More problematic than the loss of water, which can be estimated with reasonable confidence, is the loss of other labile species such as nitric acid, that are present in unknown quantities. Difficulties in measuring such species have made it difficult to quantify the composition of polar stratospheric cloud particles (Marti and Mauersberger, 1993; Molina et al., 1993; Worsnop et al., 1993).

Experts who participated in the 1991 Airborne Aerosol Inlet Workshop in Boulder, CO (Baumgardner and Huebert, 1993) concluded that more work on aircraft particle sampling inlets is required. They recommended a three-part program that includes modeling, aircraft studies, and wind tunnel studies. There has since been significant progress on this topic. Seebaugh and Lafleur (1996) have shown that it is possible to draw a portion of the flow through porous walls of conical inlets, thereby drastically reducing the development of turbulent flow within the probe. Theory suggests that particle deposition should be negligible if turbulence is negligible. Other investigators have used multiple diffusers to reduce flow speeds and shrouded inlets to both reduce speeds and align particles with the sampling inlet so as to permit isokinetic sampling.

3. Measurements of aerosol physical properties

3.1. Integral measurements

Instruments that provide totals (integrals) of specified variables over a given size range are often used for aerosol measurement. For example, condensation nucleus counters provide the total number concentration of particles larger than a minimum size, and cloud condensation nuclei counters measure the subset of particles that can form cloud droplets when exposed to water vapor at a specified supersaturation. Filter samplers are often used to measure total mass concentrations, integrated with respect to both size and time. The introduction and Fig. 1 provide graphical and mathematical explanations of integral versus size-resolved aerosol measurements.

3.1.1. Number concentration

Condensation nucleus counters (also referred to as condensation particle counters or Aitken Nuclei Counters: CNCs, CPCs, ANCs) measure the total aerosol number concentration larger than some minimum detectable size. In addition to their use in studies of aerosol climatology, CNCs are often used as detectors with other instruments such as electrical mobility classifiers (Keady et al., 1983). Recent health effects studies have suggested that particulate health effects may be more sensitive to number concentration than to mass (Oberdörster et al., 1995).

   Particles are grown by condensation in CNCs until they are sufficiently large to be detected optically. Diameter growth factors as large as 100–1000 are common, so the original particle constitutes only a minuscule fraction ($\sim 10^{-3}$–10^{-9}) of the detected droplet. CNCs can detect individual particles as small as 0.003 μm ($\sim 10^{-20}$ g), so they provide an extraordinarily sensitive means for detecting small amounts of material. A variety of substances have been used as the condensing vapor, but water and n-butyl alcohol are currently used most often. Because the supersaturation of the condensing vapor is very high, the response of CNCs is typically insensitive to the composition of the measured particles.

CNCs are often categorized according to the method used to produce the supersaturated vapor. Expansion-type CNCs were predominant until the introduction of steady-flow, forced-convection heat transfer instruments (Sinclair and Hoopes, 1975a,b; Bricard et al., 1976). In the latter instruments, supersaturation is achieved when the saturated aerosol at $\sim$35–40°C enters a laminar-flow cylindrical condenser. Supersaturation is achieved by heat transfer from the warm aerosol to the walls of the condenser, which are typically maintained at $\sim$10°C. It is advantageous to use a high molecular weight working fluid with instruments of this type to ensure that sufficient cooling (and therefore supersaturation of the vapor) occurs before the vapor is depleted by condensation on the cool walls. The working fluid should also have a vapor pressure that is high enough to cause particles to grow to $\sim$10 μm during the $\sim$0.3 s flow time through the condenser but low enough to ensure that the vapor is a small fraction of the total gas flow. n-butyl alcohol is commonly used because it meets these requirements. Steady-flow, forced-convection instruments are the most commonly used CNCs today due to their reliability and accuracy. Steady-flow instruments that achieve supersaturation by mixing cool and warm saturated air streams have also been reported and commercialized (Kousaka et al., 1982, 1992), although they have seen only limited application for atmospheric measurement.

CNCs can also be categorized as direct or indirect detection instruments. Direct (or single-particle-counting) instruments determine particle concentrations by counting individual droplets formed by condensation. The original expansion-type instruments of John Aitken (Aitken, 1890, 1891) involved the use of a microscope to manually count individual droplets collected from a known volume of air onto a grid, and today's commercially available steady-flow instruments use automated single-particle counting if concentrations are low enough ($< 10^3$–10^4 cm^{-3}, depending on the instrument design). Indirect measurement of particle concentration is achieved by measuring light attenuation through or the light scattered by the "cloud" formed by vapor condensation. Such indirect measurements require calibration with an independent concentration standard. The design of indirect-detection expansion-type CNCs culminated with the manually operated Pollak Model 1957 (Metnieks and Pollak, 1959). This instrument uses a photoelectric detector to measure the transmittance of light along the axis of a cylindrical expansion chamber. Concentrations are inferred from the measured attenuation based on a calibration scheme that involved dilution to levels that were low enough to permit measurement by manual counting techniques (Nolan and Pollak, 1946). The accuracy of measurements with the Model 1957 has been confirmed by an independent calibration (Liu et al., 1975), and the Model 1957 is still in limited use today.

The accuracy of concentrations determined by CNCs depends on the detection scheme. Uncertainties for single-particle-counting instruments are determined primarily by uncertainties in aerosol sampling rate, Poisson counting statistics, and the minimum detectable size. Uncertainties in measurements with indirect counting instruments depend on calibration accuracy, instrument stability, signal-to-noise, minimum detectable size, and sampling line losses. Under most practical situations for atmospheric sampling, accuracies of $\sim 10\%$ are typical, although discrepancies can be much greater if high concentrations of nanoparticles (particles smaller than ~ 20 nm) are present. This is because the fraction of nanoparticles that is detected varies with instrument design, and because the counting efficiency of nanoparticles can also be affected by vapor depletion due to preferential condensation on "large" particles in the condenser. For example, in comparisons of CNCs operating on NCAR's C-130 aircraft during ACE-1, Weber et al. (1999) found that when nucleation was not occurring a TSI 3760, a TSI 3025 and a University of Minnesota prototype ultrafine CNC agreed, on average, within 4%. Two other instruments were, on average, 20 and 65% lower than these. Post-campaign measurement revealed that these latter measurements were affected by flow calibration errors. When nucleation was occurring (i.e., when concentrations of particles in the 3–4 nm range were high), the average concentrations of the TSI 3760, the TSI 3025 and the University of Minnesota prototype instrument were, respectively, 1045, 1577, and 2854 cm^{-3}, illustrating the sensitivity of the lower detection limit to measured total number concentrations. Measurements of CN concentrations on the R/V Discoverer, a NOAA ship, were about 20% below values measured on the C-130 nearby, while measurements at Cape Grim, a ground station, were within 4% of C-130 values.

The minimum detectable size and the size-dependent detection efficiencies of CNCs vary significantly with the instrument design and sampling pressure (Liu and Kim, 1977; Wilson et al., 1983a, b; Bartz et al., 1985; Wiedensohler et al., 1997). The key variables that affect the minimum detectable size are transport efficiency to the condenser and the vapor supersaturations to which particles are exposed. Due to the effects of curvature on vapor pressure (Thompson, 1871), the supersaturation that is required to activate particles increases with decreasing size. Particles of ~ 3 nm require supersaturations of several hundred percent; 3 nm is near the lower detection limit of CNCs because new particles are produced by self-nucleation of the vapor when supersaturations ratios are increased beyond this. Stolzenburg and McMurry (1991) designed an instrument that provides high detection efficiencies for particles down to 3 nm. Pressure affects the performance of steady-flow CNCs through its effect on heat and mass transfer (Bricard et al., 1976; Wilson et al., 1983a, b; Zhang and Liu, 1991; Saros et al., 1996). Instruments that function at pressures down to 40 mbar (21.5 km altitude) have been developed.

3.1.2. Cloud condensation nuclei concentrations

Cloud condensation nuclei (CCN) counters measure the concentration of particles that are converted to cloud droplets by condensation of water (i.e., "activated") at a specified supersaturation. CCN concentrations depend on both the aerosol size distribution and the aerosol composition (Junge and McLaren, 1971; Fitzgerald, 1973; Pruppacher and Klett, 1980; Harrison, 1985). Particles down to ~0.04 μm diameter can serve as cloud condensation nuclei. CCN measurements are of central importance in determining the influence of anthropogenic particles on the atmosphere. For example, a significant fraction of the sulfate aerosol production in the atmosphere occurs in cloud droplets (Schwartz, 1989), and particulate pollution may increase cloud albedo, thereby decreasing the Earth's net incoming radiative energy (Twomey et al., 1984; Twomey, 1991). In order to develop valid models for such phenomena, it is necessary to understand the relationship between atmospheric aerosol properties and the number and size of cloud droplets that can be produced from them.

Unlike CNCs, which use a variety of working fluids at supersaturations of several hundred percent, CCN counters use only water and operate at supersaturations pertinent to cloud formation (~0.01 to ~1%). Therefore, the CCN concentration is a fraction of the CNC concentration and depends on the supersaturation employed. The saturation ratio that is required to activate particles increases with decreasing size. Important design parameters for CCN counters include the range of saturation ratios for which information can be obtained, the method used for determining the relationship between CCN concentrations and saturation ratio, and the particle growth time (for example, 100 s is required to achieve equilibrium for supersaturations of 0.01% (Hoppel et al., 1979)). Most CCN instruments use thermal gradient diffusion chambers to produce the desired supersaturations.

The Third International CCN Intercomparison Workshop was held in Reno, NV, in 1980 and involved nearly all CCN instruments in the world at that time including 9 static thermal gradient diffusion cloud chambers (STGDCC), 5 continuous flow diffusion cloud chambers (CFDCC), 4 isothermal haze chambers (IHC), and 2 diffusion tubes. The reader is referred to the special issues of Journal de Recherches Atmosphériques (1981, pp. 181–373) for a complete discussion of the results of this most recent workshop. A more recent review of CCN instruments is provided by Hudson (1993). The following discussion summarizes observations from these reviews.

Isothermal haze chambers (IHCs) (Fitzgerald et al., 1981) are used to measure concentrations of CCN that are activated in the low supersaturation (0.015–0.15%) range. The principle of operation of these instruments was first described by (Laktionov, 1972), who showed that there exists a unique relationship between the critical supersaturation required to activate a particle and its equilibrium size at 100% relative humidity. In IHCs, therefore, size distributions of aerosols are measured after they are exposed for an extended period of time to an atmosphere at 100% relative humidity. Optical particle counters are typically used to measure size distributions. Accurate results require accurate measurements of both droplet size and concentration. Such measurements are subject to sizing errors associated with droplet evaporation in the warm optical particle counter and differences between the refractive index of the water droplets and that of the calibration aerosols. An advantage of this technique is that a single measurement provides the spectrum of droplets that are activated over the specified range. The useful range of supersaturations for this technique is limited by the minimum size that can be detected optically. Two of the four IHCs tested during the Third International Workshop agreed to within 40% over the entire supersaturation range tested, while discrepancies for the other instruments exceeded this.

Diffusion tubes (Leaitch and Megaw, 1982) function in the 0.04–0.3% supersaturation range. These instruments involve steady flow through a heated, wetted tube. Because water vapor diffuses faster than heat, the aerosol along the centerline of the tube becomes supersaturated. The radial and axial saturation ratio profiles within the tube depend on the inlet relative humidity and on the temperature difference between the tube walls and the incoming aerosol. In practice, theory is used to calculate the saturation ratio profiles based on operating conditions. An optical particle counter is used to measure droplet distributions downstream of the diffusion tube. Theory for activation and growth of particles of known composition is then applied to these data to infer the CCN concentration at an effective average supersaturation corresponding to operating conditions. Limitations of this technique are that particle composition must be known, and an extended time ($\sim$45 min) is required to scan through a range of supersaturations. Also, the OPC measurements have limitations similar to those encountered with IHCs.

Static thermal gradient diffusion cloud chambers were the first commonly used type of CCN counter (Twomey, 1967). The vast majority of instruments used in the first two international workshops were of this type. These instruments consist of two wetted surfaces maintained at different temperatures. Water vapor is saturated at each of the surfaces but rises to a peak supersaturation at a point between the surfaces due to the nonlinear dependence of vapor pressure on temperature. Photographic or optical techniques are used to count droplet concentrations at the location of this peak supersaturation (Juisto et al., 1981). By operating over a range of temperature differentials, the relationship between CCN and supersaturation is determined. In practice, measurements are limited to the 0.1 to 1.0% supersaturation range. A limitation of this approach is that a long time is required to carry out measurements over a range of supersaturations due to the time required for thermal stabilization. Nine different static thermal gradient instruments were compared during the Third International Workshop. It was found that five of these typically agreed to within 20%; 10% agreement was achieved at 1% supersaturation. Discrepancies among the remaining instruments were greater than this, due in part to the newness and/or lack of previous calibration for these instruments.

Continuous flow diffusion cloud chambers (CFDCCs) involve flow between wetted parallel plates that are maintained at different temperatures (Hudson and Alofs, 1981). As with static diffusion chambers, the water vapor supersaturations achieve a peak value at some point between the plates. An optical counter measures the droplet concentration at the exit from the chamber along the streamline that is exposed to the peak supersaturation. Measurements are made at several temperature differentials to determine the relationship between CCN concentration and supersaturation. Again, a significant time is required to complete a single measurement. The five instruments tested during the Third International Workshop typically agreed to within 15%.

Several CCN spectrometers have been developed that permit rapid measurements of CCN-supersaturation spectra. The instrument of Radke et al. (1981) involves four continuous flow diffusion cloud chambers operated in parallel, each with a different temperature difference, thereby providing near-real-time measurements of CCN concentrations at four saturation ratios in the 0.1–1% range. This instrument was designed for aircraft use where rapid measurements are necessary. The University of North Carolina spectrometer (Fukuta and Saxena, 1979) involves flow through a rectangular channel, as in CFDCCs. In contrast to CFDCCs, however, the temperature gradient is maintained across the width of the channel rather than between the two larger plates. Thus, the supersaturation varies across the channel. An optical particle counter is moved across the width of the channel to obtain the CCN-supersaturation relationship. Measurements require about 30 s, and operation is limited to supersaturations above about 0.1%. Hudson's "instantaneous" CCN spectrometer (Hudson, 1989) provides information on the CCN-supersaturation spectrum over the 0.01 to 1% range at a rate of ~1 Hz. This instrument is similar in design to a CFDCC but is different in that CCN spectra are obtained from droplet size distributions measured with an optical particle counter at the exit of the instrument, much as is done with the IHC. The relationship between the final droplet size and the initial size is obtained by calibration with monodisperse particles of known composition. This instrument has seen a great deal of use on aircraft due to its fast time response and broad supersaturation range.

The CCN apparatus of Khlystov et al. (1996) is unique for its ability to handle large volumetric flows (30 m^3 min^{-1}), thereby it can be equipped with instrumentation designed for in-cloud studies. While not designed for routine measurements of CCN concentrations, this apparatus will provide new information on the influence of anthropogenic aerosols on the formation and microstructure of marine clouds.

CCN counters in use today are mostly laboratory prototype instruments rather than standard commercial products. Intercomparison workshops have led to higher confidence in the accuracy of particular instruments, but measurements made by different groups are not necessarily comparable. Furthermore, because the tendency of the atmosphere to produce clouds depends upon supersaturation, it is essential to carry out CCN measurements with instruments that operate in a relevant supersaturation range. Establishing the relationship between the size-resolved composition of ambient aerosols and their cloud nucleating characteristics is essential for developing valid models for the "indirect" effect of aerosols on radiative forcing and in-cloud chemical transformations.

3.1.3. Particle mass concentrations

Measurements of particulate mass concentrations are important for regulatory and scientific reasons. The current US National Ambient Air Quality Standard for particulate matter applies to mass concentrations smaller than 10 μm aerodynamic diameter, and a new standard for mass concentrations of particles smaller than 2.5 μm aerodynamic diameter has been promulgated (Fedeal Register, 1997). Federal Reference Methods for these mass measurement techniques are discussed later in the paper. While research studies tend to focus on speciation and size, it is essential to be able to reconcile measured mass concentrations with the sum of measured species. Therefore, mass concentrations are also routinely measured in aerosol research studies. In this section the various techniques that are used to measure mass concentration are discussed.

3.1.3.1. Manual methods. The most commonly used technique for measuring particulate mass concentrations involves filtration. Filters are weighed under controlled temperature and relative humidity conditions before and after sampling, and mass concentrations are determined from the increase in filter mass and the volume of air sampled. Filter samplers are most commonly equipped with inlets that eliminate particles above a specified size cut.

Fiber, membrane, granular bed and Nuclepore filters made from a wide variety of materials are used to collect aerosols (Lippmann, 1989; Lee and Ramamurthi, 1993; Chow, 1995). The physics of particle collection by filters is similar for all types of filters. Particles smaller than about 0.1 μm are collected by diffusion. Because particle diffusivities increase with decreasing size, collection efficiencies increase as size drops below ∼0.1 μm. Particles larger than about 0.5 μm are collected by interception and impaction. Collection efficiencies by these mechanisms increase with increasing size. Therefore, collection efficiencies tend to increase with increasing size above 0.5 µm. It follows the that "most penetrating particle size" typically falls between 0.1 and 0.5 µm. The value of this most penetrating particle size depends on the filter characteristics and the flow rate through the filter (Lee and Liu, 1980). Many filters that are used for aerosol measurement collect all particles with >99% efficiency. The collection efficiency of loosely woven fiber filters or membrane filters having large pore sizes tend to be less than this, however (Liu et al., 1983).

Analytical sensitivities for gravimetric analyses are currently about ±1 µg. Therefore, the analytical uncertainty for a 24-h sample obtained using the proposed EPA PM$_{2.5}$ reference method sampler, which operates at 1.0 m^3 h^{-1}, would ideally be 0.04 µg m^{-3}. Measurements have shown that actual uncertainties are substantially greater than this. Factors including water adsorption/desorption by the filter media, adsorption or volatilization of reactive species, particle losses associated with handling, etc., lead to these higher uncertainties in gravimetric measurements. The Federal Reference Method for PM$_{2.5}$ indicates that the lower detection limit for mass concentration is ~2 µg m^{-3}.

Side-by-side measurements with identical samplers and replicate measurements on a given sample permit one to establish the precision with which filter samplers can measure mass concentrations. Determining measurement accuracy is more problematic. Filter measurements are affected by vapor adsorption on substrates (McMurry and Zhang, 1989; Hering et al., 1990; McDow and Huntzicker, 1990), by evaporative losses of semivolatile compounds during or after sampling (Smith et al., 1978; Appel and Tokiwa, 1981; Dunwoody, 1986; Wang and John, 1988; Witz et al., 1990; Eatough et al., 1993), and by reactions between collected particles and substrates (Smith et al., 1978). The extent of these processes varies with location depending on the aerosol mass concentration and composition and temperature and relative humidity. It is likely that such measurement errors are substantially in excess of reported measurement precision.

For more complete information on filter gravimetric measurements, the reader is referred to the review paper of Chow (1995), in which she provided a comprehensive discussion of the relative merits and disadvantages of currently available size-selective inlets, filter sampling media, and filter holders for gravimetric analyses.

3.1.3.2. Automated methods. Automated methods for measurements of aerosol mass concentrations are discussed by Williams et al. (1993). Available methods include the beta gauge, piezoelectric crystals, and the oscillating element instruments. These techniques are briefly reviewed below.

Beta gauges. Beta gauges measure the attenuation of 0.01–0.1 MeV beta particles from a radioactive source through a particle-laden filter. Attenuation results from scattering of the beta particles by atomic electrons in the filter media and by the deposited particles and is therefore determined by the areal density of atomic electrons. Except for hydrogen, which usually constitutes a small fraction of the particulate mass, the ratio of atomic number to mass is nearly independent of element, and ranges from 0.38 to 0.50. For the elements that constitute the majority of the atmospheric particulate mass (C, Ca, Cl, Fe, Mg, N, O, K, Si, Na, S), however, this ratio ranges from 0.47 to 0.50. For ammonium sulfate, which contains a significant amount of H, the ratio increases to 0.53. Thus, errors associated with the assumed ratio of atomic number to mass may be roughly 10%. In practice, Beta gauges are calibrated with ambient aerosols to minimize this error.

Particle mass loadings are determined from the increase in attenuation that is measured as particles are added to the filter. Experimental studies of Beta gauges show that measurement precision tends to be poorer for instruments used for routine ambient monitoring than for instruments used under controlled laboratory conditions. Early studies with beta gauges show that precisions on the order of 25 $\mu g\,cm^{-2}$ can be achieved with monitoring instruments (Husar, 1974), while $\sim$5 $\mu g\,cm^{-2}$ is possible in careful laboratory experiments (Jaklevic et al., 1981b; Courtney et al., 1982). Courtney et al. concluded that mass measurements by beta attenuation are in good agreement with gravimetric mass measurements, and that the advantages of automation make beta attenuation an attractive alternative. Macias and Husar (1976) argued strongly for the utility of mass measurements by beta attenuation. Two commercially produced beta gauges have been designated by EPA as Equivalent Methods for measuring sub-10 μm particulate mass concentrations (PM-10).

Piezoelectric crystals. Piezoelectric crystals undergo mechanical deformations when an electrical potential is applied across certain crystal planes (Ward and Buttry, 1990). If a periodic potential is applied, then the crystal will expand and contract periodically. Crystals have a resonant vibrational frequency that depends on the crystalline material and thickness. This resonant frequency can be altered by adding mass to a vibrating surface. Piezoelectric crystal mass monitors determine aerosol mass loadings by measuring the change in this resonant frequency caused by the deposition of particles from a known volume of air. The use of piezoelectric crystals for monitoring ambient particulate mass loadings has been reviewed by Lundgren et al. (1976) and Williams et al. (1993).

Piezoelectric crystals used for particulate mass monitoring typically consist of quartz cut on the AT crystallographic planes and have natural resonant frequencies, of 5–10 MHz. The sensitivity of this resonant vibrational frequency to incremental mass are typically 10^3 Hz μg^{-1}, and stabilities of ± 0.5 Hz at 10 MHz can be achieved. The change in the frequency of the particle-laden crystal is determined by electronically mixing its resonant frequency with that from an identical crystal maintained at the same thermodynamic conditions. The difference, or beat, signal is proportional to the particulate mass loading.

Piezoelectric crystals can measure particulate masses as small as 1 ng, although loadings of tens of ng are typically used for measurement. Nonlinearities in the relationship between Δf and Δm typically become significant when mass loadings exceed 5 to 10 µg, at which point the deposition surface must be cleaned. The need to provide such routine maintenance is a disadvantage for instruments that are used for routine monitoring purposes. Other sources of error that have been reported include sensitivity to temperature and relative humidity and poor mechanical coupling between some particle types and the oscillating surfaces, which invalidates the relationship between frequency change and mass increment.

Several instruments that utilize piezoelectric crystals for measuring particulate mass concentrations in the 10 µg m^{-3} to 10 mg m^{-3} range are available. None of these instruments are designated by EPA as an Equivalent Method for measurements of particulate mass concentration.

Work with surface acoustic wave (SAW) mode microbalances has also been reported (Bowers and Chuan, 1989). This approach involves the creation of surface waves by a pair of electrodes on a common surface. Resonant frequencies produced in this way can be much higher than are achieved with AT-cut crystals, leading to much higher mass sensitivities. Sampling using SAW mode microbalances has been reported for situations where aerosol loadings are extremely low, such as in rockets used to sample stratospheric aerosols.

Harmonic oscillating elements. The unique component of the harmonic oscillating element instruments (Patashnick and Rupprecht, 1991) is a tapered tube, the wide end of which is mounted to a rigid base. Particles are collected on a replaceable 0.5 cm diameter filter that is mounted on the narrow end of the tapered element, which is free to oscillate. The element vibrates at a frequency that depends on its geometrical and mechanical properties and on the mass of the filter. As particles are collected on the filter, the element's natural frequency of oscillation decreases. An optical system is used to measure the natural oscillation frequency; oscillations are induced electrically.

The resonant frequency of the tapered element is affected by thermal expansion and contraction associated with temperature fluctuations. Therefore, these instruments must operate at constant temperature. For ease of operation, this temperature is fixed at a value in excess of ambient values, typically 50°C, which exacerbates the loss of semivolatile compounds. Lower temperatures (30°C) were tried in the San Joaquin Valley during the winter of 1995, however, condensation and evaporation of water vapor on the filter during high humidity events negated the measurements (Solomon 1997, personal communication, PG & E, San Ramon, CA).

Harmonic oscillating element instruments provide a very sensitive technique for particulate mass measurements. The mass resolution for 10-min samples is ± 5 μg m^{-3}. Based on comparisons with EPA's designated reference method when sampling ambient aerosols, a commercially available instrument of this design was designated an equivalent method for PM-10 monitoring.

3.1.4. Epiphaniometer

The epiphaniometer (Gäggeler et al., 1989) measures the diffusion-limited mass transfer rate of a gas to aerosol particles. These measurements provide information on the maximum possible rates of vapor condensation to or gas reaction with the aerosol. The epiphaniometer also provides a sensitive, real-time measurement of an integral aerosol property (Baltensperger et al., 1991).

Measurements involve adding the gas-phase radioactive isotope ^{211}Pb to the aerosol. The ^{211}Pb, which is produced from a ^{227}Ac source, diffuses and attaches to the aerosol particles. After exposure to the ^{211}Pb for about 2 min, particles are collected on a filter where the amount of attached ^{211}Pb is measured using an α detector.

The mass transport rate of gases to particles occurs at a rate that is proportional to particle surface area (particle diameter squared) for particles that are small compared to the mean free path of the gas ($\sim$0.067 μm for air at normal temperature and pressure). Mass transport rates vary in proportion to particle diameter for particles that are large compared to the mean free path. Atmospheric aerosols fall mostly in the "transition" regime, where neither of these simple limiting cases applies. The integral aerosol property that is proportional to the gas mass transfer rate is often referred to as the "Fuchs surface," and the Epiphaniometer directly measures the value of this integral. Theoretical expressions that enable one to calculate the "Fuchs surface" from the aerosol size distribution are given by Fuchs and Sutugin (1970) and by Davis and Ray (1978) among others.

3.1.5. Aerosol optical properties

The appearance of a distant object viewed through the atmosphere is affected by several factors: the amount and color of light emitted by the object (initial radiance); the transmittance of that light from the object to the observer; and the scattering of ambient light into the sight path by the atmosphere (path radiance) (Duntley et al., 1957; Malm, 1979; Richards, 1988). Transmittance is determined by the scattering and absorption of light as it traverses the atmosphere. Because the initial radiance, transmittance, and path radiance are sensitive to the wavelength of the light, one must know how those factors depend on wavelength before optical effects such as visibility impairment or atmospheric albedo can be characterized fully. Middleton (1952) provides a comprehensive discussion of atmospheric visibility and its measurement, and Quinn et al. (1996) and McMurry et al. (1996a, b) discuss the measurement of aerosol optical properties. We focus here on point measurements of aerosol scattering and absorption coefficients.

A significant issue in the measurement of atmospheric optical properties is the replacement of human observers by the Automated Surface Observing System (ASOS) at airports which is currently underway. Although data from human observers are qualitative, they provide a comprehensive historical record that has been useful for assessing visibility trends across the US ASOS instrumentation measures the amount of light scattered in the forward direction with a time resolution of 1 min. Because ASOS is used for flight safety, it does not record data for visibilities exceeding 10 miles. However, a recent study by Richards et al. (1997) showed that ASOS equipment is capable of measuring scattering at aerosol concentrations as low as $\sim 10\ \mu g\,m^{-3}$. It would be sensible to record ASOS data routinely at the highest time resolution and sensitivity possible. This could be done at a nominal cost, and would serve to ensure a long-term record at a consistent set of measurement sites.

3.1.5.1. Scattering coefficient. Integrating nephelometers (Beuttell and Brewer, 1949) measure the total amount of light scattered by an aerosol. The "integration" covers scattering angles from near forward to near backward. To determine the contribution of gases and electronic noise to the scattering signal, the instrument's light scattering response to filtered air is measured periodically. The contribution of particles to scattering is then determined by difference. When equipped with a photon-counting detector (Charlson et al., 1974), the integrating nephelometer can measure particle light scattering coefficients of less than $0.1\ Mm^{-1}$, a value equal to about 1% of the light scattering coefficient of particle-free air at normal atmospheric pressure. The design and applications of the integrating nephelometer were recently reviewed by Heintzenberg and Charlson (1996).

Because of its potential for high accuracy, portability, and moderate cost, the nephelometer has been used widely for measurements of light scattering coefficients. There are, however, several sources of measurement error with this instrument. First, the contributions of coarse particles (particle diameter greater than about 5 μm) to scattering are underestimated because they tend to deposit at the inlet. Such inlet losses increase strongly with size. Second, the optics do not permit measurement of light scattered in the near forward direction (between 0° and 5–10°, depending on the instrument design). The magnitude of this truncation error is typically ~ 10–15% for submicron particles (Ensor and Waggoner, 1970; Sloane et al., 1991; Anderson et al., 1996a). However, because large particles scatter strongly in the forward direction, truncation errors can be as large as 50% for particles of ~5 µm. Finally, errors caused by droplet evaporation due to heating of the aerosol can be significant, especially at high relative humidities (>90%), where water constitutes most of the particle volume.

Two new nephelometers have recently become commercially available. The Optec NGN-2 (Malm et al., 1996) is an "open air" design which minimizes errors associated with inlet losses and with heating. Scattering is measured between 5 and 175°for illuminating radiation centered at 550 nm. The TSI 3563 measures total scattering (7–170°) and back scattering (90–170°) for monochromatic radiation at 450, 550, and 700 nm. Back scattering is of particular importance in evaluating the contribution of aerosols to the Earths albedo in climate-effects studies. Intercomparisons of field measurements during the Southeastern Aerosol and Visibility Study (SEAVS) in the summer of 1995 showed that three Optec NGN-2 instruments agreed with each other to within 3%. Green light scattering (550 nm) measured with the TSI 3563 was, on average, 70% of the value measured by the Optec instruments at relative humidities less than 60%. These instruments were found to agree to within about 3% after data were corrected to account for differences in illuminating radiation and instrument truncation angles (Saxena et al., 1996). At higher relative humidities, agreement tends to deteriorate because small differences in relative humidities at the point of measurement can lead to significant differences in scattering. These results show that accurate assessments of sunlight scattering by atmospheric aerosols require that instrumental measurements be corrected for truncation, for differences in illuminating radiation, and for subtle differences between the relative humidity at the point of measurement and in the atmosphere.

Although the amount of light that an aerosol with a given mass concentration scatters depends on its size distribution, measurements have shown that that the ratio of the dry scattering coefficient to the dry fine particle mass concentration measured at various locations does not vary a great deal. For example, Charlson et al. (1968) found that this ratio (referred to as the dry fine particle mass scattering efficiency) measured at New York, San Jose, and Seattle averaged 3.3 $m^2 g^{-1}$, with a range of 1.5 to 5.6 $m^2 g^{-1}$.

Information on wavelength-dependent light scattering provided by multi-wave length nephelometers provides useful information on aerosol size distributions (Thielke et al., 1972; van de Hulst, 1981). It is often found that within the optical subrange (~0.16–1.2 µm; Junge, 1963) the light scattering coefficient, b_{sp}, and aerosol size distribution function, dN/dD_p, obey the following power-law relationships:

$$b_{sp} = C\lambda^{-\alpha}, \tag{5}$$

where α is referred to as the Angstrom exponent, and

$$\frac{\mathrm{d}N}{\mathrm{d}D_{\mathrm{p}}} = K \cdot D_p^{-\nu}. \tag{6}$$

Power-law aerosol size distribution functions of this form are referred to as Junge distributions. When these relationships apply it can be shown that

$$\alpha = \nu - 3. \tag{7}$$

Thus, if b_{sp} depends strongly on wavelength (large α), then the size distribution function decreases strongly with size. Measurements have shown that values of α tend to be higher for continental aerosols than for clean marine aerosols (Ogren, 1995).

3.1.5.2. Absorption coefficient. Absorption coefficients are most commonly inferred from measurements on particles collected on filters. However, the viability of conducting in-situ measurements using photoacoustic spectrometry has also been demonstrated. These techniques are discussed in this section.

Filter techniques are the most common methods for measuring particle absorption coefficients. Because light transmittance through filters is affected by scattering and absorption, the effects of scattering, including multiple scattering, must be accounted for. If light interacts with more than one particle as it passes through the filter, the apparent absorption coefficient will exceed the correct value. Also, filter techniques are problematic because the optical properties of deposited particles may be different from those of airborne particles, especially if the particles undergo chemical reactions on the filter.

Lin et al. (1973) developed the integrating plate technique for measuring absorption coefficients of particle deposits on filters. With this method, an opal glass plate is located between the filter and the optical detector. Because the opal glass is a diffuse reflector, light scattered by particles in the forward direction is detected with the same efficiency as light that enters the glass directly. If backward scattering is small in comparison to absorption, changes in filter transmittance before and after particle collection can be attributed to particle absorption. Lin et al. concluded that neither backward scattering nor multiple scattering contributed significantly to errors in their measurements. This technique was modified somewhat by Clarke (1982) to improve measurement accuracy. Clarke reports that with his modifications this technique can measure absorption coefficients as low as 0.005 Mm^{-1} for a 10-h sampling period.

Hänel (1987) argued that multiple scattering and backward scattering led to significant errors in absorption coefficients measured by previous investigators using filter techniques. He developed an approach for measuring absorption coefficient that permitted accounting for forward, backward, and multiple scattering from collected particles. Reported values for absorption coefficients using this approach are somewhat smaller than values determined with other techniques.

The earliest and simplest method of measuring light absorption by particles on filters is the coefficient of haze (COH) technique (Hemeon et al., 1953). The aethalometer described by Hansen et al. (1984) involves an updated and more sensitive absorption measurement that operates on a similar principle. These techniques measure the light attenuation caused by an aerosol sample on a filter; no integrating plate is used to correct for light scattering. Wolff et al. (1983) found a good correlation between COH and concentrations of elemental carbon, and Campbell et al. (1989) found good correlations between COH and absorption measurements using an integrating plate and integrating sphere. Commercially available aethalometers provide continuous, near real-time data and have been used for monitoring. Quantification is achieved by calibrating against a more accurate absorption measurement standard.

Photoacoustic spectroscopy measures the absorption coefficients of suspended particles in real time (Adams, 1988). Work on a small and very sensitive photoacoustic instrument that shows promise for field measurements was recently reported by Moosmüller and coworkers (Arnott et al., 1995, 1999; Moosmüller et al., 1997a, b). In these instruments, the air stream, from which NO_2 has been removed, is drawn into an acoustic cell where it is illuminated by light that is modulated at the resonant frequency of the cell. Light energy absorbed by the particles heats the carrier gas, which expands and then contracts according to the modulation frequency of the light. The associated pressure variation is a sound wave whose intensity can be measured with a microphone. Adams (1989) reports a sensitivity of 3 Mm^{-1} for absorption coefficients measured with her instrument, while Moosmüller reports a sensitivity of 0.4–0.5 Mm^{-1} for his. Because photoacoustic spectroscopy involves measurements on gas borne particles and is therefore not affected by errors inherent to filter-based techniques, it is conceptually the best available technique for measuring particle absorption coefficients. However, it requires skilled personnel and complex equipment that is not commercially available. Therefore, it is not yet suitable for routine monitoring.

Aerosol absorption coefficients are frequently inferred from measurements of "elemental carbon" concentrations. Obtaining an accurate value for absorption coefficientes with this approach requires (1) an accurate measurement of the elemental carbon concentration, (2) knowledge that elemental carbon is the only significantly absorbing particulate species, and (3) knowledge of the elemental carbon "mass absorption efficiency". Foot and Kilsby (1989) compared particle absorption coefficients measured with a filter technique with those measured with a photoacoustic technique. They used laboratory particles with known properties and found that agreement between the two methods was ±15%. Measurements in Los Angeles during the Southern California Air Quality Study (SCAQS) showed that absorption coefficients measured with a photoacoustic spectrometer agreed, on average, to within about 10% ($r = 0.926$) of values obtained from elemental carbon concentrations measured with a thermal-optical technique (Turpin et al., 1990a) assuming a mass absorption efficiency for elemental carbon of 10 m^2 g^{-1}. In a previous study designed to compare nine techniques for measuring elemental and organic carbon concentrations it was found that the ratio of the method mean to the grand mean for all elemental carbon measurements for these nine techniques ranged from 0.43 to 1.48 (Hering et al., 1990). Therefore, different techniques for measuring elemental carbon concentrations will lead to estimates of absorption coefficients that vary by a factor of ~3 for a given value of the mass absorption efficiency.

3.2. Size-resolved measurements

3.2.1. Optical particle counters

Single-particle optical counters (OPCs) measure the amount of light scattered by individual particles as they traverse a tightly focused beam of light. A fraction of the scattered light is collected and directed to a photodetector, where it is converted to a proportional voltage pulse. Particle size is determined from the magnitude of this voltage pulse by using a calibration curve typically obtained from measurements using spherical particles of known size and composition. Pulse height and area are commonly used measures of pulse magnitude. Size distributions are obtained by measuring the distribution of pulse magnitudes obtained from a representative population of particles. A review of aerosol measurement by light scattering is given by Gebhart (1993).

OPCs tend to heat aerosols leading to a decrease in size and an increase in refractive index for hygroscopic atmospheric particles. These perturbations make it difficult to accurately measure atmospheric aerosol size distributions with OPCs, and insufficient attention is often given to these effects. Biswas et al. (1987) showed that these errors can be substantially reduced by using a heat exchanger to control the sheath air temperature within the OPC.

Instrument design and particle optical properties both play roles in determining the relationship between size and the pulse magnitude. Important instrument design features include characteristics of the illuminating radiation and the solid angle from which scattered light is collected and focused into the photodetector. Illuminating radiation is either monochromatic (laser) or incandescent (white light), and collecting optics of most commercial OPCs can be categorized as either near forward scattering or wide angle.

Both incandescent and monochromatic forward scattering instruments exhibit a monotonic dependence of pulse magnitude on size for very small particles. For nonabsorbing particles that are somewhat greater than the wavelength of the illuminating radiation, however, responses of both types oscillate with size, leading to nonmonotonic relationships between size and response. White light instruments that collect scattered light over a wide solid angle show a monotonic dependence of response on size for nonabsorbing particles. Monochromatic wide angle instruments exhibit oscillations for particles on the order of the wavelength of light and larger. For strongly absorbing particles, wide angle incandescent and monochromatic instruments exhibit a very weak dependence of pulse height on size for particle sizes between 0.3 and 1 μm.

Lasers provide illuminating intensities several orders of magnitude higher than can be achieved with incandescent sources, thereby enabling the detection of significantly smaller particles; laser OPCs having minimum detection limits of ∼0.05 μm are available, while white light OPCs typically cannot detect particles smaller than ∼0.3 μm. Therefore, laser illumination is almost always preferable for particles smaller than the wavelength of the illuminating radiation, while white light illumination can have distinct advantages for larger particles (Gebhart et al., 1976). Nevertheless, most commercially available OPCs utilize laser illumination.

The angular distribution of scattered light for homogeneous spheres of known refractive index can be rigorously determined from theory (Mie, 1908). Numerous studies have shown reasonable agreement between the predictions of Mie theory and measured OPC responses for homogeneous spheres of known size and refractive index (Cooke and Kerker, 1975; Willeke and Liu, 1976; Garvey and Pinnick, 1983; Liu et al., 1985; Hinds and Kraske, 1986; Szymanski and Liu, 1986). For instruments that show the expected dependence of response on size, theory can be used to determine the relationship between response and size for particles having a refractive index different from the calibration aerosol. This approach is sometimes used for atmospheric measurements.

The challenge that arises when OPCs are used for atmospheric measurements is that the particle properties (shape, refractive index, and morphology) required to determine size from pulse data are typically unknown. For example, the aerosol may contain a mixture of particles consisting of homogeneous spheres, irregularly shaped solids, and solid seeds encapsulated by liquid droplets. Even for the ideal case of homogeneous spheres, uncertainties in the knowledge of particle chemical composition can lead to significant uncertainties in estimates of refractive index.

To avoid such uncertainties, OPCs can be calibrated with atmospheric aerosols. Hering and McMurry (1991) used an electrical classifier to deliver

Los Angeles aerosols of known size and found that particles of a given size often produced two distinct pulse heights, indicating that two distinct types of particles were present. When this occurs, there is no unique relationship between pulse height distribution and size distribution. In this case such calibrations provide information that can be used to quantify measurement uncertainties. It is likely that particles in urban areas are more diverse than particles in remote regions and that OPC measurement uncertainties are therefore inherently more uncertain in urban areas.

Valuable information about the shape and/or refractive index of atmospheric particles can be inferred by measuring the angular distribution of scattered light (differential light scattering (DLS)). The multiangle aerosol spectrometer probe (MASP) (Baumgardner et al., 1993) measures the light scattered by individual particles for polar angles of 30–60° and 120–150°. If the particles are homogeneous spheres, then Mie theory can be used to infer refractive indices that are consistent with measurements. This instrument is being used routinely in aircraft measurements of atmospheric aerosols (e.g., Baumgardner et al. (1996)), and information on refractive index is being inferred from these measurements. Kaye and coworkers have developed several differential light scattering (DLS) instruments that provide information on shape for particles in the 1–10 μm diameter range (Kaye et al., 1991, 1996; Hirst and Kaye, 1996). Dick and coworkers (Dick et al., 1994, 1996; Sachweh et al., 1995) have used the DAWN-A (Wyatt et al., 1988) to measure azimuthal variabilities in light scattering so as to distinguish between spherical and nonspherical particles in the 0.2–2 μm range. The nonspherical fraction was found to be reasonably well correlated with the fraction of the aerosol that was of crustal origin and with the fraction of the aerosol that was "less hygroscopic" (Dick et al., 1998). These various studies illustrate the potential of DLS to provide valuable new information about properties of atmospheric aerosols.

In summary, while optical particle sizing techniques have been widely used for about 50 years, these techniques have evolved significantly in the past decade. Recent advances permit the detection of smaller particles, the calibration of optical detectors with optically complex atmospheric particles, and the measurement of particle properties such as shape and refractive index. It is likely that such advances will continue as digital signal processing techniques and laser technology evolve.

3.2.2. Aerodynamic particle size

When an aerosol is rapidly accelerated through a nozzle, particles tend to lag behind the carrier gas due to inertia (Wilson and Liu, 1980). The difference between the particle and gas speeds increases with size and density since inertia increases with these properties. At least three commercial instruments are

available that utilize measurements of particle speed in an accelerating gas flow
to determine size (Baron et al., 1993). These measured sizes are closely related
to aerodynamic size. Because lung deposition and dry deposition of particles
larger than 0.5–1 μm depend on aerodynamic size, data from these instruments
provides direct information on such aerosol effects.

With these instruments, aerodynamic particle size is inferred from particle
velocity, which is determined by measuring the time of flight between two
illuminated volumes separated by a known distance (Dahneke, 1973; Dahneke
and Padliya, 1977; Dahneke and Cheng, 1979; Remiarz et al., 1983; Mazumder
et al., 1991). Unlike optical counters, which determine particle size from the
intensity of the scattered light, these instruments simply use the scattered light
to detect particles. This technique offers the advantage that measurements are
not compromised by Mie resonances, which introduce complications in the
interpretation of data from optical particle counters.

The aerodynamic diameter is defined as the diameter of a unit density sphere
that has the same settling velocity as the particle (Hinds, 1982). Settling velocity is determined by a balance between aerodynamic drag and gravitational
force. For most atmospheric particles, the Stokes's drag law can be used to determine the aerodynamic drag force on a settling particle. Stokes's law, however, applies only when the relative speed between the particle and the carrier
gas is quite small (i.e., particle Reynolds Number ≪1.0). Because particles are
rapidly accelerated in these instruments, particle Reynolds numbers often exceed 1.0, especially for large particles. In this case non-Stokesian corrections
must be made when determining aerodynamic size from measured particle velocities (Wang and John, 1987b; Ananth and Wilson, 1988; Cheng et al., 1990;
Lee et al., 1990; Rader et al., 1990).

During the measurement of aerodynamic particle size, particles are expanded through a nozzle to a pressure that is well below atmospheric. The
flow cooling and pressure drop associated with this expansion can lead to a
change in relative humidity and may therefore affect measurements of particle size. Sizing errors also occur due to deformations in the shapes of liquid
droplets (Baron, 1986). Insufficient attention has been given to these phenomena to permit an estimate of measurement error, but errors are likely to be
significant, especially at high humidities.

Measurement of aerodynamic particle size requires the optical detection of
individual particles. The smallest reported size that can be measured with these
instruments varies with instrument design and ranges from 0.2 to 0.5 μm. These
instruments are capable of providing high-resolution information on aerodynamic size distributions in real time. Although they have seen only limited use
for atmospheric measurements, such instruments have the potential to provide
new and useful high-quality information in the future.

3.2.3. Electrical mobility analyzers

Electrical mobility analyzers classify particles according to the electrical mobility, Z, which for spherical particles is given by (e.g., Hinds, 1982):

$$Z = \frac{neC(D_p, P)}{3\pi\mu D_p},$$

(8)

where n is the number of elementary charges carried by the particle, e is the magnitude of the elementary unit of charge, C is the slip correction factor (Rader, 1990), μ is the absolute gas viscosity, and D_p is particle diameter. Note that Z depends on gas properties, particle charge, and the geometric particle size but is independent of other particle properties such as density. Flagan (1998) has written a comprehensive review of electrical aerosol measurements.

The first practical electrical mobility analyzer was developed by Kenneth Whitby and coworkers (Whitby and Clark, 1966). A refined design of the Whitby Aerosol Analyzer became a successful commercial product (the electrical aerosol analyzer or EAA (Liu et al., 1974)) and was used in some of the first measurements of ultrafine (particle diameter down to ~10 nm) urban aerosol size distributions (Whitby et al., 1972). In the EAA, particles flow through a unipolar charger, where they are exposed to small positive ions before entering the classifier. Particles having mobilities larger than a value determined by flow rates and the precipitating voltage are removed in the coaxial cylindrical classifier; all particles that are not precipitated are collected in a Faraday cup. Currents delivered to the Faraday cup are measured as a function of the precipitating voltage to obtain size distributions. Unipolar charging is used to deliver the maximum possible charge to the particles so as to maximize the current delivered to the Faraday cup.

The EAA has been largely replaced by the differential mobility particle sizer (DMPS; Keady et al., 1983). The DMPS includes a differential mobility analyzer (DMA, also referred to as the electrostatic classifier) (Liu and Pui, 1974a, b; Knutson and Whitby, 1975) and a particle detector (typically a CNC, but aerosol electrometers are occasionally used). Systems of this type can measure size distributions in the 3–500 nm diameter range. The DMA is the heart of the DMPS. In the DMA, the aerosol is first exposed to a bipolar cloud of ions, where it achieves Boltzmann charge equilibrium (Liu and Pui, 1974a,b; Adachi et al., 1983; Wiedensohler, 1988; Reischl et al., 1996). The mean charge of particles leaving the charger is close to zero, but a fraction of the particles contain ± 1, ± 2 charges, etc. The contribution of multiply charged particles increases with increasing size. Particles in a narrow mobility range determined by the classifying voltage and flow rates are separated from the main flow and delivered to the detector. The relationship between the measured concentration in the narrow mobility slice and the inlet size distribution is well defined (Knutson, 1976; Hoppel, 1978; Fissan et al., 1983). The complete size distribution is obtained by carrying out measurements at a number of classifying voltages. The deconvolution procedure used to determine inlet size distributions requires accounting for the multiple sizes associated with singly-charged, doubly-charged, etc., particles that are obtained at each classifying voltage (e.g., Hagen and Alofs, 1983).

The DMPS typically requires about 20 min to measure size distributions. The measurement time is determined by the time required for concentrations to stabilize after the classifying voltage is changed and the time required to achieve a statistically significant sample. Flagan and coworkers (Wang and Flagan, 1990) showed that measurement times can be reduced to $\sim$2 min by ramping the classifying voltage continuously. Instrument systems that use this approach are referred to as scanning electromobility spectrometers (and also scanning mobility particle spectrometers – SEMS or SMPS). Voltage scanning is now typically used in measurements of atmospheric aerosol size distributions.

The most common geometry for DMAs involves annular flow through coaxial cylinders, as originally described by the Minnesota group (Liu and Pui, 1974a; Knutson and Whitby, 1975). An alternative cylindrical design that offers advantages in flow stability at high flow rates was developed by Reischl and coworkers in Vienna (Winklmayr et al., 1991). Electrostatic classifiers that involve radial flow between a pair of flat, parallel circular discs have been independently developed recently by two groups (Pourprix and Daval, 1990; Zhang et al., 1995).

The transport of particles through DMAs is unaffected by diffusion for particles larger than $\sim$50–100 nm. This leads to a particularly simple expression for the probability that particles in this size range will exit with the "monodisperse" exit flow (Knutson and Whitby, 1975). This size-dependent probability is referred to as the "DMA transfer function", and having an accurate expression for this transfer function is essential for determining size distributions from measured concentrations of classified particles. Diffusion leads to depositional losses during transport to, through and beyond the DMA, and it also leads to a broadening of the range of sizes that are carried by the classified aerosol flow. Quantitative investigations into the effect of particle diffusion in DMAs were first reported by Kousaka et al. (1985, 1986) and by Stolzenburg (1988). More recently, an extensive analysis of ultra-fine particle classification by DMAs has been carried out in a collaborative activity between Chen and Pui of the University of Minnesota and Fissan and coworkers at the University of Duisburg (Chen and Pui, 1995; Fissan et al., 1996) and at the University of Leipzig in preparation for ACE-1 measurements (Birmili et al., 1997). The Duisburg/Minnesota collaboration led to the development of a detailed numerical model for particle transport through DMAs and to experimental measurements of DMA transfer functions for DMAs of various designs. An objective for this work has been to extend accurate measurements of size distributions with DMAs to sizes approaching 3 nm, where diffusion has a significant effect. A newly designed nanometer DMA was recently reported to minimize particle diffusional losses and diffusional broadening (Chen et al., 1996), and Fernandez de la Mora and coworkers have shown that the detrimental effects of diffusion can be largely eliminated for the Vienna-type DMA if minor design modifications are incorporated that permit operation at high flow rates (Rosell-Llompart et al., 1996; Seto et al., 1997; de Juan and Fernández de la Mora, 1998).

The primary limitation of DMPS/SMPS systems at the small particle limit is detecting very low concentrations of very small particles. Although CNCs can measure very low concentrations, their lower detection limit is presently $\sim$3 nm. Aerosol electrometers can detect arbitrarily small particles, but cannot detect currents below $\sim 10^{-16}$ A, corresponding to a number concentration of $\sim$40 cm^{-3} at a typical sampling rate of 1 LPM. Concentrations of nanometer aerosols downstream of DMAs are often below this because the classified particles include only the charged fraction ($\sim$1% at Boltzmann charge equilibrium for 3 nm particles) of particles in a narrow size range. A strength of SMPS/DMPS systems for sub 10 nm particles is that measurements are not affected by multiple charging. Thus, there is a unique relationship between the electrical mobility of classified particles and size.

Multiple charging is a primary limitation of SMPS/DMPS systems at the high end of the size spectrum. Large particles of a given mobility contain one, two, three or more elementary charges, each charge state corresponding to a different size. This increases the difficulty of deconvoluting the data to determine the contribution of each size to measurements at a given classifying voltage.

SMPS/DMPS instruments are clear improvements over the EAAs of 20 years ago and are without question the best available technique for measurement of size distributions between $\sim$8 and 200 nm. Outside this range, these systems can still be excellent, depending on aerosol concentrations and size distributions, but other techniques that offer distinct advantages also require consideration. Measurement accuracies are likely to be size-dependent, and will be affected by particle shape. The accuracy with which these systems can measure atmospheric aerosol size distributions is difficult to quantify due, in part, to the lack of polydisperse aerosol standards.

3.2.4. Diffusion batteries

Particle diffusivities increase with decreasing size. Therefore, as particle sizes decrease, the rate at which they deposit on nearby surfaces increases. Diffusion batteries use this size-dependent deposition rate to obtain information on size distributions. They are most commonly used for particles smaller than 0.1 µm, because diffusion coefficients in this size range are high enough to lead to appreciable deposition rates. In the most commonly used diffusion batteries, the aerosol flows through a series of fine capillaries (Gormley and Kennedy, 1949; Sinclair, 1972) or fine wire-mesh screens (Sinclair and Hoopes, 1975a,b; Cheng and Yeh, 1980). Particles deposit on the inner surfaces of the capillaries or on the outer surfaces of the screens. Typically, a series of capillaries or screens is used, and the aerosol number concentration is measured downstream of each collecting element. Data for the decay in aerosol concentration through this series of collecting elements can be mathematically "inverted" to obtain the size distribution (Knutson et al., 1988; Ramamurthi and Hopke, 1989; Wu et al., 1989; Cooper and Wu, 1990; Reineking et al., 1994; Knutson, 1995).

Diffusion batteries are rugged, simple and are well suited for use in hostile environments, such as in-stack sampling. They have been used extensively for measurements of nanometer-sized radon progeny, since sensitive radioactive counting techniques can be used to measure the activity of ultrafine particles deposited on the particle collection surfaces. However, there are significant limitations to the quality of data that can be obtained with this approach. Because diffusion is a stochastic phenomenon, a wide range of sizes deposits on each collecting element. Thus, there is no simple relationship between the change in aerosol concentration across a collecting element and particle size. Furthermore, because size distributions are obtained from measurements of the change in concentration as the aerosol flows through the battery, measurements are adversely affected by other phenomena that cause change, such as shifts in the size distribution of the sampled aerosol. Finally, Cheng and Yeh (1980) showed that for screen-type diffusion batteries, two sizes of particles larger than ∼0.1 µm can be collected with the same efficiency since interception and impaction become important collection mechanisms in addition to diffusion. Thus, in this size range there is not a unique relationship between size and collection efficiency. This leads to ambiguities in measured size distributions.

Diffusional separation has been largely superceded in recent years by electrostatic classification, which provides higher sizing resolution for most measurements of sub-0.1 µm atmospheric aerosol size-distributions. Nevertheless, diffusional separation offers benefits that will ensure its use for limited applications in the foreseeable future.

3.2.5. CNC pulse height analysis (PHA)

Recent work has shown that useful information about size distributions in the 3 to 10 nm diameter range can be obtained by measuring pulse height distributions produced by a steady-flow CNC operating in the single-particle-counting mode Saros et al. (1996). While particles larger than 10–15 nm all grow to about the same final droplet size in the CNC condenser, final droplet sizes decrease with initial particle size for smaller particles. This is due to the effect of curvature on equilibrium vapor pressure (Thompson, 1871): smaller particles must travel farther into the CNC condenser before they are exposed to sufficiently high supersaturations for condensation to occur. Therefore they have less time to grow. Recent work has shown that measured pulse height distributions can be mathematically inverted to determine the size distribution (Weber et al., 1998).

The PHA technique does not provide sizing resolution comparable to that obtained with SMPS (scanning mobility particle spectrometer) systems. However, it offers the advantage that every particle entering the CNC provides a signal. In contrast, only the charged fraction of the selected mobility fraction is detected with the SMPS. Because the charged fraction can be very small ($\sim$1% for particles of 3 nm) and because measurements require scanning through a range of mobilities, the time required to acquire a statistically significant number of counts with the SMPS is significantly longer. The PHA technique offers significant benefits for studies of nucleation in the remote troposphere, where concentrations are low and changes can occur quickly (Weber et al., 1998). It is especially well suited for aircraft measurements.

3.3. Aerosol water content

Water comprises more than 50% of the fine particle mass at relative humidities exceeding 70–80% (e.g., Häncl, 1976; Zhang et al., 1993). The aerosol water content is determined by particle composition and relative humidity, and the amount of water in particles rises sharply above relative humidities of $\sim$80%. Most ionic species such as sodium chloride, sulfates, and nitrates are hygroscopic. Recent work (Saxena et al., 1995; Saxena and Hildemann, 1996; Dick, 1998) has shown that organic compounds may also significantly affect the aerosol water content.

One approach for determining particulate water content is to use thermodynamic models to calculate the aerosol water content (e.g., Pilinis et al., 1989) based on the measured composition of the major particulate species. A limitation of this approach is that current thermodynamic models do not account for water associated with organics. More quantitative information on concentrations of the major organic species and their hygroscopic properties is needed before they can be incorporated into such thermodynamic models.

Ho et al. (1974) determined the liquid water content of atmospheric particles by using microwave resonance to measure the dielectric constant of samples that were collected on glass fiber filters. Their measurements showed that water mass content ranged from $\sim$10% at 50% RH to 40% at 70% RH. Their instrument was unable to make measurements at relative humidities above 70%. An attractive feature of this approach is that it has the potential to provide information semi-continuously.

Several investigators have measured particulate water content by using a sensitive microbalance to measure the sensitivity of mass to relative humidity for particles collected on filter or impactor substrates (Winkler and Junge, 1972; Thudium, 1978; Hänel and Lehmann, 1981; Hitzenberger et al., 1997). Because relatively long times are required for deposits to equilibrate and because a specially designed relative-humidity controlled microbalance is required, this technique has seen only limited application. It does have the potential to provide accurate information.

Speer et al. (1997) used a β-gauge to infer the mass of particulate samples collected on 37 mm Teflon filters. They measured mass at relative humidities ranging from $\sim$5 to 95% and determined the incremental water mass by difference. They found excellent agreement between β-gauge measurements and gravimetric measurements. Furthermore, the water mass uptake for ammonium sulfate measured with their instrument appears to be in good agreement with thermodynamic expectations. Although this technique has not been thoroughly studied, and while it is affected by the usual problems of all filtration techniques, it appears to offer promise as a practical technique for inferring water mass of atmospheric particles.

Lee and Hsu (1998) measured water in aerosol samples by using a chermal conductivity detector to measure the amount of water released by particles deposited on a filter upon exposure to pure helium. The detection limit for water was found to be 24 µg, and good agreement was found between measured water content and thermodynamic predictions for sodium sulfate and ammonium sulfate.

The tandem differential mobility analyzer (TDMA) (originally referred to as the "aerosol mobility chromatograph" (Liu et al., 1978)) has also been used to infer water content. This instrument system involves the use of two DMAs operated in series (Rader and McMurry, 1986). The aerosol classified by the first DMA is humidified or dehumidified between the DMAs, and the second DMA measures the effect of humidity on particle size (McMurry and Stolzenburg, 1989; Covert et al., 1991; Svenningsson et al., 1992,1994). TDMA data can provide information on variations in water uptake among particles of a given size. TDMA measurements have shown that when atmospheric particles of a given size are brought to high humidity, they often separate into two distinct types, which have been termed "more" and "less" hygroscopic. Based on comparison with known materials, it is found that measured growth factors are typically accurate to within 2%.

Because number concentrations of particles larger than about 0.5 μm are too low to permit TDMA measurements, all atmospheric data reported to date applies to smaller particles. Also, while the TDMA provides accurate information on the dependence of size on relative humidity, it does not provide direct information on particulate water mass concentrations. TDMA data can be used together with size distribution data, however, to obtain estimates of humidity-dependent mass concentrations. The TDMA has provided valuable insights into hygroscopic properties of atmospheric aerosols, but due to its high cost and complexity it is likely to remain a research tool rather than a monitoring device.

In summary, water is a significant component of atmospheric aerosols, and its contribution to mass increases strongly with relative humidity above ∼70%. The tendency of water to rapidly evaporate or condense with changes in relative humidity affects measurements with most instruments discussed in this review, and measurements of aerosol water content can sometimes be used to correct data for such errors. While a variety of techniques to measure water have been proposed, most of them involve differencing and thus cannot detect bound or hydrated water. Furthermore, some of these techniques involve measurements of size change, while others involve measurement of mass. Inferring mass change from size change (or vice versa) requires information about relative humidity-dependent density, which is typically unknown. It would be ideal to have a chemical technique for measuring water, but the techniques discussed above have substantially advanced our understanding of aerosol water content.

3.4. Aerosol volatility

Aerosol volatilization has been used as an indirect method of inferring particle composition. For example, Brock et al. (1995) sampled in parallel with two CNCs, one of which was equipped with an inlet heated to 192°C. Laboratory measurements showed that the heated inlet volatilized 90% of the particles that were smaller than 0.04 μm and contained only H_2SO_4 and H_2O. Field measurements showed that in some locations significant differences were found between the two CNCs, while in other locations the measurements were similar. By coupling these observations with other information that was known about the measured aerosol, they were able to conclude that ultrafine particles containing primarily H_2SO_4 and H_2O were present at some locations. Clarke (1993) has used similar systems in aircraft to measure the effect of volatilization on measured number concentrations and size distributions.

It would obviously be far preferable to have direct measurements of aerosol composition rather than the indirect information that is provided by volatilization. However, volatilization measurements can be done in real time (it has a time response of seconds) with instruments that are compact and convenient to deploy in the field. Fast time response is especially important for aircraft sampling.

3.5. Particle density

Particle density is needed to convert aerodynamic sizes to geometric (Stokes) sizes and to establish the relationship between aerosol mass and volume concentrations. In practice, density is usually calculated from measured particle composition. However, uncertainties in aerosol composition and the thermodynamic properties of mixtures lead to uncertainties in calculated densities.

Several approaches have been used to measure densities of submicron atmospheric aerosols. Hänel and Thudium (1977) measured the mass of bulk samples with an electronic balance, and measured the volume of the same sample using a specially designed pycnometer. The reported accuracy for these density measurements was 2%. Because bulk samples were used, information about variations with size or among particles of a given size was not obtained. Measurements of aerosol chemical composition were not reported, so it was not possible to compare measured to expected values.

Stein et al. (1994) measured the density of atmospheric particles in the 0.06 to 0.18 μm diameter range using the DMA-impactor technique. A DMA was used to deliver monodisperse particles of a known electromobility equivalent size (which does not depend on density and which equals the geometric size for spherical particles) to an inertial impactor. The particles' aerodynamic size, which depends on density, was measured with the impactor. Particle densities were determined to within ~4% from the measured geometric and aerodynamic sizes. Densities calculated from measured aerosol composition were found to be ~20% lower than measured values; the reason for this discrepancy was not resolved.

In summary, only a few efforts have been made to measure the density of submicron particles. Additional work is required to ensure that density can be calculated from measured particle composition. Closure studies of this type are an essential ingredient in the development of experimentally verified thermodynamic models of aerosol properties.

4. Measurements of aerosol chemical composition

4.1. Off-line measurements

Measurements of particle composition typically involve the chemical analysis of deposited particles in a laboratory some time after sample collection. Filters are the most commonly used collection substrates, but a variety of films and foils have been used with impactors to collect size-resolved samples. Sampling times vary with ambient loadings, sampling rates, substrate blanks, and analytical sensitivities but typically vary from several hours in urban areas to a day or more under clean background conditions. In addition, several off-line techniques are available for analyzing the composition of individual particles.

A variety of sampling artifacts can affect the measured composition of the collected particle deposit relative to what was actually in the atmosphere. Volatilization of semivolatile compounds (compounds that are found in both the vapor and particulate phases) is known to be a significant source of error for species like ammonium nitrate and many organics. Volatilization can occur because of pressure drop in the sampler, which upsets the equilibrium between the deposited particles and the vapor, or due to changes in temperature, relative humidity or composition of the incoming aerosol during sampling. Artifacts associated with sampling transport and storage have also been reported (Chow, 1995). Evaporative losses of particulate nitrates have been investigated in laboratory and field experiments with filters and impactors (Wang and John, 1988). The laboratory studies involved parallel sampling of ammonium nitrate particles with a Berner impactor and a Teflon filter. Both samplers were followed by nylon filters to collect evaporated nitric acid. Losses from the impactor were 3–7% at 35°C and 18% relative humidity, and losses from the filter were 81–95% under the same conditions. This result (that evaporative losses from the filter exceeded those from the impactor) is consistent with theoretical predictions (Zhang and McMurry, 1993) and was borne out by measurements in Los Angeles where negligible losses of nitrates from the impactor were found. Because evaporative loss rates from a given substrate are determined largely by the equilibrium vapor pressure of the evaporating species (Zhang and McMurry, 1987), the relative effect of evaporative losses will be usually be higher in background areas where particulate concentrations are low or at higher temperatures where vapor pressures are higher.

The diffusion denuder method was developed for the measurement of such semivolatile compounds (Possanzini et al., 1983). With this approach, the vapor phase diffuses and sticks to a coated surface upstream of the particle filter. An adsorber is located downstream of the filter to collect material that evaporates from the deposited particles during sampling. The particle-phase concentration is determined from the loading on the filter and on the adsorber following the filter. The gas-phase concentration is determined either by measuring the amount of vapor-phase material collected on the "denuder" surface upstream of the particle filter or by subtracting the total (filter plus adsorber) loadings obtained with undenuded and denuded samplers. Diffusion denuders have been used with excellent success to distinguish inorganic gas phase species such as nitric acid and ammonia and from their particulate forms (Shaw et al., 1982; Mulawa and Cadle, 1985; Eatough et al., 1986; Knapp et al., 1986; Hering et al., 1988; Keuken et al., 1988; Koutrakis et al., 1988; Klockow et al., 1989; Koutrakis et al., 1993).

Measurements show that evaporative losses of semivolatile organic compounds can be significant (Commins and Lawther, 1957; De Wiest and Rondia, 1976; Katz and Chan, 1980; Peters and Seifert, 1980; Galasyn et al., 1984; Marty et al., 1984; Eatough et al., 1990). Application of the diffusion denuder technology to semivolatile organic compounds is an active area of research and shows promise for this difficult measurement task (Lane et al., 1988, 1992; Eatough et al., 1993; Gundel et al., 1995; Lawrence and Koutrakis, 1996). Because there are a wide variety of semivolatile organic compounds with varying adsorptive properties, finding the ideal denuder coatings is a nontrivial task. Turpin et al. (1993) demonstrated an alternative diffusion separator for semivolatile organic compounds that does not need a denuder. Definitive field testing has not been carried out.

The adsorption of organic gases on quartz filters is another source of error when sampling particulate organic carbon. Cadle et al. (1983) found that when two quartz fiber filters were used in series, the amount of carbon collected on the second filter was at least 15% of that on the first filter. Because the particulate collection efficiency exceeded 99.9%, it was concluded that the signal on the second filter was due to adsorption of carbon-containing gases. McDow and Huntzicker (1990) found that quartz backup filters collected more organic carbon when they followed Teflon pre-filters than when they followed quartz pre-filters, presumably because quartz is more effective than Teflon at removing adsorbing vapors. McMurry and coworkers (McMurry and Zhang, 1989; McMurry et al., 1996a) have found that the amount of "organic carbon" found on the quartz after-filter following an impactor can be comparable to the amount of organic carbon collected on the impactor stages. Because measurements of physical size distributions show that very little particulate mass should be found below the 0.05 μm cut point of the bottom impactor stage, and because the absence of comparable amounts of sulfate on the after-filter suggests that particle bounce is not responsible for the observed high organic carbon loadings, it was concluded that the high after-filter loadings are due to gas adsorption.

These discussions of sampling artifacts illustrate the dismal state of the art for measurement of particulate organic carbon, which can comprise nearly

50% of the fine particle aerosol in the arid southwest and in regions like Denver and Los Angeles, which are heavily impacted by vehicular emissions. An understanding of the particulate organic composition will require improved sampling methodologies and more attention to speciation (Schauer et al., 1996).

4.1.1. Filter sampling

The most common approach for determining the composition of atmospheric aerosols involves the analysis of deposits collected on filter substrates. While filter samplers are inexpensive, they require manual operation. Furthermore, the number of filters that must be analyzed in a monitoring network or in an intensive field campaign can be large. For example, 60,000 filters were collected during the 1990 NGS visibility study, and their analyses contributed significantly to the cost of the $14 million study (NRC, 1993).

Chow (1995) provides a comprehensive treatment of the use of filters to determine the chemical content of particulate matter. In this review she discusses suitable filter materials for various analytical methods, species sampling artifacts, and analytical techniques that can be used for various species. The discussion of analytical techniques includes a valuable comparison of sensitivities. The reader is referred to this paper for a discussion of this topic.

4.1.2. Impactors

Impactors are used to classify particles according to aerodynamic diameter. The aerodynamic diameter is defined as the diameter of the unit density sphere having the same settling speed as the particle. The relationship between aerodynamic diameter, D_a, and geometric diameter, D_p, for spherical particles is

$$D_a = D_p \left(\frac{\rho_p}{\rho_0} \right)^{1/2} \left(\frac{C(D_p)}{C(D_a)} \right)^{1/2}, \tag{9}$$

where the slip correction factor C (Rader, 1990) accounts for noncontinuum effects that become significant when particles sizes approach the mean free path of the gas. Thus, aerodynamic diameters exceed geometric diameters for particles with densities above $1 \, \mathrm{g \, cm^{-3}}$. Cascade impactors with a series of stages, each with a successively smaller cut point, are commonly used to collect size-resolved atmospheric samples for chemical analysis. Classifying particles according to aerodynamic diameter is ideal for health effects studies since lung deposition of particles larger than a few tenths of a micron depends on aerodynamic diameter.

The dimensionless parameter that determines whether particles are collected by an impactor is the Stokes number, St, defined as

$$St = \frac{\tau U_0}{D_{nozzle}/2} = \frac{\rho_p D_p^2 C U_0}{9\mu D_{nozzle}}, \tag{10}$$

where τ is particle relaxation time, U_0 is the mean velocity through the accelerating nozzle, D_{nozzle} is the diameter of the accelerating nozzle, ρ_p is particle density, D_p is particle diameter, C is the slip correction factor and μ is the absolute viscosity of the gas. Marple and coworkers (Marple, 1970; Marple and Willeke, 1976) developed design criteria for impactors that allow impactors to be designed with predictable cut points. In practice, impactors collect particles that have Stokes numbers larger than a critical value typically in the range 0.21–0.23.

Impactors that collect particles larger than a few tenths of a micron are straightforward to design and fabricate. Collecting particles smaller than this requires either the use of very small nozzles or low pressures (C increases as pressure decreases). Both of these approaches have been used, and impactors that collect particles down to 0.05 μm are now used routinely (Berner et al., 1979; Hering and Friedlander, 1979; Cahill and Malm, 1987; Marple et al., 1991). Because very little mass is associated with particles smaller than 0.05 μm, these impactors can collect virtually all of the particulate mass. Very small nozzle diameters are required to collect small particles, so multinozzle impactors are commonly used to achieve adequate sampling rates.

Particle bounce is an inherent problem with impactors. Coated substrates largely eliminate bounce and are commonly used for atmospheric sampling (Dzubay et al., 1976; Wesolowski et al., 1977; Lawson, 1980; Turner and Hering, 1987; Wang and John, 1987a; Pak et al., 1992). Measurements have shown that liquid oils tend to provide better bounce-prevention characteristics than do viscous greases. While coatings that do not interfere with some types of chemical analysis have been found, no available coating is compatible with measurements of the particulate organic carbon content. An alternative approach involves sampling at elevated relative humidities, where submicron atmospheric particles typically contain enough liquid water to prevent bounce (Winkler, 1974). Stein et al. (1994) showed that bounce of small (~0.2 μm) atmospheric particles is largely eliminated at relative humidities exceeding ~75%.

A variety of impaction substrates have been used for sampling ambient aerosols with impactors. Aluminum foil is often used when samples are to be analyzed for organic and elemental carbon (OC/EC), since precleaning can reduce the carbon blanks in these substrates to very low levels. Carbon-free substrates are required since OC/EC analyses involve measuring the amount of CO_2 that is released when the samples and substrates are burned. Precleaned

Teflon or Mylar film is often used for ion chromatography analyses, since ion blanks can be made very low on such surfaces. Teflon membrane filters have also been used as impaction substrates. Although these are more costly than film or foil substrates, they do not require precleaning, and they are compatible with nondestructive analytical methods such as x-ray fluorescence analysis (XRF) or proton induced x-ray emission (PIXE).

4.1.3. Laser microprobe mass spectrometry

The analysis of individual particles by mass spectrometry has been reviewed in several papers (Spurny, 1986; McKeown et al., 1991; Noble et al., 1994). Laser microprobe mass spectrometry (LAMMS) (Wieser et al., 1980; de Waele and Adams, 1986; Kaufmann, 1986; Spurny, 1986; Artaxo et al., 1992) involves the off-line analysis of particles collected on a substrate. Particles are irradiated with a high-power pulse laser, and the ejected ion fragments are analyzed by mass spectrometry. LAMMS can detect trace levels of metals in individual particles at the parts-per-million level (Otten et al., 1987; Bruynseels et al., 1988a), can speciate inorganic compounds including nitrates and sulfates (Bruynseels and van Grieken, 1984; Bruynseels et al., 1988b; Ro et al., 1991), can detect trace organic compounds (De Waele et al., 1983; Mauney and Adams, 1984; Niessner et al., 1985), and can distinguish surface species from those contained within the particle (De Waele et al., 1983; Bruynseels and Van Grieken, 1985; Niessner et al., 1985; Bruynseels and Van Grieken, 1986; Wouters et al., 1988). Because LAMMS is an off-line technique that exposes particles to a vacuum environment before they are analyzed, particle composition can be altered by chemical reactions or evaporation before analysis. Also, because collected particles must be returned to the laboratory for analysis, there is typically a significant time delay before data are available.

4.1.4. Electron microscopy

Individual particle analysis by electron microscopy can provide information on particle morphology and elemental composition. With this approach, particles are collected on a filter or impaction substrate and are irradiated by electrons under vacuum conditions. Information on elemental composition is achieved by measuring the X-ray energy spectrum produced by interactions of the electrons with the particles. Windowless or thin-window detectors can detect X-rays from elements with atomic number 11 (sodium) and greater, and the location of elements on or within particles can be determined by using electron beams that are small relative to particle size. A review of the various electron analytical techniques for particles is given by Fletcher and Small (1993).

Electron microscopy has led to important discoveries concerning atmospheric aerosol chemistry. For example, Andreae et al. (1986) found that remote marine aerosols contained internal mixtures of silicates and sea-salt, which they attributed to cloud coalescence. Sheridan and coworkers (1994) found that particles consisting mostly of crustal species or soot are coated with sulfur when found in the lower stratosphere but not in the upper troposphere. McInnes and coworkers (1994) found evidence for the substitution of sulfate for chloride in sea-salt particles in marine atmospheres, and McMurry and coworkers (1996) showed that less hygroscopic particles in urban areas tended to consist of carbon-containing chain agglomerates, while more hygroscopic particles were rich in sulfur. Other researchers have used electron microscopy to categorize individual particles into groups that provided information on source categories (Linton et al., 1980; Kim and Hopke, 1988; Van Borm and Adams, 1988; Rojas et al., 1990; van Borm et al., 1990; Katrinak et al., 1995; Anderson et al., 1996b). Because these observations require data on the composition of individual particles, bulk analysis techniques could not have provided similar information.

A limitation of microscopic techniques is that obtaining data for a statistically significant sample can be extremely time consuming. To deal with this issue, several groups have developed automated systems that can analyze large numbers of particles (Casuccio et al., 1983; Anderson et al., 1988; Schwoeble et al., 1988; Artaxo et al., 1992).

Obtaining quantitative chemical information by X-ray microanalysis can also be problematic. Several researchers have proposed standardless techniques for obtaining quantitative elemental composition (Russ, 1974; Armstrong and Buseck, 1975; Janossy et al., 1979; Aden and Buseck, 1983; Wernisch, 1985; Raeymaekers et al., 1987). These techniques account for interactions of X-rays with neighboring atoms and, in some cases, particle shape. However, in measurements with monodisperse particles of 2,6-naphthalenedisulfonic acid, disodium salt ($C_{10}H_6(SO_3Na)_2$) ranging in size from 0.207 to 1.122 μm, Huang and Turpin (Huang and Turpin, 1996) found that standardless techniques led to compositional errors exceeding 78%.

The measurement of volatile species by electron microscopy is also problematic. Volatilization occurs because particles are exposed to vacuum conditions for extended times during analysis and because samples are heated by the electron beam (Gale and Hale, 1961; Almasi et al., 1965; Watanabe and Someya, 1970; Curzon, 1991; Huang, 1997). For example, nitrates, which tend to be relatively volatile, are usually not detected by X-ray analysis even though they are often present in significant quantities (McInnes et al., 1997). Similar losses of semivolatile organic compounds are likely. Several researchers have shown that volatilization loss rates of sulfuric acid droplets are much greater than loss rates of ammonium sulfate particles (e.g., Webber, 1986; Huang and Turpin, 1996). The environmental scanning electron microscope (ESEM) (Danilatos and Postle, 1982; Danilatos, 1988) permits the analysis of particles exposed to gas pressures exceeding 5 Torr, thereby eliminating some of the volatilization losses that occur in conventional electron microscopes. Huang and coworkers (1994) used the ESEM to observe in real time the effects of condensating and evaporating liquid water on diesel chain agglomerates on a substrate.

Selecting the optimal substrate for electron microscopic analysis is also an issue. Ideally, the substrate should contain no elements that will interfere with the analysis of atmospheric particles. Due to its low atomic weight, beryllium produces no interfering X-rays and thus might seem to be an ideal substrate for scanning electron microscopy. However, beryllium is impractical to handle due to its toxicity, and laboratory measurements have shown that acid sulfate particles are not detected on beryllium due to interactions between the particles and the beryllium (Huang and Turpin, 1996). Samples are often collected on carbon-containing membranes, leading to difficulties with the measurement of particulate carbon content. McInnes et al. (1997) found that particles on TEM grids must contain at least 30% carbon in order to measure the particulate carbon. There is no ideal substrate that is suitable for analysis of all major elements found in atmospheric particles.

In summary, despite the limitations outlined above, electron microscopy has provided valuable information on the composition, sources, and atmospheric transformations of atmospheric aerosols. Electron microscopy is the only individual particle technique that provides both morphological and compositional information on ultrafine particles, and samplers that are used to collect particles for electron microscopic analysis are typically relatively inexpensive and simple to operate. It is likely that electron microscopy will continue to be an important tool in the analysis of atmospheric particles for some time to come.

4.2. Real-time measurements

There is, at present, little readily available instrumentation for measuring the composition of aerosol species in real time. Laboratory prototype instruments have been developed for measuring particulate sulfur, nitrogen and carbon concentrations in the field on time scales ranging from minutes to hours, and these instruments are discussed below. Also, prototype versions of mass spectrometers that can determine the composition of individual particles in real time have recently been developed. These mass spectrometers are, arguably, the most significant development in aerosol measurement in the past 20 years and show great promise for providing rich new insights into sources and chemical transformations of atmospheric aerosols.

4.2.1. Real-time particulate carbon analyzers

Turpin et al. (1990b) developed an automated instrument for in situ measurements of fine particle "organic" and "elemental" carbon with a detection limit of 0.2 μgC m^{-3}. Sampling intervals vary with ambient concentrations; an interval of two hours was used for measurements in Los Angeles (Turpin and Huntzicker, 1995). Sampling involves parallel trains for collecting particles and for adsorbing organic vapors. Particulate samples (train 1) are collected on a quartz filter after coarse particles are removed with a 2.5 μm impactor. Because of the affinity of quartz for organic vapors, the particle filter also collects such vapors. Adsorbing vapors (train 2) are collected on a quartz filter located downstream of a Teflon filter that removes particles but presumably not organic vapors (McDow and Huntzicker, 1990). The particulate carbon concentration is obtained by subtracting the carbon concentration on train 2 from that on train 1. Samples are analyzed by sequentially heating the filters in trains 1 and 2, first in helium at 650°C to volatilize organic carbon. The gases that evolve from the heated filters are converted to CO_2 in a MnO_2 catalyst at 1000°C and then to CH_4 in a nickel-firebrick methanator at 500°C. The methane is detected using a flame ionization detector. Elemental carbon concentrations are then obtained by measuring the carbon-containing gases that evolve when the particulate filter (train 2) is exposed to a 2% mixture of O_2 in He and ramped from 350 to 750°C.

An ambient carbon particulate monitor is available commercially (Rupprecht et al., 1995). In this instrument, sub-10 μm particles are collected at 16.7 1 m^{-1} with a muti-jet impactor having a 0.14 μm aerodynamic diameter size cut. The impactor can be operated at temperatures ranging from ambient to 150°C, and sample collection times can be adjusted from 1 to 24 h. The particulate carbon content is determined by measuring the amount of CO_2 that is produced in a 750°C afterburner when the collected particles are heated in air. "Organic" carbon is determined from the amount of CO_2 that is produced when the aerosol sample is heated to 340°C, and "soot" is determined from the amount of CO_2 that is produced when the aerosol sample is heated to 750°C. The manufacturer reports the instrument resolution to be ±0.25 μg m^{-3} at 1 σ and 1-h collection. This instrument is probably less affected by adsorption artifacts than samplers that rely on quartz filters since the surface area of the impaction substrate is much smaller than that of filters. Sources of measurement error include evaporative losses, pyrolysis of the sample which may lead to an overestimate of the "soot" concentration, and omission of particles smaller than 0.14 μm, which will lead to an underestimate of the true particulate carbon content. Measurements with MOUDI impactors in Los Angeles (Zhang et al., 1993), in Meadview, AZ (McMurry et al., 1996b), and in the Great Smoky Mountain National Park (Vasiliou and McMurry, 1997) showed that the fractions of the particulate carbon associated with particles smaller than 0.18 μm were 0.14±0.19, 0.22±0.06 and 0.18±0.08, respectively. Based on these results it can be concluded that roughly 20% of the carbon-containing particlate matter is not collected by the 0.14 μm impactor of this carbon analyzer.

4.2.2. Real-time particulate sulfur and nitrogen species analyses

Several groups have reported techniques for in situ measurements of particulate sulfur concentrations. Many of these techniques involve the use of flame photometric detectors (FPD) (Coburn et al., 1978; Huntzicker et al., 1978; Kittelson et al., 1978; Tanner et al., 1978, 1980; D'Ottavio et al., 1981; Garber et al., 1983; Allen et al., 1984; Slanina et al., 1985). The FPD detects ~394 nm light given off by excited-state S_2 molecules formed when sulfur compounds are burned in a hydrogen-rich flame. Although the FPD was originally intended for measurements of gaseous sulfur compounds, research has shown that it also detects particle-borne sulfur. The response times of the FPD instruments for particulate sulfur are as low as ~1 min, and the minimum detectable limits are typically ~1 μg m^{-3}.

Because the FPD responds to both gaseous and particulate sulfur, particles and gases are usually separated before they enter the hydrogen-rich flame. For measurements of sulfur-containing gases, the sample flows through a filter prior to entering the flame. For particulate sulfur measurements, the interfering gases are usually removed with a denuder. An alternative approach developed by Kittelson et al. (1978) and Keady (1987) permits simultaneous measurements of gases and particles. It uses an electrostatic precipitator cycling at about 0.2 Hz; only gases are measured when the precipitator is on, while both gases and particles are measured when it is off.

Measurements of sulfate speciation with the FPD have been reported. Speciation is achieved by heating the aerosol to preset temperatures upstream of the particulate sulfur monitor. Several studies have shown that temperatures of 71, 142, and 190°C, respectively volatilize sulfuric acid, ammonium sulfate salts, and refractory sulfur species (Huntzicker et al., 1978; Kittelson et al., 1978; Tanner et al., 1980). Allen et al. (1984) report that sulfuric acid can be removed from sulfate salts if the aerosol is preheated to a temperature of 120°C before entering the FPD. The volatilized components are typically removed by a denuder, and the remaining aerosol is then measured by the particulate sulfur monitor. Species concentrations are determined by difference.

Jaklevic et al. (1981a) developed an automated sampler in which particulate sulfur concentrations were determined by X-ray fluorescence (XRF) analysis. This latter system has a minimum detectable limit of 0.1 μg m^{-3} for sampling periods of ~1 hour. A comparison of particulate sulfur measurements by five FPD instruments and the XRF instrument of Jaklevic et al. was carried out in

St. Louis in August, 1979 (Camp et al., 1982). This study showed that measurements of all samplers were well correlated, and four of the six instruments in the study agreed to within ±5% on average.

Suh et al. (1994) compared measurements of sulfate concentrations obtained with their FPD instrument (the CSTS-continuous sulfate/thermal speciation system) with values obtained using the Harvard/EPA annular denuder system (HEADS) and a Micro-Orifice Impactor (MOI). They found that the CSTS was well correlated with the integrating samplers and provided concentrations that agreed with the other samplers to within experimental uncertainty.

Despite these encouraging results, such instruments are not widely used. Instrument calibrations are sensitive to relative humidity and to pressure, and obtaining accurate measurements at low ambient concentrations requires constant surveillance by a knowledgeable operator. Additional development would be required before an instrument of this type could be reliably deployed for routine monitoring.

Stolzenburg and Hering (1999) recently developed a system for measuring particulate nitrate concentrations. Particles are collected with a single stage impactor that is >95% efficient for particles larger than 0.1 μm aerodynamic diameter, and the sampled aerosol is humidified to prevent bounce in the impactor. Collected particles are analyzed by flash vaporization using a chemilluminescent NO_x analyzer. Measurements in Southern California showed good agreement with data from a denuded filter sampler. Deposits containing 5 ng of nitrate can be analyzed, corresponding to a detection limit of 0.7 μg m^{-3} for 8 min samples collected at 1 l m^{-1}. The detection limit is due to variabilities in the blank and could be reduced by if a source of more consistent blanks were developed.

Ion chromatographs (ICs) have also been adapted to semicontinuous measurements of particulate and gaseous species. Simon and Dasgupta (1995a) used an IC to analyze the effluent from a parallel plate wet denuder to determine concentrations of gas-phase nitrous (HONO) and nitric (HNO_3) acids. Detection limits for HONO and NHO_3 were 110 and 230 ppq, respectively. Particulate sulfate, nitrite, and nitrate were collected with a vapor condensation aerosol collection system and analyzed by IC. Detection limits for 8-min sampling intervals were, respectively, 2.2, 0.6 and 5.1 ng m^{-3} (Simon and Dasgupta, 1995a,b). Khlysiou et al. (1995) reported detection limits of 0.7 μg m^{-3} for ammonium, sulfate, nitrate, and chloride particles that were collected by steam condensation and analyzed with an automated IC system. Buhr et al. (1995) combined a wet denuder for nitric acid with a pyrex frit for particle collection to determine HNO_3 vapor and particulate nitrate and sulfate. Zellweger et al. (1999) modified the method of Dasgupta and coworkers to reduce interference from NO_2. Karlsson et al. (1997) used a single-stage impactor to collect particles larger than 0.4 μm onto a wetted substrate for semicontinuous analysis of sulfate and nitrate by IC; this methodology could be adapted to real-time measurements of size-resolved ionic composition. Because these approaches are amenable to the simultaneous measurement of semivolatile gases and particulate species, provide information on a number of species, and provide relatively good time resolution for typical ambient concentrations, they show promise for the future.

4.2.3. Real-time single-particle mass spectrometry

Several groups have developed real-time in-situ techniques for the analysis of individual particles in a flowing gas stream by mass spectrometry. This technique involves rapid depressurization of the aerosol, formation of a particle beam, and irradiation of particles by a high-power pulse laser to produce ions that are analyzed by mass spectrometry. Early work on this technology was reported by Stoffels and Lagergren (1981), by Sinha et al. (1982), Sinha and Friedlander (1985), Giggy et al. (1989), Marijnissen et al. (1988) and McKeown et al. (1991). Within the past five years this technology has been advancing rapidly, largely through innovations by several groups in the US and Europe (Thomson and Murphy, 1993, 1994, 1997; Hinz et al., 1994, 1996; Mansoori et al., 1994; Noble et al., 1994; Nordmeyer and Prather, 1994; Kievit, 1995; Murphy and Thomson, 1995; Reents et al., 1995; Salt et al., 1996). This approach offers most of the capabilities of LAMMS and avoids some of its limitations. Real-time instruments have been shown capable of providing information on surface coatings (Carson et al., 1997a, b), multicomponent crystallization (Ge et al., 1996), compound speciation (Neubauer et al., 1996), and oxidation state (Neubauer et al., 1995). Although particles must be brought into a vacuum before they can be analyzed, the time at reduced pressure is typically ~ 1 ms, which is short enough to avoid losses of most semivolatile compounds, although some loss of water is likely. Also, because particles are analyzed in real time, these instruments are well suited for studying time-dependent phenomena or for use on aircraft.

Within the past several years, single-particle mass spectrometers have been used in field measurements of atmospheric aerosols in Düsseldorf, Germany, in the Rocky Mountains, in Los Angeles and at Cape Grim, Australia. Each of these field campaigns has provided important new information about properties of atmospheric aerosols. For example, Hinz et al. (1994,1996) found that principal component analysis could be used to identify the major chemical components of the primary particle types, thereby providing information on particle source categories. Murphy et al. (1997a, b) found that in the Rocky Mountain aerosols, sulfate and nitrate usually (but not always) occur in different particles and that organics occur in most particles. Prather and coworkers (Noble and Prather, 1996; Liu et al., 1997) were able to identify pyrotechnically derived particles following July 4th fireworks, and they also demonstrated the ability of their instrument to resolve size/composition correlations for atmospheric particles. Finally, Murphy and coworkers found that at Cape Grim, Australia (Middlebrook et al., 1998; Murphy and Thomson, 1997c, 1998), most sulfate-containing particles down to 0.16 μm also contained some sea salt, indicating that sea salt may provide the seeds on which cloud condensation nuclei form in this region, that organics were commonly internally mixed with sea salt, and that halogens in individual particles were anti-correlated with sulfates, suggesting that the halogens were displaced by sulfates as the particles aged.

Most of the real-time instruments currently under development measure particle composition with time-of-flight mass spectrometry. Depending on particle size, however, various approaches are used to trigger the pulse laser and to measure particle size. For particles larger than about 0.2 μm, particles are usually detected as they flow through a volume illuminated by a low-power continuous wave (cw) laser. The scattered light is used to trigger the high-power pulse laser to volatilize and ionize the particles. The instruments of Hinz et al., Murphy et al., Johnston/Wexler et al., and Marijnissen et al. also use the intensity of scattered light from the cw laser to obtain an estimate of particle size. In the instrument of Prather and coworkers, the time of flight between two low-powered cw laser beams separated by a known distance is used to measure particle velocity, which is used to infer aerodynamic size. Sizes obtained from particle velocity are more accurate than those obtained from the intensity of scattered light. Particles smaller than ~0.2 μm cannot easily be detected by light scattering. Nevertheless, Reents et al. (1995) and Carson et al. (1997a,b) have had success at detecting particles as small as ~0.01 μm by firing the pulse laser at the maximum possible rate. Although particles are irradiated for only a small fraction of the laser pulses and although only a small fraction of the particles are present when the laser is fired, useful information on the composition of individual ultrafine particles has been obtained in this way. These measurements do not provide direct information on size. However, by using a differential mobility analyzer Liu and Pui (1974a, b) to deliver particles of known size, Carson et al. (1997a, b) were able to determine the size and composition of ultrafine (sub 0.2 μm) particles.

An alternative approach for real-time analysis of size-resolved aerosol composition by mass spectrometry was reported by Jayne et al. (1999). They collect size-segregated particles on a filament that is heated to vaporize and ionize the deposited particles by electron impact. Ions are detected using a quadrupole mass spectrometer. They report that their technique can be used to measure the composition of particles as small as 5 nm diameter. Because all particles in a specified size range are collected on the filament, this approach has a higher detection efficiency for ultrafine particles than other approaches that have been reported. Their instrument can detect either individual particles or an ensemble of particles.

In summary, the real-time compositional analysis of particles by mass spectrometry has advanced rapidly in the past five years. In the coming years, instruments of this type will provide fresh insights into particle sources, chemical transformations between particles and gases, and the distribution of species among, on, and within particles. Although most measurements to date do not provide quantitative information on mass concentrations, recent work has provided some encouragement that obtaining quantitative mass concentration may be possible (Jayne et al., 1996; Gross et al., 1997).

5. Calibration of atmospheric aerosol instrumentation

A review of techniques used to produce calibration aerosols is given by Chen (1993). In this section the techniques that are most commonly used to calibrate atmospheric aerosol instrumentation are discussed. Significant progress has been made since 1970 in the development of techniques for generating calibration aerosols, but the measurement of atmospheric aerosols introduces challenges that are not all resolved by these tools.

As was mentioned in the introduction, measured particle "sizes" depend on physical-chemical properties such as shape, density, refractive index, and geometric size, depending on the measurement method. The techniques described below can be used to produce laboratory calibration aerosols of known size and with known physical–chemical properties. However, atmospheric aerosols are often hygroscopic and typically contain mixtures of species and phases. Furthermore, properties such as shape, density and refractive index of atmospheric particles are typically not known with high accuracy. Therefore, it is not possible to produce synthetic particles that accurately mimic all qualities of atmospheric particles. These differences in properties of calibration and atmospheric aerosols introduce measurement uncertainties that are often difficult to quantify.

One approach that can reduce measurement uncertainties is to calibrate instruments with atmospheric particles. This can be done by using a DMA (see below) to select atmospheric particles of known electrical mobility equivalent diameter. For spherical particles (most but not all submicron atmospheric particles are spherical), the electrical mobility equivalent diameter equals the geometric size. Thus, this approach can provide a reasonably good measure of the relationship between instrument response and particle geometric diameter. Significant ambiguities remain, however, even when an instrument is calibrated directly with atmospheric particles. For example, it is known that atmospheric particles of a given size often vary in composition. Therefore, an unambiguous relationship between size and properties does not exist. Furthermore, the water content of hygroscopic particles varies with relative humidity. Therefore, refractive indices and densities also vary with relative humidity. The response of an OPC depends on refractive index, and responses of aerodynamic size classifiers depend on density, so calibrations of such instruments done at one relative humidity will be invalid when the relative humidity changes. Furthermore, the dependence of refractive index and density on relative humidity is not well known, so it is not straightforward to correct data for changes in properties that occur when relative humidities change.

The challenge of calibrating aerosol instruments is greater for instruments that significantly alter the thermodynamic environment of the aerosol. For example OPCs tend to heat the aerosol and aerodynamic particle sizers and impactors expand the aerosol through nozzles, both of which lead to changes in relative humidity and therefore particle size. It is somewhat easier to control the temperature in electrostatic classifiers, so these instruments are less prone to such errors.

Improving the accuracy of atmospheric aerosol measurements will require more attention to physical-chemical properties that affect instrument response, and this should remain a goal of atmospheric aerosol research. Only when it is possible to reconcile measurements made with different methods will we be able to refine our understanding of measurement accuracy.

5.1. Polystyrene latex spheres

Monodisperse polystyrene latex (PSL) and polyvinyl toluene (PVT) spheres ranging in size from 0.01 to 30 µm suspended in dilute liquid solutions are available from a variety of commercial sources. Monodisperse calibration aerosol particles are produced by atomizing the liquid solutions into a fine mist and evaporating the solvent. A fraction of the mist droplets contain a PSL or PVT sphere, and these spheres provide a useful calibration aerosol.

PSL and PVT spheres are well characterized with respect to size and composition and are inexpensive and relatively easily to generate in the laboratory. One difficulty is that the atomized liquid always contains a small amount of nonvolatile solute. Therefore, atomized droplets that do not contain a PSL or PVT sphere produce a small "residue" particle (Whitby and Liu, 1968). The concentration of these residue particles is typically much higher than the concentration of the larger PSL or PVT particles. Therefore, in some applications it is necessary to remove the residue particles before the atomized aerosol can be used for calibration. This is most often accomplished by electrostatic classification.

5.2. Electrostatic classification by the differential mobility analyzer (DMA)

Differential mobility analyzers (DMAs; for more information see section on electrical mobility analyzers) can be used to produce calibration aerosols of known size ($\sim$1–500 nm), concentration and composition. They can also be used to select monodisperse slices of "unknown" aerosols such as atmospheric aerosols for use in instrument calibration or for studies of aerosol properties. Size is determined from its known dependence on electrical mobility, classifying voltage and flow rates (Knutson and Whitby, 1975). Careful measurements have shown that the mean size exiting a DMA agrees with theory to within ±3% (Kinney et al., 1991). DMAs are typically operated under conditions that lead to a size range of $\sim$±10%, although this can be adjusted within limits. Concentration is determined by collecting charged particles at the exit from the DMA on a filter housed within a Faraday cage (Liu and Pui, 1974a,b). By using a sensitive electrometer to measure the current delivered to the filter, the charge delivery rate is obtained. If particles are singly charged, the concentration equals this charge delivery rate divided by the volumetric flow rate through the filter. Composition is determined by the composition of aerosols entering the DMA. It is often relatively easy to generate polydisperse aerosols of known composition by atomizing known liquid solutions or by nucleating and condensing known vapors, etc.

Multiple charging complicates the use of DMAs for producing calibration aerosols, and it becomes more problematic as particle size increases. Multiple charging leads to a multiplicity of sizes at a given classifying voltage, which is clearly a problem if ones objective is to determine the response of an instrument to particles of a known size. Multiple charging also leads to errors in measurements of aerosol concentrations using the aerosol electrometer, since multiply charged particles deliver more than a one elementary charge to the electrometer. Several schemes for eliminating multiply charged particles from DMA-produced aerosols have been published. Romay-Novas and Pui (Romay-Novas and Pui, 1988) used an impactor to remove multiply charged particles, and Gupta and McMurry (Gupta and McMurry, 1989) used a specially designed charger that produced few multiply charged particles. Despite these limitations, the DMA is one of the most versatile calibration tools due to its wide dynamic range and due to its ability to select particles of known size from any aerosol.

5.3. Vibrating orifice aerosol generator

A vibrating orifice aerosol generator (VOAG) (Berglund and Liu, 1973) can produce monodisperse particles when solutions containing a known concentration of a nonvolatile solute are forced through a very small orifice (5 to 20 μm diameter) at a known volumetric rate. The orifice is connected to a piezoelectric crystal that vibrates at a known frequency. This vibration delivers a regular instability to the liquid jet leaving the orifice, thereby causing the liquid to break into uniform droplets of known size. When the solute evaporates, a nonvolatile particle of known size remains. In principle, the particle production rate equals the crystal oscillation rate, thereby opening the possibility that the VOAG could be used to produce aerosols of known concentration as well as size. In practice, the VOAG is not often used as a concentration standard since it is difficult to avoid some deposition losses of the relatively large droplets produced by the vibrating orifice. Another problem that is encountered with VOAG aerosols is that droplets can collide, leading to particles that contain an integral number of primary droplets. The effect of collision can be minimized by adjusting the VOAG operation until the number of mulitplets detected with an aerodynamic particle sizer is negligible.

VOAGs can produce monodisperse particles ranging in diameter from 0.5 to 50 μm. Obtaining particles smaller than 1.0 μm is difficult since it is difficult to avoid plugging of the very small orifices that must be used. Particle sizes are uniform to within about ±1.4%.

6. Federal reference method

The original National Ambient Air Quality Standard (NAAQS) for particulate matter was for "total suspended particulate matter" (TSP) and was in force from 1970 to 1987, when it was replaced by a standard for particles smaller than 10 μm aerodynamic diameter (PM_{10}) (Register, 1987). More recently, an additional fine particle ($PM_{2.5}$) has been proposed (Register, 1997). These methods are briefly reviewed in light of the preceding discussion on measurement methodologies.

The PM_{10} standard defines performance specifications of PM_{10} samplers. Particles smaller than 10 μm are inertially separated from larger particles. The sampler "cut point" is defined such that the aerodynamic size for which 50% of the particles is collected falls in the 9.5–10.5 μm diameter range. The sampling efficiency increases to ∼100% for smaller particles and drops to 0% for larger particles. Particles are collected on filters, and mass concentrations are determined gravimetrically. Measurement precision for 24-h samples must be ±5 μg m^{-3} for PM_{10} concentrations below 80 μg m^{-3}, and 7% above this level. Sample volumes are adjusted to sea level pressure and 25°C.

Provision is made for "reference" and "equivalent" measurement PM_{10} methods. Reference methods meet all of performance specifications including those cited above. "Equivalent" methods may involve collection or analysis methods different from those defined by the PM_{10} standard, yet which can be shown to perform equivalently based on specified side-by-side comparisons with an approved reference method sampler. Several beta-gauge instruments and a harmonic oscillating element instrument have been certified by EPA as PM_{10} equivalent methods.

The new Federal Reference Method for determining mass concentrations of sub-2.5 µm particles ($PM_{2.5}$) was published in the Federal Register on July 18, 1997 (Register, 1997). In contrast to the PM_{10} performance specification, this $PM_{2.5}$ FRM specifies all details of the sampler design and of sample handling and analysis. The sampler consists of a PM_{10} inlet, an oil-soaked impaction substrate to remove particles larger than 2.5 µm, and a 47 mm polytetrafluoroethylene (PTFE) filter with a collection efficiency exceeding 99.7% for particle collection. The sampling duration is 24 h, and the sampler temperature is to not to exceed ambient values by more than 5°C. Filters are weighed before and after sampling at relative humidities in the 30% to 40% range and controlled to within ±5%. Mass measurements at relative humidities down to 20% are permissible if the mean ambient relative humidity during sampling is less than 30%. The description of the $PM_{2.5}$ reference method in the Federal Register acknowledges that "because the size and volatility of the particles making up ambient particulate matter vary over a wide range and the mass concentration of particles varies with particle size, it is difficult to define the accuracy of $PM_{2.5}$ measurements in an absolute sense ... Accordingly, accuracy shall be defined as the degree of agreement between a subject field $PM_{2.5}$ sampler and a collocated $PM_{2.5}$ reference method audit sampler...."

The FRM also provides for Classes I–III equivalent methods for $PM_{2.5}$. Class I equivalent methods use samplers with relatively small deviations from the sampler described in the FRM. Class II equivalent methods include "all other $PM_{2.5}$ methods that are based on a 24-h integrated filter sampler that is subjected to subsequent moisture equilibration and gravimetric mass analysis." Class III equivalent methods include filter-based methods having other than a 24-h collection interval or non-filter-based methods such as beta attenuation, harmonic oscillating elements, and nephelometry.

The strength of the $PM_{2.5}$ FRM is that because the sampler design is specified in complete detail, measurements at all locations should be comparable. In light of the previous discussion, however, several limitations can be identified. The oil-soaked impaction substrate requires maintenance to ensure that it does not become caked with deposits of coarse particles since this could lead to particle bounce and cause fine particle concentrations to be artifically inflated. It is not always safe to assume that such maintenance will be carried out, especially in routinely operated sampling networks. Furthermore, the oil-soaked impaction substrate must collect all particles between 2.5 and 10 µm. If an inlet with a smaller size cut were used, the delivery of soil dust to the 2.5 µm impactor could be substantially reduced, thereby reducing the required maintenance frequency. Another concern is that transfer of oil or organic vapors from the oiled frit could contaminate the filter, thereby precluding analysis of organics on archived filters.

Both the PM10 and PM2.5 standards are defined with consistency rather than accuracy in mind. We know, for example, that evaporative losses of species including ammonium nitrate and organics will lead to a significant underestimate of the true fine particle mass concentration in some locations. If measurement techniques were developed that eliminated such systematic errors, they could not be certified as either reference or equivalent techniques because they would not agree with data from the reference method samplers.

7. Summary of significant advances

- Instrumentation for producing laboratory calibration aerosols of known size, composition and concentration became available about 25 years ago. This instrumentation is now widely used to characterize the response of aerosol instrumentation to known aerosols. These calibration techniques have facilitated a steady advance in the quality of atmospheric aerosol measurements.
- Mass spectrometers that can measure the composition of individual atmospheric particles in real time are now available. These instruments open a new set of questions that can be addressed regarding sources and reactivity of atmospheric aerosols, and may evolve to techniques for routine, real-time measurement of aerosol chemical composition.
- Methods for routine, real-time measurement of size distributions of aerosols in the 3 nm to >10 μm are now available.

8. Summary of future aerosol measurement needs

- Gravimetric techniques that are used for regulatory compliance purposes involve filtration. While such methods are relatively simple and inexpensive to implement, they require manual operation, provide only rough time and spatial resolution, and are subject to sampling errors that cannot be quantified. Real-time techniques for accurate measurement of mass that avoid such sampling errors are needed.
- The response of aerosol instruments depends on particle properties including density, complex refractive index and shape. More information on such properties is required to improve our understanding of measurement accuracy.

- Instrumentation to measure aerosol composition in real time is needed. Most analytical measurements are done off-line and are expensive. While some effort has been made to measure particulate sulfur, carbon and nitrate concentrations in real time, the instruments that have been developed are mostly laboratory prototypes and would require further development if they were to be used for routine measurements. It is possible that individual particle mass spectrometry will evolve into instruments that can provide accurate real-time data of size resolved composition, but more work is needed to evaluate the potential of these instruments for such measurements.

- The measurement of particulate carbon-containing species is especially problematic. While aerosols include many carbon-containing species with a wide variety of chemical and thermodynamic properties, most measurements crudely identify these as either "elemental" or "organic carbon". There is no standard analytical technique that is accepted as the "correct" method for distinguishing between elemental and organic carbon, so different laboratories often obtain different results for the same sample. Furthermore, substantial sampling errors are encountered, especially for organic carbon. It is likely that significant progress on measurements of the carbon-containing portion of particles will require more attention to speciation.

- Water is a major component of the aerosol, and is the most abundant fine particle species for relative humidities exceeding ~80%. An accurate, real-time technique for measurements of aerosol water concentration is needed.

- Our understanding of particle surface composition and the distribution of chemical species and phases within individual particles is poor. Chemical interactions between particles and gases are almost certainly affected by particle surface composition, and transformations within particles must depend on details of particle makeup. The development of instrumentation to provide such information would provide fresh insights into chemical reactivities of atmospheric aerosols.

- Our understanding of chemical interactions involving atmospheric particles is primitive relative to our understanding of purely gas-phase chemistry. This is due, in large part, to the fact that chemists have historically played a secondary role to physicists and engineers in aerosol science. Many of the challenges for the future are chemical, and recent work shows a rapid increase in the sophistication with which aerosol chemistry is treated. It is likely that this trend will continue.

- Nucleation is an unsolved problem in aerosol science. Nucleation occurs in combustors that produce primary pollutant emissions and in the atmosphere as a result of chemical transformations. Understanding the chemical and physical processes that lead to the birth of new particles by nucleation will require information on the composition and concentration of the molecular clusters that serve as their precursors. "Bridging the gap" between molecules and macroscopic (>3 nm) particles remains a challenge to the scientific community.

Acknowledgements

Preparation of this review was supported in part by the Electric Power Research Institute through Grant No. EPRI W09116-08/W04105-01 and in part by the Department of Energy through Grant No. DE-FG02-91ER61205. Colleagues too numerous to mention have readily responded to my requests for information. Thank you all.

References

Adachi, M., Okuyama, K., Kousaka, Y., 1983. Electrical neutralization of charged aerosol particles by bipolar ions. Journal of Chemical Engineering of Japan 16 (3), 229–235.

Adams, K.M., 1988. Real-time in situ measurements of atmospheric optical absorption in the visible via photoacoustic spectroscopy. 1: evaluation of the photoacoustic cells. Applied Optics 27 (19), 4052–4056.

Adams, K.M., Davis, L.I.J., Japar, S.M., Pierson, W.R., 1989. Real-time in situ measurements of atmospheric optical absorption in the visible via photoacoustic spectroscopy – II. Validation for atmospheric elemental carbon aerosol. Atmospheric Environment 23 (3), 693–700.

Aden, G.D., Buseck, P.R., 1983. Aminicomputer procedure for quantitative EDS analyses of small particles. Microbeam Analysis – 1983, R. Gooley. San Francisco Press, Inc. San Francisco, CA, pp. 195–201.

Aitken, J., 1890–1891. On a simple pocket dust-counter. Proceedings of the Royal Society of Edinburgh XVIII, 39–53.

Allen, G.A., Turner, W.A., Wolfson, J.M., Spengler, J.D., 1984. Description of a continuous sulfuric acid/sulfate monitor. National Symposium on Recent Advances in Pollutant Monitoring of Ambient Air and Stationary Sources, U.S. EPA No. EZPA-600/9-84-019.

Almasi, G.S., Blair, J., Ogilvie, R.E., Schwartz, R.J., 1965. A heat-flow problem in electron-beam microprobe analysis. Journal of Applied Physics 36, 1848–1854.

Ananth, G.P., Wilson, J.C., 1988. Theoretical analysis of the performance of the TSI aerodynamic particle sizer. Journal of Aerosol Science 9, 189–199.

Anderson, J.R., Aggett, F.J., Buseck, P.R., Germani, M.S., Shattuck, T.W., 1988. Chemistry of individual aerosol particles from Chandler, Arizona, an Arid Urban environment. Environmental Science and Technology 22 (7), 811–818.

Anderson, J.R., Buseck, P.R., Patterson, T.L., Arimoto, R., 1996a. Characterization of the Bermuda tropospheric aerosol by combined individual-particle and bulk-aerosol analysis. Atmospheric Environment 30 (2), 319–338.

Anderson, T.L., Covert, D.S., Marshall, S.F., Laucks, M.L., Charlson, R.J., Waggoner, A.P., Ogren, J.A., Caldow, R., Holm, R.L., Quant, F.R., Sem, G.J., Wiedensohler, A., Ahlquist, N.A., Bates, T.S., 1996b. Performance characteristics of a high-sensitivity, three wavelength, total scatter/backscatter nephelometer. Journal of Atmospheric and Oceanic Technology 13 (5), 967–986.

Andreae, M.O., Charlson, R.J., Bruynseels, F., Storms, H., van Grieken, R., Maenhaut, W., 1986. Internal mixture of sea salt, silicates, and excess sulfate in marine aerosols. Science 222, 1620–1623.

Appel, B.R., Tokiwa, Y., 1981. Atmospheric particulate nitrate sampling errors due to reactions with particulate and gaseous strong acids. Atmospheric Enviroment 15 (6), 1087–1090.

Armstrong, J.T., Buseck, P.R., 1975. Quantitative chemical analysis of individual microparticles using the electron microprobe. Analytical Chemistry 47, 2178–2192.

Arnott, W.P., Moosmüller, H., Abbott, R.E., Ossofsky, M.D., 1995. Thermoacoustic enhancement of photoacoustic spectroscopy: theory and measurements of the signal to noise ratio. Review of Scientific Instruments 66 (10), 4827–4833.

Arnott, W.P., Moosmüller, H., Rogers, C.F., 1999. Photoacoustic spectrometer for measuring light absorption by aerosol: Instrument description. Atmospheric Environment 33, 2845–2852.

Artaxo, P., Rabello, M.L.C., Maenhaut, W., Grieken, R.v., 1992. Trace elements and individual particle analysis of atmospheric aerosols from the antarctic peninsula. Tellus 44B, 31–334.

Baltensperger, U., Gäggeler, H.W., Jost, D.T., Emmenegger, M., Nägeli, W., 1991. Continuous background aerosol monitoring with the epiphaniometer. Atmospheric Environment 25A (3/4), 629–634.

Baron, P.A., 1986. Calibration and use of the aerodynamic particle sizer (APS 3300). Aerosol Science and Technology 5, 55–67.

Baron, P.A., Mazumder, M.K., Cheng, Y.S., 1993. Direct-reading techniques using optical particle detection. Aerosol Measurement: Principles, Techniques, and Applications. K. Willeke and P.A. Baron. New York, Van Norstrand Reinhold.

Bartz, H., Fissan, H., Helsper, C., Kousaka, Y., Okuyama, K., Fukushima, N., Keady, P., Fruin, S., McMurry, P.H., Pui, D.Y.H., Stolzenburg, M.R., 1985. Response characteristics for four condensation nucleus counters to particles in 3 to 50 nm diameter range. Journal Aerosol Science Vol. 16 (No. 5), 443–456.

Baumgardner, D., Dye, J.E., Gandrud, B., Barr, K., Kelly, K., Chan, K.R., 1996. Refractive, indices of aerosols in the upper troposphere and lower stratosphere. Geophysical Research Letters 23 (7), 749–752.

Baumgardner, D., Huebert, B., 1993. The airborne aerosol inlet workshop – meeting report. Journal of Aerosol Science 24 (6), 835–846.

Baumgardner, D., Weaver, K., Gandrud, B., Dye, J.E., 1993. The Multiangle Aerosol Spectrometer Probe (MASP): A New Aerosol Probe for Airborne Stratospheric Research. American Association for Aerosol Research, (AAAR), Oak Brook, IL.

Belyaev, S.P., Levin, L.M., 1974. Techniques for collection of representative aerosol samples. Aerosol Science 5, 325–338.

Berglund, R.N., Liu, B.Y.H., 1973. Generation of monodisperse aerosol standards. Environmental Science and Technology 7 (2), 147–153.

Berner, A., Lürzer, C., Pohl, F., Preining, O., Wagner, P., 1979. The size distribution of the urban aerosol in Vienna. The Science of the Total Environment, 245–261.

Beuttell, R.G., Brewer, A.W., 1949. Instruments for the measurement of the visual Range. Journal of Science and Instruction 26, 357–359.

Birmili, W., Stratmann, F., Wiedensohler, A., Covert, D., Russell, L.M., Berg, O., 1997. Determination of differential mobility analyzer transfer functions using identical instruments in series. Aerosol Science and Technology 27 (2), 215–223.

Biswas, P., Jones, C.L., Flagan, R.C., 1987. Distortion of size distributions by condensation and evaporation in aerosol instruments. Aerosol Science and Technology 7 (2), 231–246.

Bowers, W.D., Chuan, R.L., 1989. Surface acoustic wave piezoelectric crystal aerosol mass monitor. Review of Scientific Instrument 60 (7), 1297–1302.

Bricard, J., Delattre, P., Madelaine, G., Pourprix, M., 1976. Detection of ultra-fine particles by means of a continuous flux condensation nuclei counter. In: Liu, B.Y.H. (Ed.), Fine Particles: Aerosol Generation, Measurement, Sampling, and Analysis. Academic Press, New York, pp. 565–580.

Brock, C.A., Hamill, P., Wilson, J.C., Honsson, H.H., Chan, K.R., 1995. Particle formation in the upper tropical troposphere: A source of nuclei for the stratospheric aerosol. Science 270, 1650–1653.

Bruynseels, F., Otten, P., Van Grieken, R., 1988a. Inorganic nitrogen speciation in single micrometer-size particles by laser microprobe mass analysis. Journal of Analytical Atomic Spectrometry 3, 237–240.

Bruynseels, F., Storms, H., Van Grieken, R., 1988b. Characterization of North Sea aerosols by individual particle analysis. Atmospheric Environment 22, 2593–2602.

Bruynseels, F., Van Grieken, R., 1985. Direct detection of sulfate and nitrate layers on sampled marine aerosols by laser microprobe mass analysis. Atmospheric Environment 19 (11), 1969–1970.

Bruynseels, F., Van Grieken, R., 1986. Recombination reactions and geometry effects in laser microprobe mass analysis studies with 12C/13C bilayers. International Journal of Mass Spectrometry and Ion Processes 74, 161–177.

Bruynseels, F.J., Van Grieken, R.E., 1984. Laser microprobe mass spectrometric identification of sulfur species in single micrometer-size particles. Analytical Chemistry 56, 871–873.

Buhr, S.M., Buhr, M.P., Fehsenfeld, F.C., Holloway, J.S., Karst, U., Norton, R.B., Parrish, D.D., Sievers, R.E., 1995. Development of a semi-continuous method for the measurement of nitric acid vapor and particulate nitrate and sulfate. Atmospheric Environment 29 (19), 2609–2624.

Cadle, S.H., Groblicke, P.J., Mulawa, P.A., 1983. Problems in the Sampling and Analysis of Carbon Particulate. Atmospheric Environment 17, 593–600.

Cahill, T.A., Eldred, R.A., Feeney, P.J., Beveridge, P.J., Wilkinson, L.K., 1990. The stacked filter unit revisited. In: Mathai, C.V. (Ed.), Transactions: Visibility and Fine Particles. Air and Waste Management Association, Pittsburgh, PA, pp. 213–221.

Cahill, T.A., Malm, W.C., 1987. Size/time/composition data at grand Canyon National Park and the role of ultrafine sulfur particles. In: P.S. Bhardwaja (Ed.), Visibility Protection: Research and Policy Aspects. APCA, P.O. Box 2861, Pittsburgh, PA 15230, APCA Publication No. TR-10, pp. 657–667.

Camp, D.C., Stevens, R.K., Cobourn, W.G., Husar, R.B., Collins, J.F., Huntzicker, J.J., Husar, J.D., Jaklevic, J.M., McKenzie, R.L., Tanner, R.L., Tesch, J.W., 1982. Intercomparison of concentration results from fine particle sulfur monitors. Atmospheric Environment 16, 911–916.

Campbell, D., Copeland, S., Cahill, T., Eldred, R., Cahill, C., Vesenka, J., VanCuren, T., 1989. The coefficient of optical absorption from particles deposited on filters: integrating plate, integrating sphere and coefficient of haze measurements. 82nd Annual Meeting & Exhibition Anaheim, APCA, California.

Carson, P.G., Johnston, M.V., Wexler, A.S., 1997a. Laser desporption ionization of ultrafine aerosol particles. Rapid Communications in Mass Spectrometry 11, 993–996.

Carson, P.G., Johnston, M.V., Wexler, A.S., 1997b. Real-time monitoring of the surface and total composition of aerosol particles. Aerosol Science and Technology 26 (4), 291–300.

Casuccio, G.S., Janocko, P.B., Lee, R.J., Kelly, J.F., Dattner, S.L., Mgebroff, J.S., 1983. The use of computer controlled scanning electron microscopy in environmental studies. Journal of the Air Pollution Control Association 33 (10), 937–944.

Chan, T., Lippmann, M., 1977. Particle collection efficiencies of air sampling cyclones: an empirical theory. Environmental Science and Technology 11, 377–382.

Charlson, E.J., Porch, W.M., Waggoner, A.P., Ahlquist, N.C., 1974. Background aerosol light scattering characteristics: nephelometric observations at Mauna Loa Observatory compared with results at other remote locations. Tellus XXVI (3), 345–359.

Charlson, R.J., Ahlquist, N.C., Horvath, H., 1968. On the generality of correlation of atmospheric aerosol mass concentration and light scatter. Atmospheric Environment 2, 455–464.

Chen, B.T., 1993. Instrument calibration. In: Willeke, K., Baron, P.A. (Eds.), Aerosol Measurement: Principles, Techniques, and Applications. Van Norstrand Reinhold, New York, pp. 493–520.

Chen, D., Pui, D.Y.H., 1995. Numerical modeling of the performance of differential mobility analyzer for nanometer aerosol measurements. Journal of Aerosol Science 26, S141–S142.

Chen, D.R., Pui, D.Y.H., Hummes, D., Fissan, H., Quant, F.R., Sem, G.J., 1996. Nanometer Differential Mobility Analyzer: Design and numerical modeling. AAAR, Orlando, FL.

Cheng, Y.-S., Yeh, H.-C., 1979. Particle bounce in cascade impactors. Environmental Science and Technology 13 (11), 1392–1396.

Cheng, Y.S., Chen, B.T., Yeh, H.C., 1990. A study of density effect and droplet deformation in the TSI aerodynamic particle sizer. Aerosol Science Technology 4, 89–97.

Cheng, Y.S., Yeh, H.C., 1980. Theory of a screen-type diffusion battery. Journal of Aerosol Science 11, 313–320.

Chow, J.C., 1995. Measurement methods to determine compliance with ambient air quality standards for suspended particles. Journal of the Air and Waste Management Association 45 (5), 320–382.

Clarke, A.D., 1982. Integrating sandwich: a new method of measurement of the light absorption coefficient for atmospheric particles. Applied Optics 21 (16), 3011–3020.

Clarke, A.D., 1993. Atmospheric nuclei in the Pacific Midtroposphere – their nature, concentration, and evolution. Journal of Geophysical Research – Atmospheres 98 (D11), 20633–20647.

Coburn, W.G., Husar, R.B., Husar, J.D., 1978. Continuous in situ monitoring of ambient particulate sulfur using flame photometry and thermal analysis. Atmospheric Environment 12, 89–98.

Commins, B.T., Lawther, P.J., 1957. Volatility of 3,4-benzpyrene in relation to the collection of smoke samples. British Journal of Cancer 12, 351–354.

Cooke, D.D., Kerker, M., 1975. Response calculations for light scattering aerosol particle counters. Applied Optics 14, 734–739.

Cooper, D.W., Wu, J.J., 1990. The inversion matrix and error estimation in data inversion: Application to diffusion battery measurements. Journal of Aerosol Science 21 (2), 217–226.

Courtney, W.J., Shaw, R.W., Dzubay, T.G., 1982. Precision and accuracy of a ß gauge for aerosol mass determinations. Environmental Science and Technology 16 (4), 236–238.

Covert, D.S., Hansson, H.-C., Winkler, P., Heintzenberg, J., 1991. The degree of mixing of hygroscopic properties in source and receptor locations in northern Europe. AAAR'91. Traverse City, Michigan, USA.

Curzon, A.E., 1991. A first-order theory for the temperature distribution in a thermally conducting and radiating rectangular film observed by transmission electron microscopy. Journal of Physics D: Applied Physics 24, 1616–1623.

D'Ottavio, T.D., Garber, R., Tanner, R.L., Newmann, L., 1981. Determination of ambient aerosol sulfur using a continuous flame photometric detection system II. The measurement of low-level sulfur concentrations under varying atmospheric conditions. Atmospheric Environment 15 (2), 197–204.

Dahneke, B., Padliya, D., 1977. Nozzle-inled design for aerosol beam. Instruments in Rarefied Gas Dynamics 51 Part II, 1163–1172.

Dahneke, B.E., 1973. Aerosol beam spectrometry. Nature Physical Science 244, 54–55.

Dahneke, B.E., Cheng, Y.S., 1979. Properties of continuum source particle beams. I. Calculation methods and results. Journal of Aerosol Science 10, 257–274.

Danilatos, G.D., 1988. Foundations of environmental scanning electron microscopy. Advances in Electronics and Electron Physics 71, 109–250.

Danilatos, G.D., Postle, R., 1982. The environmental scanning electron microscope and its applications. Scanning Electron Microscopy I, 1–16.

Davis, J.E., Ray, A.K., 1978. Submicron droplet evaporation in the continuum and non-continuum regimes. Journal of Aerosol Science 9, 411–422.

de Juan, L., Fernández de la Mora, J., 1998. High resolution size-analysis of nanoparticles and ions: Running a DMA of near optimal length at Reunolds numbers up to 5000. Journal of Aerosol Science 29 (7), 617–626.

De Waele, J.K., Gjbels, J.J., Vansant, E.F., Adams, F.C., 1983. Laser microprobe mass analysis of asbestos fiber surfaces for organic compounds. Analytical Chemistry 55, 671–677.

de Waele, J.K.E., Adams, F.C. (Eds.), 1986. Laser-microprobe mass analysis of fibrous dusts. Chemical Characterization if Individual Airborne Particles. Ellis Horwood Limited, Chichester, U.K.

De Wiest, F., Rondia, D., 1976. Sur la validite des determinations du benzo(a)pyrene atmospherique pendant les mois d'ete (Short Communication). Atmospheric Environment 10, 487–489.

Dick, W., McMurry, P.H., Bottiger, J.R., 1994. Size- and composition-dependent response of the DAWN-A multiangle single-particle optical detector. Aerosol Science and Technology 20 (4), 345–362.

Dick, W.D., 1998. Multiangle light scattering techniques for measuring shape and refractive index of submicron atmospheric particles. Ph.D. Thesis, Department of Mechanical Engineering. Minneapolis, MN, University of Minnesota.

Dick, W.D., Sachweh, B.A., McMurry, P.H., 1996. Distinction of coal dust particles from liquid droplets using the DAWN-A detector. Applied Occupation and Environmental Hygiene 11 (7), 637–645.

Dick, W.D., Ziemann, P.A., Huang, P.F., McMurry, P.H., 1998. Optical shape fraction measurements of submicron laboratory and atmospheric aerosols. Measurement Science and Technology 9 (2), 183–196.

Dirgo, J., Leith, D., 1985. Cyclone collection efficiency: collection efficencies of experimental results with theoretical predictions. Aerosol Science and Technology 4, 401–415.

Duntley, S.Q., Boileau, A.R., Preisendorfer, R.W., 1957. Image transmission by the troposphere I. Journal of the Optical Society of America 47, 499–506.

Dunwoody, C.L., 1986. Rapid nitrate loss from PM-10 filters. Journal of Air Pollution Control Association 36, 817–818.

Dzubay, T.H., Hines, L.E., Stevens, R.K., 1976. Particle bounce in cascade impactors. Environmental Science and Technology 13, 1392–1395.

Eatough, D.J., Aghdale, N., Cottam, M., Gammon, T., Hansen, L.D., Lewis, E.A., Farber, R.J., 1990. Loss of semi-volatile organic compounds from particles during sampling on filters. Proceedings of the AWMA Visibility Conference. October 16–19, 1989, Air and Waste Management Association, Estes Park, CO.

Eatough, D.J., Brutsch, M., Lewis, L., Eatough, N.L., Farber, R.J., 1986. Diffusion denuder sampling systems for the collection of gas and particle phase organic compounds in Visibility Protection: Research and Policy Aspects. Proceedings of APCA Specialty Conference on Visibility, APCA.

Eatough, D.J., Wadsworth, A., Eatough, D.A., Crawford, J.W., Hansen, L.D., Lewis, E.A., 1993. A multiple-system, multi-channel diffusion denuder sampler for the determination of fine-particulate organic material and the atmosphere. Atmospheric Environment 27A (8), 1213.

Ensor, D.S., Waggoner, A.P., 1970. Angular truncation error in the integrating nephelometer. Atmospheric Environment 4, 481–487.

Fahey, D.W., Kelly, K.K., Ferry, G.V., Wilson, J.C., Murphy, D.M., Lowenstein, M., Chan, K.R., 1989. In situ measurements of total reactive nitrogen, total water, and aerosol in a polar stratospheric cloud in the Antarctic. Journal of Geophysical Research 94 (D9), 11299–11315.

Fissan, H., Hummes, D., Stratmann, F., Buscher, P., Neumann, S., Pui, D.Y.H., Chen, D., 1996. Experimental comparison of four differential mobility analyzers for nanometer aerosol measurements. Aerosol Science and Technology 24 (1), 1–13.

Fissan, H.J., Helsper, C., Thielen, H.J., 1983. Determination of particle size distributions by means of an electrostatic classifier. Journal of Aerosol Science 14, 354–357.

Fitzgerald, J.W., 1973. Dependence of the supersaturation spectrum of CCN on aerosol size distribution and composition. Journal of Atmospheric Science 30, 628–634.

Fitzgerald, J.W., Rogers, C.F., Hudson, J.G., 1981. Review of Isothermal Haze Chamber Performance. Journal of Recherches Atmosphere 15, 333–346.

Flagan, R.C., 1998. History of electrical aerosol measurements. Aerosol Science and Technology 28 (4), 301–380.

Fletcher, R.A., Small, J.A., 1993. Analysis of individual collected particles. In: Willeke, K., Baron, P.A. (Eds.), Aerosol Measurement: Principles, Techniques and Applications. Van Norstrand Reinhold, New York, pp. 260–295.

Flocchini, R.G., Cahill, T.A., Pitchford, M.L., Eldred, R.A., Feeney, P.J., Ashbaugh, L.L., 1981. Characterization of particles in the arid west. Atmospheric Environment 15, 2017–2030.

Foltescu, V.L., Selin, E., Below, M., 1995. Corrections for particle losses and sizing errors during aircraft aerosol sampling using a Rosemount inlet and the PMS LAS-X. Atmospheric Environment 29 (3), 449–453.

Foot, J.S., Kilsby, C.G., 1989. Absorption of light by aerosol particles: an intercomparison of techniques and spectral observations. Atmospheric Environment 23 (2), 489–496.

Friedlander, S.K., 1970. The characterization of aerosols distributed with respect to size and chemical composition. Journal of Aerosol Science 1, 295–307.

Friedlander, S.K., 1971. The characterization of aerosols distributed with respect to size and chemical composition – II. Aerosol Science 2, 331–340.

Friedlander, S.K., 1977. Smoke, Dust and Haze: Fundamentals of Aerosol Behavior. Wiley-Interscience, New York.

Friedlander, S.K., Johnstone, H.F., 1957. Deposition of suspended particles from turbulent gas streams. Industrial and Engineering Chemistry 49, 1151–1156.

Fuchs, N.A., Sutugin, A.G., 1970. Highly dispersed aerosols. Ann Arbor Science Publishers, Ann Arbor, MI.

Fukuta, N., Saxena, V.K., 1979. A horizontal thermal gradient cloud condensation nucleus spectrometer. Journal of Applied Meteorology 18, 1352–1362.

Gäggeler, H.W., Baltensperger, U., Emmenegger, M., Jost, D.T., Schmidt-Ott, A., Haller, P., Hofmann, M., 1989. The epiphaniometer, a new device for continuous aerosol monitoring. Journal of Aerosol Science 20, 557–564.

Galasyn, J.F., Hornig, J.F., Soderberg, R.H., 1984. The loss of PAH from quartz fiber high volume filters. Journal of the Air Pollution Control Association 34 (1), 79.

Gale, B., Hale, K.F., 1961. Heating of metallic foils in an electron microscope. British Journal of Applied Physics 12, 115–117.

Garber, R.W., Daum, P.H., Doering, R.F., D'Ottavio, T., Tanner, R.L., 1983. Determination of ambient aerosol and gaseous sulfur using a continuous FPD-III. Design and characterization of a monitor for airborne applications. Atmospheric Environment 17 (7), 1381–1386.

Garvey, D.M., Pinnick, R.G., 1983. Response characteristics of the particle measuring systems active scattering aerosol spectrometer probe (ASASP-X). Aerosol Science and Technology 2, 477–488.

Ge, Z., Wexler, A.S., Johnston, M.V., 1996. Multicomponent aerosol crystallization. Journal of Colloid and Interface Science 183, 68–77.

Gebhart, J., 1993. Optical direct-reading techniques: light intensity systems. In: Willeke, K., Baron, P.A. (Eds.), Aerosol Measurement: Principles, Techniques, and Applications. Van Norstrand Reinhold, New York, pp. 313–344.

Gebhart, J., Heyder, J., Roth, C., Stahlhofen, W., 1976. Optical aerosol size spectrometry below and above the wavelength of light – a comparison. In: Liu, B.Y.H. (Ed.), Fine Particles: Aerosol Generation, Measurement, Sampling, and Analysis. Academic Press, New York.

Giggy, C.L., Friedlander, S.K., Sinha, M.P., 1989. Measurement of externally mixed sodium containing particles in ambient air by single particle mass spectrometry. Atmospheric Environment 23 (10), 2223–2230.

Gormley, P.G., Kennedy, M., 1949. Diffusion from a stream flowing through a cylindrical tube. Proceedings of the Royal Irish Academy 52, 163–169.

Gross, D.S., Gaelli, M.E., Prather, K.A., 1997. Relative response of ion signals for quantitation of species in atmospheric aerosol particles. AAAR Annual Meeting, Denver, CO.

Gundel, L.A., Lee, V.C., Mahanama, K.R.R., Stevens, R.K., Daisey, J.M., 1995. Direct determination of the phase distributions of semi-volatile polycyclic aromatic hydrocarbons using annular denuders. Atmospheric Environment 29 (14), 1719–1733.

Gupta, A., McMurry, P.H., 1989. A device for generating singly charged particles in the 0.1-1.0 μm diameter range. Aerosol Science and Technology 10, 451–462.

Hagen, D.E., Alofs, D.J., 1983. Linear inversion method to obtain aerosol size distributions from measurements with a differential mobility analyzer. Aerosol Science and Technology 2, 465–475.

Hänel, G., 1976. The properties of atmospheric aerosol particles as functions of the relative humidity at thermodynamic equilibrium with the surrounding moist air. Advaces in Geophysics 19, 73–188.

Hänel, G., 1987. Radiation budget of the boundary layer. Part II. Simultaneous measurement of mean solar volume absorption and extinction coefficients of particles. Beitraege Physikalische Atmosphere 60, 241–247.

Hänel, G., Lehmann, M., 1981. Equilibrium size of aerosol particles and relative humidity: new experimental data from various aerosol types and their treatment for cloud physics application. Contributions to Atmospheric Physics 54 (1), 57–71.

Hänel, G., Thudium, J., 1977. Mean bulk densities of samples of dry atmospheric aerosol particles: a summary of measured data. Pageoph 115, 799–803.

Hansen, A.D.A., Rosen, H., Novakov, T., 1984. The aethelometer – an instrument for the real-time measurement of optical absorption by aerosol particles. Science of the Total Environment 36, 191–196.

Harrison, L., 1985. The segregation of aerosols by cloud-nucleating activity. Part II: Observation of an urban aerosol. Journal of Climate and Applied Meteorology 24, 312–321.

Heintzenberg, J., Charlson, R.J., 1996. Design and applications of the integrating nephelometer: a review. Journal of Atmospheric and Oceanic Technology 13 (5), 987–1000.

Hemeon, W.C.L., Haines, G.F., Ide, H.M., 1953. Determination of haze and smoke concentrations by filter paper sampling. Journal of the Air Pollution Control Association 3, 22–28.

Hering, S., McMurry, P.H., 1991. Optical counter response to monodisperse atmospheric aerosols. Atmospheric Environment 25A (2), 463–468.

Hering, S.V., 1995. Impactors, cyclones and other inertial and gravitational collectors. In: Cohen, B., Hering, S.V. (Eds.), Air Sampling Instruments for Evaluation of Atmospheric Contaminants, 8th Edition. American Conference of Governmental Industrial Hygienists, Cincinnati, OH.

Hering, S.V., Appel, B.R., Cheng, W., Salaymeh, F., Cadel, S.H., Mulawa, P.A., Cahill, T.A., Eldred, R.A., Surovik, M., Fitz, D., Howes, J.E., Knapp, K.T., Stockburger, L., Turpin, B.J.,

Huntzicker, J.J., Zhang, X.Q., McMurry, P.H., 1990. Comparison of sampling methods for carbonaceous aerosols in ambient air. Aerosol Science and Technology 12, 200–213.

Hering, S.V., Friedlander, S.K., 1979. Design and evaluation of a new low-pressure impactor. 2. Environmental Science and Technology 13 (2), 184–188.

Hering, S.V., et al., 1988. The nitric acid shootout: field comparison of measurement methods. Atmospheric Environment 22 (8), 1519–1540.

Hinds, W.C., 1982. Aerosol Technology: Properties, Behavior, and Measurement of Airborne Particles. Wiley, New York.

Hinds, W.C., Kraske, G., 1986. Performance of PMS model LAS-X optical particle counter. Journal of Aerosol Science 17 (1), 67–72.

Hinz, K.P., Kaufmann, R., Spengler, B., 1994. Laser-induced mass analysis of single particles in the airborne state. Analytical Chemistry 66 (13), 2071–2076.

Hinz, K.P., Kaufmann, R., Spengler, B., 1996. Simultaneous detection of positive and negative ions from single airborne particles by real-time laser mass spectrometry. Aerosol Science and Technology 24 (4), 233–242.

Hirst, E., Kaye, P.H., 1996. Experimental and theoretical light scattering profiles from spherical and nonspherical particles. Journal of Geophysical Research – Atmospheres 101 (D14), 19231–19235.

Hitzenberger, R., Berner, A., Dusek, U., Alabashi, R., 1997. Humidity-dependent growth of size-segregated aerosol samples. Aerosol Science and Technology 27 (2), 116–130.

Ho, W., Hidy, G.M., Govan, R.M., 1974. Microwave measurements of the liquid water content of atmospheric aerosols. Journal of Applied Meteorology 13, 871–879.

Hoppel, W.A., 1978. Determination of the aerosol size distribution from the mobility distribution of the charged fraction of aerosols. Journal of Aerosol Science 9, 41–54.

Hoppel, W.A., Twomey, S., Wojciechowski, T.A., 1979. A segmented thermal diffusion chamber for continuous measurements of CN. Journal of Aerosol Science 10, 369–373.

Huang, P.-F., 1997. Electron microscopy of atmospheric particles. In: Mechanical Engineering. University of Minnesota, Minneapolis, MN, p. 213.

Huang, P.-F., Turpin, B.J., Pipho, M.J., Kittelson, D.B., McMurry, P.H., 1994. Effects of water condensation and evaporation on diesel chain-agglomerate morphology. Journal of Aerosol Science 25 (3), 447–459.

Huang, P.F., Turpin, B., 1996. Reduction of sampling and analytical errors for electron microscopic analysis of atmospheric aerosols. Atmospheric Environment 30 (24), 4137–4148.

Hudson, J.G., 1989. An instantaneous CCN spectrometer. Journal of Atmospheric and Oceanic Technology 6 (6), 1055–1065.

Hudson, J.G., 1993. Cloud condensation nuclei. Journal of Applied Meteorology 32, 596–607.

Hudson, J.G., Alofs, D.J., 1981. Performance of the continuous flow diffusion chambers. Journal of Rechereches Atmospshere 15, 321–331.

Huebert, B.J., Lee, G., Warren, W.L., 1990. Airborne aerosol inlet passing efficiency measurement. Journal of Geophysical Research 95 (D10), 16369–16381.

Huntzicker, J.J., Hoffman, R.S., Ling, C.-S., 1978. Continuous measurement and speciation of sulfur-containing aerosols by flame photometry. Atmospheric Environment 12 (1), 83–88.

Husar, R.B., 1974. Atmospheric particulate mass monitoring with a beta radiation detector. Atmospheric Environment 8, 183–188.

Jaklevic, J.M., Gatti, R.C., Goulding, F.S., Loo, B.W., 1981a. A ß-gauge method applied to aerosol samples. Environmental Science and Technology 15 (6), 680–686.

Jaklevic, J.M., Loo, B.W., Fujita, T.Y., 1981b. Automatic particulate sulfur measurements with a dichotomous sampler and on-line X-ray fluorescence analysis. Environmental Science and Technology 15 (6), 687–690.

Janossy, A.G.S., Kovacs, K., Toth, I., 1979. Parameters for ratio method by X-ray microanalysis. Analytical Chemistry 51, 491–495.

Jayne, J.T., Worsnop, D.R., Kolb, C.E., 1999. Aerosol mass spectrometer for size and composition analysis of submicron particles. Aerosol Science and Technology, in press.

John, W., Hering, S., Reischl, G., Sasaki, G., Goren, S., 1983a. Characteristics of nuclepore filters with large pore size – I, physical properties. Atmospheric Environment 17 (1), 115–120.

John, W., Hering, S., Reischl, G., Sasaki, G., Goren, S., 1983b. Characteristics of nuclepore filters with large pore size – II. filtration properties. Atmospheric Environment 17, 373–382.

Jonsson, H.H., Wilson, J.C., Brock, C.A., Knollenberg, R.G., Newton, R., Dye, J.E., Baumgardner, D., Borrmann, S., Ferry, G.V., Pueschel, R., Woods, D.C., Pitts, M.C., 1995. Performance of a focused cavity aerosol spectrometer for measurements in the stratosphere of particle size in the 0.06–2.0-µm-diameter range. Journal of Atmospheric and Oceanic Technology 12 (1), 115–129.

Juisto, J.E., Ruskin, R.E., Gagin, A., 1981. CCN comparisons of static diffusion chambers. Journal of Recherches Atmosphere 15, 291–302.

Junge, C.E., 1963. Air Chemistry and Radioactivity. Academic Press, New York.

Junge, C.E., McLaren, E., 1971. Relationship of cloud nuclei spectra to aerosol size distribution and composition. Journal of Atmospheric Science 28, 382–390.

Karlsson, A., Irgum, K., Hansson, H.-C., 1997. Single-stage flowing liquid film impactor for continuous on-line particle analysis. Journal of Aerosol Science 28 (8), 1539–1551.

Katrinak, K.A., Anderson, J.R., Buseck, P.R., 1995. Individual particle types in the aerosol of Phoenix, Arizona. Environmental Science and Technology 29 (2), 321–329.

Katz, M., Chan, C., 1980. Comparative distribution of eight polycyclic aromatic hydrocarbons in airborne particles collected by conventional high-volume sampling and by size fractionation. Environmental Science and Technology 14 (7), 838–842.

Kaufmann, R.L. (Ed.), 1986. Laser microprobe mass spectroscopy of particulate matter. Physical and Chemical Characterization of Individual Airborne Particles. Ellis Horwood Limited, Chichester, UK.

Kaye, P.H., Alexanderbuckley, K., Hirst, E., Saunders, S., Clark, J.M., 1996. A real-time monitoring system for airborne particle shape and size analysis. Journal of Geophysical Research – Atmospheres 101 (D14), 19215–19221.

Kaye, P.H., Eyles, N.A., Ludlow, I.K., Clark, J.M., 1991. An instrument for the classification of airborne particles on the basis of size, shape, and count frequency. Atmospheric Environment 25A (3/4), 645–654.

Keady, P.B., 1987. Development and field testing of a sulfur aerosol and gas monitor using a pulsed precipitator. University of Minnesota.

Keady, P.B., Quant, F.R., Sem, G.J., 1983. Differential mobility particle sizer: a new instrument for high-resolution aerosol size distribution measurement below 1 µm. TSI Quarterly 9 (2), 3–11.

Keuken, M.P., Schoonebeek, C.A.M., van Wensveen-Louter, A., Slanina, J., 1988. Simultaneous sampling of NH_3, HNO_3, HCl, SO_2, and H_2O_2 in ambient air by wet annular denuder system. Atmospheric Environment 22 (11), 2541–2548.

Khlystov, A., Kos, G.P.A., Tenbrink, H.M., 1996. A high-flow turbulent cloud chamber. Aerosol Science and Technology 24 (2), 59–68.

Khlysiou, A., Wyers, G.P., Slanina, J., 1995. The steam-jet aerosol collector. Atmospheric Environment 29 (17), 2229–2234.

Kievit, 1995. Technische Universiteit Delft.

Kim, D.S., Hopke, P.K., 1988. Multivariate-analysis of CCSEM auto emission data. Science of the Total Environment 59, 141–155.

Kinney, P.D., Pui, D.Y.H., Mulholland, G.W., Bryner, N.P., 1991. Use of electrostatic classification method to size 0.1 μm SRM particles – a feasibility study. Journal of Research of National Bureau of Standards Technology 96 (2), 147–176.

Kittelson, D.B., McKenzie, R., Wermeersch, M., Dorman, F., Piu, D., Linne, M., Liu, B.Y.H., Whitby, K.T., 1978. Total sulfur aerosol concentration with an electrostatically pulsed flame photometric detector system. Atmospheric Environment 12, 105–111.

Klockow, D., Niessner, R., Malejczyk, M., Kiendl, H., vom Berg, B., Keuken, M.P., Wayers-Ypelaan, A., Slanina, J., 1989. Determination of nitric acid and ammonium nitrate by means of a computer-controlled thermodenuder system. Atmospheric Environment 23 (5), 1131–1138.

Knapp, K.T., Durham, J.L., Ellestad, T.G., 1986. Pollutant sampler for measurements of atmospheric acidic dry deposition. Environmental Science and Technology 20 (6), 633–637.

Knutson, E.O., 1976. Extended electric mobility method for measuring aerosol particle size and concentration. In: Liu, B.Y.H. (Ed.), Fine Particles: Aerosol Generation, Measurement, Sampling, and Analysis. Academic Press, New York, pp. 739–762.

Knutson, E.O., 1995. Random and systematic errors in the graded screen technique for measuring the diffusion coefficient of radon decay products. Aerosol Science and Technology 23 (3), 301–310.

Knutson, E.O., Tu, K.W., Solomon, S.B., Strong, J., 1988. Intercomparison of three diffusion batteries for the measurement of radon decay product particle size distributions. Radiation Protection Dosimetry 24 (1/4), 261–264.

Knutson, E.O., Whitby, K.T., 1975. Aerosol classification by electric mobility: Apparatus, theory, and application. Journal of Aerosol Science 6, 443–451.

Kousaka, Y., Endo, Y., Muroya, Y., Fukushima, N., 1992. Development of a high flow rate mixing type CNC and its application to cumulative type electrostatic particle size analysis. Aerosol Research 7 (3), 219–229.

Kousaka, Y., Niida, T., Okuyama, K., Tanaka, H., 1982. Development of a mixing type condensation nucleus counter. Journal of Aerosol Science 13 (3), 231–240.

Kousaka, Y., Okuyama, K., Adachi, M., 1985. Determination of particle size distribution of ultrafine aerosols using a differential mobility analyzer. Aerosol Science and Technology 4 (2), 209–225.

Kousaka, Y., Okuyama, K., Adachi, M., Mimura, T., 1986. Effect of Brownian diffusion on electrical classification of ultafine aerosol particles in differential mobility analyzer. Journal of Chemical Engineering of Japan 19, 401–407.

Koutrakis, P., Sioutas, C., Ferguson, S.T., Wolfson, J.M., Mulik, J.D., Burton, R.M., 1993. Development and evaluation of a glass honeycomb denuder filter pack system to collect atmospheric gases and particles. Environmental Science and Technology 27 (12), 2497–2501.

Koutrakis, P., Wolfson, J.M., Slater, J.L., Brauer, M., Spengler, J.D., Stevens, R.K., Stone, C.L., 1988. Evaluation of an annular denuder/filter pack system to collect acidic aerosols and gases. Environmental Science and Technology 22 (12), 1463–1468.

Laktionov, A.G., 1972. A constant-temperature method of determining the concentrations of cloud condensation nuclei. Izveystiya Atmospheric Oceanic Physics 8, 382–385.

Lane, D.A., Johnson, N.D., Barton, S.C., Thomas, G.H.S., Schroeder, W.H., 1988. Development and evaluation of a novel gas and particle sampler for semivolatile chlorinated organic compounds in ambient air. Environmental Science and Technology 22 (8), 941–947.

Lane, D.A., Johnson, N.D., Hanley, M.-J.J., Schroeder, W.H., Ord, D.T., 1992. Gas and particle-phase concentrations of a-hexachlorocyclohexane, g-hexachlorocyclohexane, and hexachlorobenzene in Ontario air. Environmental Science and Technology 26 (1), 126–133.

Lawrence, J., Koutrakis, P., 1996. Measurement and speciation of gas and particulate phase organic acidity in an urban environment. 1. Analytical. Journal of Geophysical Research – Atmospheres 101 (D4), 9159–9169.

Lawson, D.R., 1980. Impaction surface coatings intercomparison and measurements with cascade impactors. Atmospheric Environment 14 (2), 195–200.

Leaitch, R., Megaw, W.J., 1982. The diffusion tube; a cloud condensation nucleus counter for use below 0.3% supersaturation. Journal of Aerosol Science 13 (4), 297–319.

Lee, C.T., Hsu, W.C., 1998. A novel method to measure aerosol water mass. Journal of Aerosol Science 29 (7), 827–837.

Lee, K.W., Kim, J.C., Han, D.S., 1990. Effects of gas density and viscosity on response of aerodynamic particle sizer. Aerosol Science and Technology 13, 203–212.

Lee, K.W., Liu, B.Y.H., 1980. On the minimum efficiency and the most penetrating particle size for fibrous filters. Journal of the Air Pollution Control Association 30 (4), 377–382.

Lee, K.W., Ramamurthi, M., 1993. Filter collection. In: Willeke, K., Baron, P.A. (Eds.), Aerosol Measurement: Principles, Techniques and Applications. Van Nostrand Reinhold, New York.

Leith, D., Mehta, D., 1973. Cyclone performance and design. Atmospheric Environment 7, 527–549.

Lin, C.I., Baker, M., Charlson, R.J., 1973. Absorption coefficient of atmospheric aerosol: a method for measurement. Applied Optics 12 (6), 1356–1363.

Linton, R.W., Farmer, M.E., Hopke, P.K., Natusch, D.F.S., 1980. Determination of the sources of toxic elements in invironmental particles using microscopic and statistic analysis techniques. Environment International 4, 453–461.

Lippmann, M., Hering, S.V., 1989. Sampling aerosols by filtration. In: Hering, S.V. (Ed.), Air Sampling Instruments. American Conference of Governmental Industrial Hygienists, Cincinnati, OH.

Liu, B.Y.H., Agarwal, J.K., 1974. Experimental observation of aerosol deposition in turbulent flow. Aerosol Science 5, 145–155.

Liu, B.Y.H., Kim, C.S., 1977. On the counting efficiency of condensation nuclei counters. Atmospheric Environment 11, 1097–1100.

Liu, B.Y.H., Pui, D.Y.H., 1974a. A submicron aerosol standard and the primary, absolute calibration of the condensation nuclei counter. Journal of Colloid and Interface Science 47, 155–171.

Liu, B.Y.H., Pui, Y.H., 1974b. Electrical neutralization of aerosols. Aerosol Science 5, 465–472.

Liu, B.Y.H., Pui, D.Y.H., Hogan, A.W., Rich, T.A., 1975. Calibration of the pollak counter with monodisperse aerosols. Journal of Applied Meteorology 14 (1), 46–51.

Liu, B.Y.H., Pui, D.Y.H., Rubow, K.L., 1983. Characteristics of air sampling filter media. In: Marple, V.A., Liu, B.Y.H. (Eds.), Aerosols in the Mining and Industrial Work Environments. Ann Arbor Science, Ann Arbor, MI, pp. 989–1038.

Liu, B.Y.H., Pui, D.Y.H., Whitby, K.T., Kittelson, D.B., Kousaka, Y., McKenzie, R.L., 1978. The aerosol mobility chromatograph: a new detector for sulfuric acid aerosols. Atmospheric Environment 12, 99–104.

Liu, B.Y.H., Szymanski, W.W., Ahn, K.-H., 1985. On aerosol size distribution measurement by laser and white light optical particle counters. The Journal of Environmental Sciences 28 (3), 19–24.

Liu, B.Y.H., Whitby, K.T., Pui, D.Y.H., 1974. A portable electrical analyzer for size distribution measurement of submicron aerosols. Air Pollution Control Association Journal 24 (11), 1067–1072.

Liu, D.-Y., Rutherford, D., Kinsey, M., Prather, K.A., 1997. Real-time monitoring of pyrotechnically derived aerosol particles in the troposphere. Analytical Chemistry 69 (10), 1808–1814.

Loo, B.W., Jaklevic, J.M., Goulding, F.S., 1976. Dichotomous virtual impactors for large scale monitoring of airborne particulate matter. In: Liu, B.Y.H. (Ed.), Fine Particles: Aerosol Generation, Measurement, Sampling, and Analysis. Academic Press, New York, pp. 311–349.

Lundgren, D.A., Carter, L.D., Daley, P.S., 1976. Aerosol mass measurement using piezoelectric crystal sensors. In: Liu, B.Y.H. (Ed.), Fine Particles: Aerosol Generation, Measurement, Sampling and Analysis. Academic Press, New York.

Macias, E.S., Husar, R.B., 1976. A review of atmospheric particulate mass measurement via the beta attenuation technique. In: Liu, B.Y.H. (Ed.), Fine Particles: Aerosol Generation, Measurement, Sampling, and Analysis. Academic Press, Inc, New York.

Malm, W.C., 1979. Considerations in the measurement of visibility. Journal of the Air Pollution Control Association 29 (10), 1042–1052.

Malm, W.C., Molenar, J.V., Eldred, R.A., Sisler, J.F., 1996. Examining the relationship among atmospheric aerosols and light scattering and extinction in the Grand Canyon area. Journal of Geophysical Research – Atmospheres 101 (D14), 19251–19265.

Mansoori, B.A., Johnston, M.V., Wexler, A.S., 1994. Quantitation of ionic species in single microdroplets by on-line laser desorption/ionization. Analytical Chemistry 66 (21), 3681–3687.

Marijnissen, J., Scarlett, B., Verheijen, P., 1988. Proposed on-line aerosol analysis combining size determination, laser-induced fragmentation and time-of-flight mass spectrometry. Journal of Aerosol Science 19 (7), 1307–1310.

Marple, V.A., 1970. A fundamental study of inertial impactors, Ph.D. Thesis, Department of Mechanical Engineering, University of Minnesota, Minneapolis, MN.

Marple, V.A., Chien, C.M., 1980. Virtual impactors: a theoretical study. Environmental Science and Technology 14 (8), 977–985.

Marple, V.A., Liu, B.Y.H., 1975. On fluid flow and aerosol impaction in inertial impactors. Journal of Colloid and Interface Science 53, 31–34.

Marple, V.A., Liu, B.Y.H., Burton, R.M., 1990. High-volume impactor for sampling fine and coarse particles. Journal of the Air and Waste Management Association 40 (5), 762–768.

Marple, V.A., Rubow, K.L., Behm, S.M., 1991. A microorifice uniform deposit impactor (MOUDI): Description, calibration, and use. Aerosol Science and Technology 14 (4), 434–446.

Marple, V.A., Willeke, K., 1976. Impactor design. Atmospheric Environment 10, 891–896.

Marple, V.S., Liu, B.Y.H., 1974. Characteristics of laminar jet impactors. Environmental Science and Technology 8, 648–654.

Marti, J., Mauersberger, K., 1993. Laboratory simulations of PSC particle formation. Geophysical Research Letters 20 (5), 359–362.

Marty, J.C., Tissier, M.J., Saliot, A., 1984. Gaseous and particulate polycyclic aromatic hydrocarbons (PAH) from the marine atmosphere. Atmospheric Environment 18 (10), 2183–2190.

Mauney, T., Adams, F., 1984. Laser microprobe mass spectrometry of environmental soot particles. Science of the Total Environment 36, 215–224.

Mazumder, M.K., Ware, R.E., Yokoyama, T., Rubin, B.U., Kamp, D., 1991. Measurement of particle size and electrostatic charge distributions on toners using E-SPART analyzer. IEEE Transactions on Industry Applications 27 (4), 611–619.

McDow, S.R., Huntzicker, J.J., 1990. Vapor adsorption artifact in the sampling of organic aerosol: face velocity effects. Atmospheric Environment 24A (10), 2563–2571.

McInnes, L., Covert, D., Baker, B., 1997. The number of sea-salt, sulfate, and carbonaceous particles in the marine atmosphere: EM measurements consistent with the ambient size distribution. Tellus Series B - Chemical and Physical Meteorology 49 (3), 300–313.

McInnes, L.M., Covert, D.S., Quinn, P.K., Germani, M.S., 1994. Measurements of chloride depletion and sulfur enrichment in individual sea-salt particles collected from the remote marine boundary layer. Journal of Geophysical Research – Atmospheres 99 (D4), 8257–8268.

McKeown, P.J., Johnston, M.V., Murphy, D.M., 1991. On-line single-particle analysis by laser desorption mass spectrometry. Analytical Chemistry 63, 2069–2073.

McMurry, P.H., Litchy, M., Huang, P.-F., Cai, X., Turpin, B.J., Dick, W., Hanson, A., 1996a. Elemental composition and morphology of individual particles separated by size and hygroscopicity with the TDMA. Atmospheric Environment 30, 101–108.

McMurry, P.H., Stolzenburg, M.R., 1989. On the sensitivity of particle size to relative humidity for Los Angeles aerosols. Atmospheric Environment 23 (2), 497–507.

McMurry, P.H., Zhang, X.Q., 1989. Size distributions of ambient organic and elemental carbon. Aerosol Science and Technology 10, 430–437.

McMurry, P.H., Zhang, X.Q., Lee, C.T., 1996b. Issues in aerosol measurement for optics assessments. Journal of Geophysical Research – Atmospheres 101 (D14), 19189–19197.

Metnieks, A.L., Pollak, L.W., 1959. Instruction for use of photo-electric condensation nucleus counters, Geophysical Bulletin No. 16, School of Cosmic Physica. Dublin Institute for Advanced Studies.

Middlebrook, A.M., Murphy, D.M., Thomson, D.S., 1998. Observations of organic material in individual marine particles at Cape Grim during ACE-1. Journal of Geophysics Research 103, 16475–16483.

Middleton, W.E.K., 1952. In: Vision through the Atmosphere. University of Toronto Press, Toronto, Ontario, Canada, pp. 83–102.

Mie, G., 1908. Beitraege zur Optik trueber Medien, speziell kolloidaler Metalosungen. Annales de Physik 25, 377–445.

Molina, M.J., Zhang, R., Wooldridge, P.J., McMahon, J.R., Kim, J.E., Chang, H.Y., Beyer, K.D., 1993. Physical chemistry of the $H_2SO_4/HNO_3/H_2O$ system: implications for polar stratospheric clouds. Science 261, 1418–1432.

Moosmüller, H., Arnott, W.P., Rogers, C.F., 1997a. Design and field evaluation of a photoacoustic instrument for the real time, in situ measurement of aerosol light absorption. AAAR, Denver, CO.

Moosmüller, H., Arnott, W.P., Rogers, C.F., 1997b. Methods of real-time, in situ measurement of aerosol light absorption. Journal of Air and Waste Management Association 47, 157–166.

Mulawa, P.A., Cadle, S.H., 1985. A comparison of nitric acid and particulate nitrate measurements by the penetration and denuder difference methods. Atmospheric Environment 19 (8), 1317–1324.

Murphy, D.M., Thomson, D.S., 1995. Laser ionization mass spectroscopy of single aerosol particles. Aerosol Science and Technology 22 (3), 237–249.

Murphy, D.M., Thomson, D.S., 1997a. Chemical composition of single aerosol particles at Idaho Hill: negative ion measurements. Journal of Geophysical Research 102 (D5), 6353–6368.

Murphy, D.M., Thomson, D.S., 1997b. Chemical composition of single aerosol particles at Idaho Hill: Positive ion measurements. Journal of Geophysical Research 102 (D5), 6341–6352.

Murphy, D.M., Thomson, D.S., Middlebrook, A.M., 1997c. Bromine and iodine in single aerosol particles at Cape Grim. Geophysical Research Letters 24, 3197–3200.

Murphy, D.M., Thomson, D.S., Middlebrook, A.M., Schein, M.E., 1998. In situ single particle characterization at Cape Grim. Journal of Geophysical Research 103, 16485–16491.

Neubauer, K.R., Johnston, M.V., Wexler, A.S., 1995. Chromium speciation in aerosols by rapid single-particle mass spectrometry. International Journal of Mass Spectrometry and Ion Processes 151, 77–87.

Neubauer, K.R., Sum, S.T., Johnston, M.V., Wexler, A.S., 1996. Sulfur speciation in individual aerosol particles. Journal of Geophysical Research – Atmospheres 101 (D13), 18701–18707.

Niessner, R., Klockow, D., Bruynseels, F.J., Van Grieken, R.E., 1985. Investigation of heterogeneous reactions of PAH's on particle surfaces using laser microprobe mass analysis. International Journal of Environmental and Analyticla Chemistry 22, 281–295.

Noble, C.A., Nordmeyer, T., Salt, K., Morrical, B., Prather, K.A., 1994. Aerosol characterization using mass spectrometry. Trends in Analytical Chemistry 13 (5), 218–222.

Noble, C.A., Prather, K.A., 1996. Real-time measurement of correlated size and composition profiles of individual atmospheric aerosol particles. Environmental Science and Technology 30 (9), 2667–2680.

Nolan, P.J., Pollak, L.W., 1946. The calibration of a photoelectric nucleus counter. Proceedings of the Royal Irish Academy 51A, 9–31.

Nordmeyer, T., Prather, K.A., 1994. Real-time measurement capabilities using aerosol time-of-flight mass spectrometry. Analytical Chemistry 66 (20), 3540–3542.

NRC, 1993. Protecting Visibility in National Parks and Wilderness Areas. Washington, DC, National Academy Press.

Oberdörster, G., Gelein, R.M., Ferin, J., Weiss, B., 1995. Association of particulate air pollution and acute mortality: involvement of ultrafine particles?. Inhalation Toxicology, 7, 111–124.

Ogren, J.A., 1995. In Situ observations of aerosol properties. In: Charlson, R.J., Heintzenberg, J. (Eds.), Aerosol Forcing of Climate. Wiley, Chichester, pp. 215–226.

Otten, P., Bruynseels, F., van Grieken, R., 1987. Study of inorganic ammonium compounds in individual marine aerosol particles by laser microprobe mass spectrometry. Analytica Chimica Acta 195, 117–124.

Pak, S.S., Liu, B.Y.H., Rubow, K.L., 1992. Effect of coating thickness on particle bounce in inertial impactors. Aerosol Science and Technology 16, 141–150.

Patashnick, H., Rupprecht, E.G., 1991. Continuous PM-10 measurements using the tapered element oscillatng microbalance. Journal of the Air and Waster Management Association 41 (8), 1079–1084.

Peters, J., Seifert, B., 1980. Losses of benzo(a)pyrene under the conditions of high-volume sampling. Atmospheric Environment 14 (1), 117–120.

Pilinis, C., Seinfeld, J.H., Grosjean, D., 1989. Water content of atmospheric aerosols. Atmospheric Environment 23 (7), 1601–1606.

Possanzini, M., Febo, A., Liberti, A., 1983. New design of a high-performnance denuder for the sampling of atmospheric pollutants. Atmospheric Environment 17 (12), 2605–2610.

Pourprix, M., Daval, J., 1990. Electrostatic precipitation of aerosol on wafers, a new mobility spectrometer. Third International Aerosol Conference. Pergamon Press, Oxford, Kyoto, Japan.

Pruppacher, H.R., Klett, J.D., 1980. Microphysics of Clouds and Precipitation. D. Reidel Publishing Co., Dordrecht, Holland.

Pui, D.Y.H., Romay-Novas, F., Liu, B.Y.H., 1987. Experimental study of particle deposition in bends of circular cross section. Aerosol Science and Technology 7, 301–315.

Quinn, P.K., Anderson, T.L., Bates, T.S.D., Heintzenbert, J., von Hoyningen-Huene, W., Kulmala, M., Russcll, P.B., Swietlicki, E., 1996. Closure in tropospheric aerosol-climate research: direct shortwave radiative forcing. Beitraege Physikalische Atmosphere 69 (4), 547–577.

Rader, D.J., 1990. Momentum slip correction factor for small particles in nine common gases. Journal of Aerosol Science 21 (2), 161–168.

Rader, D.J., Brockmann, J.E., Ceman, D.L., Lucero, D.A., 1990. A method to employ the APS factory calibration under different operating conditions. Aerosol Science and Technology 13 (4), 514–521.

Rader, D.J., Marple, V.A., 1985. Effect of ultra-Stokesian drag and particle interception on impaction characteristics. Aerosol Science and Technology 4, 141–156.

Rader, D.J., Marple, V.A., 1988. A study of the effects of anisokinetic sampling. Aerosol Science and Technology 8, 283–299.

Rader, D.J., McMurry, P.H., 1986. Application of the tandem differential mobility analyzer to studies of droplet growth or evaporation. Journal of Aerosol Science 17 (5), 771–787.

Radke, L.F., Domonkos, S.K., Hobbs, P.V., 1981. A cloud condensation nucleus spectrometer designed for airborne measurements. Journal of Recherches Atmosphere 15, 225–229.

Raeymaekers, B.J., Liu, X., Janssens, K.H., Espen, P.J.V., Adams, F.C., 1987. Determination of the thickness of flat particles by automated electron microprobe analysis. Analytical Chemistry 59, 930–937.

Ramachandran, G., Leith, D., Dirgo, J., Feldman, H., 1991. Cyclone optimization based on a new empirical model for pressure drop. Aerosol Science and Technology 15, 135–148.

Ramamurthi, M., Hopke, P.K., 1989. On improving the validity of wire screen unattached fraction Rn daughter measurements. Health Physics 56 (2), 189–194.

Rao, A.K., Whitby, K.T., 1977. Nonideal collection characteristics of single stage and cascade impactors. American Industrial Hygiene Association Journal 174–179.

Rao, A.K., Whitby, K.T., 1978. Nonideal collection characteristics of inertial impactors I: single stage impactors and solid particles. Journal of Aerosol Science 9, 77–86.

Reents, W.D., Downey, S.W., Emerson, A.B., Mujsce, A.M., Muller, A.J., Siconolfi, D.J., Sinclair, J.D., Swanson, A.G., 1995. Single particle characterization by time-of-flight mass spectrometry. Aerosol Science and Technology 23 (3), 263–270.

Register, F., 1987. Reference method for the determination of particulate matter as PM10 in the atmosphere. Federal Register 52, 24664.

Register, F., 1997. National ambient air quality standards for particulate matter; final rule. Federal Register 62 CFR Parts 50, 53 and 58(138).

Reineking, A., Knutson, E.A., George, A.C., Solomon, S.B., Kesten, J., Butterweck, G., Porstendorfer, J., 1994. Size distribution of unattached and aerosol-attached short-lived radon decay products: some results of intercomparison measurements. Radiation Protection Dosimetry 56 (1–4), 113–118.

Reischl, G.P., John, W., 1978. The collection efficiency of impaction surfaces. Staub Reinhalt Luft 38, 55.

Reischl, G.P., Makela, J.M., Karch, R., Necid, J., 1996. Bipolar charging of ultrafine particles in the size range below 10 nm. Journal of Aerosol Science 27 (6), 931–949.

Remiarz, R.J., Agarwal, J.K., Quant, F.R., Sem, G.J., 1983. Real-time aerodynamic particle size analyzer. In: Marple, V.A., Liu, B.Y.H. (Eds.), Aerosols in the Mining and Industrial Work Environments. Ann Arbor Science Publishing, Ann Arbor, MI.

Richards, L.W., 1988. Sight path measurements for visibility monitoring and research. Journal of the Air Pollution Control Association 38 (6), 784–792.

Richards, L.W., Dye, T.S., Hurwitt, S., Allen, G., Oh, J.A., 1997. ASOS visibility sensor data as an indicator of PM and haze. Visual Air Quality: Aerosols and Global Radiation Balance. AWMA, Bartlett, NH.

Ro, C.-U., Musselman, I.H., Linton, R.W., 1991. Molecular speciation of microparticles: application of pattern recognition techniques to laser microprobe mass spectrometric data. Analytica Chimica Acta 243, 139–147.

Rojas, C.M., Artaxo, P., van Grieken, R., 1990. Aerosols in Santiago de Chile: a study using receptor modeling with X-ray fluoroescence and single particle analysis. Atmospheric Environment 24B (2), 227–241.

Romay-Novas, F.J., Pui, D.Y.H., 1988. Generation of monodisperse aerosols in the 0.1 to 1.0 µm diameter range using a mobility classification-inertial impaction technique. Aerosol Science and Technology 9, 123–131.

Rosell-Llompart, J., Loscertales, I.G., Bingham, D., De la Mora, J.F., 1996. Sizing nanoparticles and ions with a short differential mobility analyzer. Journal of Aerosol Science 27 (5), 695–719.

Rupprecht, G., Patashnick, H., Beeson, D.E., Green, R.N., Meyer, M.B., 1995. A new automated monitor for the measurement of particulate carbon in the atmosphere. Particulate Matter: Health and Regulatory Issues. AWMA, Pittsburgh, PA.

Russ, J.G., 1974. X-ray microanalysis in the biological sciences. J. Submicr. Cytol. 6, 55–79.

Sachweh, B.A., Dick, W.D., McMurry, P.H., 1995. Distinguishing between spherical and non-spherical particles by measuring the variability in azimuthal light scattering. Aerosol Science and Technology 23 (3), 373–391.

Salt, K., Noble, C.A., Prather, K.A., 1996. Aerodynamic particle sizing versus light scattering intensity measurement as methods for real-time particle sizing coupled with time-of-flight mass spectrometry. Analytical Chemistry 68 (1), 230–234.

Saros, M.T., Weber, R.J., Marti, J.J., McMurry, P.H., 1996. Ultrafine aerosol measurement using a condensation nucleus counter with pulse height analysis. Aerosol Science and Technology 25 (2), 200–213.

Saxena, P., Hildemann, L.M., 1996. Water-soluble organics in atmospheric particles: a critical review of the literature and application of thermodynamics to identify candidate compounds. Journal of Atmospheric Chemistry 24, 57–109.

Saxena, P., Hildemann, L.M., McMurry, P.H., Seinfeld, J.H., 1995. Organics alter hygroscopic behavior of atmospheric particles. Journal of Geophysics Research 100 (D9), 18755–18770.

Saxena, P., Musarra, S., Malm, W., Day, D., Hering, S., 1996. Optical Characteristics of Southeastern Aerosol. AAAR96, Orlando, FL.

Schauer, J.J., Rogge, W.F., Hildemann, L.M., Mazurek, M.A., Cass, G.R., 1996. Source apportionment of airborne particulate matter using organic compounds as tracers. Atmospheric Environment 30 (22), 3837–3855.

Schwartz, S.E., 1989. Acid deposition: unraveling a regional phenomenon. Science 243, 753–762.

Schwoeble, A.J., Dalley, A.M., Henderson, B.C., Casuccio, G.S., 1988. Computer-controlled SEM and microimaging of fine particles. Journal of Metals 40, 11–14.

Seebaugh, W.R., Lafleur, B.G., 1996. Low turbulence inlet for aerosol sampling from aircraft. American Association for Aerosol Research, Orlando, FL.

Seinfeld, J.H., 1986. Atmospheric Chemistry and Physics of Air Pollution. Wiley Interscience, New York.

Seto, T., Okuyama, K., de Juan, L., Fernández de la Mora, J., 1997. Condensation of supersaturated vapors on monovalent and divalent ions of varying size. Journal of Chemical Physics 107, 1576–1585.

Shaw, R.W.J., Stevens, R.K., Bowermaster, J., Tesch, J.W., Tew, E., 1982. Measurements of atmospheric nitrate and nitric acid: the denuder difference experiment. Atmospheric Environment 16 (4), 845–853.

Sheridan, P.J., Brock, C.A., Wilson, J.C., 1994. Aerosol particles in the upper troposphere and lower stratosphere – elemental composition and morphology of individual particles in northern midlatitudes. Geophysical Research Letters 21 (23), 2587–2590.

Simon, P.K., Dasgupta, P.K., 1995a. Continuous automated measurement of gaseous nitrous and nitric acids and particulate nitrite and nitrate. Environmental Science and Technology 29 (6), 1534–1541.

Simon, P.K., Dasgupta, P.K., 1995b. Continuous automated measurement of the soluble fraction of atmospheric particulate matter. Analytical Chemistry 67 (1), 71–78.

Sinclair, D., 1972. A portable diffusion battery. American Industrial Hygiene Association Journal, 729–735.

Sinclair, D., Hoopes, G.S., 1975a. A continuous flow condensation nucleus counter. Aerosol Science 6, 1–7.

Sinclair, D., Hoopes, G.S., 1975b. A novel form of diffusion battery. American Industrial Hygiene Association Journal, 39–42.

Sinha, M.P., Friedlander, S.K., 1985. Real-time measurement of sodium chloride in individual aerosol particles by mass spectrometry. Analytical Chemistry 57 (9), 1880–1883.

Sinha, M.P., Giffin, C.E., Norris, D.D., Estes, T.J., Vilker, V.L., Friedlander, S.K., 1982. Particle analysis by mass spectrometry. Journal of Colloid and Interface Science 87 (1), 140–153.

Sioutas, C., Koutrakis, P., Burton, R.M., 1994a. Development of a low cutpoint slit virtual impactor for sampling ambient fine particles. Journal of Aerosol Science 25 (7), 1321–1330.

Sioutas, C., Koutrakis, P., Burton, R.M., 1994b. A high-volume small cutpoint virtual impactor for separation of atmospheric particulate from gaseous pollutants. Particulate Science and Technology 12 (3), 207–221.

Sioutas, C., Koutrakis, P., Olson, B.A., 1994c. Development and evaluation of a low cutpoint virtual impactor. Aerosol Science and Technology 21 (3), 223–235.

Slanina, J., Schoonebeek, C.A.M., Klockow, D., Niessner, R., 1985. Determination of sulfuric acid and ammonium sulfates by means of a computer-controlled thermodenuder system. Analytical Chemistry 57, 1955–1960.

Sloane, C.S., Rood, M.J., Rogers, Dr., F.C., 1991. Measurements of aerosol particle size: Improved precision by simultaneous use of optical particle counter and nephelometer. Aerosol Science and Technology 14, 289–301.

Smith, J.P., Grosjean, D., Pitts, J.N., 1978. Observation of significant losses of particulate nitrate and ammonium from high volume glass fiber filter samples stored at room temperature. Journal of Air Pollution Control Association 28, 929–933.

Solomon, P.A., Moyers, J.L., Fletcher, R.A., 1983. High-volume dichotomous virtual impactor for the fractionization and collection of particles according to aerodynamic size. Aerosol Science and Technology 2, 455–464.

Speer, R.E., Barnes, H.M., Brown, R., 1997. An instrument for measuring the liquid water content of aerosols. Aerosol Science and Technology 27, 50–61.

Spurny, K.R. (Ed.), 1986. Physical and chemical characterization of individual airborne particles. Ellis Horwood Limited, Chichester, West Sussex, England.

Stein, S.W., Turpin, B.J., Cai, X.P., Huang, C.P.F., McMurry, P.H., 1994. Measurements of relative humidity-dependent bounce and density for atmospheric particles using the DMA-impactor technique. Atmospheric Environment 28 (10), 1739–1746.

Stoffels, J.J., Lagergren, C.R., 1981. On the real-time measurement of particles in air by direct-inlet surface-ionization mass spectrometry. International Journal of Mass Spectrometry and Ion Physics 40, 243–254.

Stolzenburg, M.R., 1988. An ultrafine aerosol size distribution measuring system. Ph.D. Thesis, Department of Mechanical Engineering, University of Minnesota, Minneapolis, MN.

Stolzenburg, M.R., Hering, S.V., 1999. A method for the automated measurement of fine particle nitrate in the atmosphere, Environmental Science and Technology, submitted for publication.

Stolzenburg, M.R., McMurry, P.H., 1991. An ultrafine aerosol condensation nucleus counter. Aerosol Science and Technology 14, 48–65.

Suh, H., Allen, G.A., Aurian-Bläjeni, B., Koutrakis, P., Burton, R.M., 1994. Field method comparison for the characterization of acid aerosols and gases. Atmopsheric Environment 28 (18), 2981–2989.

Svenningsson, B., Hansson, H.C., Wiedensohler, A., Noone, K., Ogren, J., Hallberg, A., Colvile, R., 1994. Hygroscopic growth of aerosol particles and its influence on nucleation scavenging in cloud: experimental results from kleiner feldberg. Journal of Atmospheric Chemistry 19 (1-2), 129–152.

Svenningsson, I.B., Hansson, H.-C., Wiedensohler, A., Ogren, J.A., Noone, K.J., Hallberg, A., 1992. Hygroscopic growth of aerosol particles in the po valley. Tellus 44B, 556–569.

Szymanski, W.W., Liu, B.Y.H., 1986. On the sizing accuracy of laser optical particle counters. Particle Characteristics 3, 1–7.

Tanner, R.L., D'Ottavio, T., Garber, R., Newmann, L., 1980. Determination of ambient aerosol sulfur using a continuous flame photometric detection system. I. Sampling system for aerosol solfate and sulfuric acid. Atmospheric Environment 14 (1), 121–128.

Tanner, R.L., D'Ottavio, T., Garber, R.W., Newman, L., 1978. Determination of atmospheric gaseous and particulate sulfur compounds. In: Nriagu, J. (Ed.), Sulfur in the Environment. Wiley, New York.

Thielke, J.F., Charlson, R.J., Winter, J.W., Ahlquist, N.C., Whitby, K.T., Husar, R.B., Liu, B.Y.H., 1972. Multiwavelength nephelometer measurements in Los Angeles smog aerosols. Journal of Colloid and Interface Science 39 (1), 252–259.

Thompson, W., 1871. On the equilibrium of vapour at a curved surface of liquid. Philosophical Magazine 42, 448–453.

Thomson, D.S., Murphy, D.M., 1993. Laser-induced ion formation thresholds of aerosol particles in a vacuum. Applied Optics 32 (33), 6818–6826.

Thomson, D.S., Murphy, D.M., 1994. Analyzing single aerosol particles in real time. Chemtech 24, 30–35.

Thomson, D.S., Murphy, D.M., 1997. Thresholds for laser-induced ion formation from aerosols in a vacuum using ultraviolet and vacuum-ultraviolet laser wavelengths. Aerosol Science and Technology 26, 544–559.

Thudium, J., 1978. Water uptake and equilibrium sizes of aerosol particles at high relative humidities: their dependence on the composition of the water-soluble material. Pageoph 116, 130–148.

Tsai, C.J., Pui, D.Y.H., 1990. Numerical study of particle deposition in bends of a circular cross-section-laminar flow regime. Aerosol Science and Technology 12, 813–831.

Turner, J.R., Hering, S.V., 1987. Greased and oiled substrates as bounce-free impaction surfaces. Journal of Aerosol Science 18, 215–224.

Turpin, B.J., Cary, R.A., Huntzicker, J.J., 1990a. An in situ, time-resolved analyzer for aerosol organic and elemental carbon. Aerosol Science and Technology 12, 161–171.

Turpin, B.J., Huntzicker, J.J., 1995. Identification of secondary organic aerosol episodes and quantitation of primary and secondary organic aerosol concentrations during SCAQS. Atmospheric Environment 29 (23), 3527–3544.

Turpin, B.J., Huntzicker, J.J., Adams, K.M., 1990b. Intercomparison of photoacoustic and thermal-optical methods for the measurement of atmospheric elemental carbon. Atmospheric Environment 24A (7), 1831–1836.

Turpin, B.J., Liu, S.P., Podolske, K.S., Gomes, M.S.P., Eisenreich, S.J., McMurry, P.H., 1993. Design and evaluation of a novel diffusion separator for measuring gas particle distributions of semivolatile organic compounds. Environmental Science and Technology 27 (12), 2441–2449.

Twomey, S., 1967. Remarks on the photographic counting of cloud nuclei. Journal of Recherches Atmosphere 3, 85–90.

Twomey, S., 1991. Aerosols, clouds, and radiation. Atmospheric Environment 25A (11), 2435–2442.

Twomey, S.A., Piepgrass, M., Wolfe, T.L., 1984. An assessment of the impact of pollution on global cloud albedo. Tellus 36B, 356–366.

van Borm, W., Wouters, L., van Grieken, R., Adams, F., 1990. Lead particles in an urban atmosphere: an individual particle approach. The Science of the Total Environment 90, 55–66.

Van Borm, W.A., Adams, F.C., 1988. Cluster analysis of electron microprobe analysis data of individual particles for source approtionment of air particulate matter. Atmospheric Environment 22 (10), 2297–2308.

van de Hulst, H.C., 1981. Light scattering by small particles. Dover Publications, New York.

Vasiliou, J., McMurry, P.H., 1997. MOUDI Data Report: 1995 Southern Aerosol Visibility Study (SEAVS). University of Minnesota, Minneapolis, MN.

Vincent, J.H., 1989. Aerosol Sampling: Science and Practice. Wiley, Cichester.

Wang, H.-C., John, W., 1987a. Comparative bounce properties of particle materials. Aerosol Science and Technology 7, 285–299.

Wang, H.C., John, W., 1987b. Particle density correction for the aerodynamic particle sizer. Aerosol Science and Technology 6, 191–198.

Wang, H.C., John, W., 1988. Characteristics of the Berner impactor for sampling inorganic ions. Aerosol Science and Technology 8, 157–172.

Wang, S.C., Flagan, R.C., 1990. Scanning electrical mobility spectrometer. Aerosol Science and Technology 13, 230–240.

Ward, M.D., Buttry, D.A., 1990. In situ interfacial mass detection with piezoelectric transducers. Science 249, 1000–1007.

Watanabe, M., Someya, T., 1970. Temperature rise of specimen due to electron irradiation. Journal of Physics D: Applied Physics 3, 1461.

Webber, J.S., 1986. Using microparticle characterization to identify sources of acid-rain precursors reaching whiteface mountain, New York. In: Basu, S. (Ed.), Electron Microscopy in Forensic, Occupational, and Environmental Health Sciences. Plenum Press, New York, pp. 261–273.

Weber, R., McMurry, P.H., Bakes, T.S., Clarke, A.D., Couert, D.S., Brechyel, F.J., Kok, G.L., 1999. Intercomparison of airborne and surface-based measurements of condensation nuclei in the remote marine troposphere during ACE-1. Journal of Geophysical Research 104 (D17), 21673–21683.

Weber, R.J., Stolzenburg, M.R., Pandis, S.N., McMurry, P.H., 1998. Inversion of ultrafine condensation nucleus counter pulse height distributions to obtain nanoparticle (3 to 10 nm) size distributions. Journal of Aerosol Science 29 (5/6), 601–615.

Wernisch, J., 1985. Quantitative electron microprobe analysis without standard samples. X-Ray Spectrometry 14, 109–119.

Wesolowski, J.J., John, W., Devor, W., Cahill, T.A., Feeny, P.J., Wolfe, G., Flocchini, R., 1977. Collection surfaces of cascade impactors. In: Dzubay, T. (Ed.), X-Ray Fluorescence Analysis of Environmental Samples. Ann Arbor Science, Ann Arbor, MI, pp. 121–131.

Whitby, K.T., Clark, W.E., 1966. Electrical aerosol particle counting and size distribution measuring system for the 0.015 to 1.0 μm size range. Tellus 13, 573–586.

Whitby, K.T., Husar, R.B., Liu, B.Y.H., 1972. The aerosol size of distribution of Los Angeles smog. Journal of Colloid and Interface Science 39 (1), 177–204.

Whitby, K.T., Liu, B.Y.H., 1968. Polystyrene Aerosols – electrical charge and residue size distribution. Atmospheric Environment 2, 103–116.

Wiedensohler, A., 1988. An approximation of the bipolar charge distribution for particles in the submicron size range. Journal of Aerosol Science 19 (3), 387–389.

Wiedensohler, A., Orsini, D., Covert, D.S., Coffmann, D., Cantrell, W., Havlicek, M., Brechtel, F.J., Russell, L.M., Weber, R.J., Gras, J., Hudson, J.G., Litchy, M., 1997. Intercomparison study of the size-dependent counting efficiency of 26 condensation particle counters. Aerosol Science and Technology 27 (2), 224–242.

Wieser, P., Wurster, R., Seiler, H., 1980. Identification of airborne particles by laser induced mass spectroscopy. Atmospheric Environment 14 (4), 485–494.

Willeke, K., Liu, B.Y.H., 1976. Single particle optical counter: principle and application. In: Liu, B.Y.H. (Ed.), Fine Particles: Aerosol Generation, Measurement, Sampling and Analysis, Academic Press Inc., New York, pp. 698–729.

Williams, K., Fairchild, C., Jaklevic, J., 1993. Dynamic mass measurement techniques. In: Willeke, K., Baron, P.A. (Eds.), Aerosol Measurement: Principles, techniques and applications. Van Norstrand Reinhold, New York.

Wilson, J.C., Blackshear, E.D., Hyun, J.H., 1983a. An improved continuous-flow condensation nucleus counter for use in the stratosphere. Journal of Aerosol Science 14 (3), 387–391.

Wilson, J.C., Hyun, J.H., Blackshear, E.D., 1983b. The function and response of an improved stratospheric condensation nucleus counter. Journal of Geophysical Research 88 (C11), 6781–6785.

Wilson, J.C., Liu, B.Y.H., 1980. Aerodynamic particle size measurement by laser-doppler velocimetry. Journal of Aerosol Science 11 (2), 139–150.

Wilson, J.C., Stolzenburg, M.R., Clark, W.E., Loewenstein, M., Ferry, G.V., Chan, K.R., Kelly, K.K., 1992. Stratosphereic sulfate aerosol in and near the northern hemisphere polar vortex: the morphology of the sulfate layer, multimodal size distributions, and the effect of denitrification. Journal of Geophysical Research 97 (D8), 7997–8013.

Winkler, P., 1974. Relative humidity and the adhesion of atmospheric particles to the plates of impactors. Aerosol Science 5, 235–240.

Winkler, P., Junge, C., 1972. The growth of atmospheric aerosol particles as a function of the relative humidity. Journal de Recherches Atmosphériques, 617–638.

Winklmayr, W., Reischl, G.P., Linder, A.O., Berner, A., 1991. A new electromobility spectrometer for the measurement of aerosol size distribution in the size range from 1 to 1000 nm. Journal of Aerosol Science 22, 289.

Witz, S., Eden, R.W., Wadley, M.W., Dunwoody, C., Papa, R.P., Torre, K.J., 1990. Rapid loss of particulate nitrate, chloride, and ammonium on quartz fiber filters during storage. Journal of Air Waste Management Association 40, 53–61.

Wolff, G.T., Stroup, C.M., Stroup, D.P., 1983. The coefficient of haze as a measure of particulate elemental carbon. Journal of the Air Pollution Control Association 33 (8), 746–751.

Worsnop, D.R., Fox, L.E., Zahniser, M.S., Wofsy, S.C., 1993. Vapor pressures of solid hydrates of nitric acid: implications for polar stratospheric clouds. Science 259, 71–74.

Wouters, L.C., van Grieken, R.E., Linton, R.W., Bauer, C.F., 1988. Discrimination between coprecipitated and adsorbed lead on individual calcite particles using laser microprobe mass analysis. Analytical Chemistry 60, 2218–2220.

Wu, J.J., Cooper, D.W., Miller, R.J., 1989. Evaluation of aerosol deconvolution algorithms for determining submicron particle size distributions with diffusion battery and condensation nucleus counter. Journal of Aerosol Science 20 (4), 477–482.

Wyatt, P.J., Schehrer, K.L., Philips, S.D., Jackson, C., Chang, Y.-J., Parker, R.G., Phillips, D.T., Bottiger, J.R., 1988. Aerosol particle analyzer. Applied Optics 27 (2), 217–221.

Zellweger, C., Ammann, M., Hofer, P., Baltensperger, U., 1999. NO_y speciation with a combined wet effluent diffusion denuder aerosol collector coupled to ion chromatography. Atmospheric Environment 33, 1131–1140.

Zhang, S.H., Akutsu, Y., Russell, L.M., Flagan, R.C., Seinfeld, J.H., 1995. Radial differential mobility analyzer. Aerosol Science and Technology 23 (3), 357–372.

Zhang, X.Q., McMurry, P.H., 1987. Theoretical analysis of evaporative losses from impactor and filter deposits. Atmospheric Environment 21 (8), 1779–1789.

Zhang, X.Q., McMurry, P.H., 1993. Evaporative losses of fine particulate nitrates during sampling. Atmospheric Environment 26A, 3305–3312.

Zhang, X.Q., McMurry, P.H., Hering, S.V., Casuccio, G.S., 1993. Mixing characteristics and water content of submicron aerosols measured in Los-Angeles and at the grand canyon. Atmospheric Environment Part A – General Topics 27 (10), 1593–1607.

Zhang, Z., Liu, B.Y.H., 1991. Performance of TSI 3760 condensation nuclei counter at reduced pressures and flow rates. Aerosol Science and Technology 15, 228–238.

Chapter 18

Formation and cycling of aerosols in the global troposphere

Frank Raes, Rita Van Dingenen, Elisabetta Vignati, Julian Wilson,
Jean-Philippe Putaud

Environment Institute, European Commission, 21020 Ispra (VA), Italy
E-mail: frank.raes@jrc.it

John H. Seinfeld, Peter Adams

California Institute of Technology, Pasadena, CA, USA

Abstract

Aerosols are formed, evolve, and are eventually removed within the general circulation of the atmosphere. The characteristic time of many of the microphysical aerosol processes is days up to several weeks, hence longer than the residence time of the aerosol within a typical atmospheric compartment (e.g. the marine boundary layer, the free troposphere, etc.). Hence, to understand aerosol properties, one cannot confine the discussion to such compartments, but one needs to view aerosol microphysical phenomena within the context of atmospheric dynamics that connects those compartments. This paper attempts to present an integrated microphysical and dynamical picture of the global tropospheric aerosol system. It does so by reviewing the microphysical processes and those elements of the general circulation that determine the size distribution and chemical composition of the aerosol, and by implementing both types of processes in a diagnostic model, in a 3-D global Chemical Transport Model, and in a General Circulation Model. Initial results are presented regarding the formation, transformation, and cycling of aerosols within the global troposphere.

1. Introduction

Particles in the atmosphere arise from natural sources, such as wind-borne dust, sea spray, and volcanoes, and from anthropogenic activities, such as combustion of fuels (Table 1). Emitted directly as particles (primary aerosol) or formed in the atmosphere by gas-to-particle conversion processes (secondary aerosol),

First published in Atmospheric Environment 34 (2000) 4215–4240

Table 1.   Estimated global emission rates of particles into the atmosphere (Tg yr^{-1})

	Source strength (Tg yr^{-1})	Reference
Sea salt		
Total	5900	Tegen et al. (1997)
0–2 μm	82.1	Gong et al. (1997)
2–20 μm	2460	
Soil dust		
< 1 μm	250	Tegen and Fung (1995)
1–10 μm	1000	
0.2–2 μm	250	Penner, personal comm.
2–20 μm	4875	
Organic carbon		
Total	69	Liousse et al. (1996)
Biomass burning	54.3	Penner, personal comm.
Fossil fuel	28.8	
Terpene oxidation	18.5	Griffin et al. (1999)
Black carbon		
Total	12	
Biomass burning	5.6	Liousse et al. (1996)
Fossil fuel	6.6	
Sulfate (as H_2SO_4)		
Total	150	Chin and Jacob (1996)
Natural	32	Koch et al. (1999)
Anthropogenic	111	
Nitrate	11.3[a]	Adams et al. (1999)
Ammonium	33.6	Adams et al. (1999)

[a]Nitrate source strength is based on a computed burden of 0.13 Tg and an assumed lifetime of 4.2 d (same as ammonium).

atmospheric aerosols range in size from a few nanometers (nm) to tens of micrometers (μm) in diameter. Once airborne, particles evolve in size and composition through condensation of vapour species or by evaporation, by coagulating with other particles, by chemical reaction, or by activation in the presence of supersaturated water vapour to become cloud and fog droplets. Particles smaller than 1 μm diameter generally have atmospheric concentrations in the range from 10 to 10,000s per cm^3; those exceeding 1 μm diameter typically exhibit concentrations less than 10 cm^{-3}.

    There is evidence that anthropogenic particles, at concentrations typical of urban airsheds, directly affect human health. Biomass burning, especially in the tropics, leads to significant perturbations to tropospheric aerosol loadings in that region, perhaps accompanied by alterations of cloud behaviour. Aircraft exhaust particles in the upper atmosphere are a source of ice and cloud nuclei. Atmospheric particles provide surfaces for heterogeneous chemical reactions that may influence gas-phase chemistry in the troposphere. It is not possible to survey each of these aspects in a review of modest length; consequently, we focus here on aerosol processes in the global atmosphere, the dynamics that shape the size and composition of the global aerosol.

The first measurements of the aerosol number concentration in the atmosphere were performed by Aitken (1888) who used an expansion chamber to make water vapour condense on the particles and make them grow to visible droplets. Aitken proclaimed that "without aerosols there would be no clouds and no precipitation". The water vapour supersaturation (= relative humidity (%) – 100) created in the Aitken counter reached 300%, enough to activate any particle. In the atmosphere, however, supersaturations of at most 2% are reached (Pruppacher and Klett, 1980), and Köhler (1936) showed that at such low supersaturations only those particles will activate that are sufficiently hygroscopic, i.e. particles that contain sufficient amount of soluble material to reduce the equilibrium water vapour pressure above the solution droplet. Hence, aerosol chemical and physical properties do control cloud droplet formation, and accordingly cloud microphysical properties, precipitation potential and optical properties. There are now many observations that this is effectively the case (Boers et al., 1994; Cerveny and Balling, 1998; Rosenfeld, 1999; Pawlowska and Brenguier, 2000; Johnson et al., 2000; Chuang et al., 2000).

Aerosols are important players in the hydrological cycle and climate system. It is therefore necessary to understand their cycling in the atmosphere, and to be able to predict their characteristics. Within the context of global climate change, aerosol studies have focused either on descriptions of global sources and spatial distributions of aerosols, neglecting the microphysical aspects, or they have focused on the microphysics of their formation and evolution, without placing these processes in the context of atmospheric large-scale circulation. In this paper we will review progress achieved by the two approaches, and we will attempt to synthesise a combined microphysical and dynamical picture of the global tropospheric aerosol system. We will also review observations of some key aerosol characteristics in a number of environments, which have been helpful to constrain our understanding of aerosols. In the model studies, presented at the end of the paper, we draw particularly from the global sulphur cycle because much has been learned recently about this cycle, and it serves as an excellent vehicle to discuss the effect of global circulation on aerosol properties and behaviour.

2. Microphysics of aerosol formation and evolution

2.1. Processes

Fig. 1 depicts generally the microphysical processes that influence the size distribution and chemical composition of the atmospheric aerosol, highlighting the large range of sizes that are involved in the formation and evolution of aerosol particles. Traditionally, atmospheric aerosols have been divided into two size classes: coarse ($D_p > 1$ μm) and fine ($D_p < 1$ μm), reflecting the two major formation mechanisms: primary and secondary. Both populations strongly overlap, however, in the 0.1–1 μm diameter range.

Primary particles that are derived from the break-up and suspension of bulk material by the wind, such as sea salt, soil dust, and biological material, have most of their mass associated with particles of diameters exceeding 1 μm, however, their highest number concentrations occur in the 0.1–1 μm range. For such emission mechanisms, the particle number concentration increases nonlinearly with increasing wind speed (O'Dowd and Smith, 1993; Schulz et al., 1998). Because of their low concentrations and large sizes, primary particles generally do not coagulate with one another, but they can mix with other species through exchange of mass with the gas phase. A particular and impor-

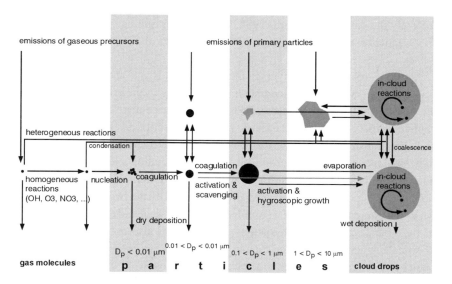

Figure 1.   Scheme of the microphysical processes that influence the size distribution and chemical composition of the atmospheric aerosol. The scheme highlights the large range of sizes that are involved in the formation and evolution of aerosol particles, and how aerosols participate in atmospheric chemical processes through homogeneous, heterogeneous and in-cloud reactions.

tant type of primary particles are soot particles formed in combustion. Initially they are formed at high concentrations within the combustion process as particles with a diameter of 5–20 nm. They coagulate however rapidly to form fractal-like aggregates, which, in turn, will collapse to more compact structures of several tens of nanometers due to capillary forces of condensing vapours.

Secondary aerosol mass is formed by transformation of gaseous compounds to the liquid or solid phase. This occurs when the concentration of the compound in the gas phase exceeds its equilibrium vapour pressure above the aerosol surface. In the atmosphere, several processes can lead to such a state of supersaturation:

(1) gas-phase chemical reactions leading to an increase in the gas-phase concentration of compounds with low equilibrium vapour pressure. Examples are:

$$SO_2 + OH \rightarrow \text{(known reaction scheme)} \rightarrow H_2SO_4,$$

$$NO_2 + OH \rightarrow HNO_3,$$

$$\alpha - \text{pinene} + O_3 \rightarrow \text{(unknown reaction scheme)} \rightarrow \text{pinic acid,}$$

(2) lowering the ambient temperature leading to a lowering of the equilibrium vapour pressure above the aerosol,

(3) formation of multicomponent aerosol, so that the equilibrium vapour pressure of the single compounds above the aerosol is lowered by the presence of other species in the aerosol (Raoult effect).

The equilibrium vapour pressure over a spherical particle increases with increasing curvature of the particle (Kelvin effect), hence the equilibrium vapour pressure above molecular clusters formed by random collisions is much larger than that above a film on a pre-existing particle or flat surface. Consequently, molecular clusters will generally evaporate. Their growth to a stable size, i.e. nucleation, will be favoured primarily by the absence of pre-existing aerosol surface, and by extreme realisations of the three processes described above. Classical nucleation theory predicts that nucleation is highly nonlinearly dependent on the concentration of the nucleating species in the gas phase.

When nucleation does occur, the new particles grow by condensation and self-coagulation. As particles reach a diameter of the order of the mean free path length of the condensing molecule (typically about 60 nm), condensation becomes diffusion limited and slows down. Also, self-coagulation, which is a second-order process, eventually quenches as number concentrations fall. Hence, under background tropospheric conditions, particles formed initially by nucleation require days to weeks to grow larger than about 0.1 μm solely by condensation and coagulation. Under polluted, urban type conditions, this growth can occur within a day (Raes et al., 1995).

One straightforward way of accumulating secondary aerosol species in the 0.1–1 μm diameter range is by condensation on primary particles emitted in that range. Another more elusive but important growth process is by chemical processing in non-precipitating clouds (Mason, 1971; Friedlander, 1977; Hoppel et al., 1986, 1994a). This process begins with the activation of aerosols, which is the uncontrolled uptake of water once water vapour becomes supersaturated above a certain critical limit. According to traditional Köhler theory, the critical supersaturation for activation depends on the amount of soluble material in the particle and its hygroscopicity, i.e. tendency of the material, once dissolved, to lower the equilibrium water vapour pressure over the solution (Pruppacher and Klett, 1980). The critical supersaturation needed to activate all particles larger than a given dry size increases with decreasing particle size. When the supersaturation in an air parcel rises, cloud activation will therefore preferentially occur on larger particles. The rapid condensation of water quenches a further increase of the supersaturation (which usually does not exceed 2%), so that activation is limited to a subset of particles (cloud condensation nuclei, CCN). For example, for pure ammonium sulphate aerosol and a maximum supersaturation of 0.2%, typical for marine stratus clouds, only particles larger than about 80 nm in diameter will activate. Once a droplet is formed, gaseous species like SO_2 can dissolve and be oxidised in the aqueous phase. When the droplets evaporate, the residue particles are larger than the original CCN upon which the droplets formed as a result of the additional oxidised material, e.g. sulphate from the following aqueous-phase reactions:

$$S(IV) + O_3 \rightarrow S(VI) + O_2,$$

$$S(IV) + H_2O_2 \rightarrow S(VI) + H_2O.$$

Reactions occurring in clouds might also occur in non-activated aerosol solution droplets, however, with different efficiencies because of the larger ionic strength in such droplets. Moreover, some gases might also react on the aerosol surface producing products that might either remain on the particle or return into the gas phase. Examples are the heterogeneous conversion of NO_x to HONO on fresh soot aerosol (Ammann et al., 1998) and halogen release from sea salt (Vogt et al., 1996).

Aerosols are removed from the atmosphere by dry and wet processes. Small particles ($D_p < 1$ μm) diffuse to the Earth's surface, a process which becomes less efficient as the particle size increases. Large particles ($D_p > 1$ μm) settle gravitationally, a process which becomes less efficient as the particle size decreases. In the range $0.1 < D_p < 1$ μm, dry removal is very slow, and the formation and growth processes discussed above tend to accumulate the aerosol in this size range. These particles, when they have the right hygroscopic properties, will be removed mainly by activation in clouds and subsequent precipitation.

It has become generally accepted to name particles with a diameter in the range $0.1 < D_p < 1$ μm *Accumulation mode particles*, particles in the range $0.01 < D_p < 0.1$ μm *Aitken mode particles*, and particles with $D_p < 0.01$ μm *Nucleation mode particles*. Particles with $D_p > 1$ μm are called *Coarse mode particles*. The idea to represent the aerosol size distribution with a number of log-normally distributed modes is supported by measurements (see Section 3) and was first ventured by Whitby (1978).

Each of the processes in Fig. 1 has a characteristic time. Table 2 presents such times for some of these processes. They are derived in Appenix A con-

Table 2. Aerosol properties and characteristic times of processes in various atmospheric compartments. As the characteristic time the "e-folding time" is used, i.e. the inverse relative rate of change of a property through a process. (See Appendix A)

	Continental boundary layer		Marine boundary layer		Free troposphere	
	Urban	Sub-urban	Polluted	Clean	Clean	Dust
Characteristic time for coagulation with Aitken and accumulation mode, and with clouds (d)						
Nucleation	0.036	0.095	0.14	1.0	9.4	2.8
($D_p = 6$ nm)						
Aitken	0.2	0.7	1.1	7.6	34	11
($D_p = 60$ nm)						
Accumulation	1.4	4.8	10	66	754	168
($D_p = 300$ nm)						
Characteristic time for volume production by condensation (d)						
Nucleation mode	0.5	1.0	2.4	1.5	0.5	
Aitken mode	1.1	2.6	42	11	1.6	94
Accumulation mode	5.0	12	97	52	5.5	206
All aerosol	4.5	12	83	49	4.9	156
Characteristic time for volume production by cloud processing (d)						
Activated aerosol	0.16	0.42	3.0	1.8		
Characteristic time for wet deposition of volume (d)						
All aerosol			0.5–50			
Compartment Properties						
Residence time	1 day	5–10 days	1–2 weeks	1–2 weeks	2–4 weeks	2–4 weeks
SO_2 (pptV)	1000	600	100	20	20	20
fraction SO_4	0.2	0.3	0.5	0.5	0.5	0.5
N_{tot} (/cm^3)	35000	6500	2600	480	300	550
N ($D_p > 80$ nm) (/cm^3)	4520	2695	1866	181	65	186

sidering conditions that are typical for various atmospheric compartments. In order to understand the behaviour of aerosols in a given compartment, it is useful to compare the process characteristic times with the compartment residence times. Loss processes (coagulation, deposition) which occur on a time scale that is small compared to the compartment residence time, indicate an unstable, i.e. decaying, property. Time scales that are long compared to compartment residence times lead to stable properties within the compartment and inter-compartment exchange of those properties. For instance, the extremely short residence time of nucleation mode particles in the boundary layer, due to coagulation with larger particles, implies that they can only be observed in the immediate vicinity of their sources. Hence, as these ultrafine particles are frequently observed in the polluted continental boundary layer (see Fig. 2), the urban and sub-urban continental compartment must contain sources for nucleation mode aerosol. The absence of a persistent nucleation mode in the MBL indicates that in this case no strong in-situ sources for this mode are available. However, occasional nucleation bursts could temporally occur leading to an unstable and rapidly decaying nucleation mode. Because of their short life time, nucleation mode particles are generally not exchanged between the atmospheric compartments, unless they happen at the boundaries of such compartments. Accumulation mode particles on the other hand decay much slower and as such they can travel from one compartment to the other, mixing their properties with the ones of newly formed aerosol within the next compartment.

A comparison of the time scales for condensation and cloud processing show that cloud processing is the major mechanism for growth. Obviously, this applies only for particles that are activated in clouds. In the urban boundary layer, enough condensing material is available to maintain the Aitken mode mass ($\tau_{cond} < \tau_{res}$). When a polluted air mass advects over the ocean, it is cut off

Figure 2. Number (N) and Volume (V) size distributions observed at (a) an urban site (Milan, Italy); (b) a con-urban site (semi-rural area 50 km from Milan, Italy); (c) a marine boundary layer site, 2 days downwind the European continent (Tenerife in the north–east Atlantic); (d) a marine boundary layer site during transport from the open Atlantic (Tenerife); (e) north–east Atlantic lower free troposphere (FT) during background conditions (Tenerife, 2360 m altitude), (f) north–east Atlantic lower free troposphere during African dust transport in the FT (Tenerife, 2360 m altitude). Black lines are data obtained with a differential mobility analyser. Gray lines are data obtained with an optical particle counter (Grimm Dustcheck®) (b), or aerodynamic particle sizer (APS®, TSI Inc.) (c–f), (data not avialable for urban site).
Mean chemical composition (C) of the sub-micron aerosol ($\mu g/m^3$). Data represent similar conditions as for the N and V size distributions, but for case (a), (b) and (f) they are not obtained concurrently. The unknown fraction (unk) was calculated by comparing the sum of the quantified aerosol components with the sub-micron aerosol mass calculated from sub-micron particle volume size distributions (data not available at the urban and rural sites).

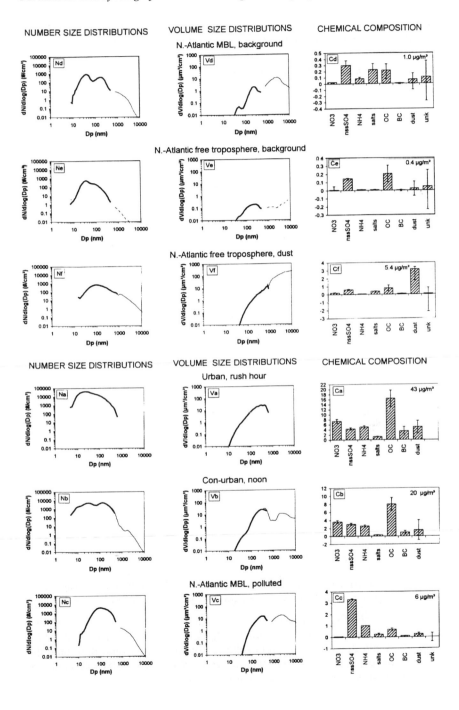

from the aerosol precursor sources, condensation becomes too slow to maintain the continental aerosol loading, so aerosol volume decays.

2.2. Recent advances and remaining issues

Fig. 1 integrates knowledge that is available, at least qualitatively, since the middle of the 1980s. This knowledge resulted from theoretical developments (e.g. Fuchs, 1964; Friedlander, 1977) as well as key observations (Whitby, 1978; Hoppel et al., 1990; Section 3). However, during the last decade major progress has been made in dealing with the various chemical species that can be involved in one or more of the processes depicted in Fig. 1, and with the multicomponent nature of the atmospheric aerosol in general.

2.2.1. Nucleation

With respect to nucleation, it has in general been presumed that the principal gas-phase species involved in atmospheric nucleation is H_2SO_4, and that, if particle formation occurs, it does so via binary nucleation of $H_2SO_4-H_2O$ (Jaecker-Voirol and Mirabel, 1989; Kulmala et al., 1998). Except for the case of cloud outflow (see below), observed rates of new particle formation significantly exceed predictions based on classical $H_2SO_4-H_2O$ nucleation theory (Clarke et al., 1998a; Weber et al., 1996). It has been suggested that NH_3, an ubiquitous molecule in the troposphere, enhances nucleation rates of $H_2SO_4-H_2O$ beyond the binary rate (Coffman and Hegg, 1995). Korhonen et al. (1999) have derived a ternary $H_2SO_4-H_2O-NH_3$ nucleation theory that predicts indeed substantial enhancement, and model studies predict an ubiquitous presence of stable $H_2SO_4-H_2O-NH_3$ clusters, which, however, remain too small to be detected ($D_p < 3$ nm) (Kulmala et al., 2000). Laboratory studies show that oxidation of volatile organic carbon (VOC) leads to new particle formation through nucleation. In the case of biogenic VOCs some oxidation products with low equilibrium vapour pressure have been identified (Christoffersen et al., 1998; Yu et al., 1999; Glasius et al., 1999). Regardless of the species involved, or the specific theoretical expression used for predicting nucleation, nucleation is a process that strongly depends on the gaseous precursor concentration, relative humidity and temperature. This implies that fluctuations in atmospheric conditions may play a major role in determining the location and magnitude of nucleation in the atmosphere (Easter and Peters, 1994). Such fluctuations might occur preferentially when mixing different air masses (Hegg et al., 1992; Nilsson and Kulmala, 1998). Finally it has been pointed out that nucleation around ions is energetically more favourable than homogeneous nucleation and that in certain atmospheric conditions it would dominate the particle formation process (Hamill et al., 1982; Raes et al., 1986). Recent theoretical developments including a whole range of ion–molecule and ion–ion interactions, promote "ion-mediated nucleation" as the explanation to close gap between theory and observations (Turco et al., 1998).

At present, however, whether or not NH_3 or organics or fluctuations or ions are responsible for enhanced nucleation has yet to be verified experimentally.

2.2.2. Condensation

Semi-volatile species, like ammonium nitrate and many organics, will tend to distribute themselves between gas and aerosol phases to achieve thermodynamic equilibrium (Wexler and Seinfeld, 1990). Condensation occurs when the equilibrium shifts towards the aerosol phase.

In the case of inorganic species, the time scale over which this equilibration takes place has been analysed (Meng and Seinfeld, 1996); depending on the particle size and ambient conditions, equilibration times can vary from seconds to hours. Inorganic thermodynamic equilibrium models have been developed over the past two decades, reaching a state of maturity in terms of thermodynamic properties, such as condensed phase activity coefficient parameterizations, and computational implementation (Zhang et al., 2000). Originally developed to simulate gas–aerosol equilibrium in urban and regional-scale atmospheric models, inorganic gas–aerosol equilibrium models have now been embedded into general circulation models to predict the global distribution of sulphate–nitrate–ammonium aerosols (Adams et al., 1999). Predictions of the phase distribution of inorganic aerosol species are in general agreement with observations (Adams et al., 1999).

In the case of organic semi-volatile species much progress still is to be made. Because of the large number of such compounds in the atmosphere and because methods to predict their thermodynamic properties in complex organic-water mixtures pose significant theoretical challenges, gas–aerosol thermodynamic models for organic atmospheric species are not yet available. Experimentally based gas–particle distribution factors for complex mixtures generated by the photo-oxidation of hydrocarbons are available (Odum et al., 1996; Griffin et al., 1999b); however, the goal is fundamentally based thermodynamic models that predict the phase partitioning of individual organic compounds between the gas phase and complex organic–inorganic–water mixtures. Such models would allow first-principles prediction of the amount of organic aerosol formed from primary or secondary organic species in the atmosphere. Such models are currently under development.

2.2.3. Degree of chemical mixing

In treating multicomponent aerosols, the issue arises whether the components are mixed within a single particle (internal mixing), or whether the various components are present as pure particles (external mixing). The manner in which different aerosol species are mixed in individual particles affects both optical and hygroscopic properties (Heintzenberg and Covert, 1990). The only processes depicted in Fig. 1 that lead to aerosols that are externally mixed from pre-existing aerosol are nucleation and primary emissions. The remainder of the processes lead to internal aerosol mixtures. The degree of "internal mixing", that is the degree to which the chemical composition of each individual particles resembles that of the bulk, will increase with the time available for interaction, hence with the residence time of the aerosol in the atmosphere. In urban and polluted continental conditions, the characteristic times of many of these interactions (see Table 2) are small, and internal mixing will occur on a short time scale. However, these regions provide generally also sources for primary and nucleated secondary aerosol, enhancing external mixing. Field observations have indeed shown that in continental areas, two distinct aerosol types with different hygroscopic properties are often present, a direct indication of externally mixed aerosol (e.g. Zhang et al., 1993; Svenningsson et al., 1994). In aged polluted air masses, such as found in continental plumes over the ocean, this feature is not observed, suggesting a chemically more homogeneous (sub-micron) aerosol (Swietlicki et al., 2000), in agreement with the long residence, hence interaction, times.

2.2.4. Aerosol activation

The understanding of activation of aerosols to cloud droplets, and, in particular, the role of organic species in this process, is rapidly evolving. Whereas before only the inorganic salts were considered to constitute the soluble aerosol fraction, it is now clear that a large fraction of aerosol organics is water soluble (Saxena et al., 1995; Saxena and Hildemann, 1996), and there is indirect evidence that these species contribute to the chemical composition of CCN (Novakov and Penner, 1993). Laboratory studies have shown that oxidation products of biogenic compounds, like terpenes, are only slightly hygroscopic; however, internally mixed with ammonium sulphate particles, these organic products do not inhibit water uptake by forming an impermeable layer, but rather mix homogeneously with the ammonium sulphate in the solution droplet (Virkkula et al., 1999).

   The formation of layers is expected when organic surfactants are mixed with solution droplets. Surfactants have been found in fog droplets in polluted air masses, however their overall effect on the aerosol activation process is complex (Li et al., 1998). On the one hand, surfactants will reduce the surface tension of the droplets (as recently observed by Facchini et al. (1999)), hence decrease the supersaturation needed to activate them. On the other hand their presence will reduce the molality of the solution because of their high molecular weight, and therefore increase the supersaturation needed to activate the droplet.

One further effect on aerosol activation is that, apart from water vapour, other soluble vapours can co-condense during the activation process, increase the solute content of the droplet and eventually decrease the supersaturation needed for activation. Theoretical work showed that in aerosol mixtures consisting of weakly soluble substances interacting with water vapour and other soluble gases present in the atmosphere (like HNO_3), particles can grow to cloud droplet sizes (~ 10 µm) at RHs below 100% (Kulmala et al., 1997). In order to describe the effects mentioned above, the traditional Köhler theory has been generalised to a multi-component multi-phase theory (Shulman et al., 1996; Kulmala et al., 1997).

All these studies indicate that the number of cloud droplets critically depends on the detailed physico-chemical properties of the initial individual aerosol particles, which in turn are determined by the complex of processes shown in Fig. 1.

3. Observations

3.1. *Size distributions*

Measurements of the atmospheric aerosol size distributions were essential in identifying the various processes involved in the formation and evolution of the atmospheric aerosol (Whitby, 1978; Hoppel et al., 1986, 1990). Jaenicke (1988) has reviewed such measurements up till the early 1980s and made a climatology of aerosol size distributions. Fig. 2 shows a similar climatology of number distributions and corresponding volume distributions as a function of particle diameter, obtained more recently by our groups with state-of-the-art aerosol counting and sizing equipment (Van Dingenen et al., 1995, 1999; Raes et al., 1997). They will be discussed in detail below. The improvement in aerosol measuring equipment is likely to be the main reason for the difference between the Jaenicke climatology and the present collection of distributions. Note that Fig. 2 shows the *dry* aerosol size, and that under ambient conditions the uptake of water may lead to a diameter growth by a factor up to 1.7. Lognormal parameters for the distributions in the sub-micron aerosol fraction (i.e. below 1 µm dry diameter) are given in Table 2.

3.1.1. *Urban environment (traffic rush hour) (Fig. 2a)*

The size distribution contains three modes. The maximum in the nucleation mode is around 15 nm diameter. This small particle mode appears consistently during traffic rush hours in the morning and the evening, and can thus be linked to emissions from cars. The total particle concentration is of the order 3×10^4 cm^{-3}. Aitken and accumulation mode are merged into a broad mode. Considering the process time scales (Section 2), the Aitken mode can be explained by coagulation and condensational growth of nucleation mode particles, whereas the accumulation mode is a more aged aerosol, advected with the regional scale circulation, and formed by cloud processing and condensation.

3.1.2. Sub-urban environment (Fig. 2b)

The distribution shown is typical for a sunny summer day at noon. Also in this case numerous small particles are often observed, in particular when the total aerosol surface area is low. This indicates that these particles result from photochemically induced nucleation. Total number concentration is of the order $0.5–10 \times 10^4$ cm^{-3}. The high Aitken mode concentration is unstable with respect to coagulation (Table 2), indicating that it is freshly formed by coagulation and growth of the nucleation mode. The third mode is again the stable accumulation mode, aged (by cloud processing and coagulation) and advected into the area. A distinct coarse mode is also observed in the number and, a fortiori, in the volume size distributions. Chemical analysis shows that this mode contains mainly dust.

3.1.3. Polluted marine boundary layer (Fig. 2c)

The distribution represents a continentally derived aerosol which has advected for two days over the ocean. A fairly unimodal size distribution is observed below 1 μm, and a coarse sea-salt mode appears above 1 μm. The ultrafine particles observed in the continental aerosol have vanished, and number concentrations are reduced to about 2500 cm^{-3}. The time scales discussed above show that this type of distribution might further evolve through cloud processing. Although the size distribution still clearly shows a pollution signature (elevated number and volume compared to marine background), the two days of ageing in the MBL have made it chemically internally mixed, as shown by hygroscopicity measurements (Swietlicki et al., 2000).

3.1.4. Clean MBL (Fig. 2d)

The size distribution shows the typical tri-modal structure, with a pronounced Aitken mode, accumulation mode, and the coarse sea-salt mode. The minimum in particle number between the Aitken and accumulation mode is a result of cloud activation of those aerosols sufficiently large to act as CCN, followed by in-cloud chemical reactions and re-evaporation. In most cases no particles smaller than 20 nm are observed, indicating that nucleation is not a major process in the remote MBL. The number is again reduced compared to the previous case and is in the range of 200–500 cm^{-3}.

3.1.5. Background lower free troposphere (2300 m asl) (Fig. 2e)

Size distributions generally have a multi-modal shape, with a dominant mode around 60 nm. Number concentrations range between 300 and 500 cm^{-3}. Recent observations (Maring et al., 2000) have shown that the mode round 60 nm correlated with Be-7, a tracer for upper troposphere air masses. Its form and number concentration have been explained by self-preserving theory (Lai et al., 1972; Raes, 1995), which, predicts that a coagulating and growing aerosol population eventually develops a (close to) lognormal size distribution and number concentrations that are independent of the initial size distribution and number concentration. The mode centred around 100–200 nm is related to anthropogenic tracers like sulphate (Maring et al., 2000). Background FT aerosol does not contain a significant coarse mode; the number of coarse particles is generally below 0.01 cm^{-3}.

3.1.6. Saharan dust layer (FT, 2300 m asl) (Fig. 2f)

This distribution illustrates a particular case of continentally derived aerosol (Saharan dust) injected into the FT and transported over long distances. The sub-micron aerosol shows a broad accumulation mode, in addition to a very pronounced coarse mode, obviously containing dust particles (some tens of coarse particles per cm^3 (Maring et al., 1999)). The accumulation mode is strongly enhanced compared to the 200 nm "pollution" mode in the background FT. The air masses into which the dust was injected, often contain anthropogenic pollution that was previously picked up. The interaction between these two aerosol types is of interest as it will determine the optical and cloud activation properties of the dust.

3.2. Chemical composition

Measurements of the chemical composition are important to identify the various sources contributing to the aerosol as well as its effect. Heintzenberg (1989) has reviewed the data until 1986 and reconstructed grand average chemical compositions for the sub-micron aerosol in a number of environments. Fig. 2 shows the average relative contributions of chemical compounds in the sub-micron fraction of the aerosols for the same environments discussed above. We will denote with PM1 the total aerosol mass in this size fraction.

3.2.1. Urban/sub-urban environment (Figs. 2a and b)

Preliminary chemical mass closure results indicate that the sub-micron urban aerosol is characterized by the highest PM1 concentration ($>40 \, \mu g \, m^{-3}$), and the highest black carbon (BC) contribution ($>8\%$). At a sub-urban site, 50 km away from a major urban centre, PM1 is already decreased by a factor of two. The aerosol sub-micron composition at the sub-urban site is close to what is observed at the urban site, except for primary aerosols, BC and dust which are significantly reduced. Organic compounds (OC) might represent the main component of the sub-micron aerosol a both urban and sub-urban sites (about 40%). However, OC data are affected by large and difficult to assess uncertainties (Putaud et al., 2000). At both sites, NO_3^- and SO_4^{2-} are fully neutralized by NH_4^+.

3.2.2. Clean and polluted MBL (Figs. 2c and d)

Chemical mass closure was achieved by comparing results of chemical analyses with size distribution measurements. The striking feature of the polluted MBL aerosol composition is the predominance of nss-SO_4. The polluted MBL nss-SO_4 concentration is indeed comparable to the rural site nss-SO_4 concentration. This could be due to the presence of specially important SO_2 sources (coal power plants?) upwind of the MBL sampling site, which was on Tenerife, Canary Islands. It is also interesting to notice that the contribution of NO_3^- to PM1 is very small. This can be explained by the displacement of the equilibrium $NH_4NO_3 \rightleftharpoons NH_3 + HNO_3$ above the ocean through the reaction of HNO_3 with sea salt:

$$NaCl + HNO_3 \rightarrow NaNO_3 + HCl.$$

Actually, the concentration of NO_3^- in the supermicron fraction averages $1.6 \, \mu g \, m^{-3}$ in the polluted MBL, i.e. 20 times more than in the fine fraction. In the background North Atlantic MBL, nss-SO_4 remains the main PM1 component. Measurements of MSA, an indicator for DMS particulate oxidation products, suggest that oceanic biogenic sources could account for up to 50% of the background nss-SO_4. The relative contribution of OC is twice as high as in the polluted MBL. Even at moderate wind speed ($5 \, m \, s^{-1}$), sea-spray contributes significantly to sub-micron aerosol mass in the background MBL (24%).

3.2.3. Background free troposphere (2300 m asl) (Fig. 2e)

PM1 is as low as $0.4 \, \mu g \, m^{-3}$. OC may be the main component of the sub-micron aerosol. Estimation of the uncertainties in OC mean concentration indicates that its contribution to the aerosol sub-micron mass is in the range 31–56%.

3.2.4. Saharan dust layer (FT, 2300 m asl) (Fig. 2f)

During transport events out of North Africa, mineral dust is the main component of sub-micron aerosol in the FT. However, the SO_4^{2-}/Ca^{2+} ratio observed during dust events is significantly higher than in fine Saharan sand grains and glacier ice cores representative of pre-industrial times (Schwikowski et al., 1995). This suggests that the desert dust plumes transported over the Atlantic Ocean tend to be mixed with pollution aerosol.

4. Modelling the clean marine boundary layer

The existence of significantly different size distributions and chemical compositions in various environments (see Section 3) has led the aerosol community to think in terms of atmospheric compartments, such as the marine boundary layer, the continental boundary layer, the free troposphere. Of those, the MBL has been studied extensively, because of its dominant role in the climate system (Charlson et al., 1987) and because of it simplicity relative to others.

The aerosol in the marine boundary layer (MBL) has traditionally been divided into two categories, that derived from sea salt and that from all non-sea salt (nss) sources. The dominant chemical component of the non-sea salt aerosol is sulphate. It has been established that the principal source of this sulphate in areas uninfluenced by anthropogenic sources, is gaseous DMS produced by phytoplankton in surface (Charlson et al., 1987; Ayers et al., 1991; Calhoun et al., 1991). A major question concerns the sensitivity of the levels of MBL aerosol number concentration to the sea-surface DMS emission rate, and correspondingly, the importance of other processes in controlling the levels of particles over the remote ocean (Bates et al., 1998).

The first studies dealing with this question, relied on box models of the MBL, and implemented descriptions of those processes in Fig. 1 dealing with sea-salt and sulphate aerosol formation (Raes and Van Dingenen, 1992; Russell et al., 1994). Model results, however, converged and were in better agreement with the observed size distributions only when entrainment of aerosol from the free troposphere was considered as a source of aerosol number in the MBL (Raes, 1995; Capaldo et al., 1999).

Indeed, whereas a few measurements have been reported of the occurrence of large numbers of ultrafine particles following rain events or strong subsidence of free tropospheric air (Covert et al., 1992, 1996; Hoppel et al., 1994b; Clarke et al., 1998a), observations generally indicate that subsidence from the free troposphere is the main process controlling MBL aerosol number concentration (Clarke et al., 1996b; Bates et al., 1998; Raes et al., 1997; Van Dingenen et al., 1999, 2000).

In a recent study, Katoshevski et al. (1999) have simulated the relative influence of sources and sinks of particles using a simplified model of the clean MBL. Based on the typical conditions considered, MBL aerosol number concentration is predicted to be dominated by free tropospheric aerosol under virtually all conditions: 89% in the average case they consider, and even 69% at a $17 \mathrm{~m\,s}^{-1}$ wind speed. MBL aerosol mass, on the other hand, is dominated by sea salt particles: 62% in the base case and 98% at a wind speed of $17 \mathrm{~m\,s}^{-1}$. Even under conditions when a high rate of nucleation is presumed to occur, still only about 5% of the total particle number, on average, is predicted to be provided by nucleation events. Cloud processing, while not a major contributor to aerosol number, does provide, except under high wind conditions, the order of 20% of the aerosol mass. Although nucleation occurs only infrequently in the MBL and does not contribute appreciably to long-term average aerosol number or mass, nucleation can replenish particles in brief, intense episodes when aerosol surface areas are substantially reduced by precipitation.

In summary, the studies of aerosol formation and evolution in the clean MBL have highlighted that the aerosol characteristics observed within a certain compartment, cannot always be explained by processes occurring within that compartment. Instead, exchanges between compartments can play a major role. This is in agreement with the result of the time scale analysis in Section 2.1. We argue that it is necessary to consider general atmospheric circulation which connects marine and continental boundary layers and the boundary layer with the free troposphere, in order to understand the different characteristics of the aerosol throughout the global troposphere.

5. Global atmospheric circulation and the life cycle of the tropospheric aerosol

5.1. Tropospheric general circulation

Tropospheric general circulation is characterised by rapid, localized upward motion due to convection (in the tropics) or slantwise ascent along frontal surfaces (in the mid-latitudes), which is compensated by relatively slow and large-scale subsidence in the sub-tropical and polar regions. Horizontal transport in the lower and upper troposphere connects areas of upward and downward transport, in what are supposed to be toroidal circulation patterns. Long-term averages of both the meridional and zonal wind fields in the tropics/sub-tropics reveal the existence of these patterns, which are called the Walker and Hadley circulations, respectively (see Fig. 3). In a snapshot of the global wind fields,

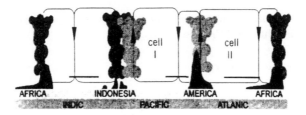

Figure 3. Global circulation in the tropics/subtropics, after Newell (1979). Convection occurs preferentially over the continents, whereas subsidence occurs over the oceans. The subsidence inversion creates a well-defined marine boundary layer, which is topped by stratiform clouds.

these toroidal circulations are less evident (Newell et al., 1996; Wang et al., 1998).

Subsidence over the sub-tropical oceans leads to the existence of a temperature inversion and the creation of a marine boundary layer, which is topped by vast stratiform clouds. Thus in the subtropics there is a clear separation between the marine boundary layer and the free troposphere aloft, whereas in convective regions this separation is less clear.

The general circulation is described by global observations of fields of winds and other meteorological parameters (Oort, 1983), or it can be reproduced by General Circulation Models (GCM's) from basic physical principles. Using the observed climatologies, or off-line versions of the GCMs, Chemical Transport models (CTMs) have been built in which the descriptions of emissions, transport, transformations and removal of chemical species have also been considered (Zimmermann, 1984; Heimann et al., 1990).

5.2. Recent progress and remaining issues

5.2.1. Global budgets and mass concentration fields of individual aerosol species

Major progress has been achieved in simulating the global distribution of tropospheric aerosol mass using global CTMs. The first simulation of the global distribution of biogenic and anthropogenic sulphur (Langner and Rodhe, 1991) lead to the recognition that anthropogenic sulphate aerosols may have a significant impact on the global radiation balance (Charlson et al., 1991). This spurred a large interest, and simulations of the global mass distributions for the aerosols types listed in Table 1 followed. Despite the simplification of considering each aerosol type independently, these studies were important to relate emissions to global distributions, to construct global and regional budgets, to estimate the contribution of anthropogenic sources to the burden of aerosol species that are also produced naturally, and to draw attention to elements of the general circulation that are important in aerosol transport, in particular deep convection (Feichter and Crutzen, 1990).

5.2.2. The role of deep convection

The high updraft velocities in convective clouds and the corresponding super-saturations up to 2% lead to activation of soluble aerosol particles with diameters as small as 0.01 µm; partly soluble particles activate at somewhat larger diameters. Soluble trace gases in these updrafts will be taken up by the cloud droplets. Precipitation, which, on a global scale, is produced mainly in convective clouds, eventually removes the activated particles and dissolved gases. During convective cloud transport a separation therefore occurs between soluble species that are rained out and insoluble species that are pumped into the free troposphere (Rodhe, 1983).

Initially it was supposed that convective transport of fairly insoluble DMS and its subsequent oxidation to SO_2 and H_2SO_4 and MSA was the main source of sulphate aerosol in the background free troposphere (Chatfield and Crutzen, 1984). Recent simulations of convective transport (Wang et al., 1995) and, in particular, measurements over the Western Pacific (Thornton et al., 1997), now suggest that DMS contributes only 1–10% to the SO_2 in the upper troposphere over the Northern Hemisphere.

Fig. 4 shows the fraction of total SO_2 derived from DMS oxidation in the upper troposphere (300 mb) as simulated by the GISS GCM II' aerosol model (Adams et al., 1999; Koch et al., 1999). The simulation shown here is based on emissions from the IPCC SRES A2 scenario for the year 2000. This scenario prescribes 26.0 $Tg\,S\,yr^{-1}$ of natural DMS emissions and 73.8 $Tg\,S\,yr^{-1}$ of SO_2 emissions. Of the SO_2 emissions, 69.0 $Tg\,S\,yr^{-1}$ stem from anthropogenic activities. The remaining 4.8 $Tg\,S\,yr^{-1}$ represent volcanic emissions.

These results show that, in the northern hemisphere, SO_2 in the upper troposphere is mostly anthropogenic in origin. North of about 15°N latitude, less than 20% of the SO_2 is the product of DMS oxidation. In the most anthropogenically perturbed areas above the industrial centres of North America, Europe, and eastern Asia, the fraction is less than 10%, in agreement with the observations of Thornton et al. (1997) and results of Wang et al. (1995). The situation is mostly reversed in the southern hemisphere, where the SO_2 in the upper troposphere above remote ocean regions, such as the south Pacific and Indian oceans, is mostly derived from DMS oxidation. An important exception is a plume of anthropogenic SO_2 emissions visible above South America and Africa stemming from biomass burning and copper smelting activities in those areas.

Therefore, through the process of deep convection, anthropogenic activities have significantly perturbed not only the continental boundary layer, but also

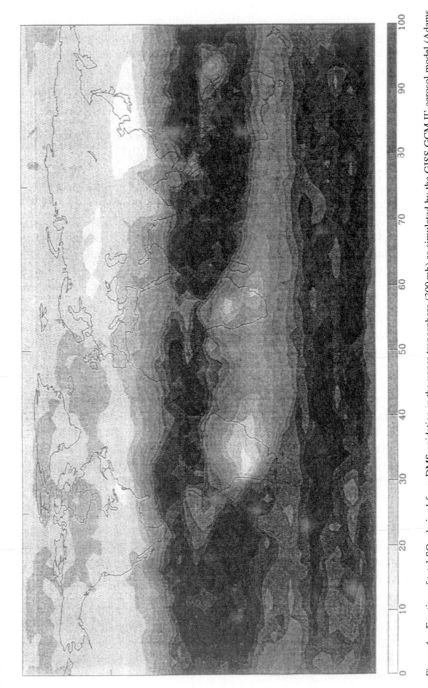

Figure 4. Fraction of total SO_2 derived from DMS oxidation in the upper troposphere (300 mb) as simulated by the GISS GCM II' aerosol model (Adams et al., 1999; Koch et al., 1999). Sulphur emissions were prescribed based on the IPCC SRES A2 scenario, including 26.0 Tg S yr^{-1} of DMS emissions, 69.0 Tg S yr^{-1} of anthropogenic SO_2 emissions, and 4.8 Tg S yr^{-1} of volcanic SO_2 emissions.

the free and upper troposphere as well. Because oxidation of SO_2 in the upper troposphere proceeds mainly by gas-phase reaction with the OH radical, anthropogenic activities have the tendency to increase gas-phase concentrations of H_2SO_4 in the upper troposphere. Given the low aerosol surface area concentrations in this part of the atmosphere, this process may result in enhanced rates of new particle formation in the upper troposphere.

5.2.3. Aerosol nucleation in the free troposphere

Deep convection leads to an increase of the photo-oxidizing capacity of the upper troposphere by pumping up nitrogen oxides produced by anthropogenic sources or biomass burning at the surface or by lightning within the convective cloud (Lelieveld and Crutzen, 1994; Jacob et al., 1996). Given the link between photochemistry and aerosol formation (see Fig. 1), as well as the increase in gas-phase concentrations of H_2SO_4 discussed in the previous section, a link between convective clouds and aerosol nucleation might be expected.

Measurements in the upper troposphere of the Pacific demonstrate an anti-correlation between aerosol number and aerosol surface area (Clarke, 1992, 1993; Andronache et al., 1997), suggesting that production of new particles by nucleation occurs when the pre-existing aerosol surface area is low, in agreement with the process understanding. These measurements show, furthermore, enhanced ultrafine particle concentrations above the convective areas of the Inter Tropical Convergence Zone. Other data indicate that conditions favourable for nucleation exist near evaporating cloud boundaries in both mid-latitude and equatorial marine environments, i.e. elevated water vapour, cold temperatures, low aerosol surface areas (about 5 $\mu m^2\,cm^{-3}$) or a combination of these (Hegg et al., 1990; Perry and Hobbs, 1994; Clarke et al., 1998b). The high actinic flux in the vicinity of clouds can furthermore increase OH production (Mauldin et al., 1997), which, in turn, enhances formation of H_2SO_4 vapour. As noted in Section 2.3, observations that include measurements of gas-phase H_2SO_4 have detected nucleation events near clouds, in agreement with the predictions of the classical theory (Clarke et al., 1999). In this way, convective clouds seem to both initialize the life cycle of part of the background aerosol and terminate it, when, after going through the Hadley/Walker circulation, particles re-enter a convective cloud and are removed by precipitation.

Measurements at various sites in the remote troposphere indicate that the rate at which freshly formed particles grow exceeds that which can be explained by condensation of H_2SO_4 and associated H_2O (Weber et al., 1998; Kulmala et al., 2000). It has been suggested that condensation of other compounds (notably organics) on the nucleated H_2SO_4–H_2O–NH_3 clusters might account for this "excess" growth.

5.2.4. Organics aerosol in the free troposphere

Organic aerosol has been observed to be ubiquitous in the upper troposphere (Novakov et al., 1997; Murphy et al., 1998b; Putaud et al., 2000, Section 3.2). Although the accuracy of organic matter determination in aerosols is in question, these studies do claim that there is relatively more organic matter in the upper troposphere than sulphate. This would be consistent with a separation between soluble and insoluble species during vertical transport.

A major question is the extent to which organic aerosol in the upper troposphere is derived from atmospheric oxidation of biogenic hydrocarbons to organic aerosol products. Such organic aerosol can reach the free troposphere by two routes. Deep convection over tropical areas could transport relatively insoluble biogenic hydrocarbons into the free troposphere where they are oxidized to produce organic aerosol or oxidation and aerosol formation can occur in the boundary layer, with the aerosol subsequently being transported to the free troposphere.

Biogenic hydrocarbons such are terpenes are among the most reactive gaseous compounds in the atmosphere, and they are important precursors of secondary organic aerosol (Went, 1960; Christoffersen et al., 1998; Griffin et al., 1999a; Glasius et al., 2000). Terpene oxidation occurs by reaction with OH, O_3, and NO_3, with OH and O_3 being generally the most important under atmospheric conditions. Laboratory chamber data on the aerosol-forming potential of individual terpenes, together with temporally and spatially resolved, compound-specific estimates of global biogenic hydrocarbon emissions and global OH and O_3 fields, can be combined to obtain an estimate of the annual global production of organic aerosol from biogenic hydrocarbons. This procedure has been employed to lead to an estimate of 18.5 Tg per year of present day biogenic organic aerosol production (Griffin et al., 1999b). Because of the high reactivity of the biogenic parent compounds with OH and O_3, most of the oxidation is predicted to occur in the lowest few kilometers of the atmosphere. One aspect of the process that the estimate of Griffin et al. (1999b) does not account for is the effect of temperature on the conversion of oxidation products to aerosol. As temperature decreases, semi-volatile oxidation products will increasingly partition to the aerosol phase; thus, all else being equal, the above estimate can be considered to be conservative, since it is based on aerosol yields measured at room temperature.

With an assumed free troposphere residence time of two weeks, this annual global formation rate leads to an estimate of the average mass concentration of tropospheric organic aerosol derived from biogenic hydrocarbons of about $0.5 \, \mu g \, m^{-3}$. Such an estimate has necessarily a considerable degree of uncertainty associated with it, easily plus or minus a factor or two. The estimate does not account for, for example, the effect of wet scavenging of terpene oxidation products in deep convection, although such products are not expected to be overly hygroscopic (Virkkula et al., 1999). Nevertheless, the estimate $0.5\ \mu g\,m^{-3}$ can be compared with the $0.4\ \mu g\,m^{-3}$ measured in the North Atlantic free troposphere (Putaud et al., 2000, Section 3.2), and does suggest that terpene oxidation is a potential contributor to organic aerosol in the free troposphere and that the route of injection by convective mixing of boundary layer oxidation products can lead to measurable levels of such aerosol.

5.2.5. Ubiquity of layers in the troposphere

Recent observations have shown that the free troposphere is chemically not homogeneous. Quasi-horizontal layers are frequently observed which are characterised by various combinations of ozone and water vapour (Newell et al., 1999), and other chemical species (Wu et al., 1997). Layers with lower O_3 and higher H_2O than the background are tentatively interpreted as due to convection from the boundary layer. Layers with higher O_3 and lower H_2O, which are the most abundant are interpreted as originating in the stratosphere. Vertical profiles of aerosols also indicate layered structures in, e.g. aerosol number concentration (e.g. Clarke et al., 1996) but a detailed study of their relationship with other gas-phase species has not been made yet. Convective transport of (insoluble) pollution aerosols or nucleation near clouds might be two ways of producing such layers in the middle and upper troposphere. Johnson et al. (2000), however, observed layers in the lower free troposphere immediately above the marine boundary layer, which they explained as originating from a deep (2–5 km) convectively driven continental aerosol layer which advects over the colder ocean. Layers of mineral dust from the Gobi and Sahara desert are frequently found over the Pacific and Atlantic oceans, respectively.

Turbulent mixing in the free troposphere is extremely slow. Following the formulation of Dürbeck and Gerz (1996), the time needed to mix a layer throughout the depth of the free troposphere is larger than the time needed to subside a layer from the upper free troposphere to the surface, the latter being about 10–14 d (Gage et al., 1991). Hence, interaction and mixing of aerosols in different free tropospheric layers, might not happen often. Evidence for this lack of mixing is given by Ostrom and Noone (2000), who measured different aerosol properties at different heights within the same Saharan dust outbreak over the N. Atlantic.

6. Aerosol microphysics in the context of the general circulation

A straightforward way to link microphysics and the general circulation and treat fully the issues discussed above is to implement the descriptions of the processes depicted in Fig. 1 in a general circulation model or global CTM, which captures the transport patterns depicted in Fig. 3. However, in order to accurately treat aerosol dynamic processes such as nucleation, coagulation, and condensation, the aerosol size distribution between 1 nm and 1 μm should be described with a high resolution in particle size (Raes and Van Dingenen, 1995). Furthermore, within each size class several chemical compounds should in principle be tracked. Hence, the model must handle a large number of extra tracers, which is computationally not possible yet. Simplified descriptions of the multi-dimensional size distributions are still in order. Another problem with GCMs or global CTMs is their low spatial resolution (presently 100×100 km at best). The discussion in the previous sections identified various processes that are likely to be important at a smaller scale; this is particularly the case with cloud processes, nucleation, and with the existence of fine horizontal layers in the vertical. Parameterization of these sub-grid processes are also needed: they might be developed with box models which are more appropriate to describe detailed microphysical processes in air parcels or layers and to probe how these processes are sensitive to elements of the global circulation.

In the following sections we give an example of both box and 3-D modelling approaches, including some initial results.

6.1. Diagnostic 0-D modelling of aerosol microphysics and global circulation

Raes et al. (1993) used a box model in which they implemented the aerosol dynamics processes of Fig. 1, related only to the H_2SO_4–SO_2 aerosol system as it results from biogenic DMS emissions over the oceans and excluding the primary emissions of seasalt. They "moved" this box through an oceanic Hadley/Walker cell (cell I in Fig. 3). This approach turned out to be adequate to study in a qualitative way the cycling of aerosols in the clean marine environment. It predicted, in particular, the occurrence of nucleation in the outflow regions of the convective clouds, and the absence of nucleation within the MBL.

This simulation is repeated here with an updated version of the aerosol dynamics model, AERO3, which considers also the primary emissions of insoluble soot particles (Vignati, 1999, see Appendix B). The inclusion of soot allows one to study also aerosol cycling in a Hadley/Walker cell that has its convective updraft over the polluted continent (cell II in Fig. 3).

In both the clean marine and polluted continental scenario the simulation starts with the aerosol that enters a convective cloud. In cloud, the fraction of SO_2 oxidised in the cloud droplets and the fraction of the aerosol that is activated at a supersaturation of 2% is completely removed by precipitation. DMS, the remainder of the SO_2 and the unactivated aerosol is injected into the Free Troposphere (FT), where an immediate dilution by a factor of 4 is assumed to account for the turbulent mixing at the exit of the cloud. This factor is in agreement with the reduction of DMS measured below and near the top of convective clouds (Ferek et al., 1986). Subsequently the aerosol plume travels for 15 d in the FT, where it is slowly dispersed due to atmospheric turbulence (Dürbeck and Gerz, 1996). In the plume DMS is oxidized by OH to SO_2 and further to H_2SO_4. If the conditions allow, nucleation can occur, in which case the aerosol further develops by coagulation and condensation. As a rough simulation of the conditions expected in the free troposphere, the plume experiences 90% relative humidity during the first day after leaving the cloud. Afterwards the relative humidity is kept at 40%. During the first 8 d of the simulation, transport is kept at constant height in the FT and the temperature is maintained constant at 239 K. During the last 7 d the plume subsides, and the temperature increases linearly to the MBL value. After 15 d, the aerosol entrains into the MBL, where it travels for another 7 d back to its starting point. In the MBL the SO_2 concentration responds to the oxidation of locally emitted DMS with OH radicals, and the MBL aerosol size distribution is governed by the aerosol dynamical processes, by cycling trough MBL clouds and by entrainment of the free tropospheric aerosol. The free tropospheric aerosol is that resulting after the initial 15 d of transport in the FT. A comprehensive list of input data to the model is given in Table 3.

Table 3.   Input data for the AERO3 diagnostic box model calculations, shown in Fig. 5[a]

	Free troposphere	Marine boundary layer
Temperature (K)	239	279
Relative humidity (%)	99 decreasing to 40	90
Background SO_2 (pptV)	20	
Initial SO_2 (pptV)		20 (CM)
		840 (PC)
Initial DMS (pptV)		80
OH 24 h average (mol cm^{-3})	2×10^6	2×10^6
DMS flux ($\mu mol\, m^{-2}\, d^{-1}$)		5
H_2O_2 (pptV)	500	500
pH of cloud	5	5
LWC ($g\, m^{-3}$)	0.3	0.3
Supersaturatio (%)	2 (convective cloud)	0.2 (stratiform cloud)
Cloud cover (%)		43
Removal efficiency by precip (%)	100	15
Boundary layer height (m)		1500

[a]CM = Clean Marine Case; PC = Polluted Continental Case.

6.1.1. Clear marine case (Cell I)

The model is initialised with the bi-modal clean MBL aerosol size distribution in Fig. 2d and with log-normal parameters given in Table 2. (Note that this bi-modality is hardly resolved in Fig. 5a.) The particles are assumed to be pure H_2SO_4–H_2O droplets. Fig. 5 shows the evolution of the number size distribution. At time zero the total particle number sharply decreases, as all particles larger than a (dry) critical diameter of 20 nm are processed and eventually removed in the convective cloud. The surface of the aerosol (at 90% rh) is reduced from 32 μm^2 cm^{-3} in the MBL to 0.0175 μm^2 cm^{-3} when exiting the cloud. This small surface area, together with the initial high relative humidity and the low temperatures, leads to a burst of nucleation of new particles. After the first day the particle number steadily decreases due to coagulation, while condensation of H_2SO_4 increases the aerosol mass. The distribution after 15 d of transport in the FT is mono-modal, in agreement with measurements at the free tropospheric station of Izana only in the cleanest conditions (Raes et al., 1997). When after 15 d this aerosol entrains into the MBL, in-situ MBL nucleation is quenched by the surface area of the entrained aerosol, despite the increase in relative humidity and DMS derived H_2SO_4 in the gas phase. Entrainment and MBL cloud processing determine the final aerosol size distribution. The mode at $D_{p,dry} = 40$ nm is due to the shape of the FT size distribution, while the second mode is formed by the cloud processing in the MBL stratiform clouds. The fact that after going through one cycle the size distribution exhibits the same basic bi-modality as the initial size distribution supports the sequence of microphysical and dynamical processes used to explain these size distributions.

According to this model, formation by nucleation in the upper troposphere in cloud outflow and subsequent coagulation during subsidence would explain the observed increase in total number concentration with increasing altitude over the remote oceans (Clarke et al., 1993, 1998; Andronache et al., 1997). Condensation increases the aerosol mass and shifts the aerosol size distribution towards larger sizes, but due to the decreasing availability of SO_2 the importance of the process diminishes during transport in the FT. Entrainment of free tropospheric particles in the MBL is the process determining the final MBL particle number, whereas cloud processing by marine stratus clouds is the main contributor of the aerosol mass increase. This is in agreement with the discussion in Section 4.

6.1.2. Polluted continental case (Cell II)

The difference with the previous simulation is that the model is initialised with an aerosol size distribution and SO_2 concentration, typical of continen-

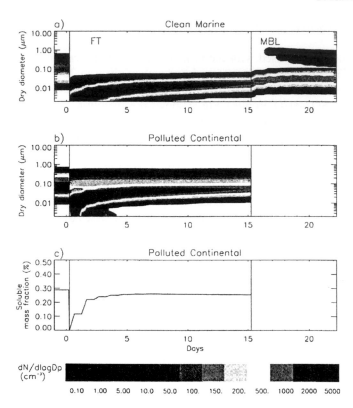

Figure 5. (a) Modelled time evolution of the number size distribution of H_2SO_4–H_2O particles in an air parcel cycling through a clean marine circulation cell (Cell I in Fig. 3). At $t = 0$, the parcel is transported through a convective cloud over the ocean. The concentration is initially reduced by activation and wet deposition. A burst of nucleation is predicted at the outflow of the cloud in the upper troposphere. The number decreases mainly by coagulation and the particles grow by coagulation and condensation during 15 d of transport in the FT. On day 15, the aerosol is entrained in the marine boundary layer to form the MBL Aitken mode. Nucleation within the MBL is quenched by the entrained aerosol surface area, but an accumulation mode develops through cloud processing. (b) Modelled time evolution of the number size distribution of internally and externally mixed Black Carbon–H_2SO_4–H_2O particles in an air parcel cycling through a polluted marine circulation cell (Cell II in Fig. 3). At $t = 0$, the parcel is transported through a convective cloud over the polluted continent. The concentration is initially reduced by activation and wet deposition. A burst of nucleation is predicted at the outflow of the cloud in the upper troposphere, despite the higher number of particles surviving wet deposition. The number decreases mainly by coagulation, and the particles grow by coagulation and condensation during 15 d of transport in the FT. After 15 d of evolution in the FT the aerosol is bi-modal (however not resolved in the figure) with the smallest particles consisting of pure H_2SO_4–H_2O particles and the larger of mixed BC–H_2SO_4–H_2O particles. The form of the distribution is comparable with the one shown in Fig. 2e. (c) Soluble mass fraction of the aerosol. At $t = 0$ the soluble mass fraction decreases as the most soluble aerosol is activated in the cloud and removed by precipitation. The soluble mass fraction subsequently increases by condensation of H_2SO_4.

Table 4. Characteristics of the input distributions for the AERO3 box model calculations shown in Fig. 5

		Clean marine	Polluted continental
Aitken mode (dry)	N (cm^3)	300	2800
	D_p (nm)	36	76
	σ	1.42	1.52
Accumulation mode (dry)	N (cm^3)	70	350
	D_p (nm)	150	230
	σ	1.47	1.36
Soluble fraction	ε (%)	100	28[a]

[a]Of the particles with diameter smaller than 0.3 µm 50% are assumed to be purely insoluble, and 50% internally mixed with ε equal to 0.52. Of the particles larger than 0.3 µm 20% are considered purely insoluble and 80% internally mixed with ε ranging between 0.52 and 0.90, to obtain an average ε equal to 0.28. Pure soluble particles are considered initially zero.

tal polluted conditions. The aerosol is an external/internal mixture consisting of water-soluble H_2SO_4 and insoluble soot. The average water-soluble mass fraction is 28%, and its distribution across the size distribution has been determined using results from various experiments looking at either the hygroscopicity (Svenningsson et al., 1994) or the solubility of the aerosol (Sprengard-Eichel et al., 1998) (see Table 4). Figs. 5b and c show the evolution of the number size distribution and soluble mass fraction of the aerosol, respectively. Only the insoluble and part of the mixed particles survive wet deposition in the convective cloud, and the aerosol surface is reduced from 144 µm^2 cm^{-3} in the boundary layer to 6.2 µm^2 cm^{-3} at the exit of the cloud. Although this offers a larger surface area for condensation than in the clean case, nucleation still does occur because of the larger concentration of SO_2 that survives through the cloud. Twelve hours after the injection in the FT, condensation of H_2SO_4 and coagulation with the nucleated soluble particles have transformed the remaining insoluble particles into mixed particles. During the FT transport the total number of new particles, which initially are mainly soluble, decreases due to coagulation, whereas the soluble mass fraction keeps increasing due to condensation. After 15 d of transport in the FT, the distribution is bimodal, with the largest mode being the boundary layer aerosol, modified by convective transport and transport in the FT, and the smallest mode the new aerosol nucleated near the convective cloud. The bimodal distribution is in qualitative agreement with most measurements at the free tropospheric site of Izana (see Fig. 2e), and the interpretation of these distributions by Maring et al. (2000) (see above) supports the sequence of events described by the model.

6.2. CTM 3-D modelling of aerosol microphysics and global circulation

A "modal" model, M^3, for H_2SO_4–H_2O aerosol dynamics has been developed, in which the aerosol size distribution is represented by overlapping log-normal size distributions representing, respectively, the nucleation mode, Aitken mode and accumulation mode (Wilson and Raes, 1996, see Appendix C). This model is used to resolve the particle size and number concentration of the sulphate mass calculated by a model of the atmospheric global sulphur cycle (Langner and Rodhe, 1991). It is further coupled to a model of the atmospheric black carbon (BC) cycle (Cooke and Wilson, 1996), and a static sea salt aerosol distribution (Blanchard and Woodcock, 1980), which together provide a background aerosol for the H_2SO_4–H_2O dynamics processes to interact with. This coupled model is implemented in the global off-line CTM TM2 (Heimann et al., 1990). Hence, in each gridbox of the CTM, the model treats the processes depicted in Fig. 1: the secondary aerosol consists of H_2SO_4–H_2O resulting from the oxidation of biogenic DMS and anthropogenic SO_2, or from in-cloud oxidation of SO_2, and the primary aerosol consist of sea salt and of black carbon from fossil fuel and biomass burning. Dust and organics are not included. Details on the global emission inventories for DMS, SO_2, H_2SO_4 and black carbon used are given in Appendix C, Table 5.

The model is run for 3 yr with meteorology and emissions representative of the mid–1980s, and the results discussed below pertain to the third year of the

Table 5. Input data for the M^3-TM3 CTM model calculations, shown in Fig. 6

Transport Model TM2	
Basic reference	Heimann et al. (1990)
Resolution	$8° \times 10° \times 9$ levels
Meteorology	ECMWF 1987
Clouds	ISCCP monthly mean cloud data
Photo chemistry	prescribed monthly average OH, O_3 and H_2O_2 from TM3 (Dentener, personal communication)
Global emissions and fields	
DMS	Bates et al. (1987)
SO_2	GEIA v1A 15% of SO_2 emitted as H_2SO_4 (70% attached to BC 30% as gas)
Black Carbon	Mass based emissions from Cooke and Wilson (1996)
	Transformed into number based emissions assuming a lognormal distribution with geometric mean diameter of 60 nm and a geometric standard deviation of 2.0
Sea salt	Mass loading after Blanchard and Woodcock (1980)
	Transformed into number concentrations assuming a lognormal distribution with geometric mean diameter of 600 nm and geometric standard deviation of 1.8

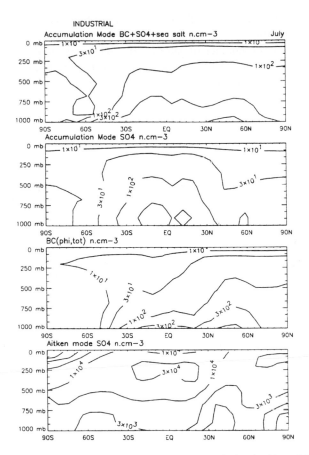

Figure 6. Zonal plots of aerosol number concentration as calculated with a global Chemical Transport Model (see text for full description): (a) total number of accumulation mode particles ($D_{p,dry} > 80$ nm) including pure BC, pure H_2SO_4–H_2O, mixed BC–H_2SO_4–H_2O particles and sea-salt particles; (b) number of pure H_2SO_4–H_2O accumulation mode particles; (c) number of internally mixed BC–H_2SO_4–H_2O accumulation mode particles; (d) number of pure H_2SO_4–H_2O particles in the Aitken mode ($10 < D_{p,dry} < 80$ nm).

simulation. Fig. 6 shows the resulting zonally averaged fields for July, of the number concentration in several aerosol classes.

Fig. 6a shows the concentration of the total number of accumulation mode particles ($0.08 < D_p$), including pure BC, pure H_2SO_4–H_2O particles, mixed BC–H_2SO_4–H_2O particles and sea-salt particles. In the clean Southern Hemisphere the number concentration is between 30 and 300 cm^{-3} at the surface. In the northern hemisphere, the zonally averaged number concentration increases to between 300 and 1000 cm^{-3} at the surface, as a result of the anthropogenic emissions. Accumulation mode number concentrations generally decrease with height.

Fig. 6b shows the number of pure H_2SO_4–H_2O accumulation mode particles. They are confined to the lower layers in the atmosphere, as they are removed by wet deposition in convective clouds, but are not affected by sub-cloud scavenging in the model.

Fig. 6c shows the mixed BC–H_2SO_4–H_2O accumulation mode particles. They typically comprise a larger fraction of the total than the pure H_2SO_4–H_2O accumulation mode particles. Because of the reduced solubilty of BC when emitted, the model transports them more effciently through convective clouds into the upper troposphere.

By differencing Fig. 6a with Figs. 6b and c, it can be seen that in the mid-lattitudes and polar regions the number of accumulation mode particles is governed by the primary emissions of sea salt.

Fig. 6d shows the pure H_2SO_4–H_2O particles in the Aitken mode ($10 < D_p < 80$ nm). In the model, these particles are only formed from the growth of nucleation mode particles. Their concentration is greatest in the upper troposphere, in particular in the outflow regions of the tropical convective clouds. A lesser maxima is found in the Northern Hemisphere at the surface, near the anthropogenic SO_2 source regions, and overall zonal average surface concentrations are in the range 300–3000 cm^{-3}. The same features are also seen in the nucleation mode concentration fields (not shown).

There is a lack of global aerosol size distribution data so that it is not possible to reconstruct decent zonal averages from measurements and test the model results presented above (Heintzenberg et al., 2000). However, a comparison with individual data sets is possible. The results of Fig. 6a, e.g., are in general agreement with the data on accumulation mode particles ($N_{D_p > 80\,nm}$) shown at the bottom of Table 2. The measured urban and sub-urban aerosol number concentrations are 4500 and 2700 accumulation mode particles cm^{-3} respectively, but these high values pertain to local situations which most likely are smeared out in a zonal average of a CTM. Observations over the ocean, downwind of polluted continental areas indicate an internal mixture of the aerosol, with the number of the accumulation mode aerosol being governed by a refractery aerosol, most likely black carbon (Clarke et al., 1996b). The predominant role of sea salt in determining the number of accumulation mode particles in certain marine areas of high wind speed such as the mid-latitutes has been documented by Murphy et al. (1998a). The decrease in accumulation mode particles and increase in Aitken mode particles with altitude are in agreement with vertical profile measurements. However, the predicted zonal average Nucleation and Aitken mode concentrations are larger throughout the model domain than observations suggest. For example in the FT, the modelled zonal average concentrations (3000–30,000 cm^{-3}) have been observed only locally near the outflow of clouds. We believe the over-prediction in Aitken mode concentrations is driven by nucleation in the free troposphere; consequently, the effect is seen most strongly in the FT and less so at the surface and in the accumulation mode zonal average concentrations. The two model layers spanning 70–255 mbar, have minimum temperatures of 190 and 210 K, respectively. At these temperatures and 30% RH, the modified formulation of Jaecker-Voirol and Mirabel used in M^3 effectively nucleates all gas phase sulphate concentrations in excess of 10^2 and 10^5 molecules cm^{-3}, respectively. While the nucleation parameterization is obviously playing a big role in the excess particle formation predicted in the FT, this docs not preclude a contribution from the chemistry of the FT yielding too much OH oxidation of SO_2. This could be due to the simplified DMS chemistry scheme, or over abundant OH concentrations, or under-prediction of cloud volumes.

7. Summary and outlook

During the past decade enormous progress has been made in the understanding of the life cycle of aerosols in the global atmosphere. In the previous sections we argued that even a basic understanding of aerosols at a global scale requires the understanding and integration of both microphysical and large-scale dynamics processes. This is primarily because the time scales of aerosol evolution are in many cases longer than the residence time in particular atmospheric compartments. Furthermore, important phenomena such as nucleation and particle wet removal are occurring at the boundary of such compartments.

At present, however, the picture of the aerosol life cycle remains fragmentary. Observational data sets are incomplete and models need to take simple approaches, favouring one aspect of the aerosol (e.g. calculation of aerosol mass) at the expense of others (e.g. calculation of aerosol number). Based on the process understanding, observations and model calculations presented in this paper, the following general statements are tentatively made.

- In areas of strong primary emissions (e.g. sea salt over the ocean at high wind speed, black carbon in industrial areas) secondary aerosol species will preferentially condense on the primary particles, and the number of the resulting mixed particles will be determined by the latter. Even if nucleation does occur, the freshly nucleated particles are expected to coagulate with the primary aerosol within 1–2 d, and eventually the primary particles will again govern the number concentration of the aerosol.
- The removal of the aerosol in the 0.01–1 μm diameter range is mainly through activation in clouds and subsequent precipitation. The efficiency of this wet removal process depends on the size and chemical composition of the aerosol, which itself results from the complex of aerosol dynamic

processes. Although essential in determining the aerosol burden, wet removal is the least understood.

- Assuming a 100% wet removal of particles activated in convective clouds, conditions at the outflow of convective clouds are favourable for nucleation both when the convection occurs over the remote ocean and when it occurs over the polluted continent.
- The convective updraft of DMS over the remote ocean and SO_2 over the polluted continent is sufficient to induce nucleation of H_2SO_4–H_2O in the upper troposphere. Measurements and global budget calculations suggest however that half of the free tropospheric aerosol *mass* might be organic in nature.
- The number concentration of the FT aerosol is governed by the initial nucleation bursts and subsequent coagulation. These processes are expected to be happening in isolated layers originating at the convective cloud outflow.
- Entrainment of FT layers in the boundary layer might constitute a source of particle number, e.g. in case of the clean marine boundary layer, or it might dilute the BL aerosol when the latter is polluted.

In this paper a discussion on the effects of changing emissions or climate change on the global aerosol has not been attempted. A complex web of chemistry–aerosol–climate interactions exists, which makes any calculation of the future aerosol system speculative. The tropospheric aerosol system is nonlinearly dependent on the meteorological and chemical variables that govern its behaviour. Emissions of sea salt, mineral dust, and DMS increase nonlinearly with increasing wind speed. Hence, a future climate characterised by altered wind speeds would experience different amounts of natural aerosols. Nucleation rates depend nonlinearly on temperature, relative humidity, and vapour concentrations. Small changes in temperature and water content can produce large changes in new particle formation rates. However, coagulation serves to damp the effect of such excursions. Aerosols are the nuclei around which clouds form, and cloud properties depend in a nonlinear way on the quantity and composition of available particles. In a future atmosphere in which particle number concentrations are different in number and composition, e.g. by changing emission of primary particles and secondary precursors in industrial areas, cloud properties are likely to be changed. Such changes could result in altered cloud prevalence and convection that, in turn, would affect the aerosol life cycle through modified patterns of deep convection and wet removal. Increase of NO_x imported into background regions can affect the abundance of OH and thus the oxidising capacity of the troposphere, which could affect the rate of oxidation of SO_2 and organics to produce condensable species. Alteration of aerosol levels can affect tropospheric actinic flux and atmospheric photolysis rates that, in turn, change the rate of generation of OH itself.

The life cycle of the tropospheric aerosol is intimately interwoven with the climate–atmospheric chemistry system. To predict how the tropospheric aerosol will respond to climate change will require deep understanding of the dynamics of the climate system itself.

Appendix A

We use as the characteristic time the "e-folding time" τ, i.e., the inverse relative rate of change of a property (aerosol number, volume, diameter, etc.) through a process (coagulation, condensation, etc.). For instance, the characteristic time for change of number by dry deposition is given by

$$\tau_{N,\mathrm{dry}} = \frac{1}{N} \frac{\mathrm{d}N^{-1}}{\mathrm{d}t} = \frac{1}{\lambda_{\mathrm{dry}}},$$

by coagulation by

$$\tau_{N,\mathrm{coa}} = \frac{1}{N} \frac{\mathrm{d}N^{-1}}{\mathrm{d}t} = \frac{1}{KN}.$$

The characteristic times for coagulation given in Table 2 are those for depletion of aerosol number in the various modes by coagulation with Aitken and accumulation mode particles, and with cloud droplets. Coagulation rates are obtained from the coagulation coefficients for monodisperse particles of typical nucleation, Aitken and accumulation mode size with Aitken, accumulation and cloud droplet size particles, respectively (Seinfeld and Pandis, 1998). For the coagulation with cloud droplets the average time an air parcel spends in and outside clouds is taken into account (Lelieveld et al., 1989). Coagulation plays a significant role in the atmosphere when particle number concentrations are high and/or residence times are long. Small particles coagulating with larger ones do not significantly increase the mass of the larger particles, but the process reduces the number of small particles.

The characteristic time for the change in volume by condensational growth is derived assuming that the rate-limiting step is the oxidation of SO_2 in the gas phase. The amount of sulphate produced per unit of time is then distributed over the aerosol modes, weighted by the surface area in each mode and $\tau_{V_i,\mathrm{cond}}$ is obtained as (Van Dingenen et al., 2000)

$$\frac{1}{V_i} \frac{\Delta V_i^{-1}}{\Delta t}.$$

The contribution of aqueous-phase chemistry is estimated by transferring all available gas-phase SO_2 to accumulation mode-SO_4 during 20% of the time, the latter being the average time an air parcel spends inside clouds (Lelieveld et al., 1989). As sulphate is not the only condensing species, the contribution of other secondary material in the growth rate has been taken into account, dividing the growth rate by the fraction of sulphate observed in the aerosol (see Fig. 2). The resulting $\tau_{V_i,\text{cond}}$ can be interpreted as the time needed to produce in the particular compartment the amount of aerosol volume observed in the compartment.

Appendix B

AERO3 is a box model and simulates the aerosol dynamics of three particle populations: pure H_2SO_4–H_2O particles, pure soot particles, and mixed H_2SO_4–H_2O–soot particles. It allows for the internal mixing of the particles by coagulation, condensation, nucleation and in-cloud SO_2 oxidation.

In the model, sulphuric acid is formed from the gas phase by oxidation of SO_2 with OH radicals and can nucleate with water vapour forming H_2SO_4–H_2O droplets, or condense on aerosol particles. The nucleation of H_2O–H_2SO_4 droplets is treated using a parameterization of the homogeneous nucleation rates (Kulmala et al., 1998).

For relative humidity below 100% a H_2O–H_2SO_4 droplet can be considered to be in equilibrium with water vapour. Therefore, the equilibrium of the droplets may be described by the generalized Kelvin equation (Raes et al., 1992). For condensational growth both kinetic and continuum regimes are considered (Fuchs, 1964). Dry deposition is parameterized using a general correlation for particle deposition from turbulent gas flows to completely rough surfaces (Schack et al., 1985).

The H_2SO_4–H_2O and soot particles are assumed to be spheres. Their geometrical diameter is discretized using $D_{p,n} = 0.002 \times 10^{(n/10)}$ μm, where $n = 0, 45$, corresponding to a range from 0.002 μm ($n = 0$) to 63 μm ($n = 45$). The mixed particles are assumed to consist of an insoluble (soot) core surrounded by a soluble (H_2SO_4–H_2O) shell. Their size is are discretisized into 46×46 classes, one dimension for the particle size and one dimension for the size of the insoluble core (Strom et al., 1992). Since the model cannot resolve particles with a water-soluble fraction less than 10%, the corresponding "less hygroscopic" particles have been assimilated into the insoluble population.

The evolution of particle number concentration is described by a system of coupled non-linear equations, one equation for each class (Vignati, 1999). The system of equations is solved using the Euler Backward Iterative method.

To describe the activation process, its assumed that all sulphuric acid is neutralized to ammonium sulphate and the Kohler equations extended to account for an insoluble core are applied (Seinfeld and Pandis, 1998).

Appendix C

The M^3 (Multi-Modal Model) model (Wilson and Raes, 1996) simulates the evolution of a sulphuric acid-water aerosol population, as three log-normally distributed modes (nucleation, Aitken and accumulation), with prescribed standard deviations, but varying number and mass (and therefore mean diameter).

The model simulates the competing processes of nucleation of new particles and condensation of gas-phase sulphate onto existing particles, together with coagulational growth of existing particles, cloud processing and dry and wet removal. M^3 has been compared with the full sectional aerosol model AERO2 (Raes et al., 1992) for a sample of 64 representative boundary layer cases, extracted at random from a climatological sulphur cycle model (Langner and Rodhe, 1991). The 24 h average total aerosol number concentrations and accumulation mode number concentrations predicted by the two models are shown in Fig. 7. The 24 h averages of the accumulation mode number concentrations predicted by M^3 are well within a factor of two of those predicted by AERO2.

In reality, sulphate is only one component of the aerosol burden. Therefore in implementing the M^3 model in the global off-line chemical transport model TM2, it is coupled to a model of the atmospheric black carbon cycle, and a static sea salt burden model. The following key assumptions are made:

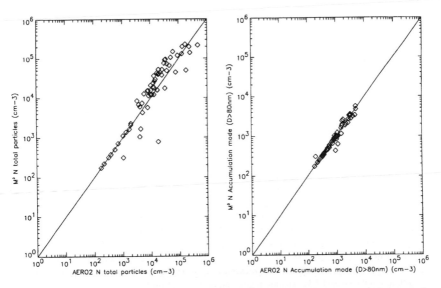

Figure 7. Comparison of the total number concentration and the number of accumulation mode particles calculated by the full sectional model AERO2 (Raes et al., 1992) and the simplified modal model M^3 (Wilson and Raes, 1996).

- Emissions of fossil fuel black carbon are assumed to be purely insoluble, and hence not subject to wet deposition. They age to a mixed and wet depositable aerosol through condensation of H_2SO_4.
- Biomass burning black carbon is a similar condensation sink, but its ageing, which is arbitrarily assumed to be 2.5% of mass per hour is independent of the sulphate condensed, as the co-emitted organic compounds are most likely to perform the ageing role rather than sulphate.

Details on the transport model and the emissions are given in Table 5.

References

Adams, P.J., Seinfeld, J.H., Koch, D.M., 1999. Global concentrations of tropospheric sulphate, nitrate, and ammonium simulated in a general circulation model. Journal of Geophysical Research 104, 13791–13823.

Aitken, J., 1888. On the number of dust particles in the atmosphere. Nature 37, 428–430.

Ammann, M., Kalberer, M., Jost, D.T., Tobler, L., Rossler, E., Piguet, D., Gaggeler, H.W., Baltensperger, U., 1998. Heterogeneous production of nitrous acid on soot in polluted air masses. Nature 395, 157–160.

Andronache, C., Chameides, W.L., Davis, D.D., Anderson, B.E., Pueschel, R.F., Bandy, A.R., Thornton, D.C., Talbot, R.W., Kasibhatla, P., Kiang, C.S., 1997. Gas-to-particle conversion of tropospheric sulphur as estimated from observations in the western North Pacific during PEM-West B. Journal of Geophysical Research 102, 28511–28538.

Ayers, G.P., Ivey, J.P., Gillet, R.W., 1991. Coherence between seasonal cycles of dimethyl sulphide, methanesulphonate and sulphate in marine air. Nature 349, 404–406.

Bates, T.S., Kapustin, V.N., Quinn, P.K., Covert, D.S., Coffman, D.J., Mari, C., Durkee, P.A., De Bruyn, W.J., Saltzman, E.S., 1998. Processes controlling the distribution of aerosol particles in the lower marine boundary layer during the First Aerosol Characterization Experiment (ACE 1). Journal of Geophysical Research 103, 16369–16383.

Bates, T.S., Cline, J.D., Gammon, R.H., Kelly-Hansen, S.R., 1987. Regional and seasonal variation in the flux of oceanic dimethylsulphide to the atmosphere. Journal of Geophysical Research 92, 2930–2938.

Blanchard, D.C., Woodcock, A.H., 1980. The production concentration and vertical distribution of the sea-salt aerosol. Annales of the New York Academy of Sciences 338, 330–347.

Boers, R., Ayers, G.P., Gras, J.L., 1994. Coherence between seasonal variation in satellite-derived cloud optical depth and boundary layer CCN concentrations at a mid-latitude Southern Hemisphere station. Tellus 46B, 123–131.

Calhoun, J.A., Bates, T.S., Charlson, R.J., 1991. Sulphur isotope measurements of submicrometer sulphate aerosol particles over the Pacific Ocean. Geophysical Research Letters 18, 1877–1880.

Capaldo, K.P., Kasibhatla, P., Pandis, S.N., 1999. Is aerosol production within the remote boundary layer sufficient to maintain observed concentrations? Journal of Geophysical Research 104, 3483–3500.

Cerveny, R.S., Balling, R.C., Jr., 1998. Weekly cycles of air pollutants, precipitation and tropical cyclones in the coastal NW Atlantic region. Nature 394, 561–663.

Charlson, R.J., Langner, J., Rodhe, H., Leovy, C.B., Warren, S.G., 1991. Perturbation of the nothern hemisphere radiative balance by backscattering from anthropogenic sulphate aerosols. Tellus 43AB, 152–163.

Charlson, R.J., Lovelock, J.E., Andreae, M.O., Warren, S.G., 1987. Oceanic phytoplankton, atmospheric sulphur, cloud albedo and climate. Nature 326, 655–661.

Chatfield, R.B., Crutzen, P.J., 1984. Sulphur dioxide in remote oceanic air: cloud transport of reactive precursors. Journal of Geophysical Research 89, 7111–7132.

Chin, M., Jacob, D., 1996. Anthropogenic and natural contributions to tropospheric sulphate: a global model analysis. Journal of Geophysical Research 101, 18691–18699.

Christoffersen, T.S., Hjorth, J., Horie, O., Jensen, N.R., Kotzias, D., Molander, L.L., Neeb, P., Ruppert, L., Winterhalter, R., Virkkula, A., Wirtz, K., Larsen, B., 1998. cis-Pinic acid, a possible precursor for organic aerosol formation from ozonolysis of alpha-pinene. Atmospheric Environment 32, 1657–1661.

Chuang, P.Y., Collins, D.R., Pawlowska, H., Snider, J.R., Jonsson, H.H., Brenguier, J.L., Flagan, R.C., Seinfeld, J.H., 2000. CCN measurements during ACE-2 and their relationship to cloud microphysical properties. Tellus 52B, 843–867.

Clarke, A.D., 1993. Atmospheric nuclei in the Pacific midtroposphere: their nature, concentration, and evolution. Journal of Geophysical Research 98, 20633–20647.

Clarke, A.D., 1992. Atmospheric nuclei in the remote free-troposphere. Journal of Atmospheric Chemistry 14, 479–488.

Clarke, A.D., Davis, D., Kapustin, V.N., Eisele, F., Chen, G., Paluch, I., Lenschow, D., Bandy, A.R., Thornton, D., Moore, K., Mauldin, L., Tanner, D., Litchy, M., Carroll, M.A., Collins, J., Albercook, G., 1998a. Particle nucleation in the tropical boundary layer and its coupling to marine sulphur sources. Science 282, 89–92.

Clarke, A.D., Porter, J.N., Valero, F.P.J., Pilewskie, P., 1996a. Vertical profiles, aerosol microphysics, and optical closure during ASTEX: measured and modeled column optical properties. Journal of Geophysical Research 101, 4443–4453.

Clarke, A.D., Uehara, T., Porter, J.N., 1996b. Lagrangian evolution of an aerosol column during the Atlantic Stratocumulus Transition Experiment. Journal of Geophysical Research 101, 4351–4362.

Clarke, A.D., Kapustin, V.N., Eisele, F.L., Weber, R.J., McMurry, P.H., 1999. Particle production near marine clouds: sulphuric acid and predictions from classical binary nucleation. Geophysical Research Letters 26, 2425–2428.

Clarke, A.D., Varner, J.L., Eisele, F., Mauldin, R.L., Tanner, D., Litchy, M., 1998b. Particle production in the remote marine atmosphere: cloud outflow and subsidence during ACE-1. Journal of Geophysical Research 103, 16397–16409.

Coffman, D.J., Hegg, D.A., 1995. A preliminary study of the effect of ammonia on particle nucleation in the marine boundary layer. Journal of Geophysical Research 100, 7147–7160.

Cooke, W., Wilson, J.N., 1996. A global black carbon aerosol model. Journal of Geophysical Research 101, 19395–19409.

Covert, D.S., Kapustin, V.N., Quinn, P.K., Bates, T.S., 1992. New particle formation in the marine boundary layer. Journal of Geophysical Research 92, 20581–20589.

Covert, D.S., Kapustin, V.N., Bates, T.S., Quinn, P.K., 1996. Physical properties of marine boundary layer aerosol particles of the Mid Pacific in relation to sources and meteorological transport. Journal of Geophysical Research 101, 6919–6930.

Dürbeck, T., Gerz, T., 1996. Dispersion of aircraft exhausts in the free atmosphere. Journal of Geophysical Research 101 (20), 26007–26015.

Easter, R.C., Peters, L.K., 1994. Binary homogeneous nucleation: temperature and relative humidity fluctuations, nonlinearity, and aspects of new particle production in the atmosphere. Journal of Applied Meteorology 33, 775–784.

Facchini, M.-C., Mircea, M., Fuzzi, S., Charlson, R.J., 1999. Cloud albedo enhancement by surface-active organic solutes in growing droplets. Nature 401, 257–259.

Feichter, J., Crutzen, P.J., 1990. Parameterization of vertical tracer transport due to deep cumulus convection in a global transport model and evaluation with radon measurements. Tellus 42B, 100–117.

Ferek, R.J., Chatfield, R.B., Andreae, M.O., 1986. Vertical distribution of dimethylsulphide in the marine atmosphere. Nature 320, 514–516.

Friedlander, S.K., 1977. Smoke, Dust and Haze: Fundamentals of Aerosol Behaviour. Wiley, New York.

Fuchs, N., 1964. The Mechanics of Aerosols. Pergamon, New York.

Gage, K.S., McAfee, J.R., Carter, D.A., Ecklund, W.L., Riddle, A.C., Reid, G.C., Balsley, B.B., 1991. Long-term mean vertical motion over the tropical pacific: wind-profiling doppler radar measurements. Science 254, 1771–1773.

Glasius, M., Lahaniati, M., Calogirou, A., Di Bella, D., Jensen, N.R., Hjorth, J., Kotzias, D., Larsen, B.R., 1999. Carboxilic acids in secondary aerosols from oxidation of cyclic monoterpenes by ozone. Environmental Science and Technology, in press.

Gong, S.L., Barrie, L.A., Blanchet, J.-P., 1997. Modeling sea-salt aerosols in the atmosphere 1. Model development. Journal of Geophysical Research 102, 3805–3818.

Griffin, R.J., Dabdub, D., Cocker III, D.R., Seinfeld, J.H., 1999a. Estimate of global atmospheric organic aerosol from oxidation of biogenic hydrocarbons. Geophysical Research Letters 26, 2721–2724.

Griffin, R.J., Cocker III, D.R., Flagan, R.C., Seinfeld, J.H., 1999b. Organic aerosol formation from the oxidation of biogenic hydrocarbons. Journal of Geophysical Research 104, 3555–3567.

Hamill, P., Turco, R.P., Kiang, C.S., Toon, O.B., Whitten, R.C., 1982. An analysis of various nucleation mechanisms for sulphate particles in the stratosphere. Journal of Aerosol Science 13, 561–585.

Hegg, D.A., Covert, D.S., Kapustin, V.N., 1992. Modeling case of particle nucleation in the marine boundary layer. Journal of Geophysical Research 97, 9851–9857.

Hegg, D.A., Radke, L.F., Hobbs, P.V., 1990. Particle production associated with marine clouds. Journal of Geophysical Research 95, 13917–13925.

Heimann, M., Monfray, P., Polian, G., 1990. Modeling the long-range transport of Rn-222 to subantarctic and antarctic areas. Tellus 42B, 83–99.

Heintzenberg, J., 1989. Fine particles in the global troposphere: a review. Tellus 41B, 149–160.

Heintzenberg, J., Covert, D.S., 1990. On the distribution of physical and chemical particle properties in the atmospheric aerosol. Journal of Atmospheric Chemistry 10, 383–397.

Heintzenberg, J., Covert, D.C., Van Dingenen R., 2000. Size distribution and chemical composition of marine aerosols: a compilation and review. Tellus, in press.

Hoppel, W.A., Frick, F.M., Larson, R.E., 1986. Effect of nonprecipitating clouds on the aerosol size distribution in the marine boundary layer. Geophysical Research Letters 13, 125–128.

Hoppel, W.A., Frick, G.M., Fitzgerald, J.W., Wattle, B.J., 1994a. A cloud chamber study of the effect that nonprecipitating water clouds have on the aerosol size distribution. Aerosol Science and Technology 20, 1–30.

Hoppel, W.A., Frick, G.M., Fitzgerald, J.W., Larson, R.E., 1994b. Marine boundary layer measurements of new particle formation and the effects nonprecipitating clouds have on aerosol size distribution. Journal of Geophysical Research 99, 14443–14459.

Hoppel, W.A., Fitzgerald, J.W., Frick, G.M., Larson, R.E., 1990. Aerosol size distributions and optical properties found in the marine boundary layer over the Atlantic Ocean. Journal of Geophysical Research 95, 3659–3686.

Jacob, D.J., Heikes, B.G., Fan, S.-M., Logan, J.A., Mauzerall, D.L., Bradshaw, J.D., Singh, H.B., Gregory, G.L., Talbot, R.W., Blake, D.E., Sachse, G.W., 1996. Origin of ozone and NO_x in the tropical troposphere: a photochemical analysis of aircraft observations over the South Atlantic basin. Journal of Geophysical Research 101, 24235–24250.

Jaenicke, R., 1988. Aerosol physics and chemistry. In: Landolt-Bornstein, Vol. 4. Springer, Berlin, pp. 391–457.

Jaecker-Voirol, A., Mirabel, P., 1989. Heteromolecular nucleation in the sulphuric acid-water system. Atmospheric Environment 23, 2053–2057.

Johnson, D., Osborne, S., Wood, R., Suhre, K., Johnson, R., Businger, S., Quinn, P.K., Wiedensohler, A., Durkee, P.A., Russell, L.M., Andreae, M.O., O'Dowd, C., Noone, K.J., Bandy, B., Rudolph, J., Rapsomanikis, S., 2000. An overview of the Lagrangian experiments undertaken during the North Atlantic regional aerosol characterization experiment (ACE-2). Tellus 52B, 290–320.

Katoshevski, D., Nenes, A., Seinfeld, J.H., 1999. A study of processes governing the maintenance of aerosols in the marine boundary laycr. Journal of Aerosol Science 30, 503–532.

Koch, D.M., Jacob, D., Tegen, I., Rind, D., Chin, M., 1999. Tropospheric sulphur simulation and sulphate direct radiative forcing in the GISS GCM. Journal of Geophysical Research 104, 23799–23822.

Kohler, H., 1936. The nucleus in and the growth of hygroscopic droplets. Transactions of the Faraday Society 32, 1152.

Korhonen, P., Kulmala, M., Laaksonen, A., Viisanen, Y., McGraw, R., Seinfeld, J.H., 1999. Ternary nucleation of H_2SO_4, NH_3, and H_2O in the atmosphere. Journal of Geophysical Research 104, 26349–26353.

Kulmala, M., Laaksonen, A., Charlson, R.J., Korhonen, P., 1997. Clouds without supersaturation. Nature 388, 336.

Kulmala, M., Laaksonen, A., Pirjola, L., 1998. Parameterizations for sulphuric acid/water nucleation rates. Journal of Geophysical Research 103, 8301–8308.

Kulmala, M., Pirjola, L., Makela, J.M., 2000. Stable sulphate clusters as a source of new atmospheric particles. Nature 404, 66–69.

Lai, F.S., Friedlander, S.K., Pich, J., Hidy, G.M., 1972. The self-preserving particle size distribution for Brownian coagulation in the free-molecular regime. Journal of Colloid and Interface Science 39, 395–405.

Langner, J., Rodhe, H., 1991. A global three-dimensional model of the tropospheric sulphur cycle. Journal of Atmospheric Chemistry 13, 225–263.

Lelieveld, J., Crutzen, P.J., 1994. Role of deep cloud convection in the ozone budget of the troposphere. Science 264, 1759–1761.

Lelieveld, J., Crutzen, P.J., Rodhe, H., 1989. Zonal average cloud characteristics for global atmospheric chemistry modelling. Report CM-76, Department of Meteorology University of Stockholm, Stockholm, Sweden.

Li, Z., Williams, A.L., Rood, M.J., 1998. Influence of soluble surfactant properties on the activation of aerosol particles containing inorganic solute. Journal of Atmospheric Sciences 55, 1859–1865.

Liousse, C., Penner, J.E., Chuang, C., Walton, J.J., Eddleman, H., Cachier, H., 1996. A global three-dimensional model study of carbonaceous aerosols. Journal of Geophysical Research 101, 19411–19432.

Maring, H., Savoie, D.L., Izaguirre, M.A., McCormick C., Arimoto, R., Prospero, J.M., Pilinis, C., 2000. Aerosol physical and optical properties and their relation to aerosol composition in the free troposphere at Izaña, Tenerife, Canary Islands during July 1995. Journal of Geophysical Research 105, 14677–14700.

Mason, B.J., 1971. The Physics of Clouds. Clarendon Press, Oxford.

Mauldin III, R.L., Madronich, S., Flocke, S.J., Eisele, F.L., Frost, G.J., Prevot, A.S.H., 1997. New insights on OH: measurements around and in clouds. Geophysical Research Letters 24, 3033–3036.

Meng, Z., Seinfeld, J.H., 1996. Time scales to achieve atmospheric gas-aerosol equilibrium for volatile species. Atmospheric Environment 30, 2889–2900.

Murphy, D.M., Anderson, J.R., Quinn, P.K., McInnes, L.M., Brechtel, F.J., Kreidenweiss, S.M., Middlebrook, A.M., Posfai, M., Thomson, D.S., Buseck, P.R., 1998a. Influence of sea-salt on aerosol radiative properties in the Southern Ocean marine boundary layer. Nature 392, 62–65.

Murphy, D.M., Thomson, D.S., Mahoney, M.J., 1998b. In situ measurements of organics, meteoritic material, mercury, and other elements in aerosols at 5 to 19 kilometers. Science 282, 1664–1669.

Newell, R.E., 1979. Climate and the ocean. American Scientist 67, 405–416.

Newell, R.E., Zhu, Y., Browell, E.V., Read, W.G., Waters, J.W., 1996. Walker circulation and tropical upper tropospheric water vapor. Journal of Geophysical Research 101, 1961–1974.

Newell, R.E., Thouret, V., Cho, J.Y.N., Stoller, P., Marenco, A., Smit, H.G., 1999. Ubiquity of quasi-horizontal layers in the tropophere. Nature 398, 316–319.

Nilsson, E.D., Kulmala, M., 1998. The potential for atmospheric mixing processes to enhance the binary nucleation rate. Journal of Geophysical Research 103, 1381–1389.

Novakov, T., Penner, J.E., 1993. Large contribution of organic aerosols to cloud-condensation-nuclei concentrations. Nature 365, 823–826.

Novakov, T., Hegg, D.A., Hobbs, P.V., 1997. Airborne measurements of carbonaceous aerosols on the East Coast of the United States. Journal of Geophysical Research 102, 30023–30030.

O'Dowd, C.D., Smith, M.H., 1993. Physicochemical properties of aerosols over the Northeast Atlantic: evidence for wind-speed-related sub-micron sea-salt aerosol production. Journal of Geophysical Research 98, 1137–1149.

Odum, J.R., Hoffman, T., Bowman, F., Collins, D., Flagan, R.C., Seinfeld, J.H., 1996. Gas/Particle partitioning and secondary aerosol yields. Environmental Science and Technology 30, 2580–2585.

Oort, A.H., 1983. Global atmospheric circulation statistics 1958–1973. NOAA Professional Paper No 14. US Government Printing Office, Washington, DC.

Ostrom, E., Noone, K.J., 2000. Vertical profiles of aerosol scattering and absorption measured in-situ during the North Atlantic aerosol characterization experiment (ACE-2). Tellus 52B, 526–545.

Pawlowska, H., Brenguier, J.-L., 2000. Microphysical properties of stratocumulus clouds during ACE-2. Tellus 52B, 868–887.

Perry, K.D., Hobbs, P.V., 1994. Further evidence for particle nucleation in clean air adjacent to marine cumulus clouds. Journal of Geophysical Research 99, 22803–22818.

Pruppacher, H.R., Klett, J.D., 1980. Microphysics of Clouds and Precipitation. Reidel, Dordrecht.

Putaud, J.-P., Van Dingenen, R., Mangoni, M., Virkkula, A., Raes, F., Maring, H., Prospero, J., Swietlicki, E., Berg, O., Hillamo, R., Makela, T., 2000. Chemical mass closure and assessment of the origin of the sub-micron aerosol in the marine boundary layer and the free troposphere at Tenerife during ACE-2. Tellus 52B, 141–168.

Raes, F., Wilson, J., Van Dingenen, R., 1995. Aerosol dynamics and its implication for the global aerosol climatology. In: Charson, R.J., Heintzenberg, J. (Eds.), Aerosol Forcing of Climate. Wiley, New York.

Raes, F., Van Dingenen, R., 1992. Simulations of condensation and cloud condensation nuclei from biogenic SO_2 in the remote marine boundary layer. Journal of Geophysical Research 97, 12901–12912.

Raes, F., Van Dingenen, R., 1995. Comment on "The relationship between DMS flux and CCN concentrations in remote marine regions" by S.N. Pandis, L.M. Russell, and J.H. Seinfeld. Journal of Geophysical Research 100, 14355–14356.

Raes, F., 1995. Entrainment of free tropospheric aerosols as a regulating mechanism for cloud condensation nuclei in the remote marine boundary layer. Journal of Geophysical Research 100, 2893–2903.

Raes, F., Janssens, A., Van Dingenen, R., 1986. The role of ion-induced aerosol formation in the lower atmosphere. Journal of Aerosol Science 17, 466–470.

Raes, F., Van Dingenen, R., Saltelli, A., 1992. Modelling the dynamics of H_2SO_4–H_2O aerosols with $AERO_2$: model description, uncertainty analysis and experimental validation. Journal of Aerosol Science 23, 759–771.

Raes, F., Van Dingenen, R., Cuevas, E., Van Velthoven, P.F.J., Prospero, J.M., 1997. Observations of aerosols in the free troposphere and marine boundary layer of the subtropical Northeast Atlantic: discussion of processes determining their size distribution. Journal of Geophysical Research 102, 21315–21328.

Raes, F., Van Dingenen, R., Wilson, J., Saltelli, A., 1993. Cloud condensation nuclei from dimethyl sulphide in the natural marine boundary layer: remote vs. in-situ production. In: Restelli, G., Angeletti, G. (Eds.), DMS: Ocean, Atmosphere and Climate. Kluwer Academic Publisher, Dordrecht, pp. 311–322.

Rodhe, H., 1983. Precipitation scavenging and tropospheric mixing. In: Pruppacher et al. (Eds.), Precipitation Scavenging, Dry Deposition, and Resuspension, pp. 719–728.

Rosenfeld, D., 1999. TRMM observed first direct evidence of smoke from forest fires inhibiting rainfall. Geophysical Research Letters 26, 3105–3108.

Russell, L.M., Pandis, S.N., Seinfeld, J.H., 1994. Aerosol production and growth in the marine boundary layer. Journal of Geophysical Research 99, 20989–21003.

Saxena, P., Hildemann, L.M., McMurry, P.H., Seinfeld, J.H., 1995. Organics alter hygroscopic behaviour of atmospheric particles. Journal of Geophysical Research 100, 18755–18770.

Saxena, P., Hildemann, L., 1996. Water-soluble organics in atmospheric particles: a critical review of the literature and application of thermodynamics to identify candidate compounds. Journal of Atmospheric Chemistry 24, 57–109.

Schack, C.J., Pratsinis, S.E., Friedlander, S.K., 1985. A general correlation for deposition of suspended particles from turbulent gases to completely rough surfaces. Atmospheric Environment 19, 953–960.

Schulz, M., Balkanski, Y.J., Guelle, W., Dulac, F., 1998. Role of aerosol size distribution and source location in a three-dimensional simulation of a Saharan dust episode tested against satellite-derived optical thickness. Journal of Geophysical Research 103, 10579–10592.

Schwikowski, M., Seibert, P., Baltensberger, U., Gaggeler, H.W., 1995. A study of an outstanding Saharan dust event at the high-alpine site Jungfraujoch. Switzerland, Atmospheric Environment 29, 1829–1842.

Seinfeld, J.H., Pandis, S.N., 1998. Atmospheric Chemistry and Physics: from Air Pollution to Climate Change. Wiley, New York.

Shulman, M.L., Jacobson, M.C., Charson, R.J., Synovec, R.E., Young, T.E., 1996. Dissolution behaviour and surface tension effects of organic compounds in nucleating cloud droplets. Geophysical Research Letters 23, 277–280.

Sprengard-Eichel, C., Krämer, M., Schütz, L., 1998. Soluble and insoluble fractions of urban, continental and marine aerosol. Journal of Aerosol Science 29, S175–S176.

Strom, J., Okada, K., Heintzenberg, J., 1992. On the state of mixing of particles due to Brownian coagulation. Journal of Aerosol Science 23, 467–480.

Svenningsson, B., Hansson, H.-C., Wiedensohler, A., Noone, K.J., Ogren, J., Hallberg, A., Colville, R., 1994. Hygroscopic growth of aerosol particles and its influence on nucleation scavenging in cloud: experimental results from Kleiner Feldberg. Journal of Atmospheric Chemistry 19, 129–152.

Swietlicki, E., Zhou, J., Covert, D.S., Hameri, K., Busch, B., Vakeva, M., Dusek, U., Berg, O.H., Wiedensohler, A., Aalto, P., Makela, J., Martinsson, B.G., Papaspiropoulos, G., Mentes, B., Frank, G., Stratmann, F., 2000. Hygroscopic properties of aerosol particles in the north-eastern Atlantic during ACE-2. Tellus 52B, 201–227.

Tegen, I., Hollrig, P., Chin, M., Fung, I., Jacob, D., Penner, J., 1997. Contribution of different aerosol species to the global aerosol extinction optical thickness: estimates from model results. Journal of Geophysical Research 102, 23895–23915.

Thornton, D.C., Bandy, A.R., Blomquist, B.W., Bradshaw, J.D., Blake, D.R., 1997. Vertical transport of sulphur dioxide and dimethyl sulphide in deep convection and its role in new particle formation. Journal of Geophysical Research 102, 28501–28509.

Turco, R.P., Zhao, J.-X., Yu, F., 1998. A new source of tropospheric aerosols: ion-ion recombination. Geophysical Research Letters 25, 635–638.

Van Dingenen, R., Raes, F., Jensen, N.R., 1995. Evidence for anthropogenic impact on number concentration and sulphate content of cloud-processed aerosol particles over the North-Atlantic. Journal of Geophysical Research 100, 21057–21067.

Van Dingenen, R., Raes, F., Putaud, J.-P., Virkkula, A., Mangoni, M., 1999. Processes determining the relationship between aerosol number and non-seasalt sulphate mass concentrations in the clean and perturbed marine boundary layer. Journal of Geophysical Research 104, 8027–8038.

Van Dingenen, R., Virkkula, A.O., Raes, F., Bates, T.S., Wiedensohler, A., 2000. A simple non-linear analytical relationship between aerosol accumulation number and sub-micron volume, explaining their observed ratio in the clean and polluted marine boundary layer. Tellus 52B, 439–451.

Vignati, E., 1999. Modelling interactions between aerosols and gaseous compounds in the polluted marine atmosphere. Ph.D. Thesis, RISOE National Laboratory, Report No. Riso-R-1163(EN), p. 133.

Virkkula, A., Van Dingenen, R., Raes, F., Hjorth, J., 1999. Hygroscopic properties of aerosol formed by oxidation of limonene, alpha-pinene and beta-pinene. Journal of Geophysical Research 104, 3569–3579.

Vogt, R., Crutzen, P.J., Sander, R., 1996. A mechanism for halogen release from sea-salt aerosol in the remote marine boundary layer. Nature 383, 327–330.

Wang, C., Crutzen, P.J., Ramanathan, V., Williams, S.F., 1995. The role of a deep convective storm over the tropical Pacific Ocean in the redistribution of atmospheric chemical species. Journal of Geophysical Research 100, 11509–11516.

Wang, P.-H., Rind, D., Trepte, C.R., Kent, G.S., Yue, G.K., Skeens, K.M., 1998. An empirical model study of the tropospheric meridional circulation based on SAGE II observations. Journal of Geophysical Research 103, 13801–13818.

Weber, R.J., Marti, J.J., McMurry, P.H., Eisele, F.L., Tanner, D.J., Jefferson, A., 1996. Measured atmospheric new particle formation rates: implications for nucleation mechanisms. Chemical Engineering Communications 151, 53–64.

Weber, R.J., McMurry, P.H., Eisele, F.L., Mauldin, L., Tanner, D., 1998. Rapid growth of freshly formed nanoparticles in the remote troposphere. Journal of Aerosol Science 29, S179–S180.

Went, F.W., 1960. Blue hazes in the atmosphere. Nature 187, 641–643.

Wexler, A.S., Seinfeld, J.H., 1990. The distribution of ammonium salts among a size and composition dispersed aerosol. Atmospheric Environment 24A, 1231–1246.

Whitby, K.T., 1978. The physical characteristics of sulphur aerosols. Atmospheric Environment 12, 135–159.

Wilson, J.J.N., Raes, F., 1996. M3 a multi modal model for aerosol dynamics. In: Kulmala, D., Wagner, D. (Eds.), Proceedings of the 14th International Conference on Nucleation and Atmospheric Aerosols. Pergamon, Oxford, pp. 458–461.

Wu, Z., Newell, R.N., Zhu, Y., Anderson, B.E., Browell, E.V., Gregory, G.L., Sachse, G.W., Collins, Jr., J.E., 1997. Atmospheric layers measured from the NASA DC-8 during PEM-West B and comparison with PEM-West A. Journal of Geophysical Research 102, 28353–28365.

Yu, J., Cocker, D.R., Griffin, R.J., Flagan, R.C., Seinfeld, J.H., 1999. Gas-phase ozone oxidation of monoterpenes: gaseous and particulate products. Journal of Atmospheric Chemistry 34, 207–258.

Zhang, Y., Seigneur, C., Seinfeld, J.H., Jacobson, M., Clegg, S.L., Binkowski, F.S., 2000. A comparative review of inorganic aerosol thermodynamics equilibrium modules: similarities, differences, and their likely causes. Atmospheric Environment 34, 117–137.

Zhang, X.Q., McMurry, P.H., Hering, S.V., Casuccio, G.S., 1993. Mixing characteristics and water content of sub-micron aerosols measured in Los Angeles and at the Grand Canyon. Atmospheric Environment 27A, 1593–1607.

Zimmermann, P.H., 1984. Ein dreidimensionales numerisches Transportmodell fur atmospharische Spurenstoffe. Thesis, Univeristy of Mainz, FRG.

Air Pollution Science for the 21st Century
J. Austin, P. Brimblecombe and W. Sturges, editors
© 2002 Elsevier Science Ltd. All rights reserved.

565

Chapter 19

New Directions: Particle air pollution down under

Lidia Morawska

School of Physical Sciences, Queensland University of Technology,
2 George Street, Brisbane, Q 4001, Australia
E-mail: l.morawska@qut.edu.au

Flying across the oceans separating the continents of the Southern Hemisphere, the traveller often develops almost a physical feeling of distance and space. It is not only nature that astonishes with its grandness: human problems and the catastrophic effects of human actions are on a grand scale as well. These include air pollution and its two major sources: fires and motor vehicles. When flying between Australia and Asia, it is not uncommon to find oneself surrounded by a large plume of smoke blocking out the sun. Blue sky is an uncommon view over the megacities of the Southern Hemisphere too.

Fires raging through forests, grasslands or savannahs of South America, Asia, Africa and Australia blanket the land with dense smoke for weeks at a time. The smoke can travel hundreds and thousands of kilometres, and be detected in neighbouring countries or even continents. Fires may be lit deliberately, sometimes maliciously, or can be spontaneous. Australian eucalyptus forests, for example, need fire to regenerate and to grow. According to the World Health Organisation's "Health Guidelines for Episodic Vegetation Fire Events" (WHO, Geneva, 1999), the very small particles generated by such fires are a significant threat to human health. But how can these fires be controlled and the particle emissions reduced? At present there is no obvious solution, and prevention measures are not always possible or effective.

Emissions from motor vehicles are another major area of concern, particularly in the megacities of South America and Asia. Visitors unused to such conditions invariably feel discomfort, and many even develop clinical symptoms. Such is the pervasive nature of pollution that judges at a children's painting competition in Mexico City were astonished to find that all the children had painted the sky grey. Huge numbers of vehicles on the streets, old technology, lack of maintenance and dirty fuels, all contribute to emission levels that are unacceptable by the standards of developed countries. An annual mean airborne particulate matter concentration on the order of 500 μg m^{-3} has been

First published in Atmospheric Environment 35 (2001) 1711–1712

reported for a number of cities in India and South and Central America (WHO, Healthy Cities Air Management Information System, AMIS, Geneva, 1998). Peak concentrations would almost certainly be several times higher than this, and would probably compare with the particulate matter and SO_2 concentrations of about 1800 $\mu g \, m^{-3}$ recorded during the "killer" smogs of London in 1952, in which 4000 people died from pollution-related causes.

Since the 1950's, however, pollution levels have dropped dramatically in the Northern Hemisphere, notably in terms of particulate matter and SO_2. Nevertheless, there is one component of this pollution that has not declined, and that is the fine/ultrafine particulate fraction arising from combustion. Whilst improvements in combustion have resulted in a decrease of the total mass of particles emitted, emissions of the smallest particles have increased in number. These particles remain in the air for long periods of time, usually carry most of the toxins, trace elements and compounds, and have the ability to penetrate to the deepest parts of the respiratory tract. They also have an affect on the radiation balance of the atmosphere, scattering the incoming sunlight and absorbing the outgoing infrared radiation. The net effect on surface temperature, however, is still a matter of debate.

Fig. 1 compares particle number density and particulate matter mass (PM_{10}) in Birmingham (England), Brisbane (Australia), and Santiago de Chile. Admittedly different instruments and different sampling periods and frequencies were used, so comparisons must be taken with caution. Despite these uncertainties, however, the picture emerging is that the two Southern Hemisphere cities have relatively low particle numbers compared with the total mass of particulate, whereas the opposite is true for Birmingham. Particulate matter in Brisbane, and especially Santiago, is plainly more dominated by coarse particles than in Birmingham, whilst the latter is characterised by large numbers of fine or ultrafine particles.

With the work that is being initiated or considered in many countries in Asia and South and Central America towards advancement in motor vehicle technologies, it can be expected that the total mass emission levels of particulate matter will decline, as it has done in the Northern Hemisphere. Expected to follow the improvements to the combustion process is, however, an increase in the number of very small particles. It might be speculated, therefore, that cities such as Santiago, with extreme particle mass concentrations now, may experience much higher particle number concentrations – with consequent health effects – some years from now.

In developed countries, regulators say that before considering any legislation they need to know whether the health effects associated with exposure to particles are really caused by the particles themselves. They want to know the mechanisms by which the reported health effects occur, not only the statistical associations. But, for a number of years now, doctors have provided

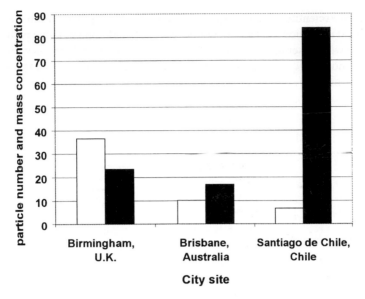

Figure 1. Particle number and mass concentrations in Birmingham, England (R.M. Harrison et al., Atmospheric Environment, Vol. 33 (1999) 309–321); Brisbane, Australia (L. Morawska et al., Atmospheric Environment, Vol. 32 (1998) 2467–2478); and Santiago de Chile (A. Trier, Atmospheric Environment, Vol. 31 (1997) 909–914). Open bars are particle number densities (10^3 particles cm^{-3}), closed bars are PM_{10} ($\mu g\, m^{-3}$).

only contradictory hypotheses. In reality the investigation of such small particles, with size spectra extending down to molecular dimensions, constitutes a particularly difficult scientific and technical challenge (R.L. Maynard, New Directions: Reducing the toxicity of vehicle exhaust, Atmospheric Environment, 34 (2000) 2667–2668).

Nevertheless, measures to curb fine particle emissions have continued unabated, with cleaner diesel fuels and experimental particle traps for vehicle exhaust, despite the costs and technological difficulties involved. We may, in fact, end up removing the particles before either understanding them or regulating them.

So what of countries in the Southern Hemisphere? Will they be willing and able to invest in technologies for removing particles? Or in the absence of knowledge on the mechanisms causing deleterious health and environmental effects, will they simply ignore them? If they ignore the issue now, will it become a problem in the future?

Given that answers are still lacking in the developed world, we can only speculate on the prospects for the Southern Hemisphere. What has to be realised, however, is that the direction taken by those countries with the resources to improve scientific understanding, to invest in technology to eliminate emissions, and introduce legislation, will have a profound impact on the path taken by those with lesser resources. It should also be remembered that, as with many other atmospheric pollutants, while health effects are normally experienced close to emission sources, effects on climate are of global significance. That is how the Southern Hemisphere shares its grand problems with the North.

Chapter 20

Review and intercomparison of operational methods for the determination of the mixing height

Petra Seibert[a], Frank Beyrich[b], Sven-Erik Gryning[c], Sylvain Joffre[d], Alix Rasmussen[e], Philippe Tercier[f]

[a] *Institute of Meteorology and Physics, University of Agricultural Sciences Vienna (BOKU), Türkenschanzstr. 18, A-1180 Wien, Austria*
E-mail: seibert@boku.ac.at

[b] *Meteorologisches Observatorium Lindenberg, Deutscher Wetterdienst, D-15864 Lindenberg, Germany*
E-mail: frank.beyrich@dwd.de

[c] *Meteorology and Wind Energy Department, Risø National Laboratory, DK-4000 Roskilde, Denmark*
E-mail: sven-erik.gryning@risoe.dk

[d] *Finnish Meteorological Institute, P.B. 503, FIN-00101 Helsinki, Finland*
E-mail: sylvain.joffre@fmi.fi

[e] *Danish Meteorological Institute, Lyngbyvej 100, DK-2100 Copenhagen, Denmark*
E-mail: ali@dmi.min.dk

[f] *Swiss Meteorological Institute, Les Invuardes, CH-1530 Payerne, Switzerland*
E-mail: pht@sap.sma.ch

Abstract

The height of the atmospheric boundary layer (ABL) or the mixing height (MH) is a fundamental parameter characterising the structure of the lower troposphere. Two basic possibilities for the practical determination of the MH are its derivation from profile data (measurements or numerical model output) and its parameterisation using simple equations or models (which only need a few measured input values). Different methods suggested in the literature are reviewed in this paper. The most important methods have been tested on data sets from three different sites in Europe (Cabauw – NL, Payerne – CH, Melpitz – D). Parcel and Richardson number methods applied to radiosonde profiles and the analysis of sodar and wind profiler data have been investigated. Modules for MH determination implemented in five currently used meteorological preprocessors for dispersion models have been tested, too. Parcel methods using a revised coefficient for the excess temperature and Richardson number methods using a surface excess

First published in Atmospheric Environment 34 (2000) 1001–1027

temperature worked well under convective conditions. Under stable conditions, the inherent difficulties call for a combination of several methods (e.g., mast and sodar). All the tested parameterisation schemes showed deficiencies under certain conditions, thus requiring more flexible algorithms able to take into account changing and non-classical conditions. Recommendations are formulated regarding both the analysis of profile measurements and the use of parameterisations and simple models, and suggestions for the preprocessor development and for future research activities are presented.

1. Introduction

Air quality assessments at the local or regional scale are required for a variety of purposes, e.g., emission control, air quality forecasts and implementation of legislation, and can be carried out using different types of models. A key input to these models are the meteorological data required to compute the transport, dispersion and removal of pollutants. The dispersion of pollutants depends mostly on atmospheric turbulence, but turbulence measurements per se are not routinely performed by meteorological services. Thus, dispersion characteristics are either inferred from basic meteorological parameters such as wind, temperature and radiation using parameterisation schemes or from the output of numerical models.

On the background of increasing internationalisation of policy and trade exchange there is a need for the harmonisation of legislation (e.g., on air quality), so that assessment methods applied in different countries are transparent and comparable. As a first step in that direction, an intercomparison of meteorological preprocessors embedded in some air quality models used in Europe was conducted within the COST[1] Action 710. This action was divided into 4 working groups: surface energy balance, mixing layer height, vertical profiles of mean and turbulent quantities, and complex terrain (Fisher et al., 1998). This paper is based on the comprehensive final report of the working group dealing with the mixing layer height (Seibert et al., 1998).

2. The concept of the mixing layer and definition of its height

Substances emitted into the atmospheric boundary layer (ABL) are gradually dispersed horizontally and vertically through the action of turbulence, and finally become completely mixed over this layer if sufficient time is given and if there are no significant sinks. Therefore, it has become customary in air pollution meteorology to use the term "mixed layer" or "mixing layer". Since

---

[1] COST: European Co-operation in the Field of Scientific and Technological Research.

under stable conditions complete mixing is often not reached, the term "mixing layer" seems preferable, because it emphasises more the process than the result. Obviously, the mixing layer coincides with the ABL if the latter is defined as the turbulent domain of the atmosphere adjacent to the ground. However, other definitions of the ABL have also been used which may, e.g., include the domain influenced by nocturnal radiative exchange processes.

The height h of the mixing layer (the mixing height MH) is a key parameter for air pollution models. It determines the volume available for the dispersion of pollutants and is involved in many predictive and diagnostic methods and/or models to assess pollutant concentrations, and it is also an important parameter in atmospheric flow models. The MH is not measured by standard meteorological practices, and moreover, it is often a rather unspecific parameter whose definition and estimation is not straightforward.

The practical and theoretical problems associated with the determination of the MH, and sometimes even its definition, are reflected in the numerous definitions found in the literature (see Stull, 1988; Garratt, 1992; Seibert et al., 1998). It seems also that the MH definitions of different authors have to be seen in the context of the data available to them.

The definition we have adopted as a general guideline for our work is: *The mixing height is the height of the layer adjacent to the ground over which pollutants or any constituents emitted within this layer or entrained into it become vertically dispersed by convection or mechanical turbulence within a time scale of about an hour.*

In order to proceed from this general definition to practical realisations, it is necessary to consider separately the structure of the stable boundary layer (SBL) and of the convective boundary layer (CBL). For the CBL, an important feature is the entrainment layer (Gryning and Batchvarova, 1994), a zone which is not well-mixed and where turbulence intensity declines towards its top. The above definition corresponds to the top of the entrainment layer. The most widespread definition, however, is the value z_i, defined as the height where the heat flux gradient reverses its sign. It is usually applied for scaling purposes and it is the definition closest to the thermodynamical CBL height definition in a zero-order jump model (i.e., where the entrainment layer thickness is neglected). We will use here this definition, but one should be aware, e.g., in the specification of turbulence parameterisations for dispersion models, that turbulence extends beyond z_i.

The SBL can be divided into two layers: a layer of continuous turbulence and an outer layer of sporadic or intermittent turbulence. Under very stable conditions the layer of sporadic turbulence may extend to the ground. Since it is notoriously difficult to measure sporadic turbulence, and even more to develop a related scaling theory, the scaling height, h, used for the SBL generally is the layer of continuous turbulence. As in the convective case, however, this does not mean that turbulence is strictly confined to the region below h.

The asymptotic case with the heat flux approaching zero from either stable or unstable stratification is often termed neutral boundary layer. It must be kept in mind, however, that even in this case stable stratification will prevail above the ABL, which limits the validitiy of idealised concepts based on an infinitely deep neutral boundary layer. In this situation, like in the stable boundary layer, wind shear is the main source of turbulence and therefore in this paper it is subsumed under the SBL.

There are situations where these definitions have to be carefully discussed and possibly modified, e.g., patches of sporadic turbulence caused by the breaking of gravity waves, regions of turbulence generated by wind shear due to low-level jets, situations with strong non-stationarity (e.g., the evening transition period), the presence of clouds (e.g., cloud venting of the CBL, or in frontal zones), situations with significant horizontal advection, and in complex terrain regions.

We encourage researchers to pay attention to which definitions of the MH or ABL height their work is based upon, and to specify it clearly.

3. Methods for the determination of the mixing height

This section contains a literature review of practical methods for the determination of the MH. At first, we describe methods based on profile measurements and assess existing comparisons among them (Section 3.1). Then we discuss methods based on parameterisations and simple models which require only operationally available input data, from measurements or from numerical weather prediction (NWP) models (Section 3.2).

3.1. Mixing height determination from profile measurements

3.1.1. Radiosoundings

Radiosoundings are the most common source of data for operational determination of the MH. They are widely distributed, and the data are continuously quality controlled. On the other hand, at most stations they are only taken twice daily at specified synoptic times (00 UTC, 12 UTC). Consequently, in Europe the soundings can often be used as a reference for comparison with modelled MHs only around midnight and noon. Other limitations of radiosoundings are the poor vertical resolution of standard aerological data with respect to boundary layer studies, the smoothing due to the sensor lag constant bounded with the high ascent rate of the sonde, and the fact that a sounding gives only a "snapshot"-view of the ABL structure.

MH estimations based on (standard) radiosonde data may result in quite high uncertainty (e.g., Russell et al., 1974; Hanna et al., 1985; Martin et al., 1988). Specific problems occur in the stable (nocturnal) boundary layer since no universal relationship seems to exist between the profiles of temperature, humidity or wind and turbulence parameters (heat or momentum fluxes, turbulent kinetic energy TKE). The interpretation of profiles thus is not straightforward and several criteria have been used (see Table 1 in Seibert ct al., 1998).

MH estimation from profiles obtained with tethered balloons or aircraft, which may include turbulence and/or trace gas concentration profiles, is, in principle, not very different from the analysis of radiosonde data. The operation of both systems, however, is very expensive and therefore is not suited for routine applications. Temperature (and trace gas) profiles taken by commercial aircraft during take-off and landing may become a useful data source in the future.

3.1.1.1. Subjective methods. Radiosonde temperature and wind profiles in the lower part of the atmosphere are often used for a subjective estimation of the MH. Under convective conditions, the MH is often identified with the base of an elevated inversion or stable layer, or as the height of a significant reduction in air moisture, often accompanied by wind shear. Some authors recommend to take the inversion base altitude increased by half of the depth of the inversion layer as the characteristic CBL height (Stull, 1988).

3.1.1.2. Objective methods. Holzworth (1964, 1967, 1972) and others have developed objective methods to simplify and homogenise the estimation of the MH under convective conditions. The basic idea of the "Holzworth" or "parcel method" is to follow the dry adiabate starting at the surface with the measured or expected (maximum) temperature up to its intersection with the temperature profile from the most recent radiosounding. It determines the MH as the equilibrium level of a hypothetical rising parcel of air representing a thermal. However, this method strongly depends on the surface temperature, and a high uncertainty in the estimated MH value may result in situations without a pronounced inversion at the CBL top. Some authors have noticed that the MH determined by the "Holzworth method" is not strongly correlated with observed trace gas concentrations (e.g., Aron, 1983; Jones, 1985).

Different refinements of this simple scheme have been suggested, to account for temperature advection, subsidence and other effects (e.g., Miller, 1967; Garrett, 1981). They differ in how the temperature of this air parcel is found, and in the thermodynamical variable used to define the equilibrium level. An advanced parcel method has been proposed by Beljaars and Betts

Table 1. Measuring platforms and their qualification for MH determination

Method	Advantages	Shortcomings
Direct measuring techniques/sensor platforms		
Radiosonde	• Routine ascents for many years all over the world, therefore especially suited for climatological studies • Measured data transmitted via international communication networks with very short time delay, therefore well suited for operational use • Compatibility with measurements in the free atmosphere	• Crossing the ABL along a slanted path within a few minutes, provides a "snapshot"-like profile • Limited height resolution of routine ascents • Operationally only 2–4 soundings per day at fixed times, even during field campaigns 1.5–3 h as closest interval • Tracking problems at low levels (site-dependent) may affect wind profiles
Tethered balloon	• Ascent velocity can be chosen according to the desired vertical resolution • Turbulence and trace gas concentration measurements possible	• Limited to field campaigns, no unmanned operation • Synchronous profile measurement difficult • Limited measurement range, usually below 500 m • Not possible in cases of high wind speed or strong convection
Mast	• Installation of a large number of different sensor types possible including detailed turbulence measurements • Continuous operation • Good resolution of the lowest layers	• Very high installation/operation costs, increasing with height • Limited range: 50 to at most 300 m • High vertical resolution requires a high number of sensors (increasing costs)
Aircraft and remote sensing techniques		
Aircraft	• Possibility to operate many different sensors, including mean meteorology, chemistry, and turbulence sensors, as well as remote sensing systems • Provides spatial information, well suited for mesoscale studies	• High costs, only for field campaigns • Operation mostly limited to daylight hours • Lowest flight level subject to restrictions (security)
Doppler weather radar/wind profiler	• Ground based and aircraft based operation possible (for radar only) • Alternating RHI/PPI-Scan for 3D-studies (for radar only) • High sampling rate and continuous operation	• Lowest range normally not below 200 m • Limited vertical resolution (50–250 m) • Expensive • Weather radars do not work well in clear air • Interpretation not always straightforward

Table 1. (Continued)

Method	Advantages	Shortcomings
Lidar	• Ground-based and aircraft-based operation possible • Alternating RHI/PPI-Scan for 3D-studies • High sampling rate • Return signals originate directly from aerosols ("pollution")	• Expensive • Unattended operation often not possible for safety reasons • Limited range resolution and lowest range gate • Tracer necessary (gas, aerosol) • Interpretation sometimes ambiguous
Sodar	• Relatively simple, not very expensive: suited for unmanned long-term operation • High temporal and vertical resolution • Minisodars allow probing of shallow SBL	• Limited sounding range (500–1000 m) • Sensitivity to environmental noise • Noise contamination to the environment • Interpretation requires experience, sometimes ambiguous

(1992) and has been also applied (with slightly different values of the constants) by Wotawa et al. (1996) and in this paper (see Section 4.1).

Another popular approach are the bulk Richardson number methods (e.g., Troen and Mahrt, 1986; Vogelezang and Holtslag, 1996). They differ mainly in the choice of the level for the near-surface temperature and wind, the parameterisation of shear production of turbulence in the surface layer, and the consideration of an excess surface temperature under convective conditions. Parcel methods can be understood as a simplification of the Ri-number methods where the shear contribution is neglected. Thus they are only suited for unstable conditions.

Methods based on conserved variables (e.g., mixing ratio, equivalent-potential temperature, etc.) permit analyses of air mass structures and vertical mixing. Betts and Albrecht (1987) proposed respective criteria on the basis of averaged and smoothed profiles. For individual profiles, a refinement of these criteria is necessary to differentiate the main features from secondary stratification (e.g., Tercier et al., 1995).

3.1.2. Remote sounding systems

Remote sounding systems (lidars, sodars, RASS, wind profiling radars; see Clifford et al. (1994) for an overview) are more and more introduced into operational application. They provide an interesting alternative for MH estimations. The basic advantages of remote sounding systems are the continuous operation and that they do not cause any modification of the investigated flow.

The sodar is one of the simpler and less expensive remote sounding systems, making it well suited for routine operation. Sodar signals are scattered by temperature inhomogeneities characterised by the structure parameter of the acoustic refractive index, C_n^2. Vertical profiles of C_n^2 and thus the sodar backscatter intensity show typical features under stable and convective conditions (e.g., a strong decrease above a region of less variable C_n^2 in a SBL with significant shear-produced turbulence, or an elevated maximum at the top of a CBL) which can be used to derive the MH (for more details, see Beyrich, 1997). In addition, sodar systems with Doppler capability allow determination of the mean wind and vertical velocity variance profiles which may also be employed for MH determination. Methods and algorithms to derive the MH from sodar data are compiled in Beyrich (1997) and Seibert et al. (1998). However, the vertical range of most sodars is limited to a maximum of about 1 km, but often to only a few hundred metres. The lowest range gate of typical sodars is around 40 m. This is lower than for the other remotes sounding systems, but it can still be too high for very shallow SBLs. In this case, minisodars (Asimakopoulos et al., 1996) are an attractive option.

Lidars allow the measurement of aerosol or trace gas concentration profiles and may therefore be considered to provide direct measurements of the MH. Since the top of a convectively mixed layer is often associated with strong gradients of the aerosol content, a simple aerosol backscatter lidar seems suited to determine the convective MH. However, interpreting data from aerosol lidars is often not straightforward, because the detected aerosol layers are not always the result of ongoing vertical mixing, but may originate from advective transport or past accumulation processes (e.g., Russell et al., 1974; Coulter, 1979; Baxter, 1991; Batchvarova et al., 1999). Under stable conditions, problems in estimating the MH from lidar data can arise from the weak vertical gradients in the aerosol content. Moreover, in the evening, it usually takes some time until a sufficiently clear discontinuity in the backscatter intensity profile develops at the top of the SBL, within the previously well-mixed layer (e.g., Russell et al., 1974).

The boundary layer wind profiler seems to be a very promising device for direct and continuous measurement of the MH in a deep CBL (Angevine et al., 1994a,b; Gaynor et al., 1994; Dye et al., 1995). The backscatter intensity of the electromagnetic signal is proportional to the structure parameter of the electromagnetic refractive index C_n^2 which depends on small-scale fluctuations of the temperature and especially the moisture fields. Vertical profiles of C_n^2 usually show a maximum at the top of a well-developed CBL. However, the moisture profile is often not as well-mixed as that of temperature which may result in some ambiguity of the derived MHs. Additional problems occur in the presence of cumulus clouds, even if only shallow, in the upper part of the ABL.

Radio-acoustic sounding systems (RASS) are extensions of either sodars or radar wind profilers, providing in addition to wind and C_n^2 profiles also virtual potential temperature profiles. They appear therefore well suited for the determination of the MH with Richardson number methods, provided that temperature is retrieved with sufficient accuracy (Görsdorf and Lehman, 2000) and that range and resolution are adequate.

The combination of different remote sounding systems (e.g., sodar + wind profiler, or sodar + lidar) offers a promising way towards the direct and continuous monitoring of the evolution of the MH throughout the complete diurnal cycle (e.g., Beyrich and Görsdorf, 1995). However, the interpretation of data measured with remote sounding systems is not always straightforward. Nevertheless, this holds true also for the direct measuring systems and may (at least partially) be attributed to the general problem of MH definition as discussed in Section 2.

3.1.3. Assessment of measurement-based methods

3.1.3.1. General considerations. The variety of empiricial methods to estimate MHs requires a critical assessment of their advantages and shortcomings (summarised in Table 1) and the agreement of their results. Strictly following our definition given in Section 2, the MH should be determined by investigating the dispersion process of non-reactive tracer gases through the analysis of concentration profiles. However, vertical mixing is not the only process determining such profiles and additional measurements would be needed for a safe interpretation. The second-best choice would be turbulence profile measurements. Both types of measurements are difficult, expensive and therefore not operational. Thus, MH determination is based in most cases on profile measurements of mean meteorological variables such as wind, temperature, humidity, and refractive index. These profiles should satisfy the following conditions:

- They should cover the layer between the earth's surface and about 2–3 km above ground, considering the typical height range over which the MH varies during its annual and diurnal cycles in defined climatic regions.
- The profile measurements should be available with a time resolution of about 1 h or less in order to properly describe the evolution of the MH, especially during the morning and evening transition phases.
- The measured profiles must have a vertical resolution of about 10–30 m to avoid relative uncertainties of more than 10–20%, especially for low MHs (< 250 m).
- The measured parameters should be linked physically to the vertical mixing of pollutants.

Table 2. Critical assessment of different methods to determine the MH

	Continuous data output	Range covered well			Determination of turbulence parameters
		10–100 m (low SBL)	100–500 m (SBL/CBL)	0.5–3 km (CBL)	
In-situ measurements					
Radiosonde	.	.	✓	✓	.
Tethered balloon	.	✓	✓	.	(✓)
Mast	✓	✓	.	.	✓
Aircraft	.	.	(✓)	✓	✓
Remote Sounding					
Mini-sodar	✓	✓	.	.	✓
Sodar	✓	.	✓	.	✓
Radar[a]	✓	.	(✓)	✓	✓
RASS[b]	(✓)	.	(✓)	(✓)	(✓)
Lidar	✓	.	(✓)	✓	✓
Numer. models	(✓)	(✓)		✓	✓

Note: ✓ means fulfilled; (✓) partly fulfilled and . not fulfilled.

[a] Electromagnetic boundary-layer wind profiler.

[b] As an extension of the electromagnetic boundary-layer wind profiler.

Table 2 indicates which of the above-mentioned requirements are fulfilled by the different sounding systems. It clearly appears that none of the systems meets all the requirements, i.e., the *"MH-meter"* does not exist. Reliable MH determination under all conditions is therefore still an unsolved problem. The best approach is to use a combination of systems.

In this section we discuss comparative studies of empirical MH determination. Complete agreement between MH values derived from different sounding systems cannot be expected a priori due to several reasons, the most important ones being:

- Different sounding systems measure different atmospheric parameters (mean temperature, humidity, wind, turbulent fluxes or structure parameters) with varying height resolution and accuracy.
- Vertical profiles of these parameters are influenced in a different way by the processes occurring on the earth's surface. In addition, a host of turbulent and non-turbulent processes (heating and cooling, convection and subsidence, radiation processes, baroclinity, advection, gravity waves, phase changes of water) interact with each other within the ABL and influence the vertical profiles. It is nearly impossible to separate the various contributions to the observed ABL structure.
- Often it is difficult to identify a clear upper boundary of the mixing layer or ABL because vertical profiles of turbulent parameters are smooth without any clear signatures and decrease asymptotically towards values close to

zero which are typical for the residual layer. This occurs especially under stable conditions with weak turbulence and near-neutral conditions.

3.1.3.2. Convective boundary layer. Field studies to compare MH values derived from different measurement systems (radiosonde, sodar, radar, lidar, aircraft) under convective conditions have been described, e.g., by Russell et al. (1974), Noonkester (1976), Coulter (1979), Kaimal et al. (1982), Baxter (1991) and Marsik et al. (1995). These studies show that the relative differences are mostly less than 10%, provided that the elevated inversion capping the well-mixed CBL is not too weak and has a well defined base. Conclusions on possible systematic deviations between different estimates of the MH are not consistent (except for the lidar – see below). This should be attributed to different criteria applied to analyse the profiles as well as to the often limited number of observations and in some cases also to spatial differences between the sites where the different systems had been operated.

In cases of a weak inversion or a non-perfectly mixed CBL, measurements from different systems and even the analysis of the same potential temperature profile by several experienced meteorologists may easily result in relative differences of 25% or even larger (e.g., Hanna et al., 1985; Martin et al., 1988).

MH values derived from lidar measurements have generally been found to be slightly but systematically higher than values derived from temperature profiles or sodar measurements (e.g., Coulter, 1979; Hanna et al., 1985; Martin et al., 1988; Dupont, 1991). This is basically explained by the fact that the most energetic convective plumes penetrate into the stable or inversion layer thereby transporting aerosols up to levels higher than the mean height of the inversion or stable layer base. Under certain conditions, pollutants trapped within the stable capping inversion or free atmosphere can cause a systematic overestimation of MHs from lidar observations (McElroy and Smith, 1991).

3.1.3.3. Stable boundary layer. (a) Methods using wind and temperature profiles: The comparison of MHs derived from different observing systems under stable conditions is much more difficult. This is due to certain features of the structure and evolution of the SBL such as the intermittent and weak turbulence, gravity waves, radiative cooling, drainage flows, and inertial oscillations. Time scales of most of the relevant processes are much longer than in the CBL, so that the SBL is often far from stationarity. Different SBL height scales derived from temperature and wind profiles are compared, e.g., in Hanna (1969), Yu (1978), Mahrt and Heald (1979), Mahrt et al. (1979, 1982), Arya (1981), and Wetzel (1982). No significant relationship exists between the height scales based on the temperature profile and the height of the low-level wind maximum. This is basically due to the different time evolution of the temperature and wind profiles during the night. The surface inversion under undisturbed meteorological conditions normally grows with time due to continuous radiative cooling (e.g., Anfossi et al., 1976; Klöppel et al., 1978; Stull, 1983a; Godowitch et al., 1985). On the contrary, the axis of the nocturnal low-level jet seems to exhibit more of a tendency to descend during the night, although the observations are sometimes contradictory (e.g., Beyrich and Klose, 1988; Mahrt et al., 1979; Godowitch et al., 1985; Smedman, 1988). Normally, a temperature-derived SBL height scale is smaller than the height of the wind maximum at the beginning of the night, whereas towards its end the opposite often holds true. Thus, the structure and the evolution stage of the SBL should be considered when deriving the stable MH from temperature or wind profiles, or when comparing MH values derived from different observing systems under stable conditions.

(b) Methods based on turbulence data: Comparing MH estimates for the SBL derived from turbulence profiles to those derived from mean temperature and wind profiles profiles is difficult due to the scarcity of data and to the variety of definitions for the SBL height. Caughey (1982) noticed that "there is no simple relationship between the SBL depth and the depth of the surface inversion layer. As this layer deepens and becomes more intense, significant turbulence exchange becomes confined to a shallow layer close to the ground". This was confirmed by Garratt (1982a) and Smedman (1991). Kurzeja et al. (1991) found a good correlation between the top of a strong surface inversion and a SBL height derived from profile measurements of the wind direction standard deviation at the beginning of the night and in general between the latter height and the height of the wind maximum. Model calculations often show an increase of the SBL height defined by the turbulence profile or a different time behaviour during different phases of the SBL evolution (e.g., Nieuwstadt and Driedonks, 1979). Model simulations using a one-dimensional ABL model with an algebraically approximated second-order closure (Dörnbrack, 1989) have shown that the turbulent MH scales derived from the vertical profiles of heat flux, momentum flux, and TKE exhibit both a different behaviour in time and a different relationship between each other, depending on the external conditions (Beyrich, 1994b). The level at which the turbulent heat flux has decreased to 5% of its surface layer value is often used as a definition of the MH under stable conditions. Mason and Derbyshire (1990) deduced from large-eddy simulations (LES) that "if radiative heat transfer is negligible, the flux of heat is roughly analogous to that of contaminants emitted from low-level sources, and so definitions of SBL depth based on the buoyancy flux profile are easy to relate to a 'mixing depth' appropriate for dispersion applications". The turbulent SBL height scale estimated from profiles by determining the level at which the gradient Richardson number Ri exceeds a critical value Ri_c, depends on the vertical resolution of the profile data used, and there is still some controversy about the numerical value of Ri_c. Thus, it is clear that the question which of the suggested SBL height scales is best suited to characterise the vertical mixing of pollutants under stable conditions has not received yet a final answer.

(c) Methods using remote sounding data: Among the remote sounding instruments, solely acoustic sounders seems capable in providing MH data under stable conditions. Radar profilers and lidars have mostly a range resolution which does not allow to resolve the SBL in detail. In addition, their first usable range gate is often at or above the SBL top, a fact which sometimes even limits the application of a conventional sodar (Garratt, 1982b; Smedman, 1988; Baxter, 1991). For very shallow stable boundary layers (below 50–100 m), minisodars – with a lowest range gate around 10 m – can overcome this deficit. Unfortunately, the interpretation of sodar data for MH determination under stable conditions is controversial (Hanna, 1992). Results from comparisons of sodar measurements with MHs derived from temperature and wind profiles are quite inconsistent (Arya, 1981; Hanna et al., 1985). No systematic differences between sodar observations and the height of the nocturnal surface inversion were found by Hicks et al. (1977), Hayashi (1980), or Fitzharris et al. (1983). Some authors concluded that the sodar-based MH is generally lower than the surface inversion height (Nieuwstadt and Driedonks, 1979; Bacci et al., 1984). Piringer (1988) found agreement with the top of the lowest strongly stable layer of a layered surface inversion. Gland (1981) and Dohrn et al. (1982) did not find any clear connection between echo intensity and inversion features. On the other hand, the rare comparisons of sodar observations with turbulent MH scales by Nieuwstadt and Driedonks (1979) and Tjemkes and Duynkerke (1989) yielded a reasonable agreement between sodar data and the height at which modelled sensible heat flux or TKE have decreased to 5% of their surface layer values. Dupont (1991) and Devara et al. (1995) found a good agreement between MH values derived from simultaneous sodar and lidar operation, though based on a small number of case in a simply structured SBL. Van Pul et al. (1994) found a high correlation between the SBL height derived from lidar measurements and radiosonde profiles. Comparing ozone profiles measured with a tethered balloon and sodar data under conditions of a complex structured SBL, Beyrich et al. (1996) reported a generally good agreement of the MH values derived from both systems, except for very low MHs. All these findings are consistent with the fact that the sodar echo intensity is more or less strongly influenced by turbulence whereas inversion heights are also influenced by radiative cooling. Beyrich and Weill (1993) have demonstrated that the relationship between a sodar-derived value for the stable MH and any other height scale strongly depends on the stage of the SBL evolution. They concluded that different criteria have to be applied to derive a MH value from sodar signal intensity profiles depending on the actual shape of these profiles. Beyrich (1994a) demonstrated that a well-developed nocturnal low-level jet governs the time evolution of the stable MH especially in the second half of the night.

3.2. *Mixing height determination from parameterisations and models*

As continuous profile measurements for the operational determination of the MH are not generally available, simple parameterisations based on standard surface observations and single profile data as well as numerical models are widely used in the practice of meteorological and environmental services. Simple diagnostic or prognostic parameterisation equations for the MH are still very attractive for operational purposes because of their simplicity and the limited number of required input data. They are also used within comprehensive parameterisation schemes for the treatment of the ABL in some numerical weather prediction and climate models.

One-dimensional prognostic models of the atmosphere with (local) turbulence closure of the order of 1.5, 2 or even higher have been used by different authors over the last 20 yr in order to estimate the height of the stable turbulent boundary layer, usually as the level where turbulence (measured by the TKE, the heat flux, or the momentum flux) decays below a certain threshold (e.g., Delage, 1974; Wyngaard, 1975; Brost and Wyngaard, 1978; Rao and Snodgrass, 1979; Nieuwstadt and Driedonks, 1979; Dörnbrack, 1989; Tjemkes and Duynkerke, 1989; Estournel and Guedalia, 1990). They could be considered an alternative to simple parameterisations, especially if coupled to observations. Over the last years, also three-dimensional numerical flow models have been increasingly used to derive MH values, and especially to produce regional MH patterns over larger areas (e.g., Sørensen and Rasmussen, 1997; Fay et al., 1997). Successful intercomparisons of daytime maximum MHs from a NWP model and observed values from windprofiler measurements over a several weeks period have been presented by Engelbart (1998). Klein Baltink and Holtslag (1997) found differences between model results and measurements up to a factor of two for the daytime maximum MHs which they attributed to a wrong soil moisture in the model. Batchvarova et al. (1999) found good agreement between a comprehensive data set on the MH in a coastal area determined from different observational devices (balloons, airborne lidar) and the MH derived from the RAMS model, where the top of the MH was identified by a critical value of the TKE. This development will certainly continue, as non-hydrostatic models with a few kilometres horizontal resolution and turbulence closures based on a prognostic equation for TKE are widespread research tools nowadays and are about to be used as operational NWP models (e.g., Schlünzen, 1994). More systematic comparisons of MHs derived from these models and from observations are desirable. It is intrinsic to them that they are prognostic, and unless four-dimensional variational data assimilation is applied, they will not be able to make full use of existing observations. Their use is justified especially in situations with strong horizontal inhomogeneities, and if geographical fields of the MH are needed.

3.2.1. Modelling and parameterisation of the MH under stable conditions

Many parameterisation expressions for the height of the turbulent SBL have been suggested in the literature (e.g., Hanna, 1969; Zilitinkevich, 1972; Etling and Wippermann, 1975; Arya, 1981; Mahrt, 1981; Nieuwstadt, 1984; Koracin and Berkowicz, 1988). Both diagnostic and prognostic relationships have been proposed and there has been a controversial debate on which type is the most suitable (Nieuwstadt, 1981, 1984; Garratt, 1982a,b).

The most popular *diagnostic equations* (based on scaling arguments) are:

$$h = a_1 L_E = a_1 u_*/f \tag{1}$$

and

$$h = a_2 (L_E L_*)^{1/2} = \frac{a_2 u_*^2}{\sqrt{-\beta \kappa Q_0 f}} \tag{2}$$

with the empirical coefficients $a_1 = 0.07$–0.3 and $a_2 = 0.3$–0.7. Nieuwstadt (1981) proposed a combination of (1) and (2), namely:

$$h = \frac{L_*}{3.8} \left(-1 + \sqrt{1 + 2.28 \frac{u_*}{fL_*}} \right). \tag{3}$$

Most of the verification studies done in the past do not seem to favour the application of more elaborated parameterisations. However, one should be cautious in using these equations because Eq. (1) assumes a neutral, stationary boundary layer, and because $1/f$ is generally too long to be a relevant time scale. The subordinate role of the Coriolis force for the turbulent fluxes is also supported by large-eddy simulation (see Fig. 11 in Andrén, 1995).

As an alternative to $1/f$ as time scale, some authors, e.g., Kitaigorodskii and Joffre (1988), have suggested to use $1/N_{BV}$, and

$$h = a_3 L_N = a_3 \frac{u_*}{N_{BV}} \tag{4}$$

with the empirical constant $a_3 = 4$–14. This model has been corroborated by measurements in the Arctic (Overland and Davidson, 1992), by lidar measurements in the Netherlands (van Pul et al., 1994), and by LES-computations (Vogelezang and Holtslag, 1996).

Classically, it is assumed that the structure of the stable ABL depends on external parameters such as the Coriolis parameter f and the surface roughness length z_0, and on internal turbulent parameters such as the friction velocity u_* and the surface heat flux $Q_0 = \langle w'\Theta' \rangle$. Then, the stable (and neutral) ABL height is assumed to be a function of the Ekman and Monin-Obukhov length scales $L_E = u_*/f$ and $L_* = -u_*^3/(\beta \kappa Q_0)$, respectively. Joffre (1981) and Kitaigorodskii and Joffre (1988) have extended these similarity theories to include the effect of the background stratification of the atmosphere through the length scale $L_N = u_*/N_{BV}$, with the Brunt-Väisälä frequency $N_{BV} = \sqrt{\beta \gamma_\theta}$, where γ_θ is the potential temperature gradient above the MH. Zilitinkevich and Mironov (1996) have proposed a multi-asymptotic expression which combines all these different scales. A comprehensive survey of the various formulations can be found in Seibert et al. (1998).

It has also been proposed to set h proportional to L_*. A wide range of constants have been found for this relationship, between 1.2 and 100 (Zilitinkevich and Mironov, 1996); values between 2 and 10 appear to be most typical (Arya, 1981).

A second controversy has been whether it is possible to parameterise the SBL height solely based on surface layer variables (mainly u_*, L_*), or whether bulk SBL parameters should be considered additionally or even exclusively (e.g., Mahrt, 1981; Garratt, 1982b; Smedman, 1991; Vogelezang and Holtslag, 1996). Diagnostic formulae considering the bulk structure of the SBL are based on the assumption that turbulence production must vanish at the top of the SBL and the Richardson number must therefore exceed its critical value Ri_c (Hanna, 1969; Mahrt et al., 1979). Thus,

$$h = Ri_c(\Delta V)^2/\beta \, \Delta\Theta. \tag{5}$$

The proposed equations differ basically in the choice of the levels over which the wind and temperature differences ΔV and $\Delta\Theta$ across the ABL are determined and in the value of Ri_c ($Ri_c = 0.33$, Hanna (1969), Wetzel (1982); $Ri_c = 0.5$, Mahrt (1981), Troen and Mahrt (1986); $Ri_c = 0.25$, Holtslag et al. (1990)). Joffre (1981) found that the value of Ri_c depends on the parameter hf/u_* with the classical value of 0.25 being relevant for small values of hf/u_* ($\leqslant 0.1$), but large values of $Ri_c \sim 7$ can be reached when $hf/u_* \geqslant 0.3$ (see also Maryon and Best, 1992). Vogelezang and Holtslag (1996); hereafter the VH-method) extended this method by incorporating shear production in the surface layer parameterised by an additional term depending on u_*.

Most *prognostic equations* proposed to describe the SBL height are based on a relaxation process during which h approaches a certain equilibrium value

h_e with a time scale τ_S, i.e.:

$$\frac{dh}{dt} = \frac{(h_e - h)}{\tau_S}. \tag{6}$$

The equilibrium height h_e is often parameterised using one of the diagnostic equations given above (Eqs. (1), (2), or (3)). The time scale δ/τ_S has been proposed to be proportional, e.g., to $1/f$, to a combination of surface layer scaling variables such as L_*/u_*, or to the inverse of a normalised cooling rate $\Delta\Theta(\partial\Theta_0/\partial t)^{-1}$ where $\Delta\Theta$ is $T_h - T_0$ (subscripts h and 0 refer to $z = h$ and surface values, respectively).

The evolution of the SBL is highly non-stationary; one would thus expect that prognostic equations taking into account relevant mechanisms should be superior to each diagnostic relationship for simulating the stable MH. However, comparisons of different diagnostic and prognostic relationships with observations have not corroborated this expectation. For practical use it is also helpful that diagnostic equations do not require an initial value of the MH as input.

The verification of such diagnostic or prognostic relationships has long been performed with the aid of radiosonde data (e.g., Hanna, 1969; Yu, 1978; Mahrt and Heald, 1979; Wetzel, 1982). Since the early 1980s, sodar observations have increasingly been used for this purpose (e.g., Arya, 1981; Nieuwstadt, 1984; Koracin and Berkowicz, 1988; Beyrich, 1993, 1994b). The prognostic equations have rarely been tested at all and only single case studies are reported in the literature. For the diagnostic equations, correlation coefficients typically range between 0.4 and 0.7, quite often with large scatter. Also, the numerical constants appearing in all the parameterisations vary significantly, and it seems difficult to recommend any universal and site-independent value. Hence, there is still a need for verification studies of the different SBL height parameterisations using comprehensive data sets.

3.2.2. Modelling and parameterisation of the MH under convective conditions

Diagnostic relations based on similarity theory have a few times been suggested to parameterise the CBL depth (e.g., Tennekes, 1970; Zilitinkevich, 1972; San José and Casanova, 1988). However, these are valid only under certain conditions (e.g., free convection) and are not of much practical relevance. The Ri-method has been used by Vogelezang and Holtslag (1996) also for unstable situations by adding an excess temperature to the near-surface temperature, as suggested by Troen and Mahrt (1986) and described in Section 4.1.

Today, the numerical integration of mixed-layer slab models is a well-established way to simulate the evolution of the convective MH. These models use surface fluxes and an initial temperature profile as basic input parameters. The latter one represents a general problem for these models since normally the network of radiosounding stations is not dense enough for boundary layer studies.

Prognostic equations describing the growth of the CBL are normally derived from a parameterisation of the TKE budget equation which is either averaged over the whole mixed layer or specified at the mixed layer top. The equations proposed by various authors mainly differ in the terms which are neglected in the TKE budget and how, the remaining terms are parameterised. The spectrum ranges from simply considering surface heating as the only relevant driving force (Betts, 1973; Carson, 1973; Tennekes, 1973) to additional consideration of mechanical turbulence production due to surface friction (Driedonks, 1981, 1982b), local changes of TKE at the level $z = h$ (so-called "*spin-up*" effect, Zilitinkevich, 1975; Gryning and Batchvarova, 1990; Batchvarova and Gryning, 1991), wind shear across the entrainment layer (Stull, 1976a; Driedonks, 1981; Manins, 1982; Rayner and Watson, 1991), explicit parameterisation of TKE dissipation (Zeman and Tennekes, 1977), and finally rather complex equations taking also into account energy losses in connection with gravity waves (Stull, 1976b) or the influences of moisture and advection (Steyn, 1990). A survey of these relationships is given in Seibert et al. (1998). Comparisons of some of them with observational data can be found in Driedonks (1981, 1982b), Arya and Byun (1987), or Batchvarova and Gryning (1991, 1994). These studies as well as sensitivity experiments carried out by Beyrich (1994b) have shown that the observed variability of the MH during daytime can in general be well described if surface heating and mechanical turbulence production due to surface friction are taken into account, with the entrainment heat flux parameterised in terms of the surface heat flux.

These effects are properly parameterised in the following two equations from Driedonks (1982a) and Batchvarova and Gryning (1991), respectively:

$$\frac{dh}{dt} = A \frac{Q_0}{\Delta\Theta} + B \frac{u_*^3}{\beta h \Delta\Theta} = \frac{A w_*^3 + B u_*^3}{\beta h \Delta\Theta}, \tag{7}$$

$$\frac{dh}{dt} = (1 + 2A) \frac{Q_0}{\gamma_\Theta h} + 2B \frac{u_*^3}{\gamma_\Theta \beta h^2} = \frac{(1 + 2A) w_*^3 + 2B u_*^3}{\gamma_\Theta \beta h^2}. \tag{8}$$

The values of the constants A and B given in the literature differ considerably, ranging between 0 and 1 for A, and 0 and >10 for B. Many authors use $A = 0.2$ as a typical value (e.g., Tennekes, 1973; Yamada and Berman, 1979; Driedonks, 1982b). However, recent comparisons with measurements from different climatic regions suggest a higher typical value of $A \approx 0.4$ (e.g., Tennekes and van Ulden, 1974; Clarke, 1990; Betts, 1992; Culf, 1992). The constant A represents the ratio $-Q_h/Q_0$ between the entrainment-layer and surface-layer heat fluxes, therefore, Carson (1973) suggested the use of different values for A corresponding to different stages of the CBL evolution. Betts and Barr (1997) found that A increases with the wind speed, but they did not include a u_*-term. For B, Tennekes (1973) proposed $B = 2.5$, a value used later also by Gryning and Batchvarova (1990). Driedonks (1981, 1982b) achieved the best agreement with observations using $B = 5$, a value also applied by, e.g., Zilitinkevich et al. (1992). The possible ranges for A and B reported in the literature significantly affect the simulated CBL growth. Beyrich (1994b) showed that a variation of their values over the typical range reported in the literature results in differences of the simulated evolution of the MH much larger than those which would originate from an application of a more complex equation. Beyrich (1995) therefore suggested to adapt the model constants to actual observations (e.g., from sodar or wind profiler data) to improve the model output results.

3.2.3. Determination of the MH from NWP model output

Dispersion models for the regional scale or for long-range transport often derive the MH from NWP model output. This may also be considered for the local scale if suitable measurements are not available at the site under consideration. It is obvious that methods and their reliability depend on the degree of sophistication of the ABL parameterisation and on the resolution of the boundary layer within the NWP model. The bulk Richardson-number method is the standard approach to derive MHs from the NWP models.

A procedure to determine h from the output of HIRLAM, the limited area model used in the Nordic and some other European countries, has been validated by Sørensen et al. (1996). It is based on a bulk Richardson number derived from the model level data:

$$Ri_b(h) = \frac{gh}{\Theta_{v1}} \frac{\Theta_v(h) - \Theta_{v1}}{U(h)^2 + V(h)^2} \tag{9}$$

where Θ_{v1} is now the virtual potential temperature at the lowest model level (about 30 m above ground). This formula is consecutively applied for $h = z_2, z_3, \ldots$, where the z_i are the model levels. The actual value of h is chosen as the height where Ri_b reaches a critical value. An optimal value of 0.25 was found after applying the method to radiosoundings with a clear convective lid. This method is used in the "Danish Emergency Response Model of the Atmosphere" (Sørensen, 1998; Sørensen et al., 1998).

Maryon and Best (1992) studied the boundary layer height h determined from vertical profiles obtained from NWPs for use in NAME (UK Met. Office's long range transport model for nuclear accidents). They compared model boundary layer heights diagnosed by different methods with radiosonde measurements. These methods, based on identifying the level at which a critical gradient Ri-number equal to 1.3 is reached, gave generally poor results, underestimating h particularly for the daytime boundary layer. A best-fit procedure yielded a value $Ri_c = 7.2$, but the improvement was limited. Also Wotawa et al. (1996) and Fay et al. (1997) used Richardson number methods to derive MHs from NWP model output.

The EMEP MSC-W model introduced another method to derive MHs from HIRLAM output (Jakobsen et al., 1995). The mechanical MH is defined as the lowest level where the turbulent diffusion coefficient K_M is less than $1 \text{ m}^2 \text{ s}^{-1}$, with K_M determined by the method of Blackadar (1979) from the Richardson number. The convective MH is the height of the adiabatic layer, after the sensible heat input of one hour has been distributed via dry-adiabatic adjustment. The MH is taken as the larger one of the two values.

4. Intercomparison of mixing height determination methods

Data sets suitable for the testing of MH routines should, according to our opinion, fulfil at least the following requirements:

  (i)   high frequency of radiosonde data (at least 4 ascents a day);
 (ii)   continuous profile information from a remote sensing instrument (sodar, lidar, wind profiler) as a second (independent), continuous data source for MH determination from measurements;
(iii)   measurements of turbulent fluxes at the surface;
(iv)   sufficiently uniform terrain without too much orographic influence.

It turned out that the number of such data sets is rather limited. We finally selected three data sets: SADE-campaign (Germany), Cabauw (The Netherlands) and Payerne (Switzerland); details are given in Appendix A. Only data sets of less than a full year were available. Some of the data sets continue to grow, and a new data set is being built up at Lindenberg, a meteorological observatory and aerological station of the German Weather Service. Future work should be based on these more extensive data sets.

4.1. Assessment of the parcel methods

We have tested two different versions of the parcel methods (Fig. 1). The simple parcel method (Holzworth, 1964) uses the measured virtual potential temperature Θ_v of a radiosounding at ground level, and the MH is taken as the equilibrium level of an air parcel with this temperature. In the advanced parcel method, the temperature of the parcel is given as the temperature near the

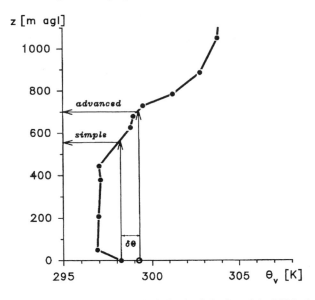

Figure 1. Illustration of the two parcel methods for the derivation of the MH in the CBL from radiosoundings. The simple parcel method uses the virtual potential temperature at ground level, while according to the advanced method an excess temperature (see Eq. (10)) is added.

ground plus an excess temperature $\delta\Theta_v$ calculated as (Troen and Mahrt, 1986; Holtslag et al., 1990; Beljaars and Betts, 1992)

$$\delta\Theta_v = \frac{C_1 \langle w' \Theta_v' \rangle}{\sqrt[3]{u_*^3 + C_2 w_*^3}}. \tag{10}$$

Values of 5 and 8.5 have been suggested for C_1, while for C_2 the value of 0.6 has been used.

Since under convective conditions a superadiabatic layer is usually found near the ground, also the simple parcel method implicitly applies an excess temperature. One may thus question whether it is justified to add another excess temperature, especially as the authors who used this concept mainly applied it to NWP model output. Therefore, we plotted both the temperature excess according to Eq. (10) and the one found in radiosoundings against the heat flux (which is the dominant term in $\delta\Theta_v$) for the Cabauw data (Fig. 2). The temperature excess in the radiosounding was defined as the difference between Θ_v at the lowest level and at the first level above which the stratification was neutral or stable. Two features are striking in Fig. 2, the much higher scatter of the observed $\delta\Theta_v$ as compared to the computed one, and its larger magnitude. This implies, first, that the simple parcel method depends on stochastic influences, since its implicit $\delta\Theta_v$ is not closely related to the heat

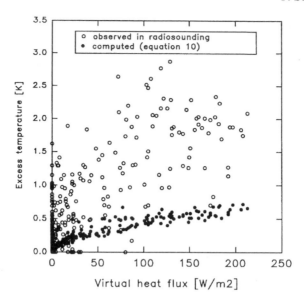

Figure 2.   Excess temperatures derived from radiosoundings and computed from Eq. (10) plotted against the virtual heat flux. Radiosonde data are from De Bilt and heat flux data from Cabauw.

flux. Secondly, adding $\delta\Theta_v$ according to the similarity formula does not make much sense, as one would add only a small quantity (typically 0.5 K) to a much larger (typically 2 K) but stochastically influenced quantity. Thus, we suggest that when the advanced parcel method is applied to radiosoundings, one should omit the superadiabatic near-surface layer and instead use a larger value of the constant C_1. In order to achieve the same order of magnitude as found in the observed $\delta\Theta_v$, $C_1 = 20$ was used in this study.

This reasoning applies, to some extent, also to model data, as normal K-type diffusion parameterisation requires superadiabatic gradients to produce an upward heat flux and thus, in general, model data will already contain a certain excess temperature. This may be the reason why such a small value of the constant C_1 has been used. However, one may wonder if this mix of two excess temperatures, one created by the model and another one from similarity theory, is really adequate.

4.2. Intercomparison of empirical methods

4.2.1. Stable situations

The sodar-derived MHs during stable situations were compared with different analyses of radiosoundings for all days of the SADE intensive observation periods. The radiosoundings were subjectively analysed with respect to the temperature profile, and with Ri-number methods (with $Ri_c = 0.2$). The relationships between MHs derived from sodar and with the Ri-methods are characterised by a lot of scatter, whereas MHs from the sodar and from the temperature profile agree much better. This demonstrates that during stable situations sodar-derived MHs are strongly influenced by the shape of the temperature profile which, however, reflects the effects of mechanical turbulence production in stable situations with moderate to strong winds. The Ri-method based on the ground-level temperature yields many cases of very shallow MHs (< 50 m) whereas the sodar gives values between 50 and 200 m. It is hard to check the validity of the value of Ri_c for these circumstances of extremely shallow MHs, because if they are real, they were overestimated by the sodar which had its lowest range gate at 50 m. On the other hand, there is some uncertainty in the Ri-number method as well: if the temperature at the 20 m level is used instead of the ground-level temperature in the Ri-number (similarly to the VH-method), many Ri-number-derived MH values exceed the sodar-derived ones. This would mean that the VH-method tends to overpredict the stable MH.

Another comparison was made with the Cabauw data set, where we compared MHs under stable conditions derived from the sodar and from Ri-number analyses (Fig. 3). In contrast to SADE, the sodar MHs were derived here with a completely automatic algorithm. Like in the SADE data, Ri-number derived MHs tend to be lower than those indicated by the sodar. The number of extremely shallow MHs is much lower with the VH-method than with the standard Ri-number method. In some cases the Ri-number meth-

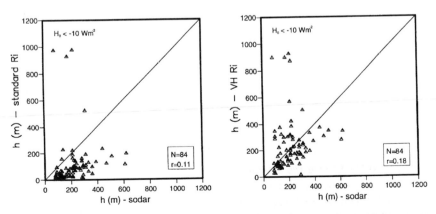

Figure 3. Scatter plots of mixing heights derived from different Ri-number methods: (Left: standard method, right: VH-method) versus sodar-derived MHs in Cabauw: for all stable hours ($H_0 < -10 \text{ W m}^{-2}$).

ods indicate MHs around 1000 m while according to the sodar it should be below 300 m; these cases belong to the evening transition period, when the Ri-methods can pick up the height of the residual layer. Overall, the agreement between sodar-derived MHs and those obtained with the VH-method is moderate, though the sodar data tend to be somewhat higher. The correlation between MHs obtained with the standard Ri-number and from the sodar does not differ much from the correlation obtained for the VH-method, but the standard Ri-number method leads to systematically lower MHs.

During stable situations, sodars give a reasonable magnitude of the MH as long as it is not lower than the lowest sodar range gate. Sometimes they may overestimate it.

It appears difficult to judge the relative performance of the standard and the VH Ri-number method. As the critical Ri-number in the VH-method has been determined from Cabauw sodar data (obtained with an automatic routine), evaluations with other, independent data sets are necessary before final conclusions can be drawn.

4.2.2. Unstable situations

Sodars usually capture the rise of the CBL very well during a period in the morning of well-developed convective days, as long as the MH is within the sodar range. However, for a large part of the convective hours the MH is higher and it cannot be covered by sodar. There are automatic routines for MH determination by sodars, but they have the problem to recognise situations when the MH is outside the sodar range and the routine's estimates will be invalid. One automatic routine, part of a commercial sodar software (which is said to use also vertical velocities) turned out to show a lot of unexplainable variability during all times of the day in our comparison at Payerne. Unfavourable results were also found by Keder (1999).

The comparison of different methods using daytime SADE-data indicates a very good agreement between the results of the parcel method with sodar-derived MHs as well as with MHs evaluated subjectively from radiosoundings, except for a small number of outliers (Fig. 4). The agreement between the parcel method and the Ri-number method is also very good. The comparison between the standard and the advanced parcel method shows a rather small scatter while the MHs obtained with the standard parcel method are slightly lower than those from the advanced method. From the other comparisons, it appears that the advanced method is unbiased.

The advanced Ri-method (or VH method) adapted by Vogelezang and Holtslag (1996) has been used also for unstable situations with an excess temperature added to the near-surface temperature (with $C_1 = 8.5$). For the Cabauw data set, all the methods (simple parcel, advanced parcel, standard Ri-number,

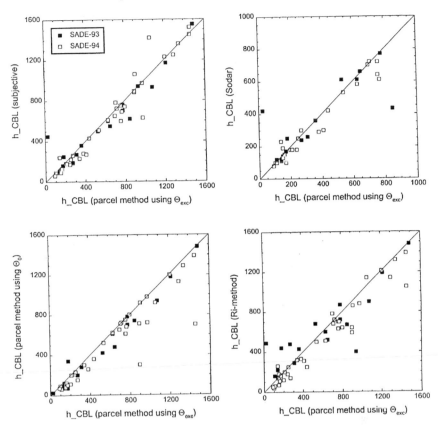

Figure 4. Scatter plot comparing a subjective evaluation of the MH based on radiosoundings (temperature, humidity, wind), a semi-objective evaluation of sodar backscatter data, the simple parcel method and the standard Ri-number method with the advanced parcel method. Data sets from SADE-93 (full symbols) and SADE-94 (open symbols), unstable hours only.

VH Ri-number) correlated well with each other ($r > 0.9$), and without significant biases (Fig. 5). The standard Ri-number method and the simple parcel methods yield almost equal results since under convective conditions, when the wind shear is usually small, Ri-number and parcel methods are almost equivalent. The comparison between the standard and the VH Ri-number method shows that the VH-method overpredicts low MHs (< 500 m, according to the standard method) as compared to the standard method, and has a weak tendency to underpredict for high MHs.

In the CBL, the Ri-number methods give very similar results to the parcel methods when the same near-ground temperature is used. Thus, we can recommend to use either the advanced parcel method or a Ri-number method

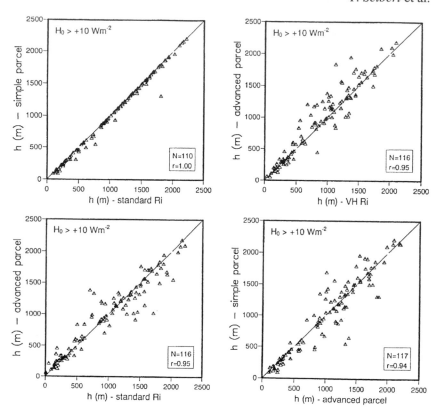

Figure 5. Scatter plot of mixing heights derived by the standard and the VH *Ri*-number methods as well as the simple and advanced parcel methods at Cabauw, unstable cases.

where the near-ground temperature contains an excess temperature as in the advanced parcel method (Vogelezang and Holtslag, 1996). For boundary layers where shear-produced turbulence is important, *Ri*-number methods should be preferred to the parcel method even if the ABL is unstable.

4.3. Intercomparison of preprocessor modules

Advanced meteorological preprocessors utilise information from both ground meteorological measurements and sounding profiles. Within the COST Action 710 we have compared five numerical preprocessors that were available to us. These were the OML meteorological preprocessor (Olesen et al., 1987), the Hybrid Plume Dispersion Model (HPDM, Hanna and Chang, 1992), the meteorological preprocessor library of Servizi Territorio (Servizi Territorio,

1994), the routine of the Finnish Meteorological Institute (Karppinen et al., 1997, 1998), and the RODOS preprocessor (Mikkelsen et al., 1997). These preprocessors are shortly described in Appendix B.

We have compared the results from the MH-modules included in these five preprocessors with each other and with the results of empirical methods for typical days of the selected data sets described in Appendix A. The empirical methods are based on radiosoundings and sodar measurements. The radiosoundings were analysed with the Ri-number method and – for hours with positive heat flux – the parcel method. Sodar-derived MHs are the only empirical method available which gave continuous information, but they are limited to night-time and shallow unstable boundary layers.

Through these intercomparisons, a number of specific problems for each preprocessor could be identified, but the cause of these problems was not always apparent. Thus, it will be a major task for the model developers to investigate them more in depth.

4.3.1. 28–29 May 1996, Payerne (Fig. 6)

OML and HPDM gave similar results during the night. However, during the convective period of the first day, OML used all the time the MH given by

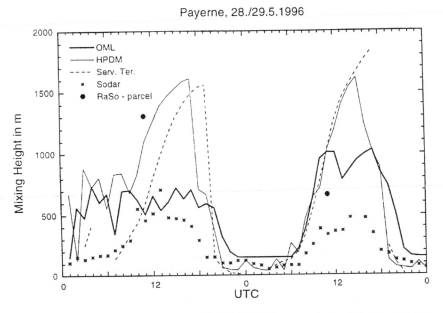

Figure 6. Evolution of the MH in Payerne, 28–29 May 1996, as computed by different preprocessors (lines) and as derived with empirical methods (symbols).

u_*/f, while HPDM produced a convective growth of the mixing layer during the day. Note the difference between the convective MHs produced by HPDM and ST on the first day, which was probably due to different initialisations. On the second day, HPDM and ST agreed well as they both start from low initial values. OML again used u_*/f, indicating that the convective MH calculated in OML was lower than the one obtained by u_*/f. During the second night, OML gave its minimum height of 150 m while other methods indicated lower MHs. During daytime, the sodar-based MHs were erroneously placed within the CBL (see discussion of sodar routines in Section 4.2).

4.3.2. 6–7 July 1995, Cabauw (Fig. 7)

On these days, OML produced a growing mixing layer during the daytime, at first determined by u_*/f (growing u_*) and then by the convective model. Compared with the sodar, the early phase of this growth was overestimated by OML (because it used the u_*/f formula which is not appropriate for this situation) while it was correct in the ST and FMI models. Later on the day, OML gave MHs which coincided with the radiosounding while the ST and RODOS model values remained lower. We cannot explain the behaviour of

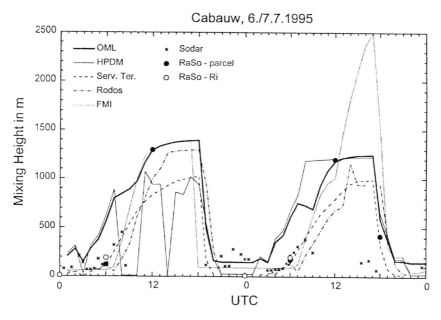

Figure 7.   Evolution of the MH in Cabauw, 6–7 July 1995, as computed by different preprocessors (lines) and as derived with empirical methods (symbols).

HPDM on the first day. The plateau of the MH given by HPDM on the second day is not shown by other models. During the second day, the FMI model produced an unexplained, unrealistic growth in the afternoon. We consider the sodar values of the night from July 6 to 7 too high, but they show nicely the growth of the mixing layer on the morning of the second day, coinciding with the ST convective slab model and the morning radiosounding. However, while OML reached the correct MH at noon, ST and RODOS MHs again remained too low.

4.3.3. 12 13 November 1995, Cabauw (Fig. 8)

The sodar was able to detect the MH throughout this period, obviously connected with a strong inversion, and also picked up by the parcel method. OML detected this inversion as a "sustained lid" and just interpolated the height of this lid linearly between the radiosoundings. There was a positive heat flux only for a few hours on each of the two days, as indicated in ST-slab model. In contrast to OML, strong winds with associated high u_* caused the HPDM to yield much too high MHs in the first 30 h. The other models fluctuated around the values observed with the sodar during the whole period. With the excep-

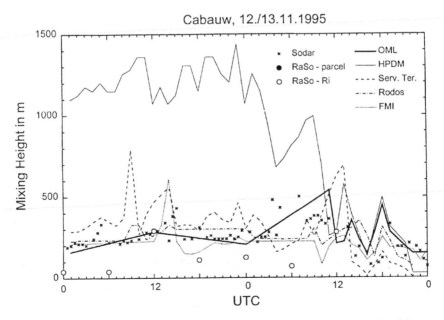

Figure 8. Evolution of the MH in Cabauw, 12–13 November 1995, as computed by different preprocessors (lines) and as derived with empirical methods (symbols).

tion of 12 UTC on the second day, the standard Ri-number method always yielded MHs considerably lower than obtained by all other methods while the VH method (not shown) gave more consistent values.

4.3.4. 7–8 October 1994, Melpitz/SADE (Fig. 9)

The OML and HPDM models showed an explosive growth of the mixing layer in the morning of the first day, while FMI, ST and RODOS started the mixing layer growth later and at a slower pace. Sodar and radiosonde data showed that HPDM and OML were not realistic. Their behaviour cannot be explained by u_* either, in contrast to some other cases. The behaviour of HPDM was even more unexpected than that of OML, which at least showed a normal convective growth phase between 1100 and 1500 m. The slab model of ST did not reach a sufficient height, due to its constant temperature gradient, while all the other models reached the correct maximum MH as indicated by the radiosoundings. On the morning of the second day, OML, FMI, RODOS and ST all created a very similar growth of the CBL which is also supported by the sodar data. However, in HPDM the MH rose until its maximum values of 3000 m. During the night between the first and the second day, all the methods and the sodar

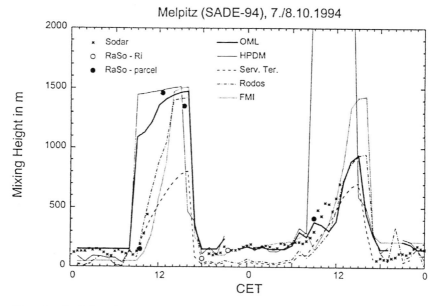

Figure 9. Evolution of the MH in Melpitz, 7–8 October 1994, as computed by different pre-processors (lines) and as derived with empirical methods (symbols).

measurements are in the same range, with exception of the ST and RODOS models which lead to considerably smaller values.

5. Conclusions and recommendations

There are two general possibilities for operational MH determination, namely the analysis of profile measurements on one hand and the application of parameterisations or models based on operationally available data on the other hand. If suitable data are available, the first option is to be preferred. Since none of the methods and models are perfect, it is recommended to have results obtained in an operational context checked by a qualified scientist, considering the basic data. It is possible to substitute NWP model output for measurements, but then the results strongly depend on the characteristics of the model, especially its ABL formulation and vertical resolution. Therefore, no general recommendations can be made here. High-resolution mesoscale models with good ABL formulation can be useful especially in complex terrain.

A flow chart for the practical determination of the MH according to available data is shown in Fig. 10. It is to be seen as a guidance but not as an algorithm which must be followed strictly. Considering the contradictions and difficulties we found, and the limited amount of data that could be used in this study, the present conclusions cannot be seen as definite ones.

5.1. Analysis of measurements

Parcel methods based on the virtual potential temperature are the most reliable ones for the detection of the convective MH. The simple parcel method already gives reasonable results. When using an excess temperature, attention has to be paid to the value to which it is added. If this is the temperature of the well-mixed CBL, a value of the constant C_1 around 20 appears to be more appropriate than the values 5 or 8.5 which have been suggested so far (see also 5.3). Methods using a bulk Richardson number can also be applied to convective boundary layers, but then they should use an excess temperature similar to the parcel methods to compute the temperature differences.

The determination of the MH in situations where mechanically produced turbulence is important is much more difficult. Given the fact that temperature and wind profiles are the most widespread information, we consider bulk Richardson number methods to be the most appropriate ones under such conditions. However, with shallow stable MHs, such methods give questionable results if the wind profile is available from radiosondes only, as such wind data may be inaccurate and/or not sufficiently well resolved near the ground. Forming a composite with wind information from other measurement systems (sodar, mast) might help to overcome this problem.

FLOWCHART FOR THE DETERMINATION OF THE MIXING HEIGHT *h*

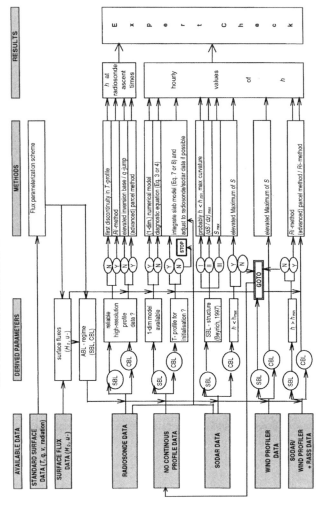

Figure 10. Flowchart to guide the practical determination of the mixing height *h*. The left column provides possible entries, depending on the available data. All paths which lead into the GOTO box continue to the "NO CONTINUES PROFILE DATA" entry box. STOP means that mixing height determination is not possible. *S* denotes the intensity of the backscattered signal of a sodar or wind profiler.

Profiles obtained with tethersondes need to be smoothed, and trends to be removed, as substantial local temperature changes may occur during the ascent or descent.

Direct determination of the MH from sodar data is possible only if the MH is well within the range of the sodar (typically, about 50 to 500 m). If it is not known whether the MH really falls into the range of the sodar, erroneous conclusions are easily possible. MH detection routines currently implemented on commercial sodars by manufacturers should be considered cautiously. End-users should have the performance of any routine carefully checked on their site, using other reliable methods for comparison.

The backscatter intensity profiles obtained from electromagnetic wind profilers (operation frequency around 1 GHz) are a promising basis for MH determination in well-developed cloudless CBLs. However, algorithms developed so far appear not to be reliable enough for operational use. In the near future, radio-acoustic sounding systems (RASS) providing temperature profiles in combination with electromagnetic or acoustic wind profilers will allow the application of Ri number methods on a continous base.

5.2. Application and improvement of preprocessors

For convective boundary layers, the numerical slab model is appropriate. Its integration should use the actual initial temperature profile, and not a predetermined value of the gradient of Θ_v above the mixing layer. Rate equations in such models should include also the mechanical contribution to mixing layer growth, parameterised by u_*.

For the mechanically-driven MH, all the current preprocessors rely on similarity formulae involving u_*, f, and partly also L_*; when using such formulae, Eq. (3) (Nieuwstadt, 1981) is to be preferred to the simple u_*/f approach for stable conditions. However, we regard this approach as not satisfactory from a physical point of view (see Section 3.2), and it is not applicable in low latitudes. Richardson-number methods appear to be better in this respect (input data may be a problem, however). Using numerical models may become a solution in the future.

Preprocessors should allow the substitution of measured data for parameterised ones at any stage, even if they do not follow idealised parameterisations: e.g., if there are periods of downward heat flux during the day or upward heat flux during the night.

Preprocessors should be able to use all soundings available, not just those at the standard hours 00 and 12 UTC. The real launching time of the sounding should be considered, as it may vary by about one hour around the standard time.

Preprocessors should be able to work with high-resolution radiosonde data (e.g., readings every 2 to 10 s) and not only with significant levels as reported in the TEMP part B. This requires appropriate dimensioning of the arrays in the code as well as taking into account the presence of small fluctuations in the high-resolution data, which may, e.g., lead to relatively strong potential temperature gradients over small layers in a well-mixed CBL. The preferred solution would be, in addition to the archival of full resolution data, to provide TEMP part B data with enhanced resolution in the lower atmospheric layers.

Constants and parameters specific to a certain climatic region (e.g., absolute maxima or minima of the MH, or criteria to find convective lids) should be clearly documented, and users should have the possibility to change them.

5.3. *Future research*

The suitability of the parcel method for the analysis of temperature profiles in the CBL calls for studies on both the practical determination of the mean temperature from an arbitrary sounding (including cases where the ABL is not really well-mixed) and the determination of the constants in the calculation of the excess temperature.

There is a general need for more work on the SBL. Formulae for the mechanical MH, especially those based on Richardson numbers, should be tested on more data sets. The effects of intermittent turbulence and of waves in the outer SBL on the dispersion of pollutants should be studied, too.

The development of routines to derive the MH in deep CBLs from the backscatter intensity profile of electromagnetic wind profilers should be continued with the aim of operational applicability. Such work would ideally include sodar data to detect MHs below the lowest profiler range gates, especially under stable conditions. Supplementing this chain of profilers with minisodars for very shallow MHs should be considered.

The spatial and temporal representativity of MHs derived from measurements (including indirect effects, such as, e.g., those of the initial profile and the heat flux in CBL slab models) should be studied. Parameterisations of the entrainment zone (Gryning and Batchvarova, 1994; Beyrich and Gryning, 1998) and the influence of waves, wind shear and synoptic-scale vertical motion at the CBL top on the MH development should be further investigated.

Numerical simulation models using a prognostic equation for the turbulent kinetic energy (closure of the order 1.5 or higher) should be considered as an alternative to simple parameterisations in the future. Where spatial inhomogeneities are relevant, three-dimensional models must be used. As long as computer resources do not allow their application for long data series, one-dimensional models could be considered in simple terrain.

If NWP models are to provide input for dispersion modelling, specific data requirements, different from those of weather forecasting, should be taken into account. For example, better temporal resolution of the NWP model output or providing additional variables such as TKE, diffusion coefficients or parameters describing deep convection (venting of the ABL) would be desirable.

It should be investigated whether subsidence velocities and horizontal advection at the top of the CBL obtained from NWP model output (presently the only practical source) are reliable enough to improve the performance of CBL models. Techniques to fit parameterisations and models to observed data, such as variational methods, should be developed and implemented. We believe that this is the most important improvement which can be made to CBL slab models.

Last but not least, we recommend to establish long-term monitoring programmes for the ABL in different climatic regions, including measurements of the surface energy budget components, turbulence parameters at several levels, at least two continuous profiling systems, and radiosoundings. This goal could be accomplished by supplementing existing field sites. Further development of remote sounding systems in order to measure turbulence parameters accurately would bring considerable benefits for MH determination.

Acknowledgements

The authors of this report acknowledge the support given by COST through the European Commission and (P. Tercier) the Swiss Government. Data for the comparisons have kindly been supplied by the KNMI, the Institute of Meteorology and Climate Research at the Research Centre of Karlsruhe, the Meteorological Institute of the University of Munich, and the Institute of Troposphere Research in Leipzig. Ari Karppinen from the Finnish Meteorological Institute ran the FMI model for the computations of Section 4.3.

Appendix A. Data sets used for testing mixing height routines

A.1. Cabauw (1995/96)

Cabauw is a boundary layer study site operated by the Royal Netherlands Meteorological Institute KNMI (Monna and van der Vliet, 1987). It is located between Utrecht and Rotterdam ($51°58'$N, $4°56'$E, 2 m), surrounded mainly by pastures and meadows with interspersed small water channels. Turbulent fluxes of sensible and latent heat are available, derived by different methods. A sodar without Doppler capabilities is operated on the site; it is used to derive

MH from the backscatter profile. The data used in this study cover the period from July 1995 until January 1996. In addition to the on-site data, regular aerological soundings from the station De Bilt, which is about 25 km north-east of Cabauw, have been used.

A.2. Payerne 1995/96

Payerne is the aerological station of the Swiss Meteorological Institute located in the western part of the Swiss Midland ($46°49'$N, $6°57'$E, 491 m). The Swiss Midland is a hilly basin surrounded to the north by the Jura mountains and to the south by the Alps. The data relevant for this study comprise routine aerological soundings, measurements of the sensible heat and the momentum flux, and data obtained from a Remtech PA1 Doppler sodar and a Radian LAP-3000 1290 MHz wind profiler. Several periods between August 1995 and June 1996 have been chosen for the investigation.

A.3. SADE (1993/94)

The SADE-93 and SADE-94 experiments were carried out at the field site Melpitz ($51°32'$N, $12°54'$E, 87 m) about 40 km NE of the city of Leipzig. Melpitz is a flat-terrain site situated within relatively large agricultural fields and pastures. The analysis within this study focussed on three intensive observation periods: 18–23 September 1993, 20–30 September 1994, and 6–11 October 1994. For the MH determination, data from the Doppler-sodar ECHO-1D and from frequent radiosoundings were used. Surface fluxes were derived from gradient measurements and from eddy-correlation measurements.

Appendix B. Computer routines to derive mixing height values

B.1. The OML meteorological preprocessor (Olesen et al., 1987)

The meteorological preprocessor of the Danish dispersion model OML contains a module for the MH calculation. Each hour, the module computes a mechanical and, during daytime, a convective MH and selects the larger one as the actual value, but a minimum value of 150 m is applied. The mechanical MH is calculated as $h = 0.25u_*/f$ (Eq. (1)). The convective MH is obtained from the integration of the (prognostic) Eq. (7) with $A = 0.2$ and $B = 5$. It starts with the observed temperature profile, and uses observed or parameterised hourly values of u_* and Q_0. If a so-called convective lid is found in the noon sounding, the calculated convective MHs before noon are multiplied by the ratio of the base height of this lid to the calculated MH for 11 UTC. If a so-called sustained lid is found also in the following midnight sounding, the interpolated lid height is used as an upper bound for the MH.

B.2. The HPDM meteorological preprocessor (Hanna and Chang, 1992)

The meteorological preprocessor of the Hybrid Plume Dispersion Model (HPDM) produces time series of hourly values of the surface heat and momentum fluxes and of the mixing depth, using observations of wind speed, cloudiness, surface roughness length, surface moisture availability and albedo. Since the preprocessor in its standard form does not accept observed fluxes, the source code was modified in the present study in order to use measured fluxes. HPDM uses Eq. (3) (Nieuwstadt, 1981) to estimate the MH at night. During daytime, two separate MHs are calculated. One is obtained with Carson's (1973) prognostic formula (first term in Eq. (8)) which parameterises the entrainment heat flux as a fraction A of the surface heat flux ($A = 0.2$). A second MH is calculated, according to a suggestion of Weil and Brower (1983), considering solely the growth rate of the CBL due to mechanical turbulence. Finally, the larger value is taken as the convective MH. For neutral conditions (defined as Pasquill stability category D), Eq. (1) is used with $a_1 = 0.3$.

B.3. The meteorological preprocessor library of Servizi Territorio (1994)

The aim of Servizi Territorio was to offer a library of subroutines as a flexible tool which can be applied in different environments. One set of subroutines is provided for the computation of the ABL parameters. For convective MH (daytime), the so-called encroachment model (first term of Eq. (8) with $A = 0$) or the full mixed-layer growth model (Eq. (8)) are offered. Both routines assume a constant lapse rate above the top of the mixing layer throughout the day, to be taken from an early morning temperature sounding. For neutral and stable MH (night-time) the library includes Nieuwstadt's (1981) Eq. (3), Zilitinkevich's (1972) Eq. (2) with $a_2 = 0.4$, and Eq. (1). If $L_* < 0$, only Eq. (1) can be used. In all the formulae, f is set to $10^{-4}\,\mathrm{s}^{-1}$, but partly absorbed into the constants.

B.4. The FMI Routine (Karppinen et al., 1997, 1998)

The module used for the calculation of the MH at the Finnish Meteorological Institute considers 3 cases.

B.4.1. Wintertime situations

During Finnish winter the boundary layer is mostly stable or near neutral, even during daytime, and h can be estimated as a function of the friction velocity (Eq. (1)), with the coefficient estimated from the 00 and 12 UTC soundings.

B.4.2. Summer night situations

A summer night is a period when the ABL is stable with unstable sections at both ends of the nocturnal period. After sunset the thickness of the developing shallow inversion h_{inv} is assumed to keep increasing through the course of the night. The MH is expressed by the bulk formulae of Stull (1983a, b) using the integrals of Q_0 over time. If the height h given by Eq. (1) with $a_1 = 0.2$ is larger than that of the previous evening's unstable boundary layer prevailing at sunset, the latter value is used.

B.4.3. Summer day situations

Daytime is defined by an upward turbulent heat flux. The evolution of the CBL is simulated by solving numerically the standard slab model (Tennekes, 1973; Driedonks and Tennekes, 1984) leading to a system of equations similar to Eq. (7) but with an additional term in the denominator containing the velocity scale $w_M = (w_*^3 + C_\tau u_*^3)^{1/3}$. After solving the equations for the whole period, the height h is compared to the 12 UTC sounding and corrected if necessary. The solution of the equations is continuously checked against the potential temperature observed at 2 m. If the sounding shows that the situation is not unstable as had been suggested by the energy budget method, u_*, Q_*, and L_* are replaced by profile estimates obtained from the lowest layer of the 12 UTC sounding data.

B.5. The RODOS preprocessor (Mikkelsen et al., 1996; Mikkelsen and Desiato, 1993)

MET-RODOS is a comprehensive atmospheric transport and diffusion module, designed for operational use within the real-time, on-line emergency management system RODOS. The MH can be calculated from NWP model output utilising the bulk Richardson number method with $Ri_c = 0.25$, or from measurements.

During daytime, the mixed-layer growth model (Eq. (8)) is used. The integration starts with an initial value of h equal to 50 m; the lapse rate is taken from the previous midnight sounding. The MH during night is calculated with Nieuwstadt's (1981) formula (Eq. (3)). In the case of an upward heat flux at night or a downward heat flux during the day, the MH is calculated using Eq. (1) with $a_1 = 0.3$. In addition, if a strong inversion (lid) is found above 150 m, the MH is calculated using a bulk Richardson number formulation (Eq. (5)) with $Ri_c = 3.0$. If this lid is lower than the MH calculated otherwise, it is taken as the MH. The lid height is not interpolated in time.

References

Andrén, A., 1995. The structure of the stably stratified atmospheric boundary layer: a large-eddy simulation study. Quarterly Journal of the Royal Meteorological Society 121, 961–986.

Anfossi, D., Bacci, P., Longhetto, A., 1976. Forecasting of vertical temperature profiles in the atmosphere during nocturnal radiation inversions from air temperature trend at screen height. Quarterly Journal of the Royal Meteorological Society 102, 173–180.

Angevine, W.M., White, A.B., Avery, S.K., 1994a. Boundary layer depth and entrainment zone characterisation with a boundary-layer profiler. Boundary-Layer Meteorology 68, 375–385.

Angevine, W.M., Trainer, M., Parrish, D.D., Buhr, M.P., Fehsenfeld, F.C., Kok, G.L., 1994b. Wind profiler mixing depth and entrainment measurements with chemical applications. Proceedings of Eigth AMS Conference on Air Pollution Meteorology and AWMA, Nashville, pp. 32–34.

Aron, R., 1983. Mixing height – an inconsistent indicator of potential air pollution concentrations. Atmospheric Environment 17, 2193–2197.

Arya, S.P.S., 1981. Parameterizing the height of the stable atmospheric boundary layer. Journal of Applied Meteorology 20, 1192–1202.

Arya, S.P.S., Byun, D.W., 1987. Rate equations for the planetary boundary layer depth (urban vs. rural). In: Modeling the Urban Boundary Layer. Amer. Meteorol. Soc., Boston, pp. 215–251.

Asimakopoulos, D.N., Helmis, C.G., Petrakis, M., 1996. Mini-acoustic sounding – a powerful tool for ABL applications: recent advances and applications of acoustic mini-sodars. Boundary-Layer Meteorology 81, 49–61.

Bacci, P., Giraud, C., Longhetto, A., Richiardone, R., 1984. Acoustic sounding of land and sea breezes. Boundary-Layer Meteorology 28, 187–192.

Batchvarova, E., Gryning, S.E., 1991. Applied model for the growth of the daytime mixed layer. Boundary-Layer Meteorology 56, 261–274.

Batchvarova, E., Gryning, S.-E., 1994. Applied model for the height of the daytime mixed layer and the entrainment zone. Boundary-Layer Meteorology 71, 311–323.

Batchvarova, E., Cai, X., Gryning, S.E., Steyn, D., 1999. Modelling internal boundary layer development in a region with complex coastline. Boundary-Layer Meteorology 90, 1–20.

Baxter, R.A., 1991. Determination of mixing heights from data collected during the 1985 SCC-CAMP field program. Journal of Applied Meteorology 30, 598–606.

Beljaars, A.C.M., Betts, A.K., 1992. Validation of the boundary layer representation in the ECMWF model. ECMWF Seminar Proceedings: Validation of Models over Europe, Vol. II. Reading, UK, 7–11 September 1992.

Betts, A.K., 1973. Non-precipitating cumulus convection and its parameterisation. Quarterly Journal of the Royal Meteorological Society 99, 178–196.

Betts, A.K., 1992. FIFE atmospheric boundary layer budget methods. Journal of Geophysical Research 97, 18,523–18,531.

Betts, A.K., Barr, A.G., 1997. First international satellite land surfcae climatology Field Experiment 1987 sonde budget revisited. Journal of Geophysical Research 101, 23,285–23,288.

Betts, A.K., Albrecht, B.A., 1987. Conserved variable analysis of the convective boundary layer thermodynamic structure over tropical oceans. Journal of the Atmospheric Sciences 44, 83–99.

Beyrich, F., 1993. On the use of sodar data to estimate mixing height. Applied Physics B 57, 27–35.

Beyrich, F., 1994a. Sodar observations of the stable boundary layer height in relation to the nocturnal low-level jet. Meteorologische Zeitschrift (N.F.) 3, 29–34.

Beyrich, F., 1994b. Bestimmung der Mischungsschichthöhe aus Sodar-Daten unter Verwendung numerischer Modellrechnungen. Frankfurt/M.: Wiss.-Verlag Dr. W. Maraun (ISBN 3-927548-67-7). Schriftenreihe des FhI für Atmosphärische Umweltforschung Garmisch-Partenkirchen, Bd. 28, 161 pp. + Appendix.

Beyrich, F., 1995. Mixing height estimation in the convective boundary layer using sodar data. Boundary-Layer Meteorology 74, 1–18.

Beyrich, F., 1997. Mixing height estimation from sodar data – a critical discussion. Atmospheric Environment 31, 3941–3954.

Beyrich, F., Görsdorf, U., 1995. Composing the diurnal cycle of mixing height from simultaneous sodar and wind profiler measurements. Boundary-Layer Meteorology 76, 387–394.

Beyrich, F., Gryning, S.-E., 1998. Estimation of the entrainment zone depth in a shallow convective boundary layer from sodar data. Journal of Applied Meteorology 37, 255–268.

Beyrich, F., Güsten, H., Sprung, D., Weisensee, U., 1996. Comparative analysis of sodar and ozone profile measurements in a complex structured boundary layer and implications for mixing height estimation. Boundary-Layer Meteorology 81, 1–9.

Beyrich, F., Klose, B., 1988. Some aspects of modeling low-level jets. Boundary-Layer Meteorology 43, 1–14.

Beyrich, F., Weill, A., 1993. Some aspects of determining the stable boundary layer depth from sodar data. Boundary-Layer Meteorology 63, 97–116.

Blackadar, A.K., 1979. High resolution models of the planetary boundary layer. In: Pfafflin, J.R., Zeigler, E.N. (Eds.), Advances in Environmental Science and Engineering, Vol. 1. Gordon and Breach, New York, pp. 50–85.

Brost, R.A., Wyngaard, J.C., 1978. A model study of the stably stratified planetary boundary layer. Journal of the Atmospheric Sciences 35, 1427–1440.

Carson, D.J., 1973. The development of a dry inversion-capped convectively unstable boundary layer. Quarterly Journal of the Royal Meteorological Society 99, 450–467.

Caughey, S.J., 1982. Observed characteristics of the atmospheric boundary layer. In: Nieuwstadt, F.T.M., van Dop, H. (Eds.), Atmospheric Turbulence and Air Pollution Modeling. Reidel Publ. Co., Dordrecht, pp. 107–158.

Clarke, R.H., 1990. Modeling mixed layer growth in the Koorin experiment. Australian Meteorological Magazine 38, 227–234.

Clifford, S.F., Kaimal, J.C., Lataitis, R.J., Strauch, R.G., 1994. Ground-based remote profiling in atmospheric studies: an overview. Proceedings of IEEE 82 (3), 313–355.

Coulter, R.L., 1979. A comparison of three methods for measuring mixing layer height. Journal of Applied Meteorology 18, 1495–1499.

Culf, A., 1992. An application of simple models to Sahelian convective boundary layer growth. Boundary-Layer Meteorology 58, 1–18.

Delage, Y., 1974. A numerical study of the nocturnal boundary layer. Quarterly Journal of the Royal Meteorological Society 100, 351–364.

Devara, P.C.S., Raj, P.E., Murthy, B.S., Pandithurai, G., Sharma, S., Vernekar, K.G., 1995. Intercomparison of nocturnal lower-atmospheric structure observed with lidar and sodar techniques at Pune, India. Journal of Applied Meteorology 34, 1375–1383.

Dohrn, R., Raschke, E., Bujnoch, A., Warmbier, G., 1982. Inversion structure heights above the city of Cologne (Germany) and a rural station nearby as measured with two sodars. Meteorologische Rundschau 35, 133–144.

Dörnbrack, A., 1989. Approximative Berechnung turbulenter Flüsse und des Tensors der turbulenten Diffusion auf der Grundlage einer Schließung 2. Ordnung. Dissertation, Humboldt-Univ. Berlin, 138 S.

Driedonks, A.G.M., 1981. Dynamics of the well-mixed atmospheric boundary layer. De Bilt: KNMI Sci. Rep. WR 81–2, 189 pp.

Driedonks, A.G.M., 1982a. Sensitivity analysis of the equations for a convective mixed layer. Boundary-Layer Meteorology 22, 475–480.

Driedonks, A.G.M., 1982b. Models and observations of the growth of the atmospheric boundary layer. Boundary-Layer Meteorology 23, 283–306.

Driedonks, A.G.M., Tennekes, H., 1984. Entrainment effects in the well-mixed atmospheric boundary layer. Boundary-Layer Meteorology 30, 75–105.

Dupont, E., 1991. Étude méthodologique et expérimentale de la couche limite atmosphérique par télédetection laser. Ph.D. Thesis, Univ. Paris VI, 220 pp.

Dye, T.S., Lindsay, C.G., Anderson, J.A., 1995. Estimates of mixing depth from boundary layer radar profilers. Proceedings of 9th AMS Symposium on Meteorological Instruments & Obs., Charlotteville, pp. 156–160.

Engelbart, D., 1998. Determination of boundary-layer parameters using windprofiler/RASS and sodar/RASS. Proceedings of 9th International Symposium on Acoustic Remote Sensing (IS-ARS'98), Vienna, Austria. Österr. Beiträge zu Meteorologie und Geophysik 17, 192–195.

Estournel, C., Guedalia, D., 1990. Improving the diagnostic relation for the nocturnal boundary layer height. Boundary-Layer Meteorology 53, 191–198.

Ftling, D., Wippermann, F., 1975. The height of the planetary boundary layer and of the surface layer. Beitraege Phys. Atmos. 48, 250–254.

Fay, B., Schrodin, R., Jacobsen, I., Engelbart, D., 1997. Validation of mixing heights derived from the operational NWP models at the German Weather Service. In: The Determination of the Mixing Height – Current Progress and Problems. EURASAP Workshop Proceedings, 1–3 Oct 1997, Report Risø-R-997(EN), ISBN 87-550-2325-8, Risø National Laboratory, Roskilde, Denmark, pp. 55–58.

Fisher, B.E.A., Erbrink, J.J., Finardi, S., Jeannet, P., Joffre, S., Morselli, M.G., Pechinger, U., Seibert, P., Thomson D.J. (Eds.), 1998. COST Action 710-Final Report. Harmonisation of the pre-processing of meteorological data for atmospheric dispersion models. L-2985 European Commission, Luxembourg, EUR 18195 EN (ISBN 92-828-3302-X).

Fitzharris, B.B., Turner, A., McKinley, W., 1983. Cold season inversion frequencies as measured with acoustic sounder in the Cromwell Basin. New Zealand Journal of Science 26, 307–313.

Garrett, A.J., 1981. Comparison of observed mixed layer depth to model estimates using observed temperature and winds, and MOS forecasts. Journal of Applied Meteorology 20, 1277–1283.

Garratt, J.R., 1982a. Observations in the nocturnal boundary layer. Boundary-Layer Meteorology 22, 21–48.

Garratt, J.R., 1982b. Surface fluxes and the nocturnal boundary layer height. Journal of Applied Meteorology 21, 725–729.

Garratt, J.R., 1992. The Atmospheric Boundary Layer. University Press, Cambridge, 316 pp.

Gaynor, J.E., Ye Jin Ping, White, A.B., 1994. Determining mixing depths in complex terrain near a power plant with radar profiler reflectivities. Proceedings of 8th AMS Conference on Air Pollution Meteorology & AWMA, Nashville, pp. 335–339.

Gland, H., 1981. Qualifying test on a three-dimensional Doppler-sodar (Satolas: July–December 1980). Electricité de France, Paris, Report EDF HE/32-81.9, 36 pp.

Godowitch, J.M., Ching, K.S., Clarke, J.F., 1985. Evolution of the nocturnal inversion layer at an urban and nonurban location. Journal of Climate and Applied Meteorology 24, 791–804.

Görsdorf, U., Lehman, V., 2000. Enhanced accuracy of RASS measured temperatures due to an improved range correction. Journal of Atmospheric and Oceanic Technology, in print.

Gryning, S.-E., Batchvarova, E., 1990. Analytical model for the growth of the coastal internal boundary layer during onshore flow. Quarterly Journal of the Royal Meteorological Society 116, 187–203.

Gryning, S.-E., Batchvarova, E., 1994. Parametrization of the depth of the entrainment zone above the daytime mixed layer. Quarterly Journal of the Royal Meteorological Society 120, 47–58.

Hanna, S.R., 1969. The thickness of the planetary boundary layer. Atmospheric Environment 3, 519–536.

Hanna, S.R., 1992. Effects of data limitations on hopes for improved short range atmospheric dispersion models. In: Olesen, H.R., Mikkelsen, T. (Eds.), Proceedings of Workshop "Objectives for next generation of practical short-range atmospheric dispersion models" (Risø, 1992). DCAR Roskilde, pp. 77–85.

Hanna, S.R., Burkhart, C.L., Paine, R.J., 1985. Mixing height uncertainties. Proceedings of 7th AMS Symposium on Turbulence and Diffusion, Boulder, pp. 82–85.

Hanna, S.R., Chang, J.C., 1992. Boundary-layer parameterizations for applied dispersion modeling over urban areas. Boundary-Layer Meteorology 58, 229–259.

Hayashi, M., 1980. Acoustic sounding of the lower atmospheric inversion layer. Journal of the Meteorology Society of Japan 58, 194–201.

Hicks, R.B., Smith, D., Irwin, P.J., Mathews, T., 1977. Preliminary studies of atmospheric acoustic sounding at Calgary. Boundary-Layer Meteorology 12, 201–212.

Holtslag, A.A.M., De Bruin, E.I.F., Pan, H.-L., 1990. A high resolution air mass transformation model for short range weather forecasting. Monthly Weather Review 118, 1561–1575.

Holzworth, C.G., 1964. Estimates of mean maximum mixing depths in the contiguous United States. Monthly Weather Review 92, 235–242.

Holzworth, C.G., 1967. Mixing depths, wind speeds and air pollution potential for selected locations in the United States. Journal of Applied Meteorology 6, 1039–1044.

Holzworth, C.G., 1972. Mixing depths, wind speeds, and potential for urban pollution throughout the contiguous United States. EPA, Office of Air Programs Publ. AP-101, 118 pp. (Can be obtained from EPA, Research Triangle Park NC 277711, USA).

Jakobsen, H.A., Berge, E., Iversen, T., Skålin, R., 1995. Status of the development of the multilayer Eulerian model. EMEP/MSC-W Note 3/95 (can be obtained from: Norwegian Meteorological Institute, P.O. Box 43, N-0313 Oslo 3, Norway).

Joffre, S.M., 1981. The physics of the mechanically driven atmospheric boundary layer as an example of air–sea ice interaction. Univ. of Helsinki, Dept. of Meteorology Rep. No. 20, 75 pp.

Jones, D.E., 1985. Mixing depth in La Trobe Valley. Clean Air in Australia 19, 49–51.

Kaimal, J.C., Abshire, N.L., Chadwick, R.B., Decker, M.T., Hooke, W.H., Kroepfli, R.A., Neff, W.D., Pasqualucci, F., Hildebrand, P.H., 1982. Estimating the depth of the daytime convective boundary layer. Journal of Applied Meteorology 21, 1123–1129.

Karppinen, A., Kukkonen, J., Nordlund, G., Rantakrans, E., Valkama, I., 1998. A dispersion modelling system for urban air pollution. Finnish Meteorological Institute: Publications on Air Quality 28. Helsinki, 58 pp.

Karppinen, A., Joffre, S., Vaajama, P., 1997. Boundary layer parametrization for Finnish regulatory dispersion models. International Journal of Environment and Pollution 8, 557–564.

Keder, J., 1999. Detection of inversions and mixing height by REMTECH PA2 sodar in comparison with collocated radiosonde measurements. Meteorology and Atmospheric Physics 71, in print.

Kitaigorodskii, S.A., Joffre, S.M., 1988. In search of a simple scaling for the height of the stratified atmospheric boundary layer. Tellus 40A, 419–433.

Klein Baltink, H., Holtslag, A.A.M., 1997. A comparison of boundary-layer heights inferred from windprofiler backscatter profiles with diagnostic calculations using regional model forecasts. Proc. EURASAP Workshop on The Determination of the Mixing Height – Current Progress and Problems, Risø National Laboratory, Denmark, 51–54.

Klöppel, M., Stilke, G., Wamser, C., 1978. Experimental investigations into variations of ground based inversions and comparison with results of simple boundary layer models. Boundary-Layer Meteorology 15, 135–145.

Koracin, D., Berkowicz, R., 1988. Nocturnal boundary layer height: observations by acoustic sounders and prediction in terms of surface layer parameters. Boundary-Layer Meteorology 43, 65–83.

Kurzeja, R.J., Berman, S., Weber, A.H., 1991. A climatological study of the nocturnal planetary boundary layer. Boundary-Layer Meteorology 54, 105–128.

Mahrt, L., 1981. Modeling the depth of the stable boundary layer. Boundary-Layer Meteorology 21, 3–19.

Mahrt, L., Heald, R.C., 1979. Comments on determining the height of the nocturnal boundary layer. Journal of Applied Meteorology 18, 383.

Mahrt, L., Heald, R.C., Lenschow, D.H., Stankov, B.B., Troen, I., 1979. An observational study of the structure of the nocturnal boundary layer. Boundary-Layer Meteorology 17, 247–264.

Mahrt, L., André, J.C., Heald, R.C., 1982. On the depth of the nocturnal boundary layer. Journal of Applied Meteorology 21, 90–92.

Manins, P.C., 1982. The daytime planetary boundary layer: A new interpretation of Wangara data. Quarterly Journal of the Royal Meteorological Society 108, 689–705.

Marsik, F.J., Fischer, K.W., McDonald, T.D., Samson, P.J., 1995. Comparison of methods for estimating mixing height used during the 1992 Atlanta Field Initiative. Journal of Applied Meteorology 34, 1802–1814.

Martin, C.L., Fitzjarrald, D., Garstang, M., Oliveira, A.P., Greco, S., Browell, E., 1988. Structure and growth of the mixing layer over the Amazonian rain forest. Journal of Geophysical Research 93, 1361–1375.

Maryon, R.H., Best, M.J., 1992. 'NAME', 'ATMES' and the boundary layer problem. Met O (APR) Turbulence and Diffusion Note No. 204 (U.K. Met. Office).

Mason, P.J., Derbyshire, S.H., 1990. Large-eddy simulation of the stably stratified atmospheric boundary layer. Boundary-Layer Meteorology 53, 117–162.

McElroy, J.L., Smith, T.B., 1991. Lidar descriptions of mixed layer thickness characteristics in a complex terrain/coastal environment. Journal of Applied Meteorology 30, 585–597.

Mikkelsen, T., Desiato, F., 1993. Atmospheric dispersion models and pre-processing of meteorological data for real-time application. Radiation Protection Dosimetry 50, 205–218.

Mikkelsen, T., Thykier-Nielsen, S., Astrup, P., Santabarbara, J.M., Sørensen, J.H., Rasmussen, A., Robertson, L., Ullerstig, A., Deme, S., Martens, R., Bartzis, J.G., Päsler-Sauer, J., 1997. METRODOS: A Comprehensive Atmospheric Dispersion Module. Radiation Protection Dosimetry 73, 45–56.

Miller, M.E., 1967. Forecasting afternoon mixing depth's and transport wind speed. Monthly Weather Review 95, 35–44.

Monna, W.A.A., van der Vliet, J.G., 1987. Facilities for research and weather observations on the 213 m tower at Cabauw and at remote locations. De Bilt: KNMI Sci. Rep. WR 87–5, 27 pp.

Nieuwstadt, F.T.M., 1981. The steady state height and resistance laws of the nocturnal boundary layer: theory compared with Cabauw observations. Boundary-Layer Meteorology 20, 3–17.

Nieuwstadt, F.T.M., 1984. Some aspects of the turbulent stable boundary layer. Boundary-Layer Meteorology 30, 31–55.

Nieuwstadt, F.T.M., Driedonks, A.G.M., 1979. The nocturnal boundary layer – a case study compared with model calculations. Journal of Applied Meteorology 18, 1397–1405.

Noonkester, V.R., 1976. The evolution of the clear air convective layer revealed by surface based remote sensors. Journal of Applied Meteorology 15, 594–606.

Olesen, H.R., Jensen, A.B., Brown, N., 1987. An operational procedure for mixing height estimation. Risø National Laboratory MST-Luft-A96. 2nd edition 1992, 182 pp.

Overland, J.E., Davidson, K.L., 1992. Geostrophic drag coefficients over sea ice. Tellus 44A, 54–66.

Piringer, M., 1988. The determination of mixing heights by sodar in an urban environment. In: Grefen, K., Löbel, J. (Eds.), Environmental Meteorology. Kluwer Academic Publishers, Dordrecht, pp. 425–444.

Rao, K.S., Snodgrass, H.F., 1979. Some parameterizations of the nocturnal boundary layer. Boundary-Layer Meteorology 17, 41–55.

Rayner, K.N., Watson, D., 1991. Operational prediction of daytime mixed layer heights for dispersion modeling. Atmospheric Environment 25A, 1427–1436.

Russell, P.B., Uthe, E.E., Ludwig, F.L., Shaw, N.A., 1974. A comparison of atmospheric structure as observed with monostatic acoustic sounder and lidar techniques. Journal of Geophysical Research 79, 5555–5566.

San José, R., Casanova, J., 1988. An empirical method to evaluate the height of the convective boundary layer by using small mast measurements. Atmospheric Research 22, 265–273.

Schlünzen, K.H., 1994. Mesoscale modelling in complex terrain – an overview on the German nonhydrostatic models. Contributions to Atmospheric Physics 67, 243–253.

Seibert, P., Beyrich, F., Gryning, S.E., Joffre, S., Rasmussen, A., Tercier, P., 1998. Mixing layer depth determination for dispersion modelling. European Commission. In: Fisher, B.E.A., Erbrink, J.J., Finardi, S., Jeannet, P., Joffre, S., Morselli, M.G., Pechinger, U., Seibert, P., Thomson, D.J. (Eds.), 1998: COST Action 710-Final Report. Harmonisation of the pre-processing of meteorological data for atmospheric dispersion models. L-2985 Luxembourg: European Commission, EUR 18195 EN (ISBN 92-828-3302-X).

Servizi Territorio, 1994. PBL_MET, a software library for advanced meteorological and air quality data processing. Cinisello Balsamo, Italy.

Smedman, A.S., 1988. Observations of multi-level turbulence structure in a very stable atmospheric boundary layer. Boundary-Layer Meteorology 44, 231–253.

Smedman, A.S., 1991. Some turbulence characteristics in stable atmospheric boundary layer flow. Journal of the Atmospheric Sciences 48, 856–868.

Sørensen, J.H., 1998. Sensitivity of the DERMA long-range Gaussian dispersion model to meteorological input and diffusion parameters. Atmospheric Environment 32, 4195–4206.

Sørensen, J.H., Rasmussen, A., Svensmark, H., 1996. Forecast of atmospheric boundary layer height utilised for ETEX real-time dispersion modelling. Physics and Chemistry of the Earth 21, 435–439.

Sørensen, J.H., Rasmussen, A., Ellermann, T., Lyck, E., 1998. Mesoscale influence on long-range transport, evidence from ETEX modelling and observations. Atmospheric Environment 32, 4207–4217.

Steyn, D.G., 1990. An advective mixed layer model for heat and moisture incorporating an analytic expression for moisture entrainment. Boundary-Layer Meteorology 53, 21–31.

Stull, R.B., 1976a. The energetics of entrainment across a density interface. Journal of the Atmospheric Sciences 33, 1260–1267.

Stull, R.B., 1976b. Internal gravity waves generated by penetrative convection. Journal of the Atmospheric Sciences 33, 1279–1286.

Stull, R.B., 1983a. A heat flux history length scale for the nocturnal boundary layer. Tellus 35A, 219–230.

Stull, R.B., 1983b. Integral scales for the nocturnal boundary layer. Part I: Empirical depth relationships. Journal of Climate Applied Meteorology 22, 673–686.

Stull, R.B., 1988. An Introduction to Boundary Layer Meteorology. Kluwer Academic Publishers, Dordrecht, 665 pp.

Tennekes, H., 1970. Free convection in the turbulent Ekman-layer of the atmosphere. Journal of the Atmospheric Sciences 27, 1027–1033.

Tennekes, H., 1973. A model for the dynamics of the inversion above a convective boundary layer. Journal of the Atmospheric Sciences 30, 558–567.

Tennekes, H., van Ulden, A.P., 1974. Short term forecasts of temperature and mixing height on sunny days. Proceedings of Second AMS Symposium Atmospheric Diffusion & Air Pollution, Santa Barbara, pp. 35–40.

Tercier, Ph., Stübi, R., Häberli, Ch., 1995. Evaluation de la hauteur de la couche limite de mélange dans le cadre du projet POLLUMET. Report of the Swiss Meteorol. Inst., 31 pp.

Tjemkes, S.A., Duynkerke, P.G., 1989. The nocturnal boundary layer: Model calculations compared with observations. Journal of Applied Meteorology 28, 161–175.

Troen, I., Mahrt, L., 1986. A simple model of the planetary boundary layer: Sensitivity to surface evaporation. Boundary-Layer Meteorology 37, 129–148.

van Pul, W.A.J., Holtslag, A.A.M., Swart, D.P.J., 1994. A comparison of ABL-heights inferred routinely from lidar and radiosonde at noontime. Boundary-Layer Meteorology 68, 173–191.

Vogelezang, D.H.P., Holtslag, A.A.M., 1996. Evolution and model impacts of alternative boundary layer formulations. Boundary-Layer Meteorology 81, 245–269.

Weil, J.C., Brower, R.P., 1983. Estimating convective boundary layer parameters for diffusion applications. Report PPSP-MP-48. Prepared by Environmental Center, Martin Marietta Corporation, for Maryland Department of Natural Resources, Annapolis, MD.

Wetzel, P.J., 1982. Toward parameterization of the stable boundary layer. Journal of Applied Meteorology 21, 7–13.

Wotawa, G., Stohl, A., Kromp-Kolb, H., 1996. Parameterization of the planetary boundary layer over Europe: A data comparison between the observation-based OML preprocessor and ECMWF model data. Contributions to Atmospheric Physics 69, 273–284.

Wyngaard, J.C., 1975. Modeling the planetary boundary layer – extension to the stable case. Boundary-Layer Meteorology 9, 441–460.

Yamada, T., Berman, S., 1979. A critical evaluation of a simple mixed layer model with penetrative convection. Journal of Applied Meteorology 18, 781–786.

Yu, T.W., 1978. Determining the height of the nocturnal boundary layer. Journal of Applied Meteorology 17, 28–33.

Zeman, O., Tennekes, H., 1977. Parameterization of the turbulent kinetic energy budget at the top of the daytime boundary layer. Journal of the Atmospheric Sciences 34, 111–123.

Zilitinkevich, S.S., 1972. On the determination of the height of the Ekman boundary layer. Boundary-Layer Meteorology 3, 141–145.

Zilitinkevich, S.S., 1975. Comments on a paper by H. Tennekes. Journal of the Atmospheric Sciences 32, 991–992.

Zilitinkevich, S.S., Fedorovich, E.E., Shabalova, M.V., 1992. Numerical model of a non-steady atmospheric planetary boundary layer, based on similarity theory. Boundary-Layer Meteorology 59, 387–411.

Zilitinkevich, S.S., Mironov, D.V., 1996. A multi-limit formulation for the equilibrium depth of a stably stratified boundary layer. Boundary-Layer Meteorology 81, 325–351.

Chapter 21

Computation, accuracy and applications of trajectories— A review and bibliography

Andreas Stohl

Lehrstuhl für Bioklimatologie und Immissionsforschung, Ludwig-Maximilians-Universität München, Am Hochanger 13, D-85354 Freising-Weihenstephan, Germany

Abstract

To the authors knowledge, there exists no recent review paper on the computation and use of trajectories. To fill this gap, this study summarizes the current knowledge on the calculation and application of trajectories. The different techniques that can be used to compute trajectories are presented and their error sources are described. The assumptions often made to account for the vertical wind velocity are explained. Most studies agree now that fully three-dimensional trajectories are the most accurate trajectory type. Methods to assess trajectory errors are outlined and a summary of the errors presented in the literature is given. Errors of 20% of the distance travelled seem to be typical for trajectories computed from analyzed wind fields. Finally, some important applications of trajectories, namely Lagrangian particle dispersion models, Lagrangian chemical box models and trajectory statistics, are discussed.

1. Introduction

Trajectory models, which describe the paths air parcels take, have been used to study dynamical processes in the atmosphere for several decades now. Applications vary from synoptic meteorology, for instance to investigate airmass flow around mountains (Steinacker, 1984), to climatology, for instance to identify pathways of water vapor transport (D'Abreton and Tyson, 1996) or desert dust (Chiapello et al., 1997), to the environmental sciences, for instance to establish *source–receptor relationships* of air pollutants (Stohl, 1996a). They may be even used to detect illegal cultivation of marihuana by combining pollen measurements in the ambient air with back trajectories (Cabezudo et al., 1997).

Various methods to compute trajectories, based on different assumptions, have been developed, and the accuracy of calculated trajectories has improved

First published in Atmospheric Environment 32 (1998) 947–966

gradually since 1940 when Petterssen (1940) introduced a graphical technique to compute trajectories. In the literature, calculated trajectories are sometimes interpreted as if they represent "ground truth", but this is never the case. Trajectories are often highly uncertain, which can eventually lead to serious misinterpretations of a flow situation if the magnitude of the errors cannot be estimated (Kahl, 1993). One objective of this paper is to describe and compare the different methods developed to calculate trajectories, and to describe the data sources usually available for trajectory calculations. It will be discussed how accurate trajectories typically are, on what factors their accuracy depends, and how it can be estimated in individual situations. This is of fundamental importance, since the resolution at which source–receptor relationships can be established is limited by the accuracy of the trajectories (Pack et al., 1978).

Another topic of this paper is to discuss how trajectories are applied in the environmental sciences, beginning with a brief discussion on Lagrangian particle dispersion models (LPDM). LPDM are an interesting extension to conventional trajectory models, since they allow a more realistic representation of transport in the planetary boundary layer where turbulence is important.

It will be analyzed how trajectories can serve as an input to Lagrangian box models. Lagrangian box models are often used to simulate complex physical and chemical processes occurring in the atmosphere. These models are very popular because they pose smaller computational demands than Eulerian chemical transport models.

Several decades ago trajectories were only used to investigate the transport processes associated with individual air pollution events, but now there exist sophisticated methods based on large sets of trajectories, to study also the air pollution climatology of a site. These methods will be presented.

2. Computation of trajectories

2.1. Definition

There exist two different ways to view air motions, namely the *Eulerian* and the *Lagrangian* perspectives (Byers, 1974; Dutton, 1986). The first one focuses on points fixed in space through which the air flows, the second one on individual air parcels as they move through time and space. The paths of these air parcels are known as *trajectories*.

Let us assume that we have a specific infinitesimally small air parcel, then its trajectory is defined by the differential *trajectory equation*

$$\frac{d\mathbf{X}}{dt} = \dot{\mathbf{X}}[\mathbf{X}(t)] \tag{1}$$

with t being time, $\mathbf{X}$ the position vector and $\dot{\mathbf{X}}$ the wind velocity vector. If we know the initial position $\mathbf{X}_0$ at time t_0 of the parcel, its path is completely determined through Eq. (1). We can write

$$\mathbf{X}(t) = \mathbf{X}(\mathbf{X}_0, t). \tag{2}$$

We can also find the inverse transformation

$$\mathbf{X}_0(t = t_0) = \mathbf{X}_0(\mathbf{X}, t) \tag{3}$$

which gives the initial coordinates of the parcel, which at time t is at position $\mathbf{X}$. Thus, air parcels may be followed either forward (*forward trajectories*) or backward (*back trajectories*) in time. The spatial coordinates $\mathbf{X}_0$ at time t_0 provide a means of identifying each air parcel for all time. These initial coordinates are called *material* or *Lagrangian coordinates* (Dutton, 1986).

An important feature of trajectories is that particles that are initially neighbors remain neighbors for all time. A line of particles at time t_0 remains an unbroken line at time t, no matter how it is distorted by the motion. This can be expressed by

$$\lim_{\Delta X_0 \to 0} |\mathbf{X}(\mathbf{X}_0 + \Delta \mathbf{X}_0, t) - \mathbf{X}(\mathbf{X}_0, t)| = 0. \tag{4}$$

The most important property of Eq. (4) is that particles that are inside a closed surface at time t_0 are forever separated from those outside. The closed surface that moves with the flow is called a *material surface*. An interesting application of this feature is contour advection (Waugh and Plumb, 1994).

It has to be noted that trajectories are different from *streamlines*. A streamline represents the direction of flow at a fixed instant of time and thus is everywhere tangent to the velocity vectors. At a certain instant of time, each particle in the flow is moving along its trajectory and so the streamlines are parallel to the trajectories, but as time passes, the streamlines must adjust to the flow and trajectories and streamlines are no longer parallel. Only during *stationary conditions* are the trajectories and the streamlines the same.

The idealized concept discussed above is not fully applicable in the real atmosphere. With the limited information available, it is not possible to pick an infinitesimal air parcel and follow its path with infinite accuracy. A real parcel of finite size may become distorted so strongly in a divergent flow that it is torn apart. Pflüger et al. (1990) remark to this problem (from German): "Real air parcels of finite dimensions are drawn asunder and are being deformed by inhomogeneities of the wind field, by turbulent and convective motions and by precipitation processes. The combined action of these factors is the reason for the fact that in general a single trajectory is not sufficient for a description of the path [of an air parcel] and that the mass center of gravity of the air parcel does not exactly follow the path of the computed trajectory." Hence, a computed trajectory is representative for the path of an air parcel only for a limited period.

2.2. *Solution of the trajectory equation*

Eq. (1) can be solved analytically only for simple flow fields; for meteorological applications, a finite-difference approximation of Eq. (1) must be used (Walmsley and Mailhot, 1983). Expanding $\mathbf{X}(t)$ in a Taylor series about $t = t_0$ and evaluating at $t_1 = t_0 + \Delta t$, one obtains

$$\mathbf{X}(t_1) = \mathbf{X}(t_0) + (\Delta t) \left. \frac{d\mathbf{X}}{dt} \right|_{t_0} + \frac{1}{2} (\Delta t)^2 \left. \frac{d^2\mathbf{X}}{dt^2} \right|_{t_0} + \cdots. \tag{5}$$

The first approximation to Eq. (5) is

$$\mathbf{X}(t_1) \approx \mathbf{X}(t_0) + (\Delta t)\dot{\mathbf{X}}(t_0) \tag{6}$$

a *"zero acceleration"* solution of Eq. (1) that is computationally cheap since it involves no iteration. It is accurate to the first order, which means that differences between the real and the numerical solution occur from the ommision of the second- and higher-order terms. If trajectories are calculated using very short integration time steps, Eq. (6) might be of sufficient accuracy. However, more accurate approximations at acceptable computational costs exist. If $\mathbf{X}(t)$ is also expanded in a Taylor series about $t = t_1$ and evaluated a $t = t_0$, this yields:

$$\mathbf{X}(t_0) = \mathbf{X}(t_1) - (\Delta t) \left. \frac{d\mathbf{X}}{dt} \right|_{t_1} + \frac{1}{2} (\Delta t)^2 \left. \frac{d^2\mathbf{X}}{dt^2} \right|_{t_1} - \cdots. \tag{7}$$

Combining Eqs. (5) and (7), we obtain

$$\mathbf{X}(t_1) = \mathbf{X}(t_0) + \frac{1}{2} (\Delta t)\left[\dot{\mathbf{X}}(t_0) + \dot{\mathbf{X}}(t_1)\right]$$
$$+ \frac{1}{4} (\Delta t)^2 \left[\left. \frac{d\dot{\mathbf{X}}}{dt} \right|_{t_0} - \left. \frac{d\dot{\mathbf{X}}}{dt} \right|_{t_1} \right] + \cdots. \tag{8}$$

If only the first two terms on the right-hand side of Eq. (8) are retained, the *"constant acceleration"* solution

$$\mathbf{X}(t_1) \approx \mathbf{X}(t_0) + \frac{1}{2} (\Delta t)\left[\dot{\mathbf{X}}(t_0) + \dot{\mathbf{X}}(t_1)\right] \tag{9}$$

results (Walmsley and Mailhot, 1983). This approximation is identical to Petterssen's (1940) scheme, originally a graphical method to construct isobaric trajectories manually from weather charts. Eq. (9) is accurate to the second order. It has to be solved by iteration starting with Eq. (6), since $\dot{\mathbf{X}}(t_1)$ is not *a priori* known:

$$\mathbf{X}^1(t_1) \approx \mathbf{X}(t_0) + (\Delta t)\dot{\mathbf{X}}(t_0)$$

$$\mathbf{X}^2(t_1) \approx \mathbf{X}(t_0) + \frac{1}{2}(\Delta t)\left[\dot{\mathbf{X}}(t_0) + \dot{\mathbf{X}}^1(t_1)\right]$$

$$\vdots$$

$$\mathbf{X}^i(t_1) \approx \mathbf{X}(t_0) + \frac{1}{2}(\Delta t)\left[\dot{\mathbf{X}}(t_0) + \dot{\mathbf{X}}^{i-1}(t_1)\right]. \qquad (10)$$

The superscripts indicate the number of iteration, and $\dot{\mathbf{X}}^i(t_1)$ is taken at position $\mathbf{X}^i(t_1)$.

Sometimes, the third term on the right-hand side of Eq. (8) is retained, too ("*variable acceleration*" *method*). In principle, this solution gives higher accuracy at the cost of increased computing time, but it has the disadvantage that the accelerations at two times must be evaluated. This can be inaccurate because wind fields are often available only at large temporal intervals. Hence, the variable acceleration method may even be less accurate than the constant acceleration method. If linear interpolation of the wind is used, the third term on the right-hand side of Eq. (8) vanishes, and the "variable acceleration" method reduces to the "constant acceleration" method.

All solutions discussed sofar are *kinematic* because they use the wind information only. Danielsen (1961) developed a technique to construct trajectories by tagging air parcels with (*quasi-*) *conservative quantities* such as potential temperature. Although two-dimensional kinematic trajectories can also be constructed on isentropic surfaces, Danielsen's (1961) method is *dynamic* because it makes use of velocity *and* mass field information and of *dynamic equations* linking the two (Merrill et al., 1986). Various different dynamic methods have appeared in the literature. The one of Petersen and Uccellini (1979) is based on the integration of the equation of motion for inviscid, adiabatic flow in isentropic coordinates,

$$\frac{d\mathbf{v}_h}{dt} + \nabla_\Theta M + f\,\mathbf{k} \times \mathbf{v}_h = 0, \qquad (11)$$

where $\mathbf{v}_h$ represents the horizontal wind vector, ∇_Θ is the gradient on isentropic surfaces, $M = c_p T + gz$ is the Montgomery potential, with c_p being the heat capacity of air at constant pressure, T the temperature and gz the geopotential, and f the Coriolis parameter. Using an estimation for the wind at the starting position and integrating Eq. (11), this method produces wind vectors along a trajectory.

For some time, dynamic methods were very popular (Merrill et al., 1986; Steinacker, 1984) because they allowed to use long intervals between the wind fields. For instance, Merrill et al. (1986) found that dynamic and two-dimensional kinematic trajectories calculated on isentropic surfaces agree very well for short intervals between the wind fields, but that dynamic trajectories calculated either with the implicit technique of Danielsen (1961) or with the explicit technique of Petersen and Uccellini (1979) are superior for wind field intervals in excess of 3 h. However, Stohl and Seibert (1997) showed that dynamic trajectories calculated with Steinacker's (1984) explicit method can perform unrealistic ageostrophic oscillations at travel times longer than 24 to 48 h that result from inaccurate determination of M. Since nowadays accurate wind fields with high space and time resolution are available, kinematic trajectories are more accurate (Stohl and Seibert, 1997).

2.3. Data sources for the computation of trajectories

In principle, trajectories can be calculated directly from wind observations by interpolating between the measurement locations. In practice, however, trajectory calculations are mostly based on the gridded output of numerical models. The most simple models are *diagnostic wind field models* (Sherman, 1978; Goodin et al., 1980; Ross et al., 1988; Mathur and Peters, 1990; Scire et al., 1990; Ludwig et al., 1991), some doing little more than interpolating radiosonde measurements, and adjusting them to fulfill mass consistency. Although their grid spacing may be much smaller than the average distance between radiosondes, due to a lack of model physics their effective resolution at upper levels is often limited by the resolution of the radiosonde network. Hence, there may be not much difference to interpolating directly from the measurement locations. If a dense network of surface wind measurements exists, the downscaling technique of Stohl et al. (1997) can yield more accurate wind fields in the planetary boundary layer (PBL) than conventional diagnostic models, but its applicability may be restricted to rather simple terrain.

On the synoptic scale, the most accurate wind data come from *numerical weather prediction* (NWP) centers. They use the most sophisticated methods currently available to produce accurate analysis fields to start their model forecasts. Hence a time series of these analyses should be used whenever possible. An additional bonus of this data source is that the data are easily accessible to many researchers. From most NWP models, data are available either on levels used internally by the model or on pressure levels that are interpolated from the model levels for synoptic purposes. For trajectory calculations, data on model levels are clearly better suited since interpolation errors are much smaller.

On the mesoscale, *prognostic mesoscale models* may produce more accurate wind fields (Pielke et al., 1992; Grell et al., 1995; Schlünzen, 1994). This requires, however, a very sophisticated and well validated modeling system. Important criteria that should be considered when the output of a mesoscale model is selected as the basis of trajectory computations are:

- Is the model physics suitable for the scale considered, for instance is the model non-hydrostatic?
- Is the terrain resolution sufficient?
- Is the soil moisture specified adequately?
- Arc the data used for model initialization and for boundary conditions of high quality?
- Are four-dimensional data assimilation techniques used to improve the forecast or to develop a better analysis archive?

Only modeling systems that are carefully selected according to the above and many other criteria can reproduce real small-scale structures in the wind fields. Otherwise, the wind fields, although containing much variation, may actually be less accurate than those of NWP models or those derived from radiosonde measurements.

2.4. Error sources for the computation of trajectories

Before discussing error sources, it is necessary to describe how differences between trajectories can be measured. Assessment of trajectory errors is often done in a statistical framework. Unfortunately, there exists some confusion in the literature regarding the statistical parameters used. Sometimes the same names have been given to different parameters, or different names have been given to identical ones. A measure that has been adopted by many authors in recent years is the absolute horizontal transport deviation (Kuo et al., 1985; Rolph and Draxler, 1990)

$$\text{AHTD}(t) = \frac{1}{N} \sum_{n=1}^{N} \{[X_n(t) - x_n(t)]^2 + [Y_n(t) - y_n(t)]^2\}^{1/2}, \quad (12)$$

where N is the number of trajectories used, X and Y are the locations of the test trajectories and x and y are the locations of some reference trajectories at travel time t. A similar measure can be defined in the vertical.

Another parameter often used is the relative horizontal transport deviation (RHTD). However, there exist several different definitions of RHTD. Some authors define RHTD as the AHTD divided by the average length of the reference trajectories (e.g. Rolph and Draxler, 1990), while others divide by the average length of test and reference trajectories (e.g. Stohl et al., 1995). In addition, the length of the trajectories is sometimes defined as the straight-line distance between the starting and ending points (e.g. Kuo et al., 1985), and sometimes as the length of the curved trajectory (e.g. Rolph and Draxler, 1990; Stohl et al., 1995). Currently, no general suggestion can be made regarding which of these definitions should be used in future work, but it is important to keep these differences in mind when comparing relative errors from different studies. Several other measures, like directional deviations, differences in length or in meandering, etc., have also been used, but it seems advisable to report them only in addition to AHTD and RHTD since they cannot be compared as readily with results of other authors.

Once large trajectory errors have occurred, trajectories separate further not only due to the occurrence of additional errors, but also due to the wind shear. An anonymous reviewer of the original version of this paper commented to this fact: "It hardly seems important to compare trajectories after they have separated some distance and are in different flow regimes—after which the errors will just amplify. It seems we need some type of approach that measures the length scale of the flow in which the trajectory is calculated. Once the error exceeds the length scale any further statistics are meaningless." Unfortunately, no such length scale has been used yet.

2.4.1. Truncation errors

The so-called *truncation errors* result when Eq. (1) is approximated by a finite-difference scheme that neglects the higher order terms of the Taylor series. Walmsley and Mailhot (1983) showed that the truncation error is proportional to Δt for the zero [Eq. (6)] and proportional to $(\Delta t)^2$ for the constant [Eq. (9)] and variable acceleration method. It can be kept below any desired limit by using sufficiently small Δt.

Walmsley and Mailhot (1983) computed trajectories with a numerical scheme of high order using very short time steps to keep truncation errors negligible. They compared these "exact" trajectories with ones computed with the operational methods discussed above using time steps of 3 h. After 42 h, position errors of the trajectories were 300 km for the zero acceleration, 100 km for the constant acceleration and 40 km for the variable acceleration method. Since an error of 300 km is significant when compared to other errors, the zero acceleration method can only be used with time steps much shorter than 3 h.

Similar results were obtained by Seibert (1993) who compared analytical solutions of the trajectory equation with numerical solutions. For a purely rotational flow she derived a stability criterion $\Delta t < 4/|\zeta|$, where ζ is the relative vorticity of the flow. For longer time steps, the constant acceleration scheme does not converge. Assuming a (large) relative vorticity $\zeta = 2 \times 10^{-4}$ s^{-1}, the time step must be smaller than 6 h. Seibert (1993) also showed that even when convergence is achieved, prohibitive truncation errors can occur. If truncation errors are to be kept below 1% of the distance travelled, the time step must be shorter than 1 h for $\zeta = 2 \times 10^{-4}$ s^{-1}. Since, on the one hand in most of the situations the time step requirements will be less demanding, while on the other hand a few situations may even be more demanding, Seibert (1993) recommended to use a scheme that automatically adjusts the time step to the actual flow situation. This was already done by Maryon and Heasman (1988), who used the difference between the constant and the variable acceleration scheme to estimate the truncation error. If the smallest features resolved by the wind data are to be reproduced in the trajectories, no grid cell must be skipped during a time step. Thus, the Courant–Friedrichs–Lewy criterion $\Delta t < \Delta x_i / |v_i|$, where Δx_i are the grid distances and v_i are the wind components, may serve as an upper limit for the flexible time step (Seibert, 1993).

2.4.2. Interpolation errors

Wind data are available only at discrete locations in space and time, either as irregularly spaced observations or as the gridded output of meteorological models. In either case, the wind speed must be estimated at the trajectory position by the trajectory model. This *interpolation* causes errors that affect the trajectory accuracy substantially.

Most trajectory models are based on gridded wind fields, but since radiosondes are the most important data for analyzing wind fields, it is interesting to discuss the errors caused by interpolation of radiosonde winds. Kahl and Samson (1986) selected three reference sites in the United States and tried several interpolation methods to estimate the wind vectors at these sites from all other radiosonde measurements. They found mean spatial interpolation errors in the horizontal wind components at an altitude of 1000 m of 3–4 m s^{-1}. The temporal interpolation errors were of similar magnitude. Kahl and Samson (1986) developed what they called a "trajectory of errors" procedure. They calculated an *error trajectory*, assuming that at each time step a normally distributed error with a standard deviation determined from the interpolation experiments occurs. Conducting a Monte Carlo simulation, Kahl and Samson (1986) found a mean trajectory position error after 72 h travel time of 400 km. However, since their approach did not account for the growth of trajectory errors in divergent wind fields, this might be an underestimation. In a different estimation, Kahl and Samson (1988a) found a lower limit of the mean trajectory error caused by interpolation of 100 km after 24 h travel time for the same data set. Kahl and Samson (1988b) repeated their study for another dataset gathered during highly convective conditions and found larger average interpolation errors of 5 m s^{-1}, yielding estimated mean trajectory position errors of 500 km after 72 h.

To examine the errors caused by interpolation from wind fields produced by prognostic meteorological models (either forecasts or initialized analyses), the usual method is to artificially degrade the grid resolution, interpolate the wind data to the original grid and compare with the undegraded data. Also, the temporal resolution may be varied. Stohl et al. (1995) evaluated the performance of several different interpolation methods. They found that linear interpolation is most accurate *in time*, but interpolation methods of higher order reduce errors *in space* as compared to linear interpolation. Similar results for spatial interpolation were obtained by Walmsley and Mailhot (1983). Stohl et al. (1995) also found that interpolation of the vertical wind component w produces larger errors than the interpolation of the horizontal components because of its high-frequency variability.

The effect of degrading the wind field resolution on trajectory accuracy was addressed in several studies (Kuo et al., 1985; Doty and Perkey, 1993; Rolph and Draxler, 1990; Stohl et al., 1995). One finding of these studies is that the growth of trajectory position errors with travel time caused by interpolation is approximately linear, but the most important result is that the spatial and temporal resolution of the wind fields must be in balance in order to limit the trajectory errors. An increase in spatial resolution alone does result in just marginally more accurate trajectories when the temporal resolution is low. On the contrary, increasing the temporal resolution alone is also not very effective when the spatial resolution is low. In any case, a minimum resolution of 6 h is necessary if any diurnal variations in the flow field are to be resolved.

As an example, the results of Rolph and Draxler (1990) are summarized in Table 1. At high spatial resolution, trajectories are more sensitive to a reduction of the temporal resolution than to a reduction of the horizontal resolution. However, at 360 km resolution, except for the 12 h case, the coarse spatial resolution becomes the dominant reason for trajectory errors.

The effect of interpolation errors on trajectory accuracy may also depend on the complexity of the flow situation. Stohl et al. (1995) found larger sensitivity to interpolation errors for trajectories crossing the Alps than for others. On the contrary, the sensitivity to interpolation errors did not depend on the starting

Table 1. Mean horizontal trajectory position deviations (km) from the high-resolution reference trajectories after 96 h travel time for varying spatial and temporal resolutions of the input data as found by Rolph and Draxler (1990)

Spatial resolution (km)	Temporal resolution			
	2 h	4 h	6 h	12 h
90	0	250	411	734
180	166	281	418	733
360	417	444	517	730

position of the trajectories (either in or outside a tropopause fold) in the study of Scheele et al. (1996).

2.4.3. *Errors resulting from certain assumptions regarding the vertical wind*

Trajectory errors are also related to different assumptions regarding the vertical wind component w. In contrast to the horizontal wind, there are no routine observations of w. Fields of w are a sole product of meteorological models, and hence they are less accurate than the fields of the horizontal wind. Sardeshmukh and Licbmann (1993) compared circulation analyses of the European Centre for Medium-Range Weather Forecasts (ECMWF) and the U.S. National Meteorological Center (NMC) for the tropics. They found a signal to noise ratio of the divergence of the horizontal wind at 200 hPa of only 2.1 : 1. Since w is balanced by the vertically integrated horizontal wind divergence, this shows the large uncertainty of w at low latitudes, mainly caused by difficulties in the parameterization of cumulus convection which is also of significance at higher latitudes during the summer. Nevertheless, if accurate fields of w are available, *three-dimensional* trajectories are more accurate than all the others (Martin et al., 1987, 1990; Stohl and Seibert, 1997).

Fuelberg et al. (1996) pointed out that it is important to obtain vertical velocities directly from a dynamically consistent numerical model. Vertical motions diagnosed from the horizontal wind components using the principle of continuity are much less accurate. Trajectories computed with these vertical motions sometimes experience unrealistically large diabatic heating or cooling rates, whereas dynamically consistent vertical motions keep the diabatic heating and cooling rates within the limits expected from theory.

The simplest alternative to three-dimensional trajectories is the neglection of w, resulting in *two-dimensional* trajectories in different coordinate systems. Often used are *isobaric* trajectories, but they are the least realistic; they may even travel below the topography. Also popular are terrain-following trajectories, e.g. the *isosigma* trajectories, for which $\sigma = (p - p_t)/(p_s - p_t)$ is kept constant (where p is the pressure, p_s the pressure at the ground and p_t the pressure at the highest model level). They may be better than isobaric trajectories in mountainous areas, but they neglect vertical motions of synoptic origin.

In *isentropic* coordinates, trajectories are truly two-dimensional under adiabatic and inviscid conditions (Danielsen, 1961; Petersen and Uccellini, 1979; Merrill et al., 1985, 1986; Artz et al., 1985). However, large problems are encountered in the PBL and in saturated moist air, where diabatic effects are most important. Draxler (1996a) demonstrated that isentropic and three-dimensional trajectories resemble each other closely during most of the time (90%), but can differ substantially when they enter baroclinic regions of the troposphere. Stohl and Seibert (1997) found that isentropic trajectories are affected more than others by *dynamical inconsistencies* between subsequent wind fields which occur when a sequence of wind *analyses* is used for the trajectory calculations.

In the PBL, the concept of a trajectory being representative for the path of an air parcel does not hold; such a parcel quickly looses its identity by turbulent mixing. Stochastic Lagrangian particle dispersion models (see Section 4.1) can be applied to simulate the transport in the PBL, but they are demanding on computer resources and not generally applicable. Several simple methods have been proposed instead to approximate transport processes in the PBL by single trajectories. Often used are trajectories that are advected with the *vertically averaged wind* in the PBL (Heffter, 1980; Rao et al., 1983). This approach was found to agree best with the dispersion of tracer material (Haagenson et al., 1987, 1990). Harris and Kahl (1994) combined it with the isentropic concept, switching between layer-averaged advection close to the ground and advection on isentropic surfaces above.

If the PBL height fluctuates, transport of tracer material is even more complex. It may be advected partly in the PBL and partly in nighttime *residual layers* aloft. One approach (Henmi, 1980; Heffter, 1983; Comrie, 1994) is to branch a trajectory between boundary and free tropospheric layer at each transition between day and night. Another approach is to average the wind not within the boundary layer, but up to the top of the pollutant reservoir layer (Stohl and Wotawa, 1995). However, all these methods give only crude approximations of the real complexities encountered in turbulent flow.

2.4.4. Wind field errors

In many cases, errors of the underlying wind fields are the largest single source of error for trajectory calculations. Wind field errors can be due to either *analysis errors* or *forecast errors*, depending on the type of wind fields used. Trajectory errors caused by erroneous forecasts are relatively simple to evaluate by comparing forecast with analysis trajectories. Maryon and Heasman (1988) studied a one-year collection of 950 hPa forward trajectories computed from wind fields generated by a NWP model. For trajectories released at $T = 0$ (at the begin of the forecast), the mean position error after 36 h travel time was 245 km, whereas for trajectories released at $T = +36$ h (36 h into the forecast period) the mean position error after 36 h was 720 km or 60% of the distance travelled.

Stunder (1996) compared a one-year set of forecast and analyzed three-dimensional 48 h trajectories starting at altitudes of 500, 1000 and 1500 m at $T = 0$ h. She found approximately linear separation rates of 200 km/day. After 12 h, the mean position error was between 30 and 40% of the distance travelled. Trajectories with minimum relative errors tended to originate within strong steady flow either ahead or behind a cold front. Trajectories with maximum relative errors originated within high pressure systems. Haagenson et al. (1990) found forecast trajectory errors of 400 km d^{-1} for a set of 20 boundary layer trajectories that were started at $T = +5$ h. They reported smaller separation rates of 200 km d^{-1} for analysis trajectories. Stohl (1996a) investigated a one year set of terrain-following 96 h *back* trajectories at a level of 800 m above ground terminating at $T = +24$, $T = +48$ and $T = +72$ h. He found relative errors at the origin of the trajectories of 16, 26 and 36% of the travel distance, respectively. Investigations of forecast trajectories also have been presented by Heffter et al. (1990) and Heffter and Stunder (1993). Summarizing the above findings, it must be concluded that forecast errors can have a large effect on trajectory accuracy. Even for calculations that extend only shortly into the forecast period, average position errors of 30% are typical. One possibility to assess the uncertainty of forecast trajectories on-line would be to use ensemble prediction products, provided for instance by the ECMWF.

The evaluation of trajectory errors caused by wind field *analysis* errors can be done by comparing trajectories calculated from different analysis data sets. Kahl et al. (1989a, b) applied an isobaric model to 850 and 700 hPa analyses provided by the ECMWF and the NMC. The median separation between the trajectories after 120 h was 1000 km. Pickering et al. (1994) made a similar comparison of isentropic trajectories based on wind analyses of the ECMWF and the NMC. Their area of study was the South Atlantic, a region with scarce observational data. The average separation of the trajectories was approximately 1500 km after 120 h travel time and nearly 2500 km after 192 h travel time, approximately 60% of the distances travelled! Pickering et al. (1996) argued that the ECMWF analyses may be slightly more accurate than the NMC analyses in that data scarce region. Isentropic trajectory separations were slightly smaller in this more recent study, probably caused by an improvement in the NMC analysis scheme. However, they also showed that three-dimensional trajectories are even more affected by analysis errors than isentropic trajectories.

2.4.5. *Starting position errors, amplification of errors, and ensemble methods*

The starting positions of the trajectories are often not exactly known. For example, estimations of the effective source height of accidentally released material are usually very inaccurate. Another uncertainty is due to the differences between the model topography and the real topography, making the selection of a starting height difficult. Although the initial trajectory position error may be rather small, it can strongly amplify in *divergent* (forward trajectories) or *convergent* (back trajectories) flow.

To account for such effects, Merrill et al. (1985) started an *ensemble* of trajectories with slightly differing initial positions. In most cases, the ensembles stayed close together, but in a few cases the trajectory end points covered a very large region, indicating large uncertainty. The use of ensemble methods was also encouraged by Seibert (1993) and their capacity for assessing trajectory uncertainty was demonstrated by Baumann and Stohl (1997) in a comparison with balloon tracks. Although the ensemble method does not produce more accurate trajectories, it gives a reliable estimation of the sensitivity of the trajectories to initial errors and other errors.

Kahl (1996) used a set of stochastic trajectories with random wind components typical for interpolation errors added at each time step to the mean wind to determine a *"meteorological complexity factor"* (MCF), defined as the average separation distance between the stochastic trajectories and a reference trajectory. He assumed that trajectory uncertainty can be predicted as a function of the MCF but since the magnitude of the MCF depends critically on the integration time step, this method might not be generally applicable.

An ensemble should ideally consist of trajectories started within a four-dimensional domain, scaled by the horizontal, vertical and temporal resolution of the wind fields and with random interpolation errors added at each time step. There remains, however, the difficulty to find the most appropriate scaling factors for the initial position displacements and for the interpolation errors. The random interpolation errors should be autocorrelated, since otherwise their effect would depend on the integration time step. Unfortunately, it is virtually unexplored how to design an ensemble method such that it gives a quantitative estimation of the trajectory uncertainty.

3. Accuracy of trajectories: tracer studies

The overall accuracy of a trajectory is determined by the integral effect of all errors discussed in the previous section. Its assessment is difficult because it requires the determination of a *"true"* reference trajectory. Except for laboratory studies (Tajima et al., 1997), this is only possible by tagging an air parcel by a *tracer* that is conserved along the trajectory. Many different tracers have been used, but none of them is ideally suited, either because it is not conserved well enough, because its determination is difficult, or because it is not normally available. Studies based on three different classes of tracers, namely balloons, material and dynamical tracers, can be found in the literature. Table 2 summarizes some of the results.

3.1. Balloons

Different types of balloons can be used for trajectory evaluations, but most studies have been done with *constant level balloons* which are intended to remain on constant density surfaces (Angell and Pack, 1960; Angell et al., 1972;

Table 2. Absolute and relative trajectory errors reported in the literature

Type of errors	Evaluated against	Comment	Travel time (h)	Errors	Reference
Truncation	Traj. computed with short integration time steps	Errors resulting from time step of 3 h using zero (constant) [variable] acceleration method	42	300 (100) [40] km	Walmsley and Mailhaut (1983)
Interpolation	Zero-interpolation error traj.	Superposition of stochastic interpolation errors occurring along traj.	72	400 km	Kahl and Samson (1986)
Interpolation	Zero-interpolation error traj.	Same as above, but for more convective conditions	72	500 km	Kahl and Samson (1988b)
Temporal interpolation	Calculated traj.	3-month set of 3-D traj. calculated from wind fields of 12 h (6 h) [4 h] time resolution *vs.* 2 h time resolution	96	730 km (410 km) [250 km]	Rolph and Draxler (1990)
Temporal interpolation	Calculated traj.	86 3-D traj. in an intense cyclone calculated from wind fields of 6 h (3 h) [1 h] time resolution *vs* 15 minutes time resolution	36	250 km (170 km) [30 km]	Doty and Perkey (1993)
Temporal interpolation	Calculated traj.	1-yr set of 3-D (2-D) traj. calculated from wind fields of 6 h time resolution *vs* 3 h time resolution	96	590 km, 20% (280 km, 9%)	Stohl et al. (1995)
Horizontal interpolation	Calculated traj.	3-month set of traj. calculated from wind fields of 360 km (180 km) resolution *vs* 90 km resolution	96	420 km (170 km)	Rolph and Draxler (1990)
Horizontal interpolation	Calculated traj.	1-yr set of 3-D (2-D) traj. calculated from wind fields of 1° resolution *vs* 0.5° resolution	96	411 km, 14% (111 km, 4%)	Stohl et al. (1995)
Forecast	Analysis traj.	1-yr set of 950 hPa forward traj. started at $T = 0$ h ($T = +36$ h)	36	245 km, 25% (720 km, 60%)	Maryon and Heasman (1988)
Forecast	Analysis traj.	1-yr set of forward 3-D traj. started 500, 1000, 1500 m above ground	> 12	200 km/day	Stunder (1996)

Table 2. (Continued)

Type of errors	Evaluated against	Comment	Travel time (h)	Errors	Reference
Forecast	Analysis traj.	1-yr set of back traj. travelling 800 m above ground terminating at $T = +24$ h ($T = +48$ h) [$T = +72$ h]	96	16% (26%) [36%]	Stohl (1996a)
Wind field analysis	ECMWF traj. compared to NMC traj.	Isobaric 850 and 700 hPa traj.	120	1000 km	Kahl et al. (1989a, b)
Wind field analysis	ECMWF traj. compared to NMC traj.	Isentropic traj. over the south Atlantic	120 (192)	1500 km, 60% (2500 km, 60%)	Pickering et al. (1994)
Total	Constant level balloon	26 cases, diagnostic wind field model used	< 24	25–30%	Clarke et al. (1983)
Total	Constant level balloon	16 cases in and immediately above the PBL	1–3	5–40%	Koffi et al. (1997a, b)
Total	Constant level balloon	Stratospheric traj.	12–144	≈ 20%	Knudsen and Carver (1994) Knudsen et al. (1996)
Total	Manned balloon	Single flight at a typical height of 500 hPa	100	10%	Draxler (1996b)
Total	Manned balloon	4 flights at a typical height of 2000 m	46	< 20%	Baumann and Stohl (1997)
Total	Tracer (CAPTEX)	6 cases, different types of traj.	24	≈ 200 km	Haagenson et al. (1987)
Total	Tracer (CAPTEX)	6 cases	24–42	150–180 km	Draxler (1987)
Total	Tracer (ANATEX)	30 cases	< 30	20–30%	Draxler (1991)
Total	Tracer (ANATEX)	23 boundary layer traj.	24–72	≈ 100 km/d^{-1}	Haagenson et al. (1990)
Total	Smoke plumes	112 traj. based on a fine-scale (global) analysis	< 60	10% (14%)	McQueen and Draxler (1994)
Total	Saharan dust	Single case, 3-D traj.	3000 km	200 km, 7%, vertical error 50 hPa	Reiff et al. (1986)
Total	Potential vorticity	1-yr set of 3-D traj. based on ECMWF data	120	< 20%, < 400 km, vertical error < 1300 m	Stohl and Seibert (1997)

Note. The table summarizes not only total errors, but also errors caused by single error sources, such as interpolation. Different errors reported by the same author are put in parantheses.

Hoecker, 1977; Pack et al., 1978; Kahl et al., 1991). Reisinger and Mueller (1983) compared 45 tetroon flights (altitudes 300–1500 m, mean travel distance 90 km) with trajectories calculated from interpolated radiosonde measurements. They found that computed trajectories underestimated the meandering of the tetroons due to the smoothing effect of interpolation. The mean direction difference between computed and tetroon trajectories was 28°. Warner et al. (1983) highlighted the importance of prognostic models for providing high-resolution input to trajectory models, because trajectories calculated from wind analyses were smoother than tetroon trajectories. In an evaluation of several different trajectory models using 26 tetroon flights, Clarke et al. (1983) found mean errors of approximately 25–30% of the travel distance. Koffi et al. (1997a, b) found relative errors of approximately 5–40% of the travel distance for constant level balloon flights undertaken during the ETEX tracer experiment.

Knudsen and Carver (1994) and Knudsen et al. (1996) looked at stratospheric long-duration (0.5–6 d) constant level balloon flights. They found typical differences of 20% of the travel distance between trajectories calculated from ECMWF analyses and balloon trajectories. Under special circumstances, 24 and 48 h trajectories had errors exceeding 40%.

The tracks of *manned balloons* may also be used for comparison with calculated trajectories, if changes in the balloon heights are accounted for (Wetzel et al., 1995). Draxler (1996b) found an error of 10% of the travel distance for a 100 h transpacific balloon flight at a level of approximately 500 hPa. Baumann and Stohl (1997) evaluated trajectories calculated from ECMWF analyses using four gas balloon flights (typical height 2000 m) of 46 to 92 h duration and found errors of 4–45% of the distance travelled. The average error after 46 h was less than 20%.

Stocker et al. (1990) used 190 000 tagged helium-filled latex balloons launched from 800 sites around the United States of which 4.5% were later found and returned. Balloon and calculated boundary layer trajectory motion were often in complete disagreement, especially under light wind conditions. Similar difficulties were reported by Baumann et al. (1996) based on a study of hot-air balloon flights in complex terrain.

The obvious disadvantage of all balloon-based studies is that balloons do *not* follow the real three-dimensional air motions, but tend to stay on pressure surfaces or actively change height (manned balloons). Therefore, errors in the vertical wind are not detected.

3.2. Material tracers

There are two groups of material tracers, *material tracers of opportunity*, such as smoke plumes or geochemical tracers, and *inert tracer gases* released during specifically designed experiments. Tracers of opportunity may be available without the high costs and efforts of planned tracer experiments, but usually the source position, strength and/or release time of the tracer material are not accurately known. This problem is not encountered within tracer experiments, but only a few well-documented data sets on the regional to continental scale are available.

3.2.1. Tracer experiments

Different chemical compounds have been used as tracers, basic requirements being that the material must be chemically stable, not subjected to wash-out, rain-out or dry deposition, non-toxic, environmentally safe, detectable at low concentrations, and must have near-zero background concentrations. Sulfur hexafluoride (SF_6) on the local scale (Lamb et al., 1978a, b) and perfluoro-carbons on the regional to continental scale (Draxler, 1991) have often been used. We will focus on regional to continental scale tracer experiments, be-cause short-range experiments are usually conducted to study turbulent flows for which the accuracy of single trajectories can hardly be judged.

Three tracer experiments received the most interest until now. The first one was the *Cross-Appalachian Tracer Experiment* (CAPTEX), conducted during September and October 1983 (Ferber et al., 1986). A total of seven releases of a perfluorocarbon was made when winds were predicted to carry it over the ground-level sampling network (80 sites) which covered the northeast United States and Canada.

Haagenson et al. (1987) analyzed the tracer data to a grid and derived *tracer trajectories* that tracked the location of the tracer plume centroid. They pointed out that this definition of the tracer trajectories is somewhat arbitrary, and alter-native definitions might be used as well. With the tracer trajectories, they evalu-ated several types of trajectories and came to the conclusions that surface winds should not be used to simulate PBL transport, that wind flow corresponding to the low to middle PBL is much more appropriate, that isentropic, layer-averaged and isosigma trajectories are more realistic than isobaric trajectories, and that average errors after 24 h travel time were approximately 200 km.

Draxler (1987), using the same dataset, found smaller trajectory errors (150–180 km as the average for 24–42 h travel time); isentropic trajectories were more accurate than isosigma trajectories; increasing the spatial *and* tem-poral resolution of the wind fields (using additional radiosondes) improved the trajectory accuracy significantly, whereas increasing spatial or temporal reso-lution alone did not improve it. Chock and Kuo (1990) compared trajectories computed from model predicted wind fields with trajectories computed from objective wind analyses. They stated that the former were more accurate, espe-cially in the presence of fronts, but gave no quantitative measure of accuracy.

Determining a tracer trajectory from the surface concentration pattern is difficult, because the spatial structure of the tracer concentrations depends also on other factors than the horizontal transport. Shi et al. (1990) used a Lagrangian particle model to simulate three CAPTEX releases. Since the calculated tracer plumes at higher levels looked very different from the calculated ground-level plumes, they concluded that comparisons with the surface tracer concentration footprint are not ideally suited to assess the accuracy of trajectories, because differences in vertical mixing can lead to substantially different horizontal trajectories due to the vertical shear of the horizontal wind.

The *Across North America Tracer Experiment* (ANATEX) was conducted from January to March 1987 (Draxler et al., 1991). 33 releases of different tracer gases were made simultaneously from each of two locations every 2.5 days, alternating between daytime and nighttime releases. Ground-level air samples of 24 h duration were taken at 77 sites situated throughout North America and air samples of 6 h duration were taken at 5 towers.

Haagenson et al. (1990) used tracer-derived trajectories to validate calculated trajectories. Layer-averaged trajectories performed best (separation rate 180 km d^{-1}), but isosigma trajectories were nearly equally good. The rate of increase of the trajectory error decreased slightly with time. Although three-dimensional trajectories were generally similar to isosigma trajectories, large upward displacement of the trajectories was associated with low tracer concentrations at the surface. Dispersion models based on two-dimensional trajectories may thus overpredict surface tracer concentrations. Haagenson et al. (1990) also used different types of meteorological data (objective analyses, forecasts, four-dimensional data assimilation) to drive the trajectory model. They found that the quality of the forecasts (96 h forecasts initiated 5 h prior to the tracer release) was much lower than the quality of the analyses. This is in direct contradiction to Chock and Kuo's (1990) study based on the CAPTEX data. Possible reasons for this are the longer forecast period used for the ANATEX data and the more complex meteorological conditions.

Draxler (1991) used aircraft measurements available for 30 ANATEX releases and calculated *back trajectories* from the tracer centroid positions to minimize the complications introduced in comparing trajectories while also having to model the vertical diffusion. He defined the trajectory error to be the nearest distance of a trajectory to the tracer origin point. He compared short-term forecasts of the Nested Grid Model (NGM) that were initialized every 12 h with analyses of radiosonde observations. Both for the NGM forecasts and for the analyses, transport errors were approximately 20–30% of the distance travelled.

The most recent tracer experiment on a continental scale was the *European Tracer Experiment* (ETEX) in October to November 1994 with two releases of a perfluorocarbon tracer from a site in western France. The sampling network provided high spatial (168 stations) and temporal (3 h) resolution (Mosca et al., 1997). The major task of ETEX was to evaluate real time forecasting of the tracer concentration field (Archer et al., 1996) as well as subsequent model analyses of the transport patterns, but the data set could be also used to explore the accuracy of trajectories.

3.2.2. Tracers of opportunity

Any detectable material that is conserved along trajectories may serve as a tracer; for instance pollen (Raynor et al., 1983), volcanic ash (Heffter et al., 1990; Heffter and Stunder, 1993) or the radioactive emissions from the Chernobyl accident (Kolb et al., 1989; Klug et al., 1992). Reiff et al. (1986) studied an episode of African dust transport to northwestern Europe. They used satellite imagery, upper-air soundings, surface observations, X-ray analyses of the dust composition and low-level dust-concentration measurements to determine the place of origin and the path of the desert dust. Independently calculated trajectories based upon wind analyses produced at the ECMWF tracked the dust cloud very accurately (error less than 200 km after a travel distance of 3000 km). Also the vertical displacements calculated by the model could be confirmed by the measurements (error less than 50 hPa). Martin et al. (1990) investigated several episodes of Saharan dust transport to the Mediterranean. They also suggested that the vertical wind component should be used for the trajectory computations.

McQueen and Draxler (1994) used the smoke plumes from the Kuwait oil fires during the Gulf war to evaluate the accuracy of trajectories. They determined the plume centerline from satellite data and the vertical plume position by comparing back trajectories released at different heights. They assumed that the plume centroid height is identical to that of the trajectory that most accurately tracked the plume back to the burning oil fields. Using a global analysis and a fine-scale analysis, they found trajectory errors of 10% for the fine grid and 14% for the coarse grid. Since the plume height was determined by the "best" trajectory, this is likely to be an underestimation of the "true" error.

3.3. Dynamical tracers

One dynamical tracer is *potential temperature* which is often used to constrain the vertical motion of trajectories (isentropic assumption). Another tracer is *isentropic potential vorticity* (PV) that is conserved for inviscid adiabatic motions (Davis, 1996). It is defined by

$$PV = -g\eta_\Theta \frac{\partial \Theta}{\partial p} \tag{13}$$

with $\eta_\Theta = \zeta_\Theta + f$, ζ_Θ being the relative vorticity on an isentropic surface, and g the gravitational acceleration. PV has been used sometimes to study the accuracy of individual trajectories (Artz et al., 1985; Jäger, 1992). However, Knudsen and Carver (1994) found that due to the high-frequency variability of PV, difficulties in its analysis and its non-conservation under diabatic conditions, only large PV changes can be taken as an indication of individual trajectories being wrong.

Nevertheless, PV is very useful in a statistical framework. Recently, Stohl and Seibert (1997) studied PV conservation to determine the average accuracy of a large set of trajectories calculated from ECMWF analyses. They found that three-dimensional trajectories were the most accurate trajectory type, followed by kinematic isentropic trajectories. Isobaric trajectories were clearly less accurate. Average relative horizontal position errors for three-dimensional trajectories were estimated to be less than 20% for travel times longer than 24 h in the free troposphere. Tentative upper bounds for average absolute horizontal and vertical errors after 120 h travel time were 400 km and 1300 m, respectively.

4. Applications of trajectories

4.1. Lagrangian particle dispersion models

Although trajectory models have been used successfully to study complex transport processes such as recirculation of pollutants (Tyson et al., 1996), it is virtually impossible to describe transport phenomena in *turbulent flows* by calculating single trajectories. More sophisticated models are needed, both in the PBL, where an air parcel quickly looses its identity due to strong mixing (Lyons et al., 1995), and at higher levels of the atmosphere when longer time scales are considered (Sutton, 1994). These models can be either Eulerian transport models or Lagrangian particle dispersion models (LPDM). LPDM have no artificial numerical diffusion like Eulerian models (Nguyen et al., 1997) and hence have a greater potential to resolve fine-scale structures of the flow.

LPDM numerically simulate the transport and diffusion of a passive scalar tracer by calculating the Lagrangian trajectories of tens or hundreds of thousands of tagged "*particles.*" These trajectories are calculated according to

$$\mathbf{X}(t + \Delta t) = \mathbf{X}(t) + \Delta t \big[\bar{\mathbf{v}}(t) + \mathbf{v}'(t) \big] \tag{14}$$

where $\bar{\mathbf{v}}$ is the *resolvable scale wind vector* obtained directly from a meteorological model, and $\mathbf{v}'$ is the *turbulent wind vector* that describes the turbulent diffusion of the tracer in the PBL. The concentration of the tracer at a specific location at a given time is linearly proportional to the number of particles per unit volume. It can be evaluated simply by counting all particles that reside within a certain volume or, more favorably, by using a kernel method (Lorimer, 1986; Uliasz, 1994).

Different alternative names are given to LPDM; e.g. Langevin, Markov chain, Monte Carlo, random walk, and stochastic Lagrangian (Rodean, 1996), each referring to a certain attribute of the models. See Thomson (1984, 1987), Wilson and Sawford (1996) and Rodean (1996) for detailed treatments of the mathematical and physical background of LPDM, and Uliasz (1994) and Zannetti (1992) for discussions of more practical aspects of LPDM simulations.

The core problem of LPDM is the determination of the turbulent velocities $\mathbf{v}'$. Turbulent diffusion, like molecular diffusion (*Brownian motion*), can be described as a *Markov process*, a stochastic process that has a future that depends only on its present state and a transition rule. The particles forget their current velocity state after some time characterized by the *Lagrangian time scale*. This is written as the *Langevin equation* (Thomson, 1987)

$$dv_i' = a_i(\mathbf{X}, \mathbf{v}', t)dt + b_{ij}(\mathbf{X}, \mathbf{v}', t)dW_j(t) \tag{15}$$

where a and b are functions of $\mathbf{X}$, $\mathbf{v}'$ and t and the dW_j are the increments of a vector-valued *Wiener process* with independent components. The dW_j represent Gaussian white noise with mean zero and variance dt; increments dW_i and dW_j occurring at different times, or at the same time with $i \neq j$, are independent.

For instance, given a state n of the turbulent vertical velocity w' of a particle at time t, a future state $n+1$ at time $t + \Delta t$ can be determined using the autocorrelation coefficient $r = \exp(-\Delta t/\tau_L)$ with τ_L being the Lagrangian time scale for w' (Wilson et al., 1983)

$$\frac{w'^{n+1}}{\sigma_w^{n+1}} = r\frac{w'^n}{\sigma_w^n} + \chi\sqrt{(1-r^2)} + (1-r)\tau_L\frac{d\sigma_w}{dz} \tag{16}$$

where χ is a normally distributed random number and σ_w is the root-mean-square Eulerian turbulent vertical velocity. Both τ_L and σ_w can be calculated diagnostically based on certain boundary layer characteristics (Hanna, 1982) or can be related to the turbulent kinetic energy (Fay et al., 1995). Using sufficiently small time steps, Eq. (16) fulfills the most important criterion for an LPDM, the "*well-mixed*" criterion, which states that if the particles of a passive tracer are initially mixed uniformly in a turbulent flow, they will remain so (Thomson, 1987). Only models that fulfill this criterion are physically correct.

The last term on the right-hand side of Eq. (16) is the *drift correction velocity*, introduced by Legg and Raupach (1982) in a similar, albeit, as was found later, not correct form in order to avoid the accumulation of particles in regions of low turbulence which would violate the well-mixed criterion (McNider et al., 1988). Special attention must be paid to flow boundaries (i.e. the ground and the top of the PBL), where particles are reflected (Wilson and Flesch, 1993; Thomson and Montgomery, 1994).

Usually, Gaussian turbulence is assumed in an LPDM, but under convective conditions vertical tracer transport occurs primarily in *updrafts* and *downdrafts*. Updrafts have higher velocities but occupy less area than downdrafts leading to a skewed vertical velocity distribution (Luhar et al., 1996). Baerentsen and Berkowicz (1984) approximated this by the sum of two Gaussian distributions, one for the updrafts and the other for the downdrafts.

There exist several approaches to deal with the turbulent horizontal velocities. The simplest solution is to solve an independent Langevin equation for all three wind components, but measurements of wind fluctuations indicated that there exist cross-correlation terms between the individual wind components (Zannetti, 1992). These cross-correlations can be very important near the source, but Uliasz (1994), who compared an LPDM simulation that accounted for the cross-correlations with another that did not, found that in mesoscale applications the cross-correlations are not important. Actually, the horizontal turbulent velocities had little effect at all, even when they were completely neglected.

Another simplification to save computing time is to neglect the autocorrelations of the turbulent wind components or to increase the model time step above the Lagrangian timescale. This is equivalent to converting the Markov process for the velocity *and* position of the particles into one for their position only. Uliasz (1994) showed that even this gross simplification had no large effects on the simulated tracer concentrations in regional applications. The reason for this is that the most important process affecting the regional-scale tracer dispersion is the evolution of the PBL height along the particles path and the formation of tracer reservoir layers above the PBL. The details of the transport within the PBL are not so important since the temporal scale of vertical mixing is much shorter than the transport times. However, Maryon and Buckland (1994) argued that errors near the source caused by an oversimplified dispersion algorithm could translate into much larger errors at later times. A compromise to save computation time while maintaining sufficient accuracy would therefore be to use short time steps close to the source, but longer ones when the tracer is already well-mixed within the PBL.

Validations of LPDM are usually based on tracer experiments (Shi et al., 1990; Eastman et al., 1995; Fay et al., 1995; Moran and Pielke, 1996) or on tracers of opportunity (Draxler et al., 1994; Uliasz et al., 1997). The best way to validate LPDM is to compare modeled concentrations at the locations of the measurement sites to the measured concentrations. This method does not require to interpolate the measurements to a regular grid, which introduces substantial uncertainties. This validation strategy was followed systematically in the model evaluation study following the ETEX experiment (Mosca et al., 1997). The major remaining difficulty is the definition of the statistical measures most suitable for model evaluation.

LPDM can simulate, in addition to the dispersion, all linear processes, such as dry and wet deposition, radioactive decay, and linear chemical transformations, but a major drawback is that, currently, non-linear chemical reactions cannot be accounted for. There exist experiments to compute the chemical transformations on a grid based on the concentrations predicted by the LPDM (Chock and Winkler, 1994a, b) and to subsequently re-transform the gridded concentrations to particle masses, but these models are not yet operational, an exception being the study of Stevenson et al. (1997).

Traditionally, LPDM are used to obtain a time- and space-varying concentration field for a given emission scenario. However, because all processes simulated by an LPDM are linear, the inverse modeling is simple. Computing back trajectories from a given receptor location, a time- and space-varying *influence function* can be obtained (Uliasz, 1994). It is independent of a certain emission scenario and can therefore be used to calculate the pollutant concentration at the receptor for multiple emission scenarios without having to re-run the model. Since this receptor-oriented approach is computationally efficient, it is also possible to compute long-term concentration time series (Uliasz et al., 1997). It would be rewarding to examine these influence functions for long-term simulations using methods similar to the trajectory statistics presented in Section 4.3 to construct emission fields. A different method to estimate emissions from concentration measurements combined with backward LPDM simulations was presented by Flesch et al. (1995) and Flesch (1996), but this approach requires that the geometry of the source is known beforehand.

Finally, some attention shall be drawn to a problem of current LPDM that— to the author's knowledge—has not yet been solved: all procedures available to calculate $\mathbf{v}'$ assume constant density flows. This might be a justified approximation for shallow stable PBLs but certainly not for deep convective PBLs. There, density differences of 20% between the bottom and the top are typical. Neglect of these differences—as it is current practice—leads to an underestimation of ground-level concentrations and an overestimation of concentrations at the PBL top, which also translate into transport errors. Therefore, it would be a rewarding task to derive a Langevin equation that accounts for density fluctuations.

4.2. Lagrangian box models

LPDM have not yet been combined with nonlinear chemical reaction schemes because of the conceptual difficulties mentioned above, but Lagrangian *box models* have been very popular because of their computational efficiency. If the concentrations of some chemical species at a specific location are to be modeled, first a *back* trajectory is calculated from this location. Then, a box is moved *forward* along this trajectory and changes in the concentrations in the box caused by chemical reactions and deposition are calculated. Compared to *zero dimensional* Eulerian models, Lagrangian box models are more practical because no advection from outside the box occurs and hence no *boundary conditions* are required. However, such models are fully applicable only at higher levels of the atmosphere (Sparling et al., 1995) where turbulence is weak. For models used in the PBL (Eliassen et al., 1982), the height of the box must be adjusted to the variations of the PBL height, allowing air to be entrained from above which requires the specification of boundary conditions.

Lagrangian column models calculate the exchange of air between boxes arranged in a column. The most important boundary layer processes, such as the formation of nighttime reservoir layers or the rapid growth of the mixed layer depth in the morning, can be described with such models (Hertel et al., 1995), while their computational requirements are still moderate. For the calculation of the turbulent vertical exchange it is necessary that the boxes of the column remain exactly above each other. In reality, however, a vertical shear of the horizontal wind would separate the boxes. To avoid this *grid tangling*, the wind shear must be neglected and the whole column of boxes must be advected along a single trajectory. It is for this reason Lagrangian column models are less accurate than three-dimensional Eulerian models (Peters et al., 1995). Therefore, Lagrangian models are used mainly when computational costs are too high for Eulerian models, such as for long-term studies (e.g. De Leeuw et al., 1990) or when very complex chemical reaction schemes are applied (e.g. Derwent, 1990). They might be useful also for assimilating observations to produce accurate analysis maps of chemical species (Fisher and Lary, 1995), an application that is still extremely costly using Eulerian models.

4.3. Statistical analyses of trajectories

Trajectories have been used to interpret individual flow situations for several decades now, but statistical analysis methods for large sets of trajectories have been developed more recently. An early bibliography of such methods was prepared by Miller (1987), but more sophisticated methods have evolved since then.

4.3.1. Flow climatologies

The first statistical trajectory analyses were *flow climatologies*: back trajectories were calculated over a time span of several years and their transport directions and travel speeds were classified according to some criteria. These criteria were defined to discriminate, for instance, between oceanic, clean continental and polluted continental air masses.

An early example of this technique is the study of Miller (1981). He calculated more than 7000 back trajectories and classified them into five transport sectors. Many authors used variations of this technique, mostly to group air and precipitation chemistry data to identify roughly the source areas of air pollutants, namely those transport sectors associated with high pollutant concentrations at the receptor site (Henderson et al., 1982; Colin et al., 1989; Miller et al., 1993; White et al., 1994). For a bibliography see Miller (1987).

4.3.2. Cluster analysis

Cluster analysis is a *multivariate* statistical technique that splits a data set into a number of groups. Cluster analysis is often described as an *objective* classification method, but this is not true since the selection of the clustering algorithm, the specification of the distance measure and the number of clusters used are *subjective*. Kalkstein et al. (1987), who evaluated three different clustering procedures for use in synoptic climatological classification, remarked to the problem of subjectivity: "The selection of the proper clustering procedure to use in the development of an objective synoptic methodology may have far-reaching implications on the composition of the final 'homogeneous' groupings."

Cluster analysis has only recently been applied to meteorological data (Kalkstein et al., 1987; Fernau and Samson, 1990a,b; Eder et al., 1994). Moody and Galloway (1988) were the first to consider trajectory coordinates as the clustering variables. In principle, the result of a cluster analysis is similar to a flow climatology (which means trajectories are classified into some groups), but cluster analysis is *more* objective, and it accounts for variations in transport speed and direction simultaneously, yielding clusters of trajectories which have similar length and curvature (Moody and Samson, 1989; Harris and Kahl, 1990). Harris (1992) based a flow climatology for the South Pole on cluster analysis of trajectories and Moody et al. (1995) used cluster analysis to interpret ozone concentrations. Sirois and Bottenheim (1995) applied both cluster analysis and residence time analysis (see below) to interpret a 5-yr record of PAN and O_3. Kahl et al. (1997) used cluster analysis to investigate whether clear air tends to arrive at the Grand Canyon National Park over certain

a) Cluster 1, n = 18   b) Cluster 2, n = 60   c) Cluster 3, n = 26

d) Cluster 4, n = 29   e) Cluster 5, n = 36   f) Cluster 6, n = 44

g) Cluster 7, n = 21   h) Cluster 8, n = 17   i) Cluster 9, n = 67

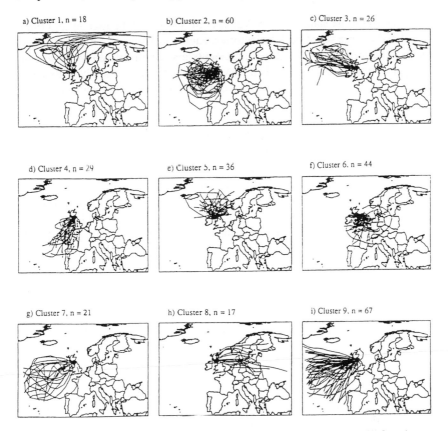

Figure 1. An example, originally presented by Dorling et al. (1992a), for the result of a trajectory cluster analysis. Shown are nine clusters identified with their trajectory members.

preferred pathways. Dorling et al. (1992b) and Dorling and Davies (1995) presented applications of trajectory clustering; the paper of Dorling et al. (1992a) contains a detailed description of their procedure which works very well in discriminating distinct flow patterns and large-scale circulation features (Stohl and Scheifinger, 1994). Fig. 1 shows as an example the result of the cluster analysis by Dorling et al. (1992a).

4.3.3. Residence time analysis and conditional probability fields

The methods discussed so far (flow climatologies, cluster analysis) classify trajectories without directly involving air pollution data. Ashbaugh (1983) and Ashbaugh et al. (1985) developed a method, *residence time analysis*, to identify source areas of air pollutants. A similar, albeit simpler, approach was published at about the same time by Munn et al. (1984). Ashbaugh et al. (1985) calculated a large set of back trajectories, each consisting of a number of segments separated by specific time increments and characterized by their positions and time, respectively. Then they covered the area of study with a grid and defined an event A to occur if an air parcel at time t happens to be inside a certain grid cell.

If N is the total number of trajectory segments and n_{ij} is the number of points falling in the ijth grid cell during a time interval T, then the probability that a randomly selected air parcel resides in the ijth grid cell is $P[A_{ij}] = n_{ij}/N$. Next, let m_{ij} be the number of trajectory segment points in the ijth grid cell, but only for those trajectories which arrive at the receptor when a certain criterion value for the pollutant concentration is exceeded. Then, the probability $P[B_{ij}] = m_{ij}/N$ represents the residence time of high pollutant concentration air parcels in the ijth grid cell (events B_{ij}) relative to the total time period considered. First knowledge of possible source regions can be gained from the differences in the probability fields $P[B_{ij}]$ and $P[A_{ij}]$ (Poirot and Wishinski, 1986), but both probability fields have their peak values in the grid cell where the receptor site is situated because all trajectories have to pass through this grid cell. Ashbaugh et al. (1985) normalized $P[B_{ij}]$ with $P[A_{ij}]$ to eliminate the redundant information. This can be written as

$$P[B_{ij} \mid A_{ij}] = \frac{P[B_{ij}]}{P[A_{ij}]} = \frac{m_{ij}}{n_{ij}} \tag{17}$$

with $P[B_{ij} \mid A_{ij}]$ being the *conditional probability* of the event B_{ij} given that event A_{ij} occurs. Regions with high conditional probability have a large potential to adversely affect the air quality at the receptor site when they are crossed by a trajectory. They do not necessarily make a large contribution to long-term air pollutant concentrations, since this also depends on the frequency at which air parcels actually travel over that region.

Ashbaugh's method was adopted by Zeng and Hopke (1989), Hopke et al. (1993), Cheng et al. (1993a, b) and Gao et al. (1993) who used it for studies of acid precipitation, sulfate, sulfur dioxide and heavy metals. Comrie (1994) and Stohl and Kromp-Kolb (1994a) used the technique to track ozone, and Sirois and Bottenheim (1995) combined it with cluster analysis to investigate ozone and PAN concentrations. Vasconcelos et al. (1996b) investigated the spatial resolution of the method. They found that the angular resolution is good, but that the radial resolution is poor because of the convergence of all trajectories toward the receptor.

It is clear that the certainty with which the conditional probability of a grid cell is known depends on the number of events A_{ij} in that cell. Recently, Vasconcelos et al. (1996a) proposed two statistical tests to examine the significance of conditional probability maps, one using a bootstrapping technique, the other based on a binomial distribution.

4.3.4. Concentration fields

Seibert et al. (1994a, b) computed *concentration fields* to identify source areas of air pollutants. Like Ashbaugh, they superimposed a grid to the domain of trajectory computations. Then they calculated a logarithmic mean concentration for each grid cell according to

$$\overline{C}_{ij} = \frac{1}{\sum_{l=1}^{M} \tau_{ijl}} \sum_{l=1}^{M} \log(c_l) \tau_{ijl} \tag{18}$$

where i, j are the indices of the horizontal grid, l the index of the trajectory, M the total number of trajectories, c_l the concentration observed on arrival of trajectory l and τ_{ijl} the time spent in grid cell (i, j) by trajectory l. A high value of $\overline{C}_{ij}$ means that, on average, air parcels passing over cell (i, j) result in high concentrations at the receptor site.

The fields exhibit small-scale variations which are not necessarily statistically significant. Simple smoothing of the concentration field, however, is not justified, because this would also remove many significant structures. Therefore, a confidence interval for the mean concentration of each grid cell is calculated using t-statistics based on the number of trajectories passing through each cell. Then, the concentration field is smoothed with a 9-point filter, imposing the restriction that the values must be kept within their confidence interval. The smoothing is repeated until the change in the concentration field is less than a prespecified value. This procedure assures that significant variations are preserved while most of the insignificant ones are removed. Stohl and Kromp-Kolb (1994b) used Seiberts method for an analysis of the ozone plume of Vienna.

4.3.5. Redistributed concentration fields

Sources of air pollutants are often concentrated in "hot spots", but Ashbaugh's and Seibert's procedures underestimate the spatial gradients of the "true" source fields because a measured concentration is attributed equally to all segments of its related trajectory. Let us imagine some trajectories which differ from each other, except for the fact that they all pass over one specific grid cell. Let all but one be "clean" trajectories, associated with low concentrations at the receptor site. Thus, no major pollutant source is located along their paths and specifically not in the grid cell which they share with the one "polluted" trajectory. Therefore, the latter one must have taken up its pollutant load some-where else. This information can be used for an iterative redistribution of the concentrations along the trajectories (Stohl, 1996b).

The major assumption of the redistribution method is that the measured species is directly emitted or produced by linear chemistry. Thus, problems are likely to occur for species produced by non-linear chemistry such as ozone. Wet deposition processes may also have a negative impact on the results. Stohl (1996b) tested the capacity of his method with particulate sulfate data mea-sured at 14 sites in Europe and found that the concentration field was in good agreement with an emission inventory. Virkkula et al. (1995) applied the redis-tribution method to the concentrations of non-sea-salt sulfate, ammonium and sodium measured in the Finnish subarctic and were able to identify the very different sources of these species.

4.3.6. Inverse modeling

Recently, Seibert (1997) presented a new approach towards establishing trajectory-derived source–receptor relationships. It is not based on statistics, but on inverse modeling, viewing trajectories as the output of a primitive La-grangian dispersion model. This is an ill-posed inversion problem, since the dimensions of the receptor-concentration-vector and of the source-vector are not equal. Seibert (1997) overcomes this difficulty by introducing additional constraints, but the effect of these constraints has to be explored further. Nev-ertheless, inverse modeling could be very useful in future work, especially be-cause it can be extended more easily than trajectory statistics to the results of more sophisticated dispersion models including, for instance, wet deposition processes.

5. Summary and conclusions

In this paper, it was shown how trajectories are being calculated, how accurate they typically are and how their accuracy can be assessed. There exists some evidence (see Table 2) that the accuracy of trajectory calculations has been improved in recent years, especially since three-dimensional wind fields are available from numerical weather prediction models. There is consensus now in the literature that three-dimensional trajectories are more accurate than any other type of trajectories, including isentropic trajectories.

Nevertheless, position errors of 20% of the travel distance can still be con-sidered as typical, limiting the general applicability of calcuated trajectories. Slightly smaller errors are often being reported for trajectories calculated from high-quality analysis fields in data-rich regions, but much larger errors, on the order of 30% or more, are typical for forecast trajectories. Since trajectory errors vary considerably from case to case, ranging from practically zero error to trajectories heading into the opposite direction of the real trajectories, it is essential to assess the uncertainty of trajectories on an individual case basis. This can be done either by using on-line tracers, such as potential vorticity, or by using trajectory ensemble methods.

Due to turbulent mixing, single trajectories are hardly sufficient to describe transport processes in the boundary layer. LPDM provide a more adequate representation of transport in turbulent flows. Therefore, studies that are currently based on the interpretation of forward or back trajectories (for instance interpretations of the measurements of trace constituents of the atmosphere), should in the future be based on the simulation results of LPDM.

Lagrangian chemical box or column models often suffer from large errors introduced by the vertical shear of the horizontal wind that cannot be accounted for with these models. Therefore, they should only be applied when other models (Eulerian chemical transport models) cannot be used due to a lack of computer capacity. However, it would be interesting to design LPDM with nonlinear chemistry.

New methods of trajectory statistics to interpret long-term air pollutant time series have been developed in the last few years, but these methods still need considerable improvement if they shall be applied to reconstruct emission fields. A large improvement could probably be achieved if the output of LPDM could be used in the statistics instead of single trajectories.

Acknowledgements

I am grateful to Jonathan Kahl, Petra Seibert and an anonymous reviewer for their very valuable comments.

References

Angell, J.K., Pack, D.H., 1960. Analysis of some preliminary low-level constant level balloon (tetroon) flights. Monthly Weather Review 7, 235–248.

Angell, J.K., Pack, D.H., Machta, L., Dickson, C.R., Hoecker, W.H., 1972. Three-dimensional air trajectories determined from tetroon flights in the planetary boundary layer of the Los Angeles basin. Journal of Applied Meteorology 11, 451–471.

Archer, G., Girardi, F., Graziani, G., Klug, W., Mosca, S., Nodop, K., 1996. The European long range tracer experiment (ETEX). Preliminary evaluation of model intercomparison exercise. In: Gryning, S.E., Schiermeier, F.A. (Eds.), Air Pollution Modeling and its Application XI, Vol. 21. Plenum Press, New York, pp. 181–190.

Artz, R., Pielke, R.A., Galloway, J., 1985. Comparison of the ARL/ATAD constant level and the NCAR isentropic trajectory analyses for selected case studies. Atmospheric Environment 19, 47–63.

Ashbaugh, L.L., 1983. A statistical trajectory technique for determining air pollution source regions. Journal of Air Pollution Control Association 33, 1096–1098.

Ashbaugh, L.L., Malm, W.C., Sadeh, W.Z., 1985. A residence time probability analysis of sulfur concentrations at Grand Canyon National Park. Atmospheric Environment 19, 1263–1270.

Baerentsen, J.H., Berkowicz, R., 1984. Monte Carlo simulation of plume dispersion in the convective boundary layer. Atmospheric Environment 18, 701–712.

Baumann, K., Langer, M., Stohl, A., 1996. Hot-air balloon tracks used to analyze air flow in alpine valleys. Proceedings of the 24th Conference on Alpine Meteorology, Bled, Slovenia, pp. 60–66.

Baumann, K., Stohl, A., 1997. Validation of a long-range trajectory model using gas balloon tracks from the Gordon Bennett Cup 95. Journal of Applied Meteorology 36, 711–720.

Byers, H.R., 1974. General Meteorology, 4th ed., McGraw–Hill, New York, USA.

Cabezudo, B., Recio, M., Sánchez-Laulhé, J.M., Trigo, M., Toro, F.J., Polvorinos, F., 1997. Atmospheric transportation of marihuana pollen from North Africa to the southwest of Europe. Atmospheric Environment 20, 3323–3328.

Cheng, M.-D., Hopke, P.K., Barrie, L., Rippe, A., Olson, M., Landsberger, S., 1993a. Qualitative determination of source regions of aerosol in Canadian high arctic. Environment Science and Technology 27, 2063–2071.

Cheng, M.-D., Hopke, P.K., Zeng, Y., 1993b. A receptor-oriented methodology for determining source regions of particulate sulfate at Dorset, Ontario. Journal of Geophysical Research 98, 16,839–16,849.

Chiapello, I., Bergametti, G., Chatenet, B., Bousquet, P., Dulac, F., Santos Soares, E., 1997. Origins of African dust transported over the northeastern tropical Atlantic. Journal of Geophysical Research 102, 13,701–13,709.

Chock, D.P., Kuo, Y.H., 1990. Comparison of wind-field models using the CAPTEX data. Journal of Applied Meteorology 29, 76–91.

Chock, D.P., Winkler, S.L., 1994a. A particle grid air quality modeling approach. 1. The dispersion aspect. Journal of Geophysical Research 99, 1019–1031.

Chock, D.P., Winkler, S.L., 1994b. A particle grid air quality modeling approach. 2. Coupling with chemistry. Journal of Geophysical Research 99, 1033–1041.

Clarke, J.F., Clark, T.L., Ching, J.K.S., Haagenson, P.L., Husar, R.B., Patterson, D.E., 1983. Assessment of model simulation of long-distance transport. Atmospheric Environment 12, 2449–2462.

Colin, J.L., Renard, D., Lescoat, V., Jaffrezo, J.L., Gros, M.J., Strauss, B., 1989. Relationship between rain and snow acidity and air mass trajectory in Eastern France. Atmospheric Environment 23, 1487–1498.

Comrie, A.C., 1994. Tracking ozone: air-mass trajectories and pollutant source regions influencing ozone in Pennsylvania forests. Annals of the Assocociation of American Geographers 84, 635–651.

D'Abreton, P.C., Tyson, P.D., 1996. Three-dimensional kinematic trajectory modelling of water vapour transport over Southern Africa. Water SA 22, 297–306.

Danielsen, E.F., 1961. Trajectories: isobaric, isentropic and actual. Journal of Meteorology 18, 479–486.

Davis, C.A., 1996. Potential vorticity. In: Schneider, S.H. (Ed.), Encyclopedia of Climate and Weather. Oxford University Press, Oxford.

De Leeuw, F.A.A.M., Van Rheineck Leyssius, H.J., Builtjes, P.J.H., 1990. Calculation of long term averaged ground level ozone concentrations. Atmospheric Environment 24A, 185–193.

Derwent, R.G., 1990. Evaluation of a number of chemical mechanisms for their application in models describing the formation of photochemical ozone in Europe. Atmospheric Environment 24A, 2615–2624.

Dorling, S.R., Davies, T.D., 1995. Extending cluster analysis—synoptic meteorology links to characterise chemical climates at six northwest European monitoring stations. Atmospheric Environment 29, 145–167.

Dorling, S.R., Davies, T.D., Pierce, C.E., 1992a. Cluster analysis: a technique for estimating the synoptic meteorological controls on air and precipitation chemistry—method and applications. Atmospheric Environment 26A, 2575–2581.

Dorling, S.R., Davies, T.D., Pierce, C.E., 1992b. Cluster analysis: a technique for estimating the synoptic meteorological controls on air and precipitation chemistry—results from Eskdalemuir, south Scotland. Atmospheric Environment 26A, 2583–2602.

Doty, K.G., Perkey, D.J., 1993. Sensitivity of trajectory calculations to the temporal frequency of wind data. Monthly Weather Review 121, 387–401.

Draxler, R.R., 1987 Sensitivity of a trajectory model to the spatial and temporal resolution of the meteorological data during CAPTEX. Journal of Climate and Applied Meteorology 26, 1577–1588.

Draxler, R.R., 1991. The accuracy of trajectories during ANATEX calculated using dynamic model analyses versus rawinsonde observations. Journal of Applied Meteorology 30, 1446–1467.

Draxler, R.R., 1996a. Boundary layer isentropic and kinematic trajectories during the August 1993 North Atlantic Regional Experiment Intensive. Journal of Geophysical Research 101, 29,255–29,268.

Draxler, R.R., 1996b. Trajectory optimization for balloon flight planning. Weather and Forecasting 11, 111–114.

Draxler, R.R., Dietz, R., Lagomarsino, R.J., Start, G., 1991. Across North America Tracer Experiment (ANATEX) sampling and analysis. Atmospheric Environment 25A, 2815–2836.

Draxler, R.R., McQueen, J.T., Stunder, B.J.B., 1994. An evaluation of air pollutant exposures due to the 1991 Kuwait oil fires using a Lagrangian model. Atmospheric Environment 28, 2197–2210.

Dutton, J.A., 1986. The Ceaseless Wind. An Introduction to the Theory of Atmospheric Motion. Dover, New York.

Eastman, J.L., Pielke, R.A., Lyons, W.A., 1995. Comparison of lake-breeze model simulations with tracer data. Journal of Applied Meteorology 34, 1398–1418.

Eder, B.K., Davis, J.M., Bloomfield, P., 1994. An automated classification scheme designed to better elucidate the dependence of ozone on meteorology. Journal of Applied Meteorology 33, 1182–1199.

Eliassen, A., Hov, Ø., Isaksen, I.S.A., Saltbones, J., Stordal, F., 1982. A Lagrangian long-range transport model with atmospheric boundary layer chemistry. Journal of Applied Meteorology 21, 1645–1661.

Fay, B., Glaab, H., Jacobsen, I., Schrodin, R., 1995. Evaluation of Eulerian and Lagrangian atmospheric transport models at the Deutscher Wetterdienst using ANATEX surface tracer data. Atmospheric Environment 29, 2485–2497.

Ferber, G.J., Heffter, J.L., Draxler, R.R., Lagomarsino, R.J., Thomas, F.L., Dietz, R.N., Benkovitz, C.M., 1986. Cross-Appalachian Tracer Experiment (CAPTEX 83). Final Rep. NOAA Techn. Memo. ERL ARL-142. Air Resources Laboratory, NOAA Environmental Research Laboratories. Silver Spring, Maryland 20910, p. 60.

Fernau, M.E., Samson, P.J., 1990a. Use of cluster analysis to define periods of similar meteorology and precipitation chemistry in Eastern North America. Part I: Transport patterns. Journal of Applied Meteorology 29, 735–750.

Fernau, M.E., Samson, P.J., 1990b. Use of cluster analysis to define periods of similar meteorology and precipitation chemistry in Eastern North America. Part II: Precipitation patterns and pollutant deposition. Journal of Applied Meteorology 29, 751–761.

Fisher, M., Lary, D.J., 1995. Lagrangian four-dimensional variational data assimilation of chemical species. Quarterly Journal of the Royal Meteorological Society 121, 1681–1704.

Flesch, T.K., 1996. The footprint for flux measurements, from backward Lagrangian stochastic models. Boundary-Layer Meteorology 78, 399–404.

Flesch, T.K., Wilson, J.D., Yee, E., 1995. Backward-time Lagrangian stochastic dispersion models, and their application to estimate gaseous emissions. Journal of Applied Meteorology 34, 1320–1332.

Fuelberg, H.E., Loring, R.O., Jr, Watson, M.V., Sinha, M.C., Pickering, K.E., Thompson, A.M., Sachse, G.W., Blake, D.R., Schoeberl, M.R., 1996. TRACE A trajectory intercomparison. 2. Isentropic and kinematic methods. Journal of Geophysical Research 101, 23,927–23,939.

Gao, N., Cheng, M.-D., Hopke, P.K., 1993. Potential source contribution function analysis and source apportionment of sulfur species measured at Rubidoux, CA during the Southern California Air Quality Study, 1987. Analytica Chimica Acta 277, 369–380.

Goodin, W.R., McRae, G., Seinfeld, J.H., 1980. An objective analysis technique for constructing three-dimensional urban-scale wind fields. Journal of Applied Meteorology 19, 98–108.

Grell, G.A., Dudhia, J., Stauffer, D.R., 1995. A description of the fifth-generation Penn State/NCAR mesoscale model (MM5). NCAR Technical Note, NCAR/TN-398 + STR, p. 122.

Haagenson, P.L., Gao, K., Kuo, Y.-H., 1990. Evaluation of meteorological analyses, simulations, and long-range transport using ANATEX surface tracer data. Journal of Applied Meteorology 29, 1268–1283.

Haagenson, P.L., Kuo, Y.H., Skumanich, M., Seaman, N.L., 1987. Tracer verification of trajectory models. Journal of Climate and Applied Meteorology 26, 410–426.

Hanna, S.R., 1982. Applications in air pollution modeling. In: Nieuwstadt, F.T.M., van Dop, H.D. (Eds.), Atmospheric Turbulence and Air Pollution Modelling. Reidel, Dordrecht, Holland.

Harris, J.M., 1992. An analysis of 5-day midtropospheric flow patterns for the South Pole: 1985–1989. Tellus 44B, 409–421.

Harris, J.M., Kahl, J.D., 1990. A descriptive atmospheric transport climatology for the Mauna Loa Observatory, using clustered trajectories. Journal of Geophysical Research 95, 13,651–13,667.

Harris, J.M., Kahl, J.D.W., 1994. Analysis of 10-day isentropic flow patterns for Barrow, Alaska: 1985–1992. Journal of Geophysical Research 99, 25,845–25,855.

Heffter, J.L., 1980. Air Resources Laboratories Atmospheric Transport and Dispersion Model (ARL-ATAD). NOAA Techn. Memo. ERL ARL-81. NOAA Environmental Research Laboratories. Boulder, Colorado, USA, p. 17.

Heffter, J.L., 1983. Branching Atmospheric Trajectory (BAT) Model. NOAA Techn. Memo. ERL ARL-121, Air Resources Laboratory. Silver Springs, Maryland, USA, p. 19.

Heffter, J.L., Stunder, B.J.B., 1993. Volcanic ash forecast transport and dispersion (VAFTAD) model. Weather and Forecasting 8, 533–541.

Heffter, J.L., Stunder, B.J.B., Rolph, G.D., 1990. Long-range forecast trajectories of volcanic ash from Redoubt Volcano eruptions. Bulletin of the American Meteorological Society 71, 1731–1738.

Henderson, R.G., Weingartner, K., 1982. Trajectory analysis of MAP3S precipitation chemistry data at Ithaca, New York. Atmospheric Environment 16, 1657–1665.

Hertel, O., Christensen, J., Runge, E.H., Asman, W.A.H., Berkowicz, R., Hovmand, M.F., Hov, Ø., 1995. Development and testing of a new variable scale air pollution model—ACDEP. Atmospheric Environment 29, 1267–1290.

Henmi, T., 1980. Long-range transport of SO_2 and sulfate and its application to the eastern United States. Journal of Geophysical Research 85, 4436–4442.

Hoecker, W.H., 1977. Accuracy of various techniques for estimating boundary-layer trajectories. Journal of Applied Meteorology 16, 374–383.

Hopke, P.K., Gao, N., Cheng, M.-D., 1993. Combining chemical and meteorological data to infer source areas of airborne pollutants. Chemom. Intell. Lab. Syst. 19, 187–199.

Jäger, A., 1992. Isentrope Trajektorien und ihre Anwendung auf das Konzept der potentiellen Vorticity bei orographisch induzierten Lee-Zyklogenesen. Diploma Thesis. University of Innsbruck, Austria.

Kahl, J.D., 1993. A cautionary note on the use of air trajectories in interpreting atmospheric chemistry measurements. Atmospheric Environment 27A, 3037–3038.

Kahl, J.D.W., 1996. On the prediction of trajectory model error. Atmospheric Environment 30, 2945–2957.

Kahl, J.D., Harris, J.M., Herbert, G.A., Olson, M.P., 1989a. Intercomparison of long-range trajectory models applied to arctic haze. Proceedings of the 17th NATO/CCMS ITM on Air Pollution Model and its Applications. Plenum Press, New York, pp. 175–185.

Kahl, J.D., Harris, J.M., Herbert, G.A., Olson, M.P., 1989b. Intercomparison of three long-range trajectory models applied to Arctic haze. Tellus 41B, 524–536.

Kahl, J.D., Liu, D., White, W.H., Macias, E.S., Vasconcelos, L., 1997. The relationship between atmospheric transport and the particle scattering coefficient at the Grand Canyon. Journal of Air and Waste Management Association 47, 419–425.

Kahl, J.D., Samson, P.J., 1986. Uncertainty in trajectory calculations due to low resolution meteorological data. Journal of Climate and Applied Meteorology 25, 1816–1831.

Kahl, J.D., Samson, P.J., 1988a. Trajectory sensitivity to rawinsonde data resolution. Atmospheric Environment 22, 1291–1299.

Kahl, J.D., Samson, P.J., 1988b. Uncertainty in estimating boundary-layer transport during highly convective conditions. Journal of Applied Meteorology 27, 1024–1035.

Kahl, J.D., Schnell, R.C., Sheridan, P.J., Zak, B.D., Church, H.W., Mason, A., Heffter, J.L., Harris, J.M., 1991. Predicting atmospheric debris transport in real-time using a trajectory forecast model. Atmospheric Environment 25A, 1705–1713.

Kalkstein, L.S., Tan, G., Skindlov, J.A., 1987. An evaluation of three clustering procedures for use in synoptic climatological classification. Journal of Climate and Applied Meteorology 26, 717–730.

Klug, W., Graziani, G., Grippa, G., Pierce, D., Tassone, C.W., 1992. Evaluation of Long Range Atmospheric Transport Models Using Environmental Radioactivity Data from the Chernobyl Release. Elsevier, Amsterdam.

Knudsen, B.M., Carver, G.D., 1994. Accuracy of the isentropic trajectories calculated for the EASOE campaign. Geophysical Research Letters 21, 1199–1202.

Knudsen, B.M., Rosen, J.M., Kjome, N.T., Whitten, A.T., 1996. Comparison of analyzed stratospheric temperatures and calculated trajectories with long-duration balloon data. Journal of Geophysical Research 101, 19,137–19,145.

Koffi, N.E., Nodop, K., Benech, B., 1997a. Constant volume balloon model used to derive tracer plume trajectories (ETEX experiment first release). In: Nodop, K. (Ed.), ETEX Symposium on Long-Range Atmospheric Transport, Model Verification and Emergency Response. European Commission EUR 17346, pp. 75–78.

Koffi, N.E., Nodop, K., Benech, B., 1997b. Constant volume balloon model used to derive tracer plume trajectories (ETEX experiment second release). In: Nodop, K. (Ed.), ETEX Symposium on Long-Range Atmospheric Transport, Model Verification and Emergency Response. European Commission EUR 17346, pp. 79–82.

Kolb, H., Seibert, P., Zwatz-Meise, V., Mahringer, G., 1989. Comparison of trajectories calculated during and after the Chernobyl accident. In: Evaluation of atmospheric dispersion models applied to the release from Chernobyl. Österreichische Beiträge zu Meteorologie und Geophysik 1, 61–69.

Kuo, Y.-H., Skumanich, M., Haagenson, P.L., Chang, J.S., 1985. The accuracy of trajectory models as revealed by the observing system simulation experiments. Monthly Weather Review 113, 1852–1867.

Lamb, B.K., Lorenzen, A., Shair, F.H., 1978a. Atmospheric dispersion and transport within coastal regions—part I. Tracer study of power plant emissions from the Oxnard plain. Atmospheric Environment 12, 2089–2100.

Lamb, B.K., Shair, F.H., Smith, T.B., 1978b. Atmospheric dispersion and transport within coastal regions—part II. Tracer study of industrial emissions in the California delta region. Atmospheric Environment 12, 2101–2118.

Legg, B.J., Raupach, M.R., 1982. Markov-chain simulation of particle dispersion in inhomogeneous flows: the mean drift velocity induced by a gradient in Eulerian velocity variance. Boundary-Layer Meteorology 24, 3–13.

Lorimer, G.S., 1986. The kernel method for air quality modelling—I. Mathematical foundation. Atmospheric Environment 20, 1447–1452.

Luhar, A.K., Hibberd, M.F., Hurley, P.J., 1996. Comparison of closure schemes used to specify the velocity PDF in Lagrangian stochastic dispersion models for convective conditions. Atmospheric Environment 30, 1407–1418.

Ludwig, F.L., Livingston, J.M., Endlich, R.M., 1991. Use of mass conservation and critical dividing streamline concepts for efficient objective analysis of wind fields in complex terrain. Journal of Applied Meteorology 30, 1490–1499.

Lyons, W.A., Pielke, R.A., Tremback, C.J., Walko, R.L., Moon, D.A., Keen, C.S., 1995. Modeling impacts of mesoscale vertical motions upon coastal zone air pollution dispersion. Atmospheric Environment 29, 283–301.

Martin, D., Bergametti, G., Strauss, B., 1990. On the use of the synoptic vertical velocity in trajectory model: validation by geochemical tracers. Atmospheric Environment 24A, 2059–2069.

Martin, D., Mithieux, C., Strauss, B., 1987. On the use of the synoptic vertical wind component in a transport trajectory model. Atmospheric Environment 21, 45–52.

Maryon, R.H., Buckland, A.T., 1994. Diffusion in a Lagrangian multiple particle model: a sensitivity study. Atmospheric Environment 28, 2019–2038.

Maryon, R.H., Heasman, C.C., 1988. The accuracy of plume trajectories forecast using the U.K. meteorological office operational forecasting models and their sensitivity to calculation schemes. Atmospheric Environment 22, 259–272.

Mathur, R., Peters, L.K., 1990. Adjustment of wind fields for application in air pollution modeling. Atmospheric Environment 24A, 1095–1106.

McNider, R.T., Moran, M.D., Pielke, R.A., 1988. Influence of diurnal and inertial boundary layer oscillations on long-range dispersion. Atmospheric Environment 22, 2445–2462.

McQueen, J.T., Draxler, R.R., 1994. Evaluation of model back trajectories of the Kuwait oil fires smoke plume using digital satellite data. Atmospheric Environment 28, 2159–2174.

Merrill, J.T., Bleck, R., Avila, L., 1985. Modeling atmospheric transport to the Marshall islands. Journal of Geophysical Research 90, 12,927–12,936.

Merrill, J.T., Bleck, R., Boudra, D., 1986. Techniques of Lagrangian trajectory analysis in isentropic coordinates. Monthly Weather Review 114, 571–581.

Miller, J.M., 1981. A five-year climatology of back trajectories from the Mauna Loa observatory, Hawaii. Atmospheric Environment 15, 1553–1558.

Miller, J.M., 1987. The use of back air trajectories in interpreting atmospheric chemistry data: a review and bibliography. NOAA Technical Memo. ERL ARL-155. Air Resources Laboratory, NOAA Environmental Research Laboratories. Silver Spring, Maryland, USA, p. 60.

Miller, J.M., Moody, J.L., Harris, J.M., Gaudry, A., 1993. A 10-year trajectory flow climatology for Amsterdam island, 1980–1989. Atmospheric Environment 27A, 1909–1916.

Moody, J.L., Galloway, J.N., 1988. Quantifying the relationship between atmospheric transport and the chemical composition of precipitation on Bermuda. Tellus 40B, 463–479.

Moody, J.L., Oltmans, S.J., Levy II, H., Merrill, J.T., 1995. Transport climatology of tropospheric ozone: Bermuda, 1988–1991. Journal of Geophysical Research 100, 7179–7194.

Moody, J.L., Samson, P.J., 1989. The influence of atmospheric transport on precipitation chemistry at two sites in the midwestern United States. Atmospheric Environment 23, 2117–2132.

Moran, M.D., Pielke, R.A., 1996. Evaluation of a mesoscale atmospheric dispersion modeling system with observations from the 1980 Great Plains mesoscale tracer field experiment. Part II: Dispersion simulations. Journal of Applied Meteorology 35, 308–329.

Mosca, S., Graziani, G., Klug, W., Bellasio, R., Bianconi, R., 1997. ATMES-II – Evaluation of Long-Range Dispersion Models Using 1st ETEX Release Data, Vol. I (in press).

Munn, R.E., Likens, G.E., Weisman, B., Hornbeck, J.W., Martin, C.W., Bormann, F.H., 1984. A meteorological analysis of the precipitation chemistry event samples at Hubbard Brook (N.H.). Atmospheric Environment 18, 2775–2779.

Nguyen, K.C., Noonan, J.A., Galbally, I.E., Physick, W.L., 1997. Predictions of plume dispersion in complex terrain: Eulerian versus Lagrangian models. Atmospheric Environment 31, 947–958.

Pack, D.H., Ferber, G.J., Heffter, J.L., Telegadas, K., Angell, J.K., Hoecker, W.H., Machta, L., 1978. Meteorology of long-range transport. Atmospheric Environment 12, 425–444.

Peters, L.K., Berkowicz, C.M., Carmichael, G.R., Easter, R.C., Fairweather, G., Ghan, S.J., Hales, J.M., Leung, L.R., Pennell, W.R., Potra, F.A., Saylor, R.D., Tsang, T.T., 1995. The current state and future direction of Eulerian models in simulating the tropospheric chemistry and transport of trace species: a review. Atmospheric Environment 29, 189–222.

Petersen, R.A., Uccellini, L.W., 1979. The computation of isentropic atmospheric trajectories using a "Discrete Model" formulation. Monthly Weather Review 107, 566–574.

Petterssen, S., 1940. Weather Analysis and Forecasting. McGraw–Hill, New York, pp. 221–223.

Pflüger, U., Roos, M., Jacobsen, I., 1990. Trajektorienberechnungen auf Basis des neuen Wettervorhersagesystems des Deutschen Wetterdienstes und ihre Anwendung für die Ausbreitungsrechnung. Deutscher Wetterdienst.

Pickering, K.E., Thompson, A.M., McNamara, D.P., Schoeberl, M.R., 1994. An intercomparison of isentropic trajectories over the South Atlantic. Monthly Weather Review 122, 864–879.

Pickering, K.E., Thompson, A.M., McNamara, D.P., Schoeberl, M.R., Fuelberg, H.E., Loring, R.O., Jr., Watson, M.V., Fakhruzzaman, K., Bachmeier, A.S., 1996. TRACE A trajectory intercomparison. 1. Effects of different input analyses. Journal of Geophysical Research 101, 23,903–23,925.

Pielke, R.A., Cotton, W.R., Walko, R.L., Trembach, C.J., Lyons, W.A., Grasso, L.D., Nicholls, M.E., Moran, M.D., Wesely, D.A., Lee, T.J., Copeland, J.H., 1992. A comprehensive meteorological modeling system – RAMS. Meteorology Atmospheric Physics 49, 69–91.

Poirot, R.L., Wishinski, P.R., 1986. Visibility, sulfate and air mass history associated with the summertime aerosol in Northern Vermont. Atmospheric Environment 20, 1457–1469.

Rao, S.T., Pleim, J., Czapaski, J., 1983. A comparative study of two trajectory models of long-range transport. Journal of Air Pollution Control Association 33, 32–41.

Raynor, G.S., Hayes, J.V., Lewis, D.M., 1983. Testing of the Air Resources Laboratories trajectory model on cases of pollen wet deposition after long-distance transport from known source regions. Atmospheric Environment 17, 213–220.

Reiff, J., Forbes, G.S., Spieksma, F.T.M., Reynders, J.J., 1986. African dust reaching northwestern Europe: a case study to verify trajectory calculations. Journal of Climate and Applied Meteorology 25, 1543–1567.

Reisinger, L.M., Mueller, S.F., 1983. Comparisons of tetroon and computed trajectories. Journal of Climate and Applied Meteorology 22, 664–672.

Rodean, H., 1996. Stochastic Lagrangian models of turbulent diffusion. Meteorological Monographs 26 (48). American Meteorological Society, Boston, USA.

Rolph, G.D., Draxler, R.R., 1990. Sensitivity of three-dimensional trajectories to the spatial and temporal densities of the wind field. Journal of Applied Meteorology 29, 1043–1054.

Ross, D.G., Smith, I.N., Manins, P.C., Fox, D.G., 1988. Diagnostic wind field modeling for complex terrain: model development and testing. Journal of Applied Meteorology 27, 785–796.

Sardeshmukh, P.D., Liebmann, B., 1993. An assessment of low-frequency variability in the tropics as indicated by some proxies of tropical convection. Journal of Climate 6, 569–575.

Scire, J., Insley, E., Yamartino, R., 1990. Model formulation and user's guide for the CALMET meteorological model. Report No. A025-1, prepared for the State of California Air Resources Board.

Scheele, M.P., Siegmund, P.C., Velthoven, P.F.J., 1996. Sensitivity of trajectories to data resolution and its dependence on the starting point: in or outside a tropopause fold. Meteorological Applications 3, 267–273.

Schlünzen, K.H., 1994. Mesoscale modeling in complex terrain—an overview on the German nonhydrostatic models. Beiträge zur Physik der Atmosphäre 67, 243–253.

Seibert, P., 1993. Convergence and accuracy of numerical methods for trajectory calculations. Journal of Applied Meteorology 32, 558–566.

Seibert, P., 1997. Inverse dispersion modelling based on trajectory-derived source-receptor relationships. Proceedings of the 22nd International Technical Meeting on Air Pollution Modelling and its Applications, Clermont-Ferrand.

Seibert, P., Kromp-Kolb, H., Baltensperger, U., Jost, D.T., Schwikowski, M., 1994a. Trajectory analysis of high-alpine air pollution data. In: Gryning, S.-E., Millan, M.M. (Eds.), Air Pollution Modelling and its Application X. Plenum Press, New York, pp. 595–596.

Seibert, P., Kromp-Kolb, H., Baltensperger, U., Jost, D.T., Schwikowski, M., Kasper, A., Puxbaum, H., 1994b. Trajectory analysis of aerosol measurements at high alpine sites. In: Borrell, P.M., Borrell, P., Cvitaš, T., Seiler, W. (Eds.), Transport and Transformation of Pollutants in the Troposphere. Academic Publishing, Den Haag, pp. 689–693.

Sherman, C.A., 1978. A mass-consistent model for wind fields over complex terrain. Journal of Applied Meteorology 17, 312–319.

Shi, B., Kahl, J.D., Christidis, Z.D., Samson, P.J., 1990. Simulation of the three-dimensional distribution of tracer during the Cross-Appalachian Tracer Experiment. Journal of Geophysical Research 95, 3693–3703.

Sirois, A., Bottenheim, J.W., 1995. Use of backward trajectories to interpret the 5-year record of PAN and O_3 ambient air concentrations at Kejimkujik National Park, Nova Scotia. Journal of Geophysical Research 100, 2867–2881.

Sparling, L.C., Schoeberl, M.R., Douglass, A.R., Weaver, C.J., Newman, P.A., Lait, L.R., 1995. Trajectory modeling of emissions from lower stratospheric aircraft. Journal of Geophysical Research 100, 1427–1438.

Steinacker, R., 1984. Airmass and frontal movement around the Alps. Rivista di Meteorologia Aeronautica 44, 85–93.

Stevenson, D.S., Collins, W.J., Johnson, C.E., Derwent, R.G., 1997. The impact of aircraft nitrogen oxide emissions on tropospheric ozone studied with a 3D Lagrangian model including fully diurnal chemistry. Atmospheric Environment 31, 1837–1850.

Stocker, R.A., Pielke, R.A., Verdon, A.J., Snow, J.T., 1990. Characteristics of plume releases as depicted by balloon launchings and model simulations. Journal of Applied Meteorology 29, 53–62.

Stohl, A., 1996a. On the use of trajectories for establishing source-receptor relationships of air pollutants. Ph.D. Thesis. University of Vienna.

Stohl, A., 1996b. Trajectory statistics—a new method to establish source–receptor relationships of air pollutants and its application to the transport of particulate sulfate in Europe. Atmospheric Environment 30, 579–587.

Stohl, A., Baumann, K., Wotawa, G., Langer, M., Neininger, B., Piringer, M., Formayer, H., 1997. Diagnostic downscaling of large scale wind fields to compute local scale trajectories. Journal of Applied Meteorology 36, 931–942.

Stohl, A., Kromp-Kolb, H., 1994a. Origin of ozone in Vienna and surroundings, Austria. Atmospheric Environment 28, 1255–1266.

Stohl, A., Kromp-Kolb, H., 1994b. Frequency of ozone formation in the plume of Vienna. In: Baldasano, J.M., Brebbia, C.A., Power, H., Zannetti, P. (Eds.), Air Pollution 2, Vol. 2—Pollution Control and Monitoring. Computational Mechanics Publications, Southampton, Boston, UK, pp. 449–456.

Stohl, A., Scheifinger, H., 1994. A weather pattern classification by trajectory clustering. Meteorologische Zeitschrift N.F. 6, 333–336.

Stohl, A., Seibert, P., 1997. Accuracy of trajectories as determined from the conservation of meteorological tracers. Quarterly Journal of the Royal Meteorological Society, in press.

Stohl, A., Wotawa, G., 1995. A method for computing single trajectories representing boundary layer transport. Atmospheric Environment 29, 3235–3239.

Stohl, A., Wotawa, G., Seibert, P., Kromp-Kolb, H., 1995. Interpolation errors in wind fields as a function of spatial and temporal resolution and their impact on different types of kinematic trajectories. Journal of Applied Meteorology 34, 2149–2165.

Stunder, B.J.B., 1996. An assessment of the quality of forecast trajectories. Journal of Applied Meteorology 35, 1319–1331.

Sutton, R., 1994. Lagrangian flow in the middle atmosphere. Quarterly Journal of the Royal Meteorological Society 120, 1299–1321.

Tajima, T., Nakamura, T., Kurokawa, K., 1997. Experimental observations of 3-D Lagrangian motion in a steady baroclinic wave. Journal of the Meteorological Society of Japan 75, 101–109.

Thomson, D.J., 1984. Random walk modelling of diffusion in inhomogeneous turbulence. Quarterly Journal of the Royal Meteorological Society 110, 1107–1120.

Thomson, D.J., 1987. Criteria for the selection of stochastic models of particle trajectories in turbulent flows. Journal of Fluid Mechanics 180, 529–556.

Thomson, D.J., Montgomery, M.R., 1994. Reflection boundary conditions for random walk models of dispersion in non-Gaussian turbulence. Atmospheric Environment 28, 1981–1987.

Tyson, P.D., Garstang, M., Swap, R., 1996. Large-scale recirculation of air over southern Africa. Journal of Applied Meteorology 35, 2218–2236.

Uliasz, M., 1994. Lagrangian particle dispersion modeling in mesoscale applications. In: Zannetti, P. (Ed.), Environmental Modeling, Vol. II. Computational Mechanics Publications, Southampton, UK, pp. 71–101.

Uliasz, M., Stocker, R.A., Pielke, R.A., 1997. Regional modeling of air pollution transport in the southwestern United States. To be published in: Zannetti, P. (Ed.), Environmental Modeling, Vol. III. Computational Mechanics Publications, Southampton, U.K.

Vasconcelos, L.A.P., Kahl, J.D.W., Liu, D., Macias, E.S., White, W.H., 1996a. A tracer calibration of back trajectory analysis at the Grand Canyon. Journal of Geophysical Research 101, 19329–19335.

Vasconcelos, L.A.P., Kahl, J.D.W., Liu, D., Macias, E.S., White, W.H., 1996b. Spatial resolution of a transport inversion technique. Journal of Geophysical Research 101, 19,337–19,342.

Virkkula, A., Mäkinen, M., Hillamo, R., Stohl, A., 1995. Atmospheric aerosol in the Finnish Arctic: particle number concentrations, chemical characteristics, and source analysis. Water, Air and Soil Pollution 85, 1997–2002.

Walmsley, J.L., Mailhot, J., 1983. On the numerical accuracy of trajectory models for long-range transport of atmospheric pollutants. Atmos.-Ocean 21, 14–39.

Warner, T.T., Fizz, R.R., Seaman, N.L., 1983. A comparison of two types of atmospheric transport models—use of observed winds versus dynamically predicted winds. Journal of Climate and Applied Meteorology 22, 394–406.

Waugh, D.W., Plumb, R.A., 1994. Contour advection with surgery: a technique for investigating finescale structures in tracer transport. Journal of Atmospheric Science 51, 530–540.

Wetzel, M., Borys, R., Lowenthal, D., Brown, S., 1995. Meteorological support to the Earthwinds Transglobal Balloon project. Bulletin of the American Meteorological Society 76, 477–487.

White, W.H., Macias, E.S., Kahl, J.D., Samson, P.J., Molenar, J.V., Malm, W.C., 1994. On the potential of regional-scale emissions zoning as an air quality management tool for the Grand Canyon. Atmospheric Environment 28, 1035–1045.

Wilson, J.D., Flesch, T.K., 1993. Flow boundaries in random-flight dispersion models: enforcing the well-mixed condition. Journal of Applied Meteorology 32, 1695–1707.

Wilson, J.D., Legg, B.J., Thomson, D.J., 1983. Calculation of particle trajectories in the presence of a gradient in turbulent-velocity scale. Boundary-Layer Meteorology 27, 163–169.

Wilson, J.D., Sawford, B.L., 1996. Review of Lagrangian stochastic models for trajectories in the turbulent atmosphere. Boundary-Layer Meteorology 78, 191–210.

Zeng, Y., Hopke, P.K., 1989. A study of the sources of acid precipitation in Ontario, Canada. Atmospheric Environment 23, 1499–1509.

Zannetti, P., 1992. Particle modeling and its application for simulating air pollution phenomena. In: Melli, P., Zannetti, P. (Eds.), Environmental Modelling. Computational Mechanics Publications, Southampton, UK, pp. 211–241.

Air Pollution Science for the 21st Century
J. Austin, P. Brimblecombe and W. Sturges, editors
© 2002 Elsevier Science Ltd. All rights reserved.

Chapter 22

Future Directions: Could transgenic mice hear pollution?

J.G.T. Hill

Department of Chemistry, University of Cambridge, Lensfield Road, Cambridge, CB2 1EW, UK

Matthew Lythe

Department of Civil Engineering, University of Surrey, Guildford, Surrey, GU2 7XH, UK

In this world of Genetically Modified food that gives cheap tomato puree, cloned pigs that are going to guarantee us all immortality and biological weapons that scare us all witless, will we soon see GM instrumentation that will cost us all our jobs? With a near infinite supply of inexpensive and accurate biosensors on the horizon there will soon be a sonde for everything. Whereas currently we search for an organism that could be a biosensor, we are proposing to engineer them. This will allow sensors to be built for the task rather than found and then employed, eliminating much fruitless work as well as creating a new field of instrumentation.

Biosensors can be defined as any sensing device that employs a biological system, be it a "simple" enzyme through to a miners' canary. Currently many different systeris are used, moss and lichen, enzymes, bacteria and devices using antibodies. All based on naturally occurring organisms, these can be split into two rough groups, those where the pollutant acts as a messenger triggering further reactions and those where it is stored in the sensor ready for later analysis.

In moss and lichen the skill lies in selecting the size and species of the sampler for a specific target material (Gailey, Lloyd, Environ. Pollut. Ser. B, Vol. 12 (1986) pp. 41–59). Enzymes, however, are typically used in potentiometric devices such as sulphite biosensors in the wine industry (Situmorang et al., Analyst, Vol. 124 (1999) pp. 1775–1779). There has been suggestion of using antibodies to alter the physical properties of an optical system, allowing *in situ* measurements to be made. Measuring the toxic effect of the pollutant on bacteria growth rates is another useful method, though bacteria also act as enzymatic biosensors. This however only touches on the potential of biological systems to be used for detecting pollution.

The intrinsic specificity of biosensors drives work in this field, as it enables the sensors to differentiate between related compounds, or even enantiomers. For repetitive analyses this offers significant advantages over current laboratory methods, both in speed and cost. Biological sensors can also benefit from an increased sensitivity due to chemical amplification. This comes about from the pollution acting as a messenger, causing large changes in the chemistry across a membrane. Sensitivity can also be enhanced for enzymatic systems by paying careful attention to the medium (e.g. controlling pH or ionic strength). Because of the nature of the chemistry involved it is possible to get an absolute measurement of the "pollution messenger". Further gains in sensitivity can be made by the stabilising of reactive compounds by binding to fats and proteins, or storing in a less hostile environment. The technology has the potential to reduce costs by removing the regular need for expensive machine time. Since bacteriological systems can be self-replicating, it would make large networks of instrumentation cheaper to produce. All this will lead to a less expensive, more sensitive, accurate system of pollution detection.

Moss and lichen work primarily by storing the pollutant in the cells – this means in order to get the data they are rendered down and the contents of their cells analysed. This is true of other systems such as bio-accumulation in estuarine fish. The enzymatic systems work by using the messenger to activate or inhibit the enzymes catalytic activity. What is measured in these systems is usually the product of enzyme reaction or (if NAD^+/NADH is required by the reaction (Stryer, Biochemistry, Freeman, 1995) the optical absorbance. The same can be said for antibody systems where binding of the pollutant molecule will cause a conformational change which will either drive an enzyme catalysed chemical response, or a change in the physical properties (e.g. the optics of a wave guide on gold leaf). The bacterial systems can work either as bioaccumulators or similarly to enzymatic systems.

The potential for using genetic manipulation to engineer sensors brings about many intriguing possibilities. Initially these manipulations could be used to understand and fine tune existing sensors to improve their functionality. There is also tie potential for creating entirely new sensors and families of sensors. As a starting point for this we should turn to bacteria.

The reason for starting off with bacteria is that the genetics of the host can be easily and well defined (Alberts et al., The Cell, Garland, 1994). The host itself does not have to be "genetically engineered" as such, what is used is a designed plasmid that will encode for the receptor. This receptor should use an already present signal transduction method, such as regulating an ion channel. If we choose a host so that the response can be something that can be measured (such as pumping protons), then the binding of a pollutant to the receptor can be measured as a change in pH of the growth medium. The skill is then in designing the receptor to bind the molecule of interest. Once this is done then the building of plasmid libraries will be a relatively simple task allowing the mass production of a sensor for any molecule that is in the library.

The recognition of hydrocarbon pollutants may well be a matter of finding relevant receptors or attempting protein engineering to alter parts of the receptor protein. The latter process is more desirable, though much harder as secondary and tertiary interactions are difficult to model (Smith, Regan, Science, Vol. 270 (1995) pp. 980–982).

Once a sensor has been designed then how it is implemented is then an issue. For bacteria there is the problem of how the cellular signal is resolved and what form an instrument will take. An associated problem is deployment, how do we get the bacteria to where we want the measurement? The most obvious extension is to move on to "higher lifeforms", and to some extent this happens already. The correlation between traffic noise and carbon monoxide emissions have been investigated by Tirabassi in an article "Listening out for Urban Air Pollution (Atmospheric Environment 33 (1999) pp. 4219–4220). Effectively a redesign of the miners' canary is needed. Although this would represent a movement from an enzymatic assay style sensor to a bioaccumulation sensor, the organism would still have to be altered to preferentially retain the pollutant. "Higher lifeforms" could well be self-deploying and then have their pollutant levels measured when recaptured. Releasing tagged, sterile transgenic mice could well be the future for monitoring environmental pollution in air conditioning vents or similar confined areas.

Chapter 23

Future Directions: The case for a "Law of the Atmosphere"

Adil Najam

*Department of International Relations and Center for Energy and Environmental Studies,
Boston University, 152 Bay State Road, Boston, MA 02215, USA
E-mail: anajam@bu.edu*

International policies for managing the global atmosphere have evolved in an ad hoc and piecemeal manner. It may now be time to adopt a more thought-out and holistic approach. While some might consider it outrageous heresy, there is a case to be made for moving towards a comprehensive "Law of the Atmosphere" (LoA).

1. A legacy of ad hoc-ism

The atmosphere is one of the very few examples of a true global commons in that it is a "domain that is beyond the exclusive jurisdiction of any one nation, but one that all nations may use for their own purposes" (Soroos, M.S., 1998, "The thin blue line: preserving the atmosphere as a global commons", *Environment* 40, 32–35). According to Garrett Hardin's (1968) now famous postulate ("The tragedy of the Commons", *Science* 162, 1234), commons need to be regulated because they have a perverse tendency to degenerate into ruin and 'tragedy' since everyone has an incentive for exploiting them, while no one has the incentive, or responsibility, for maintaining their integrity. Much of global environmental policy is intellectually premised on the desire to avert Hardin's foretold tragedy.

The Law of the Sea (which came into force in 1994) was underpinned, in part, by this concern. Although wrought out of long drawn, painstaking, and often acrimonious negotiation, and mired in debate even today, it finally declared the seabed a 'common heritage of mankind'. Proposals to apply this phrase – which has significant legal implications about jurisdiction and responsibility – to the atmosphere have been repelled, and the best that international policymakers have to offer is to recognise that climate change is a subject of "common concern for humankind" (*UN Framework Convention on*

First published in Atmospheric Environment 34 (2000) 4047–4049

Climate Change). Instead, they have adopted an approach of individually and reactively applying 'Band-Aid' solutions to each individual problem (acid rain, stratospheric ozone depletion, climate change, etc.) as it arises, with little (or belated) attention to how it might relate to other issues. Instead of focussing on the global atmosphere as a vital planetary organ, we are reduced to 'fire-fighting' skirmishes that, vital as they are, are ultimately symptomatic of more fundamental threats to the essential services provided by the atmosphere.

This legacy of ad hoc-ism has its basis in certain substantive and procedural conveniences. Substantively, the atmosphere is an extremely complex system; and one that is still much less than well understood. Trying to tackle sub-systems that relate to long-range air pollution, stratospheric ozone, or global climate – and the myriad scientific uncertainties associated with them – is difficult enough in itself. Expanding the scope of global negotiations to the entire atmospheric system, it is justifiably argued, would simply be too complex to handle. There is a prevailing sense that "the ad hoc, problem-specific approach to regulating pollution and protecting the atmosphere has proven to be quite flexible and adaptable" and therefore pursuing a comprehensive treaty would be "an ill-advised use of limited diplomatic resources" (Soroos, M.S., 1997, *"The Endangered Atmosphere*: *Preserving a Global Commons"*, University of South Carolina Press).

The root of such fears, however, is not just substantive complexity but the procedural nightmare it would entail. This is brought into sharp relief by the history of the Law of the Sea which took nearly a decade of intense interna-tional bickering to negotiate and another decade to be ratified. The scars of this traumatic experience run deep, particularly in the United States. These are compounded by fears that such a comprehensive treaty would somehow im-pinge upon the almighty principle of territorial sovereignty that national poli-cymakers hold so dear.

2. Toward a Law of the Atmosphere

Valid as these concerns are, they can, in fact, be addressed by adopting an in-cremental (as opposed to ad hoc) approach. Such an approach would begin with a general 'framework' law that deals with basic declaratory principles. In the aforementioned article by Marvin Soroos (34, 1998) the author sug-gests taking "the first step toward developing a comprehensive law of the at-mosphere" based on "a declaration on the atmosphere, etc. couched in terms of general principles rather than specific obligations". However, to have a mean-ingful impact, one would need to do more. We will also have to take the sub-sequent steps towards treaty harmonisation by imbedding existing agreements and ongoing negotiations, such as those on transboundary air pollution, ozone depletion, and climate change, within the umbrella of the principles espoused in the framework law. The proposed incremental approach would, over time, systematically mature into binding legal provisions encompassing the general declaratory principles, as well as any imbedded conventions. Where necessary, new issue-specific agreements would be negotiated as imbedded conventions to the LoA.

The ultimate goal is more than simply putting existing agreements under one umbrella. It is to consciously move towards an internally consistent Law of the Atmosphere focussed on the challenge of maintaining the health and vitality of the atmosphere as a whole rather than the management of a few selected pollutants. There are important reasons, both substantive and procedural, why doing so is a good idea.

Substantively, science clearly calls for a comprehensive policy response. It is evident to anyone who regularly reads this journal that the science of ozone depletion and the science of climate change do not simply add up to become the science of the atmospheric environment. The science of the atmosphere is much more than just the sum of its parts. Integral is how these parts interact. The existing approach to atmospheric policy provides little space for dealing with these interactions. Consider, for example, that many ozone-depleting substances are also greenhouse gases, and some substances with comparatively low ozone-depleting potentials might have high greenhouse potentials. Already, much of the discussion at recent negotiations on the Vienna Convention (on ozone depletion) revolved around climate change issues while some of the more spirited debates on ozone-depleting substances are now happening at meetings related to the climate convention. There remains insufficient interaction between these two treaties, and therefore between the scientists or the policymakers working on them. A comprehensive approach would provide the common forum where such interaction can be nurtured, where early warning of potential problems can be sounded, and where links between the different components of the atmosphere can be dealt with.

Procedurally, the proposed approach provides for significant gains in negotiation efficiency. Currently, there is a serious negotiation glut with simply too many negotiations, on too many issues, in too many places. Policymakers have a near impossible task in keeping track of how all the other negotiations impact their sub-issue and how their myriad provisions might fit together. The problem is even more confounding for developing countries, with their much more limited human and financial resources. Placing all agreements related to the atmosphere within one umbrella LoA would go a long way in streamlining the process by building on synergies and reducing the level of negotiation fatigue, particularly for developing countries.

Even more importantly, a harmonised platform for negotiating atmospheric policy would allow negotiators to 'trade across differences'. This famous dictum of negotiation theory postulates that where different parties value different issues differently, the net negotiation gain will be higher if they trade across these differences (Susskind, 1994, *Environmental Diplomacy: Negotiating More Effective Global Agreements*, Oxford University Press). Assume that country A holds issue x as its most important priority but is not particularly interested in issue y. Assume, also, that country B considers y to be extremely important but is less interested in x. If issues x and y were to be negotiated separately, the likely result would be sub-optimal, if not a stalemate. However, if both issues are negotiated together it is possible that an optimal 'trade' can be made, with country A giving in on issue y in return for country B doing so on issue x. Presently, there are significant barriers to such trades (particularly between developing and industrialised countries) because different atmospheric issues are negotiated separately. In negotiation parlance, we continue to leave value on the table.

Finally, let us return to the legacy of the Law of the Sea experience. Let it be recalled that the most significant problems were not of issue or format, but of the larger politics of the time: solidarity tactics by the developing world in the 1970s, and the US White House's environmental pull-back in the 1980s. As to the sovereignty argument, that remains a sceptre over all global environmental policy. It is at present being confronted by the Kyoto Protocol which, according to some, is already assigning de facto property rights to the global atmosphere (Najam and Page, 1998, The Climate Convention: deciphering the Kyoto Convention. *Environmental Conservation* 25, 187–194).

Having said the above, it is not surprising that national decision-makers, still mired in traditional notions of territorial sovereignty, remain hesitant to declare and treat the Earth's atmosphere as a global commons. More worrisome is that environmental activists and scholars seem equally lukewarm to the idea. Activists seem afraid that negotiating such a comprehensive treaty would take too long, and many scholars are themselves so specialised in minute elements that they lose focus on the larger atmospheric challenge. The single biggest hurdle, however, is the force of inertia. Inertia, after all, is a principle that is as potent in politics than in physics. Inefficient as the current approach is, no one has the incentive to invest the effort in devising a better approach.

The Law of the Atmosphere is an idea whose time has come. What is needed now is a champion. Where better to seek the leadership than in the ranks of scholars and scientists who study the atmospheric environment?

Author Index

Subject Index